DOOR COUNTY FLORA

A *Field Guide to the Vascular Plants of Wisconsin's Door Peninsula*

Ram's-Head Orchid
Cypripedium arietinum

STEVE W. CHADDE

WISCONSIN'S DOOR COUNTY holds a special interest for anyone interested in plants and interesting habitats. From its limestone lakeshores and cliffs, to its inland lakes, wetlands, and forests, the Door Peninsula has a unique and rich biological diversity. Personally, I first visited Door County more than 50 years ago as a child and have remained fascinated by the Peninsula ever since. My hope is that this detailed look at the area's flora will be of use to researchers and students, and to those working to preserve Door County's natural heritage for future generations.

DOOR COUNTY FLORA
A Field Guide to the Vascular Plants of Wisconsin's Door Peninsula

STEVE W. CHADDE

ISBN 978-1-951682-29-3

The author can be reached via email: steve@chadde.net
VERSION 1.2 (04/10/2020)

CONTENTS

4 CONTENTS

Northern Lake Michigan Coastal Ecological Landscape

Central Lake Michigan Coastal Ecological Landscape.

The Door Peninsula (and Door County, arrows) encompass two "ecological landscapes" as defined by Wisconsin Department of Natural Resources (2015). The majority of the Peninsula lies within the Northern Lake Michigan Coastal Ecological Landscape; a portion of southwestern Door County lies within the Central Lake Michigan Coastal Ecological Landscape.

INTRODUCTION

DOOR COUNTY FLORA is a comprehensive treatment of the vascular plant species in Wisconsin's unique and diverse Door Peninsula. The book attempts to treat all known plant species verified to occur in Door County, based primarily on data provided by the *Biota of North America Program,* and the *Online Virtual Flora of Wisconsin* (see References, page 432). Included are descriptions for approximately 1,100 species in 500 genera and 116 plant families. It is also inevitable that additional species will be discovered in the county as future studies are conducted.

Arrangement of the Taxa

All plants treated in this text belong to three informal groups, presented in order: (1) ferns and fern relatives, (2) gymnosperms (conifers), and, by far the largest group, (3) angiosperms. The angiosperms are subdivided into two classes, the dicotyledons or "dicots" (sometimes termed Magnoliopsida) and the monocotyledons or "monocots" (sometimes termed Liliopsida). These subdivision names derive from the observation that the dicots most often have two cotyledons, or embryonic leaves, within each seed. The monocots usually have only one, but the rule is not absolute either way. From a diagnostic point of view, the number of cotyledons is neither a particularly handy nor a reliable character, but provides a simple way to organize plant families into smaller groups. Dicots include many familiar trees, shrubs, and "wildflowers," such as those of the Aster Family; the monocots include the large grass and sedge families, and also smaller familes such as Juncaceae, Orchidaceae, and Typhaceae.

Within each of the divisions, families are listed alphabetically. Under each family, genera and species are also listed alphabetically. If there is more than one genus in a family, a key is provided to the genera. Likewise, if there is more than one species in a genus, a key to the species is provided.

Placement of monocots and dicots genera within families, with several exceptions, follows that of the Angiosperm Phylogeny Group IV system (AGP IV) of 2016. The APG was formed in the late 1990s, when researchers from major institutions around the world gathered with the goal of providing a modern, widely accepted classification of angiosperms. Their first attempt at a new system was published in 1998 (the AGP system). To date, three revisions have been published, in 2003 (AGP II), 2009 (AGP III), and 2016 (AGP IV), each superseding the previous system.

The major exception made here to APG IV is the retention of the traditional Liliaceae (Lily Family) to facilitate field use of the *Flora;* however the APG IV families segregated from the traditional Liliaceae are noted for each genera. Another exception is the retention of the Dipsacaceae (Teasel Family), and not including *Dipsacus* within Caprifoliaceae (Honeysuckle Family).

Similarly, fern families have been updated to reflect recent realignment of some families and genera. In the past, ferns had been loosely grouped with other spore-bearing vascular plants, these often called "fern allies" or "lycophytes." However, recent studies suggest an important dichotomy within vascular plants, separating the fern relatives or lycophytes (less than 1% of all vascular plant species) from a group termed the euphyllophytes. Euphyllophytes comprise two major groups: the spermatophytes (seed plants), which number more than 260,000 species, and the monilophytes (ferns), with over 9,200 species, including horsetails, whisk ferns, and all "true ferns." Genetic studies also reveal surprises about the relationships among true ferns and fern allies. True ferns appear to be closely related to horsetails, and in fact these plants are now often grouped within the true ferns. Also, plants commonly called fern relatives (club-mosses, spike-mosses and quillworts) do not appear to be closely related to the true ferns, but that term is retained for convenience.

Nomenclature of genera and species is not based on a single source, but in general, conforms to that of the published volumes of *The Flora of North America* series (www.eflo-ras.org), the BONAP database (Biota of North America Program, www.bonap.org), and The Plant List, a collaboration between the Royal Botanic Gardens (Kew), and the Missouri Botanical Garden (www.theplantlist.org). Common names largely reflect those of the BONAP database, or sometimes are names in popular use locally.

Nativity

Native Species
Native plants are those assumed to have been present in some part of Wisconsin prior to European settlement. Also included are plants believed to have arrived in the state more recently via natural migration from areas adjoining Wisconsin and in which they are clearly native.

Introduced Species
Several different terms can be used to indicate the status of introduced (or non-native) species in Wisconsin's flora. Each introduced plant is different in terms of its origin, persistence, rate of spread, etc. *Introduced species* include those that were deliberately planted and which may escape and reproduce locally. *Waifs* are species normally found in cultivation, such as tomatoes or watermelons, that are occasionally found sprouted from seeds in waste areas and yards; these usually persist for only a single season. Some ornamental species such as lilacs and day-lilies are planted and usually remain in place but continue to live indefinitely, often growing at old home sites. *Adventive species* are those that appear here and there, usually from accidental introductions or as escapes from cultivation, but are apparently not firmly established nor spreading in the state. *Naturalized species* are those non-native species, such as timothy (*Phleum pratense*) that have arrived deliberately or accidentally in Wisconsin, and have become firmly established as part of the flora; populations may be large or cover extensive areas. Most introduced species in the *Flora* are of Asian or European origin. These species from other continents are also referred to as exotic species.

Coefficients of Conservatism

The *Flora* includes a coefficient of conservatism for the majority of native species (abbreviated "cons" in the species descriptions). The coefficient (or C value) is used to evaluate the floristic quality of a natural area, and is based on that species' tolerance for disturbance and fidelity to a particular pre-settlement plant community type.

C values range from 0 to 10 and represent an estimated probability that a plant is likely to occur in a landscape relatively unaltered from what is believed to be a pre-settlement condition. For example, a C value of 0 is given to plants such as box elder

(*Acer negundo*), that have demonstrated little fidelity to any remnant natural community (i.e. it may be found almost anywhere). Conversely, a C value of 10 is applied to plants like sage willow (*Salix candida*) that are almost always restricted to a high quality natural area. Introduced plants, by definition, were not part of the pre-settlement flora, so no C value is given.

The aggregate conservatism of all the plants inhabiting a site determine its floristic quality. See Swink and Wilhelm (1994) for a thorough discussion of the method and how to calculate a Floristic Quality Index. A list of C values for Wisconsin plants is available at: wisplants.uwsp.edu/Checklist.html.

Landscape and Vegetation

The Door Peninsula (and Door County) encompass two "ecological landscapes" as defined by Wisconsin Department of Natural Resources (see maps, page 4). The majority of the Peninsula lies within the **Northern Lake Michigan Coastal Ecological Landscape**; a portion of southwestern Door County lies within the **Central Lake Michigan Coastal Ecological Landscape**. The website of the Wisconsin DNR provides detailed descriptions and maps (see References, page 432); and is the basis for the following description of Door County's vegetation and flora.

Historically, forests covered the uplands throughout almost all of Door County, with mesic hardwood or hemlock-hardwood types the most common. In low-lying areas, acidic conifer swamps of black spruce and tamarack were present, while in areas underlain by limestone, "rich" conifer swamps of northern white-cedar, and ash-dominated hardwood swamps occured.

Today, much of the Door Peninsula has been converted to agricultural and residential uses. The formerly extensive upland forests have been largely reduced to farm woodlots, and also altered by timber harvests and livestock grazing. Larger remnants of upland forest are typically found where the dolomite bedrock is close to the surface, which made those locations unproductive for agriculture. The largest remaining forest tracts are mostly on wet soils—either as conifer swamps composed of northern white-cedar, balsam fir, tamarack, and black spruce, or as hardwood swamps of ashes, elms, and "soft" maples (*Acer saccharinum, A. rubrum*).

The east side of the Door Peninsula near Lake Michigan remains heavily wooded, though development continues. Here, "mesic" (i.e., moderate amounts of moisture) forests predominate, with localized areas of northern boreal forest (with white spruce, northern white-cedar, balsam fir, and eastern white pine), dry-mesic oak and pine forest, and wet forests dominated by conifers or ashes.

A single, small example of **Great Lakes Barrens** has been described from the coast of 2,800-acre Chambers Island in Green Bay. Several highly degraded examples of **Alvar** (open limestone plains with little or no soil and sparse vegetation) have been noted on the eastern Door Peninsula.

Sedge meadows occur in some of the coastal wetlands of the Door Peninsula and in small areas on several of the Grand Traverse Islands. Some of these meadows are alkaline and fen-like. Fens in Door County lack many of the prairie components of the calcareous fens of southeastern Wisconsin, but, like them, may harbor rare, calcium-loving plants.

Beaches and dunes occur along parts of the Door Peninsula's Lake Michigan coast and on several of the Grand Traverse Islands. They provide habitat for rare species, including plants endemic to Great Lakes shorelines such as dune thistle (*Cirsium pitcheri*) and dune goldenrod (*Solidago simplex*). These areas are, however, subject to heavy recreational use and residential development.

At several sites adjacent to Lake Michigan on the eastern side of the Door Peninsula, past lake levels have created alternating series of **ridges and swales** parallel to the shoreline. Sandy ridges support forest communities; the swales feature wetland communities or shallow water. The flora associated with these ridge-and-swale complexes is extremely diverse. A notable example is the Ridges Sanctuary, a 1,600-acre nature preserve near Baileys Harbor.

Dolomite cliffs are a prominent landscape feature associated with some stretches of the Niagara Escarpment, which runs along the western edge of the Door Peninsula and also forms headlands with cliffs on some of the Grand Traverse Islands. In some areas, especially along the east side of the northern Door Peninsula, exposures of level dolomite occur as Great Lakes Alkaline Rockshore. Cave Point County Park, south of Jacksonport, features a long series of low dolomite cliffs and ledges ranging in height from 1–3 meters high.

Rare Flora

Because of their rarity, primarily as a result of habitat loss, approximately 73 plant species are currently listed as endangered by the state of Wisconsin; a further 59 species are listed as threatened. Those species found in Door County are noted in the *Flora* by their status:

• *Endangered* – A species is considered endangered if the species is threatened with extinction throughout all or a significant portion of its range within Wisconsin.

• *Threatened* – A species is considered threatened if the species is likely to become endangered within the foreseeable future throughout all or a significant portion of its range within Wisconsin.

The Northern Lake Michigan Coastal Ecological Landscape has a rich flora and supports many rarities, including 14 species listed as Wisconsin Endangered, 22 listed as Wisconsin Threatened, and 66 listed as Wisconsin Special Concern on the Wisconsin Natural Heritage Working List. Two plants, dune thistle and dwarf lake iris (*Iris lacustris*), listed as Wisconsin Threatened, are also listed as Federally Threatened by the U.S. Fish and Wildlife Service.

The website of the Wisconsin Dept. of Natural Resources (http://dnr.wi.gov) maintains a list of species listed as endangered, threatened, or of special concern (an additional category for plants but not afforded legal protection).

Forested habitats of the Door Peninsula support many rare plant species, including plants more common in boreal regions such as ram's-head lady's-slipper (*Cypripedium arietinum*), false toadflax (*Geocaulon lividum*), and the unusual saprophyte, giant pinedrops (*Pterospora andromedea*). Mesic forests of the Door Peninsula support most of the state's known populations of long-spur violet (*Viola rostrata*). A population of the rare Wisconsin Endangered foam-flower (*Tiarella cordifolia*) occurs in a mesic forest near Lake Michigan. Mountain sweet-cicely (*Osmorhiza berteroi*) is known in Wisconsin from a few locations on the Door Peninsula (and several others near Lake Superior).

The marshes, meadows, and fens of the Door Peninsula often show the influence of alkaline substrates and groundwater by the presence of calciphilic plants. Examples include shrubby cinquefoil (*Potentilla fruticosa*), Kalm's St. John's-wort (*Hypericum kalmianum*), and brook lobelia (*Lobelia kalmii*) as well as rarities such as slender-leaf sundew (*Drosera linearis*), livid sedge (*Carex livida*), tufted leafless-bulrush (*Trichophorum cespitosum*), seaside arrow-grass (*Triglochin maritima*), and marsh arrow-grass (*Triglochin palustre*). Marsh habitats also support unusual wetland and aquatic plants such as mare's-tail (*Hippuris vulgaris*) and mermaid-weed (*Proserpinaca palustris*).

Beach and dune habitats support dune thistle, a Great Lakes endemic occuring in no other habitat. The Wisconsin Endangered Lake Huron tansy (*Tanacetum bipinnatum*) is an extremely rare inhabitant of open dunes and shoreline bedrock habitats. Two Wisconsin Threatened plants, sand-reed grass (*Calamovilfa longifolia* var. *magna*) and dune goldenrod, also inhabit dunes and sometimes the upper beach zone, along with other uncommon beach plants such as seaside sandmat (*Euphorbia polygonifolia*), and sea-rocket (*Cakile edentula*). Wisconsin's only known population of the globally rare spoon-leaf moonwort (*Botrychium spathulatum*), occurs in an open habitat along Lake Michigan.

In Wisconsin, the Great Lakes Alkaline Rockshore community occurs only on the Door Peninsula and in the Grand Traverse Islands. The extensive alkaline rockshores on the northeast side of the Door Peninsula support rare plants having restricted distributions; these include sand beach species such as silverweed (*Potentilla anserina*), endemics such as dwarf lake iris, disjuncts such as Mistassini primrose (*Primula mistassinica*), and a group of species that also occur in the calcareous fens and alkaline prairies of southeastern Wisconsin, including lesser fringed-gentian (*Gentianopsis virgata*), Ohio flat-topped goldenrod (*Solidago ohioensis*), and sticky false asphodel (*Triantha glutinosa*).

The Niagara Escarpment, with its exposures along the west side of the Door Peninsula and on several of the Grand Traverse Islands, also provides habitat for specialists, some of them rare in Wisconsin. These include hoary whitlow-grass (*Draba cana*), rock whitlow-grass (*Draba arabisans*), Allegheny-vine (*Adlumia fungosa*), and broad-leaf sedge (*Carex platyphylla*), found nowhere else in the state.

Newport State Park, Door County, looking north-northeast toward Plum Island and Washington Island. Europe Lake is visible at the upper left. (DOOR COUNTY VISITOR BUREAU.)

FERNS & FERN RELATIVES

Aspleniaceae

SPLEENWORT FAMILY

Asplenium

SPLEENWORT

Mostly small ferns, with short rootstocks covered by old petiole-bases, and a tuft of small to medium-sized leaves ("fronds"); the blades firm, simple, pinnate, or 2-pinnate, and often evergreen. Veins free or forking. Sori (clusters of spore containers) elongate, occurring along the veinlets. Indusium (covering over sori) usually membranous, attached lengthwise along one side of the sorus. Species usually of rock crevices and where shaded and mossy.

1   Blades simple or deeply pinnately lobed .... *A. rhizophyllum*
1   Blades once-pinnate 2
2   Leaves of two types, the fertile long and upright, the sterile shorter and spreading, pinnae with conspicuous basal lobes overlapping rachis .... *A. platyneuron*
2   Leaves all alike; pinnae bases not overlapping rachis 3
3   Rachis purple-brown throughout entire length .... *A. trichomanes*
3   Rachis green, dark only at base .... *A. viride*

Asplenium platyneuron (L.) B.S.P.

EBONY-SPLEENWORT   *native / cons* **4**

Tufted fern with 2 types of leaves; the fertile leaves stiff and upright, 20–40 cm long and 2.5–4 cm wide, gradually tapered to the base; pinnae linear-oblong or basal pinnae triangular, widely separated; rachis satiny chestnut purple; sterile leaves shorter and widely spreading; sori linear-oblong and on the veins. — Crevices of shady sandstone cliffs, in moss or in shallow soil over rocks, also in partial shade in open woods. The stiff, upright fertile blades are distinctive.

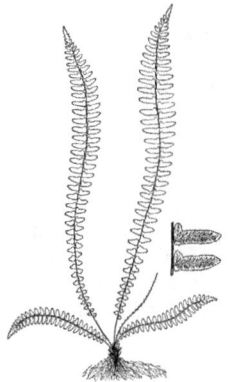

Asplenium rhizophyllum L.

WALKING FERN   *native / cons* **10**

Camptosorus rhizophyllus (L.) Link

Tufted fern with fronds 5–30 cm long or longer, clustered at the end of an erect or ascending scaly rhizome. Leaves evergreen, entire, 1–3 cm wide at the cordate or auriculate base, usually tapering to a long caudate tip; veins reticulate; **sori** elongate, scattered along the veins; indusium attached on one side of the sorus. — Shaded rocks, usually of limestone, less commonly on sandstone.

The tips of the arching blades often root to form new plants, hence the name walking fern.

Asplenium trichomanes L.

MAIDENHAIR-SPLEENWORT   *native / cons* **10**

Fern with leaves 6–20 cm long or longer, forming a dense tuft from a compact rhizome. Petiole and rachis purple brown; old rachises persistent; **blades** linear, pinnate; pinnae usually opposite or subopposite, oval, rounded to cuneate at the inequilateral base and slightly toothed on the sides and at the blunt apex; sori linear, situated on the veins between the midrib and the margin. — Sheltered rock crevices on sandstone or limestone, where moist or dry.

Asplenium platyneuron

Asplenium rhizophyllum

Asplenium trichomanes

Asplenium viride Huds.

GREEN SPLEENWORT   *endangered / native / cons* **10**
   Asplenium trichomanes-ramosum L.

Fern with leaves 2-14 cm long, tufted from a short rhizome. Petioles darkened below, bright green above; rachis green; blades linear to linear-lance-shaped, pinnate; pinnae rounded or rhomboid-ovate, crenate; sori elongate, borne near the indistinct midrib, becoming confluent at maturity. — In crevices of shaded, wet limestone cliffs and talus.

   A distinctive species less common than *Asplenium trichomanes,* with which it is sometimes associated. The green rachis (stem of the blade) distinguishes it from our other spleenworts. In Door County, collected from Washington Island in 1926; current status unknown.

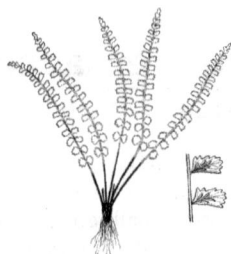

Asplenium viride

Athyriaceae

LADY FERN FAMILY

Medium to large ferns. *Deparia* formerly considered as species of *Athyrium,* now treated as separate genus. In contrast to *Deparia, Athyrium* has a more deeply grooved rachis, which is continuous from rachis to costa (vs. discontinuous in *Deparia*).

1 Blades 2-pinnate (with the pinnae again divided), the pinnules also sometimes deeply lobed. . . . . . .
  . . . . *Athyrium*
1 Blades with deeply lobed pinnae (1-pinnate-pinnatifid) . . . . *Deparia*

Athyrium
LADY FERN

Athyrium angustum (Willd.) K. Presl

NORTHERN LADY FERN   *native / cons* **5**
   Athyrium filix-femina L.

Clumped fern, rhizomes short and ascending. Leaves deciduous, sterile and fertile leaves similar; petioles with brown, linear scales; blades elliptic, 2-pinnate, broadest at middle or slightly below middle; pinnae short-stalked or stalkless; sori generally somewhat curved to hook-shaped, less often straight; indusia elongate, laterally attached. — Moist deciduous woods, thickets, streambanks, wetland margins, shaded rock outcrops.

Deparia
SILVERY GLADE FERN

Deparia acrostichioides (Swartz) M. Kato

SILVERY GLADE FERN   *native / cons* **10**
   Athyrium thelypterioides (Michx.) Desv.

Large fern from creeping rhizomes. Leaves deciduous, 50-100 cm long, sterile and fertile leaves alike; petioles straw-colored (but dark red-brown at base), with brown lance-shaped scales; blades lance-shaped to oblong in outline, tapered at tip and distinctly narrowed toward base; deeply lobed, the segments blunt to somewhat tapered at their tip, margins entire to slightly lobed. Sori crowded, elongate, straight or sometimes curved, the indusia silvery and shiny when young. — Moist, rich deciduous or mixed woods, especially in swales, ravines, depressions; streambanks.

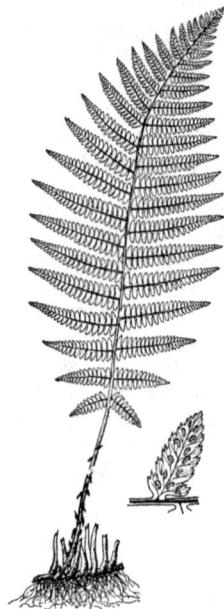

Athyrium angustum

Cystopteridaceae

BLADDER FERN FAMILY

Small to medium ferns; two genera in Door County: *Cystopteris* and *Gymnocarpium.*

1   Blades ternate (divided into 3 more or less equal parts); indusium absent . . . . *Gymnocarpium*
1   Blades 1-pinnate; indusium present . . . . *Cystopteris*

Cystopteris
BLADDER FERN
Delicate, medium-sized ferns, with 2-3-pinnate blades arising from short creeping rhizomes. Veins free. Indusium hood-shaped, thin, and withering, attached at one side and arching over the rounded sori.

1   Blades elliptic to lance-shaped, typically widest at or slightly below middle of blade; rachis and pinnule midribs without glandular hairs 2
1   Blades elliptic to triangle-shaped, usually widest at base; rachis and pinnule midribs sparsely to densely covered with glandular hairs 4
2   Stems covered with yellow hairs; leaves clustered 1–4 cm below protruding apex of stem *C. protrusa*
2   Stems without hairs; leaves clustered at apex of stem 3
3   Pinnae usually at acute angle to rachis and often curving toward apex of blade; pinnae margins rounded-toothed . . . . *C. tenuis*
3   Pinnae usually perpendicular to rachis and not curving toward apex of blade; pinnae margins sharp-toothed . . . . *C. fragilis*
4   Rachis often with bulblets, rachis and midribs usually densely glandular-hairy; blades narrowly to broadly triangle-shaped, apex of blade long-tapered . . . . *C. bulbifera*
4   Rachis occasionally with bulblets, rachis and midribs usually only sparsely glandular-hairy; blades ovate lance-shaped, apex of blade short-tapered . . . . *C. laurentiana*

Cystopteris bulbifera (L.) Bernh.
BULBLET FERN, BLADDER FERN   *native / cons* **8**
Clumped fern, rhizomes short and thick. Leaves deciduous, 30-100 cm long, sterile and fertile leaves similar but sterile blades usually shorter than fertile; petioles much shorter than blades; blades lance-shaped, 6-15 cm wide at base, long tapered to tip, with 20-30 pairs of pinnae; the veins ending in a notch (sinus). Sori round, on a small vein; indusia hood-like and attached at its base, covered with scattered, short-stalked glands. Green bulblets, 4-5 mm wide, are produced on lower side of rachis (main stem of leaf) toward upper end of blade, these falling and forming new plants. — Rocky streambanks, ravines, seepy slopes, cedar swamps, and moist, shaded, often calcium-rich rocks and cliffs.

Distinguished from *Cystopteris fragilis,* a common fern of moist woods, by the blade broadest at base, most veins ending in a notch, and the small bulblets on underside of rachis. In fragile fern, blade broadest above its base, most veins end in a tooth, and bulblets are absent.

Cystopteris bulbifera

Cystopteris fragilis (L.) Bernh.
BRITTLE BLADDER FERN   *native / cons* **6**
Fern with leaves 10-35 cm long or longer, tufted from short creeping rhizomes. Blades lance-shaped, 3-8 cm wide or wider near the base, bipinnate; pinnae pinnatifid to lobed, and at least the basal pinnules varying from orbicular to triangular and rounded to the base; veins mostly ending in a tooth or on the unnotched margin. Indusium up to 1 mm long and more or less cleft at the apex. — Sheltered crevices in cliffs, moist banks, and wooded talus slopes.

Petioles translucent (when held to a light), with veins of the blade extending to the very tips of the teeth, and a smooth rachis.

Cystopteris laurentiana (Weatherby) Blasdell
ST. LAWRENCE BLADDER FERN   *native / cons* **9**
   Cystopteris fragilis var. *laurentiana* Weath.
Fern with tufted leaves from a short creeping rhizome, to about 60 cm long. Petioles light brown to red-tinged. Blades ovate, to about 30 cm long

Cystopteris fragilis

and 12 cm wide; sterile blades usually shorter than the fertile. Indusium to 1 mm wide, very finely glandular. — Calcareous rock or slopes.

This species combines the attributes of its presumed parents, *C. fragilis* var. *fragilis* and *C. bulbifera*. It is usually an upright, vigorous plant larger than typical *C. fragilis*. The veins extend both to the teeth-tips and to the sinuses.

Cystopteris protrusa (Weatherby) Blasdell
LOWLAND BLADDER FERN   native / cons 6
Cystopteris fragilis var. *protrusa* Weatherby
Fern with leaves 20–50 cm long, scattered along a creeping rhizome. Petioles greenish, straw-colored, or pale brown. Blades lance-shaped, to 25 cm long, 5–10 cm wide; sterile blades usually shorter, bipinnate; pinnules sharply toothed, ovate-lance-shaped; lower pinnules tapered to a stalk-like base; veins mostly ending in a tooth or on the unnotched margin. Indusium to 0.5 mm long, shallowly toothed or entire at its tip. — Under deciduous trees on moist soil.

Distinguished from *C. fragilis* by the long internodes on the rhizome, the greenish or straw-colored petioles, the softer, larger blades, and the lower pinnules, which taper to a stalk-like base.

Cystopteris protrusa

Cystopteris tenuis (Michx.) Desv.
UPLAND BRITTLE BLADDER FERN   native / cons 10
Cystopteris fragilis var. *mackayi* Lawson
Similar to *C. fragilis*, but the pinnules oblong to lance-shaped and evenly wedge-shaped at the base; the indusium about 0.5 mm long and shallowly toothed or entire at its tip. — In habitats similar to *Cystopteris fragilis*, but more often on streambanks, rotted logs, and moist openings.

Cystopteris tenuis

Gymnocarpium
OAK FERN
Small ferns with 3-parted delicate blades, the blades glabrous or glandular, arising singly from slender rootstocks. Sori round. Indusium absent. Veins free, simple, or forking.

1 The two lower divisions of the blade nearly as long as the terminal division; blades membranous and thin; rachis glabrous . . . . *G. dryopteris*
1 The two lower divisions of the blade about half the length of terminal division; blades firm and somewhat stiff; rachis in part densely glandular . . . . *G. robertianum*

Gymnocarpium dryopteris (L.) Newman
NORTHERN OAK FERN   native / cons 7
Small, delicate fern with leaves to 30 cm long or longer, arising singly from a slender blackish rhizome. Blades glabrous or nearly so, triangular in outline, 3-parted; the pinnae pinnate-pinnatifid. Sori small, located near the margin. — Cool, moist coniferous and mixed woods; base of talus slopes; swamp margins.

The small, delicate, triangular blades oriented parallel to the ground and yellow-green in color are distinctive.

Gymnocarpium robertianum (Hoffmann) Newman
LIMESTONE OAK FERN   native / cons 10
Small to medium fern with leaves to 40 cm long, arising singly from a slender blackish rhizome. Blades triangular in outline, 3-parted, bipinnate-pinnatifid, glandular, the innermost basal pinnules of the lowermost pair of pinnae usually much longer than the corresponding upper pinnules. Sori small, located near the margin. — Limestone cliffs, outcrops and pavements (alvars).

The long-triangular, glandular blades (including glands on the upper surface), with pinnules at right angles are distinctive.

Gymnocarpium dryopteris

Dennstaedtiaceae

BRACKEN FERN FAMILY

Ferns with leaves all alike, arising singly from creeping rhizomes.

Pteridium
BRACKEN FERN

Pteridium aquilinum (L.) Kuhn
BRACKEN FERN   *native / cons* **2**
Coarse fern with leaves 30-70 cm or more long, often forming large
colonies from the creeping rhizomes. Blades triangular in outline, usually
3-parted, 30-50 cm wide; lower pinnules more or less pinnatifid; upper
pinnules entire, glabrous or slightly hairy on underside, with revolute
margins. Sporangia borne in marginal sori on the underside of the pinnules; sporangia covered by a nearly continuous false outer indusium
formed by the revolute pinnae margin. Plants growing in shade tend to
have more or less horizontal blades; blades of plants growing in sun tend
to be upright and stiff. — Ubiquitous in open drier woods, pine plantations,
old fields, and sandy openings.

Pteridium aquilinum

Dryopteridaceae

WOOD-FERN FAMILY

Medium to large ferns; rhizomes short, stout and scaly. Leaves dark green, sometimes evergreen;
petioles shorter than blades, straw-colored or green, with chaffy scales near base. Sterile and
fertile leaves alike or slightly different; sterile leaves sometimes persisting over winter; blades
1-3 pinnate, the smallest segments commonly toothed or lobed, veins simple to 1- or 2-branched.
Sori round, on underside veins of pinnae; indusia round to kidney-shaped.

1 Fronds 1-pinnate-pinnatifid to more divided, the pinnae pinnatifid or themselves fully divided, lacking
  a prominent basal lobe, light green to dark green, herbaceous to nearly leathery; indusia kidney-
  shaped . . . . ***Dryopteris***
1 Fronds 1-pinnate, the pinnae toothed and each with a slight to prominent lobe near the base on the
  side towards the leaf tip, dark green, leathery or nearly so; indusia peltate (umbrella-like) . . . .
  . . . . ***Polystichum***

Dryopteris
WOOD-FERN

Medium to large ferns; rhizomes short, stout and scaly, often covered with old petiole bases.
Leaves dark green, sometimes evergreen; petioles shorter than blades, straw-colored or green,
with chaffy scales near base. Sterile and fertile leaves alike or slightly different; sterile leaves
sometimes persisting over winter; blades 1-3 pinnate, the smallest segments commonly toothed
or lobed, veins simple to 1- or 2-branched. Sori round, on underside veins of pinnae; indusia
round to kidney-shaped.

HYBRIDS
Hybrids may be recognized by an appearance intermediate between the parent species, and the pres-
ence of abortive spores. Three *Dryopteris* hybrids are fairly common in Wisconsin and reported for
Door County:
  Dryopteris × *boottii* (Tuckerman) Underwood: *D. cristata* × *D. intermedia;* leaves more dissected than
D. cristata.
  Dryopteris × *triploidea* Wherry: *D. carthusiana* × *D. intermedia;* leaves similar to parents but often
somewhat larger.
  Dryopteris × *uliginosa* (A. Braun ex Dowell) Druce: *D. carthusiana* × *D. cristata;* usually in swamps and
wet woods.

1 Sori on margins of blade segments; blades leathery gray-green and paler on underside **D. marginalis**
1 Sori near middle of smallest blade segments 2
2 Lowest pinnules on lowest pinnae stalkless . . . . **D. cristata**
2 Lowest pinnules on lowest pinnae stalked 3
3 Lowermost inner pinnule shorter than adjacent lower pinnule . . . . **D. intermedia**
3 Lowermost inner pinnule longer than next outer one 4
4   Lower basal pinnule on basal pinna closer to the second upper pinnule than to the inner or first
upper pinnule . . . . **D. expansa**
4   Lower basal pinnule on basal pinna closer to the inner upper pinnule than to the second upper
pinnule . . . . **D. carthusiana**

Dryopteris carthusiana (Villars) H.P.Fuchs
SPINULOSE WOOD-FERN   *native / cons* 7
Dryopteris austriaca var. *spinulosa* (O.F. Müll.) Fisch.
Dryopteris spinulosa (O.F. Müll.) Watt
Clumped fern, rhizomes short-creeping. Leaves all alike, deciduous, smooth except for chaffy, pale brown scales near base of petioles; blades 2- to nearly 3-pinnate, 2-6 dm long and 1-4 dm wide, tapered to tip, slightly narrowed at base; pinnae usually 10-15 pairs, alternate to nearly opposite, narrowly lance-shaped; pinnules toothed to deeply lobed, mostly 5-40 mm long and 3-10 mm wide, the teeth tipped with a small spine; innermost lower pinnule longer than next outer one and 2-3x longer than opposite upper pinnule. Sori halfway between midvein and margin; indusia 1 mm wide, without stalked glands. — Moist to wet woods, hummocks in swamps, thickets; also drier sand dunes and ridges.

Dryopteris carthusiana

Dryopteris cristata (L.) A.Gray
CRESTED WOOD-FERN   *native / cons* 7
Clumped fern, rhizomes short-creeping with ascending tips. Sterile and fertile leaves somewhat different, the outer sterile leaves waxy, persistent and smaller than inner fertile leaves; fertile leaves deciduous, 3-8 dm long. Blades 1-pinnate to nearly 2-pinnate, narrowly lance-shaped, 2-6 dm long and 7-15 cm wide, tapered to tip, narrowed at base; pinnae 5-9 cm long and to 4 cm wide, typically twisted to a nearly horizontal position, giving a "venetian blind" appearance to blades; pinnae segments to 20 mm long and 8 mm wide, with small spine-tipped teeth; petioles with sparse, pale brown, long-tapered scales. Sori round, midway between midvein and margin; indusia smooth, 1 mm wide. — Swamps, thickets, open bogs, fens and seeps.

Dryopteris expansa (K.Presl) Fraser-Jenkins & Jermy
SPREADING WOOD-FERN   *native / cons* 8
Dryopteris assimilis S.Walker
Fern with leaves to 1 m long, forming a large, more or less upright crown at the end of the upright, chaffy rhizome. Petioles usually shorter than the blades, with brown-tinged, often dark-centered, ovate scales. Blades broadly triangular to ovate, abruptly tapering to the tip, twice pinnate to tripinnate; pinnae short-stalked; basal pinnae triangular, inequilateral; the inner, lowermost pinnule on each lowermost pinna closer to the second upper pinnule than to the first upper pinnule. Indusia glabrous, or rarely finely glandular. — Cool moist woods and thickets.

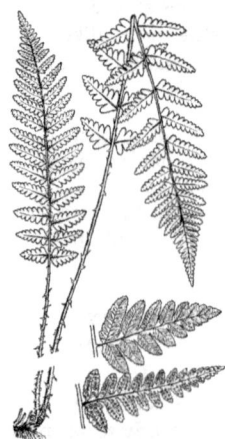
Dryopteris cristata

Dryopteris intermedia (Muhl.) A.Gray
FANCY WOOD-FERN   *native / cons* 7
Dryopteris spinulosa var. *intermedia* (Muhl. ex Willd.) Underw.
Clumped fern, rhizomes ascending. Leaves in an open vaselike cluster of evergreen leaves; blades broadest just above base and abruptly tapered near tip, 2-5 dm long and 1-2 dm wide, 2-pinnate; pinnae at right angles to stem, lowermost inner pinnule usually shorter than next outer pinnule,

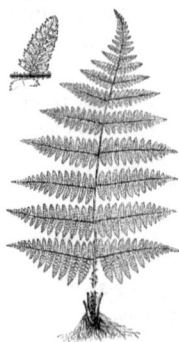
Dryopteris intermedia

pinnules toothed and tipped with small spines; petioles 1/3 as long as blade, with pale brown scales with a darker center, petioles and stems with small, gland-tipped hairs. Sori midway between midvein and margin, the indusia 1 mm wide, covered with stalked glands. — Moist hardwood and mixed hardwood-conifer forests, hummocks in swamps; soils rich in humus, slightly acid to neutral.

Dryopteris marginalis (L.) A. Gray
MARGINAL WOOD-FERN   *native / cons* **9**
Fern with leaves mostly 25-60 cm long, crowded to form a crown; lower part of the petiole covered with light brown lance-shaped scales. Blades 10-20 cm wide, dark green above, gray-green below, leathery, lance-shaped to ovate, 2-pinnate; pinnae lance-shaped; pinnules oblong, entire to deeply lobed. Sori located near the margins; indusia smooth, whitish, becoming light brown. — Rocky woods and ravines.

The leathery or spongy character of the nearly evergreen blades and the nearly marginal sori are characteristic.

Dryopteris marginalis

Polystichum
HOLLY FERN
Large, tufted ferns with mostly evergreen, leathery blades; the petioles usually scaly, arising from short, stout, chaffy rhizomes. Sori round; indusia round, attached at the center.

Polystichum acrostichoides (Michx.) Schott
CHRISTMAS FERN   *native / cons* **9**
Fern with dark green, nearly evergreen leaves 40-70 cm long. Petiole and rachis chaffy. Blades lance-shaped, 7-12 cm wide or wider, pinnate; pinnae oblong to lance-shaped, auricled at the base on the upper side; margins finely toothed. Sori borne on smaller upper pinnae. — Rich woods and shaded rocky slopes.

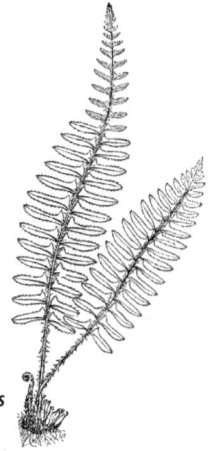

Polystichum acrostichoides

Equisetaceae
HORSETAIL FAMILY

Equisetum
HORSETAIL, SCOURING-RUSH
Rushlike herbs with dark rhizomes. Stems annual or perennial, grooved, usually with large central cavity and smaller outer cavities, unbranched or with whorls of branches at nodes. Leaves reduced to scales, united into a sheath at each node; top of sheath divided into dark-colored teeth. Spores in cones at tips of green or brown fertile stems.

HYBRIDS
> *Equisetum* × *ferrissii* Clute: *E. hyemale* × *E. laevigatum.*
> *Equisetum* × *litorale* Kuehl ex Rupr.: *E. arvense* × *E. fluviatile.*
> *Equisetum* × *mackaii* (Newm.) Brichan (A.A. Eat) Schaffn.: *E. hymale* × *E. variegatum.*

1 Stems evergreen (annual in *E. laevigatum*); unbranched or with a few scattered branches, branches not in regular whorls (scouring rushes) 2
1 Stems annual; usually with regular whorls of branches, sometimes unbranched (horsetails) 5
2 Stems solid (central cavity absent); stems small, slender and sprawling . . . . *E. scirpoides*
2 Stems hollow (central cavity present); stems larger, usually upright. 3
3 Stems 1–3 dm tall, with 5–12 ridges, central cavity to 1/3 diameter of stem . . . . *E. variegatum*
3 Stems usually taller, with 16–50 ridges, central cavity more than half diameter of stem 4
4 Cones with a distinct, small sharp tip; stem sheaths with a black band at tip and base. . . . *E. hyemale*
4 Cones blunt-tipped, sheaths with black band at tip only . . . . *E. laevigatum*

5 Stems unbranched 6
5 Stems with regular whorls of branches 10
6 Stems green 7
6 Stems brown or flesh-colored 8
7 Stems with 9–25 shallow ridges; central cavity more than half diameter of stem; sheath teeth entirely black or with narrow white margins .... *E. fluviatile*
7 Stems with 5–10 strongly angled ridges; central cavity less than 1/3 diameter of stem; sheath teeth with white margins and dark centers .... *E. palustre*
8 Sheath teeth papery and red-brown, teeth joined and forming several broad lobes .... *E. sylvaticum*
8 Sheath teeth black or brown, not papery, separate or joined in more than 4 small groups 9
9 Stems withering after spores mature, remaining unbranched. .... *E. arvense*
9 Stems persistent, becoming branched and green .... *E. pratense*
10 First internode of each branch shorter than the subtending sheath of the main stem 11
10 First internode of each branch equal or longer than the subtending sheath of the main stem 12
11 Stems with 9–25 shallow ridges; central cavity more than half diameter of stem; sheath teeth more than 12, entirely black or with narrow white margins .... *E. fluviatile*
11 Stems with 5–10 strongly angled ridges; central cavity about same size as outer cavities; sheath teeth 5–6, with white margins and dark centers .... *E. palustre*
12 Stem branches themselves branched; sheath teeth papery and red-brown, teeth joined and forming several broad lobes.. *E. sylvaticum*
12 Stem branches unbranched; sheath teeth black or brown, not papery, separate or joined in more than 4 small groups 13
13 Stem branches ascending; teeth of branch sheaths gradually tapering to a slender tip .... *E. arvense*
13 Stem branches spreading; teeth of branch sheaths broadly triangular .... *E. pratense*

Equisetum arvense L.
COMMON OR FIELD HORSETAIL   *native / cons* **1**

Stems annual, upright from creeping, branched, tuber-bearing rhizomes covered with dark hairs. Sterile and fertile stems unalike; sterile stems appearing in spring as fertile wither, green, regularly branched, 1-6 dm tall and 2-5 mm wide, with 10-14 shallow ridges, the ridges usually rough-to-touch; central cavity 1/3-2/3 stem diameter; sheaths with 6-14 persistent, black-brown teeth 1-2 mm long; branches numerous in dense whorls, usually without branchlets, upright or spreading, 3-5-angled, solid. Fertile stems flesh-colored, shorter than sterile stems and with larger sheaths, maturing in early spring and soon withering, unbranched, to 3 dm tall and 8 mm wide; sheaths with 8-12 dark brown teeth. Cones blunt-tipped, long-stalked at end of stem, 0.5-3 cm long.

Common; streambanks, meadows, moist woods, ditches, roadsides and along railroads; calcareous fens.

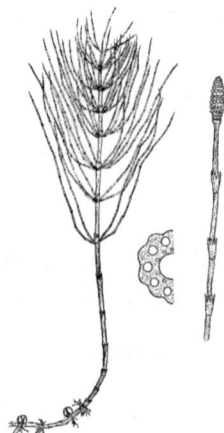

Equisetum arvense

Equisetum fluviatile L.
WATER-HORSETAIL   *native / cons* **7**

Stems annual, fertile and sterile stems alike, to 1 m or more tall, from smooth, shiny, light brown, creeping rhizomes; stems with 9-25 shallow, smooth ridges; central cavity large, about 4/5 stem diameter; stem sheaths green, 6-10 mm long; teeth 12-24, persistent, 2-3 mm long, dark brown to black, sometimes with narrow white margins; branches none or few, to many and regularly whorled from middle nodes, spreading, without branchlets, 4-6-angled, hollow. Cones 1-2 cm long at tips of stems, long-stalked, blunt-tipped, deciduous, maturing in summer. — In standing water of marshes, ponds, peatlands, ditches and swales.

Equisetum hyemale L.
COMMON SCOURING-RUSH   *native / cons* **3**
Equisetum affine Engelm.

Stems evergreen, persisting for more than 1 year, fertile and sterile stems alike, from black, slender rhizomes; stems mostly unbranched or with few, short, upright branches from upper nodes, to 15 dm tall but usually

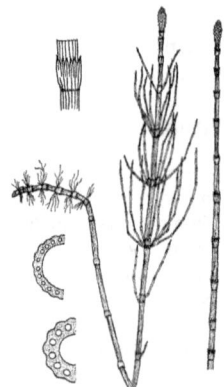

Equisetum fluviatile

shorter, 4-14 mm wide, with 14-50 rounded, very rough ridges; central cavity at least 3/4 stem diameter; stem sheaths 5-15 mm long, with a dark band at tip and usually also at base, the teeth dark brown to black with chaffy margin, 2-4 mm long, deciduous or persistent. Cones stalkless or short-stalked at tips of stems, sharp-pointed, eventually deciduous, 1-2.5 cm long, maturing in summer, or old stems sometimes developing branches with cones in the following spring. — Often forming dense colonies in seeps, wet to moist meadows, shores and streambanks, ditches, roadsides and along railroads; usually where sandy or gravelly.

Equisetum laevigatum A. Braun
SMOOTH SCOURING-RUSH native / cons 2
Stems mostly annual, fertile and sterile stems alike, from brown or black rhizomes; stems mostly unbranched or with a few upright branches, 3-10 dm tall and 3-8 mm wide, smooth and rather soft, with 10-32 ridges; central cavity 2/3-3/4 stem diameter; stem sheaths with a single dark band at tip, or rarely lowest sheaths with a dark band at base or entirely black; teeth dark brown or black with chaffy margins, free or partly joined in pairs, 1-4 mm long, soon deciduous. Cones short-stalked at tips of stems, rounded with a small sharp point, maturing in early summer and eventually deciduous. — Wet meadows, low prairie, streambanks, floodplains, seeps, and ditches, often where sandy or gravelly.

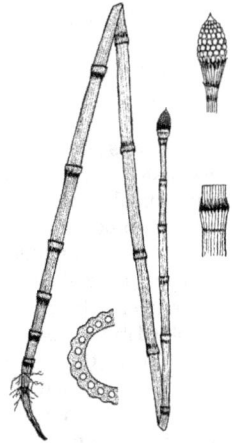

Equisetum hyemale

Equisetum palustre L.
MARSH-HORSETAIL native / cons 10
Stems annual, erect, fertile and sterile stems alike, from creeping, branched, shiny black rhizomes; stems 2-8 dm tall, with 5-10 pronounced ridges, the ridges mostly smooth; central cavity small, 1/6-1/3 stem diameter; sheaths green, loose and flared upward; teeth 5-6, free or partly joined, persistent, 3-7 mm long, brown to black, with pale, translucent margins; branches few and irregular, to many and whorled at upper nodes, upright, without branchlets, 5-6-angled, hollow. Cones long-stalked at tips of stems, 1-3 cm long, blunt-tipped, maturing in summer, deciduous. — Wetland margins, streambanks, alder thickets, fens; often in shallow water.

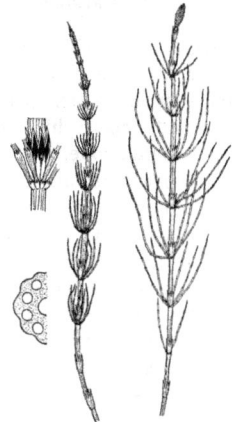

Equisetum palustre

Equisetum pratense Ehrh.
MEADOW-HORSETAIL native / cons 9
Stems annual and erect, sterile and fertile stems unalike, from creeping, dull black rhizomes. Sterile stems regularly branched, 2-5 dm tall and 1-3 mm wide; 8-18-ridged, the ridges roughened by silica on middle and upper stem; central cavity 1/3-1/2 stem diameter; main stem sheaths 2-6 mm long, the teeth persistent, 1-2 mm long, free or partly joined in pairs, brown with white margins and a dark midstripe; branches slender, many in regular whorls from middle and upper nodes, without branchlets, horizontal or drooping, mostly 3-angled, solid. Fertile stems uncommon, appearing in early spring before sterile stems and persisting, at first unbranched, fleshy and brown (without chlorophyll), later becoming green at nodes and producing many small green branches, mostly 1-3 dm tall; sheaths and teeth about twice as long as on sterile stems. Cones long-stalked at tips of stems, to 2.5 cm long, blunt-tipped, deciduous. — Moist woods, streambanks, meadows.

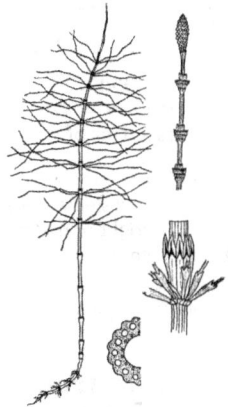

Equisetum pratense

Equisetum scirpoides Michx.
DWARF SCOURING-RUSH native / cons 7
Stems evergreen, very slender, fertile and sterile stems alike, from widely branching rhizomes; stems 5-30 cm long and only 0.5-1 mm wide, in dense clusters, usually unbranched and zigzagged, upright or trailing;

central cavity absent, 3 small outer cavities present; sheaths green with broad black band at tip, loose and flared above, with 3-4 teeth; teeth with white, chaffy margin, ± persistent, but tips usually soon deciduous. Cones black, small, 3-5 mm long, sharp-tipped. — Mossy places and moist, shaded woods, the stems often partly buried in humus.

Equisetum sylvaticum L.
WOODLAND-HORSETAIL   native / *cons* 7
Stems annual, erect, sterile and fertile stems unalike, from creeping, shiny light brown rhizomes, tubers occasionally present. Sterile stems green, 3-7 dm tall and 1.5-3 mm wide, with 10-18 ridges, rough-to-touch with sharp, hooked silica spines; central cavity 1/2-2/3 stem diameter; sheaths green at base, red-brown and flaring at tip; teeth brown, 3-5 mm long, joined in 3-5 broad lobes. Stems densely branched in regular whorls from the nodes, the branches themselves branched, often curving downward, 4-5-angled, solid. Fertile stems at first pink-brown (without chlorophyll), fleshy, unbranched, becoming green and branched as in sterile stems; sheaths and teeth larger than in sterile stems. Cones 1.5-3 cm long, stalked, blunt-tipped, deciduous. — Wet or swampy woods, thickets, usually in partial shade.

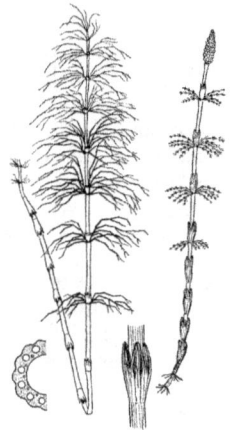
Equisetum sylvaticum

Equisetum variegatum Schleicher
VARIEGATED SCOURING-RUSH   native / *cons* 7
Stems evergreen, fertile and sterile stems alike, from creeping, much-branched, smooth rhizomes; may form thick colonies; stems 1-3 dm tall and 1-2.5 mm wide, with 5-12 shallow, rough ridges, branched near base and otherwise usually unbranched; central cavity 1/4-1/3 stem diameter, smaller outer cavities present; sheaths green at base with a broad black band above; teeth persistent, with a dark brown or black midstripe and wide white margins, abruptly narrowed to a hairlike, deciduous tip 0.5-1 mm long. Cones to 1 cm long, strongly sharp-tipped, maturing in summer or persisting unopened until following spring. — Wet calcareous open areas such as shores, low places in dunes, borrow pits and ditches.

Equisetum variegatum commonly forms hybrids with *E. hyemale* and *E. laevigatum*, sometimes making identification of this species difficult.

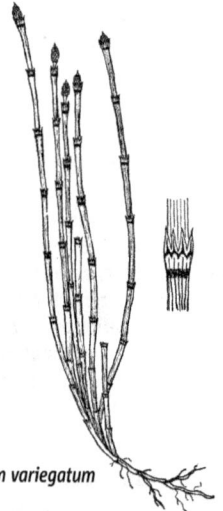
Equisetum variegatum

Lycopodiaceae
CLUBMOSS FAMILY
Low, trailing, evergreen herbs resembling large mosses. Leaves needlelike or scalelike, alternate or opposite on stem. Spore-bearing leaves (sporophylls) similar to vegetative leaves or in conelike clusters at tips of upright stems.

1 Horizontal stems absent; sporangia in axils of unmodified leaves .... *Huperzia*
1 Horizontal stems present; sporangia in axils of modified, reduced sporophylls, the sporophylls grouped into upright or nodding strobili 2
2 Ultimate shoots and their leaves 5–12 mm wide, rounded in cross-section; leaves not strongly overlapping .... *Lycopodium*
2 Ultimate shoots and their leaves to 6 mm wide, 4-angled or flattened in cross-section; leaves overlapping .... *Diphasiastrum*

Diphasiastrum
GROUND-PINE
Small plants of drier habitats resembling miniature trees; branches flattened or 4-angled in cross-section; leaves 4-ranked, neither spine- nor hair-tipped. Strobili (cones) stalked, the stalks branched into segments of equal length.

1 Stem branchlets cordlike, nearly square in cross-section, usually waxy blue-green color . . . .
. . . . ***D. tristachyum***
1 Stem branchlets flat in cross-section, usually green 2
2 Branchlets regularly fan-shaped and arching, without conspicuous constrictions between seasonal
growth; most strobili with sterile tips . . . . ***D. digitatum***
2 Branchlets irregular, with conspicuous constrictions; most strobili without sterile tips . . . .
. . . . ***D. complanatum***

Diphasiastrum complanatum (L.) Rothm.

NORTHERN RUNNING-PINE   *native / cons 7*
  Lycopodium complanatum L.

Horizontal stems mostly below the surface of the ground; leaves scale-like. Upright stems to about 30 cm tall, with forking branchlets. Branchlets flattened, often strongly constricted between yearly growths, 2-4 mm wide. Leaves 4-ranked. Strobili 1 or 2 on peduncles. — Woodlands and clearings.

Conspicuous annual constrictions present, giving plants a somewhat irregular appearance, in contrast to the regularity of fan ground-pine (*D. digitatum*). The strobili are also irregular in number per peduncle (varying from 1-4), and the naked peduncles are very slender.

Diphasiastrum digitatum (Dill.) Holub

FAN GROUND-PINE   *native / cons 6*
  Lycopodium digitatum Dill.

Horizontal stems mostly on or near the surface of the ground; leaves distant, scale-like. Upright stems to about 30 cm high, the branchlets of the branches arched and fan-like; constrictions between yearly growth absent or only slightly evident. Branchlets 2-3 mm wide. Leaves 4-ranked. Strobili mostly 3 or 4 on peduncles; peduncle branched at one point. — Dry woods and clearings. The branchlets are very regular and fan-like, annual constrictions are lacking, and the strobili are usually in groups of 4 on long, naked peduncles.

Diphasiastrum complanatum

Diphasiastrum tristachyum (Pursh) Holub

DEEP-ROOT GROUND-PINE   *native / cons 7*
  Lycopodium tristachyum Pursh

Horizontal stems usually deeply buried; leaves scalelike. Upright stems to 30 cm tall. Sterile branches ascending to loosely divergent, flattened, 1-1.5 cm wide. Leaves 4-ranked, blue-green, lance-shaped. Strobili 2-6 on leafy-bracted peduncles. — Dry, sometimes sandy woods and clearings.

The branches are vase-shaped and crowded, bluish green and white waxy on their underside; annual constrictions are present along the branches; the peduncles often branch and then branch again, resulting in 4 strobili.

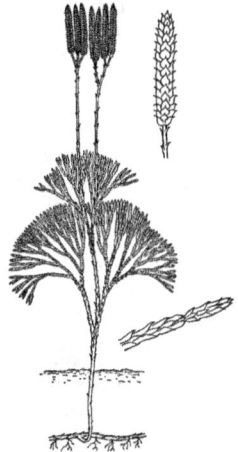

Diphasiastrum tristachyum

Huperzia

FIR-MOSS

Low evergreen perennials with erect shoots; leaves spreading or appressed and upright. Spores borne at base of upper leaves.

Huperzia lucidula (Michaux) Trev.

SHINING FIR-MOSS   *native / cons 7*
  Lycopodium lucidulum Michx.

Stems light green, creeping and rooting, upcurving stems forked several times, to 25 cm high, crowded with shiny dark green leaves which persist for more than one season. Leaves in mostly 6 rows, spreading or curved downward, in alternating groups of longer sterile and shorter fertile leaves, giving shoots a ragged look. Sterile leaves 6-12 mm long, toothed

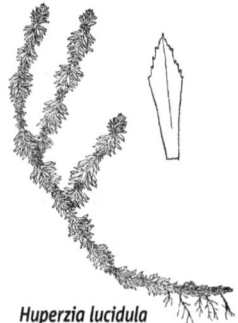

Huperzia lucidula

and broadest above middle; sporophylls barely widened and with small teeth or entire at tip. Small two-lobed buds (gemmae) produced in some upper leaf-axils; these may sprout into new plants after falling onto moist humus. — Moist to wet conifer and hardwood forests.

Lycopodium
GROUND-PINE

Plants mainly trailing on ground. Roots emerging from point of origin on underside of main stems. Horizontal stems on substrate surface or subterranean, long-creeping. Upright shoots scattered along horizontal stem, 5-16 mm diameter, round or flat in cross section, unbranched or with 1-4 lateral branchlets. Leaves not imbricate, linear to linear lance-shaped; leaves on horizontal stems scattered, appressed; leaves on lateral branchlets mostly 6-ranked or more, monomorphic with few exceptions, appressed, ascending to spreading, margins entire to dentate. Gemmae absent. Strobili single and sessile or multiple and pedunculate; peduncle, when present, conspicuously leafy; sporophylls extremely reduced, much shorter than peduncle or stem leaves. Sporangia kidney-shaped.

1 Stroboli stalked; upright stems with 2–5 branches, not forming tree-like shapes; leaves tipped with hairs . . . . *L. clavatum*
1 Stroboli not stalked 2
2 Stems creeping and horizontal; stroboli single at end of upright, mostly unbranched shoot . . . . *L. annotinum*
2 Stems upright, much-branched and tree-like; stroboli 1–7 at end of shoot 3
3 Branches flat in cross-section; leaves of unequal sizes . . . . *L. obscurum*
3 Branches round in cross-section; leaves of equal sizes 4
4 Leaves on main stem below branches dark green and appressed to stem, soft to touch . . . . *L. hickeyi*
4 Leaves on stem below branches pale green and spreading, prickly . . . . *L. dendroideum*

Lycopodium annotinum L.
STIFF GROUND-PINE   native / cons 7
Spinulum annotinum (L.) Haines

Stems elongated, prostrate, mostly unbranched, rooting at intervals; leaves uniform but the lower leaves turned upward. Erect stems simple to forked several times, increasing annually to 20 cm or more in height. Leaves 8-ranked, more or less stiff and hard, linear-subulate to linear-oblance-shaped, with a sharp spinule. Strobili sessile at the ends of leafy stems. — Moist woods and clearings, subalpine forests, and exposed rocky and peaty habitats.

Lycopodium clavatum L.
RUNNING GROUND-PINE   native / cons 6

Stems elongated, horizontal on the surface of the ground, forking, rooting at intervals; leaves uniform, but lower leaves turned upward. Erect branches at first simple, becoming dichotomous; fertile branches with a leafy-bracted peduncle bearing 2 to several sessile or short-stalked strobiles. Leaves linear-subulate, incurved-spreading, usually tipped with a soft white hair-like bristle. Bracts of strobili yellow, fimbriate-erose, at least the lower with white filiform tips. — Dry woods and clearings.

Mature fruiting plants present no problems in identification; young or sterile plants sometimes confused with *L. annotinum*. The extended, soft, hair-like bristles on the leaf tips are useful for identification.

Lycopodium dendroideum Michx.
TREE GROUND-PINE   native / cons 7
Dendrolycopodium dendroideum (Michx.) Haines
Lycopodium obscurum L. var. *dendroideum* (Michx.) D.C. Eat.

Subterranean stems creeping, branching, and rhizome-like, with broad scale-like leaves; aerial stems upright, 10-30 cm high, simple below, fork-

Lycopodium annotinum

ing above, constricted between the seasonal growth. Lower leaves strongly divergent; leaves of lateral branchlets in 2 dorsal, 2 ventral, and 2 lateral ranks; leaves strongly decurrent, the free part linear-attenuate. Strobili sessile and terminal on the main axis, or dominant branches and produced in the second, third, or fourth growing season. — Woods and clearings.

Quickly identified by grasping the base of an aerial stem; this will feel distinctly prickly because of the stiff divergent leaves.

Lycopodium hickeyi W.H. Wagner, Beitel & Moran
PENNSYLVANIA GROUND-PINE　　　　　　　　　　*native / cons* 7
　Dendrolycopodium hickeyi (W.H. Wagner, Beitel & R.C. Moran) A. Haines
　Lycopodium obscurum L. var. *isophyllum* Hickey
Similar to *L. obscurum* in that leaves of lower portion of stem are strongly appressed to slightly divergent; leaves of branchlets are all of equal size and linear-attenuate; all leaves lie in planes tangential to the branchlet axis. — Woodlands.

Lycopodium dendroideum

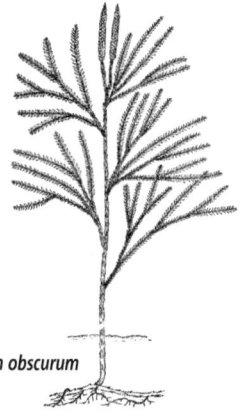

Lycopodium obscurum L.
PRINCESS-PINE　　　　　　　　　　　　　　　*native / cons* 7
　Dendrolycopodium obscurum (L.) Haines
Similar to L. dendroideum, from which it may be distinguished by the strongly appressed to slightly divergent leaves on the lower portion of the aerial shoot. Leaves of the lateral branchlets arranged in 1 dorsal, 1 ventral, and 4 lateral ranks; leaves of ventral rank linear-attenuate to long triangular, smaller than leaves of other ranks; leaves of other ranks linear-acuminate to linear-acute. — Woods.

Lycopodium obscurum

Onocleaceae

SENSITIVE FERN FAMILY

Large coarse ferns with creeping hairy rhizomes (*Onoclea*) or with stolons on ground surface (*Matteuccia*); sterile and fertile fronds strongly different, the sterile fronds deciduous, pinnatifid to 1-pinnate-pinnatifid; fertile fronds persistent. Sori enclosed under recurved margin of pinna segment (outer false indusium) and a tiny true inner indusium (membranous or of hairs).

1　Sterile blades solitary from creeping rhizomes, deeply divided into lobes (or the lowermost divisions pinnae). . . . . *Onoclea*
1　Sterile blades in a circle from a thick crown; pinnate with lobed pinnules. . . . . *Matteuccia*

Matteuccia
OSTRICH FERN

Matteuccia struthiopteris (L.) Todaro
OSTRICH FERN　　　　　　　　　　　　　　　*native / cons* 5
Large, colony-forming fern; rhizomes deep and long-creeping, black, scaly, producing erect leafy crowns. Sterile leaves upright, 1-pinnate, to 2 m tall and 15-50 cm wide; blades much longer than petioles, abruptly narrowed to tip, gradually tapered to base, stems ± hairy; each pinnae deeply divided into 20 or more pairs of pinnules, these 3-6 mm wide at base and rounded at tip; veins not netlike. Fertile leaves stiff and erect within a circle of sterile leaves, green at first, turning brown or black, much shorter than sterile leaves (to 6 dm tall), produced in mid to late summer and often persisting into following year; fertile blades 1-pinnate, pinnae upright or appressed, 2-6 cm long and 2-4 mm wide, the margins inrolled and covering the sori; indusia with a jagged margin. — Wet and swampy woods, streambanks, seeps, ditches.

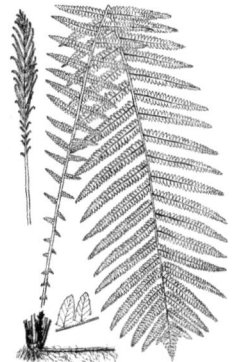

Matteuccia struthiopteris

Onoclea
SENSITIVE FERN

Onoclea sensibilis L.

SENSITIVE FERN   *native / cons* 5

Medium fern, in clumps of several leaves, spreading by branching rhizomes and forming large patches. Leaves upright, with petioles about as long as blades. Sterile leaves deciduous, 1-pinnate at base, deeply cleft upward; the stem broader-winged toward the tip; blades 15-40 cm long and 15-35 cm wide, with 8-12 pairs of opposite pinnae, these deeply wavy-margined or coarsely toothed, 1-5 cm wide, with scattered white hairs on underside veins, the veins joined and netlike. Fertile leaves produced in late summer and persisting over winter, shorter than sterile leaves; fertile blades 1-pinnate, pinnae upright, divided into beadlike pinnules with inrolled margins covering the sori; veins not joined. Sori round and covered by a hoodlike indusia, becoming dry and hard. — Swampy woods and low places in forests, wet meadows, calcareous fens, roadside ditches, wet or moist wheel ruts; sometimes weedy.

Leaves of *Onoclea* susceptible to damage from even light frosts, hence the common name of sensitive fern.

Onoclea sensibilis

Ophioglossaceae

ADDER'S-TONGUE FAMILY

Perennial herbs from short, erect rhizomes having several fleshy roots. Plants produce one leaf each year on a single stalk (stipe), with bud for next year's leaf at base of stipe. Leaves divided into a fertile segment (sporophyll) and a sterile expanded blade. Sterile blades entire (*Ophioglossum*), or lobed or 1-3x pinnately divided (*Botrychium, Botrypus, Sceptridium*). Spores in numerous round sporangia borne on simple or branched fertile blades.

1 Sterile blades simple, entire; veins netlike; sporangia embedded in rachis of spike. . . . *Ophioglossum*
1 Sterile blades pinnately lobed or dissected; veins forked; sporangia exposed, often on a branched structure 2
2 Sterile blades somewhat leathery, persisting over winter; blades with distinct stalk, usually 5-25 cm long in fertile plants; fertile portion of frond joining sterile portion at or near ground level . . . .
   . . . . *Sceptridium*
2 Plants deciduous, withering in fall; blades usually unstalked, large (more than 5 cm long) or much smaller (in some species of *Botrychium*), herbaceous or sometimes fleshy; fertile portion of frond joining sterile portion well above ground level. 3
3 Sterile blades triangular, 3–4x pinnate, stalkless and mostly 5–25 cm wide; fertile portion erect, appearing to be a continuation of stipe. . . . . *Botrypus*
3 Sterile blades short-triangular, oblong, or linear; lobed (simple) to 3-pinnate (usually 1-pinnate to 2-pinnate-pinnatifid), mostly 1-5 cm wide; fertile portion of blade upright or spreading . . . *Botrychium*

Botrychium
GRAPE-FERN, MOONWORT

Mostly small plants with one leaf, the blade divided into sterile and fetile segments. Sterile portion of blade pinnately divided or lobed, fertile portion branched to form a panicle bearing the sporangia.

1 Sterile blade (trophophore) simple to lobed, lobes rounded to square and angular, stalks usually 1/2 to 2/3 length of sterile blade; often in open grassy fields . . . . *B. simplex*
1 Sterile blade pinnately lobed (either if actual pinnae or simply lobed), lobes of varying shapes, stalk usually less than 1/4 length of sterile blade; plants often in open sunny places; dunes, streambanks, roadsides, on trails and openings in forests, etc. 2
2 Basal pinnae or segments of sterile blade with venation like the ribs of a fan; midrib absent 3
2 Basal pinnae or segments of sterile blade with pinnate venation; midrib present . *B. matricariifolium*

3　Basal pinnae broadly fan-shaped (almost perfect half moons) with narrow stalks......... *B. lunaria*
3　Basal pinnae narrowly fan- or wedge-shaped to nearly linear 4
4　Sterile blade at least partially folded longitudinally when alive (conduplicate), usually not more than 4 cm long by 1 cm wide; pinnae up to 5 pairs; basal pinnae usually 2-parted .... *B. campestre*
4　Sterile blade flat or folded only at base when alive, usually up to 10 long by 2.5 cm wide; pinnae up to 10 pairs; basal pinnae unlobed, or if lobed, not usually 2-parted 5
5　Sterile blade narrowly oblong (sterile blade widest above base), firm to herbaceous; pinnae fan-shaped, margins shallowly crenate... *B. minganense*
5　Sterile blade narrowly deltate (sterile blade widest at lowest pinna pair); pinnae spatulate to linear spatulate, margins entire to very coarsely and irregularly dentate................ *B. spathulatum*

Botrychium campestre W.H. Wagner & Farrar

IOWA MOONWORT　　　　　　　*endangered / native / cons* **10**

Small plants, sometimes less than 5 cm tall and hard to see if surrounded by taller vegetation. Plants succulent, developing early in growing season, with aboveground portion drying by mid-summer. Sterile blade fleshy, once pinnate, sessile to the common stalk, oblong in outline, longitudinally folded; to about 4 cmlong and 1.3 cm wide; usually divided into 5 pairs of linear or linear-spatulate segments; margins crenate or dentate and also usually notched into 2 or several smaller segments. Common stalk short, less than 3 cm long. Sporophore often large relative to size of sterile blade. — Inconspicuous in prairies, dunes, and fields over limestone. Leaves appear in early spring and wither in late spring and early summer, long before those of other moonworts.

Botrychium lunaria (L.) Sw.

COMMON MOONWORT　　　　　　*endangered / native / cons* **9**
　Botrychium neolunaria Stensvold & Farrar, sp. nov. ined.

Plants 3-20 cm tall, rubbery-textured, appearing in late spring and withering in summer. Leaf blades 1.5-7 cm long and 1-3 cm wide, stalkless or on a short stalk to 5 mm long; pinnately divided into 3-6 pairs of stalkless pinnae, the pinnae fan-shaped, wider than long and without a midrib; petioles 1.5-3 cm long. Fertile segments 0.5-7 cm long, on stalks about as long as the segments. — Cool, moist sandy soils in woods.

　Moonwort has a long and illustrious history in early herbals; the "seeds" could reputedly make one invisible or could be used to unlock doors. Fully mature plants are distinctive, but plants somewhat similar to *B. minganense*.

Botrychium matricariifolium (A. Braun ex Dowell) A. Braun ex Koch

DAISY-LEAF MOONWORT　　　　　　　　*native / cons* **5**

Plants to about 30 cm tall, membranous to fleshy. Leaf blades narrowly deltoid to ovate, short-stalked, inserted above the middle, pinnatifid to bipinnate-pinnatifid; segments of blade blunt and usually toothed. Fertile segment paniculate. Spores mature in June and July. — Acidic soil in old sandy and sterile fields, dry wooded slopes, rocky woods, moist cedar woods, and rich swamps.

　This species is somewhat larger than *B. simplex*. The shape of the blade is variable (deltoid to ovate) but it is stalked, and the toothed segments are distinctive.

Botrychium minganense Victorin

MINGAN MOONWORT　　　　　　　　　*native / cons* **8**

Plants to 30 cm long, somewhat membranous. Leaf blades narrowly oblong, sessile or nearly so, inserted below the middle, pinnate or occasionally pinnate-pinnatifid at the base; segments of blades opposite, obovate, rhomboidal or oblong, frequently incised, remote. Fertile segment paniculate. Spores mature in July and August. — Moist hardwood forests, aspen-balsam fir woods, and old clearings; soils mostly circumneutral.

Botrychium campestre

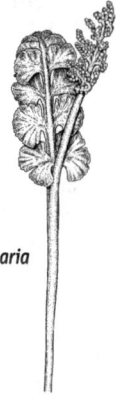

Botrychium lunaria

Botrychium minganense

Distinguished from *B. lunaria* by its yellowish green hue and by its trough-shaped sterile segments, which are ascending rather than at right angles to the stalk and which rarely overlap with each other.

Botrychium simplex E. Hitchc.
LEAST MOONWORT *native / cons* **6**
Plants to ca. 15 cm tall, rather fleshy. Blades simple, lobed or pinnately divided, inserted at the base or towards the middle; segments of blade oblong, rhomboid or kidney-shaped, and usually overlapping, with the basal segments occasionally pinnatifid. Fertile segment simple or compound. Spores mature in late May and June. — Pastures, meadows, lakeshores, and gravelly slopes; easily overlooked in the field due to its small size and grassy habitat.

Botrychium spathulatum W.H. Wagner
SPOON-LEAF MOONWORT *native / cons* **10**
Leaf single, erect, to 12 cm long; shiny yellowish-green, leathery. Sterile blade sessile or short-stalked (less than 1 mm long); pinna pairs mostly 4-5 (7), spoon- or fan-shaped, widest at tip; lowest pinnae largest, commonly folded over rachis; pinnae mostly widely spaced and not overlapping, outer pinna margins entire or lobed. Sporophore 1-2x length of the trophophore; 1-2 times pinnately divided into segments bearing the sporangia. Leaves appearing late spring through summer. — Sand dunes, beaches; in Wisconsin, known only from Door County.

Botrychium simplex

Botrypus
RATTLESNAKE FERN

Botrypus virginianus (L.) Holub
RATTLESNAKE FERN *native / cons* **6**
  Botrychium virginianum (L.) Sw.
Plants 40-75 cm tall, appearing in spring, withering in autumn, not overwintering. Blade (trophophore) broadly triangular, sessile, to 25 cm long and to 1.5x as wide, 3-4x pinnate, thin and herbaceous. Pinnae to 12 pairs, usually somewhat overlapping and slightly ascending; pinnules lance-shaped and deeply lobed, the lobes linear, sharply toothed and pointed at tip. Spore-bearing portion (sporophore) 2-pinnate, 0.5-1.5x length of trophophore. — Occasional in swamps of cedar and black spruce; more common in moist to fairly dry deciduous woods. Rattlesnake fern is widespread in North America, occurring across Canada and most of the USA.

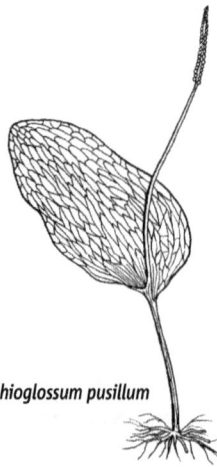

Botrypus virginianus

Ophioglossum
ADDER'S-TONGUE

Ophioglossum pusillum Raf.
NORTHERN ADDER'S-TONGUE *native / cons* **8**
Plants erect, 7-30 cm tall, from slender rhizomes. Leaves 1, entire, on a stalk 3-15 cm long; blades upright, oval to ovate, rounded to acute at tip, 3-8 cm long and 1-4 cm wide, conspicuously net-veined. Sporangia in 2 rows in a terminal, unbranched fertile segment, 1-5 cm long and 2-4 mm wide, on a stalk 6-15 cm long. — Wet sandy meadows and prairies, moist depressions, wetland margins, sandy beaches.

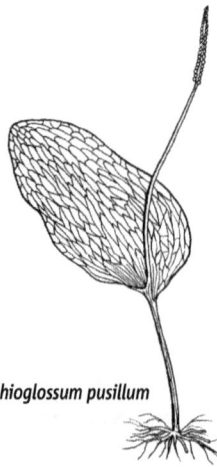

Ophioglossum pusillum

Sceptridium
GRAPE FERN
Small to medium leathery ferns found in a variety of moist to dry, open to shaded habitats, often where sandy. Sterile blades dissected, winter-green; sporophore short-lived, withering by late summer. Sterile blade and sporophore joined near or below ground level.

1 Segments of sterile blade rounded at base; symmetrically tapered to an often ± blunt or even rounded apex; larger segments mostly 9–17 mm long; margins nearly entire or finely and inconspicuously toothed . . . . ***S. multifidum***
1 Segments of sterile blade usually (obliquely) asymmetrical and angular, cuneate to the apex; larger segments mostly 4–9 mm long; margins clearly finely dentate, especially visible in immature leaves . . . . ***S. rugulosum***

Sceptridium multifidum (Gmel.) Nishida ex Tagawa
LEATHERY GRAPE FERN   *native / cons* **4**
   Botrychium multifidum (Gmel.) Trev.

Leaves to 20 cm long; stem and blade leathery; blades evergreen, long-petioled, 3-parted, attached near the base of the plant; ultimate segments of blade crowded, sometimes imbricate, ovate, more or less the same size, obtuse or somewhat acute. Fertile segment paniculate. Spores mature in Aug and Sept. — Grassy hillsides, sterile fields, exposed meadows, and sandy open places.

Sceptridium rugulosum (W.H. Wagner) Skoda & Holub
TERNATE GRAPE FERN   *native / cons* **8**
Botrychium rugulosum W.H. Wagner

Leaves 25 cm long or longer, thin and membranous; blades inserted at the base, 3-parted, with the three major divisions stalked; ultimate segments of blade all about the same size, ovate to oblong, acutish, serrate or entire, and concave. Fertile segment paniculate. Spores mature Aug-Oct. — Swampy woods, brushy fields, and wooded streambanks.

Sceptridium multifidum

Osmundaceae

ROYAL FERN FAMILY

Osmunda

ROYAL FERN

Perennial ferns with large rootstocks and exposed crowns covered with old roots and stalks, sending up tufts of coarse leaves. Leaves 1-2-pinnate, differentiated into sterile and fertile segments. Sporangia in round clusters, spores green.

The fibrous roots (osmunda fibre) were formerly used as a medium for growing orchids and bromeliads.

1 Leaves 2-pinnate, pinnae entire; sporangia on upper half of fertile leaves . . . . ***O. regalis***
1 Leaves 1-pinnate, sterile pinnae deeply cleft; sporangia only near middle of fertile leaves, or fertile and sterile leaves separate. 2
2 Fertile and sterile leaves separate, fertile leaves cinnamon-colored, sterile leaves with a tuft of wool in axil of pinnae . . . . ***O. cinnamomea***
2 Fertile pinnae near middle of vegetative leaves, with sterile pinnae above and below fertile portion, fertile portion green-black, pinnae mostly without tuft of wool in axil. . . . . ***O. claytoniana***

Osmunda cinnamomea L.
CINNAMON-FERN   *native / cons* **7**
   Osmundastrum cinnamomeum (L.) K. Presl

Large clumped fern, to 1 m or more tall. Blades of sterile leaves to 30 cm wide, gradually tapered to tip, 1-pinnate, with conspicuous tuft of white or brown woolly hairs at base of each pinna, pinnae stalkless and deeply cleft into segments, with fringe of short hairs on margins; petioles densely hairy when young. Fertile leaves at center of crown, surrounded by taller sterile leaves, without leafy tissue, arising in spring or early summer and turning cinnamon brown, withering and inconspicuous by midsummer. — Swamps, bog-margins, wooded stream-banks, and low wet places; soils acid.

Osmunda cinnamomea

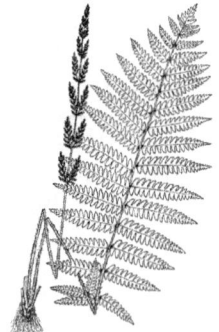

Osmunda claytoniana L.
INTERRUPTED FERN   *native / cons* **6**

Clumped fern to 1 m or more tall; often forming large colonies. Outer leaves usually sterile, inner leaves larger and with 2-5 pairs of fertile pinnae in middle of blade; fertile segments to 6 cm long and 2 cm wide and much smaller than vegetative segments above and below them; sporangia clusters at first green-black, turning dark brown and withering. Blades 4-10 dm long and 15-30 cm wide; pinnae stalkless and deeply cut into segments, with smooth or slightly hairy margins. Petioles covered with tufts of woolly hairs when young, becoming smooth or sparsely hairy with age, the hairs not forming tufts at pinna-bases (as in *O. cinnamomea*). — Moist or seasonally wet depressions in forests, hummocks in swamps, low prairie, wet roadsides; often in drier places than *Osmunda cinnamomea* or *O. regalis*.

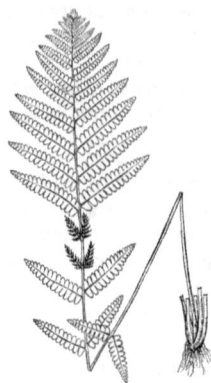

Osmunda claytoniana

Osmunda regalis L.
ROYAL FERN   *native / cons* **7**
  Osmunda spectabilis Willd.

Large fern to 1 m or more tall. Blades broadly ovate in outline, 4-8 dm long and to 3-5 dm wide, 2-pinnate into ± opposite divisions (pinnules), these well-spaced, oblong, rounded at tips, with entire or finely toothed margins. Fertile leaves with uppermost several pinnae replaced by sporangia clusters. Petioles smooth, green or red-green, to 3/4 length of blade. — Bogs, swamps, alder thickets and shallow pools; soils usually acidic.

Osmunda regalis

Polypodiaceae
POLYPODY FERN FAMILY

Polypodium
POLYPODY

Polypodium virginianum L.
ROCK POLYPODY   *native / cons* **9**
  Polypodium vulgare auct. non L. p.p.

Evergreen, colony-forming fern. Leaf blades to 50 cm long from a creeping rhizome; oblong lance-shaped; pinnatifid, leathery; veins free. Sori round, midway between the midvein and margin, and occurring on the upper segments; indusia absent. — In shallow humus on rocks, in crevices, on woodland banks, and rarely on mossy stumps and in crotches of trees.

  Easily identified by the small evergreen blades and its colony-forming habit on rocky slopes, talus, boulders, and ledges.

Polypodium virginianum

Pteridaceae
MAIDENHAIR FERN FAMILY

Delicate to coarse ferns, deciduous, or evergreen. Blades pinnate to decompound. Sori marginal, protected by the indusium, which opens toward the margin, or by the reflexed margins of the pinnae, or borne along the veins and lacking an indusium.

1 Blade segments separate from one another . . . . *Adiantum*
1 Blade segments not separate and distinct 2

2 Leaves of two types, the fertile much longer than the sterile; petioles dark brown near base, green above . . . . *Cryptogramma*
2 Fertile and sterile leaves mostly similar; petioles dark brown to black . . . . *Pellaea*

Adiantum
MAIDENHAIR FERN
Adiantum pedatum L.
NORTHERN MAIDENHAIR FERN   *native / cons* 7

Leaves 30-60 cm tall, in colonies arising from horizontal rhizomes. Petioles lustrous purple-brown, forking at the summit into two arching rachises, each of which is divided several times, thus forming a semicircular blade 15-35 cm wide or wider. Pinnules short-stalked, obliquely triangular oblong; terminal pinnule fan-shaped; main vein along the lower margin; upper margin cleft, with lobes thus formed blunt. Sori elongate, borne on the upper margins of the lobes of the pinnules; indusium formed by the inrolled margin. — Wooded, sometimes rocky slopes in humus-rich soil. The usually arching and palmately divided lustrous purple brown rachises and fan-shaped pinnules with the main vein along the lower margin are distinctive.

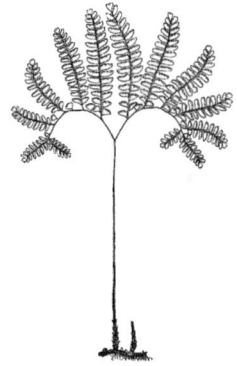
Adiantum pedatum

Cryptogramma
ROCKBRAKE
Cryptogramma stelleri (Gmel.) Prantl
FRAGILE ROCKBRAKE   *native / cons* 10

Small rock fern with dimorphic leaves, from short much branched rhizomes, or the rhizomes sometimes elongate. Leaves glabrous, deciduous, dimorphic, scattered along the horizontal rhizome. Sterile leaves almost flaccid, 3-10 cm long; blades ovate to ovate-deltoid, bipinnate; pinnules oblong or ovate; petioles purplish. Fertile leaves stiffer than sterile blades, 9-21 cm long; pinnules lance-shaped to oblong. Sori marginal, covered by a continuous indusium formed by the reflexed margin. — Moist, shaded, usually calcareous crevices and cliffs. Plants may be easily overlooked as they turn brown later in the season.

Cryptogramma stelleri

Pellaea
CLIFFBRAKE
Small tufted plants from compact rootstocks. Blades firm; petioles and rachises wiry; pinnae gray green; veins free. Sori marginal and confluent under the inrolled and altered margin of the fertile pinnules. Plants gray-green in color, blending well with the limestone rock crevices and ledges with which they are associated.

Pellaea glabella Mett.
SMOOTH CLIFFBRAKE   *native / cons* 10

Leaves similar, 10-25 cm long or longer, usually shorter than those of *P. atropurpurea*, open and spreading out beyond the rock face. Petioles and rachis dark reddish brown, smooth, and lustrous. Pinnae rigid, evergreen, bluish green, simple above, pinnate below; basal pinnae persistent; pinnules sessile or nearly so, oblong lance-shaped. Sori situated around the margins of the fertile pinnules; indusium formed by the inrolled margin of pinnule. — Crevices of dry, sometimes partly shaded, limestone cliffs.

Smooth cliffbrake is a distinctive species of high, steep limestone cliffs; it grows from small crevices and blends well with the background; often there are no other plants associated with it.

Pellaea glabella

Selaginellaceae

SELAGINELLA FAMILY

Selaginella
SPIKEMOSS
Trailing, evergreen herbs with branched, leafy stems, rooting at branching points. Leaves small and overlapping. Spore-bearing leaves similar to vegetative leaves and clustered in cones at ends of branches. Megaspores 4 in each sporangium, yellow or white; microspores numerous and very small, red or yellow, covered with small spines.

1 Leaves in four rows, of two kinds: large and spreading, and small and appressed to stem; cones 0.5–1 cm long . . . . *S. eclipes*
1 Leaves in many rows, all alike, with hairs on margins; fertile branches upright, cones 2–4 cm long . . . . *S. selaginoides*

Selaginella eclipes W.R. Buck.
HIDDEN SPIKEMOSS   *native / cons 7*
   Selaginella apoda subsp. *eclipes* (W.R.Buck) Skoda
Plants forming large, yellow-green mats of branching, trailing stems with upright tips. Stems slender, to 0.4 mm wide. Leaves scalelike, in 4 rows, of 2 types, the larger leaves spreading, 1-2 mm long and 1 mm wide; the smaller leaves appressed to stem, up to 1 mm long and 0.5 mm wide. Cones 0.5-2 cm long, cylindric in 4 rows, the sporophylls similar to lateral leaves and slightly larger. Megaspores white, with a netlike surface. — Open fens, wet meadows, sandy or marly lakeshores and riverbanks; especially where calcium-rich.

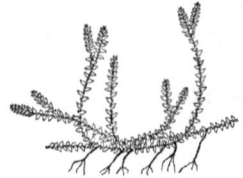

Selaginella eclipes

Selaginella selaginoides (L.) Link
NORTHERN SPIKEMOSS   *endangered / native / cons* **10**
Trailing, evergreen plants forming small mats. Stems branched, leafy, rooting at branching points. Sterile stems prostrate, 2-5 cm long; fertile stems upright, deciduous, 5-10 cm high and 0.5 mm wide (stem only), changing upward into broader sporophylls. Leaves overlapping in multiple spiral rows, all alike, 2-4 mm long and 1 mm wide, with sharp tips and sparsely hairy margins; spore-bearing leaves similar to vegetative leaves and clustered in cones at ends of branches; cones ± cylindric but with 4 rounded angles, 1.5-3 cm long and to 5 mm wide. Megaspores 4 in each sporangium, yellow-white, with low rounded projections on the 3 flat surfaces. — Calcium-rich fens and fen-like shores of Door County; mossy hummocks in cedar swamps.

Selaginella selaginoides

Thelypteridaceae

MARSH FERN FAMILY
Medium-sized deciduous ferns, spreading by rhizomes to form colonies; sterile and fertile fronds usually alike, 1-pinnate to pinnate-pinnatifid, with transparent needle-like hairs; sori usually on veins (but not marginal) on pinna underside; indusia present and often soon withering, or absent (*Phegopteris*).

1 Leaf blades broadly triangular in outline, broadest at base, lowermost pinnae directed downward; indusia absent . . . . *Phegopteris*
1 Blades lance-shaped in outline, broadest above base; indusia present. . . . . *Thelypteris*

Phegopteris
BEECH-FERN
Deciduous ferns with creeping rhizomes. Leaf blades triangular in outline.

Phegopteris connectilis (Michaux) Watt
NORTHERN BEECH-FERN   *native / cons* **7**
   Dryopteris phegopteris (L.) C. Chr.
   Thelypteris phegopteris (L.) Sloss
Fern with long, slender, scaly and densely hairy rhizomes. Leaves triangular, 15-25 cm long and 6-15 cm wide; blades 1-pinnate, the pinnalike divisions joined by a wing along rachis, except for lowermost pair which are free and angled downward; pinnules oblong, rounded at tip, and usually hairy; petioles longer than blades, hairy, with narrow, brown scales. — Cool moist woods, thickets, streambanks, sphagnum moss hummocks, shaded rock crevices.

Phegopteris connectilis

Thelypteris
MARSH-FERN
Small to medium ferns from slender rhizomes.

Thelypteris palustris Schott
MARSH-FERN   *native / cons* **7**
Leaves deciduous, ± hairy, erect, 20-60 cm long and to 15 cm wide; blades broadly lance-shaped, short-hairy on rachis and midveins, tapered to tip and only slightly narrowed at base; 1-pinnate, pinnae in 10-25 pairs, mostly alternate, narrowly lance-shaped, to 2 cm wide. Sterile and fertile leaves only slightly different; sterile leaves thin and delicate, pinnules blunt-tipped, 3-5 mm wide, veins once-forked. Fertile leaves longer than sterile leaves; pinnules oblong, 2-4 mm wide, the margins rolled under, veins mostly 1-forked; petioles longer than blades, black at base, hairless and without scales. Sori round, located halfway between midvein and margin, sometimes partly covered by the rolled under margin; indusia irregular in shape, usually with a fringe of hairs. — Swamps, low areas in forests, sedge meadows, open bogs, calcareous fens, marshes.

Thelypteris palustris

CONIFERS

Cupressaceae
CYPRESS FAMILY
Trees or shrubs; leaves opposite or whorled, sometimes dimorphic, separated by short internodes or overlapping. Flowers monoecious or dioecious, solitary, axillary or terminal. Staminate flowers of several stamens with short filaments. Pistillate flowers with opposite or whorled carpels or with 1-3 terminal ovules. Fruit a cone, or becoming fleshy and berry-like.

1 Shrubs; cones berry-like and fleshy . . . .*Juniperus*
1 Trees; cones woody or leathery . . . .*Thuja*

Juniperus
JUNIPER
Trees or shrubs; leaves evergreen, scale-like or subulate, opposite or in whorls of 3. Flowers monoecious or dioecious, axillary and short-stalked, or terminal. Staminate flowers catkin-like; stamens numerous, opposite or whorled. Pistillate flowers of several scales, the lower sterile, the terminal sometimes fertile, or sometimes sterile and the ovules terminal. Cone-scales at maturity becoming fleshy and coalescent, forming an indehiscent, usually colored, berry-like fruit with 1-10 seeds; seeds plump, wingless.

1 Leaves in clusters of three, linear and sharp-pointed, jointed at base; cones in leaf axils . *J. communis*
1 Leaves scale-like, not jointed; cones at end of branches . . . . *J. horizontalis*

Juniperus communis L.

COMMON JUNIPER   *native / cons* 3

Evergreen shrub. Leaves in whorls of 3, crowded, linear, sharp-pointed, 6-15 mm long, marked with a median white stripe above. Fruit bluish or black, 6-13 mm in diameter, normally 3-seeded. Ours var. depressa Pursh., with branches soon becoming decumbent and forming large circular patches eventually several meters in diameter, usually flat-topped and 0.5-2 m tall; leaves spreading or ascending. — Dry woods, old fields, dried bogs, rocky bluffs.

Juniperus horizontalis Moench

CREEPING JUNIPER   *native / cons* 9

Evergreen shrub; branches prostrate, often greatly elongate, bearing numerous erect branchlets 1-3 dm tall. Leaves mostly scale-like and appressed, varying from ovate, 1-2 mm long, to oblong and to 4 mm long. Fruit blue, 5-8 mm in diameter, on short recurved pedicels; seeds 3-5 mm long, not pitted. — On rocks, sandy openings, sandy or gravelly shores, sand dunes and beach ridges along Lake Michigan.

Juniperus communis

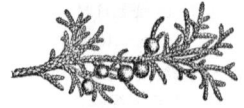

Juniperus horizontalis

Thuja

ARBOR-VITAE

Thuja occidentalis L.

NORTHERN WHITE CEDAR, ARBOR-VITAE   *native / cons* 9

Shade-tolerant tree to 20 m tall, cone-shaped with widely spreading branches, sometimes layered at base, trunk to 1 m wide or more, bark reddish or gray-brown, in long shreddy strips; twigs flattened, in fanlike sprays. Leaves scale-like and overlapping, 3-6 mm long and 1-2 mm wide, yellow-green, aromatic, persisting for 1-2 years. Seed cones small, brown, 1 cm long, maturing in fall and persisting over winter. — Cold, poorly drained swamps where *Thuja* may form dense stands; soils neutral or basic, usually highly organic, water not stagnant; also along streams, on gravelly and sandy shores of Lake Michigan, and dry soils over limestone.

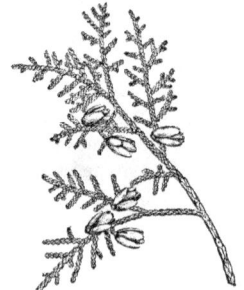

Thuja occidentalis

Pinaceae

PINE FAMILY

Resinous trees with evergreen or deciduous, needlelike leaves. Male and female cones separate but borne on same tree. Male cones small and soft, falling after pollen is shed. Female cones larger, with woody scales arranged in a spiral. Seeds on upper surface of scales.

1 Leaves grouped into clusters 2
1 Leaves not in clusters, alternate on branches 3
2 Leaves evergreen, in clusters of 2–5 needles . . . . *Pinus*
2 Leaves deciduous, with many leaves in each cluster . . . . *Larix*
3 Cones upright; leaves attached directly to branch, not leaving a bump when shed . . . . *Abies*
3 Cones drooping; leaves attached to a persistent short stalk 4
4 Leaves flat in cross-section, soft . . . . *Tsuga*
4 Leaves four-sided in cross-section, stiff . . . . *Picea*

Abies

FIR

Abies balsamea (L.) Miller

BALSAM FIR   *native / cons* 5

Shade-tolerant tree to 25 m tall, crown spirelike, trunk to 6 dm wide; bark thin, smooth and gray, becoming brown and scaly with age; lower branches often drooping; twigs sparsely short-hairy. Leaves evergreen, linear, 12-25 mm long and 1-2 mm wide, blunt or with a small notch at tip, flat in cross-section, twisted at base and arranged in 1 plane (especially on lower branches), or spiraled on twigs. Seed cones 5-10 cm long and 1.5-3 cm wide, with broadly rounded scales. — Cold boreal forests, swamps, and moist forests.

Abies balsamea

Larix
LARCH

Larix laricina (Duroi) K. Koch
TAMARACK, EASTERN LARCH   *native / cons* **8**
Shade-intolerant tree to 20 m tall, crown narrow, trunk to 6 dm wide; bark smooth and gray when young, becoming scaly and red-brown; twigs yellow-brown, ± horizontal or with upright tips. Leaves deciduous, in clusters of 10-20, linear, 1-2.5 cm long and less than 1 mm wide, soft, blunt-tipped, bright green, turning yellow in fall. Seed cones 1-2 cm long and 0.5-1 cm wide, ripening in fall and persisting on trees for 1 year. — Cold, poorly drained swamps, bogs and wet lakeshores, wet depressions.

Larix laricina

Picea
SPRUCE
Evergreen trees; bark thin and scaly, resin blisters common in white spruce (Picea glauca). Leaves linear, square in cross-section, stiff, spreading in all directions around twig, jointed at the base to a short projecting sterigma which persists on the leafless branches. Cones borne on last year's branches, drooping, with persistent scales much exceeding the bracts. Seeds wing-margined.

Several species are commonly cultivated, especially Norway spruce (*Picea abies*).

1 Leaves mostly 0.5–1.2 cm long, blunt-tipped; twigs with rust-colored hairs; cones ovoid, less than twice as long as wide when open . . . . *P. mariana*
1 Leaves mostly 1–2 cm long, sharp-tipped; twigs glabrous; cones more than twice as long as wide . . . . *P. glauca*

Picea glauca (Moench) Voss
WHITE SPRUCE   *native / cons* **7**
Moderately shade-tolerant tree to 30 m tall (often smaller), crown conelike, trunk to 60 cm or more wide; bark thin, gray-brown; branches slightly drooping, hairless. Leaves evergreen, linear, 1.5-2 cm long, 4-angled in cross-section, stiff, waxy blue-green, sharp-tipped. Cones 2.5-6 cm long, scales fan-shaped, rounded at tip, the tip entire.
— Moist to sometimes wet forests; absent from wetlands where water is stagnant.

Picea glauca

Picea mariana (Miller) BSP.
BLACK SPRUCE   *native / cons* **8**
Moderately shade-tolerant tree to 25 m tall (often smaller), crown narrow, often clublike at top, trunk to 25 cm wide; bark thin, scaly, gray-brown; branches short and drooping, often layered at base. Leaves evergreen, linear, 6-18 mm long, 4-angled in cross-section, stiff, waxy blue-green, mostly blunt-tipped. Seed cones 1.5-3 cm long, scales irregularly toothed, persisting for many years. — Cold, acid, sphagnum bogs, swamps, and lakeshores; often where water is slow-moving and low in oxygen; less common in calcium-rich, well-aerated swamps dominated by northern white cedar (*Thuja occidentalis*).

Picea mariana

Black spruce (*Picea mariana*) can be distinguished from white spruce (*Picea glauca*) by its shorter needles, the branches with fine, white to red-brown hairs, the smaller, rounded seed cones with toothed scale margins, and its occurrence in generally wetter (and sometimes stagnant) habitats.

Pinus
PINE

Trees (ours) with dimorphic branches and leaves, the foliage leaves borne on dwarf branches only, solitary or in clusters of 2-5. Staminate flowers catkin-like, in fascicles at the base of the current year's growth, each composed of numerous spirally imbricate stamens. Pistillate flowers forming a cone consisting of numerous spirally imbricate cone-scales, each subtended by a bract and bearing 2 inverted ovules at the base. Fruit a hard woody cone, maturing at the end of the second or third season and often long persistent on the tree; seeds winged.

1 Leaves in clusters of 5; cones 10–25 cm long . . . . *P. strobus*
1 Leaves in clusters of 2; cones less than 10 cm long 2
2 Leaves 10–15 cm long . . . . *P. resinosa*
2 Leaves 2.5–10 cm long . . . . *P. banksiana*

Pinus banksiana Lamb.
JACK PINE   *native / cons* **5**

Usually a small tree, but occasionally to 20 m or more tall, with spreading branches. Leaves in pairs, usually somewhat curved, 2-3.5 cm long, 1-1.5 mm wide. Cones erect or strongly ascending, usually somewhat curved or unsymmetrical, conic, yellowish-brown, 3-5 cm long; seeds about 1.5 cm long. — Dry or sterile, sandy or rocky soil.

Pinus resinosa Soland.
RED PINE   *native / cons* **7**

Tree, occasionally 40 m tall. Leaves in clusters of 2, slender, soft and flexible, 10-15 cm long, dark green. Cones spreading, conic-ovoid, 4-8 cm long; seeds 1.5-2 cm long. — Dry sandy or rocky soil.

Pinus strobus L.
EASTERN WHITE PINE   *native / cons* **5**

Tall tree, occasionally as much as 70 m tall, with thick furrowed bark. Leaves in clusters of 5, very slender, pale green and glaucous, 8-13 cm long. Cones cylindric, 10-15 cm long; seeds, including the wing, 2-3 cm long. — In many different habitats, but preferring well-drained, sandy soil.

Tsuga
HEMLOCK

Tsuga canadensis (L.) Carr.
EASTERN HEMLOCK   *native / cons* **8**

Tree to 30 m tall, the bark often purple-brown; twigs pubescent. Leaves linear, 8-15 mm long, blunt, marked beneath with two white strips of stomata, usually minutely spinulose on the margin (best detected by touch), on short (about 1 mm) petioles, spirally disposed on the twigs but forming a flat spray by twisting of the petioles. Cones at maturity pendulous on short peduncles, their scales much larger than the minute bracts; thickly ellipsoid, 12-20 mm long; seeds winged.
— Moist soil, especially on rocky ridges and hillsides. An important component of our northern forests, useful for lumber and tanbark; often cultivated in a number of horticultural forms.

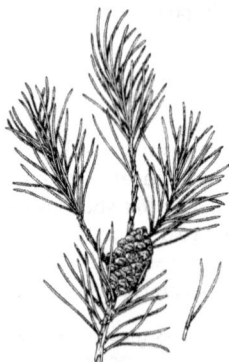
Pinus banksiana

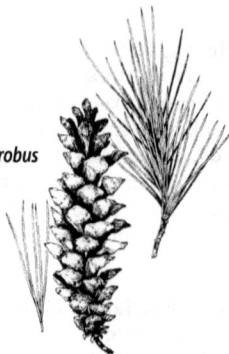
Pinus strobus

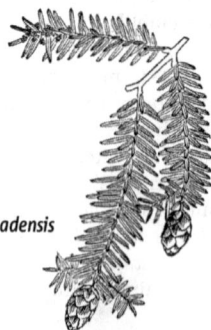
Tsuga canadensis

Taxaceae
YEW FAMILY

Taxus
YEW

Taxus canadensis Marsh.
CANADA YEW   *native / cons* **10**
Straggling evergreen shrub; stems ascending, to 2 m tall. Leaves spirally arranged on the stem, linear, 1-2 cm long, 1-2 mm wide, abruptly narrowed to a sharp point, tapering at base to a poorly defined petiole. Staminate flowers solitary in the axils; pistillate flowers in pairs, each subtended by its pair of scales. Fruit a fleshy red aril, about 5 mm long, open at the top. — Coniferous and mixed woods. A favored winter browse for deer.

Taxus canadensis

DICOTS

Adoxaceae
MUSKROOT FAMILY

Our 2 shrubby genera, *Sambucus* and *Viburnum*, were previously included in Caprifoliaceae.

1 Leaves pinnately compound; fruit with 3 (or more) seed-like pits . . . . *Sambucus*
1 Leaves simple; fruit with only 1 pit . . . . *Viburnum*

Sambucus
ELDER

Shrubs or small trees. Stems pithy, the bark with wartlike lenticels. Leaves pinnately divided. Flowers in large, rounded terminal clusters; small, perfect; calyx lobes tiny or none; corolla spreading, united at base, 5-lobed, white; stamens 5; stigmas 3. Fruit a red or dark purple, berrylike drupe with 3 nutlets.

1 Flowers opening in summer after leaves developed, in broad, nearly flat clusters; fruit purple-black, edible; leaflets usually 7 . . . . *S. canadensis*
1 Flowers opening in late spring with unfolding leaves, in pyramid-shaped or rounded clusters; fruit red, inedible; leaflets usually 5 . . . . *S. racemosa*

Sambucus canadensis L.
COMMON ELDER   *native / cons* **3**
   Sambucus nigra subsp. *canadensis* (L.) R. Bolli
Shrub to 3 m tall; spreading underground and forming thickets. Young stems soft or barely woody, smooth; older stems with warty gray-brown bark; inner pith white. Leaves large, opposite, pinnately divided into 5-11 (usually 7) leaflets, the lower pair of leaflets sometimes divided into 2-3 segments; leaflets lance-shaped to oval, tapered to a long sharp tip, base often asymmetrical, smooth or hairy on underside, especially along veins; margins with sharp, forward-pointing teeth. Flowers small, white, 5-parted, 3-5 mm wide, numerous, in flat or slightly rounded clusters 10-15 cm wide at ends of stems. Fruit a round, purple-black, berrylike drupe, edible. July-Aug (blooming when fruit of *Sambucus racemosa* is about ripe). — Floodplain forests, swamps, wet forest depressions, thickets, shores, meadows, roadsides, fencerows.

Sambucus canadensis

Sambucus racemosa L.

RED-BERRIED ELDER   *native / cons* **5**

  Sambucus pubens Michx.

Shrub to 3 m tall. Stems soft or barely woody; twigs yellow-brown and hairy, branches with warty gray-brown bark; inner pith red-brown. Leaves large, opposite, pinnately divided into 5-7 (usually 5) leaflets, the leaflets lance-shaped to ovate, tapered to a long sharp tip, smooth or hairy on underside; margins with small, sharp, forward-pointing teeth. Flowers small, white, 5-parted, 3-4 mm wide, many, in elongate, pyramidal or rounded clusters at ends of stems, the clusters 5-12 cm long and usually longer than wide. Fruit a round, red, berrylike drupe, inedible. May-June (flowers opening with developing leaves). — Occasional in swamps and thickets; more common in moist deciduous forests, roadsides and fencerows.

Sambucus racemosa

Viburnum

SQUASHBERRY, ARROW-WOOD

Shrubs or small trees. Leaves simple, entire, toothed, often palmately lobed. Flowers white or pink, in rounded clusters at ends of stems, sometimes outer florets larger and sterile. Calyx lobes and corolla lobes each 5; stamens 5; style 1; stigmas 1-3. Fruit a fleshy drupe with a single large seed; white, yellow, pink, or orange at first, maturing to orange, red, or blue-black.

1 Leaves not lobed; pinnately veined 2
2 Leaf underside with branched hairs .... *V. lantana*
2 Leaf underside glabrous or scurfy, without branched hairs .... *V. lentago*
1 Leaves 3-lobed; palmately veined from base of leaf 3
3 Outer flowers large and sterile, much larger than inner flowers .... *V. opulus*
3 Flowers all similar .... *V. acerifolium*

Viburnum acerifolium L.

MAPLE-LEAF ARROW-WOOD   *native / cons* **7**

Shrub 1-2 m tall. Younger stems, petioles, lower leaf surface and inflorescence finely stellate-pubescent. Leaf blades 6-12 cm long and about as wide, 3-lobed above the middle, coarsely toothed except in the sinuses and toward the broadly rounded or subcordate base; petioles slender, 1-2 cm long. Flowers in cymes on terminal peduncles, with usually 7 rays; corollas 4-5 mm wide; stamens long-exsert, on filaments 3-4 mm long. Fruit a drupe, purple-black when ripe, ellipsoid or subglobose, about 8 mm long; seed a lenticular stone, with 3 shallow grooves on one side and 2 on the other. May-June. — Moist or dry woods.

Viburnum acerifolium

Viburnum lantana L.

WAYFARING-TREE   *introduced*

Tall shrub. Young stems, naked winter buds, petioles, and lower leaf-surface gray-pubescent with stellate hairs. Leaf blades oblong to ovate, 5-10 cm long, acute or obtuse, finely serrate, rounded or cordate at base, pinnately veined; petioles 1-3 cm long. Flowers in short-stalked cymes, about 7-rayed; flowers all alike, about 4 mm wide. Fruit a red drupe, 8-10 mm long; seed a stone, furrowed on both sides. June. — Native of Eurasia, occasionally escaped.

Viburnum lentago L.

NANNY-BERRY   *native / cons* **4**

Tall shrub or small tree, glabrous throughout or minutely scurfy on the inflorescence or petiole. Leaf blades ovate, varying to oblong or orbicular, 5-8 cm long, all or the uppermost abruptly and sharply acuminate, sharply and finely serrate, the teeth often incurved and callous-tipped. Flowers in sessile cymes, 5-10 cm wide, with 3-5 (rarely 7) rays; flowers 4-8 mm

Viburnum lentago

wide. Fruit a drupe blue-black with a whitish bloom, ellipsoid to subglobose, 8-14 mm long; seed a flat, oval stone, scarcely grooved. May–June.
— Woods, roadsides.

Viburnum opulus L.
HIGH-BUSH CRANBERRY *native-introduced / cons 6*
Shrub, 3-4 m tall. Young stems smooth. Leaves opposite, maple-like, sharply 3-lobed and palmately veined, 5-10 cm long and about as wide, the lobes tapered to sharp tips; smooth or hairy beneath, especially on the veins; margins entire or coarsely toothed, petioles grooved, 1-3 cm long, with several club-shaped glands present near base of blade. Flowers white, in large, flat-topped clusters 5-15 cm wide at ends of stems; outer flowers sterile with large petals, surrounding the inner, smaller fertile flowers. Fruit an orange to red, round or oval drupe, 10-15 mm long. June.
— Swamps, fens, streambanks, shores, ditches.

Viburnum opulus

To date, only the native var. *americanum* is reported from Door County. However, the introduced var. *opulus* is present in s Wisconsin as a cultivated plant, and separated as follows:

1 Larger petiolar glands less than 1 mm long (rarely absent), usually stalked and flat topped . . . .
. . . . var. *americanum*
1 Larger petiolar glands 0.9–1.5 (–2) mm long, usually sessile and with the apex indented. . var. *opulus*

Amaranthaceae
AMARANTH FAMILY
Our species annual or perennial herbs. Leaves simple, alternate. Flowers small, often aggregated into large spikes, panicles, or heads, in some species with conspicuous colored bracts. Flowers perfect or unisexual; sepals usually 5; petals absent; ovary superior, 1-chambered. Fruit a 1-seeded utricle; seeds lenticular.
The Amaranthaceae now includes former members of the Chenopodiaceae.

1 Leaf tips with a sharp spine over 0.5 mm (usually ca. 1 mm, even longer on bracts subtending flowers); leaves filiform, terete; fruit horizontal, 1–1.3 mm long, slightly broader, covered by the perianth; tepals with transverse keel or wing sometimes longer than body of tepal . . . . *Salsola*
1 Leaf tips at most with mucro less than 0.5 mm long; leaves various in width, flat; fruit and perianth various 2
2 Flowers unisexual (plants monoecious or dioecious); tepals and bracts acute, scarious or fruit in most if not all flowers enveloped by a pair of bracteoles (perianth absent) 3
2 Flowers mostly bisexual; fruit not enveloped by bracts but perianth may cover it; bracts herbaceous or firm and hardened, not scarious 4
3 Bracts and tepals all acute, scarious . . . . *Amaranthus*
3 Bracts beneath pistillate flowers broad and usually tuberculate and toothed with margins partly fused, obtuse to acute but herbaceous in texture, tepals herbaceous . . . . *Atriplex*
4 Leaves linear to narrowly lanceolate, less than 4 (–6) mm broad, entire, 1 (–3)-nerved . *Corispermum*
4 Leaves usually at least 4 mm broad, toothed to sinuate or crenulate on the margin (if entire, then pinnate- or 3-nerved and not linear) 5
5 Fruit horizontal, completely encircled by the connate wing of the perianth; styles 3 . . . . . . *Cycloloma*
5 Fruit horizontal or vertical, but the perianth without connate wing; styles usually 2 . . . *Chenopodium*

Amaranthus
AMARANTH
Annual herbs; stems erect, ascending, or prostrate, usually much branched. Leaves alternate, petiolate, entire or sinuate, stipules absent. Flowers in small clusters in the axils, or aggregated into axillary or terminal, simple or panicled spikes; flowers small, each subtended by bracts, the bracts sometimes colored and showy; stamens and pistils in different flowers on the same or different plants; calyx of 3-5 scarious or membranous sepals separate to the base; stamens 2-

5; ovary short and broad, compressed; style short or none; stigmas 2 or commonly 3, pubescent. Fruit a thin-walled or leathery utricle, indehiscent or commonly opening at the middle, crowned by the persistent stigmas; seed flattened or lenticular.

1   Plants dioecious (with staminate and pistillate flowers on separate plants) . . . . ***A. tuberculatus***
1   Plants monoecious (with staminate and pistillate flowers separate but on same plants); the flowers intermixed or in separate inflorescences 2
2   Flowers all or nearly all in small clusters from the leaf axils (a small terminal panicle may also be present) 3
3   Plants bushy tumbleweeds . . . . ***A. albus***
3   Plants prostrate . . . . ***A. blitoides***
2   Flowers mainly in elongate, spike-like, terminal clusters (small axillary clusters may be present) . . . 4
4   Sepals obtuse, upper portion curved outward . . . . ***A. retroflexus***
4   Sepals acute, straight or nearly so . . . . ***A. powellii***

Amaranthus albus L.

TUMBLEWEED   *native / cons* O

Plants bushy-branched, to 1 m high and wide; stems whitish. Leaves of the flowering branches elliptic to oblong or obovate, 5-30 mm long, pale green, obtuse or rounded, attenuate at base to a long petiole; early leaves often up to 8 cm long. Flowers in short dense axillary clusters; bracts rigid, subulate, about twice as long as the flowers; sepals of the pistillate flowers commonly 3, uneven, the longest about equaling the utricle. Fruit a lenticular utricle, 1-2 mm long, opening at the middle, wrinkled when dry; seeds lenticular, to 1 mm wide. — Disturbed areas such as roadsides and railways; also sandy lakeshores and streambanks.

Amaranthus blitoides S. Wats.

MAT AMARANTH   *introduced*

Stems prostrate, much branched, 2-6 dm long. Leaves numerous, often crowded, pale green, oblong to obovate, 14 cm long, obtuse or rounded, attenuate into a long petiole. Flowers in short dense axillary clusters; bracts about equaling the sepals, acuminate, scarcely aristate; sepals of the pistillate flowers normally 5, occasionally 4, ovate to oblong, unequal in length. Utricle thick-lenticular, 2-2.5 mm long, about equaling the longest sepal, smooth or nearly so, circumscissile at the middle; seed nearly circular, 1.4-1.7 mm wide. — Disturbed areas such as yards and along roads and railways. Native of the western states, and a common weed throughout Wisconsin.

Amaranthus powellii S. Wats.

GREEN AMARANTH   *introduced*

Stems to 2 m tall, freely branched, glabrous or finely hairy. Leaves long-petioled, lance-ovate, mostly to 10 cm long in well developed plants. Inflorescence terminal, stiff, dense and spike-like, unbranched or with a few widely spaced long branches, dull greenish, not showy; bracts about 5 mm long, much longer than the sepals and fruits, with a very thick, ex-current midrib; sepals 3-5, with simple midvein, those of the pistillate flowers sharply acute, unequal, 2-3 mm long, the longer (outer) ones generally surpassing the fruit; stamens as many as the sepals; Fruit slightly rugose; seeds dark brown, 1-1.3 mm wide.

Weedy in cultivated fields and on roadsides.

Amaranthus retroflexus L.

RED-ROOT AMARANTH   *introduced*

Stems stout, erect, usually branched, finely villous, up to 2 in. tall. Leaves long-petioled, ovate or rhombic-ovate, up to 1 dm long. Terminal panicle of several or many, short, densely crowded, ovoid, obtuse spikes, the whole

Amaranthus albus

Amaranthus blitoides

5-20 cm long; similar but smaller panicles produced from the upper axils; bracts rigid, subulate, much longer than the calyx, 4-8 mm long; sepals of the pistillate flowers 5, oblong lance-shaped, rounded or truncate, mucronate, much exceeding the utricle, 3-4 mm long. Utricle compressed, 1.5-2 mm long, circuinscissile at the middle, the upper part rugulose; seeds round-obovate, dark red-brown, 1-1.2 mm long. — Weedy along roadsides and in fields and gardens, rarely along sandy lakeshores.

A. *powellii* resembles A. *retroflexus* in general habit but is nearly glabrous, with sharply acute sepals.

Amaranthus retroflexus

Amaranthus tuberculatus (Moq.) Sauer
ROUGH-FRUIT AMARANTH   *native / cons* **3**
   Acnida altissima (Riddell) Moq. ex Standl.
Annual herb. Stems erect to spreading, usually much-branched, 2-15 dm tall, usually hairless. Leaves alternate, ovate to lance-shaped, variable in size, larger leaves 4-10 cm long, smaller leaves 1-4 cm long. Flowers either staminate or pistillate flowers and on different plants, in spikes from leaf axils and at ends of stems; staminate flowers with 5 sepals, 2-3 mm long and 5 stamens; pistillate flowers without sepals or petals (rarely with 1-2 small sepals). Fruit a utricle 1-2 mm long; seeds red-brown, 1 mm wide. July-Sept. — Exposed sandy or muddy shores, streambanks, wet meadows and ditches.

Amaranthus tuberculatus

Atriplex
SPEARSCALE, ORACHE
Annual herbs (ours), usually mealy or with bran-like scales; flowers minute, sessile or short-pediceled in glomerules at the nodes, in the upper axils, or in terminal spikes; stamens and pistil in separate flowers on the same or different plants; sepals 3-5; stamens 3-5. Pistillate flowers all or mostly enclosed by 2 broad bracteoles, perianth absent; styles 2. Fruit an utricle.

Atriplex prostrata Bouchér
HASTATE ORACHE   *introduced*
   Atriplex hastata L.
   Atriplex triangularis Willd.
Plants erect and to 1 m tall, or prostrate. Leaves green, the principal ones 2-10 cm long by 2-9 cm wide, hastate to triangular or rhomboidal, with sharp basal angles or lobes. Inflorescence leafless except at base; fruiting bracteoles foliaceous and somewhat spongy-thickened toward the base, obscurely to more or less evidently veined, 3-10 mm long, triangular-ovate, the lateral angles rounded, entire or toothed. Seeds dimorphic: brown seeds 1.5-3 mm wide; black seeds 1-2 mm wide. — A weed of disturbed places and where salted in winter.

Atriplex prostrata

Chenopodium
GOOSEFOOT
Taprooted annual herbs. Stems erect to spreading. Leaves alternate, mostly lance-shaped to broadly triangular, somewhat fleshy and often mealy on lower surface. Flowers perfect, small and numerous, green or red-tinged, in dense spike-like clusters from leaf axils or at ends of stems, the spikes with small leafy bracts; sepals often curved over the fruit; petals absent; stamens 1-5; styles 2-3. Fruit a 1-seeded utricle; seeds with edge vertical or horizontal.

1 Seeds erect; sepals mostly 3 . . . . *C. glaucum*
1 Seeds horizontal; sepals 5 2
2 Mature sepals rounded to conform with fruit, the midvein not much raised . . . . *C. simplex*
2 Mature sepals raised, folded, or hood-like, the calyx appearing somewhat star-shaped . . . . *C. album*

Chenopodium album L.

LAMB'S QUARTERS, PIGWEED   *native / cons* O

Annual; leaves and inflorescence often red or reddish late in the season.
Stems stout, erect, usually much branched, to 1 m or more tall. Leaves
green or more or less white-mealy, broadly rhombic-ovate to lance-
shaped, 3-10 cm long, broadly cuneate at base, the larger almost always
toothed. Flowers in dense glomerules, these forming interrupted or continuous spikes grouped into a terminal panicle; calyx more or less white-
mealy, its segments covering the fruit. Pericarp thin and delicate, when
dry minutely rugulose-reticulate; seeds black, shining, usually 1-1.5 mm
wide, smooth or sculptured. Highly variable. — Fields, gardens, roadsides,
waste ground, dry woods, and barrens.

Chenopodium glaucum L.

OAK-LEAF GOOSEFOOT   *introduced*

Oxybasis glauca (L.) S. Fuentes, Uotila & Borsch

Annual herb. Stems upright to sprawling, 1-6 dm long, usually branched
from base, sometimes red-tinged. Leaves lance-shaped to ovate, 1-4 cm
long and to 2 cm wide, dull green above, densely white-mealy on underside
(especially when young); margins entire, wavy, or with few rounded
teeth; petioles slender, shorter on upper leaves. Flowers in small, often
branched, spike-like clusters from leaf axils, the spikes often shorter
than leaves; sepals mostly 3; petals absent; seeds dark brown, shiny, 1
mm wide. Aug-Oct. — Shores, streambanks, and disturbed areas such as
railroad ballast and barnyards, soils often brackish. Introduced from
Eurasia.

Chenopodium simplex (Torr.) Raf.

MAPLE-LEAF GOOSEFOOT   *native / cons* 1

Chenopodiastrum simplex (Torr.) S. Fuentes, Uotila & Borsch

Annual herb. Stems erect, bright green, to 1.5 m tall. Leaves long-petioled,
broadly ovate to deltoid, 5-20 cm long, truncate to rounded or cordate at
base, bearing on each side 1-4 large teeth separated by broadly rounded
sinuses. Inflorescence a loose, sparsely flowered, terminal panicle of short,
interrupted spikes, the branches often white-mealy; calyx sparsely or
not at all mealy; seeds horizontal, loosely or tightly enclosed in the readily
separable pericarp, shiny-black, 1.5-2.5 mm wide, with a bluntly keeled
margin. — Disturbed ground and moist woods.

Corispermum
BUGSEED

Corispermum americanum (Nutt.) Nutt.

BUGSEED   *native / cons* –

Corispermum orientale Lam.

Annual herb. Stems slender, much branched, 16 dm tall, often pubescent
when young, especially about the inflorescence. Leaves often deciduous
early, linear, 1-6 cm long, 1-3 mm wide, glabrous or sparsely pubescent.
Spikes densely to loosely flowered, 2-10 cm long, 3-8 mm wide; bracts
ovate, 4-10 mm long, long-acuminate, concealing the fruits, the lowest
often approximating the leaves in shape and size. Fruit obovate, 2-4 mm
long, with a pale firm wing to 0.5 mm wide. — Sandy shores and soils, occasionally adventive in waste places.

Chenopodium album

Chenopodium glaucum

Chenopodium simplex

Cycloloma
WINGED-PIGWEED

Cycloloma atriplicifolium (Spreng.) Coult.
WINGED-PIGWEED   *native / cons* **3**
Annual branched herb. Stems 1-8 dm tall, pubescent when young, soon glabrescent. Leaves pale green, early deciduous, lance-shaped in outline, coarsely and irregularly sinuate-toothed, the lower up to 8 cm long, the upper progressively reduced. All terminal branchlets bearing flowers, forming spikes 2-6 cm long. Flowers closely sessile, subtended by tiny bracts, perfect or pistillate; calyx persistent, 5-lobed to about the middle, the segments usually keeled, incurved over the ovary; stamens 5, flattened; styles 2 or commonly 3. Fruit plano-convex, purple-black, puberulent; seeds about 1.5 mm wide. — Dry or sandy ground, weedy.

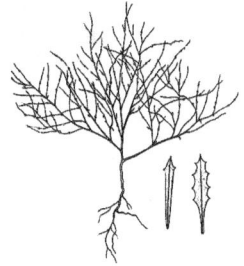

Cycloloma atriplicifolium

Salsola
RUSSIAN-THISTLE

Salsola tragus L.
PRICKLY RUSSIAN-THISTLE   *introduced*
     Kali tragus (L.) Scop.
     Salsola kali L. subsp. *tenuifolia* Moq.
Annual herb. Stems much branched, 3-8 dm tall, glabrous or pubescent. Lower leaves cylindric; upper leaves shorter, stiff, dilated at base, long-spined at tip, each subtending a solitary flower or a short spike of 2 or 3 flowers. Flowers perfect, single or few in axils of the shorter and spinier upper leaves, each subtended by a pair of bractlets; calyx deeply 5-lobed, the segments at maturity incurved over the fruit; stamens usually 5; styles 2. Variable. Late summer.
     Disturbed places.

Salsola tragus

Anacardiaceae
SUMAC FAMILY
Woody plants, juice often milky. Leaves alternate, chiefly compound. Flowers small, regular, perfect or unisexual, 5-merous. Stamens 5, inserted beneath a disk surrounding the ovary. Pistil 1, 3-carpellary. Ovary 1-celled, sessile on the disk; styles 3. Fruit a 1-seeded, dry or fleshy drupe.

1   Flowers in dense inflorescences, these terminal or lateral on previous year's twigs; fruit red, glandular-hairy .... ***Rhus***
1   Flowers in loose clusters from leaf axils; fruit whitish, nearly smooth .... ***Toxicodendron***

Rhus
SUMAC
Trees or shrubs. Leaves pinnately compound, of 3 to many leaflets. Flowers lateral or terminal, polygamo-dioecious. Calyx 5-lobed. Petals 5, white or greenish. Stamens 5. Ovary 1-celled. Fruit a drupe.

1   Bushy shrubs with 3 sessile leaflets .... *R. aromatica*
1   Sparsely branched shrubs or small trees; leaflets several to many 2
2   Twigs and leaf petioles glabrous .... *R. glabra*
2   Twigs and petioles densely hairy .... *R. typhina*

Rhus aromatica Ait.
SQUAW-BUSH   *native / cons* **10**
Bushy shrub, often forming thickets. Leaflets 3, all sessile or nearly so, the terminal elliptic to rhombic-ovate, 4-8 cm long, with usually 3-6 coarse rounded teeth on each side in the distal half, the lateral smaller,

elliptic to ovate lance-shaped, with similar teeth, at least the outer margin rounded to the petiole. Flowers in several short (1-2 cm) spike-like clusters, forming a panicle about 1 dm long, opening before or with the leaves, sessile or on pedicels no longer than the calyx. Bracts glabrescent just below the strongly ciliate apex. Drupes bright red, densely pubescent. Apr-May. — Dry woods, hills, sand dunes, and rocky soil.

Rhus glabra L.
SMOOTH SUMAC   *native / cons* 2
  Rhus × *borealis* Greene
Usually a sparsely branched shrub, but sometimes to 6 m tall, the younger branches and petioles glabrous and somewhat glaucous. Leaflets 11-31, lance-shaped to narrowly oblong, 5-10 cm long, commonly serrate, much paler beneath. Panicle dense, often 2 dm long. Fruit bright red, densely beset with minute obovoid hairs about 0.2 mm long. June-July. — Dry soil, old fields, roadsides, and margins of woods.

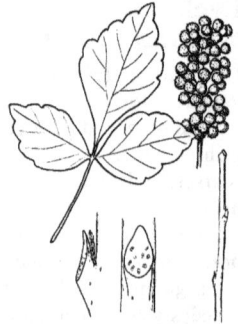

Rhus aromatica

Rhus typhina L.
STAGHORN SUMAC   *native / cons* 2
  Rhus hirta (L.) Sudw.
Tall shrub or small tree to 10 m tall. Younger branches, petioles, and leaf-rachis densely and softly hirsute. Leaflets 9-29, lance-shaped to narrowly oblong, 5-12 cm long, finely or coarsely serrate, paler beneath. Fruit red, densely covered with slender hairs 1-2 mm long. June-July. — Dry soil.

Rhus typhina

Toxicodendron
POISON-IVY
Shrubs or vines, with axillary, rather loose inflorescences often drooping in fruit; otherwise much like *Rhus,* and sometimes included in that genus. Fruit a white or yellowish drupe, shining and glabrous or inconspicuously short-hairy. All parts of these plants may cause an allergic skin reaction.

1 Leaflets 7–13, margins entire .... *T. vernix*
1 Leaflets 3, margins entire, toothed or lobed .... *T. rydbergii*

Toxicodendron rydbergii (Small) Greene
WESTERN POISON-IVY   *native / cons* 2
  Rhus radicans L. var. *rydbergii* (Small ex Rydb.) Rehder
Strongly rhizomatous shrub, forming colonies. Stems to 1(-3) m tall, nearly erect, simple or sparingly branched. Leaflets broadly ovate, tending to be openly folded along the midrib rather than flat, glabrous on both sides or strigose beneath and often with a line of minute, curly hairs along the midrib above. Inflorescence unbranched or sparingly branched, usually with fewer than 25 flowers. Fruit 4-7 mm thick, smooth, sessile or sub-sessile and crowded in an erect inflorescence — Dunes, shores, open sandy or rocky places.

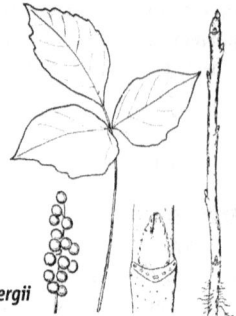

Toxicodendron rydbergii

Toxicodendron vernix (L.) Kuntze
POISON-SUMAC   *native / cons* 7
  Rhus vernix L.
Shrub or small tree to 5 m tall, often branched from base. Leaves alternate, divided into 7-13 leaflets, the leaflets oblong to oval, 4-6 cm long, tapered to a pointed tip; margins entire, smooth. Flowers small, white or green, in panicles to 2 dm long; sepals 5, joined at base; petals 5, not joined; stamens 5. Fruit a round, gray-white drupe, 4-5 mm wide. June-July. — Tamarack swamps, thickets, floating bog mats and bog margins, often in partial shade.

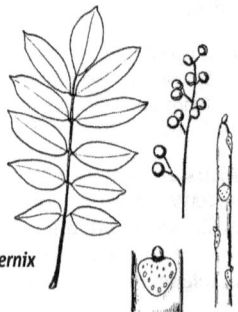

Toxicodendron vernix

Apiaceae

CARROT FAMILY

Biennial or perennial aromatic herbs with hollow stems, some very toxic. Leaves alternate and sometimes also from base of plant, mostly compound; petioles sheathing stems. Flowers small, perfect (with both staminate and pistillate parts), regular, in flat-topped or rounded umbrella-like clusters (umbels); sepals 5 or absent; petals 5, white or greenish. Fruit 2-chambered, separating into 2, 1-seeded fruit when mature.

Key to Apiaceae Groups

1	Ovary and fruit pubescent, tuberculate, bristly, or prickly	*Group A*
1	Ovary and fruit glabrous	2
2	Leaves divided into distinct and separate leaflets of about uniform shape, these often more than 2 cm wide	*Group B*
2	Leaves much dissected or 2 or more times compound, the segments ovate, oblong, linear, or thread-like and less than 1 cm wide	3
3	Plants flowering	*Group C*
3	Plants fruiting	*Group D*

Group A

Inflorescence a true umbel or compound umbel; most leaves compound, dissected, or deeply divided; fruit and ovary pubescent, covered with small bumps, or bristly or prickly.

1	Principal leaves palmately or once-pinnately compound or divided, the leaflets sometimes again divided	2
1	Principal leaves twice or more compound	3
2	Leaflets large, mostly 1 dm wide or more; fruit pubescent	*Heracleum*
2	Leaflets less than 1 dm wide; fruit bristly or spiny	*Sanicula*
3	Leaves with sharply toothed leaflets, the leaflets 1 cm wide or more	*Osmorhiza*
3	Leaves highly dissected into segments less than 1 cm wide	*Daucus*

Group B

Inflorescence a true umbel or a compound umbel; fruit and ovary glabrous; leaves divided into distinct leaflets of uniform shape, these often more than 2 cm wide.

1	Principal leaves once-compound	2
1	Main leaves twice or three-times compound	11
2	Upper leaf-sheaths expanded, 1 cm or more wide when flattened; flowers white; fruit flattened and wing-margined	*Heracleum*
2	Upper leaf sheaths not expanded, less than 1 cm wide; flowers and fruit various	3
3	Taprooted introduced weeds of waste places and disturbed areas	4
3	Native species with fibrous or tuberous-thickened roots, most common in woods or wetlands	5
4	Flowers yellow; fruit wing-margined	*Pastinaca*
4	Flowers white; fruit not winged	*Pimpinella*
5	Leaves with 3 leaflets (or the basal ones simple and toothed)	*Oxypolis*
5	Leaves with 5 or more leaflets; flowers white	*Sium*
6	Leaflets entire; flowers yellow	*Taenidia*
6	Leaflets toothed or lobed; flowers yellow or white	7
7	Plants flowering	8
7	Plants fruiting	9
8	Flowers yellow or cream-colored	*Pastinaca*
8	Flowers white	*Cicuta*
9	Fruit evidently winged	*Pastinaca*
9	Fruit not winged, or only slightly so	*Cicuta*

Group C
Inflorescence a true umbel or a compound umbel; fruit and ovary glabrous; leaves dissected or 2 or more times compound, the segments ovate, oblong, linear, or thread-like and less than 1 cm wide; plants flowering.

1 Plants annual or biennial .... *Carum*
1 Plants perennial .... *Cicuta*

Group D
Inflorescence a true umbel or a compound umbel; fruit and ovary glabrous; leaves dissected or 2 or more times compound, the leaf segments ovate, oblong, linear, or thread-like and less than 1 cm wide; plants fruiting.

1 Plants perennial .... *Cicuta*
1 Plants annual or biennial .... *Carum*

Carum
CARAWAY
Carum carvi L.
CARAWAY   *introduced*
Glabrous biennial herb, from a taproot. Stems to 1 m tall. Leaflets pinnately dissected into linear segments 5-15 mm long. Flowers in terminal and lateral compound umbels; primary rays several to many; peduncles 5-13 cm long; primary rays 7-14, commonly 2-4 cm long; umbellets small; involucel of a few minute bracts or none; pedicels very unequal; sepals absent; petals white or rarely pink. Fruit elliptic to oblong, 3-4 mm long, about half as wide, prominently ribbed. June-Aug. Native of Eurasia; sometimes cultivated and often weedy in waste places.

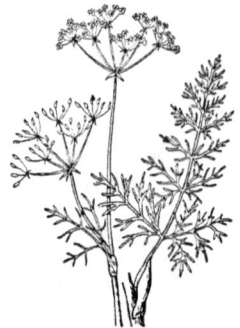

Carum carvi

Cicuta
WATER-HEMLOCK
Biennial or perennial toxic herbs. The tuberous roots, chambered stem base and young shoots of common water-hemlock (*Cicuta maculata*) are especially toxic. Leaves alternate, 2-3-pinnate; leaflets narrow or lance-shaped, entire or toothed; leaf veins ending in the lobes (sinuses) and not at teeth as in other members of this family. Flowers white or green, in few to many umbels; umbels usually without bracts, umbellets bracted. Fruit oval or round, flattened, ribbed.

1 Upper leaflet axils usually with bulblets; leaflets to 5 mm wide .... *C. bulbifera*
1 Bulblets absent; leaflets usually much more than 5 mm wide .... *C. maculata*

Cicuta bulbifera L.
BULBLET-BEARING WATER-HEMLOCK   *native / cons 7*
Biennial or perennial herb, toxic; fibrous-rooted or with a few thickened, tuberlike roots. Stems slender, upright, 3-10 dm tall, not thickened at base. Leaves alternate along stem, to 15 cm long and 10 cm wide, pinnately divided; leaflets mostly linear, 1-5 mm wide, margins sparsely toothed to entire; upper leaves reduced in size, undivided or with few segments, with 1 to several bulblets 1-3 mm long, in axils. Flowers white, in umbels 2-4 cm wide. Fruit round, 1-2 mm wide, but rarely maturing. Aug.-Sept. Streambanks, lake and pond shores, marshes, swamps, open bogs, thickets, springs, ditches.

Cicuta maculata L.
COMMON WATER-HEMLOCK   *native / cons 6*
Biennial or perennial herb. Stems single or several together, often branched, 1-2 m long, distinctly hollow above the chambered and tuberous-thickened base. Leaves from base of plant and alternate on stem, mostly 10-30 cm long and 5-20 cm wide; basal leaves larger and longer

Cicuta bulbifera

stalked than stem leaves; leaflets linear to lance-shaped, 3-10 cm long and 5-35 mm wide; margins toothed. Flowers white, in several to many umbels, these 6-12 cm wide in fruit, on stout stalks 5-15 cm long. Fruit round to ovate, 2-4 mm long, with prominent ribs. June-Sept. — Wet meadows, marshes, swamps, moist to wet forests, thickets, shores, streambanks, springs. Considered the most toxic plant in North America.

Cryptotaenia
HONEWORT

Cryptotaenia canadensis (L.) DC.
CANADIAN HONEWORT   *native / cons* **4**
Perennial glabrous herb. Stems branched, 3-8 dm tall. Leaves 3-foliate, lower leaves long-petioled, the upper on short petioles dilated as far as the leaflets. Leaflets lance-shaped to obovate, 4-15 cm long, irregularly often doubly serrate or sometimes lobed. Flowers in numerous loose, irregular, compound umbels arising terminally and from the upper axils; primary rays 2-7, ascending, 1-5 cm long, somewhat unequal; umbellets few-flowered; involucel none or of 1-3 minute bractlets; pedicels very unequal; sepals low or obsolete; corolla white. Fruit dark, slightly flattened, tipped by the slender stylopodium, 5-8 mm long; the ribs evident but low and obtuse. June-July. — Moist rich woods, swamps.

Cicuta maculata

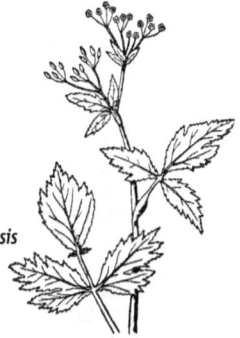

Cryptotaenia canadensis

Daucus
CARROT

Daucus carota L.
QUEEN ANNE'S-LACE   *introduced (invasive)*
Biennial herb, with a stout taproot. Stems 5-10 dm tall, glabrous, scabrous, or commonly rough-hairy. Leaves pinnately compound, the ultimate divisions linear or lance-shaped. Umbels compound, terminal and from the upper axils, long-peduncled, usually many-rayed; terminal umbel erect, commonly 7-15 cm wide, the lateral ones usually smaller; the outer primary rays curve inward after anthesis, producing a congested cluster. Flowers white or rarely pinkish, the central one of each umbellet often purple. Fruit 3-4 mm long, flattened dorsally, the primary ribs low and inconspicuous, bearing a row of short bristles, the four secondary ribs prominently winged, divided into a row of hooked or straight spines. June-Sept. — Native of Eurasia; established as a weed in fields, roadsides, waste ground, and open woods. The cultivated carrot is a race of this species.

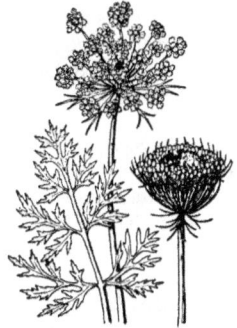

Daucus carota

Heracleum
COW-PARSNIP

Heracleum maximum Bartr.
AMERICAN COW-PARSNIP   *native / cons* **3**
  Heracleum lanatum Michx.
  Heracleum sphondylium L. subsp. *montanum*
Large perennial herb. Stems stout, hairy, 1-2 m long. Leaves alternate, nearly round in outline, divided into 3 leaflets; leaflets 1-4 dm long and as wide, margins coarsely toothed. Flowers white, in large umbels, the terminal umbel 1-2 dm wide. Fruit obovate, 8-12 mm long and nearly as wide, often hairy. May-July. — Streambanks, thickets, wet meadows, moist forest openings and disturbed areas.

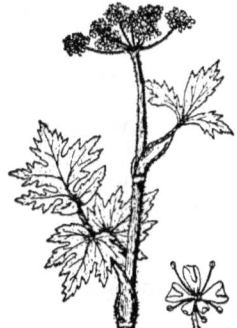

Heracleum maximum

Osmorhiza
SWEET-CICELY
Erect perennial herbs from thickened roots and glabrous to pubescent stems 4-8 dm tall; our 3 species similar in general appearance and foliage. Leaves ternate; leaflets several, the lower petioled, the upper subsessile, the ultimate segments ovate to lance-shaped, serrate or lobed. Flowers in terminal and lateral umbels, these usually surpassing the leaves; primary rays of the umbel mostly 3-6, widely ascending; involucre present or absent. Umbellets few-flowered; involucel present or lacking. Sepals none; petals white or greenish white. Fruit elongate, slightly flattened, ribbed, the base prolonged into bristly tails.

1 Umbels without bracts at base of umbel branches .... *O. berteroi*
1 Umbels with bracts at base of umbel branches 2
2 Plants anise-scented; styles 2 mm long, becoming 3–4 mm long in fruit .... *O. longistylis*
2 Plants unscented; styles less than 1.5 mm long (even in fruit) .... *O. claytonii*

Osmorhiza berteroi DC.
MOUNTAIN SWEET-CICELY   *native / cons* **8**
   Osmorhiza chilensis Hook. & Arn.
Involucel lacking. Fruit concavely narrowed to the acute summit, 14-18 mm long. Stylopodium ovoid-conic, commonly longer than thick; styles at maturity outwardly curved, 0.4-0.7 mm long including the stylopodium. June.
   Moist woods.

Osmorhiza claytonii (Michx.) C.B. Clarke
HAIRY SWEET-CICELY   *native / cons* **5**
Stems commonly sparsely villous, but vary from densely villous to nearly glabrous. Styles at anthesis distinctly shorter than the petals, in fruit nearly straight and parallel, 1.2-1.5 mm long including the stylopodium. Mericarps 2-2.5 cm long. May-June.
   Moist woods.

Osmorhiza longistylis (Torr.) DC.
ANISEROOT   *native / cons* **4**
Styles at anthesis about 2 mm long, much exceeding the petals, in fruit nearly straight and parallel, 3-3.5 mm long including the stylopodium. Mericarps about 2 cm long. May-June.   Moist woods.

Osmorhiza claytonii

Pastinaca
PARSNIP

Pastinaca sativa L.
WILD PARSNIP   *introduced (invasive)*
Stout biennial herb, to 1.5 m tall. Lower leaves long-petioled, the upper on shorter, wholly sheathing petioles, all typically 1-pinnate; leaflets 5-15, usually oblong to ovate, 5-10 cm long, variously serrate or lobed, or in vigorous plants sometimes completely divided into 2-5 segments. Umbels large, 1-2 dm wide, compound, the terminal soon overtopped by the lateral ones; primary rays unequal, 15-25; involucre and involucel usually lacking; sepals minute or none; petals yellow. Fruit broadly elliptic or obovate, strongly flattened, 5-7 mm long, with low ribs, the lateral ribs broadly and thinly winged.— Native of Eurasia; long in cultivation and thoroughly established as a weed in waste places, fields, and roadsides.
   Skin irritant if handled.

Pastinaca sativa

Pimpinella
BURNET SAXIFRAGE

Pimpinella saxifraga L.
BURNET SAXIFRAGE   *introduced*

Perennial herb. Stems 3-6 dm tall, filled with pith. Lower stem leaves 1-pinnate, the leaflets varying from ovate or subrotund and merely serrate to deeply pinnately dissected. Upper leaves much reduced, the uppermost consisting of sheaths only or of sheaths with a few small linear leaflets at the summit. Umbels peduncled, terminal and lateral, compound; involucre none or rarely of 1-few bracts; primary rays 8-20; sepals minute or lacking; petals white. Fruit glabrous, ovoid, 2-2.5 mm long, the ribs 5, narrow.

    Native of Eurasia; escaped or adventive in waste places.

Pimpinella saxifraga

Sanicula
BLACK-SNAKEROOT

Biennial or perennial herbs; stems arising from a cluster of fibrous or sometimes tuberous roots. Leaves palmately divided into 3-5 segments, the basal long-petioled, the cauline progressively reduced, and the uppermost short-petioled to nearly sessile. Umbels irregular, with spreading primary branches of unequal length, the umbellets dense or almost capitate, commonly with 3 sessile or short-pediceled perfect flowers, their hypanthium bristly, and several staminate flowers with smooth hypanthium, all or mostly on much longer pedicels; sepals narrow, joined at base, persistent; petals greenish white to greenish yellow. Fruit ovoid to subglobose, slightly flattened laterally, ribs absent, densely covered with hooked bristles.

1   Styles shorter than bristles of the fruit; staminate flowers 2–7 . . . . *S. canadensis*
1   Styles longer than bristles of the fruit; staminate flowers 12–25 in each umbellet 2
2   Staminate flowers longer than the fruit; sepals awl-shaped, 1–2 mm long . . . . *S. marilandica*
2   Staminate flowers shorter than the fruit; sepals lance-shaped or ovate, 0.5 mm long . . . . . *S. odorata*

Sanicula canadensis L.
CANADA SANICLE   *native / cons 6*

Leaves 3-parted, or 5-parted by division of the lateral leaflets. Bractlets of the involucel resembling the bracts but smaller. Flowers white, the fertile on pedicels 0.5-1 mm long, the sterile few, mostly concealed by the fertile; calyx lobes subulate, surpassing the petals. Anthers white. Fruit subglobose, 2-5 mm long, on pedicels 1-1.5 mm long, the bristles exceeding the inconspicuous styles and about equaling the sepals. June-Aug.

    Moist or dry woods.

Sanicula canadensis

Sanicula marilandica L.
MARYLAND BLACK-SNAKEROOT   *native / cons 5*

Leaves 5-parted, often appearing 7-parted, serrate, doubly serrate, or toward the apex incised. Bractlets of the involucel resembling the bracts but smaller. Flowers greenish white, the fertile ones sessile; calyx lobes lance-subulate, 1-1.5 mm long, equaling or slightly shorter than the petals. Anthers greenish white. Fruit nearly sessile, 4-6 mm long, narrowed and with shorter bristles toward the base. Styles recurved, exceeding the bristles. June-Aug. — Moist or dry woods.

Sanicula odorata (Raf.) Pryer & Phillippe
CLUSTERED BLACK-SNAKEROOT   *native / cons 3*
   Sanicula gregaria Bickn.

Leaves 3-5-parted, the segments sharply serrate to incised. Bractlets of the involucel small, subscarious. Flowers greenish yellow, the fertile on pedicels 0.5-1 mm long; calyx lobes ovatelance-shaped to ovate, obtuse or subacute, much shorter than the petals. Anthers bright yellow. Fruit

Sanicula marilandica

subglobose, about 3 mm long. Styles conspicuous, recurved, exceeding the bristles. June–Aug. — Moist or dry woods.

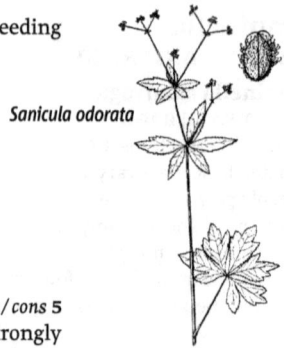

Sanicula odorata

Sium
WATER-PARSNIP

Sium suave Walt.
HEMLOCK WATER-PARSNIP   native / cons 5
Perennial emergent herb. Stems single, smooth, 5-20 dm long, strongly ribbed upward; stem base thickened and hollow with cross-partitions. Leaves once-pinnate, on long, hollow stalks (shorter stalked above); leaflets 7-17 per leaf, linear to lance-shaped, 5-10 cm long and 3-15 mm wide; margins with fine, sharp, forward-pointing teeth; finely dissected underwater leaves often present from spring to midsummer. Flowers white or green-white, 1-2 mm wide, in stalked umbels 4-12 cm wide at ends of stems and from side branches. Fruit oval, 2-3 mm long, with prominent ribs. July–Sept. — Wet forest depressions, marshes, swamps, streambanks, lakeshores, ditches; usually in shallow water.

Taenidia   *Sium suave*
PIMPERNEL

Taenidia integerrima (L.) Drude
YELLOW-PIMPERNEL   native / cons 7
Perennial herb. Stems branched, 4-11 dm tall, glabrous and somewhat glaucous. Lower leaves long-petioled, commonly 3x compound, the upper 1-2x compound, with short, wholly sheathing petioles; leaflets normally entire, ovate to oblong or elliptic. Umbels terminal and lateral, loose and irregular, primary rays numerous, the outer elongate, to 9 cm long, the inner often much shorter; involucre none; umbellets many-flowered, involucel none. Inner flowers of each umbellet staminate and short-pediceled, the marginal long-pediceled and fertile; calyx teeth tiny or none; petals yellow. Fruit elliptic to broadly ovate-oblong, 3-4 mm long, flattened; ribs faint. May–June. — Dry woods and rocky hillsides.

Taenidia integerrima

Apocynaceae
DOGBANE FAMILY
Our species herbs or twining woody vines; most species have milky juice. Leaves opposite, alternate, or sometimes whorled. Flowers 5-merous, regular, perfect. Fruit a capsule or follicle; seeds often bearing long hairs. Family now includes former members of Asclepiadaceae; *Apocynum* differs by having corolla lobes overlapping and twisted in bud, and stamens without a crown.

1 Plants trailing, subwoody, evergreen; flowers solitary in leaf axils; corolla blue; seeds glabrous **Vinca**
1 Plants erect or twining, herbaceous and not evergreen; flowers in terminal or axillary cymes or umbels; corolla various colors, not blue; seeds with tuft of silky hairs 2
2 Plant a climbing vine, strongly twining at least apically; corolla lobes spreading or ascending, dark purple to nearly black . . . . **Vincetoxicum**
2 Plant erect or ascending, not twining; corolla lobes strongly reflexed at maturity, except in Apocynum, white, pink, purple, yellow, orange, or greenish 3
3 Corolla lobes erect to spreading; flowers in small terminal (and sometimes axillary) cymes; mature fruits 3–5 mm wide. . . . . **Apocynum**
3 Corolla lobes strongly reflexed at maturity; flowers in umbels; mature fruits 6–35 mm in diameter . . . . **Asclepias**

Apocynum
DOGBANE
Perennial herbs with tough fibrous stems. Leaves opposite, mucronate. Flowers small, white or pink, in branched terminal cymes. Calyx deeply divided into triangular or lance-shaped lobes. Corolla white or pinkish, campanulate or short-cylindric, with 5 short lobes, bearing within a tooth or scale near the base of the tube opposite each lobe. Anthers lance-shaped, joined, adherent to the stigma and prolonged into a cone beyond it. Ovaries 2, subtended by 5 nectaries; style none; stigma large, 2-lobed. Fruit a cylindric follicle, pendulous; seeds numerous, bearing long soft hairs (coma).

1 Corolla pink, 5–8 mm long; leaves widely spreading or drooping . . . . *A. androsaemifolium*
1 Corolla white, 3–4 mm long; leaves ascending . . . . *A. cannabinum*

Apocynum androsaemifolium L.
SPREADING DOGBANE   *native / cons* **2**
Perennial herb. Stems more or less inclined from tbe vertical, the branches chiefly alternate. Leaves petiolate, more or less drooping, oblong lance-shaped to ovate, commonly 3-8 cm long, pilose beneath. Principal cymes terminal; secondary cymes of smaller size in the upper axils; calyx lobes triangular, a third to half as long as the corolla tube; corolla campanulate, 6–10 mm long, pink to nearly white, marked with red within, the lobes spreading or recurved. May-Aug. — Upland woods, occasionally in fields and roadsides.

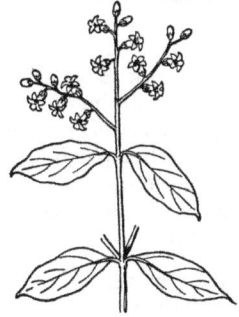

Apocynum cannabinum L.
INDIAN-HEMP   *native / cons* **3**
Perennial herb. Stems erect, branched above, 1-1.5 m tall. Leaves varying from oblong lance-shaped to ovate or broadly elliptic, acute to rounded at the mucronate apex and base, glabrous or pubescent beneath, regularly on conspicuous petioles commonly 5-10 mm long. Flowers white or greenish white. Follicles usually 10-15 cm long; seed-coma 2-2.5 cm long. June-Sept. — Dry or moist open places.

Apocynum androsaemifolium

Asclepias
MILKWEED
Perennial herbs from a thick root or deep rhizome and with milky juice (except in *A. tuberosa*). Stems usually simple. Leaves opposite (in some species whorled or rarely alternate), entire. Flowers small or medium-sized, in peduncled, terminal or axillary umbels; calyx lobes, corolla lobes, and stamens each 5; corolla lobes usually meeting in bud without overlapping; corolla deeply divided, at anthesis reflexed and concealing the calyx; anthers united with stigma forming an organ known as the gynostegium; pollen of each anther-sac united into a waxy mass known as a pollinium; ovaries 2; styles 2; stigma 1. Fruit a pod-like follicle, normally produced in pairs, commonly erect, lance-shaped or linear lance-shaped, acuminate; seeds with long silky hairs (coma).

1 Leaves linear, less than 4 mm wide, mostly whorled . . . . *A. verticillata*
1 Leaves more than 5 mm wide, alternate or opposite 2
2 Leaves mostly alternate, linear to narrowly oblong . . . . *A. tuberosa*
2 Leaves opposite, oblong to ovate 3
3 Reflexed corolla lobes red-purple, mostly less than 5 mm long . . . . *A. incarnata*
3 Reflexed corolla lobes green suffused with purple, 5 mm or more long . . . . *A. syriaca*

Asclepias incarnata L.
SWAMP MILKWEED   *native / cons* **5**
Perennial herb, from thick rhizomes; plants with milky juice. Stems stout, to 1.5 m long, branched above, smooth except for short, appressed hairs on upper stem. Leaves opposite, simple, mostly lance-shaped, 6-15

cm long and 1-5 cm wide, tapered to a sharp tip, margins entire, petioles short. Flowers pink to purple-red, numerous in umbels at ends of stems and from upper leaf axils, perfect, regular; sepals 5, spreading; petals 5, 4-6 mm long and curved downward; stamens 5; flowers with 5 petal-like "hoods", each with an awl-shaped "horn" projecting from the opening. Fruit a follicle (1-chambered and opening on 1 side only) with many seeds, the seeds having tufts of white hairs. June-Aug. — Openings in conifer swamps, marshes, streambanks, ditches, open bogs and fens; often in shallow water.

Asclepias incarnata

Asclepias syriaca L.
COMMON MILKWEED   *native / cons* 1

Stems tall and stout, mostly simple, pubescent. Leaves thick, narrowly or broadly elliptic to ovate or oblong, 10-15 cm long, acute or apiculate, softly pubescent beneath, on distinct petioles 5-15 mm long. Umbels often numerous, terminal and in the upper axils, compactly many-flowered; peduncles stout, 3-10 cm long; corolla green suffused with purple, varying from almost purple to almost green, 8-10 mm long. Hoods pale purple, somewhat divergent, 6-8 mm long, surpassing the gynostegium, the lateral margins bearing a prominent, sharp, triangular lobe at or near the middle. Horns short, inflexed. Pods erect on deflexed pedicels about 1 dm long, tomentose and beset with soft filiform to conic processes. Fields, meadows, and roadsides; often weedy.

Asclepias syriaca

Asclepias tuberosa L.
BUTTERFLY WEED   *native / cons* 6

Stems ascending or erect, 3-7 dm tall, villous or hirsute, simple to much branched above, the branches ascending or widely spreading, often flexuous. Leaves alternate or on the branches opposite, linear to lance-shaped or oblong lance-shaped, 5-10 cm long, pubescent, cuneate to truncate or subcordate at base. Umbels varying from solitary and terminal to numerous, often from most of the axils of divergent branches; corolla yellow to orange-red, 7-10 mm long. Hoods yellow to orange, 5-7 mm long, greatly exceeding the gynostegium, nearly straight, erect, the lateral margins bearing an obscure tooth below the middle. Pods 8-12 cm long, erect. June-Aug. — Dry or moist prairies and upland woods, especially in sandy soil. Our only *Asclepias* without milky juice. Plants variable in habit and shades of flower color.

Asclepias tuberosa

Asclepias verticillata L.
WHORLED MILKWEED   *native / cons* 2

Stems slender, erect, 2-5 dm tall, simple to the inflorescence, pubescent in lines. Leaves very numerous in whorls of 3-6, narrowly linear, 2-5 cm long, 1-2 mm wide, revolute. Umbels several from the upper nodes; peduncles 1-3 cm long; petals white or greenish, 4-5 mm long. Hoods white or greenish white, somewhat divergent, 1.5-2 mm long, about equaling the gynostegium, their margins entire. Horns subulate, much surpassing the hoods, slightly narrowed over the stamens. Pods slender, erect on erect pedicels, 4-5 cm long. June-Aug. — Dry or moist fields, roadsides, upland woods, and prairies.

Vinca
PERIWINKLE

Asclepias verticillata

Vinca minor L.
LESSER PERIWINKLE   *introduced (invasive)*

Perennial trailing herbs. Stems trailing or scrambling, to 1 m long, forming mats. Leaves leathery, opposite, lance-elliptic, 3-5 cm long, entire, petiolate. Flowers blue or rarely white, solitary in 1 axil only of a pair of leaves;

calyx deeply 5-parted; corolla large, salverform, corolla tube 8-12 mm long, the limb 2-3 cm wide; ovaries 2, accompanied by 2 nectaries. Fruit a linear, few-seeded follicle; seeds naked. April-May. — Native of s Europe, planted as a groundcover and escaping to roadsides and open woods.

Vincetoxicum
Vinca minor
SWALLOW-WORT

Vincetoxicum nigrum (L.) Moench
BLACK SWALLOW-WORT   *introduced (invasive)*
  Cynanchum louiseae Kartesz & Gandhi
  Cynanchum nigrum (L.) Pers. non Cav.
Perennial twining herbs, climbing 1-2 m tall. Leaves short-petioled, oblong to ovate, 5-10 cm long, acuminate, rounded to subcordate at base. Flowers in peduncled umbel-like clusters from the axils of the leaves; corolla purple-brown and dark purple, its lobes about 3 mm long, much exceeding the cuplike corona. Follicles slender, 4-6 cm long. June-Sept. — Native of s Europe; occasionally cultivated and locally established in woods and on roadsides.

Vincetoxicum nigrum

Aquifoliaceae
HOLLY FAMILY

Ilex
HOLLY
Shrubs. Leaves usually alternate, toothed or entire, not lobed. Flowers from leaf axils, 4-8-parted, usually either staminate or pistillate, sometimes perfect, on same or different plants. Fruit a fleshy berrylike drupe with 4-9 stones.

1 Leaves tipped with a short, sharp point, margins mostly entire or with a few scattered teeth; petals linear; sepals tiny or absent . . . .*I. mucronata*
1 Leaves not tipped with a short, sharp point, margins toothed; petals oblong; sepals evident . . . .
   . . . .*I. verticillata*

Ilex mucronata (L.) Powell, Savolainen & Andrews
MOUNTAIN HOLLY, CATBERRY   *native / cons* **8**
  Nemopanthus mucronatus (L.) Loes.
Much-branched shrub to 3 m tall; young twigs purple-tinged. Leaves deciduous, alternate, oval or ovate, 3-6 cm long and 2-3 cm wide, bright green above, dull and paler below, tip of leaf with a small, sharp point; margins entire or with small scattered teeth, on purple-red stalks 1 cm long. Flowers very small, yellow-white, on threadlike stalks from leaf axils; staminate flowers usually in small groups, pistillate flowers single. Fruit a purple-red berrylike drupe, 5-6 mm wide. May-June. — Open bogs (especially along outer moat), swamps, thickets, wet depressions in forests, lakeshores.

Ilex mucronata

Ilex verticillata (L.) Gray
WINTERBERRY   *native / cons* **7**
Shrub to 5 m tall; twigs smooth, finely ridged. Leaves deciduous, alternate, obovate to oval, tapered to a tip, dull green above, paler below; margins with incurved teeth. Flowers small, green-white, on short stalks from leaf axils, opening before leaves fully expanded in spring; staminate flowers in crowded clusters, pistillate flowers 1 or several in a group. Fruit a berrylike drupe, orange or red, 5-6 mm wide and persisting into winter. June. — Swamps, open bogs, thickets, shores and streambanks.

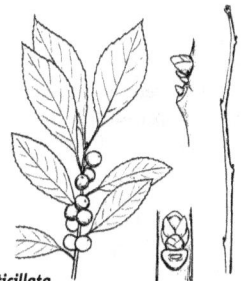

Ilex verticillata

Araliaceae

GINSENG FAMILY

Shrubs or herbs, rarely trees. Leaves usually alternate, compound or rarely simple, the petiole not sheathing at base and usually adnate to the stipules; flowers small, umbellate. Flowers regular, epigynous, perfect or unisexual, 5-10-merous. Calyx small, its limb truncate to denticulate. Petals valvate or scarcely imbricate, usually distinct, deciduous at maturity. Stamens usually as many as the petals, rarely more. inserted on a tusk within the calyx; anthers short, longitudinally dehiscent. Ovary inferior, 2-12-celled, with one pendulous ovule in each cell. Styles as many as the cells of the ovary, distinct or more or less connate. Fruit a berry or a leathery drupe.

1 Leaves alternate or basal, mostly 2–3 times compound; carpels 5 . . . . *Aralia*
1 Leaves in 1 whorl, once-palmately compound; carpels 2 or 3 . . . . *Panax*

Aralia

SARSAPARILLA

Herbs or shrubs (or rarely trees). Stems herbaceous or slightly woody at the base only, rarely thorny (*A. hispida* bristly at the base). Leaves pinnately or 3-partedly compound. Flowers white or greenish, in 2-many umbels in each inflorescence. Petals and stamens each 5. Cells of the ovary 4-6, usually 5. Styles 4-6, usually 5, free or somewhat connate at base. Fruit a berry, tipped by the persistent styles; seeds usually 5.

1 Plants with flowers on a leafless scape . . . . *A. nudicaulis*
1 Plants with leafy stems 2
2 Lower stems bristly; umbels several (3–13) in a loose cluster . . . . *A. hispida*
2 Stems smooth; umbels very many, in a large terminal panicle . . . . *A. racemosa*

Aralia hispida Vent.

BRISTLY SARSAPARILLA *native / cons 5*
Perennial herb from a stout rhizome. Stems to 1 m tall, bristly near the base with sharp slender spines and often decreasingly so above. Leaves few, on petioles usually shorter than the blade, bipinnate; leaflets oblong to ovate or lance-shaped, up to 10 cm long but usually much smaller, acute or short-acuminate, sharply serrate. Umbels several, in a loose, open, terminal inflorescence; styles connate about half their length. Berry globose, nearly black. June-July. — Dry woods, especially in sandy or sterile soil.

Aralia hispida

Aralia nudicaulis L.

WILD SARSAPARILLA *native / cons 6*
Acaulescent perennial herb, the leaves and peduncle arising from a long rhizome. Petiole erect, to 5 dm tall. Leaves 3-parted, each division pinnately 3-5-foliolate; leaflets lance-elliptic to obovate, up to 15 cm long and 8 cm wide, acuminate, finely serrate, the lateral ones asymmetric at base. Peduncles usually much shorter than the petioles, bearing 2-7 (commonly 3) umbels; styles distinct to the base. Fruit nearly black. May-June. — Moist or dry woods.

Aralia racemosa L.

SPIKENARD *native / cons 7*
Stout perennial herb to 2 m tall, lacking thorns or bristles. Leaves few, widely spreading, up to 8 dm long, the three primary divisions pinnately compound; leaflets ovate, variable in size in the same leaf, the larger up to 15 cm long, sharply and often doubly serrate, acuminate, obliquely cordate at base. Inflorescence a large panicle with numerous umbels; styles connate at base only. Fruit dark purple. July. — Rich woods.

Aralia nudicaulis

Panax
GINSENG

Perennial herbs, the unbranched stems rising from a deep-seated, thickened or tuber-like root, bearing a single whorl of once palmately compound leaves, usually 3 in number. Flowers in usually a single long-peduncled terminal umbel. Petals and stamens each 5; petals white or greenish. Fruit a small berry.

Panax trifolius L.
DWARF GINSENG   *native / cons* **8**

Stems 1-2 dm tall; root globose. Leaflets 3-5, sessile or nearly so, lance-shaped to elliptic or oblong lance-shaped, 4-8 cm long, finely serrate. Flowers white or tinged with pink, often unisexual; peduncle 2-8 cm long; styles usually 3. Fruit a yellow berry, about 5 mm wide. April-May. — Rich woods.

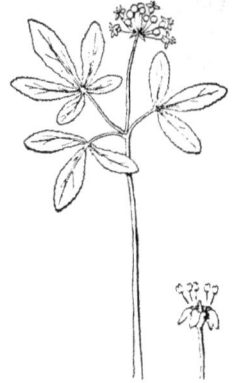

Panax trifolius

Aristolochiaceae

BIRTHWORT FAMILY

Asarum
WILD GINGER

Asarum canadense L.
CANADIAN WILD GINGER   *native / cons* **7**

Perennial herb; rhizome slender, branched, pubescent; producing annually a pair of leaves, between which arises the solitary, short-peduncled flower. Leaves 2, cordate, entire, at anthesis commonly 8-12 cm wide, larger at maturity, pubescent, especially on the long petiole. Flowers axillary, red-brown, 2-4 cm long, on a stout, pubescent pedicel 2-5 cm long; calyx tubular at base, deeply 3-lobed, the lobes spreading to reflexed, purple inside; petals absent or tiny and awl-shaped; ovary inferior, 6-celled. Fruit a capsule, bursting irregularly; seeds large, ovoid, wrinkled. April-May. — Rich woods, usually in small colonies.

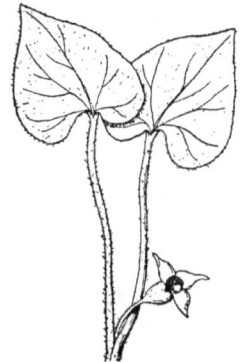

Asarum canadense

Asteraceae

ASTER FAMILY

Annual, biennial or perennial herbs. Leaves simple or compound, opposite, alternate, or whorled. Flowers perfect (with both staminate and pistillate parts) or single-sexed (sometimes sterile) and of 2 types: ray (or ligulate) and disk (or tubular). Ray flowers joined at base and have a long, flat, segment above (the ray); disk flowers tube-shaped with 5 lobes or teeth at tip.

Flowers are clustered in 1 of 3 types of heads resembling a single flower and attached to a common surface (receptacle): ray flowers only (as in dandelion, *Taraxacum*); disk flowers only (discoid, as in tansy, *Tanacetum*); and heads with both ray and disk flowers (radiate), the ray flowers surrounding the disk flowers (as in sunflower, *Helianthus*).

In addition to flowers, the receptacle may also have scales called chaff; if no scales present, the receptacle is termed naked. Each head is surrounded by involucral bracts (sometimes called phyllaries); collectively, the bracts are termed the involucre, comparable to the group of sepals (calyx) subtending an individual flower. Fertile flowers have 1 pistil tipped by a 2-cleft style (undivided in sterile flowers); stamens 5; ovary (and achene) often topped by several to many scales, awns or hairs (the pappus). Fruit a seedlike achene (sometimes termed cypsela in Asteraceae).

KEY TO ASTERACEAE TRIBES

Because of its large size, the Asteraceae is often divided into smaller groups of related species called tribes. In Door County, 10 tribes (and one subtribe) occur, and are identified in the following key. The tribal key is adapted from Kowal (2007).

1 Flowers wind-pollinated, not showy; rays absent; florets and most heads unisexual; anthers not united; involucral bracts typically connate, at least basally . . . . *HELIANTHEAE*
. . . . (Subtribe **AMBROSIINAE**, Ragweed subtribe)
1 Flowers insect-pollinated, usually showy; rays present or absent; heads and most florets bisexual; anthers united; involucral bracts free; if wind-pollinated, involucral bracts free, scarious and more than 5 (*Artemisia*) 2
2 Plants with milky juice; heads ligulate (florets bisexual and with a 5-toothed ligule) . . . .
. . . . *CICHORIEAE* (Lettuce Tribe)
2 Plants with watery juice; heads radiate (with disk florets surrounded by ray florets), or disciform but with ray florets without rays, or discoid (only disk florets) 3
3 Plants and/or heads usually prickly; heads discoid and corolla lobes of the disk florets at least 4 times longer than wide; receptacle densely bristly; leaves alternate; style with a ring of hairs (sometimes merely with a thickened ring) below the branches . . . . *CARDUEAE* (Thistle Tribe)
3 Plants and heads not prickly; heads various, corolla lobes of the disk florets less than 4 times longer than wide; receptacle various, rarely bristly; leaves various; style otherwise 4
4 Heads discoid and corollas never yellow; style-branches long and slender (thread-like), conspicuously protruding from the corolla and often attractive; receptacle naked 5
4 Heads various, but if discoid, corollas yellow (or at least creamy); style- branches much shorter (relative to their widths); receptacle various. 6
5 Style-branches hispidulous, acute or acuminate at tip; corollas purple; inflorescence corymbose; leaves alternate . . . . *VERNONIEAE* (Ironweed Tribe)
5 Style-branches merely papillate, blunt (to acutish) and sometimes thickened (clavate) towards the tip; corollas white, pink, rose or blue-violet; inflorescence various; leaves alternate, opposite or whorled
. . . . *EUPATORIEAE* (Boneset Tribe)
6 Pappus of hairs or bristles; leaves alternate (in some Senecioneae directly from a rhizome) 7
6 Pappus absent or of awns, scales, or teeth; leaves alternate or opposite. 9
7 Involucral bracts equal and in 1 row; rays yellow or absent; style-branches with a tuft of hairs at the end . . . . *SENECIONEAE* (Groundsel Tribe)
7 Involucral bracts in 2-5 rows, equal or unequal, if (rarely) in 1 row, with conspicuous white, pink, purple, or blue rays 8
8 Heads discoid or disciform; involucral bracts scarious, either virtually entirely or at least at the tip for a third of their lengths . . . . *GNAPHALIEAE* (Pussytoes Tribe)
8 Heads radiate; involucral bracts not scarious or scarious only on the margins. *ASTEREAE* (Aster Tribe)
9 Involucral bracts with scarious or hyaline margins; leaves alternate. 11
9 Involucral bracts not scarious or hyaline, or if so, leaves opposite; leaves alternate or opposite . . 11
10 Leaves entire, not aromatic; receptacle naked. . . . . *ASTEREAE* (Aster Tribe)
10 Leaves toothed, lobed, or finely divided, often aromatic; receptacle chaffy or naked; style-branches mostly truncate, with a tuft of hairs at the end (like Senecioneae). *ANTHEMIDEAE* (Chamomile Tribe)
11 Receptacle naked; rays present, widest at the prominently 3-lobed apex; leaves alternate, linear to lanceolate to ovate . . . . *HELENIEAE* (Sneezeweed Tribe)
11 Receptacle chaffy (absent in Dyssodia, with unremarkable rays and opposite pinnatisect leaves); rays present or absent, but when present usually not as above, but if so, then the leaves opposite and either lobed or pinnatifid . . . . *HELIANTHEAE* (Sunflower Tribe)

ANTHEMIDEAE (CHAMOMILE TRIBE)

GENERA: *Achillea, Anthemis, Artemisia, Leucanthemum, Matricaria, Tanacetum*

1 Receptacle chaffy; heads radiate 2
1 Receptacle naked or villous; heads radiate, disciform or discoid 4
2 Heads small, 5 mm or less in diameter, densely corymbose; receptacle flat; achenes compressed . . .
. . . . *Achillea*
2 Heads rather large, 1-4 cm in diameter, solitary and terminal on long peduncles; receptacle conic at maturity; achenes terete or angled. 3
3 Ray florets white; disk 0.5-1.2 cm in diameter . . . . *Anthemis*
3 Ray florets yellow; disk 1-2 cm in diameter . . . . *Anthemis tinctoria*

4 Inflorescence paniculate, racemose or spike-like with inconspicuous discoid heads; florets green . . .
. . . . *Artemisia*
4 Inflorescence corymbose or heads terminal on long peduncles; ray florets showy, yellow or white
(sometimes obsolete). 5
5 Receptacle conic at maturity; leaves pinnatisect . . . . *Matricaria*
5 Receptacle flat or low-convex 6
6 Heads several or many, in corymbs, disk 4-9 mm wide, with or without rays; leaves often highly lobed
. . . . *Tanacetum*
6 Heads solitary at tips of stem or long branches, large, disk 1-2.5 cm wide, with conspicuous white
rays; leaves toothed to lobed . . . . *Leucanthemum*

ASTEREAE (ASTER TRIBE)
GENERA: *Bellis, Doellingeria, Erigeron, Eurybia, Euthamia, Grindelia, Solidago, Symphyotrichum*

1 Ray corollas yellow, conspicuous; disk corollas yellow 2
1 Ray corollas white, pink, violet, bluish or purple; disk corollas various. 4
2 Pappus of 2-8 caducous awns; involucre more or less glutinous . . . . *Grindelia*
2 Pappus of numerous capillary bristles or hairs; involucre not glutinous. 3
3 Inflorescence corymbiform; leaves glandular punctate, linear to narrowly oblong, only slightly reduced
upwards on stem; ray florets more numerous than the disk florets . . . . *Euthamia*
3 Plants not with both inflorescence corymbiform and ray florets more numerous than the disk florets;
leaves usually broader, not glandular punctate . . . . *Solidago*
4 Pappus absent or inconspicuous (2-4 awns up to 2 mm long and several minute bristles); receptacle
conic, low-conical or hemispherical. . . . . *Bellis*
4 Pappus of long capillary bristles or hairs; receptacles flat. 5
5 Involucral bracts approximately in one series, neither chartaceous at base nor with herbaceous green
tip; style appendages roundish or obtuse, no longer than 0.3 mm; rays very numerous and narrowly
linear (mostly 1.3 mm or less wide); plants blooming chiefly in spring and early summer (when later,
plants also with heads past fruiting) . . . . *Erigeron*
5 Involucral bracts clearly imbricated or with a foliaceous outer series; style appendages longer and
more acute; rays in one or two series and relatively broader; plants blooming in late summer and fall
. . . . 6
6 Middle and lower stem leaves distinctly petioled, most of the petioles more than 1 cm long, wingless,
or winged but less than 1/4 as wide as the blades; blades (except the uppermost) more than 12 mm
wide, abruptly narrowed to a truncate or cordate base 7
6 Middle and lower stem leaves not distinctly petioled; sessile or subsessile on petioles less than 0.5
cm long, or apparently on broad-winged petioles wider than the blades; or with long narrow tapering
petiole-like bases, but the blades not more than 12 mm wide. 8
7 Involucral bracts narrowly to broadly ovate-lanceolate, outer ones 1.0-2.5 mm wide, less than 2.5x as
longer than wide; inflorescence corymbiform. . . . . *Eurybia*
7 Involucral bracts linear-deltoid to lanceolate, outer ones 0.2-1.0 mm wide, more than 2.5x longer
than wide; inflorescence elongate (paniculate or racemose) . . . . *Symphyotrichum*
8 Pappus double, the inner of long capillary bristles, the outer of short bristles, 1 mm long or less (very
obscure); middle and upper involucral bracts with scarious margins extending to tip, central green
line not or only slightly expanded towards tip; inflorescence corymbose . . . . *Doellingeria*
8 Pappus not double; involucral bracts various, but in most species with the central green line conspicuously dilated at tip; inflorescence various, but in most species not corymbose (*Symphyotrichum*). . 9
9 Rays much reduced or absent (heads "disciform"); pappus conspicuous at anthesis; plants annual,
with taproots . . . . *Symphyotrichum ciliatum*
9 Rays present, heads conspicuously radiate; pappus inconspicuous at anthesis or at most barely overtopping disk corollas; plants perennial, forming clumps or with rhizomes. . . . . *Symphyotrichum*

CARDUEAE (THISTLE TRIBE)
GENERA: *Arctium, Centaurea, Cirsium*

1 Neither plants nor heads prickly or involucral bracts spine-tipped and corollas yellow; achenes obliquely
attached to the receptacle; marginal disk florets often enlarged and showy; involucral bracts often
with margins scarious and deeply cleft at tip (laciniate); pappus hairs mostly less than 3 mm long or
lacking . . . . *Centaurea*
1 Plants and/or heads prickly; corollas not yellow; achenes attached by the base to the receptacle;

florets all alike; involucral bracts not laciniate at tip; pappus hairs usually more than 5 mm long . . . 2
2 Leaves unarmed, broadly rounded at base; tip of phyllary a hook . . . . **Arctium**
2 Leaves prickly, lanceolate to ovate; tip of phyllary a straight spine or merely mucronate . . . . . **Cirsium**

CICHORIEAE (LETTUCE TRIBE)

GENERA: *Cichorium, Hieracium, Hypochaeris, Lactuca, Lapsana, Leontodon, Prenanthes, Sonchus, Taraxacum, Tragopogon*

1 Pappus absent . . . . *Lapsana*
1 Pappus present 2
2 Pappus of numerous simple hairlike (capillary) bristles only 3
2 Pappus otherwise (plumose bristles, scales, scales mixed with bristles, or a ring of numerous minute
   bristles) 7
3 Achenes flattened or compressed 4
3 Achenes cylindrical, fusiform or terete, not flattened 5
4 Achenes not beaked, not enlarged at the tip; heads yellow with many florets (80 or more) . **Sonchus**
4 Achenes beaked or unbeaked, but constricted below enlarged tip; heads yellow or blue, with relatively
   few florets (5-56) . . . . **Lactuca**
5 Plants scapose; achenes beaked; pappus white; involucral bracts in more than one series **Taraxacum**
5 Stems branched or unbranched and leafy or subscapose; achenes truncate or tapered, rarely short-
   beaked; pappus pale yellow, red-brown, tannish or white; involucral bracts in 1 or 2 series 6
6 Leaves lanceolate to palmately lobed; heads cylindrical, nodding; corolla pink, purplish to yellow or
   white; pappus pale yellow to red-brown; plants sometimes tomentose, not glandular . . . **Prenanthes**
6 Leaves spatulate to oblanceolate, not lobed; heads campanulate, erect; corolla yellow to red-orange;
   pappus tannish; plants usually glandular-pubescent . . . . **Hieracium**
7 Pappus of minute scales and/or bristles . . . . **Cichorium**
7 Pappus of plumose (feathery) bristles only 8
8 Plants leafy stemmed, branched, not scaly-bracted above; leaves cauline, grasslike . . . . . **Tragopogon**
8 Plants scapose, scaly bracted above; leaves basal, coarsely dentate 9
9 Inner and outer achenes uniform, not slender-beaked; receptacle . . . . **Leontodon**
9 Inner achenes with long, slender beaks; receptacle chaffy . . . . **Hypochaeris**

EUPATORIEAE (BONESET TRIBE)

GENERA: *Ageratina, Eupatorium, Eutrochium, Liatris*

1 Leaves alternate; plants from an enlarged corm; achenes 10-ribbed; pappus of plumose or barbellate
   bristles; involucral bracts weakly ribbed. . . . . **Liatris**
1 Leaves opposite or whorled; roots fibrous; achenes 5-angled; pappus of capillary bristles; involucral
   bracts not ribbed 2
2 Leaves in whorls of 3, 4 or 5; heads purple or dull rose; involucral bracts in 5-6 series . . . **Eutrochium**
2 Leaves opposite (rarely in 3s in *E. perfoliatum*); heads white (rarely purple in *E. perfoliatum*); involucral
   bracts in 2-3 series 3
3 Leaves long-petioled, ovate; involucral bracts nearly uniseriate, narrowly linear, any basal ones usually
   much less than half the length of the longest; heads with 15-30 florets; amber resin glands absent .
   . . . . **Ageratina**
3 Leaves sessile (except *E. serotinum*), narrowly ovate or lanceolate; involucral bracts in 2-3 series, not
   narrowly linear, many roughly half the length of the longest; heads with 15 or fewer florets; tiny
   amber resin glands on leaf undersides, involucral bracts, corollas, and achenes . . . . **Eupatorium**

GNAPHALIEAE (PUSSYTOES TRIBE)

GENERA: *Anaphalis, Antennaria, Pseudognaphalium*

1 Stem leaves few, much smaller than those of the persistent basal rosette, strongly ascending; stolons
   present; plants either staminate or pistillate, populations dioecious . . . . **Antennaria**
1 Stem leaves many, about the same size as the basal leaves, which soon wither; stolons absent 2
2 Involucral bracts pure white, with conspicuous, longitudinal creases creating the appearance of wrinkled tissue paper; populations dioecious, although pistillate plants often with heads having a few staminate florets in the center; dried plants without a strong odor . . . . **Anaphalis**
2 Involucral bracts grayish white, yellow or brown, scarious, with very small longitudinal ridges but no

conspicuous creases; heads bisexual, with pistillate florets marginally and staminate heads in center; dried plants with strong tobacco-like odor .... *Pseudognaphalium*

HELENIEAE (SNEEZEWEED TRIBE)
ONE GENUS: *Helenium*

HELIANTHEAE (SUNFLOWER TRIBE)
GENERA: *Bidens, Coreopsis, Helianthus, Heliopsis, Ratibida, Rudbeckia*

1 Involucre distinctly double, the outer larger (or minute, 2 mm or less long), foliaceous, somewhat spreading, the inner broader and appressed, nearly membranous 2
1 Involucre not double, involucral bracts all about equal in length, the inner and outer similar in texture 3
2 Pappus absent or of a few teeth .... *Coreopsis*
2 Pappus of 2 to 4 barbed awns... *Bidens*
3 At least some of the leaves opposite or all basal..4
3 Leaves all alternate...5
4 Outer involucral bracts shorter than the inner; ray florets neuter, their rays thin and easily wilting, deciduous .... *Helianthus*
4 Outer involucral bracts longer than the inner; ray florets pistillate, their rays marcescent (thickish and persistent after flowering) .... *Heliopsis*
5 Disk flat or convex; leaves neither lobed nor divided .... *Helianthus*
5 Disk conical, hemispheric or columnar; leaves simple in *Rudbeckia hirta,* otherwise lobed, cleft, laciniate or pinnately parted 6
6 Leaves simple, 3-lobed, or -cleft, or laciniate; rays not subtended by receptacular bracts; achenes 4-sided .... *Rudbeckia*
6 Leaves pinnately divided; rays subtended by receptacular bracts; achenes laterally flattened *Ratibida*

HELIANTHEAE (AMBROSIINAE, RAGWEED SUBTRIBE)
GENERA: *Ambrosia, Xanthium*

1 Pistillate heads 2-flowered, with many, sharp-hooked spines; staminate heads lacking involucral bracts .... *Xanthium*
1 Pistillate heads 1 (-2) -flowered with a few vestigial spines or none; staminate heads with involucres of connate involucral bracts .... *Ambrosia*

SENECIONEAE (GROUNDSEL TRIBE)
GENERA: *Erechtites, Hasteola, Packera, Petasites, Senecio*

1 Perennials with green leaves arising individually from the ground from an underground rhizome; aerial stems consisting of scaly bracted flowering scapes arising before or as the leaves develop in early spring .... *Petasites*
1 Habit various but with well developed cauline leaves (though these may differ from the basal leaves) 2
2 Corollas yellow to orange; heads usually with rays 3
2 Corollas whitish or creamy; heads without rays 4
3 Stem leaves progressively reduced upward and lobed (unlike the basal leaves); perennials, usually with obvious vegetative reproduction .... *Packera*
3 Leaves more or less equal in size up the stem; annuals (perhaps rarely biennials) .... *Senecio*
4 Annuals; heads disciform, with 2 to several marginal rows of pistillate florets with filiform corollas; leaves roughly the same size up the stem .... *Erechtites*
4 Perennials; heads discoid, containing only bisexual florets with 5-lobed corollas......... *Hasteola*

VERNONIEAE (IRONWEED TRIBE)
ONE GENUS: *Vernonia*

Achillea
YARROW

Perennial herbs. Leaves alternate, subentire to pinnately dissected. Inflorescence more or less corymbiform, of several to many relatively small heads. Heads radiate or rarely discoid, the rays mostly 5-12, pistillate and fertile, rarely neutral, white, sometimes pink or rarely yellow. Involucral bracts imbricate in 3-4 series, dry, with scarious or hyaline margins and often greenish midrib. Receptacle conic or convex, chaffy throughout. Disk flowers about 10-75, perfect and fertile. Fruit a compressed achene; pappus none.

1 Leaves finely dissected; plants tomentose; ubiquitous . . . . *A. millefolium*
1 Leaves nearly entire; plants nearly hairless; uncommon adventive species . . . . *A. ptarmica*

Achillea millefolium L.
COMMON YARROW   *native / cons* 1

Aromatic rhizomatous perennial, sparsely to rather densely villous throughout. Stems about 2-10 dm tall. Leaves pinnately dissected, the blade about 3-15 cm long and to 2.5 cm wide, the basal petiolate, all but the lowermost stem leaves sessile. Heads numerous in a flat or round-topped, short and broad, paniculate-corymbiform inflorescence, the disk about 2-4 mm wide; involucre in ours mostly 4-5 mm high; rays about 5, white or occasionally pink, 2-3 mm long; disk flowers about 10-30. June-Oct. — Common in fields, prairies, lawns, beaches, and waste places.

Achillea ptarmica L.
SNEEZEWEED   *introduced*

Rhizomatous perennial. Stems 3-6 dm tall, villous above, often nearly glabrous below. Leaves glabrous or nearly so, linear or lance-linear, about 3-10 cm long and 2-6 mm wide, sessile, closely and rather shallowly serrate to subentire. Heads several or numerous in an open corymbiform inflorescence, the disk about 4-8 mm wide; involucre about 4-5 mm high; rays commonly 8-10, white, 3-5 mm long; disk flowers about 50-75. July-Sept. — Beaches, roadsides, and waste places; native to n Europe and Asia.

Forms escaped from cultivation are often "double," with more than the usual number of ray flowers.

Achillea millefolium

Achillea ptarmica

Ageratina
SNAKEROOT

Ageratina altissima (L.) R. M. King & H. Rob.
WHITE SNAKEROOT   *native / cons* 4
Eupatorium rugosum Houtt.

Perennial herb. Stems mostly 5-8 dm tall, finely pubescent. Leaves oppo-site, ovate to broadly ovate, base subcordate, 4-11 cm long and 3-8 cm wide, smaller upward; petioles 1-3 cm long; margins coarsely serrate. In-florescence flat-topped or flat dome-shaped; involucre 4-5 mm long, in-volucral bracts acuminate to obtuse, all about the same length; florets 12-24, corolla white; pappus of bristles. Aug-Oct. — Floodplain forests, cedar swamps, thickets, streambanks, wooded ravines, sometimes where disturbed.

A toxic substance in this plant can cause "trembles," a fatal disease of cattle which have browsed on it and transmittable to humans by their milk, in whom the consequent "milk sickness" caused many deaths in the 19th century.

Ageratina altissima

Ambrosia
RAGWEED

Coarse annual or perennial herbs. Leaves opposite or alternate, mostly lobed or dissected. Heads unisexual, small. Staminate heads in a spike-like or raceme-like bractless inflorescence; involucre

5-12-lobed; receptacle flat, its bracts slender. Pistillate heads borne below the staminate ones, in the axils of leaves or bracts; involucre closed, nut-like, usually with a single series of tubercles or short erect spines near the apex; pistil solitary, without corolla; pappus absent.

The pollen is wind-borne, and some species are among the most important causes of hay-fever in the USA.

1 Leaves palmately 3–5 lobed or unlobed; large annual plant to 2 m or more tall . . . . ***A. trifida***
1 Leaves 1–2 times pinnately lobed or divided; plants usually less than 1 m tall. 2
2 Plants perennial, forming colonies from creeping underground roots; leaves usually coarsely lobed
. . . . ***A. psilostachya***
2 Plants taprooted annuals; leaves finely divided . . . . ***A. artemisiifolia***

Ambrosia artemisiifolia L.
COMMON RAGWEED　　　　　　　　　　　　　　　　*native / cons* **0**
Annual weed. Stems branching at least above, variously hairy or sub-glabrous, mostly 3-10 dm tall. Leaves opposite below, alternate above, petiolate, 1-2x pinnatifid, ovate or elliptic in outline, commonly 4-10 cm long. Sterile heads short-pedunculate. Fruiting involucre short-beaked, about 3-5 mm long, with several short sharp spines. Aug-Oct. — Waste places.

Ambrosia artemisiifolia

Ambrosia psilostachya DC.
PERENNIAL RAGWEED　　　　　　　　　　　　　　　*native / cons* **2**
Similar to *A. artemisiifolia*. Perennial from a creeping rhizome. Leaves thicker, short-petiolate or subsessile, usually only once pinnatifid, averaging narrower in outline, sometimes 10 cm long. Fruiting involucre tuberculate above, sometimes obscurely so. July-Oct. — Waste places, usually in dry or sandy soil.

Ambrosia trifida L.
GIANT RAGWEED　　　　　　　　　　　　　　　　*native (invasive)*
Annual weed, of various heights to sometimes 5 m tall. Stems spreading-hirsute or hispid above, often glabrous or glabrate below. Leaves opposite, petiolate, broadly elliptic to more commonly ovate or suhorbicular, serrate, palmately 3-5-lobed, or, especially in depauperate specimens, lobeless, often 2 dm long or more, more or less scabrous on both sides. Sterile involucres unilaterally 3-nerved. Fertile involucres about 5-10 mm long in fruit, several-ribbed, each rib bearing a short spine at the tip. July-Oct. — Moist soil and waste places.

Ambrosia trifida

Anaphalis
PEARLY-EVERLASTING

Anaphalis margaritacea (L.) Benth.
PEARLY-EVERLASTING　　　　　　　　　　　　　　*native / cons* **3**
White-woolly perennial herbs. Stems erect, simple or branched, commonly 3-9 dm tall, leafy, loosely white-woolly. Leaves alternate, lance-shaped or linear, to about 12 cm long and 1.5 cm wide, sessile, commonly less pubescent above than beneath, or green and glabrous above, the margins entire and often revolute, basal leaves soon deciduous. Heads 1 cm wide or less, numerous and crowded in a short broad inflorescence; some flowers bearing both stamens and pistils. Involucre about 5-7 mm high, the bracts pearly white. Achenes papillate; pappus, in both staminate and pistillate flowers, of distinct capillary bristles. Variable. July-Aug. — Chiefly in dry woods and clearings.

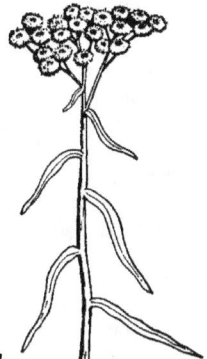

Anaphalis margaritacea

Antennaria
PUSSYTOES

Perennial woolly herbs. Leaves basal and alternate on the stem. Flowers dioecious, rarely incompletely so. Heads many-flowered, disciform or discoid, solitary to many in a crowded inflorescence. Involucral bracts imbricate in several series, scarious at least at the tip, often colored. Receptacle naked, flat or convex. Staminate flowers with scanty pappus, the bristles commonly barbellate or clavate. Pistillate flowers with filiform-tubular corolla, bifid style, and copious bristles slightly united at the base. Achenes terete or slightly compressed. Most of our species are partly or wholly apomictic, producing seeds without fertilization.

1 Rosette leaves small, 1-nerved or obscurely 3-nerved (best viewed on old basal leaves from the previous year) 2
1 Rosette leaves larger, 3- or 5-nerved 3
2 New basal leaves of the season essentially glabrous above or very soon becoming so (may appear hairy along the margin from tomentum of underside) . . . . *A. howellii*
2 New basal leaves pubescent above when young (becoming glabrous only in age) . . . . . . *A. neglecta*
3 Basal leaves glabrous or tomentose on upper surface, underside green-glabrous; pistillate involucres 8–13 mm; staminate corollas 3.5–5 mm; pistillate corollas 4–7 mm; young stolons mostly decumbent . . . . *A. parlinii*
3 Underside of basal leaves tomentose; pistillate involucres 5–7 mm; staminate corollas 2–3.5 mm; pistillate corollas 3–4 mm; young stolons mostly ascending . . . . *A. plantaginifolia*

Antennaria howellii Greene
SMALL PUSSYTOES   *native / cons* 2
Antennaria neglecta var. *howellii* (Greene) Cronquist
Plants mostly with pistillate flowers only, staminate plants rare; spreading by short stolons. Stems to 35 cm tall, sometimes with gland-tipped hairs. Basal leaves oblanceolate to ovate, 2-5 cm long, tips mucronate, upper surface tomentose, underside green-glabrous or gray-pubescent. Stem leaves linear, 1-4 cm long. Heads 3-15 in corymbiform clusters. Involucral bracts white, cream, or light brown, sometimes rose at base. — Many types of dry, open places: rock ledges and outcrops, openings in sandy or rocky woods; sometimes on moist shores, roadsides, and in fields and lawns.

Antennaria howellii

Antennaria neglecta Greene
FIELD PUSSYTOES   *native / cons* 3
Antennaria neodioica Greene
Plants 1-4 dm tall, with short and leafy or longer and merely bracteate stolons. Basal leaves and those at the ends of the stolons densely and persistently tomentose beneath, only sparsely so (and eventually glabrate) above, or the upper side glabrous from the first, relatively small, mostly under 1.5 cm wide, 1-nerved or obscurely 3-nerved. Pistillate involucres 7-10 mm long; variously sexual or apomictic. April-June. — Dry woods and open places.

Antennaria parlinii Fernald
PARLIN'S PUSSYTOES   *native / cons* 3
Antennaria plantaginifolia var. *parlinii* Cronquist
Similar to *A. plantaginifolia* and sometimes included in it. See key for distinctions between the 2 taxa. Stems usually with purple glandular hairs (especially near summits of young flowering stems).

Dry open places, including rock outcrops, banks, grassy roadsides, hillsides, and open woods; sometimes in shaded forests.

Antennaria plantaginifolia (L.) Richards.
PLANTAIN-PUSSYTOES   *native / cons* 3
Stoloniferous perennial, the stolons sparsely leafy or sometimes merely bracteate. Stems about 10-40 cm tall. Basal leaves and those at the ends

Antennaria plantaginifolia

of the stolons sooner or later glabrate above, relatively large, 3-5-nerved, evidently petiolate, the blade ovate to elliptic or obovate, mucronate, the larger ones 2-6 cm long and 1.5-5 cm wide; stem leaves reduced, mostly linear or lance-shaped. Heads several in a small cyme; some purplish glandular hairs often present in the inflorescence. Pistillate involucre 5-10 mm high, the bracts white-tipped, often pinkish toward the base; staminate involucres generally a little smaller than the pistillate, with broader and more conspicuous white tips to the bracts. Styles often crimson. Spring-early summer. — Dry, open, woodlands, banks, bluff tops. Similar to *A. parlinii* except for smaller heads, and underside of basal leaves gray-pubescent.

Anthemis
CHAMOMILE
Annual or perennial, usually aromatic herbs. Leaves alternate, dissected. Flowers in campanulate or nearly hemispheric heads terminating the branches. Heads radiate or rarely discoid, the rays elongate, white or yellow, pistillate or neutral; involucral bracts subequal or more commonly imbricate in several series, the margins more or less scarious or hyaline; receptacle convex to conic or hemispheric, chaffy at least toward the middle. Disk flowers numerous, perfect, yellow. Fruit a terete, angled, or somewhat compressed achene; pappus a short crown, or more commonly none. *Anthemis* and *Matricaria* are very similar, separated on the receptacle chaffy in *Anthemis* and naked in *Matricaria*.

1   Ray flowers yellow . . . . *A. tinctoria*
1   Ray flowers white . . . . *A. cotula*

Anthemis cotula L.
STINKING CHAMOMILE, DOGFENNEL   *introduced*
More or less branched, usually subglabrous, ill-smelling annual 1-6 dm tall. Leaves about 2-6 cm long, 2 or 3x pinnatifid, with very narrow segments. Heads more or less numerous, short-pedunculate at the ends of the branches, the disk about 5-10 mm wide, becoming ovoid or short-cylindric at maturity; involucre sparsely villous; rays about 10-20, white, neutral, 5-11 mm long; receptacle chaffy only toward the middle, its bracts narrow, tapering to the apex, scarcely awned. Achenes subterete, about 10-ribbed, glandular-tuberculate. Pappus none. May-Oct. — Native of Europe; fields and waste places.

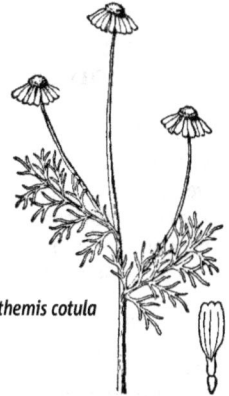

Anthemis cotula

Anthemis tinctoria L.
GOLDEN CHAMOMILE   *introduced*
  Cota tinctoria (L.) J. Gay ex Guss.
Short-lived perennial. Stems 3-7 dm tall, sparingly branched above or simple, finely hairy at least above. Leaves pinnatifid, about 2-5 cm long, with winged rachis and deeply toothed or pinnatifid segments, villous or almost floccose beneath. Heads solitary and long-pedunculate at the ends of the branches, the disk about 12-18 mm wide; involucre thinly tomentose; rays about 20-30, pistillate, yellow, about 7-15 mm long receptacle chaffy throughout, its bracts narrow, with firm yellow awn-tips equaling the disk-flowers. Achenes compressed-quadrangular, more or less striate-nerved. Pappus a very short crown. June-July. — Fields and waste places; native of Europe.

Arctium
Anthemis tinctoria
BURDOCK
Coarse biennial herbs. Leaves large, alternate, heart-shaped, entire or toothed. Flowers all tubular and perfect, the corolla pink or purplish, with long slender lobes; involucre subglobose, its

bracts multiseriate, narrow, appressed at the base, with a spreading, subulate, inwardly hooked tip. Receptacle flat, densely bristly. Achenes oblong, slightly compressed, few-angled, many-nerved, truncate at the apex, glabrous. Pappus of numerous short, separately deciduous bristles. Any of our introduced species will hybridize with any of the others.

1 Heads 1–1.5 cm long, sessile or on short peduncles; common weed . . . . **A. minus**
1 Heads about 1.5 cm long, on long peduncles; uncommmon . . . . **A. tomentosum**

Arctium minus Bernh.

LESSER BURDOCK   *introduced (invasive)*
Biennial, to 1.5 m tall or rarely taller. Leaves petiolate, the lower petioles mostly hollow; blade narrowly to very broadly ovate, to about 5 dm long and 4 dm wide, thinly tomentose and often eventually glabrate beneath, nearly glabrous above. Branches of the inflorescence ascending to widely spreading, raceme-like, the heads mostly short-pedunculate or subsessile, 1.5-3 cm thick, glabrous or slightly glandular to sometimes tomentose, usually a little shorter than the flowers, stramineous or purplish, the inner bracts often more flattened than the others and scarcely hooked. — Roadsides, fencerows, farmyards, around old buildings, disturbed places.

Arctium tomentosum P. Mill.

WOOLLY BURDOCK   *introduced*
Similar to *A. lappa* (known from a number of Wisconsin locations), but smaller, seldom over 1.3 m tall; lower petioles mostly hollow; involucre mostly 2-3 cm thick, more or less strongly arachnoid-tomentose, and only weakly or scarcely hooked at the tip. June-Oct. — Native of Eurasia, reported from a sunny creekside in Door County.

Arctium minus

Artemisia

WORMWOOD, SAGE

Annual, biennial, or perennial herbs, or shrubs, usually aromatic, with alternate entire to dissected leaves and few to numerous small, ovoid to campanulate or hemispheric heads in a spiciform, raceme-like, or panicle-like inflorescence. Heads discoid, sometimes with only perfect flowers, sometimes the outer pistillate, the central ones then sometimes sterile. Involucral bracts dry, imbricate, at least the inner scarious or with scarious margins. Receptacle flat to convex or hemispheric, naked or densely beset with long hairs. Achenes ellipsoid or obovoid to nearly prismatic, scarcely compressed, usually glabrous. Pappus none.

1 Plants perennial and somewhat woody at base; leaves covered with silky hairs; receptacle hairy . . . .
   . . . . **A. absinthium**
1 Plants annual, biennial, or perennial; leaves hairy to glabrous; receptacle naked 2
2 Disk flowers sterile; mature plants usually glabrous . . . . **A. campestris**
2 Disk flowers fertile 3
3 Leaves glabrous or nearly so, pinnately divided or dissected . . . . **A. biennis**
3 Leaves densely hairy at least on one surface, simple or dissected 4
4 Leaves green and nearly glabrous above, white hairy below; uncommon weed . . . . **A. vulgaris**
4 Leaves hairy on upper and lower sides . . . . **A. stelleriana**

Artemisia absinthium L.

COMMON WORMWOOD   *introduced*
Fragrant perennial herb or shrub. Stems 4-10 dm tall, finely sericeous or eventually glabrate. Leaves silvery-sericeous on both sides, or eventually nearly glabrous above, the lower long-petiolate and 2-3 times pinnatifid, with mostly oblong segments about 1.5-4 mm wide, the blade rounded-ovate in outline, about 3-8 cm long; upper leaves progressively less divided and shorter-petiolate, the divisions often more acute. Inflorescence ample, leafy; involucre about 2-3 mm high, finely and densely sericeous; flowers all fertile, the marginal pistillate; receptacle beset with numerous long

Artemisia absinthium

white hairs between the flowers. Achenes glabrous, nearly cylindric, but narrowed to the base and rounded at the summit. July-Sept. — Native of Europe; fields and waste places.

Artemisia biennis Willd.
BIENNIAL WORMWOOD   *introduced*

Taprooted, annual or biennial herb. Stems erect, to 1 m or more long, often branched, smooth, only faintly scented. Leaves alternate, pinnately dissected nearly to middle, 5-12 cm long and 2-5 cm wide, the segments linear and toothed. Flowers in stalkless heads from upper leaf axils; the heads composed of many small green disk flowers, grouped into spike-like inflorescences, with leafy bracts much longer than the clusters of heads; pappus none. Fruit a small oblong achene. Aug-Sept. — Sandy lakeshores, streambanks, ditches, mud flats, disturbed areas; often where seasonally flooded. Native to nw USA, throughout Wisconsin as a weed.

Artemisia campestris L.
FIELD SAGEWORT   *native / cons* **4**

Scarcely odorous perennial with a taproot and generally several glabrous to villous stems 1-10 dm tall from a branching caudex. Basal leaves crowded, about 2-10 cm long including the petiole, 0.7-4 cm wide, 2x or 3x pinnatifid or 3-parted, with mostly linear-filiform divisions seldom more than 2 mm wide, glabrous to sericeous, persistent, or, especially in the larger forms, sometimes deciduous; cauline leaves similar but smaller and less divided, the uppermost often 3-parted or simple. Inflorescence small and spike-like to diffuse and panicle-like; involucre glabrous to densely villous-tomentose, 2-4.5 mm high; outer flowers pistillate and fertile; disk-flowers sterile, with abortive ovary; receptacle glabrous. Achenes subcylindric, glabrous, those of the disk flowers abortive. July-Sept. — Open places, often in sandy soil.

Artemisia biennis

Artemisia stelleriana Bess.
DUSTY MILLER   *introduced*

Perennial from a creeping rhizome, inodorous. Stems 3-7 dm tall, simple to the inflorescence, densely white-tomentose. Leaves white-tomentose on both sides more densely so beneath, obovate, 3-10 cm long, including the petiole, and 1-5 cm wide, with a few rounded relatively broad lobes, which may be again slightly lobed. Inflorescence narrow and often dense, elongate; heads relatively large, the involucre 6-7.5 mm high, the disk corollas 3-4 mm long; receptacle glabrous; flowers all fertile, the outer pistillate. Achenes glabrous, subterete, but narrowed to the base and rounded at the summit. May-Sept. Sandy beaches; native of Asia, escaped from cultivation.

Artemisia campestris

Artemisia vulgaris L.
MUGWORT   *introduced*

Aromatic perennial herb with a stout rhizome. Stems 0.5-1.5 m tall, simple or branched above, glabrous or nearly so below the inflorescence. Leaves green and glabrous or nearly so above, densely white-tomentose beneath, chiefly obovate or ovate in outline, about 5-10 cm long and 3-7 cm wide, the principal ones cleft nearly to the midrib into ascending, unequal segments which are again toothed or more deeply cleft, and ordinarily with one or two pairs of stipule-like lobes at the base. Inflorescence generally ample and leafy; involucre 3.5-4.5 mm high, more or less tomentose; receptacle glabrous; flowers all fertile, the outer pistillate; disk corollas about 2.0-2.8 mm long. Achenes ellipsoid, not nerved or angled, essentially glabrous. July-Oct. — Fields, roadsides, and waste places; Old World native, now established throughout most of e North America.

Artemisia vulgaris

Bellis
ENGLISH DAISY

Bellis perennis L.
LAWN DAISY, ENGLISH DAISY   *introduced*

Perennial, more or less spreading-hairy. Leaves basal, elliptic or obovate to orbicular, the blades dentate or denticulate, to 3.5 cm long and 2 cm wide, narrowed to margined petioles of equal or greater length. Heads solitary atop a scape 5-15 cm high; involucral bracts herbaceous, equal; receptacle conic, naked. Rays many, pistillate, white to pink or purple; disk flowers yellow. Achenes compressed, mostly 2-nerved; pappus absent. April-Nov. — Weedy in lawns or waste places.

Bellis perennis

Bidens
BEGGARTICKS

Weedy annual herbs. Leaves opposite (ours), simple, lobed, or pinnately divided. Flower heads with both disk and ray flowers, or with disk flowers only; ray flowers often about 8, yellow; involucral bracts in 2 series, the outer row leaflike and spreading, the inner row much shorter and erect; receptacle more or less flat and chaffy. Fruit a flattened achene; pappus of 2-5 barbed awns which persist atop the achene; the body of achene barbed or with stiff hairs (at least on the angles), the "stick-tights" facilitating dispersal of seed by animals.

1 Leaves simple and toothed, or sometimes lobed; achenes 3-4-awned 2
1 Leaves all (or mostly) pinnately divided or compound; achenes 2-awned 3
2 Leaves mostly sessile 4
2 Leaves with a petiole 1-4 cm long .... *B. tripartita*
3 Heads nodding when mature; outer involucral bracts widely spreading .... *B. cernua*
3 Heads mostly upright; outer involucral bracts erect or nearly so .... *B. tripartita*
4 Heads with disk flowers only, or with short rays less than 5 mm long 5
4 Heads with both disk and ray flowers, the rays over 1 cm long .... *B. trichosperma*
5 Disk flowers orange; outer involucral bracts mostly 6-8 .... *B. frondosa*
5 Disk flowers yellow; outer involucral bracts 10 or more .... *B. vulgata*

Bidens cernua L.
NODDING BUR-MARIGOLD   *native / cons* **4**

Annual herb. Stems often branched, to 1 m long, smooth or with spreading hairs. Leaves opposite, smooth, lance-shaped to oblong lance-shaped, 3-16 cm long and 0.5-5 cm wide; margins with sharp, forward-pointing teeth and often rough-to-touch; petioles absent, the leaves usually clasping at base. Flower heads many, globe-shaped, 1.5-3 cm wide, usually nodding after flowering; rays yellow, 6-8, to 1.5 cm long, or absent; outer involucral bracts 4-8, unequal in length, the margins often fringed with hairs. Fruit a more or less straight-sided achene, 5-7 mm long, with downward-pointing barbs on margins; pappus with 4 (sometimes 2) awns, the awns with downward-pointing barbs. July-Oct. — Exposed, sandy or muddy shores, streambanks, marshes, forest depressions, wet meadows, ditches and other wet places.

Bidens frondosa L.
DEVIL'S-PITCHFORK   *native / cons* **1**

Annual herb. Stems erect, 2-10 dm tall, branched, purple-tinged, more or less smooth. Leaves pinnately divided into 3-5 segments, the segments lance-shaped, to 10 cm long and 3 cm wide, underside sometimes with short hairs; margins with coarse, forward-pointing teeth; petioles slender, 1-6 cm long. Flower heads many on long, leafless stalks; disk flowers orange, the disk to 1 cm wide; rays absent or very small; the outer involucral bracts usually 8, green and leaflike, longer than disk, fringed with hairs on margins. Fruit a flattened, nearly black achene, 5-10 mm long; pappus

Bidens cernua

of 2 slender awns with downward-pointing barbs. July-Oct. — Wet, sandy or gravelly shores, forest depressions, streambanks, pond margins; weedy in wet disturbed areas.

Bidens trichosperma (Michx.) Britton
CROWNED BEGGARTICKS   *native / cons* **7**
  Bidens coronata (L.) Britt.
Annual or biennial herb. Stems branched, 3-15 dm tall, smooth, often purple. Leaves opposite, smooth, to 15 cm long, pinnately divided into 3-7 narrow leaflets; margins coarsely toothed or deeply lobed to sometimes entire; petioles 3-15 mm long. Flower heads with both disk and ray flowers, large and numerous on slender stalks; rays about 8, gold-yellow, 1-2.5 cm long; outer involucral bracts 6-10, to 1 cm long, short-hairy on margins, inner bracts shorter. Fruit a flattened achene, 5-9 mm long, with long, stiff hairs on margins; pappus of 2 short, scale-like awns, 1-2 mm long. July-Oct. — Open bogs, fens, tamarack swamps, shores, streambanks, marshes, sand bars.

Bidens trichosperma

Bidens tripartita L.
THREE-LOBE BEGGARTICKS   *native / cons* **5**
  Bidens acuta (Wieg.) Britt.
  Bidens comosa (Gray) Wieg.
Annual herb. Stems yellow, 1-12 dm tall, branched, smooth. Leaves opposite, lance-shaped to oval, 3-15 cm long and 0.5-5 cm wide, margins with coarse, forward-pointing teeth, rough-to-touch; petioles absent, or leaves tapered to a short, winged petiole. Flower heads 1-2.5 cm wide, several to many, remaining erect after flowering; disk flowers yellow-green; rays absent; outer involucral bracts leaflike, 5-10 or more, 2-4x longer than head. Fruit an achene, 3-7 mm long, downwardly barbed on the margins; pappus of 3 downward-pointing barbed awns, the awns shorter than the achenes. Aug-Oct. — Exposed shores, streambanks, mudflats, forest depressions, pond, wet meadows, ditches and other wet places.

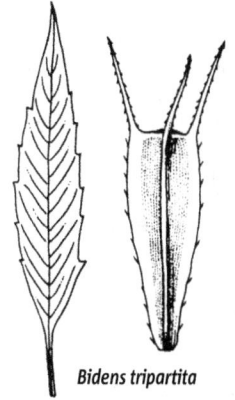

Bidens tripartita

Bidens vulgata Greene
TALL BEGGARTICKS   *native / cons* **1**
  Bidens puberula Wieg.
Annual herb. Stems to 2 m tall, smooth or upper stem and leaves short-hairy. Leaves opposite, pinnately divided into 3-5 segments, the segments lance-shaped, to 15 cm long and 5 cm wide, with prominent veins; margins with sharp, forward-pointing teeth; petioles present. Flower heads on stout, leafless stalks, disk flowers yellow; ray flowers usually present, small, yellow; outer involucral bracts about 13, leaflike. Fruit a flattened, olive-green or brown achene, 10-12 mm long; pappus of 2 awns with downward-pointing barbs. Aug-Oct. — Streambanks, wet meadows, wet forests; weedy in moist disturbed areas. Similar to devil's pitchfork (*Bidens frondosa*), but usually larger.

Bidens vulgata

Centaurea
KNAPWEED, STAR-THISTLE
Annual, biennial, or perennial herbs with alternate or all basal, entire to pinnatifid leaves, and solitary to numerous, small to large heads. Heads discoid, the flowers sometimes all tubular and perfect, or more commonly the marginal ones sterile, with enlarged, irregular, falsely radiate corolla. Involucral bracts imbricate in several series, either spine-tipped or more often some of them with enlarged appendages. Receptacle nearly flat, densely bristly. Corollas purple or blue to yellow or white, with slender tube and long narrow lobes. Achenes obliquely or laterally attached to the receptacle. Pappus of several series of graduated bristles or narrow scales, often much reduced, or wanting.

ADDITIONAL SPECIES
Centaurea scabiosa L.(Greater knapweed),introduced.

1 Leaves pinnately divided into linear-elliptic lobes; common weed .... *C. stoebe*
1 Leaves entire or toothed, sometimes few-lobed .... *C. jacea*

Centaurea jacea L.
BROWN-RAY KNAPWEED   *introduced*
Perennial, to 12 dm tall, glabrous or somewhat arachnoid. Leaves toothed
or shallowly lobed to entire, the basal ovate or lance-shaped to elliptic,
long-petiolate, the cauline reduced upwards and becoming sessile. Heads
terminating the often numerous branches; involucre mostly 12-18 mm
high, a little narrower to a little broader than high; appendages of the
involucral bracts well developed, broad, tan to dark brown, the middle
and outer ones rather irregularly lacerate, the inner less so and often
deeply bifid; marginal flowers almost always enlarged. Pappus none.
June-Sept. Fields, roadsides, and waste places; native of Europe.

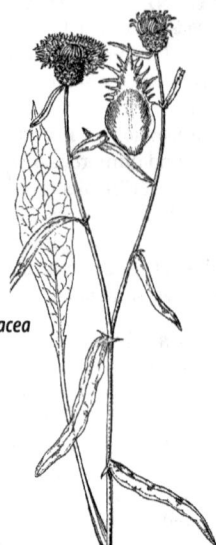

Centaurea jacea

Centaurea stoebe L.
SPOTTED KNAPWEED   *introduced (invasive)*
  Centaurea maculosa Lam.
Biennial or short-lived perennial, mostly 3-12 dm tall. Herbage with a
thin and loose, soon deciduous tomenturn, also sparsely scabrous-puberulent. Leaves obscurely to evidently glandular-punctate, pinnatifid
with narrow lobes, or the reduced ones of the inflorescence entire. Heads
terminating the numerous branches, constricted upwards; involucre
mostly 10-13 mm high, its bracts striate, the middle and outer ones with
short, dark, pectinate tips; flowers pink-purple, the marginal ones enlarged. Pappus to 2 mm long, or rarely wanting. June-Oct. — Native of
Europe; aggressive weed of fields, roadsides, waste places.

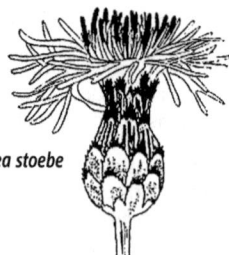

Cichorium
CHICORY
Centaurea stoebe

Cichorium intybus L.
CHICORY   *introduced*
Perennial with milky juice, from a long deep taproot. Stems branching,
hirsute or glabrous, 3-17 dm tall. Leaves alternate, lower leaves oblong
lance-shaped, petiolate, toothed or pinnatifid, 8-25 cm long and 1-7 cm
wide, becoming reduced, sessile, and entire or merely toothed upwards.
Heads sessile or short-pedunculate, borne 1-3 together in the axils of the
much reduced upper leaves. Flowers all ligulate and perfect, blue or occasionally white; involucral bracts biseriate, the outer shorter. Achenes
glabrous, 2-3 mm long, 5-angled, or the outer slightly compressed; pappus
of 2-3 series of scales, sometimes minute. July-Oct. — Roadsides, fields,
and waste places; native of Eurasia, now a cosmopolitan weed. The root
is used as an adulterant or substitute for coffee.

Cichorium intybus

Cirsium
THISTLE
Biennial or perennial herbs. Stems and leaves often spiny. Leaves from base of plant or alternate
on stem. Flower heads of pink to purple disk flowers only; involucral bracts tipped with spines.
Fruit a smooth achene; pappus of many slender bristles.

1 Involucral bracts tipped by spines mostly more than 2 mm long 2
1 Involucral bracts tipped by short spines to only 1 mm long 3
2 Leaves coarsely hairy, with cobwebby hairs on underside; introduced weed .... *C. vulgare*

2 Leaves densely white-hairy on both sides, especially on underside; rare native species in sandy soil .
. . . . *C. pitcheri*
3 Colony-forming perennial herb from deep, creeping rhizomes; common weed of dry to moist places
. . . . *C. arvense*
3 Biennial herbs; moist to wet habitats 4
4 Leaf bases not decurrent; stem not winged; involucral bracts usually with cobwebby hairs; flowers
deep rose-purple; native and not weedy . . . . *C. muticum*
4 Leaf bases decurrent, forming spiny wings on stem; involucral bracts usually without cobwebby hairs;
flowers pale pink-purple; introduced and weedy . . . . *C. palustre*

Cirsium arvense (L.) Scop.

CANADIAN THISTLE   *introduced (invasive)*
Perennial with deep-seated rhizomes, mostly 3-15 (or 20) dm tall, subglabrous, or the leaves more or less white-tomentose beneath. Heads
more or less numerous, polygamo-dioecious, the pappus of the pistillate
heads surpassing the corollas, that of the starminate heads surpassed by
the corollas; involucre 1-2 cm high, its bracts all innocuous, or the outer
with weak spine-tips about 1 mm long; flowers pink-purple or occasionally white. Achenes about 4 mm long. July-Aug. — A noxious weed of
fields and waste places; native of Eurasia, now statewide.

Cirsium arvense

Cirsium muticum Michx.

SWAMP THISTLE   *native / cons* **8**
Stout biennial herb. Stems 0.5-2 m long, branched in head, with long,
soft hairs when young, becoming more or less smooth. Leaves deeply
lobed into pinnate segments, 1-2 dm long, underside often with matted,
cobwebby hairs, becoming more or less smooth with age; margins toothed
and often tipped with spines; petioles present on lower leaves, stem
leaves sessile. Flower heads of purple or pink disk flowers only, single on
leafless stalks over 1 cm long at ends of stems; involucre 2-3.5 cm high;
the involucral bracts overlapping, densely hairy with cottony hairs (especially on margins), sometimes tipped with a short spine 0.5 mm long.
Achenes 5 mm long; pappus of long, slender bristles. Aug-Oct. — Swamps,
thickets, calcareous fens, sedge meadows, streambanks, shores.

Cirsium palustre (L.) Scop.

EUROPEAN SWAMP THISTLE   *introduced*
Biennial herb. Stems 0.5-2 m tall, spiny. Leaves to 20 cm long, deeply
lobed into pinnate segments, covered with loosely matted hairs or more
or less smooth, tapered at base and continued downward on stem as
spiny wings; margin teeth spine-tipped. Flower heads of purple disk
flowers only, on short stalks mostly less than 1 cm long; involucre 1-2 cm
high; the involucral bracts overlapping, not spine-tipped. Achenes 3 mm
long; pappus of slender bristles to 1 cm long. June-Aug. — Roadside
ditches and adjacent wetlands, including swamps, thickets and fens; resembling the native *C. muticum* in these habitats. Introduced and spreading into wetlands, especially where disturbed.

Cirsium muticum

Cirsium pitcheri (Torr. ex Eat.) Torr. & Gray

DUNE THISTLE   *threatened / native / cons* **10**
Plants biennial or short-lived perennials, commonly 5-10 dm tall; stems
and lower surfaces of the leaves densely and persistently white-tomentose, the upper surfaces of the leaves thinly so. Leaves deeply pinnatifid,
with narrow rachis and long, remote, linear, entire or few-toothed, weakly
spine-tipped lobes to about 8 cm long and 7 mm wide. Heads several; involucre mostly 2.5-3 cm high, its bracts slightly tornentose especially
marginally, well imbricate, the inner long-acuminate, the others with
weak spine-tips mostly 1-2 mm long; flowers commonly ochroleucous.

Cirsium pitcheri

Achenes mostly 5-7.5 mm long. June-Aug. — Uncommon plants of sand dunes along Lake Michigan. In their first year plants form a rosette of leaves flattened against the dune sands; the following year (or after several years) plants bloom and then die.

Cirsium vulgare (Savi) Ten.
BULL THISTLE   *introduced (invasive)*
Biennial weed mostly 5-15 dm tall. Stems conspicuously spiny-winged by the decurrent leaf-bases, copiously spreading-hirsute to sometimes arachnoid. Leaves pinnatifid, the larger ones with the lobes again toothed or lobed, scabrous-hispid above, thinly white-tomentose to sometimes green and merely hirsute beneath. Heads several, purple; involucre 2.5-4 cm high, its bracts all spine-tipped, without any well developed glutinous dorsal ridge. Achenes less than 4 mm long. June-Oct. — Pastures, fields, roadsides, and waste places; native of Eurasia, now widely established as a weed in North America.

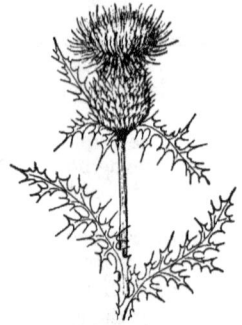

Cirsium vulgare

Coreopsis
TICKSEED

Coreopsis lanceolata L.
LANCE-LEAF TICKSEED   *native / cons* **8**
Perennial herb, with a short woody caudex. Stems usually several, 2-6 dm tall, glabrous, or, especially near the base, spreading-villous, leafy below, subnaked and elongate above. Leaves opposite or rarely alternate, spatulate to linear or lance-linear, simple or with 1 or 2 pairs of small lateral lobes, glabrous to villous or hirsute, the lower long-petiolate, to 20 cm long (including petioles) and 17 mm wide, the others reduced and sessile or nearly so. Heads radiate, few or solitary on long naked peduncles, the disk about 1-2 cm wide; outer involucral bracts about 8-10, lance-shaped to oblong-ovate, glabrous except sometimes near the tip, more or less scarious-margined, about 5-10 mm long; inner involucral bracts longer and broader than the outer; rays about 1.5-3 cm long, often over 1 cm broad, yellow; involucral bracts biseriate and dimorphic, all joined at the base, flat and chaffy below, somewhat awn-like above. Achenes with thin flat wings, orbicular, about 2-3 mm long, black. Pappus of 2 short chaffy teeth. May-July. — Dry, often sandy places.

Coreopsis lanceolata

Doellingeria
FLAT-TOPPED WHITE ASTER

Doellingeria umbellata (P. Mill.) Nees
TALL FLAT-TOPPED WHITE ASTER   *native / cons* **6**
   Aster pubentior Cronq.
   Aster umbellatus P. Mill.
Perennial herb, from thick rhizomes. Stems 0.5-2 m long, upper stem with appressed, short hairs. Leaves alternate, lance-shaped to oblong lance-shaped, 4-15 cm long and 1-4 cm wide, rough-to-touch above, densely short-hairy below; margins entire; petioles short, or absent on upper leaves. Flower heads usually many, 1-1.5 cm wide, in a ± flat-topped inflorescence; involucre 3-5 mm high, the involucral bracts short-hairy and overlapping; rays 5-10, white, 5-8 mm long. Fruit a nerved achene; pappus whitish. July-Sept. — Openings in swamps and moist forests, thickets, streambanks, sedge meadows, calcareous fens, roadside ditches.

Doellingeria umbellata

Erechtites
FIREWEED

Erechtites hieraciifolius (L.) Raf. ex DC.
FIREWEED   *native / cons* 2
Senecio hieraciifolius L.

Fibrous-rooted annual herb. Stems erect, 0.1-2.5 m tall, slightly succulent. Leaves alternate, of various sizes to sometimes 20 cm long and 8 cm wide, sharply serrate with callous-tipped teeth, sometimes also irregularly lobed. Heads cylindric to ovoid, several to many in a flat-topped or elongate inflorescence, or in depauperate plants often solitary; discoid, whitish; involucre about 1-1.5 cm high, the bracts glabrous or finely strigose, green with pale margins, 0.5-2 mm wide. Achenes about 2-3 mm long, finely strigose between the mostly 10-12 ribs, with a white annular ring at the tip; pappus of numerous bright white bristles, eventually deciduous. Aug, Sept. Various habitats, including dry woods, marshes, and waste places, often abundant after fires.

Erechtites hieraciifolius

Erigeron
DAISY, FLEABANE

Biennial to perennial herbs with simple, alternate leaves. Flower heads with both disk and ray flowers; disk flowers yellow; rays white to pink, very narrow, only to about 0.5 mm wide; involucral bracts in 1-2 series, linear, about equal in length, green in middle and at base, translucent at tip and on upper margins. Fruit a flattened achene; pappus of 20-30 slender, rough bristles.

1 Rays not over 3 mm long; taprooted annual.	*E. canadensis*
1 Rays more than 3 mm long; annuals, biennials, or perennials	2
2 Pappus of the ray flowers short, less than 1 mm long; weedy annual herbs	2
3 Plants 6 dm or more tall; stems leafy; pubescence on middle of stem long and spreading	*E. annuus*
3 Plants to 7 dm tall; stem leaves few; pubescence mostly short and appressed	*E. strigosus*
2 Pappus of long bristles; biennial or perennial herb	*E. philadelphicus*

Erigeron annuus (L.) Pers.
EASTERN DAISY FLEABANE   *native / cons* O

Annual or rarely biennial. Stems 6-15 dm tall, amply leafy, more or less hirsute, the hairs spreading except near the top. Basal leaves elliptic to suborbicular, coarsely toothed, to 10 cm long and 7 cm wide, more or less abruptly long-petiolate; stem leaves numerous, broadly lance-shaped, all except sometimes the uppermost sharply toothed, or rarely nearly entire. Heads several to very numerous; involucre 3-5 mm high, finely glandular, and sparsely hairy with long, flattened, transparent hairs; disk 6-10 mm broad. Rays about 80-125, white or rarely pinkish or bluish, about 4-10 mm long and 0.5-1.0 mm wide; disk corollas 2.0-2.8 mm long. Achenes 2-nerved. Pappus of the disk-flowers double, of 10-15 fragile bristles and several very short slender scales less than 1 mm long; pappus of the ray flowers of short scales only, lacking the longer bristles. Early and middle summer. — A weed over most of n USA and s Canada.

Erigeron annuus

Erigeron canadensis L.
HORSEWEED   *introduced*
Conyza canadensis (L.) Cronq.
Leptilon canadense (L.) Britt.

Coarse annual, 1-15 dm tall, stems simple or nearly so to the inflorescence. Leaves more or less pubescent, gradually reduced upwards, the stem leaves to about 8 cm long and 8 mm wide, the basal leaves larger and relatively wider, but generally deciduous before flowering time. Heads, except in depauperate plants, numerous in a long and open inflorescence. Involucre about 3-4 mm high, glabrous or nearly so, the bracts strongly

Erigeron canadensis

imbricate, brown or with distinct brown midvein. Rays white or sometimes pinkish, about equaling the pappus. Late summer and autumn. —
A weed in waste places.

Erigeron philadelphicus L.
PHILADELPHIA DAISY   *native / cons* 2

Biennial or short-lived perennial herb. Stems 1 to several, branched in
head, 2-7 dm long, usually long-hairy. Leaves alternate, lower leaves spatula-shaped, 5-15 cm long and 1-4 cm wide, tapered to a short petiole;
upper leaves smaller, lance-shaped, clasping at base, hairy to nearly
smooth, rounded at tip; margins entire or with rounded teeth. Flower
heads few to many, with both disk and ray flowers, 1.5-2.5 cm wide; involucre 3-6 mm high, the involucral bracts hairy; rays many, white to
deep pink, 5-10 mm long and to 0.5 mm wide. Fruit a short-hairy achene;
pappus of long rough bristles. May-Aug. — Wet meadows, shores, wet
woods, floodplains, springs; also weedy in open disturbed areas and lawns.

Erigeron strigosus Muhl.
ROUGH FLEABANE   *native / cons* 2
   Erigeron ramosus (Walt.) B.S.P.

Annual or rarely biennial. Stems 3-7 dm tall, sparsely leafy, more or less
hairy, the hairs spreading or usually appressed. Basal leaves mostly oblong
lance-shaped to elliptic, entire or toothed, the blade and petiole together
not more than 15 cm long and 2.5 cm wide; stem leaves linear to lance-
shaped, entire, or the lower ones slightly toothed, rarely the middle ones
slightly toothed also. Heads several to very numerous, involucre 2-5 mm
high, obscurely glandular and more or less hairy, the hairs long or short;
disk about 5-12 mm broad. Rays about 50-100, white, or sometimes pinkish
or bluish, to 6 mm long, 0.4-1.0 mm wide; disk corollas 1.5-2.6 mm long.
Achenes 2-nerved; pappus as in *E. annuus*. Early and mid-summer. — A
weed in much of the USA and s Canada.

Erigeron philadelphicus

Erigeron strigosus

Eupatorium
JOE-PYE-WEED, BONESET

Eupatorium perfoliatum L.
BONESET   *native / cons* 6

Perennial herb from a thick rhizome. Stems stout, erect, 3-15 dm tall,
with long, spreading hairs. Leaves opposite, mostly joined at the broad
base and perforated by the stem (upper leaves sometimes separate), lance-
shaped, 6-20 cm long and 1.5-5 cm wide, both sides dotted with yellow
glands; margins finely toothed and rough-to-touch; petioles absent.
Flower heads of dull white disk flowers only, in a flat-topped inflorescence;
involucre 3-6 mm high, the involucral bracts green with white margins,
hairy, overlapping in 3 series. Fruit a black achene, 1-2 mm long; pappus
of long slender bristles. July-Sept. — Marshes, wet meadows, low prairie,
shores, streambanks, ditches, cedar swamps, thickets, calcareous fens.
Often growing with spotted joe-pye-weed (*Eutrochium maculatum*).

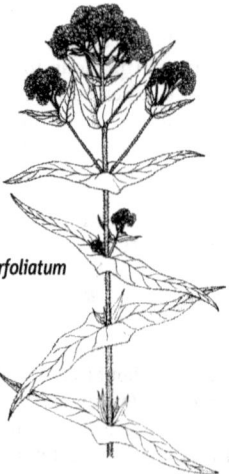

Eurybia
WOOD-ASTER

Eurybia macrophylla (L.) Cass.
LARGE-LEAF WOOD-ASTER   *native / cons* 4
   Aster macrophyllus L.

Perennial with creeping rhizomes, sometimes also with a short branched
caudex, producing abundant clusters of basal leaves on short sterile
shoots. Stems 2-12 dm tall, glandular in the inflorescence or sometimes
throughout, often also spreading-hairy. Leaves basal and alternate on

Eupatorium perfoliatum

the stem, thick and firm, varying from essentially glabrous on both sides to scabrous above and hairy beneath, and sometimes glandular; margins crenate or serrate; basal and lower stem leaves cordate, 4-20 cm long and 3-15 cm wide, long-petiolate, the middle and upper leaves gradually or abruptly reduced, becoming sessile. Inflorescence corymbiform, flat or round-topped, its bracts commonly few and broad; involucre 7-11 mm high, usually glandular and sometimes also short-hairy, its bracts firm, imbricated, the green tips sometimes obscure; rays commonly 9-20, lilac- or purple-tinged, 7-15 mm long. Pappus bristly. July-Oct. — Woodlands.

Eurybia macrophylla

Euthamia
FLAT-TOPPED GOLDENROD

Euthamia graminifolia (L.) Greene
COMMON FLAT-TOPPED GOLDENROD   *native / cons* **4**
  Solidago graminifolia (L.) Salisb.

Perennial herb, spreading by rhizomes. Stems erect, 5-15 dm tall, smooth to hairy, usually branched in head. Leaves alternate, linear to narrowly lance-shaped or oval, 3-15 cm long and 3-10 mm wide, 3-veined, with small glandular dots; margins entire, smooth or rough-to-touch; petioles absent or very short. Flower heads small, in flat-topped clusters at ends of stems; with yellow disk and ray flowers, the rays small, to 1 mm long; involucre 3-5 mm high, somewhat sticky, the involucral bracts overlapping in several series, yellow or green-tipped. Fruit a finely hairy achene, 1 mm long; pappus of many white, slender bristles. Aug-Sept. — Shores, wet meadows, low prairie, springs, fens, swamps, interdunal wetlands, often where sandy or gravelly; also weedy in abandoned fields.

Euthamia graminifolia

Eutrochium
JOE-PYE-WEED

Perennial herbs from a thick rhizome. Stems stout, erect. Leaves whorled, lower leaves smaller; margins toothed. Flower heads of pink, purple or white disk flowers only, usually many in a more or less flat-topped head at ends of stems; involucral bracts overlapping or nearly equal length. Fruit an angled achene; pappus of many slender bristles.

1 Stems green, purple only at nodes, not purple-spotted; heads with usually 3–6 flowers; dry woods . ... *E. purpureum*
1 Stems purple throughout or purple-spotted; heads with more than 8 flowers; common and widespread in wet habitats .... *E. maculatum*

Eutrochium maculatum (L.) E. Lamont
SPOTTED JOE-PYE-WEED   *native / cons* **4**
  Eupatorium maculatum L.

Perennial herb. Stems 5-20 dm long, spotted or tinged with purple, short-hairy above, especially on branches of head. Leaves in whorls of mostly 4-5, lance-shaped to ovate, 5-20 cm long and 2-7 cm wide, upper surface with sparse short hairs, underside often densely short-hairy; margins with sharp, forward-pointing teeth; petioles to 2 cm long. Flower heads of light pink to purple disk flowers only, the inflorescence more or less flat-topped; involucres 6-9 mm high, purple-tinged, the involucral bracts overlapping. Fruit a black, angled achene, 2-4 mm long; pappus of long, slender bristles. July-Sept. — Wet meadows, marshes, low prairie, shores, streambanks, ditches, cedar swamps, bogs, calcareous fens.

Eutrochium purpureum (L.) E. Lamont
PURPLE-NODE JOE-PYE-WEED   *native / cons* **6**
  Eupatorium purpureum L.

Fibrous-rooted perennial. Stems mostly 6-20 dm tall, slightly glaucous, usually purple only at the nodes, otherwise greenish, the pith usually

Eutrochium maculatum

remaining intact. Leaves mostly in 3s or 4s, lance-shaped or ovate to elliptic, mostly 8-30 cm long and 2.5-15 cm wide, gradually or sometimes
rather abruptly narrowed to the short petiole, pinnately veined, usually
sharply and coarsely toothed, loosely soft-pubescent to subglabrous beneath, usually minutely glandular as well. Inflorescence convex; involucre
imbricate, 6.5-9 mm high, mostly 4-7-flowered; corolla generally very
pale pinkish or purplish, but variable, 4.5-7.5 mm long. July-Sept. — Thickets and open woods, often in drier habitats than *E. maculatum.*

Grindelia
GUMWEED

Grindelia squarrosa (Pursh) Dunal
CURLY-TOP GUMWEED   *introduced*
Biennial or sometimes perennial. Stems branched above and sometimes
also at base, 1-10 dm tall, sometimes woody at base. Leaves alternate,
punctate and resinous, finely serrulate to entire, or, sometimes coarsely
toothed ; middle and upper leaves ovate or oblong, 3-7 cm long, 4-20 mm
wide. Heads several to many, radiate or occasionally discoid, the rays
mostly 15-45, yellow, pistillate and fertile; the disk about 1-2 cm wide;
disk-flowers yellow, the inner and often also the outer sterile. Receptacle
flat or convex, naked. Involucral bracts strongly sticky-resinous, imbricate
in several series, the green tips reflexed, especially the outer. Rays 20-35,
7-15 mm long, or absent. Fruit a compressed to 4-angled achene, scarcely
nerved, 2-3 mm long; pappus awns 2-8. July-Sept. — Open or waste places.

Hasteola
FALSE INDIAN PLANTAIN

Hasteola suaveolens (L.) Pojark.
FALSE INDIAN PLANTAIN   *native / cons* **8**
    Cacalia suaveolens L.
    Senecio suaveolens (L.) Elliott
Perennial herb, from fleshy roots. Stems more or less smooth, grooved, 1-
2.5 m tall, leafy to the inflorescence. Leaves alternate, smooth; lower
leaves triangular with a pair of outward-pointing lobes at base, 5-20 cm
long and nearly as wide; upper leaves smaller and often not lobed; margins sharply and irregularly toothed; petioles winged. Flower heads of
disk flowers only, in a more or less flat-topped inflorescence, the disk
about 1 cm wide; disk flowers white or light pink; involucre 1 cm long,
the main involucral bracts 10-15. Fruit an achene; pappus of many soft,
white bristles. July-Sept. — Riverbanks, shores, calcareous fens, wet low
areas.

Helenium
SNEEZEWEED

Helenium autumnale L.
COMMON SNEEZEWEED   *native / cons* **4**
Perennial herb. Stems single or clustered, erect, 3-13 dm tall, smooth or
finely hairy, branched in head. Leaves alternate, bright green, lance-
shaped to oval, 4-12 cm long and 0.5-3.5 cm wide, glandular-dotted, usually
short-hairy; margins entire to shallowly toothed; petioles absent, the
blades tapered to a narrow base extending downward as wings on stem.
Flower heads more or less round, 1.5-4 cm wide; few to many on slender
stalks in a leafy inflorescence, with both disk and ray flowers, the disk
flowers yellow to brown, the rays yellow and drooping, 1.5-2.5 cm long;
involucral bracts in 2-3 series, linear, short-hairy, bent downward with

Eutrochium purpureum

Grindelia squarrosa

Hasteola suaveolens

Helenium autumnale

age. Fruit a finely hairy, 4-5-angled achene, 1-2 mm long; pappus of several
translucent, awn-tipped scales. July-Sept. — Wet meadows, shores, streambanks, marshes, fens, tamarack swamps.

Helianthus
SUNFLOWER
Large perennial herbs (annual in several species), with fibrous or fleshy roots and short to long
rhizomes. Stems unbranched or branched above. Leaves usually opposite on lower part of stem
and alternate above, lance-shaped, margins entire or with forward-pointing teeth; petioles present. Flower heads large, mostly 1 to several (rarely many), at ends of stems and branches, with
yellow disk and ray flowers, the rays large and showy; involucre of several series of narrow,
overlapping bracts; receptacle chaffy. Fruit a flattened achene; pappus of 2 deciduous, awn-
tipped scales.

1 Leaves linear, less than 5 mm wide, crowded on stems .... *H. salicifolius*
1 Leaves lance-shaped to ovate, more than 5 mm wide 2
2 Stems glabrous or nearly so, sometimes glaucous; fine hairs may be present within the inflorescence
   3
2 Stems pubescent 4
3 Leaves lance-shaped, thick and coarse, the underside pale and densely hairy; petioles less than 3 cm
   long .... *H. strumosus*
3 Leaves ovate, thin and membranous, the underside glabrous or only finely hairy; petioles often more
   than 3 cm long .... *H. decapetalus*
4 Leaves lance-shaped, less than 3.5 cm wide, mostly alternate .... *H. maximiliani*
4 Leaves ovate, often more than 3.5 cm wide, the upper leaves opposite or alternate .... *H. tuberosus*

Helianthus decapetalus L.
THIN-LEAF SUNFLOWER   *native / cons 7*
Perennial herb with fibrous roots and well developed stout rhizomes.
Stems 0.5-1.5 m tall, short-hairy in the inflorescence, otherwise glabrous.
Leaves broadly lance-shaped to ovate, mostly 8-20 cm long and 3-8 cm
wide, long-acuminate, more or less abruptly contracted near the base
and with a short or fairly long tapering decurrence on the mostly 1.5-6
cm petiole, 3-nerved at the base, pinnately veined above, serrate, generally
sharply and conspicuously so, thin, moderately to very sparsely scabrous
on the upper surface, moderately rough hairy to nearly glabrous on the
paler lower surface. Heads generally several, the disk 1-2 cm wide, yellow;
involucral bracts very loose, thin, green, conspicuously ciliate on the
margins and occasionally hispidulous on the back, attenuate-acuminate,
at least some of them usually conspicuously exceeding the disk, often
some of the outer enlarged and leafy; rays 8-15, 1.5-3.5 cm long. Aug-Oct.
— Woodlands and along streams.

Helianthus decapetalus

Helianthus maximiliani Schrad.
MAXIMILIAN SUNFLOWER   *introduced*
Perennial from short rhizomes and thickened, often fleshy roots. Stems
0.5-3 m tall, conspicuously pubescent, especially upwards, with mostly
short, white, appressed hairs. Leaves lance-shaped, gradually narrowed
to the short winged petiole, commonly 7-15 cm long and 1-3 cm wide,
subentire or occasionally evidently toothed, pinnately veined, not 3-
nerved, strongly scabrous on both sides, usually some of them falcate,
the upper mostly alternate (all opposite in depauperate forms). Heads
several or occasionally solitary, the disk 1.5-2.5 cm wide, yellow; involucral
bracts narrow, often much exceeding the disk, canescent with short
white hairs; rays 10-25, 1.5-4 cm long. June-Sept. — Prairies and waste
ground, often in sandy soil.

Helianthus maximiliani

Helianthus strumosus L.

PALE-LEAF WOODLAND SUNFLOWER   native / cons 4

Perennial with rather woody roots and well developed rhizomes. Stems 1-2 m tall, short-hairy in the inflorescence, otherwise glabrous or with only a few scattered long hairs, often glaucous. Leaves opposite, or the uppermost alternate, broadly lance-shaped, 8-20 cm long and 2.5-9 cm wide, shallowly toothed or subentire, more or less abruptly contracted or sometimes broadly rounded at base, commonly with a short decurrence on the 6-30 mm petiole, scabrous-hispidulous above, some of the hairs with broad, white, slightly raised base (strumose), lower surface green and moderately short-hairy to nearly glabrous and glaucous, 3-nerved near the base, pinnately veined above. Heads several or solitary; disk yellow, 1-2 cm wide; involucral bracts lance-shaped, somewhat loose, especially the long acuminate tips, ciliolate on the margins; rays 8-15, 1.5-4 cm long. July-Sept. — Chiefly in woodlands.

Helianthus strumosus

Helianthus tuberosus L.

JERUSALEM-ARTICHOKE   native / cons 2

Perennial with tuber-bearing rhizomes. Stems stout, 1-3 m tall, pubescent with mostly spreading hairs. Leaves alternate, or sometimes all but the uppermost opposite, broadly lance-shaped, 10-25 cm long and 4-12 cm wide, on winged petioles 2-8 cm long; margins serrate; densely and coarsely rough-hairy on the upper surface, sparsely to densely velvety-hairy on the lower surface, 3-nerved near the base, pinnately veined above. Heads usually several or numerous; disk 1.5-2.5 cm wide, yellow; involucral bracts usually rather dark, especially near the base, narrowly lance-shaped, loose especially above the middle, ciliate on the margins; rays 10-20, 2-4 cm long. Aug.-Oct. — Moist soil and waste places; escaped from cultivation and also native. Cultivated since pre-Columbian times for its edible tubers.

Helianthus tuberosus

Heliopsis

SUNFLOWER-EVERLASTING

Heliopsis helianthoides (L.) Sweet

SUNFLOWER-EVERLASTING   native / cons 5

Short-lived perennial, generally with a short caudex and fibrous roots. Stems 5-15 dm tall, glabrous or more or less scabrous. Leaves opposite, ovate, serrate, often subtruncate at base, 5-15 cm long and 2.5-8 cm wide, borne on petioles about 5-35 mm long. Heads solitary or several, sometimes numerous, naked-pedunculate; radiate, the rays yellow, pistillate, fertile or rarely sterile, persistent on the achenes and becoming papery; receptacle conic, chaffy throughout, its bracts concave and clasping, subtending the rays as well as the disk-flowers; disk-flowers perfect and fertile; the disk about 1-2.5 cm wide; rays 8-15, pale yellow, 1.5-4 cm long. Achenes quadrangular, glabrous; pappus none, or of a short irregular crown or a few teeth. June-Oct. Dry woodlands, prairies, and waste places.

Heliopsis helianthoides

Hieracium

HAWKWEED

Fibrous-rooted perennial herbs with milky juice, with a rhizome which may be elongate or shortened into a caudex. Leaves alternate or all basal, entire or more or less toothed. Heads solitary to numerous, small or large, in a corymbiform or panicle-like inflorescence. Flowers all ligulate and perfect, yellow to red-orange. Involucre cylindric to hemispheric, its bracts imbricate. Achenes terete, mostly narrowed toward the base, truncate or occasionally narrowed toward the summit, strongly ribbed. Pappus of numerous whitish to brownish capillary bristles.

ADDITIONAL SPECIES

Hieracium murorum L., known in Wisconsin from a single Door County collection in 1990; plants in crevice in dolomitic limestone along lakeshore at old resort.

1　Leaves spatula-shaped, lower leaves with petioles, upper leaves sessile; involucres and peduncles with black glands .... *H. scabrum*
1　Leaves lance-shaped to oblong lance-shaped, sessile, toothed; involucres and peduncles without glands .... *H. umbellatum*

Hieracium scabrum Michx.

ROUGH HAWKWEED　　　　　　　　　　　　　　*native / cons* 6

Perennial from a short mostly simple caudex. Stems 2-14 dm tall, setose with spreading hairs seldom as much as 5 mm long, at least near the base, becoming stellate and glandular upwards, densely so in the inflorescence. Leaves sparsely or moderately setose on both sides, more densely so on the petiole and midrib beneath; basal and often also the lowermost stem leaves ordinarily deciduous, the lower leaves oblong lance-shaped to elliptic, 5-20 cm long (including the usually short petiole), 1-4.5 cm wide, the others progressively reduced upwards, soon becoming sessile. Inflorescence open-corymbiform (especially in smaller specimens) to more elongate and cylindric; heads mostly 40-100-flowered; involucre 6-9 mm high, hispid with blackish mostly gland-tipped hairs, especially toward the base. Achenes 2-3 mm long; pappus tawny. July-Sept. — Open ground and dry woods, especially in sandy soil.

Hieracium scabrum

Hieracium umbellatum L.

NARROW-LEAF HAWKWEED　　　　　　　　　　　*native / cons* 6
　Hieracium canadense Michx.
　Hieracium kalmii L.
　Hieracium scabriusculum Schwein.

Perennial from a short caudex. Stems 1.5-15 dm tall, often spreading-hairy below, sometimes stellate-puberulent above. Leaves stellate-puberulent to subglabrous, and often long-hairy beneath, the basal and lowermost stem ones small and soon deciduous, the others, except for the strongly reduced upper ones, mostly rather numerous, nearly alike in size and shape, sessile and tending to be broadly rounded and somewhat clasping at the base, elliptic to ovate, mostly 3-12 cm long and 7-40 mm wide, usually with a few irregularly spaced sharp teeth. Inflorescence loosely corymbiform to often umbel-like, or the heads occasionally solitary, the peduncles stellate-puberulent, occasionally with some longer spreading hairs as well; heads mostly 40-110-flowered; involucre 6-13 mm high, its bracts imbricate in several series, glabrous or obscurely puberulent, occasionally with a few longer hairs. Achenes 2.5-3.5 mm long; pappus tawny or yellowish. July-Sept. — Woodlands, beaches, and fields, especially in sandy soil.

Hieracium umbellatum

Hypochaeris
CAT'S-EAR

Hypochaeris radicata L.

HAIRY CAT'S-EAR　　　　　　　　　　　　　　*introduced*

Perennial from a caudex, fibrous-rooted, or more often several of the roots en- larged. Stems 1.5-6 dm tall, striate-angled, branched above or in small plants simple, often spreading-hispid below. Basal leaves hispid, oblong lance-shaped, toothed or pinnatifid, 3-3.5 cm long and 0.5-7 cm wide. Heads usually several, terminating the branches; involucre 1-1.5 cm high at anthesis, sometimes nearly 2.5 cm in fruit, its bracts imbricate, glabrous or hispid. Body of the achene 4-5 mm long, from a little longer

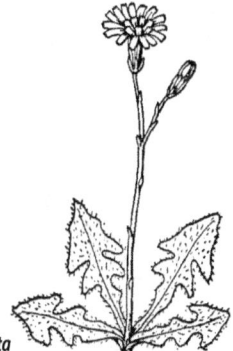

Hypochaeris radicata

to more often much shorter than the slender beak, the prominent nerves and the lower part of the beak scabrous. Some of the outer pappus-bristles often shorter than the inner. May-Sept. — Roadsides, pastures, fields, and waste places; native of Eurasia; now widely established. Similar to *Leontodon,* from which it is distinguished primarily by its chaffy-bracted receptacle.

Lactuca
LETTUCE

Annual, biennial, or perennial, herbs with milky juice. Leaves alternate, entire to pinnatifid. Heads usually numerous in a panicle-like inflorescence. Flowers all ligulate and perfect, yellow, blue, or white, the corolla tube generally more than half as long as the ligule. Involucre cylindric, often broadening at the base in fruit, generally imbricate. Achenes compressed, winged or strongly nerved marginally, with 1-several lesser nerves on each face, expanded at the summit where the pappus is attached. Pappus of capillary bristles.

1   Achenes with one central nerve on each side . . . . *L. canadensis*
1   Achenes with several prominent nerves on each face . . . . *L. biennis*

Lactuca biennis (Moench) Fern.
TALL BLUE LETTUCE   *native / cons* 3

Robust annual or biennial. Stems glabrous, 6-20 dm tall. Leaves glabrous, or hairy on the main veins beneath, sometimes sagittate at the base, commonly 10-40 cm long and 4-20 cm wide. Heads numerous in an elongate, rather narrow, panicle-like inflorescence, often crowded, mostly 15-34-flowered (rarely to 54); fruiting involucre 10-14 mm high. Achenes 4-5.5 mm long, thin-edged, prominently several-nerved on each face, tapering to the beakless or shortly beaked tip; pappus light brown. July-Sept. — Moist places.

Lactuca biennis

Lactuca canadensis L.
TALL LETTUCE   *native / cons* 2

Annual, or usually biennial. Stems 3-25 dm tall, the herbage glabrous or occasionally coarsely hirsute, often more or less glaucous. Leaves entire or toothed to pinnately lobed, sagittate or sometimes narrowed to the base, mostly 10-35 cm long and 1.5-12 cm wide. Heads numerous, relatively small, mostly 13-22-flowered, the flowers yellow; fruiting involucre mostly 10-15 mm long. Achenes blackish, flat, with a median nerve on each face, rugulose, the body 3-4 mm long, beaked; pappus 5-7 mm long. July-Sept. — Fields, waste places, woods.

Lactuca canadensis

Lapsana
NIPPLEWORT

Lapsana communis L.
COMMON NIPPLEWORT   *introduced*

Branching annual herb with milky juice, hirsute to nearly glabrous. Stems 1.5-15 dm tall. Leaves alternate, thin, petiolate; blade ovate, toothed or occasionally basally lobed, 2.5-10 cm long and 2-7 cm wide, progressively less petiolate and narrowed upwards. Heads several or numerous in a corymbiform or panicle-like inflorescence, naked-pedunculate, mostly 8-15-flowered. Flowers all ligulate and perfect, yellow; involucre cylindric-campanulate, 5-8 mm high; bracts nearly equal, uniseriate, keeled. Achenes curved, narrowed to both ends, 3-5 mm long, glabrous; pappus none. June-Sept. — Native of Eurasia; now established in woods, fields, and waste ground.

Lapsana communis

Leontodon
HAWKBIT

Leontodon saxatilis Lam.
LITTLE HAWKBIT   *introduced*
Leontodon taraxacoides (Vill.) Mérat

Fibrous-rooted perennial herb with milky juice. Stems 1-3.5 dm tall, the scapes simple and ordinarily naked. Leaves basal, oblong lance-shaped, hispid-hirsute, 4-15 cm long, 6-25 mm wide, usually shallowly lobed. Heads solitary; involucre 6-11 mm high, glabrous or hairy; receptacle not chaffy-bracted Flowers all ligulate and perfect, yellow. Achenes fusiform, nerved, short-beaked, 3-6 mm long, scabrous; pappus of the inner flowers partly of plumose bristles, partly of shorter outer scales which may be tipped with a scabrous bristle. June-Sept. — Native of Europe; a weed in lawns and waste places.

Leontodon saxatilis

Leucanthemum
OX-EYE DAISY

Leucanthemum vulgare Lam.
OX-EYE DAISY   *introduced (invasive)*
Chrysanthemum leucanthemum L.

Rhizomatous perennial. Stems 2-8 dm tall, simple or nearly so, glabrous or sparsely hairy. Leaves alternate, glabrous or hairy; basal leaves oblong lance-shaped or spatulate, petiolate, 4-15 cm long, crenate and often also lobed or cleft; stem leaves reduced and becoming sessile, pinnatifid or nearly entire. Heads hemispheric or flattened, solitary at the ends of the branches, naked-pedunculate, radiate; involucral bracts imbricate in 2-4 series, dry, scarious or hyaline at least at the margins and tips, the midrib sometimes greenish; receptacle flat or convex, naked; disk about 1-2 cm wide; rays about 15-30, white, 10-20 mm long; disk flowers tubular and perfect, the corolla with 4 or more commonly 5 lobes. Achenes terete, about 10-ribbed; pappus none. May-Oct. Fields, roadsides, and waste places; native of Europe and Asia, naturalized throughout most of temperate North America.

Leucanthemum vulgare

Liatris
BLAZING STAR, GAY FEATHER

Perennial herbs, mostly with an evident corm, rarely from a more elongate caudex or stout rhizome. Leaves alternate, entire, more or less punctate, the basal ones usually the largest. Inflorescence spike-like. Heads discoid, the flowers all tubular and perfect, 3-100 or more in each head. Involucral bracts imbricate in several series. Receptacle naked. Corollas pink-purple or occasionally white. Achenes about 10-ribbed, pubescent. Pappus of 1 or 2 series of barbellate or plumose capillary bristles.

1 Inflorescence a dense spike, heads with mostly 5–10 flowers; leaves mostly less than 1 cm wide . . . . *L. pycnostachya*
1 Inflorescence an open spike or raceme; heads with 14 or more flowers; larger leaves 1–4 cm wide . . . . *L. aspera*

Liatris aspera Michx.
TALL GAYFEATHER   *native / cons 5*

Plants 4-12 dm tall, the herbage short-hairy, or glabrous throughout. Leaves 2590 below the inflorescence, the lowermost ones 5-40 cm long (including the long petiole) and 7-45 mm wide, the middle and upper ones gradually reduced and becoming sessile. Heads generally numerous in an elongate spiciform inflorescence, or the peduncles occasionally more elongate and to 5 cm long; terminal head not evidently enlarged;

involucre 8-15 mm high, campanulate or subhemispheric, glabrous, its bracts loosely spreading, often purplish upwards, with conspicuous, lacerate, often crisped margins; flowers 16-35 in each head, the corolla hairy within toward the base; pappus barbellate. Aug-Oct. — Dry open places and thin woods, especially in sandy soil.

Liatris pycnostachya Michx.
THICK-SPIKE BLAZING STAR   *native / cons 7*

Perennial from a woody corm or rootstock, 6-15 dm tall, more or less hirsute in the inflorescence or throughout. Leaves numerous, linear or nearly so, the lowermost ones 10-50 cm long and 3-13 mm wide, reduced upwards. Heads sessile, crowded in an elongate, densely spiciform inflorescence; involucre subcylindric or narrowly turbinate, 8-11 mm high, its bracts tapering to an acuminate, conspicuously squarrose tip, or the inner ones sometimes merely loosely erect; flowers mostly 5-7, or reputedly to 12, the corolla glabrous or nearly so within; pappus strongly barbellate. July-Sept. — Moist or dry prairies and open woods.

Liatris aspera

Matricaria
MAYWEED

Matricaria discoidea DC.
PINEAPPLE-WEED   *introduced*
  Matricaria matricarioides (Less.) Porter

Pineapple-scented glabrous annual herb. Stems leafy, branching, 5-40 cm tall. Leaves alternate, 1-5 cm long, 1-3x pinnatifid, the ultimate segments short, linear or filiform. Inflorescence corymbiform, terminating the branches; heads several or numerous, rayless, the disk about 5-9 mm wide; involucral bracts with broad hyaline margins; disk corollas 4-toothed; receptacle conic, pointed. Achenes with 2 marginal and 1 or several weak nerves; pappus a short crown. May-Sept. — Roadsides and waste places.

Matricaria discoidea

Packera
GROUNDSEL

Erect perennial, biennial, or annual herbs. Leaves alternate or from base of plant, stalked near base, stalkless and usually smaller upward. Flower heads with both disk and ray flowers, few to many in clusters at ends of stems; disk flowers perfect and yellow, the rays yellow; involucral bracts in 1 series and not overlapping, of equal lengths; receptacle flat or convex, not chaffy. Fruit an achene, nearly round in section; pappus of slender bristles.

1 Basal leaves mostly oblong lance-shaped or elliptic, tapered at base to petiole . . . . *P. paupercula*
1 Basal leaves heart-shaped at base . . . . *P. aurea*

Packera aurea (L.) Á.Löve & D.Löve
HEART-LEAVED GROUNDSEL   *native / cons 6*
  Senecio aureus L.

Perennial herb, from a spreading crown or rhizome. Stems single or clumped, 3-8 dm long, slightly hairy when young, soon becoming smooth. Basal leaves heart-shaped, 5-10 cm long and to as wide, often purple-tinged, on long petioles, the margins with rounded teeth; stem leaves much smaller and more or less pinnately lobed, becoming sessile. Flower heads several to many, the disk 5-10 mm wide, rays gold-yellow, 6-13 mm long involucre 5-8 mm high, the involucral bracts often purple-tipped. Fruit a smooth achene; pappus of slender white bristles. May-July. — Floodplain forests, wet forest depressions, swamp openings and hummocks, sedge meadows, thickets, fens, ditches.

Packera aurea

Packera paupercula (Pursh) Á.Löve & D.Löve
RAYLESS ALPINE GROUNDSEL   *native / cons* 4
Senecio pauperculus Michx.
Fibrous-rooted perennial with a short, simple or slightly branched crown,
occasionally also with very short slender stolons. Stems 1-5 dm tall, the
herbage lightly floccose-tomentose when young, generally soon glabrate,
except frequently at the very base and in the leaf-axils. Basal leaves
mostly oblong lance-shaped to elliptic, occasionally suborbicular, generally tapering to the petiolar base, crenate or serrate to subentire, seldom
over 12 cm long and 2 cm wide, generally much smaller; stem leaves
more or less pinnatifid, the lower sometimes larger than the basal, the
others reduced and becoming sessile. Heads relatively few, seldom more
than 20, the disk 5-12 mm wide; involucre 4-7 mm high, its bracts cari-
nate-thickened or thin and flat, often purple-tipped; rays 5-10 mm long,
yellow, rarely wanting. Achenes glabrous or hispidulous. May-July. —
Meadows, prairies, streambanks, beaches, and cliffs.

Petasites
Packera paupercula

SWEET COLT'S-FOOT
Petasites frigidus (L.) Fries
NORTHERN SWEET COLT'S-FOOT   *native / cons* 7
Petasites palmatus (Ait.) Gray
Perennial herb, spreading by rhizomes. Stems 1-6 dm long, smooth or
short-hairy in the head. Leaves mostly from base of plant, triangular to
nearly round in outline, palmately lobed, 5-30 cm wide, upper surface
green and smooth, underside densely white-hairy, sometimes becoming
smooth with age; margins coarsely toothed; petioles of basal leaves 1-3
dm long; stem leaves small and bractlike, 2-6 cm long. Flower heads
nearly white, staminate and pistillate flowers mostly on separate plants;
rays of pistillate heads to 7 mm long; involucral bracts in a single series;
receptacle not chaffy. Fruit a linear, ribbed achene; pappus of many white,
slender bristles. May-June, flowering before or as leaves expand in spring.
— Wet conifer forests and swamps, wet trails and clearings, aspen woods.

Pilosella
Petasites frigidus

MOUSE-EAR HAWKWEED
Fibrous-rooted perennial herbs with milky juice, with a rhizome which may be elongate or
shortened into a caudex. Leaves alternate or all basal, entire or more or less toothed. Heads
solitary to numerous, small or large, in a corymbiform or panicle-like inflorescence. Flowers all
ligulate and perfect, yellow to red-orange. Involucre cylindric to hemispheric, its bracts imbricate.
Achenes terete, mostly narrowed toward the base, truncate or occasionally narrowed toward
the summit, strongly ribbed. Pappus of numerous whitish to brownish capillary bristles.

1   Ray flowers red-orange. . . . . *P. aurantiaca*
1   Ray flowers yellow 2
2   Leaves hairy above and not glaucous; mature involucres ca. 7-8 mm long; pedicels densely stellate-
    pubescent. . . . . *P. caespitosa*
2   Leaves glabrous or unevenly and sparsely hairy above, glaucous; mature involucres usually ca. 5-7
    mm long; pedicels with stellate hairs sparse (or sometimes dense) . . . . *P. piloselloides*

Pilosella piloselloides (Vill.) Soják
TALL HAWKWEED   *introduced (invasive)*
Hieracium florentinum All.
Hieracium piloselloides Vill.
Perennial from a usually rather short praemorse rhizome. Stems 2-10
dm tall, naked or with 1 or 2, rarely as many as 5, reduced leaves. Herbage
glaucous, sparsely long-setose or subglabrous, the peduncles becoming

Pilosella piloselloides

stipitate-glandular and somewhat stellate. Basal leaves oblong lance-shaped, mostly 3-18 cm long (including the petiole) and 5-18 mm wide, 5-12 times as long as wide. Heads mostly 3-75 in a corymbiform inflorescence, the involucre 6-8 mm high, hispid with blackish, mostly gland-tipped hairs and somewhat stellate. Achenes 1.5-2 mm long; pappus slightly sordid. June-Sept. Mostly in fields, meadows, pastures, roadsides, and waste places; native of Europe.

Pilosella aurantiaca (L.) F.W. Schultz & Schultz-Bip.
  Hieracium aurantiacum L.
ORANGE KING-DEVIL   *introduced (invasive)*
Perennial with slender stolons and normally with a slender elongate rhizome, commonly 1-6 dm tall, the stem naked or with a single (rarely 2) more or less reduced leaf, conspicuously long-setose, also becoming stellate-tomentose and hispid with gland-tipped hairs above. Basal leaves oblong lance-shaped or narrowly elliptic, blunt, 4-20 cm long (including the petiole), 1-3.5 cm wide, long-setose on both sides, or nearly glabrous above; leaves of the stolons few, similar but much smaller. Heads 5-25 in a compact corymbiform inflorescence; flowers red-orange, becoming deeper red in drying; involucre 5-8 mm high, long-setose, hispid with blackish gland-tipped hairs. Achenes about 2 mm long; pappus slightly sordid. June-Sept. — Native of Europe; fields, roadsides, and meadows.

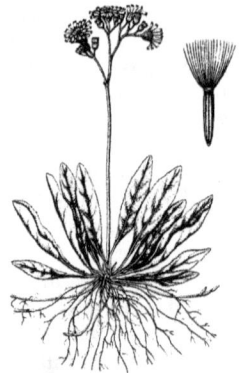

Pilosella aurantiaca

Pilosella caespitosa (Dumort.) P.D. Sell & C. West
YELLOW KING-DEVIL, MEADOW HAWKWEED   *introduced*
  Hieracium caespitosum Dumort.
  Hieracium pratense Tausch
Perennial with a short or more often elongate rhizome and commonly with short stout stolons, the stem 2.5-9 dm tall, sparsely to rather densely long-setose, becoming stellate-tomentose and hispid with blackish gland-tipped hairs above, naked or with only one or two (rarely 3) reduced leaves. Basal leaves oblong lance-shaped or narrowly elliptic, 4-25 cm long (including the petiole), 1-3 cm wide, long-setose on both sides, sometimes sparsely so above, commonly slightly stellate beneath. Heads several or rather numerous in a compact corymbiform inflorescence, the involucre 6-8 mm high, hispid with blackish, gland-tipped hairs, commonly also sparsely long-setose and slightly stellate. Achenes 1.5-2 mm long; pappus slightly sordid. May-Sept. — A weed in fields, pastures, and along roadsides, occasionally in dry woods; native of Europe.

Pilosella caespitosa

Prenanthes
RATTLESNAKE-ROOT

Prenanthes alba L.
WHITE RATTLESNAKE-ROOT   *native / cons 5*
  Nabalus albus (L.) Hook.
Perennial herb. with milky juice and tuberous-thickened roots. Stems stout, commonly 4-15 dm tall, the herbage more or less glaucous. Leaves alternate, glabrous above, paler and often hairy beneath, very variable in size and shape, the lower ones long-petioled, palmately few-lobed to sagittate and merely coarsely toothed, becoming smaller, less cut, and less petiolate upwards, the upper leaves often entire. Inflorescence elongate panicle-like, the heads nodding, 10-15-flowered. Flowers all ligulate and perfect, fragrant, greenish or yellowish white; involucre 11-14 mm long, generally somewhat purplish, its principal bracts 8, glabrous, but more or less densely papillate with white, waxy-appearing cells; pappus cinnamon-brown. Aug-Sept. — Woodlands.

Prenanthes alba

Pseudognaphalium
RABBIT-TOBACCO
Biennial glandular herbs (ours), sometimes aromatic. Stems usually erect, woolly-tomentose, sometimes glandular. Leaves basal and along the stem or mostly cauline, alternate, usually sessile; blades mostly narrowly lance-shaped, bases often clasping the stem, margins entire. Inflorescence corymbiform or panicle-like, sometimes a terminal cluster; heads disciform. Corollas yellowish. Fruit a glabrous achene. Pappus of 10-12 barbellate bristles.

1 Leaves decurrent at base; stems glandular-hairy, sometimes also woolly hairy . . . . *P. macounii*
1 Leaves not decurrent at base . . . . *P. obtusifolium*

Pseudognaphalium macounii (Greene) Kartesz
CLAMMY RABBIT-TOBACCO   *native / cons* **2**
    Gnaphalium macounii Greene
Similar to *P. obtusifolium*. Stems glandular-hairy, becoming woolly in the inflorescence, rarely somewhat woolly to near the base, as well as glandular. Leaves distinctly decurrent at the base, the upper surface glandular-hairy, the lower surface usually woolly, or sometimes glandular-hairy. July-Sept. — Open places.

Pseudognaphalium obtusifolium (L.) Hilliard & Burtt
FRAGRANT RABBIT-TOBACCO   *native / cons* –
    Gnaphalium obtusifolium L.
Annual or perhaps sometimes biennial, fragrant, 1-8 dm tall, erect. Stems thinly white-woolly, commonly becoming subglabrous or sometimes a little glandular toward the base. Leaves numerous, linear lance-shaped, up to about 10 cm long and 1 cm wide, sessile, white-woolly beneath, green and from glabrous to slightly glandular or slightly woolly above. Inflorescence branched and many-headed except in depauperate plants, flat or round-topped and often elongate, the final clusters with the heads somewhat glomerate. Involucre yellowish white or somewhat dingy, campanulate, woolly only near the base, 5-7 mm high; pappus bristles distinct, falling separately. Achenes glabrous. July-Oct. — Open places, often where sandy . When crushed, plants have a characteristic maple syrup scent.

Pseudognaphalium obtusifolium

Ratibida
CONEFLOWER
Perennial herbs (ours). Leaves alternate, pinnatifid. Heads naked-pedunculate, radiate, the rays yellow. Involucre a single serics of green, subherbaceous, linear or lance-linear bracts. Receptacle columnar. Disk flowers perfect and fertile.

Ratibida pinnata (Vent.) Barnh.
GLOBULAR CONEFLOWER   *native / cons* **4**
Perennial from a stout woody rhizome or sometimes a short caudex. Stems mostly 4-12 dm tall, strigose above, strigose or more commonly spreading-hirsute below. Leaves loosely hirsute, the segments lance-shaped, coarsely toothed to entire. Heads several or occasionally solitary, the disk about 12-20 mm high and 0.6-0.9 times as thick, much shorter than the rays; rays about 5-10, pale yellow, mostly 2.5-6 cm long, spreading or reflexed; style-appendages elongate, acuminate. Achenes smooth; pappus none. June-Aug. — Prairies and dry woods.

Rudbeckia
CONEFLOWER

Ratibida pinnata

Perennial herbs. Stems and leaves rough-hairy. Leaves alternate. Flower heads with both disk and ray flowers, the rays yellow to orange; involucral bracts green, overlapping; receptacle rounded, chaffy. Fruit a smooth, 4-angled achene; pappus none or a short crown. The genus includes the well-known black-eyed Susan (*Rudbeckia hirta*), widespread on dry sites.

1 Main leaves deeply lobed 2
1 Leaves unlobed 3
2 Disk yellow; stems glabrous or nearly so; largest leaves 5–7 lobed .... *R. laciniata*
2 Disk dark purple-red; stems pubescent; largest leaves 3-lobed .... *R. triloba*
3 Leaves ovate, margins toothed; chaff of receptacle glabrous, tapered to a sharp prolonged point ...
   .... *R. triloba*
3 Leaves and margins various; chaff of receptacle not tapered to a sharp point .... *R. hirta*

Rudbeckia hirta L.

BLACK-EYED SUSAN   native / cons **4**

Biennial or short-lived perennial, sometimes flowering the first year. Stems 3-10 dm tall, more or less hirsute throughout. Leaves variable in size and shape, toothed or subentire, the basal and lower stem leaves mostly oblong lance-shaped to elliptic and long-petiolate, the others lance-linear to oblong or ovate, mostly sessile. Heads several or solitary, mostly long-pedunculate, the disk hemispheric or ovoid, 12-20 mm wide, dark purple or brown, rarely yellow; involucral bracts copiously hirsute, nearly equal, green, spreading, sometimes elongate and equaling the rays; rays about 8-20, orange or orange-yellow, sometimes darker or marked with purple near the base, commonly 2-4 cm long; receptacular bracts more or less hispid near the tip, often also ciliate on the margins. Achenes quadrangular; pappus none. June-Oct. — Various habitats: chiefly in disturbed or waste places, meadows, roadsides.

Rudbeckia hirta

Rudbeckia laciniata L.

CUTLEAF CONEFLOWER   native / cons **6**

Perennial herb, from a woody base. Stems branched, 5-30 cm tall, smooth and often waxy. Leaves alternate, to 30 cm wide, deeply lobed, nearly smooth to hairy on underside; margins coarsely toothed as well as lobed, or entire on upper leaves; petioles long on lower leaves, becoming short above. Flower heads several to many at ends of stems, with both disk and ray flowers, disk flowers green-yellow, rays lemon-yellow, drooping, 3-6 cm long; involucral bracts of unequal lengths; receptacle round at first, becoming cylindric. Fruit a 4-angled achene; pappus a short toothed crown. July-Sept. — Floodplain forests, swamps, streambanks, thickets, ditches; usually in partial or full shade.

Rudbeckia laciniata

Rudbeckia triloba L.

THREE-LOBED CONEFLOWER   native / cons **4**

Short-lived perennial. Stems mostly 5-15 dm tall, moderately spreading-hirsute to subglabrous. Leaves thin, sharply toothed to subentire, moderately appressed hairy or nearly glabrous, the basal leaves broadly ovate or subcordate and long-petiolate, the stem leaves mostly narrower and short-petiolate or sessile, usually some of the larger leaves deeply 3-lobed. Heads several or numerous, the disk dark purple, hemispheric or ovoid, 8-15 mm wide; involucral bracts narrow, nearly equal, green and more or less leafy, spreading or reflexed, mostly strigose and ciliate-margined; rays 6-12, yellow, or partly or wholly orange, 1-3 cm long; receptacular bracts glabrous, equaling or usually a little exceeding the disk corollas, abruptly narrowed to a distinct awn-point. Achenes equably quadrangular; pappus a minute crown. July-Oct. Woodlands and moist soil. Our form is var. triloba, with the larger leaves merely 3-lobed.

Rudbeckia triloba

Senecio
GROUNDSEL, RAGWORT
Erect annual herbs (ours). Flower heads with both disk and ray flowers, few to many in clusters at ends of stems; disk flowers perfect and yellow, the rays yellow; involucral bracts in 1 series and of equal lengths. Several former members of this genus now placed in *Packera*.

Senecio vulgaris L.
OLD-MAN-IN-THE-SPRING   *introduced*
Simple or strongly branched annual with a more or less evident taproot. Stems 1-4 dm tall, leafy throughout; herbage sparsely crisp-hairy or subglabrous. Leaves alternate, coarsely and irregularly toothed or more often pinnatifid, 2-10 cm long and 5-45 mm wide, the lower tapering to the petiole, the upper sessile and clasping. Heads several or numerous, strictly discoid, the flowers all tubular and perfect; disk usually 5-10 mm wide; involucre about 5-8 mm high; bracteoles well developed, black-tipped; pappus very copious, equaling or generally surpassing the corollas. Achenes strigillose, chiefly along the angles. May-Oct. — Native of the Old World; waste places.

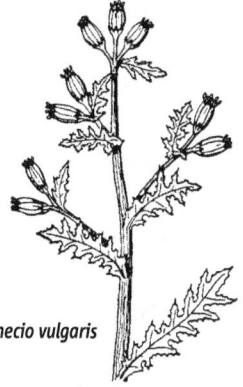

Senecio vulgaris

Solidago
GOLDENROD
Erect perennials, spreading by rhizomes or from a crown. Leaves alternate, margins entire or toothed. Flower heads small, many, in flat-topped (corymb-like), rounded (panicle-like) or spike-like clusters at ends of stems; the flowers sometimes mostly on 1 side of inflorescence branches (secund) in species with panicle-like heads; the heads with yellow disk and ray flowers; involucral bracts in several overlapping series, papery at base and tipped with green; receptacle flat or convex, not chaffy. Fruit an achene, angled or nearly round in cross-section; pappus of many slender white bristles.

1 Heads in a more or less flat-topped cluster at end of stem 2
1 Heads in an elongate or pyramid-shaped cluster 4
2 Leaf blades of middle and upper stem ovate to elliptic; stems and leaves densely hairy; common species of dry to mesic habitats .... *S. rigida*
2 Leaf blades linear to lance-shaped or oblong lance-shaped, glabrous apart from rough leaf margins; stems glabrous or nearly so, or slightly hairy below inflorescence 3
3 Pedicels rough-hairy; leaves folded inward along midrib, with 3 or more veins from base . *S. riddellii*
3 Pedicels smooth or nearly so; leaves flat, not 3-veined from base .... *S. ohioensis*
4 Inflorescence terminal, usually more or less pyramid-shaped and slightly nodding at top; inflorescence branches curving; the heads mostly on upper side of the branches 5
4 Flower heads spiraled around branches of inflorescence and not all on one side of branch 13
5 Stem leaves with 3 prominent veins (midrib plus 2 distinct lateral veins) 6
5 Stem leaves with prominent midrib and weaker lateral veins 10
6 Stem pubescent for all or most of its length 7
6 Stem glabrous (or nearly so) below inflorescence 9
7 Involucres 2-3 mm long .... *S. canadensis*
7 Involucres 3-6 mm long 8
8 Mid to upper leaves serrate, glabrous or scabrous above, pubescent on the veins beneath; stem pilose chiefly above the middle .... *S. canadensis*
8 Mid to upper leaves minutely serrate to entire, scabrous above, densely pubescent beneath; stem grayish with close puberulence throughout, except sometimes near the base .... *S. altissima*
9 Basal leaves absent; stem leaves elliptic, withering by flowering time, numerous, not reduced in size; distinctly 3-nerved; inflorescence branches densely hairy; flowering in Aug–Sept .... *S. gigantea*
9 Basal leaves present; basal and lower stem leaves oblong lance-shaped to elliptic, with long petioles, persistent, middle and upper stem leaves few, smaller than basal leaves; leaves obscurely 3-nerved; inflorescence branches glabrous or nearly so; plants begin flowering in July .... *S. juncea*
10 Stems pubescent, at least on upper half .... *S. nemoralis*
10 Stems glabrous (or sometimes with fine hairs in inflorescence)............................ 11
11 Bases of lowest stem leaves clasping stem; wet habitats .... *S. uliginosa*

11 Lower leaf bases not clasping stem 12
12 Basal and lower stem leaves much larger than leaves of middle stem, persistent; flowering begins early (July) .... *S. juncea*
12 Basal and lower stem leaves not much larger than leaves of middle stem, withered by flowering time; flowering late (August–September) .... *S. ulmifolia*
13 Basal and lower stem leaves smaller than mid-stem leaves; middle and upper stem leaves with sharp teeth, longer than the axillary inflorescences .... *S. flexicaulis*
13 Basal and lower stem leaves larger than mid-stem leaves; middle and upper stem leaves entire or with rounded teeth, not longer than inflorescences 14
14 Stem and leaves pubescent 15
14 Stems glabrous or nearly so, sparse fine hairs may be present in inflorescence .... *S. uliginosa*
15 Achenes glabrous .... *S. hispida*
15 Achenes densely covered with upward-pointing appressed hairs .... *S. simplex*

Solidago altissima L.

TALL GOLDENROD   *native / cons* **7**
Solidago canadensis var. *scabra* Torr. & Gray
Perennial from creeping rhizomes. Stems 5-20 dm tall, usually short-hairy throughout. Lower stem leaves usually withered by flowering; sessile or subpetiolate; blades oblong lance-shaped, 5-15 cm long and 7-20 mm wide, relatively thick and firm, entire to finely serrate, strongly 3-nerved, upper surface finely strigose, underside scabrous; much reduced upwards. Heads many, in a secund, pyramidal, panicle-like inflorescence, branches divergent and recurved; involucres narrowly campanulate, 2.5-4.5 mm long, the bracts in 3 series, unequal; ray florets 8-13; disk florets 3-6; corollas 2-4 mm long. Achenes sparsely to moderately hairy. Many types of wet to dry habitats.

    S. altissima is sometimes treated as *S. canadensis* var. *scabra*. The short hairs on the leaves give fresh plants a gray-green color not seen in *S. canadensis*. Subject to insect galls on the stems.

Solidago altissima

Solidago canadensis L.

COMMON GOLDENROD   *native / cons* **1**
Perennial from creeping rhizomes, without a well-developed caudex. Stems 3-13 dm tall, more or less puberulent at least above the middle. Leaves thin, sharply serrate to subentire, glabrous or slightly scabrous above, commonly finely hairy on the midrib and main veins beneath; basal leaves absent, or, like the lower stem leaves, reduced and soon deciduous; mid- and upper stem leaves numerous and crowded, only gradually reduced upwards, lance-linear, tapering to the sessile base, 3-nerved, 5-13 cm long and 5-18 mm wide. Inflorescence terminal, panicle-like, with conspicuously recurved-secund branches; involucre about 2-3 mm high, its bracts imbricate in several series, yellowish, without well defined green tips; rays mostly 10-17,, only about 1-1.5 mm long. Achenes short-hairy. July-Sept. — Open, moist or dry places.

Solidago canadensis

Solidago flexicaulis L.

ZIGZAG GOLDENROD   *native / cons* **6**
Perennial with creeping rhizomes. Stems 3-12 dm tall, grooved, glabrous below the inflorescence. Leaves sharply and often coarsely serrate or dentate, hirsute beneath, at least on the midrib and main veins, or rarely glabrous, glabrous or sparsely hairy above; basal and lowermost stem leaves deciduous by flowering time; upper leaves ovate to elliptic, 7-15 cm long and 3-10 cm wide, abruptly contracted to the winged petiole. Inflorescence a series of short raceme-like clusters, the lower in the axils of ordinary scarcely reduced foliage leaves, these progressively reduced upwards, the uppermost becoming inconspicuous and shorter than their axillary clusters; involucre 4-6 mm. high, its bracts strongly imbricate, glabrous; rays 3-4. Achenes short-hairy. Aug-Oct. — Woodlands.

Solidago flexicaulis

Solidago gigantea Ait.
SMOOTH GOLDENROD   *native / cons* 3
Solidago serotina Ait. non Retz.

Perennial herb, from stout rhizomes, often forming colonies. Stems 0.5-2 m tall, mostly smooth, sometimes waxy, short-hairy on upper branches. Leaves lance-shaped to oval, 6-15 cm long and 1-4 cm wide, prominently 3-veined, tapered to a stalkless or short petiolelike base, glabrous or sparsely hairy on underside veins; margins with sharp, forward-pointing teeth. Flower heads many, in large panicle-like clusters, on 1 side of the spreading branches (secund), with yellow disk and ray flowers, the rays 2-3 mm long; involucre 2-5 mm high, the involucral bracts linear. Achenes 1-2 mm long. July-Sept. — Wet meadows, streambanks, swamps, floodplain forests, thickets, marshes, calcareous fens, ditches; also open woods and roadsides. Canada goldenrod (*Solidago canadensis*) similar but generally smaller and densely short-hairy on leaf undersides and upper stem.

Solidago gigantea

Solidago hispida Muhl.
HAIRY GOLDENROD   *native / cons* 6

Perennial with a stout branched caudex and fibrous roots. Stems 1-10 dm tall, the herbage generally spreading-hirsute throughout. Basal and lowermost stem leaves well developed and generally persistent, broadly oblong lance-shaped, crenate or serrate to entire, petiolate, the blade and petiole 3-20 cm long and 1-5 cm wide; stem leaves reduced upwards and becoming sessile. Inflorescence terminal, elongate and narrow, generally more or less leafy-bracteate toward the base, the lower clusters often elongate and stiffly ascending, but not secund; involucre 4-6 mm high, the bracts imbricate in several series, yellowish; rays about 7-14, usually deep yellow. Achenes glabrous, at least when mature. July-Oct. Dry woodlands and rocky shores.

Solidago hispida

Solidago juncea Ait.
EARLY GOLDENROD   *native / cons* 4

Perennial with a stout branched caudex and fibrous roots, frequently with long creeping rhizomes as well. Stems 3-12 dm tall, essentially glabrous throughout except for the scabrous or ciliate leaf-margins, but sometimes short-hairy on one or both surfaces or in the inflorescence. Basal leaves tufted and persistent, 15-40 cm long and 2-7.5 cm wide, with narrowly elliptic, serrate blades tapering to the long petiole; stem leaves progressively reduced, becoming sessile. Inflorescence terminal, panicle-like, dense, generally about as wide as long or even wider, with recurved-secund branches; involucre glabrous, 3-5 mm high, its bracts imbricate; rays minute, usually 7-12. Achenes persistently short-hairy. June-Oct. — Dry open places and open woods, especially in sandy soil. One of the earliest goldenrods to flower.

Solidago juncea

Solidago nemoralis Ait.
GRAY GOLDENROD   *native / cons* 4

Perennial with a branchmg caudex and fibrous roots. Stems 1-10 dm tall, the herbage densely puberulent with loosely spreading hairs. Leaves weakly 3-nerved; basal leaves well-developed, tufted and persistent, oblong lance-shaped, long-petiolate, 5-25 cm long and 8-40 mm wide, toothed; stem leaves progressively reduced, less petiolate, and less toothed upwards, the lowermost similar to the basal, but often deciduous. Inflorescence terminal, panicle-like, sometimes elongate and nodding at the apex; sometimes larger with long, recurved, secund branches; involucre 3-6 mm high, its bracts imbricate in several series, glabrous except for the ciliolate margins; rays short, 5-9. Achenes pubescent. Aug-Oct. — Dry woods and open places, especially in sandy soil.

Solidago nemoralis

Solidago ohioensis Frank

OHIO FLAT-TOPPED GOLDENROD   *native / cons* 9
Oligoneuron ohioense (Frank) G.N. Jones

Perennial herb, from a crown. Stems 5-10 dm tall, smooth. Leaves largest at base of plant and becoming smaller upward, lance-shaped to oblong lance-shaped, to 2 dm long and 1-5 cm wide, pinnately-veined, margins entire or slightly toothed near tip, rough-to-touch; tapered to a long petiole on lower leaves, upper leaves sessile. Flower heads many in a branched, flat-topped to rounded inflorescence at ends of stems, with yellow disk and ray flowers; involucre smooth, 4-5 mm high, the involucral bracts rounded at tip. Achenes glabrous. July-Sept. — Wet, sandy or gravelly shores, streambanks, sedge meadows, calcareous fens; soils often calcium-rich.

Solidago ohioensis

Solidago riddellii Frank

RIDDELL'S FLAT-TOPPED GOLDENROD   *native / cons* 7
Oligoneuron riddellii (Frank) Rydb.

Perennial herb, from a crown and sometimes also with rhizomes. Stems 2-10 dm tall, smooth but sometimes sparsely hairy in head. Leaves glabrous, largest at base of plant, these often early-deciduous, lance-shaped to linear, 10-20 cm long and 5-30 mm wide, becoming smaller upward, the upper leaves sickle-shaped and folded along midrib; margins entire; petioles of lower leaves long and winged, upper leaves sessile and clasping stem. Flower heads many, crowded in a branched, rounded to flat-topped inflorescence, the heads not confined to 1 side of the branches, with yellow disk and ray flowers, the rays 1-2 mm long; involucre 5-6 mm high, its bracts rounded at tip. Achenes glabrous, 1-2 mm long. Aug-Oct. — Wet meadows, calcareous fens, low prairie, lakeshores, streambanks.

Solidago riddellii

Solidago rigida L.

STIFF GOLDENROD   *native / cons* 5
Oligoneuron rigidum (L.) Small

Perennial from a stout branched caudex. Stems 2.5-15 dm tall. Herbage densely pubescent with short spreading hairs, but sometimes nearly glabrous. Leaves firm, slightly toothed or entire, the basal and lowermost cauline leaves well developed and usually persistent, with elliptic, or broadly lance-shaped blades 6-25 cm long and 2-10 cm wide, often exceeded by the long petiole; stem leaves progressively reduced and less petiolate upwards, the middle ones sessile or nearly so. Inflorescence terminal, dense, corymbiform; heads large and many-flowered; involucre 5-9 mm high, its bracts conspicuously striate; rays 8-14. Achenes 10-15 nerved, glabrous, or hairy at the tip. July-Oct. — Prairies and other dry open places, especially in sandy soil.

Solidago rigida

Solidago simplex Kunth

DUNE GOLDENROD   *native / cons* 9
Solidago gillmanii (Gray) Steele
Solidago spathulata DC.

Perennial with a short usually branched caudex and fibrous roots. Stems 1-9 dm tall, glabrous or nearly so except for fine hairs in the inflorescence (or sometimes more hairy throughout in var. *gillmanii*). Basal and lower stem leaves narrowly oblong lance-shaped, 2-30 cm long (including the petiole), 4-40 mm wide, toothed to nearly entire, often irregularly ciliolate-margined, usually persistent; stem leaves progressively reduced, becoming sessile upwards. Inflorescence terminal, varying from few-headed to dense, but not secund, its leaves reduced and inconspicuous. Involucre 3-9 mm high, often glutinous, its bracts imbricate in several series; rays mostly 9-10. Achenes short-hairy. Aug-Oct. — Sand dunes and beaches. Ours mostly subsp. *randii* (Porter). Plants of var. *gillmanii* (state threatened, known from Door and Sheboygan counties) are robust, mostly 3-9 dm tall, often more hairy than the other varieties, with large heads (involucre 6-9 mm high) in a long, often branched inflorescence, and mostly found on sand dunes and sand beaches along Lake Michigan.

Solidago uliginosa Nutt.

NORTHERN BOG-GOLDENROD   *native / cons* 8

Perennial herb, from a branched crown. Stems stout, 5-15 dm long, glabrous but finely hairy in the inflorescence. Leaves largest at base of plant, 5-35 cm long and 1-5 cm wide, becoming smaller upwards, lance-shaped to oblong lance-shaped, glabrous; lower leaves tapered to long petioles, somewhat clasping stem, upper leaves stalkless; margins finely toothed, or entire on upper leaves, rough-to-touch. Inflorescence long, crowded, and spike-like, the branches ascending, straight or curved downward at tip, the heads sometimes secund; involucre 3-5 mm high. Achenes more or less glabrous. Aug-Sept. — Conifer swamps, fens, open bogs, low prairie, wet meadows, interdunal wetlands.

Solidago uliginosa

Solidago ulmifolia Muhl.

ELM-LEAVED GOLDENROD   *native / cons* 5

Perennial with a branching caudex and fibrous roots. Stems 4-12 dm tall, glabrous or nearly so below the inflorescence. Leaves thin, sharply and usually rather coarsely serrate, glabrous to more commonly hirsute above, loosely hirsute on the midrib and main veins beneath, and usually rather sparsely so over the surface as well; basal leaves elliptic-ovate, abruptly contracted to the petiole, but usually deciduous by flowering time; lowermost stem leaves soon deciduous and smaller than the persistent ones just above, these 6-12 cm long and 1.2-5.5 cm wide, becoming reduced and nearly sessile upwards. Inflorescence terminal, panicle-like, with recurved-secund branches, these generally few, long, and divergent; involucre 2.5-4.5 mm high, its bracts imbricate; rays 3-5, minute. Achenes persistently short-hairy. Aug-Oct. — Woodlands.

Solidago ulmifolia

Sonchus

SOW-THISTLE

Annual or perennial herbs with milky juice. Leaves alternate or all basal, entire to pinnatifid or dissected, mostly auriculate, often prickly-margined. Heads solitary to usually several or many in an irregular corymb-like inflorescence. Flowers all ligulate and perfect, yellow, few to more often numerous (our species with usually ca. 120-160 flowers in each head). Involucre ovoid or campanulate, its bracts generally imbricate in several series. Achenes flattened, about 6-20-ribbed, beakless, glabrous. Pappus of numerous white capillary bristles which tend to fall connected, sometimes with a few stouter ones which fall separately. The perennial species are troublesome farm weeds.

1   Perennial with creeping rhizomes; leaf bases auriculate and clasping stem, the auricles small . . . . . .
. . . . *S. arvensis*
1   Taprooted annuals; leaf bases auriculate and clasping stem, the auricles large and conspicuous . . . 2
2   Leaf margins sparsely prickly . . . . *S. oleraceus*
2   Leaf margins with numerous spine-tipped teeth . . . . *S. asper*

Sonchus arvensis L.

PERENNIAL SOW-THISTLE   *introduced*

Perennial with long vertical roots, and extensively spreading by horizontal, rhizomelike, often deep-seated roots. Stems 4-20 dm tall, glabrous below the inflorescence and often somewhat glaucous. Leaves prickly-margined, the lower and middle ones usually pinnately lobed or pinnatifid, commonly 6-40 cm long and 2-15 cm wide, becoming less lobed and often more strongly auriculate upwards, the upper reduced and distant. Heads several in an open corymbiform inflorescence, relatively large, commonly 3-5 cm wide in flower, the fruiting involucre 15-22 mm long; involucre and peduncles more or less copiously provided with coarse, spreading, gland-tipped hairs, the involucre with some small and obscure tufts of tomentum as well. Achenes about 2.5-3.5 mm long, with 5 or more prominent longitudinal ribs on each face, strongly rugulose. July-Oct. — A cosmopolitan weed of European origin.

Sonchus asper (L.) Hill
SPINY-LEAF SOW-THISTLE   *introduced*
Similar to *S. oleraceus*, but usually more prickly. Leaves pinnatifid, or frequently obovate and lobeless, with rounded, not acute auricles. Achenes with 3 or rarely 4-5 evident longitudinal ribs on each face, not rugulose, although there may be minute projections from the marginal ribs. July-Oct. A cosmopolitan weed; native of Europe.

Sonchus arvensis

Sonchus oleraceus L.
COMMON SOW-THISTLE   *introduced*
Annual with a short taproot. Stems 1-10 dm tall, glabrous except sometimes for a few spreading gland-tipped hairs on the involucre and peduncle. Leaves pinnatifid to occasionally merely toothed, the margins rather weakly or scarcely prickly, 6-30 cm long and 1-15 cm wide, all but the lowermost prominently auriculate, the auricles with well rounded margins but eventually sharply acute; leaves progressively less divided upwards, and more or less reduced. Heads several in a corymbiform inflorescence, relatively small, only about 1.5-2.5 cm wide in flower; receptacle expanding and becoming conspicuously pale and indurate in fruit; fruiting involucre mostly 9-13 mm high. Achenes 2.5-3 mm long, rugulose and 3-5-ribbed on each face. July-Oct. — A cosmopolitan weed; native of Europe.

Sonchus oleraceus

Symphyotrichum
WILD ASTER
Mostly perennial herbs (annual in S. ciliatum). Leaves simple, alternate. Flower heads with both ray and disk flowers (disk flowers only in S. ciliatum); ray flowers white, pink, blue or purple, usually more than 0.5 mm wide (in contrast to the very narrow rays in Erigeron); disk flowers red, purple or yellow; involucral bracts in 2 or more series, usually overlapping; receptacle naked (not chaffy), flat or nearly so; pappus of numerous hairlike bristles.

The traditional genus *Aster* has been split into several segregate genera to reflect differences with European species. Most species native to e USA are now placed within the genus *Symphyotrichum*, with the following exception for Door County species: *A. umbellatus* in the genus *Doellingeria*. Initially controversial, this classification is now widely accepted and is followed here.

1 Leaves, at least the lower ones, heart-shaped at base and with petioles 2
1 Leaves not both heart-shaped and petioled 6
2 Leaves entire or nearly so; involucral bracts with a short, diamond-shaped green tip . . . .
  . . . . *S. oolentangiense*
2 Leaves toothed; involucral bracts various 3
3 Inflorescence with relatively few heads, often less than 50; peduncles and inflorescence branches with only a few bracts . . . . *S. ciliolatum*
3 Inflorescence with many heads, often over 100; peduncles and inflorescence branches with many bracts 4
4 Plants glabrous or nearly so (sometimes slightly finely hairy in the inflorescence) . . . . . *S. urophyllum*
4 Plants hairy to rough-hairy, at least in part 5

5 Rays bright blue; plants often densely hairy . . . . *S. drummondii*
5 Rays pale-blue, pale-purple, or white; plants thinly hairy . . . . *S. urophyllum*
6 Base of leaf strongly clasping stem 7
6 Base of leaf not clasping stem (or only slightly so) 8
7 Involucral bracts (at least the inner), long-tapered to a slender tip . . . . *S. puniceum*
7 Involucral bracts rounded or short-tapered to the tip. . . . . *S. laeve*
8 Taprooted annual herb; rays absent . . . . *S. ciliatum*
8 Perennial herbs with fibrous roots and from rhizomes or crowns; rays well developed 9
9 Most involucral bracts with a slender green tip, the margins inrolled . . . . *S. pilosum*
9 Involucral bracts flat, the margins not inrolled 10
10 Tips of outer involucral bracts loose or recurved, tapered to a very small spine-tip; leaves entire . . . .
. . . . *S. ericoides*
10 Involucral bracts appressed or only slightly loose, the tips not spine-tipped; leaves various. 11
11 Leaves hairy on underside, at least along the midvein 12
11 Leaf underside glabrous 13
12 Leaf underside hairy; plants with creeping rhizomes . . . . *S. ontarionis*
12 Leaf underside glabrous except for hairs on the midvein; plants without creeping rhizomes . . . .
. . . . *S. lateriflorum*
13 Heads very small and numerous, the rays 3–6 mm long; heads often arranged on 1-side of the inflo-
rescence branches . . . . *S. lanceolatum*
13 Heads larger or few in number, not arranged on 1 side of the branches 14
14 Slender plants of bogs and other wetlands; inflorescence short-stalked and wide in outline . . . .
. . . . *S. boreale*
14 Plants stouter, not in bogs; inflorescence elongate . . . . *S. lanceolatum*

Symphyotrichum boreale (Torr. & Gray) A.& D. Löve

NORTHERN BOG-ASTER   *native / cons* **10**
    Aster borealis (Torr. & Gray) Prov.
    Aster junciformis Rydb.
Perennial herb, from rhizomes 1-2 mm wide. Stems erect, slender, 3-8
dm tall and to 2 mm wide, unbranched below, usually branched in the
head; smooth except for lines of short, appressed hairs below base of
upper leaves. Leaves alternate, linear, 4-12 cm long and 2-6 mm wide,
sometimes slightly clasping at base, margins rough-to-touch, petioles
absent. Flower heads usually few to rarely many, in an open, broad inflorescence; the heads 1.5-2 cm wide; involucre 5-7 mm high, the involucral
bracts overlapping, often purple at tips and on margins; ray flowers 20-
50, white to light blue or lavender, 1-1.5 cm long. Fruit an achene; pappus
of pale hairs. Aug-Sept. — Conifer swamps, calcareous fens, open bogs,
wet meadows, shores and seeps.

Symphyotrichum boreale

Symphyotrichum ciliatum (Ledeb.) Nesom

WESTERN ANNUAL ASTER   *introduced*
    Aster brachyactis Blake
Taprooted annual herb. Stems unbranched and erect, to branched and
spreading, 2-6 dm long, smooth. Leaves alternate, linear, 2-10 cm long
and mostly 2-5 mm wide, margins fringed with scattered hairs, petioles
absent. Flower heads several to many, in an open inflorescence which
forms much of plant; flower heads bell-shaped, 1-2 cm wide, involucre 5-
10 mm high, the involucral bracts mostly green, linear, of equal length
or slightly overlapping; ray flowers absent. Achenes flattened, 1-2 mm
long; pappus of many long, soft hairs. Aug-Sept. — Shores (including
along Lake Michigan), streambanks, wet meadows, roadside ditches, usually where brackish. Native of w North America; considered adventive in
Wisconsin.

Symphyotrichum ciliatum

Symphyotrichum ciliolatum (Lindl.) A.& D. Löve
NORTHERN HEART-LEAVED ASTER   *native / cons* 4
  Aster ciliolatus Lindl.
  Aster lindleyanus Torr. & Gray
Perennial with long creeping rhizomes, sometimes also with a short branched caudex. Stems 2-12 dm tall, hirsute, especially in the inflorescence and on the lower leaf surfaces, sometimes glabrous throughout. Basal and lower stem leaves petiolate and cordate or subcordate, 4-12 cm long and 2-6 cm wide, sharply serrate, often deciduous, those above abruptly narrowed to the broadly winged petiole, and often less toothed, or the upper leaves sessile and entire. Inflorescence open, relatively few-headed, the heads often less than 50, rarely more than 100, the branches and peduncles sparsely or scarcely bracteate, the bracts narrow, the peduncles of very unequal length, generally some of them over 1 cm long; involucre 5-8 mm high, its slender bracts slightly or moderately imbricate, glabrous except for the sometimes ciliolate margins, their green tips relatively narrow and elongate; rays mostly 12-25, blue, 8-15 mm long. Achenes glabrous or nearly so, gray or stramineous, 3-6-nerved. July-Oct. — Woods and clearings.

Symphyotrichum ciliolatum

Symphyotrichum drummondii (Lindl.) Nesom
HAIRY HEART-LEAVED ASTER   *native / cons* 4
  Aster drummondii Lindl.
Perennial with a branched caudex or short rhizome and numerous fibrous roots. Stems 4-12 dm tall, usually densely pubescent at least above the middle with minute, stiffly spreading hairs. Leaves relatively firm, shallowly toothed, scabrous above, densely pubescent with short spreading hairs beneath, the lowermost ones ovate or lance-ovate, cordate, 6-14 cm long and 2.5-6.5 cm wide, long-petiolate, those above progressively less cordate (or the upper merely broadly rounded) at the base and with shorter, usually broadly winged petioles. Inflorescence panicle-like, with spreadmg or ascending bracteate branches, the heads often numerous, on bracteate peduncles usually less than 1 cm long; involucre glabrous or puberulent, 4.5-7 mm high, its bracts firm, imbricate, the green tips rhombic; rays mostly 10-20, bright blue, 5-10 mm long. Achenes pale, minutely hairy especially toward the tip, or glabrous, with several weak nerves. Sept-Oct. — Clearings and open woodland.

Symphyotrichum ericoides

Symphyotrichum ericoides (L.) Nesom
WHITE HEATH ASTER   *native / cons* 4
  Aster ericoides L.
Perennial from well developed creeping rhizomes, pubescent with appressed or spreading hairs, or the leaves sometimes subglabrous. Stems 3-10 dm tall, occasionally more. Leaves numerous, linear, sessile, rarely as much as 6 cm long and 7 mm wide, the lower and often also the middle ones soon deciduous, those of the branches reduced and divaricate, often becoming mere bracts. Heads numerous, small, commonly somewhat secund on the divergent or recurved branches; involucre about 3-5 mm high, its bracts more or less strongly imbricate in several series, the outer spinulose and more or less squarrose, some or all of the bracts coarsely ciliolate-margined; rays 8-20, white, rarely blue or pink, 3-5 mm long. Achenes hairy. — Dry, open places.

Symphyotrichum laeve (L.) A.& D. Löve
SMOOTH BLUE ASTER   *native / cons* 6
  Aster laevis L.
Perennial from a short stout rhizome or branched caudex, occasionally with short creeping red rhizomes as well. Stems 3-10 dm tall. Herbage

Symphyotrichum laeve

glabrous throughout, except occasionally for some puberulent lines in the inflorescence, commonly somewhat glaucous. Leaves thick and firm, variable in size and shape but the larger ones over 1 cm wide, entire or sometimes toothed, sessile and more or less strongly clasping, or the lower tapering to winged petioles and scarcely clasping; leaves of the inflorescence reduced and often bractlike, clasping at their base. Heads several or numerous in an open inflorescence; involucre 5-9 mm high, its appressed bracts imbricate in several series, with short, commonly with rhombic green tips; rays mostly 15-25, blue or purple, 8-15 mm long. Achenes nearly glabrous; pappus reddish or sometimes white. Aug–Oct. Open, usually dry places.

Symphyotrichum lanceolatum (Willd.) Nesom
EASTERN LINED ASTER   *native / cons* 4
Aster lanceolatus Willd.
Aster hesperius Gray
Aster interior Wieg.
Symphyotrichum simplex (Willd.) A. & D. Löve
Perennial herb, forming colonies from long rhizomes. Stems 0.5-1.5 m long, upper stems with lines of hairs. Leaves alternate, all on stem, lance-shaped to linear, 8-15 cm long and 3-30 mm wide, upper surface smooth or slightly rough-to-touch, margins toothed or sometimes entire; petioles absent or blades tapered to petiolelike base, sometimes slightly clasping stem. Flower heads many in an elongate leafy inflorescence; the involucre 3-6 mm high, the involucral bracts tapered to a green tip, smooth or margins fringed with hairs, strongly overlapping; ray flowers 20-40, usually white, sometimes lavender or blue, 4-12 mm long. Fruit an achene; pappus white. Aug–Oct. — Marshes, wet meadows, fens, swamp openings, low prairie, streambanks and shores. One of our most common asters.

Symphyotrichum lanceolatum

Symphyotrichum lateriflorum (L.) A. & D. Löve
GOBLET-ASTER, FAREWELL-SUMMER   *native / cons* 3
Aster lateriflorus (L.) Britt.
Perennial from a branching caudex or short stout rhizome, with numerous fibrous roots. Stems several, 3-12 dm tall, curly-villous to glabrous. Leaves scabrous or nearly glabrous above, glabrous beneath except for the usually puberulent midrib; basal and lower stem leaves soon deciduous, or the basal occasionally persistent, obovate to lance-shaped, tending to taper from the middle to both ends, entire or serrate, mostly 5-15 cm long and 5-30 mm wide, petiolate; upper leaves sessile or nearly so. Heads numerous in a widely branched or sometimes more simple inflorescence; involucre glabrous, mostly 4-5.5 mm high, its bracts imbricate in few series, with broad green tips, often suffused with purple upwards; rays 9-14, white or slightly purple-tinged, 4-6.5 mm long; lobes of the disk corollas recurved. Achenes few-nerved, somewhat hairy. Aug.–Oct. — Various habitats, most commonly in open woodlands, dry open places, and on beaches.

Symphyotrichum lateriflorum

Symphyotrichum ontarionis (Wieg.) Nesom
ONTARIO ASTER   *native / cons* 6
Aster ontarionis Wieg.
Perennial herb, from long creeping rhizomes. Stems branched, 3-8 dm long, upper stems with short spreading hairs. Leaves alternate, thin, oblong lance-shaped, 5-10 cm long and 1-3 cm wide (upper leaves smaller), upper surface rough-hairy to nearly smooth, underside finely to densely hairy; margins with sharp, forward-pointing teeth above middle of blade; petioles absent. Flower heads 1-2 cm wide, on short stalks from short leafy branches; involucre smooth to finely hairy, 5-7 mm high, the in-

Symphyotrichum ontarionis

volucral bracts overlapping; ray flowers white, 9 or more. Fruit an achene.
Sept-Oct. — Floodplain forests, river terraces, thickets. Similar to *S. later-iflorum*, but with long rhizomes rather than a crown or short rhizomes.

Symphyotrichum oolentangiense (Riddell) Nesom
PRAIRIE HEART-LEAVED ASTER   *native / cons* 5
Aster oolentangiensis Riddell

Perennial from a branched caudex or short rhizomes, with numerous fibrous roots. Stems 2-15 dm tall, scabrous-puberulent to occasionally nearly glabrous. Leaves thick and firm, entire or occasionally shallowly serrate, scabrous-hispid above, the hairs on the lower surface softer, and usually longer and looser than those on the upper; basal and usually also the lower stem leaves long-petiolate, cordate, lance-shaped or ovate, 4-13 cm long and 1.2-6 cm wide, those above abruptly smaller, narrower, less petiolate, and generally not at all cordate, the upper leaves sessile and lance-shaped or linear. Inflorescence open, panicle-like, with narrow and usually numerous bracts, the peduncles often very long; involucre 4.5-8 mm high, its bracts imbricate in several series, with a diamond-shaped green tip, glabrous except for the often ciliolate margins; rays commonly 10-25, blue, or rarely pink, 5-12 mm long. Achenes glabrous or nearly so, 3-5-nerved, usually pale. Aug-Oct. — Prairies and dry open woods.

Symphyotrichum oolentangiense

Symphyotrichum pilosum (Willd.) Nesom
WHITE OLDFIELD ASTER   *native / cons* 1
Aster pilosus Willd.

Perennial herb, from a large crown. Stems to 1.5 m long, more or less smooth (var. pringlei) or stems and leaves with spreading hairs (var. pilosum). Lower leaves oblong lance-shaped, 5-10 cm long and 1-2 cm wide, petioled; upper leaves smaller, linear, stalkless; margins entire or slightly toothed; petioles fringed with hairs; basal leaves and lower stem leaves soon deciduous (or basal leaves persistent). Flower heads at ends of small branches, forming an open inflorescence; involucre urn-shaped, narrowed near middle and flared upward, 3-5 mm high, smooth, involucral bracts overlapping to nearly equal in length, green-tipped; ray flowers 15-35, white. Fruit an achene. Summer. — Sandy and gravelly shores, interdunal swales, wet meadows; often where calcium-rich; sometimes weedy in disturbed fields and roadsides.

Symphyotrichum pilosum

Symphyotrichum puniceum (L.) A. & D. Löve
PURPLE-STEM ASTER   *native / cons* 5
Aster firmus Nees
Aster puniceus L.

Large perennial herb, from a short rhizome or crown, sometimes also with short stolons. Stems stout, red-purple, 0.5-2 m long, unbranched, or branched in head, with long stiff hairs or sometimes nearly smooth. Leaves alternate, lance-shaped to oblong lance-shaped, 6-18 cm long and 1-4 cm wide, rough-to-touch to nearly smooth above, underside smooth or with long hairs on midvein; margins with scattered sharp teeth or sometimes entire; petioles absent, base of leaf clasping. Flower heads numerous, 1.5-2.5 cm wide; involucre 6-10 mm high, involucral bracts about equal in length, smooth or fringed with hairs, green and spreading; ray flowers 20-50, blue (rarely white). Fruit a smooth achene; pappus more or less white. Aug-Sept. Swamps, sedge meadows, thickets, calcareous fens, streambanks, shores, springs, roadside ditches.

Symphyotrichum puniceum

Symphyotrichum urophyllum (Lindl.) Nesom

ARROW-LEAVED ASTER *native / cons* **3**
   Aster sagittifolius Willd.

Perennial with a branched caudex or short rhizome and numerous fibrous roots. Stems 4-12 dm tall, glabrous or nearly so below the inflorescence, or the upper part occasionally puberulent in lines. Leaves rather thick, shallowly toothed, glabrous or scabrous above, glabrous or hirsute beneath; lowermost leaves lance-ovate, cordate, 6-15 cm long and 2-6 cm wide, long-petiolate; upper leaves narrowed to the often broadly winged petiole, or sessile. Inflorescence panicle-like, elongate, with ascending bracteate branches; the heads often very numerous, borne on branches rarely more than 1 cm long, thus appearing crowded; involucre 4-6 mm high, its imbricate bracts glabrous except for the sometimes ciliolate margins, slender with elongate green tips; rays 8-20, usually pale blue or lilac, sometimes white, 4-8 mm long. Achenes pale, glabrous, 4-5-nerved. Aug-Oct. — Streambanks, woodlands, and less often in open places.

Symphyotrichum urophyllum

Tanacetum

TANSY

Annual or perennial herbs, sometimes somewhat woody at the base. Leaves alternate, pinnately dissected. Heads small or medium-sized, corymbiform or solitary, hemispheric to campanulate; discoid or nearly so, the outer flowers pistillate, with a short tubular corolla in some species expanded into a short yellow ray, or the pistillate flowers rarely wanting. Involucral bracts imbricate, dry, the margins and tips commonly scarious. Receptacle flat or convex, naked. Disk-flowers perfect, with 5-toothed tubular yellow corolla. Achenes mostly 5-ribbed, commonly glandular. Pappus a short crown, or none.

1 Leaves undivided (though regularly toothed) . . . . *T. balsamita*
1 Leaves pinnatifid or bipinnatifid 2
2 Heads 13–20 mm wide; leaves hairy; rare along Lake Michigan beaches . . . . *T. bipinnatum*
2 Heads 5–10 mm wide; leaves glabrous or nearly so; common . . . . *T. vulgare*

Tanacetum balsamita L.

COSTMARY *introduced*
   Balsamita major Desf.
   Chrysanthemum balsamita L.

Coarse fragrant perennial. Stems 5-12 dm tall, strigose above, glabrous below. Leaves silvery-strigose when young, more or less glabrate in age, crenate, sometimes with a few reduced basal pinnae, the basal with elliptic blades 10-25 cm long and 2.5-8 cm wide, on a petiole of about equal length; stem leaves smaller, sessile or nearly so, seldom over 10 cm long, numerous. Heads numerous in a corymbiform inflorescence, the disk 4-7 mm wide; rays usually wanting, if present white and less than 1 cm long; involucral bracts narrow, with conspicuous, expanded, hyaline tip. Achenes subterete, about 10-ribbed; pappus a minute border or crown. Aug-Oct. — Native of s Europe and the Orient, escaped from cultivation to roadsides and other waste places.

Tanacetum bipinnatum (L.) Sch.Bip.

EASTERN TANSY *endangered / native / cons* **10**
   Tanacetum huronense Nutt.

Rhizomatous perennial, villous throughout. Stems 1-8 dm tall. Leaves few or fairly numerous, sessile or short-petiolate, scarcely punctate; stem leaves 5-20 cm long and 1.5-8 cm wide, the basal often longer and persistent, all dissected, with narrow ultimate segments, the main rachis scarcely or not at all winged, that of each pinna only narrowly so. Heads 1-15, rarely more, the disk about 10-18 mm wide; rays inconspicuous or

Tanacetum bipinnatum

more commonly evident, sometimes as much as 4 mm long; pappus a short toothed crown. July-Aug. — Sandy beaches, dunes, and cracks in limestone pavement (alvars); in Wisconsin, known only from Door County.

Tanacetum vulgare L.
COMMON TANSY   *introduced (invasive)*

Coarse aromatic perennial with a stout rhizome, glabrous or nearly so throughout. Stems about 4-15 dm tall. Leaves numerous, 1-2 dm long and nearly half as wide, sessile or short-petiolate, punctate, pinnatifid, with evidently winged rachis, the pinnae again pinnatifid or deeply lobed, with broadly winged rachis, the pinnules often again toothed. Heads discoid, numerous, commonly about 20-200, the disk about 5-10 mm wide; pappus a minute crown, almost obsolete. Aug-Oct. — Native of the Old World, escaped from cultivation to roadsides, fields and waste places.

Tanacetum vulgare

Taraxacum
DANDELION

Perennial, scapose, taprooted herbs with milky juice. Leaves all basal, forming a rosette, entire to pinnatifid. Flowers all ligulate and perfect, mostly numerous. Heads solitary, erect; involucral bracts biseriate, the outer usually shorter than the inner and often reflexed. Fruit a columnar achene, longitudinally ribbed. Pappus of numerous capillary bristles.

1 Leaves generally deeply lobed or cut to midrib; mature achenes red-brown ..... *T. erythrospermum*
1 Leaves various, deeply lobed to entire; mature achenes tan or olive-green .... *T. officinale*

Taraxacum erythrospermum Andrz. ex Besser
RED-SEED DANDELION   *introduced (naturalized)*
  Taraxacum laevigatum (Willd.) DC.

Similar to T. officinale, often more slender. Leaves generally very deeply cut for their whole length, the lobes narrow, the terminal one seldom much larger than the lateral ones. Heads a little smaller, the involucre mostly 1-2 cm high, its inner bracts mostly 11-13, often somewhat corniculate, the outer bracts appressed to reflexed, a third to a little more than half as long as the inner. Body of the achene becoming bright red or reddish purple at maturity, commonly somewhat rugulose below as well as muricate above; beak usually stramineous, from more than twice as long to occasionally only half as long as the body. April-June. — Native of Eurasia, now established throughout Wisconsin in fields, pastures, lawns, and other disturbed places, but less common than *T. officinale.*

Taraxacum officinale G.H. Weber
COMMON DANDELION   *introduced (naturalized)*

Leaves commonly sparsely hairy beneath and on the midrib, otherwise generally glabrous, or sometimes completely so, oblong lance-shaped, mostly 6-40 cm long and 0.7-15 cm wide, pinnatifid or lobed, the terminal lobe tending to be larger than the others, tapering to a narrow, scarcely or obscurely winged petiolar base. Scape 5-50 cm tall, glabrous or more or less villous, especially upwards. Heads usually large, the involucre mostly 1.5-2.5 cm high, the inner bracts mostly 13-20, these at first erect, finally reflexed, the mature achenes and pappus then forming a conspicuous ball easily disintegrated by the wind; outer bracts a little shorter and scarcely wider than the inner, reflexed. Body of the achene 3-4 mm long, pale gray-brown to olive-brown, muriculate above or sometimes to near the base, about half or a third as long as the slender beak; pappus white. March-Dec. — Native of Europe and adjacent Asia, now a cosmopolitan weed of lawns and disturbed sites.

Taraxacum officinale

Tragopogon
GOAT'S-BEARD

Biennial or perennial lactiferous herbs with a taproot. Leaves alternate, linear, entire, clasping, commonly grass-like. Heads solitary at the ends of the branches. Flowers all ligulate and perfect, yellow or purple. Involucre cylindric or campanulate, the bracts uniseriate, equal. Achenes linear, terete or angled, 5-10-nerved, narrowed at the base, slender-beaked, or the outer occasionally beakless. Pappus of a single series of plumose bristles, united at the base, the plume-branches interwebbed, several of the bristles commonly longer than the others and naked at the apex.

1   Peduncle enlarged or inflated below the head; leaf tips not recurved . . . . *T. dubius*
1   Peduncle not enlarged; leaf tips recurved . . . . *T. pratensis*

Tragopogon dubius Scop.
MEADOW GOAT'S-BEARD   *introduced*
   Tragopogon major Jacq.
Similar to *T. porrifolius* (salsify, sometimes cultivated in gardens and escaping; reported from elsewhere in Wisconsin) but *T. dubius* with yellow flowers and often smaller and less robust. Leaves averaging narrower. Involucral bracts sometimes more numerous. Achenes (including beak) seldom over 3.5 cm long. May-July. — Native of Europe; roadsides and waste places.

Tragopogon pratensis L.
JACK-GO-TO-BED-AT-NOON   *introduced*
Glabrous perennial 1.5-8 dm tall. Leaves to 30 cm long and nearly 2 cm wide, often much narrower. Peduncles not at all enlarged in flower, scarcely so in fruit. Involucral bracts most commonly 8, mostly 12-24 mm long in flower, equaling or shorter than the yellow rays, elongating to 18-38 mm in fruit. Achenes 12-24 mm long, relatively shorter beaked, the body nearly or quite as long as in the other two species. May-Aug. — Native of Europe; roadsides, fields, and waste places.

Tragopogon pratensis

Vernonia
IRONWEED

Vernonia fasciculata Michx.
SMOOTH IRONWEED   *native / cons 5*
Stout perennial herb, from a thick rootstock. Stems erect, single or clumped, 5-12 dm long, red or purple, smooth but short-hairy on branches of the head. Leaves alternate, lance-shaped, 5-15 cm long and 1-4 cm wide, smooth above, underside finely pitted, margins sharp-toothed, petioles short. Flower heads usually many, crowded in flat-topped clusters to 10 cm wide, with purple disk flowers only; involucre 6-9 mm high, the involucral bracts overlapping, green with purple tips; receptacle flat, not chaffy. Fruit a ribbed achene, 3-4 mm long; pappus of purple to brown, slender bristles. July-Sept. — Marshes, low prairie, streambanks.

Vernonia fasciculata

Xanthium
COCKLEBUR

Xanthium strumarium L.
COMMON COCKLEBUR   *native / cons 1*
Weedy taprooted annual herb; plants variable in size and habit, rough-to-touch or sometimes nearly smooth. Stems 2-15 dm long, often brown-spotted. Leaves alternate, ovate to nearly round, sometimes with 3-5 shallow lobes, 3-15 cm long and 2-20 cm wide, margins with blunt teeth; petioles 3-10 cm long. Flower heads either staminate or pistillate, the staminate flowers brown, in clusters of small round heads at ends of

Xanthium strumarium

stems above the larger pistillate heads; pistillate heads in several to many clusters from leaf axils, each head with 2 flowers, with a spiny involucre enclosing the head; petals absent. Fruit a brown bur formed by the involucre, 1.5-3 cm long, covered with hooked prickles; achenes thick, 1 in each of the 2 chambers of the bur. Aug-Sept. — Shores, streambanks, wet meadows, sand bars, dried depressions, often where disturbed; also in cultivated and abandoned fields, roadsides and waste places.

Balsaminaceae

TOUCH-ME-NOT FAMILY

Impatiens
TOUCH-ME-NOT

Impatiens capensis Meerb.
SPOTTED TOUCH-ME-NOT, JEWELWEED   *native / cons* **2**
   Impatiens biflora Walt.
Smooth annual herb. Stems hollow, succulent, 3-10 dm tall, usually branched above. Leaves simple, alternate, ovate to oval, 3-9 cm long and 1.5-4 cm wide, tapered to tip or rounded and tipped with a short slender point, margins shallowly and irregularly toothed; petioles longest on lower leaves, shorter upward, 0.5-5 cm long. Flowers irregular, orange-yellow, 1.5-3 cm long, usually mottled with red-brown spots, with a spur recurved parallel to the sac and to half its length, hanging from the petioles in few-flowered racemes from upper leaf axils; sepals 3, petal-like; petals 3; stamens 5. Fruit a 5-valved capsule; the mature capsules splitting when jarred or touched, scattering the seeds away from parent plants. July-Sept. — Swamps, low areas in woods, floodplain forests, thickets, streambanks, shores, marshes, fens, springs; often where disturbed.

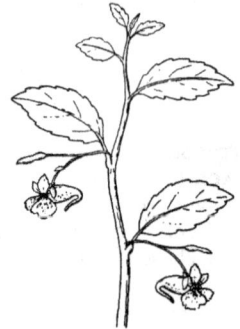

Impatiens capensis

Berberidaceae

BARBERRY FAMILY

Herbs or shrubs. Leaves alternate or basal, simple, lobed, or compound. Flowers solitary, racemose or cymose; perfect, all parts free and distinct. Sepals 4 or 6, sometimes early deciduous, in some genera petal-like. Petals as many as or more than the sepals, petaloid or reduced to nectaries. Stamens as many as the petals. Ovary 1-celled. Fruit a berry or capsule.

1 Plants spiny shrubs . . . . *Berberis*
1 Plants smooth perennial herbs 2
2 Flowers in a small panicle-like cyme . . . . *Caulophyllum*
2 Flowers single . . . . *Podophyllum*

Berberis
BARBERRY
Spiny shrubs. Leaves of the shoots reduced to alternate, simple or 3-branched spines, with clusters of small foliage leaves in their axils. Flowers yellow, in elongate racemes, umbel-like clusters, or sometimes solitary. Sepals 6, petal-like, subtended by 2 or 3 small bracts. Petals 6, usually smaller than the sepals and with 2 glands at their base. Stamens 6, appressed to the sepals until irritated, when they rapidly bend toward the center. Fruit a red one to few-seeded berry.

1 Leaves entire; flowers single or in clusters of 2–4 . . . . *B. thunbergii*
1 Leaves tipped by a small spine; flowers in racemes of 10–20 flowers . . . . *B. vulgaris*

Berberis thunbergii DC.
JAPANESE BARBERRY   *introduced (invasive)*
Densely and divaricately branched shrub to 2 m tall; spines usually simple. Leaves obovate to spatulate, usually obtuse, entire, narrowed at base to a short petiole. Flowers solitary or in small clusters of 2-4, about 8 mm wide. Fruit about 1 cm long. May. — Native of Japan; commonly planted for low hedges and frequently escaped along roadsides and in thickets.

Berberis thunbergii

Berberis vulgaris L.
EUROPEAN BARBERRY   *introduced (invasive)*
Freely branched shrub to 3 m tall. Leaves obovate to obovate-oblong, 2-5 cm long, obtuse or acute, finely spinulose-denticulate, the veinlets prominently reticulate beneath. Racemes usually 3-6 cm long, with 10-20 flowers on pedicels 5-10 mm long; petals entire. Fruit about 1 cm long. — Native of Europe; formerly widely planted and frequently escaped along roadsides and fences and in open woods; now largely purposefully exterminated as the alternate host of black rust of wheat.

Caulophyllum
BLUE COHOSH

Berberis vulgaris

Caulophyllum thalictroides (L.) Michx.
BLUE COHOSH   *native / cons* 8
Smooth perennial herbs. Stems erect, 3-8 dm tall, glaucous when young, bearing above the middle a single large, sessile, 3-parted leaf, and another smaller leaf just below the panicle. Leaflets obovate-oblong, 2-5-lobed above the middle, 5-8 cm long when fully grown. Flowers yellowish green or greenish purple, nearly 1 cm wide, in a panicle 3-6 cm long; sepals 6, petal-like, subtended by 3 or 4 sepal-like bracts; petals 6, reduced to small gland-like bodies much shorter than the sepals and opposite them; stamens 6; ovary soon ruptured by the enlarging seeds, which ripen exposed on short stout stalks and resembling drupes; seeds dark blue, 5-8 mm long, on stalks of nearly the same length. April-May. — Rich moist woods.

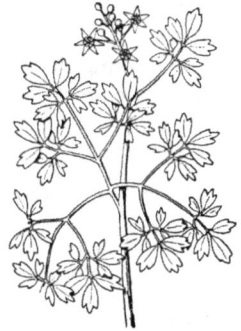

Caulophyllum thalictroides

Podophyllum
MAY-APPLE

Podophyllum peltatum L.
MAY-APPLE   *native / cons* 4
Herb, from a perennial rhizome, usually colony-forming. Flowering stem 3-5 dm tall, bearing a pair of leaves and a short-peduncled, solitary, terminal flower. Sterile plants bearing a single, large, peltate, deeply radially lobed, terminal leaf; fertile plants bearing 2 half-round, similarly lobed leaves. Flowers 3-5 cm wide, on a short nodding peduncle; sepals 6, falling early; petals 6-9, white; stamens 2x as many as the petals; ovary ovoid, with a large sessile stigma. Fruit a yellow, ovoid, fleshy, many-seeded berry 4-5 cm long. May; fruit ripe in Aug. — Moist, preferably open woods. The ripened fruit is edible in small amounts, toxic if consumed in large quantities; rhizome, leaves and seeds toxic.

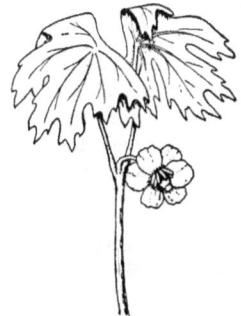

Podophyllum peltatum

Betulaceae
BIRCH FAMILY
Medium to large trees, or shrubs. Leaves deciduous, simple, alternate, with toothed margins and pinnate veins. Flowers small, staminate and pistillate flowers separate on same plant, crowded into catkins (aments) that open in spring before leaves fully open; staminate catkins

hang downward; conelike pistillate catkins erect or drooping. Fruit a small, 1-seeded, winged nutlet.

1 Plants in flower 2
1 Plants in fruit 6
2 Pistillate flowers 1 or several in a cluster .... *Corylus*
2 Pistillate flowers in catkins 3
3 Each bract of staminate catkin with 1 flower, this without sepals 4
3 Each bract of staminate catkin with 3–6 flowers, each with sepals 5
4 Staminate catkins in groups of 1 .... *Carpinus*
4 Staminate catkins usually in clusters of 3 .... *Ostrya*
5 Pistillate bracts 3-lobed; stamens 2 .... *Betula*
5 Pistillate bracts 5-lobed; stamens 3–5 .... *Alnus*
6 Each fruit (nut) subtended by leaf-like bracts 7
6 Fruit without leafy bracts, in the axil of a small scaly bract 9
7 Shrubs; nut 1 cm long or more .... *Corylus*
7 Trees; nut to 6 mm long 8
8 Bark furrowed and shredding, gray-brown, bracts saclike, enclosing the nut .... *Ostrya*
8 Bark smooth and gray; bracts not enclosing the nut .... *Carpinus*
9 Bracts woody, widely spreading from rachis of cone .... *Alnus*
9 Bracts papery, ascending .... *Betula*

Alnus
ALDER

Thicket-forming shrubs, or an introduced tree. Leaves deciduous, ovate, toothed on margins. Staminate and pistillate flowers separate on same plant, staminate flowers in long, drooping catkins which fall after shedding pollen; pistillate flowers in short, persistent conelike clusters. Fruit a flattened achene with winged or thin margins.

1 Twigs and young leaves sticky, leaves with small, sharp teeth; catkins on long stalks; fruit broadly winged .... *A. viridis*
1 Twigs and young leaves not sticky, leaves unevenly double-toothed; catkins stalkless or on short stalks; fruit narrowly winged .... *A. incana*

Alnus incana (L.) Moench
SPECKLED ALDER, TAG ALDER   native / cons 4
Alnus rugosa (Du Roi) Spreng.

Thicket-forming shrub to 5 m tall; twigs red-brown, waxy, with conspicuous pale lenticels. Leaves ovate to oval, broadest near or below middle, 6-14 cm long and 4-7 cm wide, dark green and smooth above, paler and hairy below; margins sharply toothed and shallowly lobed; petioles 1-2.5 cm long. Flowers in catkins clustered at ends of branches; staminate catkins developing in late summer, short-stalked, elongate, 4-9 cm long; pistillate catkins appear in late summer, stalkless, rounded, 1-2 cm long and to 1 cm wide, the scales unlobed, becoming conelike, persistent. Fruit a flat nutlet, narrowly winged on margin, 2-4 mm long. April-June. — Swamps, thickets, bog margins, shores and streambanks.

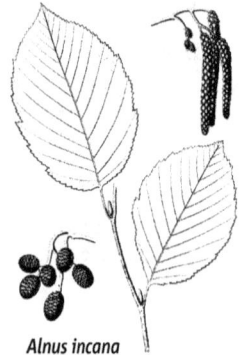

Alnus incana

Alnus viridis (Vill.) Lam. & DC.
SITKA ALDER   native / cons 8
Alnus alnobetula (Ehrh.) K. Koch p.p.
Alnus crispa (Ait.) Pursh

Thicket-forming shrub to 4 m tall; bark red-brown to gray; twigs brown, sticky, somewhat hairy, lenticels pale and scattered. Leaves round-oval, bright green above, slightly paler and shiny below, sticky when young, margins wavy with small, sharp teeth; petioles 6-12 mm long. Flowers in catkins; staminate catkins stalked, slender, developing in late summer and expanding in spring; pistillate catkins appear in spring, becoming long-stalked, blunt and conelike, persistent, 1-2 cm long. Fruit a nutlet,

Alnus viridis

2-3 mm long, with a pale, thin wing. — Lakeshores, wet depressions in woods, rock outcrops.

Betula
BIRCH

Trees or shrubs, often with multiple stems from base; bark sometimes peeling in thin layers. Leaves deciduous, alternate, sharply toothed. Staminate and pistillate flowers separate on same plant, catkins appearing in fall, opening the following spring, staminate flowers in drooping slender catkins; pistillate flowers in erect conelike catkins. Fruit a wing-margined achene (samara).

1 Shrub to 2 m tall; bark not shredding; leaves to 5 cm long .... *B. pumila*
1 Small to large trees; bark shredding with age 2
2 Bark white; samara wings as wide or wider than body .... *B. papyrifera*
2 Bark yellow-gray; samara wings narrower than body .... *B. alleghaniensis*

Betula alleghaniensis Britt.
YELLOW BIRCH   *native / cons* 7
  Betula lutea Michx. f.

Medium to large tree to 25 m tall; bark on young trees thin and smooth with conspicuous horizontal lenticels, becoming yellow-gray and shredding into thin, shaggy horizontal strips; bark of old trees breaking into large plates; twigs hairy when young, becoming smooth and shiny, wintergreen-scented when crushed. Leaves alternate, simple, ovate, tapered to a short, sharp tip, dark green above, paler yellow-green below, 6-12 cm long, margins coarsely double-toothed, petioles grooved and hairy. Staminate and pistillate flowers in catkins, separate on same tree, appearing before leaves in spring; staminate catkins drooping, yellow-purple, 7-10 cm long; pistillate catkins erect, green, 2-4 cm long, more or less stalkless. Fruit a winged nutlet, 3-5 mm wide. April-May. — Moist forests with sugar maple; also occasional in swamps, thickets, and forest depressions with red maple, black ash, black spruce, eastern hemlock and *Alnus incana*.

Betula alleghaniensis

Betula papyrifera Marsh.
WHITE BIRCH, PAPER BIRCH   *native / cons* 3

Trees, usually 20 m or shorter; trunks single or sometimes 2 or more. Bark of young trunks and branches dark reddish brown, smooth; in maturity creamy to chalky white, peeling in paper-thin sheets; lenticels pale, horizontal, in maturity dark, much expanded. Twigs without strong odor and taste of wintergreen, slightly to moderately pubescent, infrequently with small, scattered, resinous glands. Leaf blade ovate, with 9 or fewer pairs of lateral veins, 5-9 cm long and 4-7 cm wide, base rounded or truncate; lower surface pubescent, often velvety-hairy along major veins and in vein axils; margins coarsely or irregularly doubly serrate. Flowers in pendulous, cylindric catkins, 2.5-5 cm long, readily shattering with fruits in late fall; scales pubescent to glabrous. Samaras with wings as wide as or slightly wider than body. Late spring. — Moist, open, upland forest, especially where rocky; also on sand dunes swamps and sometimes in swampy woods; especially characteristic after fire or timber harvests, when seedlings are often abundant.

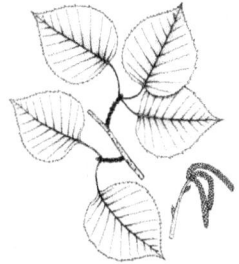
Betula papyrifera

  The bark, which has a high oil content making it waterproof, was used for a wide variety of building and clothing purposes by Native Americans.

  Includes *B. papyrifera* Marsh. var. *cordifolia* (Regel) Fern., sometimes considered a separate species (*Betula cordifolia* Regel).

Betula pumila L.
BOG BIRCH   native / cons 7
Betula glandulosa var. *glandulifera* (Regel) Gleason
Shrub 1-3 m tall; bark dull gray or brown; twigs gray, short-hairy and
dotted with resin glands, becoming red-brown and waxy with age. Leaves
leathery, rounded to obovate, 2-4 cm long and 1-3 cm wide, dark green
above, paler and often waxy below; margins coarsely toothed, the teeth
blunt or sharp; petioles 3-6 mm long. Flowers in catkins; staminate
catkins stalkless, cylindric, 15-20 mm long and 2-3 mm wide; pistillate
catkins stalked, cylindric, 1-2 cm long and 5 mm wide; scales 3-lobed.
Fruit a flat, winged, rounded nutlet, 2-3 mm long and 2-4 mm wide. May.
— Swamps, bogs, fens, seeps; often where calcium-rich.

Betula pumila

Carpinus
HORNBEAM

Carpinus caroliniana Walt.
HORNBEAM, IRONWOOD   native / cons 6
Tall shrub or small tree up to 10 m tall, with fluted trunk and smooth,
blue-gray or ashy gray bark. Leaves oblong to oblong-ovate, 5-12 cm long;
margins sharply and often doubly serrate. Staminate catkins slender,
pendulous; scales ovate, each subtending a single naked flower composed
of several stamens. Pistillate catkins slender, 2-5 cm long; scales ovate,
deciduous; pistillate flowers in pairs, each subtended by a minute bract
adnate at base to 2 minute bractlets; calyx minute. Fruit a small ribbed
nutlet. — Moist woods.

Carpinus caroliniana

Corylus
HAZELNUT

Corylus cornuta Marsh.
BEAKED HAZELNUT   native / cons 5
Shrub 1-3 m tall, the young twigs villous, later nearly glabrous. Leaves
oblong or obovate, broadly rounded to subcordate at base, pale green beneath, pubescent, especially on the veins and in the vein-axils beneath;
margins coarsely doubly serrate. Staminate catkins elongate, cylindric,
emerging in autumn, reaching anthesis in early spring; stamens 4, the
filaments deeply bipartite. Pistillate catkins small, ovoid, resembling a
leaf-bud, the few closely imbricate scales concealing the flowers, except
the elongate protruding stigmas. Involucre usually densely bristly toward
the base, closely surrounding the nut and prolonged beyond it into a
long slender beak cut at the summit into narrowly triangular lobes, the
whole 4-7 cm long. Nut short-ovoid, scarcely compressed, 1-1.5 cm long.
— Moist woods and thickets.

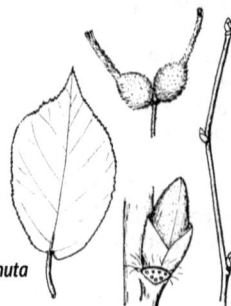

Corylus cornuta

Ostrya
HOP-HORNBEAM

Ostrya virginiana (P. Mill.) K. Koch
HOP-HORNBEAM   native / cons 5
Tree or tall shrubs to 20 m tall, with light brown scaly bark; twigs and
petioles at first pilose, eventually nearly glabrous, occasionally also stipitate-glandular. Leaves alternate, narrowly to broadly oblong or ovate;
margins sharply and often doubly serrate. Catkins opening with the
leaves in spring; staminate catkins elongate, densely flowered, composed
of spirally arranged scales, tipped with a sharp point and each subtending
a cluster of several stamens; filaments shortdivided at the summit, each
branch bearing a half-anther; pistillate catkins short-cylindric, 3-5 cm

Ostrya virginiana

long, loosely flowered, the ovate, hairy bracts early deciduous; calyx minute. Fruit a flattened-ovoid nutlet about 5 mm long. — Moist or dry woods and banks.

Boraginaceae

BORAGE FAMILY

Annual or perennial herbs with usually bristly stems and alternate, bristly leaves; plants glabrous in eastern bluebells (Mertensia virginica). Flowers typically in a spirally coiled, spike-like head that uncurls as flowers mature; flowers perfect (with both staminate and pistillate parts), with 5 petals, 4-5 sepals, and 5 stamens. Fruit a dry capsule with 4 nutlets.

Hydrophyllum previously included in Hydrophyllaceae and lacks the deeply 4-lobed ovary of other Boraginaceae.

1 Leaves shallowly palmately lobed to deeply pinnately divided . . . . *Hydrophyllum*
1 Leaves simple, entire 2
2 Plants in flower 3
2 Plants in fruit 11
3 Flowers irregular; stamens conspicuously exserted . . . . *Echium*
3 Flowers regular; stamens not longer than corolla 4
4 Corolla blue or purple, or leaf base extending downward along stem (decurrent) 5
4 Corolla not blue or purple; the leaf bases not decurrent along the stem 8
5 Flowers more than 1 cm long . . . . *Mertensia*
5 Flowers less than 1 cm long; leaf bases not decurrent 6
6 Leaves 2 cm or more wide, or calyx lobes 5 mm long or more . . . . *Cynoglossum*
6 Leaves less than 2 cm wide and calyx lobes less than 5 mm long 7
7 Flowers all subtended by bracts . . . . *Lappula*
7 Only lowest flowers with bracts . . . . *Myosotis*
8 Style 2-lobed, stigmas 2 . . . . *Lithospermum*
8 Style not lobed, stigma 1 9
9 Leaves 2 cm or more wide . . . . *Hackelia*
9 Leaves less than 2 cm wide 10
10 Flowers with subtending bracts . . . . *Lappula*
10 Most flowers without subtending bracts . . . . *Myosotis*
11 Nutlets covered with bristly hairs, the hairs hooked at tip 12
11 Nutlets not covered with bristly hairs 14
12 Leaves less than 1 cm wide . . . . *Lappula*
12 Leaves more than 1 cm wide 13
13 Sepals when mature more than 5 mm long . . . . *Cynoglossum*
13 Sepals when mature less than 5 mm long . . . . *Hackelia*
14 All flowers subtended by bracts 15
14 All, or at least upper, flowers without subtending bracts 17
15 Lateral veins on leaves conspicuous . . . . *Lithospermum*
15 Lateral veins absent or very faint 16
16 Plants covered with coarse, stiff hairs, the hairs 2–3 mm long . . . . *Echium*
16 Pubescence various, the hairs less than 2 mm long . . . . *Lithospermum*
17 Nutlets smooth and shiny . . . . *Myosotis*
17 Nutlets wrinkled and dull . . . . *Mertensia*

Cynoglossum

HOUND'S-TONGUE

Biennial or perennial herbs. Leaves large, usually pubescent. Flowers pediceled in elongating, bractless, axillary and terminal racemes. Calyx deeply parted, in fruit reflexed by the growth of the nutlets. Corolla broadly funnelform, the short tube closed by 5 appendages at the throat. Stamens included in the corolla tube. Style slender; stigma 1. Nutlets with conspicuous, stout, hooked bristles.

1 Flowers red-purple; leaves many, continuing upward on stem into inflorescence, not clasping stem .
. . . . *C. officinale*
1 Flowers blue; leaves few, not in inflorescence, the upper leaves clasping at base. . . . . *C. boreale*

Cynoglossum boreale Fern.
WILD COMFREY   *native / cons* **8**
  Andersonglossum virginianum (L.) J. I. Cohen
  Cynoglossum virginianum subsp. *boreale* (Fern). A. Haines
Perennial herb. Stems erect, unbranched, 4-8 dm tall. Basal leaves elliptic-oblong, the blades 1-2 dm long, tapering at base and decurrent upon
the long petiole; stem leaves sessile, progressively smaller, some broadly
clasping at base, some often narrowed below and more or less expanded
at the very base. Racemes 1-4, usually 3, at maturity 1-2 dm long, terminating a long, erect, terminal peduncle; calyx at anthesis 3-4 mm long;
corolla blue, 8-12 mm wide, its broadly rounded lobes more or less overlapping; fruiting pedicels 5-15 mm long, recurved. Nutlets 6-8 mm long,
uniformly bristly over the exterior surface. May-June. — Upland woods.

Cynoglossum boreale

Cynoglossum officinale L.
HOUND'S-TONGUE   *introduced*
Biennial herb, the stem and foliage finely and usually softly pubescent.
Stems erect, branched above, 6-10 dm tall. Basal and lower leaves oblong
or oblong lance-shaped, to 3 dm long, tapering to a long petiole-like base;
upper leaves progressively shorter and proportionately narrower, the uppermost sessile. Racemes numerous, divaricate, eventually 1-2 dm long;
mature pedicels spreading; corolla dull red or red-purple, rarely white,
about 8 mm wide. Nutlets 5-8 mm long, uniformly bristly, surrounded
by a low ridge with crowded bristles. May-Aug. — Native of Eurasia; established in fields, meadows, and open woods.

Echium
VIPER'S-BUGLOSS   *Cynoglossum officinale*

Echium vulgare L.
COMMON VIPER'S-BUGLOSS   *introduced*
Biennial, very hispid herb. Stems erect, simple or branched, 4-8 dm tall.
Leaves linear-oblong to oblong lance-shaped, the basal to 15 cm long, the
upper progressively smaller, the uppermost bract-like. Cymes numerous,
in the axils of the upper foliage leaves, at first coiled, straightening with
age. Flowers sessile, crowded, subtended by linear bracts; calyx deeply
parted; corolla blue, rarely white, pubescent, narrowly campanulate,
somewhat curved, conspicuously longer on the upper side, 12-20 mm
long; stamens inserted near the middle of the corolla tube, the slender
filaments unequal in length, long-exsert; style elongate, usually pubescent, shortly 2-cleft at the summit. Nutlets ovoid, 3-angled, rough. —
Native of s Europe, weedy in waste places, roadsides, and meadows, usually
where sandy or gravelly.

Hackelia
Echium vulgare
STICKSEED, BEGGAR'S-LICE
Perennial herbs with numerous, usually paired racemes terminating the axillary branches;
racemes bracteate for at least part of their length; flowers small, blue or white. Corolla deeply 5-
cleft into narrow lobes. Corolla salverform or broadly funnelform, the tube shorter than to
scarcely surpassing the calyx, the throat nearly closed by the small appendages. Stamens and
style short, included in the corolla tube. Fruiting pedicels short, recurved or reflexed. Nutlets attached by a lance-shaped to ovate area occupying the middle third only, the terminal third free,
the basal third free and often with 2 low divergent keels; dorsal area lance-shaped, bordered by
a row of hooked bristles and in some species bearing similar bristles on the surface.

1   Corolla blue; widest stem leaves to 2.5 cm broad . . . . *H. deflexa*
1   Corolla white; widest stem leaves to 3–5 cm broad . . . . *H. virginiana*

Hackelia deflexa (Wahlenb.) Opiz
NODDING STICKSEED   *native / cons* 3
Stems to 1 m tall, freely branched above, the branches all terminated by
racemes (usually paired). Lower leaves long-tapering to a petiole-like
base; middle and upper leaves oblong-elliptic, to 10 cm long, sharply
pointed at both ends. Racemes eventually 5-10 cm long, spreading; bracts
linear or lance-shaped, reduced above and often absent beyond the middle
of the raceme, opposite the flowers or alternate with them; fruiting
pedicels 2-4 mm long, abruptly deflexed at base; corolla white or pale
blue, about 2 mm wide. Nutlets with an ovate dorsal area 2-3 mm long,
with a few short bristles on the back, bearing a marginal row of flat
hooked bristles. May-Aug. — Moist woods, and thickets.

Hackelia virginiana (L.) I. M. Johnston
BEGGAR'S-LICE   *native / cons* 3
Stems to 1 m tall, freely branched above and bearing numerous racemes.
Lower leaves narrowed to a petiole, to 2 dm long; middle and upper leaves
oblong-elliptic, 5-10 cm long, about 1/4 to 1/3 as wide, sharply narrowed
to both ends, sessile; uppermost leaves, above the lowest flowering branch,
progressively reduced and passing into the small, lance-shaped to linear
bracts. Racemes eventually 5-15 cm long, spreading; bracts often alternate
with the flowers, those beyond the middle of the raceme minute or lacking; fruiting pedicels 2-5 mm long, reflexed or recurved; corolla white or
pale blue, about 2 mm wide. Nutlets forming a globose cluster of 4, bearing
about 10-15 erect bristles as long as the marginal ones. July-Sept. — Dry
or moist upland woods.

Hackelia virginiana

Hydrophyllum
WATERLEAF
Perennial herbs from horizontal rhizomes. Leaves large, lobed or divided. Flowers several to
many in a repeatedly forked cyme. Sepals separate to below the middle or nearly to the base.
Corolla campanulate to tubular, lobed to or below the middle, the lobes erect or somewhat
spreading, white to purple. Stamens equaling the corolla or exsert, the slender filaments usually
villous. Style 1, shortly bifid at the summit. Ovary 1-celled. Capsule globose, hispid or pubescent.

1   Leaves pinnately compound, divided into 5–7 leaflets or lobes . . . . *H. virginianum*
1   Leaves palmately lobed . . . . *H. appendiculatum*

Hydrophyllum appendiculatum Michx.
GREAT WATERLEAF   *native / cons* 8
Stems at anthesis 3-6 dm tall, the upper portion and inflorescence densely
pubescent with short hairs 0.3-0.5 mm long and also conspicuously hirsute with longer hairs 2-3 mm long. Stem leaves mostly overtopped by
the cymes, orbicular in outline, 6-15 cm wide at anthesis, shallowly 5-7
lobed; sepals separate nearly to the base, lance-shaped, densely hirsute,
alternating with small but conspicuous reflexed appendages; corolla
lavender or pink-purple, 9-13 mm long, the lobes about equaling the tube;
stamens as long as or slightly longer than the corolla. May-June. — Rich
moist woods.

Hydrophyllum appendiculatum

Hydrophyllum virginianum L.
EASTERN WATERLEAF   *native / cons* 4
Stems 3-8 dm tall at anthesis, the upper portion, cymes, pedicels, and
back of the sepals strigose with short hairs rarely to 0.5 mm long. Stem
leaves broadly ovate triangular in outline, 1-2 dm long and usually some-

what wider, pinnately lobed almost to the midvein, the segments usually 5, occasionally 7 or 9, the terminal one and the basal pair often 2-3-lobed, all with sharply acute apex and similar, strongly ascending teeth. Cymes very dense at anthesis; sepals sparsely hirsute; corolla white to pale pink-purple, 7-10 mm long; stamens long-exsert. May-June. — Moist or wet woods, or open wet places.

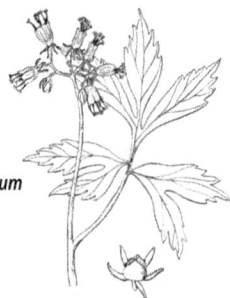

Hydrophyllum virginianum

Lappula
STICKSEED, BEGGAR'S-LICE
Lappula squarrosa (Retz.) Dumort.
TWO-ROW STICKSEED   *introduced*
Roughly pubescent, erect annual herbs. Stems usually simple to above the middle, thence freely branched, 2-8 dm tall, tending to be appressed-pubescent. Leaves linear or linear-oblong lance-shaped, 2-5 cm long, usually ascending, acute or obtuse, narrowed to a sessile base, roughly hirsute. Flowers in numerous terminal racemes, eventually 5-10 cm long; pedicels at maturity erect or ascending, 5-10 mm apart, 1-2 mm long; bracts deeply 5-cleft, linear or lance-shaped, 3-10 mm long; corolla blue, 2-3 mm wide; stamens and style short, included in the calyx. Nutlets 3-4 mm long, the lance-shaped face surrounded by 2 rows of bristles, those of the inner row usually the longer. May-Sept. — Native of Asia and the Mediterranean region; established as a weed in waste places.

Lappula squarrosa

Lithospermum
GROMWELL, PUCCOON, STONESEED
Perennial herbs (ours) with pubescent stem and foliage. Leaves narrow. Flowers solitary in the axils or crowded into a terminal leafy-bracted cyme. Calyx lobes narrow, separate nearly to the base. Corolla funnelform or salverform; the tube slender or wide, appendaged at the summit. Stamens inserted in the corolla tube, the anthers included or partly exsert. Style shortly 2-lobed. Nutlets bony, ovoid to nearly globose, smooth or pitted, usually only 1 or 2 ripening in each flower.

1 Flowers yellow or yellow-orange in a terminal inflorescence . . . . *L. caroliniense*
1 Flowers white to pale yellow from leaf axils 2
2 Largest leaves 2–4 cm wide; stem leaves below inflorescence number 20 or fewer . . . . *L. latifolium*
2 Largest leaves less than 2 cm wide; stem leaves 25 or more . . . . *L. officinale*

Lithospermum caroliniense (Walt.) MacM.
PLAINS PUCCOON   *native / cons* **10**
Perennial herb. Stems erect from a stout woody root, at anthesis 3-6 dm tall, very leafy, simple or branched above, villous or birsute. Leaves linear to lance-shaped, 3-6 cm long, roughly hirsute, the hairs often papillate at base. Cymes at first dense, leafy-bracted, becoming elongate and racemi-form after anthesis; calyx lobes linear, 9-11 mm long, hirsute; corolla tube 10-14 mm long, pubescent at the base within; corolla-limb bright orange-yellow, 15-25 mm wide. Nutlets ivory-white, smooth and shining. May-July. — In dry, moist or preferably sandy soil, upland woods, shores, and prairies.

Lithospermum caroliniense

Lithospermum latifolium Michx.
AMERICAN GROMWELL   *native / cons* **8**
Perennial from a thick root. Stems erect, simple or branched above, 4-8 dm tall, strigose, the principal internodes commonly 3-6 cm long. Leaves nearly sessile, lance-shaped to ovate lance-shaped, usually 2-4 cm wide, distinctly acuminate, scabrous above, with 2 or 3 prominent lateral veins on each side of the midvein. Flowers solitary in the axils of the upper leaves, becoming distant at maturity, yellowish-white, 5-7 mm long and

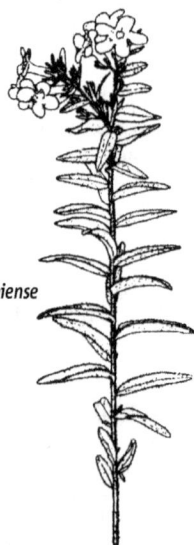

nearly as wide; calyx lobes nearly as long as the corolla. Nutlets ovoid, white, shining, 3.5-5 mm long, smooth or remotely pitted. May-June. — Dry woods and thickets.

Lithospermum officinale L.
EUROPEAN GROMWELL *introduced*
Perennial from a thick root. Stems erect, usually much branched above, to 1 m tall, strigose, the principal internodes usually less than 2 cm long, often only 5 mm. Leaves nearly sessile, lance-shaped to oblong, mostly 6-15 mm wide, scabrous above, with 2 or 3 conspicuous veins on each side of the midvein. Flowers solitary in the axils of the crowded upper leaves, 3-15 mm apart at maturity, white or nearly so, 4-5 mm long; calyx lobes nearly as long as the corolla. Nutlets ovoid, 3-3.5 mm long, white to pale brown, shining, smooth or sparsely pitted. May-Aug. — Native of Eurasia; introduced as a weed of waste places.

Lithospermum latifolium

Mertensia
BLUEBELLS

Mertensia paniculata (Ait.) G. Don
NORTHERN BLUEBELLS *native / cons 9*
Perennial herb. Stems erect, 3-10 dm long, branched above, smooth or with sparse hairs. Basal leaves ovate, rounded at base; stem leaves alternate, lance-shaped to ovate, 5-15 cm long, tapered to a tip, hairy, entire; petioles short on lower leaves, upper leaves more or less stalkless. Flowers blue-purple, narrowly bell-shaped, 10-15 mm long, on slender stalks, in few-flowered racemes at ends of stems and branches; sepal lobes lance-shaped, 3-6 mm long, with dense, short hairs. Fruit a nutlet. June-July. — Conifer swamps, streambanks, seeps.

Mertensia paniculata

Myosotis
FORGET-ME-NOT, SCORPION GRASS
Perennial (sometimes annual) herbs; plants with short, appressed hairs. Leaves alternate and entire. Flowers blue, tube-shaped and abruptly flared outward at tip, in a 1-sided raceme. Fruit a nutlet.

1 Calyx hairs all straight-tipped, appressed .... *M. scorpioides*
1 Calyx hairs mostly hooked at tip, spreading 2
2 Pedicels shorter than calyx; corolla 1–2 mm wide .... *M. stricta*
2 Pedicels equal to length of calyx (when in full-flower or fruit); corolla 3 mm or more wide 3
3 Expanded part of petal cupped, less than 4 mm wide .... *M. arvensis*
3 Expanded part of petal flat, 5–10 mm wide .... *M. sylvatica*

Myosotis arvensis (L.) Hill
ROUGH FORGET-ME-NOT *introduced*
Annual or biennial. Stems simple or branched, eventually 3-5 dm long. Leaves firm, oblong, varying to lance-shaped or oblong lance-shaped, the larger 2-5 cm long. Racemes becoming 1-2 dm long, usually completely bractless; fruiting pedicels 5-15 mm apart, divergent, 5-9 mm long; mature calyx 3-4.5 mm long, pubescent with both hooked and appressed hairs, the lobes slightly longer than the tube; corolla blue or white, broadly funnelform, the limb 2-3 mm wide. Nutlets 1.3-1.7 mm long. Summer. — Native of Eurasia; established in fields and roadsides.

Myosotis scorpioides L.
TRUE FORGET-ME-NOT *introduced (invasive)*
Myosotis palustris (L.) Hill
Perennial herb. Stems 2-6 dm long, with short, appressed hairs, often

Myosotis arvensis

creeping at base and producing stolons. Leaves 3-8 cm long and 0.5-2 cm wide, lower leaves oblong lance-shaped, upper leaves oblong or oval; stalkless or the lower leaves on short petioles. Flowers blue with a yellow center, tube-shaped, abruptly flared at tip, in a 1-sided raceme at ends of stems, becoming open; flower stalks spreading in fruit; sepals with short, appressed hairs, sepal lobes equal or shorter than the tube. Fruit a nutlet shorter than the style. May-Sept. — Streambanks, shores, ditches, swamps, wet depressions in forests.

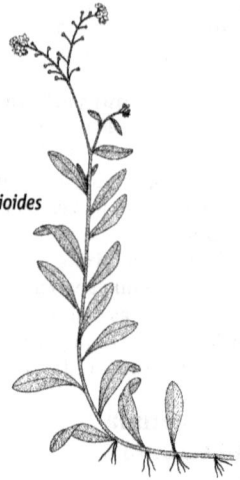

Myosotis scorpioides

Myosotis stricta Link
BLUE SCORPION-GRASS   *introduced*

Annual. Stems slender, branched from the base, 1-2 dm tall, forking at the second or third node into the primary racemes. Leaves oblong lance-shaped, 8-20 mm long, all except the very lowest bearing flowers in their axils. Racemes constituting three-fourths of the height of the plant, or even more, the lower flowers subtended by foliaceous bracts; fruiting pedicels 1-1.5 mm long, more or less hirsute with obtuse hooked hairs; mature calyx about 4 mm long, narrowed at the base, densely hirsute with hooked hairs, the lobes about as long as the tube; corolla blue, funnelform, about 1.5 mm wide, its tube not exceeding the calyx. Nutlets 1-1.2 mm long. April-July. — Native of Eurasia; locally introduced in dry waste places.

Myosotis sylvatica Ehrh.
GARDEN FORGET-ME-NOT   *introduced (invasive)*

Perennial. Stems eventually to 5 dm long, bearing several racemes. Leaves oblong to lance-shaped or spatulate, thin and soft, the larger 3-7 cm long. Racemes bractless, seldom more than 1 dm long; fruiting pedicels 5-15 mm apart, ascending or spreading, to 9 mm long; mature calyx 4-5 mm long, much shorter than the pedicel, densely pubescent with hooked hairs, the narrowly triangular lobes much longer than the tube; corolla blue or rarely white, salverform, the limb 5-8 mm wide. Nutlets 1.5-2 mm long. April-Sept. — Native of Eurasia; commonly cultivated for ornament and sometimes escaped near gardens.

Myosotis sylvatica

Brassicaceae

MUSTARD FAMILY

Annual, biennial or perennial herbs. Leaves simple or compound, alternate on stems or basal, smooth or hairy, some species with branched or star-shaped hairs. Flowers in terminal or lateral clusters (racemes), the lower portion often fruiting while tip in flower, the stalks elongating in fruit. Flowers perfect (with both staminate and pistillate parts), cross-shaped, with 4 sepals and 4 yellow, white, pink or purple petals; stamens 6, the outer 2 stamens shorter than the inner 4; pistil 1, style 1, ovary superior. Fruit a cylindrical or round pod (silique or silicle) with 2 chambers and 1 to many seeds in 1 or 2 rows in each chamber.

1 Petals yellow, yellow-tinged, or orange 2
1 Petals white, greenish, pink, purple, or absent 16
2 Leaves simple, not deeply lobed 3
2 At least the lower leaves lobed, pinnately lobed, or pinnately compound 9
3 Plants glabrous throughout 4
3 Plants pubescent, at least near the base 5
4 Stem leaves not clasping at their base . . . . ***Rorippa***
4 Stem leaves lobed at base and clasping the stem . . . . ***Sinapis***
5 At least some of the stem leaves clasping the stem 6
5 None of the leaves clasping stem 7
6 Upper stem glabrous or nearly so; fruit with many seeds and more than 3 mm wide . . . . . . ***Camelina***

6 Upper stem densely hairy; fruit with only 1 or 2 seeds, up to 3 mm wide . . . . *Neslia*
7 Ovary and fruit soon becoming much longer than wide . . . . *Erysimum*
7 Ovary and fruit rounded in outline and usually to 1–2 times longer than wide 8
8 Fruit about as long as wide; pubescence of well-branched hairs . . . . *Alyssum*
8 Fruit much longer than wide; leaves loosely pubescent, the hairs with few branches . . . . *Draba*
9 Pedicels in the lower portion of the raceme subtended by leafy bracts . . . . *Erucastrum*
9 Pedicels not subtended by bracts 10
10 Petals up to 5 mm long (including the slender or tapered claw), or if 6 mm long then plant a creeping
   or rhizomatous perennial 11
10 Petals more than 5 mm long; plants never creeping perennials 14
11 Lobes of leaves rounded, nearly round in outline to broadly oval or obovate, the terminal lobe much
   larger than the lateral lobes; plants glabrous throughout . . . . *Barbarea*
11 Lobes of leaves pointed, mostly distinctly longer than wide, the terminal lobe similar in size to the lateral lobes; plants glabrous or pubescent 12
12 Leaves 2–3 times pinnately dissected; plants with tiny stalked glands, or pubescent with branched
   hairs, or both . . . . *Descurainia*
12 Leaves 1–2 times pinnately dissected; plants without glands, glabrous or pubescent, the hairs un-
   branched 13
13 Fruits linear, more than 5x longer than wide; style very short; plants taprooted annuals or winter an-
   nuals . . . . *Sisymbrium*
13 Fruits spherical to oblong, less than 5x longer than wide (if longer, then the plant a rhizomatous
   perennial); style stout to elongate; plants annuals or perennials . . . . *Rorippa*
14 Basal rosette leaves often present at flowering time; plants glabrous throughout or nearly so; uppermost leaves usually with at least one pair of lobes . . . . *Barbarea*
14 Basal rosette leaves usually absent or mostly withered at flowering time; plants glabrous or pubescent;
   uppermost leaves variously toothed or wavy-margined, but not distinctly lobed 15
15 Petals pale yellow with dark veins; fruit strongly twisted and long-beaked . . . . *Raphanus*
15 Petals pale to deep yellow but without dark veins; fruit not twisted (or only slightly so), beak present
   or absent . . . . *Sisymbrium*
16 Leaves 3-parted or deeply palmately divided . . . . *Cardamine*
16 Leaves neither 3-parted nor palmately divided 17
17 Petals pink or purple 18
17 Petals white, greenish, or absent 21
18 At least the lower and middle stem leaves pinnately lobed or dissected . . . . *Raphanus*
18 None of the stem leaves pinnately lobed or dissected 19
19 Plants glabrous or nearly so . . . . *Cakile*
19 Plants pubescent, at least near the base or on the leaves 20
20 Stem leaves numbering less than 10, ovate or lance-ovate, entire to wavy-margined or wavy-toothed;
   petals less than 1.5 cm long . . . . *Cardamine*
20 Stem leaves more than 10, lance-shaped to ovate or oblong lance-shaped, entire to finely toothed;
   petals mostly 1.5 cm or more long . . . . *Hesperis*
21 At least the upper leaves sessile and clasping at their base 22
21 Leaves not clasping the stem 32
22 Stem leaves pinnately compound; plants aquatic or subaquatic . . . . *Nasturtium*
22 Leaves not all pinnately compound; plants never aquatic 23
23 Ovaries and fruit becoming linear, more than 5 times longer than wide 24
23 Ovaries and fruit up to 2 times longer than wide 27
24 Fruiting pedicels spreading, divaricate, or reflexed; fruit straight or somewhat curved and clearly
   spreading from the axis or even pendent . . . . *Boechera*
24 Fruiting pedicels strongly ascending to appressed; fruit straight, erect and closely appressed to the
   stem 25
25 Stem and leaves entirely glabrous or with a very few scattered hairs at the very base of the plant (es-
   pecially on leaf margins and petioles); sepals ca. half as long as the petals; mature fruit 1.4–2.5 (–3.3)
   mm wide, with seeds in 2 rows in each locule . . . . *Boechera*
25 Stem and leaves pubescent, at least at the base, with spreading simple or stellate hairs; sepals ca. 2/3
   as long as the petals; mature fruit less than 1.3 mm wide, with seeds crowded into 1 row in each
   locule 26
26 Fruit rather strongly flattened; style-beak clearly narrower than mature fruit; stem pubescent with
   simple and/or forked (or stellate) hairs on at least the lower half or third, and leaves on the same portion pubescent (often stellate) . . . . *Arabis*

26 Fruit terete or 4-angled, slightly if at all flattened at maturity; style-beak nearly or quite as wide as the fruit; stem pubescence only on the lowermost 1–3 full-grown internodes, and only the lowermost leaves pubescent.. ***Turritis***

27 Plants glabrous throughout 28
27 Plants pubescent, at least below the middle 30

28 Upper leaves nearly round in outline, less than 1.5 times longer than wide, clasping, the stem appearing to perforate the leaf .... ***Lepidium***
28 Upper leaves lance-shaped to narrowly ovate, more than 2 times longer than wide, clasping the stem 29

29 Inflorescence of corymb-like racemes; fruit widest at their base, less than 5 mm long; style present . ... ***Lepidium***
29 Inflorescence of unbranched racemes, or branched 1–2 times but not corymb-like; fruit oval or obovoid, more than 5 mm long; style absent .... ***Thlaspi***

30 Pedicels densely hairy; stem leaves unlobed .... ***Lepidium***
30 Pedicels glabrous or nearly so; leaves lobed or unlobed 31

31 Basal leaves deeply incised to pinnate, present at flowering time; stem leaves strongly reduced in size, pinnate (or if entire then less than 1 cm wide); fruit triangle-shaped, truncate at the tip .... ***Capsella***
31 Basal leaves absent or shriveled at flowering time; stem leaves well developed, entire to finely toothed, mostly more than 1 cm wide; fruit nearly round in outline........................... ***Lepidium***

32 Nearly all stem leaves deeply lobed, or pinnately divided or compound, or stem leaves absent ... 33
32 All or nearly all of the stem leaves simple, not lobed or pinnately divided 38

33 Lowest leaves 2–3 times pinnately divided or pinnately compound 34
33 Lowest leaves only once-pinnate (the segments sometimes with a few teeth) 35

34 Upper leaves simple to lobed or once-pinnate; fruit nearly round in outline .... ***Lepidium***
34 All leaves 2–3 times pinnately divided or pinnately compound; fruit linear .... ***Descurainia***

35 Petals more than 6 mm long 36
35 Petals to 5 mm long 37

36 Leaves pinnately compound, the lobes entire and all similar, the lobes of the upper leaves linear, the lobes of the lower leaves round in outline or nearly so .... ***Cardamine***
36 Leaves pinnately lobed, the terminal lobe much larger than the smaller lateral lobes; upper leaves similar to the lower leaves, though usually smaller .... ***Raphanus***

37 Plants aquatic, rooting at the nodes; stems succulent; petals about 5 mm long .... ***Nasturtium***
37 Plants not aquatic; stems slender and firm; petals to 4 mm long .... ***Cardamine***

38 Basal and lower leaves with distinct petioles, rounded to heart-shaped at their base 39
38 Leaves narrowed to the sessile or nearly sessile base 42

39 Stem leaves deeply toothed to pinnately divided; fruit to 6 mm long, but usually soon falling .... ***Armoracia***
39 Stem leaves shallowly toothed or wavy-margined, or nearly entire; fruit more than 6 mm long, persistent 40

40 Stem leaves truncate to heart-shaped at base, petioles 5 mm or more long; petals less than 7 mm long; fruiting pedicels stout, about 5 mm long .... ***Alliaria***
40 Stem leaves not both heart-shaped at base and on petioles as long as 5 mm; petals more than 7 mm long; fruiting pedicels slender, or if stout then much longer than 5 mm 41

41 Stem leaves fewer than 10, ovate, the margins entire to wavy or wavy-toothed; petals less than 1.5 cm long .... ***Cardamine***
41 Stem leaves more than 10, lance-shaped to ovate or oblong lance-shaped, the margins finely toothed; petals usually 1.5 cm or more long .... ***Hesperis***

42 Ovaries and fruit less than 2 times longer than wide 43
42 Ovaries and fruit more than 2 times longer than wide 45

43 Fruit ascending or erect; the fruit and stems densely woolly hairy .... ***Berteroa***
43 Fruit widely spreading; plants glabrous to pubescent 44

44 Plants glabrous or with tiny hairs, the hairs unbranched .... ***Lepidium***
44 Plants pubescent, the hairs branched .... ***Alyssum***

45 Fruit linear, more or less round in cross-section, to 1 mm wide; plants blooming and then withering by late spring .... ***Arabidopsis***
45 Fruit linear, flat, 1 mm or more wide; plants withering or persisting 46

46 Fruit to 18 mm long; plants usually not leafy above the middle; early blooming and withering by late spring .... ***Draba***
46 Fruit longer than 18 mm; plants usually leafy throughout; plants persisting to late summer or early fall .... ***Boechera***

Alliaria
GARLIC-MUSTARD

Alliaria petiolata (Bieb.) Cavara & Grande
GARLIC-MUSTARD   *introduced (invasive)*
   Alliaria alliaria (L.) Britt.
   Alliaria officinalis Andrz.
Biennial, garlic-scented herb, glabrous or nearly so, the hairs simple. Stems erect, to 1 m tall, simple or little branched. Basal leaves in more or less evergreen rosettes; lower leaves kidney-shaped; stem leaves deltoid, 3-6 cm long and wide, coarsely toothed. Petals white, spatulate, 5-6 mm long; pedicels at maturity stout, about 5 mm long. Fruit widely divergent, 4-6 cm long; seeds black, nearly cylindric, about 3 mm long. May-June. — Native of Europe; invasive in rich, moist, shaded soil; also on roadsides or rarely in swamps.

Alliaria petiolata

Alyssum
MADWORT

Alyssum alyssoides (L.) L.
PALE MADWORT   *introduced*
Annual herb. Stem, leaves, inflorescence, and fruits stellate-pubescent. Stems 5-25 cm tall, simple and erect, or branched from the base only. Leaves oblong lance-shaped, 6-15 mm long, entire, obtuse. Flowers pale yellow or nearly white, about 2 mm wide; petals narrowly oblong. Fruit on widely divergent pedicels, circular, 3-4 mm long, flat at the margin, convex toward the center; seeds 2 in each cell. May-June. — Native of Europe; weedy in waste places.

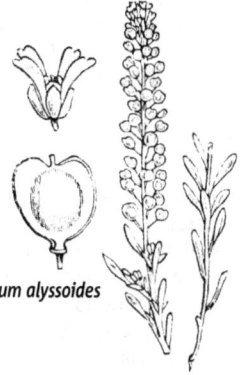

Alyssum alyssoides

Arabidopsis
THALECRESS

Arabidopsis lyrata (L.) O'Kane & Al-Shehbaz
LYRE-LEAF ROCKCRESS   *native / cons –*
   Arabis lyrata L.
Biennial herb. Stems erect or ascending, branched from the base, 1-4 dm tall, hirsute below (very rarely glabrous), glabrous or glabrescent above. Leaves mostly in a basal rosette; basal leaves spatulate, 2-4 cm long, entire to pinnately lobed; stem leaves linear to narrowly spatulate, the lowest sometimes with a few teeth or shallow lobes. Petals white, spatulate, 3-8 mm long; pedicels at maturity widely ascending, 6-15 mm long. Fruit maintaining about the same direction as the pedicel, linear, nearly terete, 2-4.5 cm long, occasionally shorter, about 1 mm wide; seeds many, oblong to elliptic, wingless, about 1 mm long. April-June. — Dry woods and fields, especially in sandy soil; sand dunes.

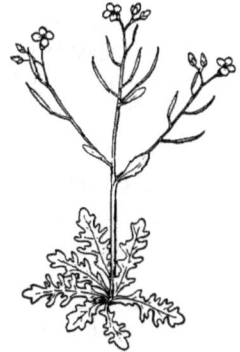

Arabis
ROCKCRESS

Arabidopsis lyrata

Most of our native species formerly in genus *Arabis* are now placed in genus *Boechera*, with either broader fruit (if tightly appressed) or the fruit spreading, reflexed or pendent; *Arabis glabra* (L.) Bernh. now placed in genus *Turritis*.

ADDITIONAL SPECIES
Arabis caucasica Willd. (Gray rockcress), garden escape, reported from Door County.

Arabis pycnocarpa M. Hopkins
HAIRY ROCK CRESS   *native / cons 6*
   Arabis hirsuta (L.) Scop.
Biennial herb. Stems erect, 2-8 dm tall, pubescent at least at base with

simple or branched hairs. Stem leaves oblong to linear lance-shaped, 1-3 cm long, sessile and more or less clasping, at least the lower pubescent. Mature pedicels erect, slender, 7-10 mm long; petals white. Fruit erect, flat, linear, 3-5 cm long, about 1 mm wide; seeds in one row, flattened and narrowly winged, 1-1.2 mm long. May-June. — Woods, often where calcareous.

Arabis pycnocarpa

Armoracia
HORSE-RADISH

Armoracia rusticana P. G. Gaertn. B. Mey. & Scherb.
HORSE-RADISH   *introduced*
Glabrous perennial herb from thick roots. Stems erect, to 1 m tall. Lower leaves long-petioled, the blade oblong, 1-3 dm long, cordate at base; upper leaves smaller, short-petioled to sessile, lance-shaped. Racemes several, terminal and from the upper axils; petals white, obovate, 6-8 mm long; pedicels after anthesis ascending, 8-12 mm long. Fruit obovoid, inflated, 2-celled, eventually up to 6 mm long but usually falling early; seeds apparently never maturing. May-July. — Native of se Europe and w Asia; commonly cultivated and escaped into moist soil of ditches, shores, roadsides, and disturbed places.

Barbarea
YELLOW-ROCKET

Barbarea vulgaris Ait. f.
GARDEN YELLOW-ROCKET   *introduced*
Biennial herbs, smooth or with a few simple hairs. Stems erect, branched above, 2-8 dm tall. Basal leaves petioled, with 1-4 pairs of small, elliptic to ovate, lateral lobes and a large ovate to rotund terminal lobe; stem leaves progressively shorter petioled and with fewer lobes, the upper sessile or clasping, angulately toothed, repand, or entire. Flowers yellow, about 8 mm wide, crowded at anthesis; pedicels at maturity 3-6 mm long, about 0.5 mm thick. Fruit linear, terete or obscurely 4-angled, several-seeded, tipped by the persistent style 1.5-3 cm long. April-June. — Native of Europe; naturalized as a weed in damp soil of fields, roadsides, and gardens.

Barbarea vulgaris

Berteroa
HOARY ALYSSUM

Berteroa incana (L.) DC.
HOARY ALYSSUM   *introduced*
Annual herb. Stem, foliage, and inflorescence finely canescent, the hairs stellate with radiating branches. Stems stiffly erect, usually branched above, to 7 dm tall. Leaves oblong lance-shaped, 2-5 cm long, acute, entire. Sepals ascending; petals white, 2-lobed, about 3 mm wide. Fruit elliptic, thinly pubescent, 5-8 mm long, 3-4 mm wide; seeds 3-6 in each cell. May-Sept. — Native of Europe; now established as a weed.

Boechera
ROCKCRESS

Biennial or perennial herbs. Basal leaves petioled, the stem leaves smaller and usually sessile; pubescence usually present, of simple, forked, or stellate hairs. Sepals erect or spreading, the outer pair sometimes saccate at base. Petals white, yellowish, or pink, spatulate to oblong or obovate. Ovary cylindric. Fruit linear, elongate, flat or subterete, many-seeded. Seeds flattened, often with a marginal wing.

Berteroa incana

Includes former native members of genus *Arabis*.

1 Pedicels becoming distinctly reflexed before the petals wither, the fruit pendent; sepals ca. half as long as mature petals or a little shorter . . . . ***B. grahamii***
1 Pedicels spreading or ascending to strongly appressed, even after anthesis, the fruit spreading to erect; sepals various **2**
2 Fruiting pedicels strongly ascending to appressed, the fruit straight, erect and closely appressed to the stem . . . . ***B. stricta***
2 Fruiting pedicels ± spreading, the fruit straight or somewhat curved and clearly spreading from the axis. **3**
3 Stem leaves ca. 25–35 or more below the inflorescence, the longest 2.5–5 cm; stem at the base and both surfaces of basal leaves stellate-pubescent; petals pink or pale purple; sepals at most barely more than half as long as the petals . . . . ***B. grahamii***
3 Stem leaves ca. 10–15 (–20) below the inflorescence, the longest 9–15 cm; stem and leaves completely glabrous at base of plant (and elsewhere); petals white; sepals much more than half as long as petals . . . . ***B. laevigata***

Boechera grahamii (Lehmann) Windham & Al-Shehbaz
SPREADING ROCKCRESS   *native / cons* **6**
   Arabis divaricarpa A. Nels.
Biennial herb. Stems erect, to 1 m tall, glabrous except at the very base. Basal leaves oblong lance-shaped, finely stellate-pubescent on both sides; stem leaves linear lance-shaped, erect or nearly so, 2-5 cm long, sessile, auriculate at base, entire, glabrous on both sides. Petals pinkish or white, 5-8 mm long; pedicels at maturity widely spreading, 6-12 mm long. Fruit at first erect, soon widely spreading, linear, straight or nearly so, 3-9 cm long, 1.2-2.2 mm wide, the valves 1-nerved to or beyond the middle; seeds in 1 row and broadly quadrate to orbicular, or in 2 rows and oblong. June-July. — Sandy or rocky soil.

Boechera laevigata (Muhl. ex Willd.) Al-Shehbaz
SMOOTH BANK CRESS   *native / cons* **5**
   Arabis laevigata (Muhl.) Poir.
   Borodinia laevigata (Muhl. ex Willd.) P. J. Alexander & Windham
Biennial herb. Glabrous and glaucous throughout, except the sparsely hirsute, spatulate, basal leaves. Stems to 1 m tall. Stem leaves narrowly lance-shaped, usually 5-10 cm long, serrate to entire, usually sagittate at the sessile base. Petals white, 3-5 mm long, equaling to or to one-fourth longer than the sepals; pedicels at maturity widely spreading, 7-12 mm long. Fruit widely spreading, horizontal, or somewhat decurved, rarely straight, linear, flat, 5-10 cm long, 1.2-2 mm wide; seeds in 1 row, oblong, narrowly winged. May-June. — Moist or dry woods.

Boechera laevigata

Boechera stricta (Graham) Al-Shehbaz
DRUMMOND'S ROCKCRESS   *native / cons* **6**
   Arabis drummondii A. Gray
Biennial herb. Stems erect, 3-9 dm tall, glabrous, or at base very thinly pubescent with 2-pronged hairs, often glaucous. Stem leaves sessile, lance-shaped to narrowly oblong, 2-8 cm long, acute, entire or with a few teeth, auriculate at base, usually wholly glabrous. Petals pink to purple (often drying whitish)5-9 mm long; fruiting pedicels erect, 10-15 mm long. Fruit straight, erect, flat, 4-7 cm long or rarely longer, 1.5-2.5 mm wide or occasionally wider; seeds in 2 rows. May-Aug. — Moist or dry places, often where calcareous.

Boechera stricta

Cakile
SEA-ROCKET

Cakile edentula (Bigelow) Hook.
SEA-ROCKET   *native / cons* 9
   Cakile lacustris (Fern.) Pobed.
Succulent annual. Stems much branched and bushy, 2–3 dm tall. Leaves
oblong lance-shaped or spatulate, sinuately toothed, varying to pinnately
lobed or nearly entire. Petals obovate, pale purple, about 5 mm wide;
pedicels at maturity 2–5 mm long; short stamens subtended by a minute
gland; long stamens separated by a larger gland. Fruit corky at maturity
when dry, divided into 2 joints of different shape, the lower persistent, 1-
seeded or seedless, the upper always conspicuously longer, eventually
deciduous, fertile, usually 1-seeded, ovoid or lance-ovoid; seed suspended
in the lower joint, erect in the upper. — Atlantic and Great Lakes coastal
species, in Wisconsin mostly along sandy beaches and low dunes near
Lake Michigan, often spreading by pieces of the floating fruit.

Cakile edentula

Camelina
FALSE FLAX

Annual or winter-annual herbs, bearing both simple and branched hairs; stems branched above.
Basal leaves narrowly spatulate; stem leaves linear to lance-shaped, clasping by a sagittate-au-
riculate base. Sepals erect, obtuse, the outer slightly saccate at base. Petals yellow, spatulate.
Short stamens flanked at base by a pair of semicircular glands. Style slender, persistent. Fruit
obovoid or pyriform, somewhat keeled, narrowed to the base and short-stipitate.

1 Fruit to 7 mm long and 5 mm wide; lower stem pubescent with both spreading and appressed hairs
   . . . . *C. microcarpa*
1 Fruit more than 7 mm long and more than 5 mm wide; stem glabrous or with tiny hairs . . . *C. sativa*

Camelina microcarpa DC.
LITTLE-POD FALSE FLAX   *introduced*
Stems erect, 3–7 dm tall, rough-pubescent, as is also the foliage, with both
simple and branched hairs, the former 1–2 mm long. Fruit erect, 5.5–8
mm long, 3–4.5 mm wide, obscurely rugulose. April–June. — Fields and
waste places, usually in sandy soil.

Camelina sativa (L.) Crantz
LARGE-SEED FALSE FLAX   *introduced*
Similar to *C. microcarpa* in foliage and habit. Stems and leaves glabrous to
sparsely pubescent, the simple hairs not projecting beyond the stellate.
Fruit commonly 7–10 mm long, 5–7 mm wide, inconspicuously veiny. May–
June. — Fields and waste places, usually where sandy.

Camelina sativa

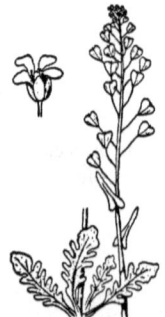

Capsella
SHEPHERD'S-PURSE

Capsella bursa-pastoris (L.) Medik.
SHEPHERD'S-PURSE   *introduced*
Annual herb, pubescent with stellate hairs. Stems 1–6 dm tall, sparingly
branched. Basal leaves oblong, 5–10 cm long, pinnately lobed; stem leaves
much smaller, lance-shaped to linear, entire or denticulate, auriculate at
base. Racemes at anthesis congested, at maturity greatly elongate, often
forming half the total height of the plant; pedicels at maturity widely
spreading, 1–2 cm long; sepals short-oblong, ascending; petals white, obovate, about 2 mm wide and about 2x as long as the sepals. Fruit oblong
cordate, flattened, 5–8 mm long, truncate to notched at the tip. Spring. —
Weedy in lawns, gardens, and waste places. Where sheltered, this is one
of the first plants to bloom in spring.

Capsella bursa-pastoris

Cardamine
BITTERCRESS, TOOTHWORT

Annual, biennial or perennial herbs, smooth or with short hairs near base of stem. Leaves simple to pinnately divided, the basal leaves often different in shape than stem leaves. Flowers in racemes or umbel-like clusters; sepals green to yellow, early deciduous; petals usually white. Fruit a 2-chambered, linear pod (silique), the seeds in a single row in each chamber.

1 Leaves simple to pinnately compound 2
1 Leaves palmately 3–5 parted or compound 4
2 Leaves simple; plants from a shallow tuber-like rhizome .... *C. bulbosa*
2 Stem leaves pinnately dissected, with 2 or more deep lobes; plants without a tuber-like base...... 3
3 Petals 8 mm or more long .... *C. pratensis*
3 Petals to 4 mm long .... *C. pensylvanica*
4 Leaves divided into 4–7 linear segments .... *C. concatenata*
4 Leaves divided into 3 ovate segments .... *C. diphylla*

Cardamine bulbosa (Schreb.) B.S.P.
BULBOUS BITTERCRESS   *native / cons* **6**
  Cardamine rhomboidea (Pers.) DC.
Perennial herb. Stems 1 to several from a short thick tuber, 2-6 dm long, unbranched or with a few branches above, smooth or with short hairs on lower stems. Leaves simple, sparsely to densely covered with short hairs; basal leaves round or heart-shaped, on long petioles, withering before anthesis; stem leaves 4-8, oblong to oval, 2-7 cm long and 0.5-2.5 cm wide; petioles shorter upward on stem. Flowers in racemes; sepals green, turning yellow after flowering, 2-4 mm long; petals white (rarely pink), 6-15 mm long. Fruit a silique, 1-2.5 cm long and 1-2 mm wide, with a style beak 2-4 mm long, on spreading stalks 1-3 cm long, the pod often falling before mature. May-June. — Wet forest depressions, floodplain forests, streambanks, wet meadows, swamps, calcareous fens.

Cardamine concatenata (Michx.) Sw.
CUT-LEAF TOOTHWORT   *native / cons* **6**
  Dentaria laciniata Muhl.
Perennial herb; rhizome constricted at intervals, the segments 2-3 cm long. Stems 2-4 dm tall, pubescent above and on the rachis. Basal and stem leaves similar, basal leaves usually absent at anthesis, stem leaves typically in a whorl of 3 above the middle of the stem, deeply 3-parted, the segments linear or lance-shaped, nearly entire to laciniately toothed, the segments often deeply bifid, the whole leaf appearing 5-parted. Sepals 5-8 mm long; petals 12-19 mm long. April-May. — Moist rich woods.

Cardamine diphylla (Michx.) Wood
BROAD-LEAF TOOTHWORT   *native / cons* **8**
  Dentaria diphylla Michx.
Perennial herb, rhizome continuous. Stems glabrous, 2-4 dm tall. Stem leaves commonly 2, opposite or nearly so; leaflets 3, coarsely crenately toothed, about half as long as wide; basal leaves similar. Sepals 5-8 mm long; petals 11-17 mm long. April-May. Rich woods.

Cardamine pensylvanica Muhl.
PENNSYLVANIA BITTERCRESS   *native / cons* **3**
Biennial herb. Stems erect or spreading, to 6 dm long, usually hairy on lower stem. Leaves pinnately divided into 2-5 pairs of lateral leaflets and a single terminal segment, 4-8 cm long and 1-4 cm wide, the leaflets entire or with a few teeth or lobes; the terminal leaflet largest, 1-4 cm long and 1-2 cm wide; petioles shorter than blades, becoming shorter upward. Flowers in a raceme; sepals 1-2 mm long; petals white, 2-4 mm

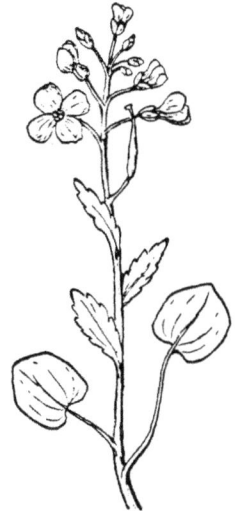
Cardamine bulbosa

Cardamine concatenata

long. Fruit an upright silique, 2-3 cm long and to 1 mm wide, with a style-beak to 2 mm long, on stalks 5-15 mm long. May-Sept. — Streambanks, swamps, and wet forests (often where seasonally flooded); wet, disturbed areas.

Cardamine pratensis L.
CUCKOO-FLOWER   *native / cons* 10
Perennial upright herb. Stems 2-5 dm long. Basal leaves on long petioles, divided into 3-8 broad leaflets, 5-20 mm long, the terminal segment largest and more or less entire; lower stem leaves similar to basal ones, becoming shorter and with shorter petioles upward on stem; stem leaves with 7-17 oval to linear leaflets. Flowers in a crowded raceme; petals white, 8-15 mm long. Fruit an upright silique, 2.5-4 cm long, with a style-beak 1-2 mm long, on stalks 8-15 mm long. May-June. — Peatlands, tamarack and cedar swamps, wet depressions in forests.

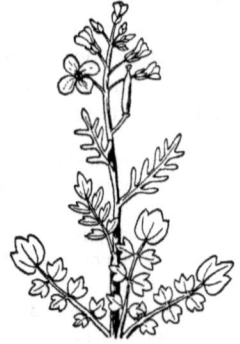

Cardamine pratensis

Descurainia
TANSY-MUSTARD
Annual or biennial herbs, more or less pubescent or canescent with wholly or partly branched hairs; leaves 1-3-pinnate with very numerous small segments. Sepals ovate, obtuse. Petals yellow or pale yellow, small, sometimes barely surpassing the calyx, obovate or spatulate. Filaments slender; anthers ovate or oblong. Staminal glands minute or none. Ovary cylindric; style very short, as thick as the ovary; stigma capitate; ovules numerous. Fruits linear or clavate, terete or slightly 4-angled, tipped with the very short persistent style; valves with a prominent midnerve. Seeds elliptic or oblong, in one or two rows.

1 Plants green, with glandular hairs; fruit less than 13 mm long .... *D. pinnata*
1 Plants gray-green, with stellate hairs, the hairs not glandular; fruit 13 mm or more long ... *D. sophia*

Descurainia pinnata (Walt.) Britt.
TANSY-MUSTARD   *native / cons* 0
Annual herb. Stems erect, simple, or abundantly branched below, or branched at the infloreseence, 2-7 dm tall. Leaves oblong in outline, the lower leaves largest, bipinnate, or pinnate with deeply pinnatifid segments; the upper leaves progressively reduced, less divided, the uppermost 1-pinnate. Flowers 2-4 mm wide.; raceme after anthesis elongate, up to 3 dm long; pedicels at maturity widely divergent, 5-20 mm long. Fruit narrowly clavate, 5-13 mm long, 1-2 mm wide; seeds in 2 rows. — Usually in disturbed places; roadsides, railroads, fields, gravel pits, shores.

Descurainia sophia (L.) Webb
HERB-SOPHIA   *introduced*
Annual herb, plants canescent throughout. Stems erect, usually much branched, 3-8 dm tall. Leaves ovate to obovate in outline, or the upper narrower, 2 to 3x pinnate into linear segments. Flowers about 3 mm wide; raceme after anthesis loose and open; pedicels widely ascending, 8-14 mm long. Fruit narrowly linear, 15-25 mm. long, 0.5-1 mm wide; seeds in 1 row. — Disturbed places, roadsides, railroads.

Draba
WHITLOW-GRASS
Annual, biennial, or perennial herbs, in some species woody at base. Leaves entire or dentate, more or less pubescent with simple, branched, or stellate hairs, or with 2 types of hairs together. Sepals ascending or erect, blunt. Petals yellow or white, rounded, sometimes bifid, narrowed below to a claw, or in certain species sometimes reduced or absent. Fruit a 2-valved silicle, rarely as much as 5x longer than wide.

Descurainia pinnata

1   Plants annuals or winter-annuals . . . . *D. verna*
1   Plants perennial 2
2   Fruit densely covered with stellate hairs . . . . *D. cana*
2   Fruit glabrous or only sparsely hairy . . . . *D. arabisans*

Draba arabisans Michx.
ROCK WHITLOW-GRASS   *native / cons* **10**

Perennial. Stems 1-4 dm tall, simple to branched above, glabrous to stellate-pubescent. Basal leaves narrowly oblong lance-shaped to spatulate, up to 6 cm long, often sharply toothed, uniformly stellate-pubescent; stem leaves few to several, oblong to obovate, narrowed or acute at base, often dentate. Mature racemes loose, up to 10 cm long, commonly glabrous or nearly so; petals white, 4-6 mm long. Fruit glabrous, lance-shaped to narrowly oblong, 7-12 mm long, about one-fourth as wide, soon twisted, in poorly grown examples shorter, ovate, and straight. May-June. — Rocks and cliffs.

Draba cana

Draba cana Rydb.
HOARY WHITLOW-GRASS   *endangered / native / cons* **10**

Perennial. Stems 1-3 dm tall. Basal rosettes very dense, the numerous spatulate, densely stellate-pubescent leaves commonly less than 3 cm long; stem leaves lance-shaped to ovate, usually 5-20 mm long. Raceme elongate, at maturity constituting 1/2 to 3/4 of the total plant height; lower flowers remote, often axillary, only the upper crowded; petals white, about 4 mm long. Fruit oblong, 5-12 mm long, usually 1.5-2 mm wide, stellate-pubescent. May-July. — Rocky limestone ledges, cliffs, and gravelly or rocky soil; known from Door County; more common in Rocky Mtns.

Draba verna L.
SPRING WHITLOW-GRASS   *introduced*

Annual or winter-annual. Leaves in a crowded basal rosette, oblong lance-shaped, 1-2 cm long, pubescent with simple and branched hairs. Scapes several, leafless, very slender, usually 10-15 cm tall; pedicels ascending, elongating in fruit and the lowest to 3 cm long. Flowers white, 2-3 mm wide; petals bifid nearly to the middle. Fruit narrowly to broadly elliptic, usually 4-10 mm long, 2-3.5 mm wide, glabrous. April-May. — Native of Europe; naturalized in fields and roadsides.

Draba verna

Erucastrum
DOG-MUSTARD

Erucastrum gallicum (Willd.) O.E. Schulz
COMMON DOG-MUSTARD   *introduced*

Annual or biennial, with the general aspect of a Brassica; pubescence of simple hairs or none. Stems erect or ascending, 3-6 dm tall, branched from the lower nodes. Basal and lower leaves oblong lance-shaped in outline, to 15 cm long, sparsely pubescent, deeply pinnatifid, the segments dentate, the terminal segment largest; stem leaves progressively reduced, the uppermost only 1-2 cm long; Mature racemes greatly elongate; pedicels slender, ascending; sepals ascending; petals yellow, spatulate, about 7 mm long. Fruit 4-angled, usually upwardly curved, the body 2-2.5 cm long, the beak about 3 mm long; seeds numerous, in 1 row. May-Sept. — Waste places.

Erucastrum gallicum

Erysimum
WALLFLOWER

Annual to perennial herbs, with narrow, entire, dentate, or pinnatifid leaves, more or less pubescent with appressed, 2-4-pronged hairs. Petals yellow to orange (in our species), obovate or spatulate, abruptly or gradually narrowed to a long claw. Ovary linear-cylindric, pubescent;

style very short; stigma capitate, 2-lobed. Fruits elongate, more or less 4-angled, thinly to densely pubescent; valves with a prominent midnerve. Seeds numerous, in one row. All our species are more or less densely pubescent on the stem, leaves, sepals, and fruit, and more or less so on the back of the petals, especially at the base of the blade. The 2-pronged hairs lie lengthwise; V-shaped hairs have the single prong directed backward.

1 Petals less than 6 mm long; fruit to 3 cm long . . . . *E. cheiranthoides*
1 Petals 6 mm or more long . . . . *E. inconspicuum*

Erysimum cheiranthoides L.
WORM-SEED WALLFLOWER *introduced*
Annual herb. Stems erect, simple or sparingly branched, 2-10 dm tall. Leaves linear to oblong lance-shaped, entire or barely sinuate, thinly pubescent but bright green, tapering to the base. Mature racemes elongate, the rachis straight, the pedicels very slender, widely divergent, 6-14 mm, commonly 8-12 mm long; sepals 2-3.5 mm long; petals bright yellow, 3.5-5.5 mm long. Fruit ascending to erect, 12-25 mm, commonly 15-20 mm long. June-Aug. — Usually in wet soil, but also appearing as a weed in fields and roadsides.

Erysimum inconspicuum (S. Wats.) MacM.
SHY WALLFLOWER *introduced*
Perennial herb. Stems erect, commonly simple, occasionally sparingly branched, 3-8 dm tall. Leaves mostly erect or ascending, linear to oblong lance-shaped, entire or obscurely and remotely sinuate-dentate, the stem leaves rarely more than 5 mm wide, canescent. Mature racemes elongate and wand-like, the stout ascending pedicels 3-9 mm long; sepals densely stellate, 5-7 mm long; petals pale yellow, 6-10 mm long. Fruit erect or nearly so, 1.5-4 cm long. May-Aug. — Dry soil of prairies, upland woods.

Erysimum cheiranthoides

Hesperis
DAME'S ROCKET

Hesperis matronalis L.
DAME'S ROCKET *introduced (invasive)*
Perennial herb. Stems erect, 5-10 dm tall, simple or branched above. Leaves lance-shaped, short-petioled or sessile, remotely and sharply denticulate, pubescent above with simple hairs, below chiefly with branched hairs. Flowers fragrant; sepals erect, the outer narrow; the inner broad, saccate at base; petals purple, varying to pink or white, 2-2.5 cm long, the blade obovate; ovary cylindric; stigma 2-lobed. Fruit widely spreading on stout pedicels, linear, terete or nearly so, 5-10 cm long, somewhat constricted between the seeds; seeds numerous, large, 3-4 mm long, angularly fusiform, in 1 row. May-June. — Formerly cultivated for ornament; frequently escaped along roads and fencerows and in open woods.

Hesperis matronalis

Lepidium
PEPPERWORT
Annual, biennial, or perennial herbs. Leaves linear to elliptic, entire, toothed, or pinnatifid. Sepals blunt. Petals small, white (rarely yellowish), linear to spatulate, sometimes notched at tip. Stamens 6, or by abortion 4 or 2. Ovary flat. Fruit a flattened silicle, thin or somewhat distended over the seeds, ovate to circular or obovate, often winged, commonly notched at tip, tipped by the persistent style or stigma.

1 At least upper stem leaves sessile and auriculate, sagittate, or clasping at base . . . . *L. campestre*
1 Stem leaves petiolate or subsessile, never auriculate, sagittate, or clasping at base 2
2 Fruit obovate, widest above middle; petals absent or often rudimentary; rachis of raceme puberulent with cylindrical or clavate hairs . . . . *L. densiflorum*

2   Fruit orbicular, widest at middle; petals present or rarely rudimentary; rachis of raceme puberulent
with curved hairs, rarely glabrous . . . . *L. virginicum*

Lepidium campestre (L.) Ait.f.
FIELD-CRESS   *introduced*
Biennial herb; densely short-pubescent. Stems erect, 2-5 dm tall, simple
to much branched. Basal leaves elongate, oblong lance-shaped, entire to
shallowly lobed; stem leaves erect or ascending, lance-shaped to narrowly
oblong, 2-4 cm long, entire or denticulate, sessile, clasping by acute auricles. Racemes dense, up to 15 cm long at maturity; fruiting pedicels 4-8
mm long, widely divergent. Fruit oblong-ovate, 5-6 mm long, about three-
fourths as wide, convex below, concave above, broadly winged distally,
the short style barely exsert. May-June. — Native of Europe; a weed of
sandy waste ground, fields, and roadsides.

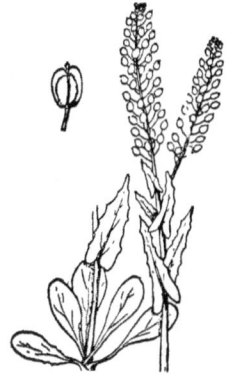

Lepidium densiflorum Schrad.
PRAIRIE PEPPERWORT   *introduced*
Annual herb. Stems 2-5 dm tall, thinly short-pubescent. Basal leaves 4-7
cm long or rarely longer, coarsely dentate or pinnatifid; stem leaves
shorter, linear or narrowly oblance-shaped, mostly entire, sharply acute.
Racemes at maturity erect, 5-10 cm long; petals none, or shorter than
the sepals, linear to narrowly spatulate. Fruit broadly oval to obovate, 2-
3.3 mm long, narrowly winged distally; stigma included in the notch.
May-June. — Dry, sandy or gravelly disturbed places; perhaps native to
w USA, considered adentive in Wisconsin and e USA.

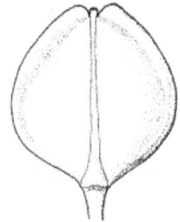

Lepidium virginicum L.
POOR-MAN'S PEPPER   *native / cons O*
Annual herb. Stems erect, 1-5 dm tall, simple to much branched. Basal
leaves oblong lance-shaped in outline, sharply toothed to pinnatifid, occasionally bipinnatifid. Upper leaves smaller, oblong lance-shaped to linear, dentate to entire, acute, narrowed to the base. Racemes numerous,
many-flowered, up to 1 dm long; petals present, equaling to twice as long
as the sepals. Fruit broadly elliptic to circular, 2.5-4 mm long, narrowly
winged across the top; style included in the notch. May-June. — Dry
fields, gardens, roadsides, and waste places.

Nasturtium
WATERCRESS

Nasturtium officinale R.Br.
WATERCRESS   *introduced (invasive)*
  Rorippa nasturtium-aquaticum (L.) Hayek
Perennial herb; plants smooth, often forming large, tangled colonies.
Stems underwater, floating, or trailing on mud; rooting from lower nodes.
Leaves 4-12 cm long and 2-5 cm wide, pinnately divided into 3-9 segments,
the lateral segments round to ovate in outline, the terminal segment
largest; margins entire or with a few shallow rounded teeth; petioles
present. Flowers in 1 to several racemes per stem, flat-topped and elongating in fruit; flowers 5 mm wide, sepals green-white, oblong, 1-3 mm
long; petals white, sometimes purple-tinged, obovate, 4-5 mm long. Fruit
a linear, often curved pod (silique), 1-2.5 cm long and 2 mm wide, tipped
with a short style beak to 1 mm long. May-Sept. — Seeps, slow-moving
streams, ditches, cedar swamps, especially in cold spring-fed waters. Naturalized throughout most of USA and s Canada.

Lepidium campestre

Lepidium densiflorum (fruit)

Lepidium virginicum

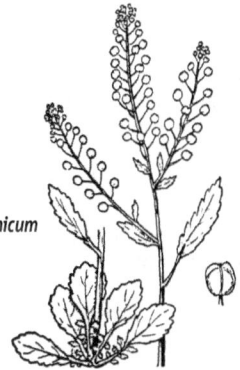

Nasturtium officinale

Neslia
BALL-MUSTARD

Neslia paniculata (L.) Desv.
YELLOW BALL-MUSTARD   *introduced*
Annual herb. Stems much branched, to 8 dm tall, pubescent with
branched hairs. Stem leaves lance-shaped, 3-6 cm long, entire or nearly
so, sessile, clasping, scabrous with 2-pronged hairs. Flowers small, to
about 1.5 mm wide; sepals oblong, obtuse; petals yellow, spatulate; fruiting
pedicels 10-15 mm long. Fruit obliquely ovoid, 7-10 mm long, slightly
compressed, reticulate and pitted. June-July. — Introduced and weedy,
sometimes in cultivated fields.

Raphanus
RADISH

Neslia paniculata

Raphanus raphanistrum L.
WILD RADISH   *introduced*
Coarse annual herb from a stout taproot, pubescence of simple hairs.
Stems stout, 3-8 dm tall, usually sparsely hispid. Lower leaves obovate in
outline, 1-2 dm long, pinnatifid into 5-15 oblong segments, the lower very
small, the upper progressively larger; upper leaves much smaller, oblong
to lance-shaped, entire, dentate, or few-lobed. Sepals obtuse, somewhat
saccate at base; petals yellow, becoming white in age, 10-15 mm long;
mature pedicels ascending, 8-15 mm long. Fruit nearly cylindrical when
fresh, when dry becoming prominently several-ribbed and constricted
between the 4-10 seeds, the body 2-4 cm long, the beak 1-3 cm long. June-
Aug. — Native of Eurasia; weedy on roadsides, fields, waste places.

Rorippa
YELLOWCRESS

Rorippa palustris (L.) Bess.
COMMON YELLOWCRESS   *native / cons* **3**
Annual or biennial herb. Stems erect, usually 1, to 1 m long, unbranched
or branched upward. Leaves lance-shaped to obovate, mostly pinnately
divided; the blades oblong to oblong lance-shaped, 5-30 cm long and 2-6
cm wide, middle stem leaves usually with basal lobes and clasping stem,
smooth to densely hairy on lower surface; margins deeply lobed and
slightly wavy; petioles short or absent. Flowers in racemes at ends of
stems and from leaf axils, the terminal raceme flowering and fruiting
first, the oldest siliques on lowest portions of raceme; sepals green, 1-3
mm long, early deciduous; petals yellow, drying white, 2-3 mm long. Fruit
a round to short-cylindric pod (silique), 3-10 mm long and 1-3 mm wide,
straight-sided or slightly tapered to tip, on stalks 3-10 mm long. June-
Sept. — Marshes, wet meadows, shores, streambanks, ditches and other
wet places.

Sinapis
WHITE-MUSTARD

Similar to *Brassica* (and previously included in that genus) but the
flowers tend to be larger.

Sinapis arvensis L.
CORN-MUSTARD   *introduced*
   Brassica arvensis Rabenh. non L.
   Brassica kaber (DC.) L.C. Wheeler
Annual herb. Stems 2-8 dm tall, usually sparsely hirsute. Leaves obovate
in outline, the lower sometimes lobed but more often merely coarsely

Rorippa palustris

toothed; upper leaves progressively smaller, coarsely toothed, roughly pubescent to nearly glabrous. Flowers about 15 mm wide; pedicels at maturity ascending, about 5 mm long. Fruit ascending, linear, nearly terete, the body 1-2 cm long, 1.5-2.5 mm thick, smooth or somewhat pubescent; beak commonly about half as long as the body, flattened-quadrangular; seeds dark brown, smooth, 1-1.5 mm wide. May-July. — Common weed of fields, gardens, and waste ground.

Sinapis arvensis

Sisymbrium
HEDGE-MUSTARD
Our species annual or winter-annual herbs, with simple hairs. Leaves, (at least the lower) deeply pinnatifid. Sepals obtuse, ascending. Petals small, yellow, obovate. Ovary cylindric; style short, scarcely differentiated. Fruits elongate, linear, terete or slightly quadrangular, tipped with the minute persistent style. Seeds in one row, oblong, nearly or quite smooth.

1   Fruit erect, appressed; pedicels erect, 2–3 mm long .... *S. officinale*
1   Fruit widely spreading; pedicels spreading, 5 mm long or more .... *S. altissimum*

Sisymbrium altissimum L.
TUMBLING MUSTARD   *introduced*
Stems erect, usually much branched, to 1 m tall, glabrous or sparsely pilose. Leaves petioled, deeply pinnately parted, the lower into 5-8 pairs, the upper into 2-5 pairs of segments; segments varying from linear and entire to lance-shaped and serrate. Racemes greatly elongate after anthesis; pedicels nearly or quite as thick as the fruit, ascending, 5-10 mm long; petals pale yellow, 6-8 mm long. Fruit ascending or spreading, slender, linear, 5-10 cm long, 1-1.5 mm wide. June-July. — Native of Eurasia; established as a weed in fields and waste ground.

Sisymbrium altissimum

Sisymbrium officinale (L.) Scop.
HEDGE-MUSTARD   *introduced*
Stems erect, 3-8 dm tall, branched above or simple. Lower leaves petioled, deeply pinnatifid; segments oblong to ovate or the terminal sometimes rotund, angularly toothed; upper leaves sessile or nearly so, few-lobed or 3-lobed or entire, the lateral lobes widely divergent. Flowers bright yellow, about 3 mm wide. Racemes erect, simple or with straight, widely divergent branches; pedicels at maturity 2-3 mm long, closely appressed, distally thickened and as wide as the fruit at the summit. Fruit closely appressed, subulate, 10-15 mm long, 1-1.5 mm wide at base. May-Oct. — Native of Europe; established as a weed in gardens, roadsides, and waste ground.

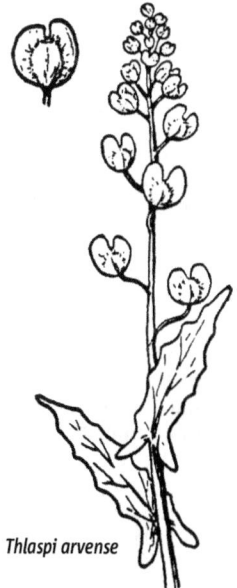

Thlaspi
PENNYCRESS
Thlaspi arvense L.
FIELD PENNYCRESS   *introduced*
Glabrous annual herb. Stems 1-5 dm tall, simple to much branched. Stem leaves sessile, oblong to lance-shaped, entire or few-toothed, with 2 narrow auricles at base 1-5 mm long. Petals white, spatulate to obovate, about 3 mm wide, 2x as long as the sepals; pedicels at maturity ascending. Fruit circular to broadly elliptic, 10-14 mm long, strongly flattened and distended over the seeds, keeled or winged at the margin, tip deeply (2-3 mm) notched. April-June. — Native of Europe; roadsides and waste ground.

Thlaspi arvense

Turritis
TOWER-MUSTARD

Turritis glabra L.
TOWER-MUSTARD   *native / cons* 5
  Arabis glabra (L.) Bernh.
Biennial herb. Stems commonly hirsute near the base with mostly simple hairs, occasionally with some or many branched hairs, glabrous and glaucous above. Lower leaves more or less pubescent, usually with Y-shaped hairs; stem leaves overlapping in the lower part of the stem, more remote above, all lance-shaped or lance-oblong, sessile and auriculate-clasping, usually glabrous and glaucous. Petals 3-6 mm long; mature pedicels erect, 7-16 mm long. Fruit erect, nearly terete, overlapping, 5-9 cm long, 0.8-1.3 mm wide; seeds angular, very narrowly winged. May-June. — Dry sandy fields, gravel pits, roadsides, gravelly shores.

Turritis glabra

Campanulaceae
BELLFLOWER FAMILY
Perennial herbs. Stems usually with milky juice. Leaves simple, alternate. Flowers in racemes at ends of stems or single from upper leaf axils, perfect (with both staminate and pistillate parts), 5-parted, regular and funnel-shaped (*Campanula*) or irregular (*Lobelia*); petals blue, white or scarlet; stamens separate or joined into a tube around style. Fruit a many-seeded capsule. *Lobelia* is sometimes placed in the Lobeliaceae, but that family discontinued under APG III.

1 Flowers irregular; stamens joined to form a tube around the style . . . . *Lobelia*
1 Flowers regular, bell-shaped or funnel-shaped; stamens separate . . . . *Campanula*

Campanula
BELLFLOWER, HAREBELL
Annual, biennial, or perennial herbs. Leaves alternate. Flowers conspicuous, solitary or in various types of inflorescence. Sepals 5, triangular to linear. Corolla rotate, campanulate, or funnelform, deeply or shallowly 5-lobed, in our species blue or violet to white. Stamens attached at the very base of the corolla; filaments widened at the base; anthers distinct. Ovary 3-5-celled with elongate style. Capsule short, conspicuously ribbed, opening by 3 or 5 lateral pores; seeds numerous.

ADDITIONAL SPECIES
Two introduced species, cultivated for their handsome flowers, are occasionally reported as escaped in Door County: *Campanula glomerata* L. and *C. persicifolia* L.

1 Stems weak, reclining on other plants . . . . *C. aparinoides*
1 Stems upright 2
2 Flowers on short pedicles in an erect, 1-sided raceme . . . . *C. rapunculoides*
2 Flowers solitary or in loose, open clusters on slender pedicels . . . . *C. rotundifolia*

Campanula aparinoides Pursh
MARSH BELLFLOWER   *native / cons* 7
Perennial herb, spreading by slender rhizomes. Stems slender, weak, usually reclining on other plants, 2-6 dm long, 3-angled, rough-to-touch. Leaves linear or narrowly lance-shaped, larger below and smaller upward on stem, 2-8 cm long and 2-8 mm wide, tapered to a sharp tip; margins and midvein on leaf underside often rough; petioles absent. Flowers single on long slender stalks from upper leaf axils, funnel-shaped, sepals triangular to lance-shaped, 2-5 mm long; petals pale blue to white, 5-12 mm long. Fruit a capsule, opening near its base to release seeds. July-Sept. — Sedge meadows, marshes, calcareous fens, conifer swamps (cedar, tamarack), thickets, open bogs; soils often calcium-rich.

Campanula aparinoides

Campanula rapunculoides L.

CREEPING BELLFLOWER   *introduced (invasive)*

Erect perennial herb. Stems 4-10 dm tall from a creeping rhizome, usually unbranched, smooth or sparsely pubescent. Leaves coarse, irregularly serrate, usually sparsely pubescent beneath, the lower long-petioled, ovate, rounded to subcordate at base, the upper progressively narrower, short-petioled or subsessile. Inflorescence strict, unbranched, secund, forming a 1-sided raceme with much reduced bracts; lower pedicels sometimes elongate to 5 cm long, the upper 1-5 mm long; calyx lobes reflexed; corolla blue, somewhat nodding, 2-3 cm long. Capsule opening from pores near base. July-Aug. — Native of Eurasia, forming persistent weedy colonies on roadsides, railroads, and in disturbed places.

Campanula rapunculoides

Campanula rotundifolia L.

BLUEBELL-OF-SCOTLAND   *native / cons* **5**

Perennial herb. Stems simple or branched from the base, 1-5 dm tall, smooth or nearly so, bearing 1 to several spreading or drooping flowers. Basal leaves broadly ovate or subcordate, seldom persistent in the flowering season; stem leaves linear, 2-10 cm long, decreasing above; pedicels slender. Sepals linear, 4-10 mm long; corolla blue (rarely white), bell-shaped, 12-20 mm long or occasionally somewhat larger, its lobes much shorter than the tube. Capsule turbinate or short-cylindric, opening by pores near the base. June-Sept. — Dry woods, meadows, cliffs and beaches. Variable in habit, stature, and number and size of flowers.

Campanula rotundifolia

Lobelia

LOBELIA

Mostly perennial herbs. Stems single, usually with milky juice. Leaves alternate. Flowers irregular, in racemes at ends of stems; white, bright red, or pale to dark blue, often with white or yellow markings; 2-lipped, the 3 lobes of lower lip spreading, the 2 lobes of upper lip erect or pointing forward, divided to base, the anthers projecting through the split; stamens 5, joined to form a tube around style, the lower 2 anthers hairy at tip and shorter than other 3. Fruit a capsule. Most species toxic if taken internally.

1 Stem leaves narrow, to 4 mm wide, margins entire or with a few small teeth; or leaves all from base of plant . . . . ***L. kalmii***
1 Stem leaves broader, 1–5 cm wide, margins toothed 2
2 Flowers small, to 1.5 cm long 3
2 Flowers larger, 2–4 cm long 4
3 Inflorescence usually branched; hypanthium equalling the corolla or nearly so, inflated in fruit . . . . .
   . . . . ***L. inflata***
3 Inflorescence unbranched; hypanthium shorter than corolla, not much inflated in fruit . . . . ***L. spicata***
4 Flowers bright red (rarely white), 3 cm or more long . . . . ***L. cardinalis***
4 Flowers blue with white stripes on lower lip, less than 2.5 cm long . . . . ***L. siphilitica***

Lobelia cardinalis L.

CARDINAL-FLOWER   *native / cons* **7**

Perennial herb. Stems erect, usually unbranched, 5-15 dm long, hairy to smooth. Leaves lance-shaped to oblong, 10-15 cm long and 3-5 cm wide, margins toothed; lower leaves on short petioles, upper leaves more or less stalkless. Flowers bright scarlet (rarely white), in racemes 1-4 dm long, the racemes with small, leafy, linear bracts; flowers 2-4 cm long, on hairy pedicels 5-15 mm long. July-Sept. — Floodplain forests, swamps, thickets, streambanks, shores and ditches; sometimes in shallow water.

Lobelia inflata L.

INDIAN-TOBACCO *native / cons* 2

Annual or biennial herb. Stems erect, usually branched, villous, to 1 m tall. Leaves sessile or subsessile, obovate, 5-8 cm long, 1.5-3.5 cm wide, more or less serrate, usually pubescent. Racemes terminating the branches, 1-2 dm long; lower bracts foliaceous, the upper gradually reduced; pedicels 3-8 mm long, glabrous or puberulent, bracteolate at the base. Flowers 7-10 mm long; sepals linear, 3-5 mm long; corolla blue or white, the lower lip pubescent; hypanthium much inflated in fruit. July-Oct. — Open woods in moist or dry soil, disturbed places such as roadsides, ditches, borrow pits, trails, utility line clearings.

The long, irregular hairs at base of stem are distinctive. Long used in herbal medicine.

Lobelia inflata

Lobelia kalmii L.

BROOK LOBELIA *native / cons* 9

Small perennial herb. Stems erect, smooth, 1-4 dm long, unbranched or with a few branches above, sometimes with a rosette of small, obovate leaves at base of plant. Stem leaves linear, 1-5 cm long and 1-5 mm wide, blunt to sharp-tipped, margins with a few small teeth. Flowers blue with a white center, 6-10 mm long, in an open raceme, the flowers on pedicels 4-10 mm long. July-Oct. — Wet, sandy or gravelly shores, wet meadows, interdunal wetlands, conifer swamps (cedar, tamarack), rock ledges and crevices; usually where calcium-rich.

Lobelia kalmii

Lobelia siphilitica L.

GREAT BLUE LOBELIA *native / cons* 5

Perennial herb. Stems stout, erect, 3-12 dm long. Leaves oblong or oval, smaller upward, 6-12 cm long and 1-3 cm wide, tip sharp or blunt, margins irregularly toothed, petioles absent. Flowers dark blue, in crowded racemes 1-3 dm long; the lower lip blue and white-striped, 1.5-2.5 cm long, on ascending stalks 4-10 mm long; sepals triangular to lance-shaped, 5-20 mm long, usually with narrow lobes near base. Aug-Sept. — Swamps, streambanks, calcareous fens, wet meadows.

Lobelia spicata Lam.

SPIKED LOBELIA *native / cons* 6

Perennial herb. Stems unbranched, 3-10 dm long, hairy toward base. Leaves obovate to lance-shaped, 5-10 cm long, hairy, becoming smaller above. Flowers pale blue to white, 6-10 mm long, on stalks 2-4 mm long, in a slender, crowded raceme; base of sepals often with distinct, curved lobes (auricles), 1-2 mm long. May-Aug. — Moist to wet prairies (sometimes where disturbed), swamp margins.

Lobelia cardinalis

Lobelia siphilitica

Lobelia spicata

Cannabaceae

HEMP FAMILY

Humulus
HOPS

Humulus lupulus L.
COMMON HOPS　　　　　　　　　　　　*native / cons* **3**

Perennial vine to 10 m long. Stems rough. Principal leaves opposite, as wide as long, cordate at base, 3-lobed to below the middle, the lateral lobes obliquely ovate-oblong, the terminal lance-shaped, constricted at base; upper leaves commonly broadly ovate, unlobed. Dioecious; flowers small, in axillary clusters; staminate panicles 5-15 cm long, stamens 5; pistillate spikes about 1 cm long at anthesis, with conspicuous slender stigmas. Fruit (hops) an achene enclosed within the persistent calyx and covered by the expanded bracts, 3-6 cm long, the bracts entire and very glandular at base. July-Aug. — Moist, rich soil, often where shaded.

Yellow glands secreting a bitter substance, lupulin, occur on many parts of the plant but are most numerous on the fruit, which is important in beer-making.

Humulus lupulus

Caprifoliaceae

HONEYSUCKLE FAMILY

Shrubs or vines, with opposite, mostly simple leaves. Flowers perfect (with both staminate and pistillate parts), mostly 5-parted. Fruit a fleshy berry or dry capsule. Family now includes members of the former Valerianaceae (*Valeriana*); herbs with opposite, simple or divided leaves, and numerous small flowers in terminal, panicled or capitate cymes.

1 Flowers numerous, in rather dense terminal inflorescences (at ends of stem and branches) *Valeriana*
1 Flowers axillary or on paired pedicels on a peduncle 2
2 Plants small, creeping, evergreen; flowers paired and nodding at tips of slender stalks . . . . . *Linnaea*
2 Plants larger shrubs or coarse herbs, upright, deciduous 3
3 Plants herbaceous; flowers from leaf axils. . . . . *Triosteum*
3 Plants woody vines or shrubs; flowers various 4
4 Corolla bell-shaped, less than 1 cm long . . . . *Symphoricarpos*
4 Corolla tube-shaped, mostly more than 1 cm long . . . . *Lonicera*

Linnaea
TWINFLOWER

Linnaea borealis L.
TWINFLOWER　　　　　　　　　　　　*native / cons* **9**

Low, evergreen, trailing vine. Stems slightly woody, to 1-2 m long, with numerous short, erect, leafy branches to 10 cm long; branches green to red-brown, finely hairy; older stems woody, 2-4 mm wide. Leaves opposite, simple, evergreen, oval to round, 1-2 cm long, blunt at tip, upper surface and margins with short, straight hairs; margins rolled under, with a few rounded teeth near tip; petiole short, short-hairy. Flowers small, pink to white, bell-shaped, shallowly 5-lobed, slightly fragrant, in nodding pairs atop a Y-shaped stalk to 10 cm long, the stalk with gland-tipped hairs and 2 small bracts at the fork and a pair of smaller bracts at base of each flower. Fruit a small, dry, 1-seeded capsule. June-Aug. — Hummocks in cedar swamps and thickets, moist conifer woods, on rotten logs and mossy boulders.

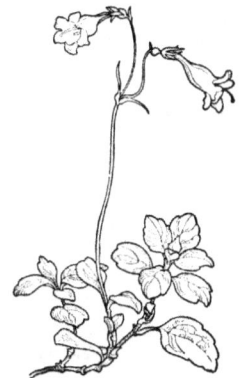

Linnaea borealis

Lonicera
HONEYSUCKLE
Shrubs or woody vines. Leaves opposite, simple, entire. Flowers long and tubular or funnel-shaped, in pairs from leaf axils. Fruit a few-seeded, blue or red berry.

1 Flowers in opposite, sessile, 3-flowered clusters, producing a whorl of 6 flowers; plants woody, climbing vines 2
1 Flowers paired at ends of peduncles from leaf axils; shrubs or vines 3
2 Leaves hairy on upper surface . . . . *L. hirsuta*
2 Leaves glabrous on upper surface . . . . *L. dioica*
3 Style glabrous; fruit red; native species . . . . *L. canadensis*
3 Style with coarse, stiff hairs; fruit red; introduced species (except *L. oblongifolia*) 4
4 Corolla strongly 2-lipped; the upper lip lobed to half its length 5
4 Corolla only weakly 2-lipped; the upper lip lobed to base or nearly so 6
5 Ovaries separate, gland-covered . . . . *L. xylosteum*
5 Ovaries partly or entirely joined, glabrous . . . . *L. oblongifolia*
6 Leaves glabrous on underside; flower peduncles 1.5 cm long or more . . . . *L. tatarica*
6 Leaves hairy on underside; flower peduncles to 1.5 cm long 7
7 Corolla glabrous on outside, barely swollen at base . . . . *L. x bella*
7 Corolla pubescent on outside, ± swollen at base . . . . *L. morrowii*

Lonicera × bella Zabel
HONEYSUCKLE   *introduced (invasive)*
Hybrid shrub (*Lonicera morrowii* x *L. tatarica*), invasive and known from a number of Door County locations; plants are often intermediate between the two parents, making identification difficult. *Lonicera x bella* tends to be taller than either of its parents, sometimes to 6 m tall; young stems hollow, sparsely pubescent. Leaves opposite, oval, entire, to ca. 7 cm long; leaf underside slightly pubescent. Flowers paired in leaf axils of the leaves, usually pink, often turning yellow with age. Fruit a red, paired berry. May–June. — Native of Asia, invasive and found in a wide range of habitats: forest edges, roadsides, old fields, disturbed places, utility right-of-ways, vacant lots. The berries are eaten by birds which helps spread the seeds.

Lonicera canadensis Bartr.
FLY-HONEYSUCKLE   *native / cons* **8**
Shrub to 2 m tall, with straggling branches. Leaves ovate to oblong, 3-12 cm long, acute or obtuse, ciliate, glabrous to sparsely pubescent beneath. Peduncles 2-3 cm long; bracts linear or subulate, from much shorter than to slightly exceeding the ovaries; bractlets to 0.5 mm long, or obsolete; corolla yellowish, 12-22 mm long, distinctly saccate at base, glabrous, its lobes a third to a half as long as the tube; ovaries glabrous, strongly divergent at anthesis; style glabrous. Berries red. May–June. — Dry or moist woods, occasionally swamps.

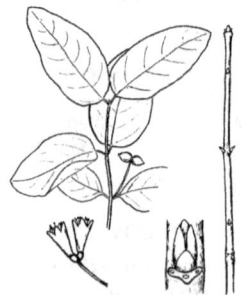

Lonicera canadensis

Lonicera dioica L.
LIMBER HONEYSUCKLE   *native / cons* **7**
Climbing woody vine with glabrous branches. Leaves ovate or obovate, 5-12 cm long, rounded or narrowed at base, glaucous beneath; uppermost one or two pairs united into a rhombic or doubly ovate disk narrowed to an obtuse or acute tip or rounded and mucronate. Spike short-peduncled (usually 5-20 mm); corolla yellow to deep maroon, 1.5-2.5 cm long, abruptly bulging just above the base, hairy inside. Berries red, 8-12 mm long, in clusters surrounded by leafy cup-like bracts, inedible and extremely bitter. May–June. — Moist woods and thickets, occasionally on dunes or in swamps.

Lonicera dioica

Lonicera hirsuta Eat.

HAIRY HONEYSUCKLE   *native / cons 7*

Twining woody vine, the younger stems pubescent and glandular. Leaves dull green, sessile or short-petioled, broadly oval, 6-12 cm long, more or less pubescent on both sides, underside also glaucous; upper 1 or 2 pairs acuminate, connate into a rhombic-elliptic to subrotund disk. Spikes bearing 1-4 crowded whorls of flowers; corolla usually yellow, corolla tube 10-18 mm long, the pubescent tube slightly bulging near the base. Berries orange to red, sessile, subtended by a pair of saucer-shaped bracts. June-July. — Moist woods, particularly on margins and in clearings, often where sandy or rocky; occasionally in white cedar swamps. The pubescent upper surface of the leaves distinguish this species from *L. dioica* which also has connate terminal leaves.

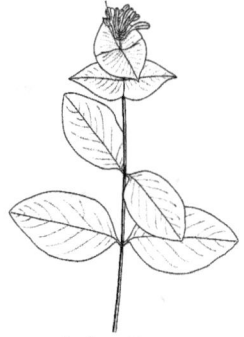

Lonicera hirsuta

Lonicera morrowii Gray

MORROW'S HONEYSUCKLE   *introduced (invasive)*

Tall shrub, the older branches hollow. Leaves ovate to oblong, 3-6 cm long, rounded at base to a short petiole, softly pubescent beneath. Peduncles 5-15 mm long; bracts linear lance-shaped, 1-3 times as long as the ovary; bractlets half as long as to equaling the ovary; sepals ciliate; corolla white, fading yellow, pubescent, saccate at base, the lips equaling or longer than the tube; upper lip 4-lobed to its base; style hirsute. Fruit a dark red berry 7-8 mm wide; seeds numerous. May-June. — Native of Asia; escaping and invasive.

Lonicera morrowii

Lonicera oblongifolia (Goldie) Hook.

SWAMP FLY-HONEYSUCKLE   *native / cons 9*

Thicket-forming shrub 1-1.5 m tall; branches upright, with shredding bark and solid pith; twigs green to purple, smooth. Leaves opposite, oblong or oval, 3-8 cm long and 1-4 cm wide, rounded or blunt at tip, underside hairy when young, becoming smooth; margins entire, not fringed with hairs; petioles absent or to 1-2 mm long. Flowers yellow-white, tube-shaped with 2 spreading lips, 10-15 mm long, in pairs at ends of slender stalks up to 4 cm long from leaf axils. Fruit an orange-red to red (or sometimes purple), few-seeded berry composed of the 2 joined ovaries. May-June. — Cedar and tamarack swamps, fens, open bogs, wet streambanks and shores; often over limestone.

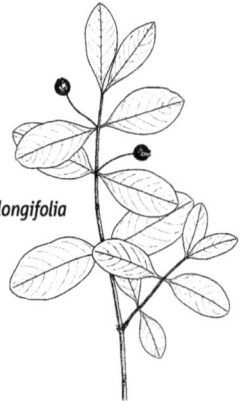

Lonicera oblongifolia

Lonicera tatarica L.

TARTARIAN HONEYSUCKLE   *introduced (invasive)*

Tall shrub, the older branches hollow. Leaves ovate to oblong, 3-6 cm long, rounded to subcordate at base, glabrous. Peduncles 15-25 mm long; bracts subulate, shorter than to exceeding the ovary; bractlets broadly ovate, about a third as long as the ovary; sepals entire; corolla white to pink, glabrous, barely gibbous at base, the lips equaling or exceeding the tube; upper lip 4-lobed to its base; style hirsute. Fruit a shiny orange or red berry to 1 cm wide. May-June. — Native of e Europe and Asia, an old favorite in cultivation and escaping and invasive.

Lonicera tatarica

Lonicera xylosteum L.

EUROPEAN FLY-HONEYSUCKLE   *introduced*

Shrub to 3 m tall, the young stems pubescent. Leaves oval, 2-6 cm long, usually more than half as broad, pubescent beneath. Peduncles 6-12 mm long; bracts subulate, equaling or surpassing the ovaries; bractlets rotund, shorter than the separate glandular ovaries; corolla yellowish, pubescent, 7-12 mm long, the lips more than twice as long as the tube; lateral clefts of the upper lip twice as long as the central one; style hirsute. Berries red. May-June. — Native of Eurasia; cultivated and sometimes escaping.

Symphoricarpos
SNOWBERRY

Symphoricarpos albus (L.) Blake
SNOWBERRY   *native / cons* 7
Bushy shrub to 1 m tall, or sometimes dwarf, the younger branches pubescent or glabrous. Leaves ovate or oval, usually 2-3 cm long. Flowers in pairs on short pedicels or in short, few-flowered, interrupted spikes; corolla campanulate, regular or nearly so, white or pink, slightly ventricose, 6-8 mm long, the lobes from half as long as to sometimes equaling the tube; style glabrous, 2-3 mm long. Fruit a 2-seeded white berry; seeds 5 mm long. — Dry or rocky soil.

Symphoricarpos albus

Triosteum
FEVERWORT, HORSE-GENTIAN

Coarse, erect, pubescent, perennial herbs. Leaves large, connate or united by a ridge around the stem. Flowers greenish yellow to dull red, solitary or in small clusters in their axils. Corolla narrowly campanulate, gibbous at the base, unequally 5-lobed. Sepals 5, linear, elongate. Stamens 5. Ovary 3-5-celled; style slender, included or exserted; stigma capitate. Fruit a yellow, red, or greenish dry berry, crowned by the persistent sepals, enclosing a few hard oblong seeds.

Triosteum aurantiacum Bickn.
HORSE-GENTIAN   *native / cons* 5
Much like *T. perfoliatum* (found mostly in s Wisconsin) but the leaves distinct, tapering to a narrow base (or seldom 1-3 pairs with connate base 1-2 cm wide), the hairs of the stem mostly over 0.5 mm long, the corolla purplish-red, the style about equaling the corolla or shortly included, and the fruit bright orange-red. May-July. — Rich woods and thickets.

Triosteum aurantiacum

Valeriana
VALERIAN

Valeriana officinalis L.
ALLHEAL   *introduced*
Perennial, strongly scented herb. Stems stout, erect, 6-15 dm tall, usually pubescent at the nodes. Leaves from base of plant and opposite along stem; basal and stem leaves similar, pinnately divided into 11-21 lance-shaped dentate segments; petioles of the upper leaves progressively shorter. Panicle large and open, its lower branches often remote (to 10 cm) from the upper, to 10 cm long; corolla obconic, 5-lobed, the tube about 4 mm long, the lobes about 1 mm long; stamens 3. Fruit a 1-chambered glabrous achene, 4-5 mm long by about half as wide. May-Aug. — Native of Eurasia; escaped from gardens where cultivated, to roadsides, ditches, fields, shores, and forest margins.

Valeriana officinalis

Caryophyllaceae
PINK FAMILY

Annual or perennial herbs. Leaves simple, entire, mostly opposite but sometimes alternate or whorled. Stems often swollen at nodes. Flowers perfect (with both staminate and pistillate parts) or imperfect, in open or compact heads at ends of stems or from leaf axils; sepals usually 5, separate or joined into a tube; petals 5 (sometimes 4), separate, often lobed or toothed, sometimes absent; stamens 3-10, anthers often distinctly colored. Fruit a few- to many-seeded capsule.

1 Leaves with chaffy or membranous stipules . . . . *Spergula*
1 Leaves without stipules 2
2 Sepals joined to form a lobed tube 3
2 Sepals not joined, distinct from one another 6
3 Styles 3–5 . . . . *Silene*
3 Styles 2 4
4 Calyx not subtended by bracts . . . . *Saponaria*
4 Calyx subtended by 1–3 pairs of bracts 5
5 Calyx with 20 or more nerves . . . . *Dianthus*
5 Calyx 5-nerved . . . . *Petrorhagia*
6 Petals entire 7
6 Petals deeply cleft at tip into 2 segments 8
7 Leaves linear-subulate, the principal stem leaves subtending dense axillary clusters; plant entirely glabrous; capsule dehiscing into 3 valves . . . . *Minuartia*
7 Leaves ovate to elliptic or lanceolate, mostly without axillary tufts; plants puberulent at least on the stem; capsule with the 3 valves again split, resulting in a total of 6 teeth . . . . *Arenaria*
8 Fruit cylindric; styles usually 5 (3 in *Cerastium nutans*) . . . . *Cerastium*
8 Fruit ovoid or oblong; styles usually 3 (5 in *Stellaria aquatica*) . . . . *Stellaria*

Arenaria
SANDWORT
Arenaria serpyllifolia L.
THYME-LEAF SANDWORT   *introduced*
Finely hairy annual herb. Stems diffuse, 5-20 cm long, the internodes usually much longer than the leaves. Leaves usually 8-10 pairs, 3-5 mm long, 3-4 mm wide, ovate, sparsely rough hairy, often pustulate, 3-5-nerved. Inflorescence a terminal cyme, short or extending to the middle of the stem; bracts leaf-like; pedicels subcapillary, 4-8 mm long; sepals 5, 2.5-3.5 mm long, ovate lance-shaped, frequently glandular; petals 5, white, obovate, usually shorter than the sepals; stamens normally 10. Capsule ovoid, exceeding the sepals, dehiscent by 6 nearly equal teeth; seeds numerous, about 0.5 mm wide, kidney-shaped, gray-black. — Native of Eurasia; sandy or rocky places.

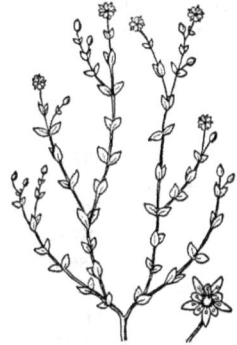

Arenaria serpyllifolia

Cerastium
MOUSE-EAR CHICKWEED
Low annual or perennial herbs. Leaves opposite. Flowers solitary or more commonly in terminal cymes. Sepals 5. Petals 5, notched at tip to bifid, seldom entire. Stamens normally 10, sometimes 5. Styles normally 5, sometimes 4 or 3. Capsule usually exceeding the sepals, cylindric, membranous, dehiscent by 10 short teeth, frequently curved. Ovary 1-celled. Seeds numerous, kidney-shaped, papillate-tuberculate.

ADDITIONAL SPECIES
Cerastium tomentosum L. (Snow-in-summer), is an ornamental, occasionally escaping or persisting near old homesteads (see key).

1 Petals large and showy, longer than the sepals 2
2 Plants covered with long, soft, straight, glandular hairs . . . . *C. arvense*
2 Plants covered with tangled hairs, these not glandular . . . . *C. tomentosum**
1 Petals about equal to the sepals 3
3 Bracts of inflorescence green and herbaceous . . . . *C. nutans*
3 Bracts of upper inflorescence with papery, non-green margins and tips 4
4 Matted perennial; stamens 10 . . . . *C. fontanum*
4 Small annual; stamens 4–5 . . . . *C. semidecandrum*

Cerastium arvense L.
FIELD CHICKWEED   *native / cons* **4**
Matted perennial. Stems ascending or erect, glabrous to densely villous, glandular or nonglandular, mostly 1.5-4 dm tall. Leaves linear to lance-

shaped, 2-7 cm long, 1-15 mm wide, glabrous or pubescent, glandular or nonglandular; primary leaves usually subtending conspicuous axillary fascicles or short sterile shoots. Inflorescence with few to many flowers, the pedicels very slender; sepals lance-shaped, mostly 5-8 mm long, 2-3x exceeded by the conspicuous white petals. Capsule cylindric, somewhat to much exceeding the sepals; seeds papillate-tuberculate, reddish brown. April-Aug. — Rocky, gravelly, or sandy areas, chiefly in calcium- or magnesium-rich soils, weedy in abandoned fields and meadows.

Cerastium fontanum Baumg.
MOUSE-EAR CHICKWEED   *introduced*
Cerastium vulgatum L.
Perennial. Stems tufted, spreading or erect, viscid-pubescent, 1.5-5 dm long. Leaves oblong to lance-shaped, 1-2 cm long, 3-12 mm wide, 1-nerved, sparingly pubescent, the lower leaves oblong lance-shaped. Inflorescence at length rather open, the mature pedicels 5-12 mm long; sepals 4.5-6 mm long, oblong lance-shaped, pubescent, rather strongly 1-nerved toward the base, scarious-margined; petals about as long as the sepals. Capsule 8-10 mm long, 2-3 mm wide, cylindric, sometimes curved; seeds rugose-papillate, reddish brown. April-Oct. — Native of Eurasia; widely naturalized in fields, woods, and waste places and frequently a troublesome weed, especially in lawns.

Cerastium arvense

Cerastium nutans Raf.
NODDING MOUSE-EAR CHICKWEED   *native / cons* O
Cerastium brachypodum (Engelm.) B.L. Robins.
Annual. Stems weak or ascending, usually branched, viscid-pubescent, 1-4.5 dm long. Leaves narrowly oblong lance-shaped or oblong lance-shaped, commonly 1.5-5 cm long, 5-10 mm wide. Inflorescence open and loosely cymose; sepals lance-shaped, 4-5 mm long, acute or obtusish, scarious-margined; petals more or less equaling, frequently conspicuously exceeding the sepals; occasionally lacking. Capsule 8-15 mm long, cylindric, straight or curved; seeds coarsely papillose, pale reddish brown. April-June. — Moist or dry woodlands or open places.

Cerastium semidecandrum L.
FIVE-STAMEN MOUSE-EAR CHICKWEED   *introduced*

Cerastium fontanum

Viscid-pubescent annual. Stems 0.5-2 dm tall. Leaves relatively small, mostly 0.5-1.5 cm long; basal leaves oblong lance-shaped; stem leaves ovate. Iinflorescence often compact, varying to moderately open, the pedicels usually deflexed in fruit: bracts scarious-margined; sepals 3-5 mm long, lance-shaped, stipitate-glandular, scarious-margined; petals shorter than the sepals, only shallowly notched; stamens 5 (sometimes 10). Capsule 4.5-7 mm long; seeds smooth. — Native of Eurasia; dry, sandy disturbed places.

Dianthus
PINK
Biennial or perennial, usually glaucous herbs. Leaves narrow. Flowers solitary or in paniculate or capitate cymes. Calyx subtended by 1-3 pairs of bracts, cylindric, with 30 or more nerves. Petals 5, without auricles or appendages. Stamens 10. Styles 2. Capsule dehiscent by 4 or 5 teeth. Seeds disc-shaped, apiculate.

Many species are well known in cultivation; the carnation is *D. caryophyltus* L.

1 Leaves not linear, more than 9 mm wide .... *D. barbatus*
1 Leaves linear, less than 9 mm wide 2
2 Annual; calyx and bracts hairy .... *D. armeria*
2 Perennial; calyx and bracts glabrous or only very finely hairy 3

3   Leaves hard and stiff; flowers clove-scented . . . .*D. plumarius*
3   Leaves soft and lax; flowers not clove-scented . . . .*D. deltoides*

Dianthus armeria L.
DEPTFORD PINK   *introduced*
Biennial. Stems 2-6 dm tall, usually strigose below the nodes, otherwise
glabrous. Basal leaves numerous; stem leaves 5-10 pairs, linear to linear
lance-shaped, 3-8 cm long, puberulent, ciliate. Cymes congested, 3-9-
flowered, the lower, lance-subulate, erect bracts frequently surpassing
the flowers; calyx 12-18 mm long, 20-25-nerved; petals 2-2.5 cm long, the
blades 4-5 mm long, pink or rose, dentate. Capsule equaling the calyx,
dehiscent by 4 recurved teeth. Summer. — Native of Europe, established
as a weed.

Dianthus barbatus L.
SWEETWILLIAM   *introduced*
Glabrous perennial. Stems stout, 3-6 dm tall. Stem leaves 5-19 pairs, lance-
shaped to oblong lance-shaped, 6-10 cm long, 1-1.8 cm wide; basal leaves
somewhat wider. Cymes densely corymbose; primary bracts leaflike; secondary bractlets awn-tipped, ciliolate; calyx 15-18 mm long, about 40-
nerved; petals 15-25 mm long, the blades dark red, pink, or whitish, cre-
nate-denticulate. Capsule about 1 cm long. Summer. — Native of the Old
World; escaped from cultivation and locally established.

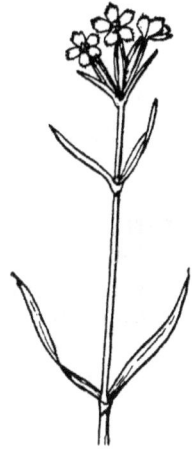

Dianthus armeria

Dianthus deltoides L.
MAIDEN PINK   *introduced*
Perennial with a very slender creeping rootstalk. Stems slender, branched,
1-4 dm tall, glabrous or hispidulous-puberulent. Basal leaves oblong lance-
shaped, 1.5-3 cm long, 1.5-3 mm wide; stem leaves 5-10 pairs, linear lance-
shaped, acute, 2-4 cm long. Flowers solitary on pedicels 1-4 cm long; calyx
12-15 mm long, with 30-40 nerves, subtended by 1 or 2 pairs of obovate,
abruptly acuminate or awned bracts; petals purple-red, lavender, or white,
1.5-2 cm long, the blades 4-8 mm long, sharply denticulate. Capsule narrowly ellipsoid-lance-shaped, about equaling the calyx. Summer. — Native
of Europe; often cultivated and locally escaped into waste places.

Dianthus plumarius L.
FEATHERED PINK   *introduced*
Perennial. Stems simple, 1-3 dm tall, glabrous, frequently glaucous. Leaves
linear, 2-8 cm long, 1.5-3 mm wide, 3-nerved, the 3-8 pairs of stem leaves
somewhat shorter than the basal ones. Inflorescence 1-5-flowered, the
pedicels 1-3 cm long; calyx 15-20 mm long, 40-nerved, the lobes 4-5 mm
long. Bracts 1 or 2 pairs, 5-10 mm long, abruptly mucronate or short-
awned; petals red to white, 1.5-2 cm long, fringed. Capsule somewhat exceeding the calyx, dehiscent by 4 subobtuse teeth. Summer. — Native of
Europe; escaped from cultivation and locally established.

Dianthus deltoides

Minuartia
STITCHWORT
Previously included in genus *Arenaria*.

Minuartia michauxii (Fenzl) Farw.
MICHAUX'S STITCHWORT   *native / cons –*
   Arenaria michauxii (Fenzl) Hook. f.
   Sabulina michauxii (Fenzl) Dillenb. & Kadereit
Perennial or annual herb, diffuse, completely glabrous or sometimes pu-
bescent, the prostrate branches with numerous short sterile shoots or
the plants merely cespitose. Stems 1-4 dm long, leafy for 1/3 to 2/3 of their

Minuartia michauxii

length, the primary leaves with short, leafy, fascicle-like sterile shoots in their axils. Primary leaves 8-30 mm, usually 10-20 mm long, subulate, somewhat involute, 3-nerved. Inflorescence open, forked; pedicels slender, 5-20 mm long; sepals 3.5-6.5 mm long, broadly lance-shaped, scarious-margined, 3-nerved or rarely 1-nerved; petals oblong lance-shaped, 5-8 mm long, entire. Capsule equaling or somewhat exceeding the sepals, the 3 valves dehiscent from near the middle to the base; seed kidney-shaped, brown-black, low-tuberculate. July-Sept. — Dry woods, sand ridges and dunes.

Petrorhagia
SAXIFRAGE-PINK
Petrorhagia saxifraga (L.) Link
SAXIFRAGE-PINK   *introduced*
Cespitose perennial. Stems decumbent at base, 1-4 dm tall, glabrous or sparsely hispidulous. Leaves linear-subulate, 5-10 mm long, 1 mm wide. Flowers solitary or in cymes; calyx 4-5 mm long, 5-ribbed, closely sub-tended by 2 or 3 pairs of lance-shaped scarious bracts; petals purple to pink, 5-6 mm long, the blades broadly notched; stamens 10; styles 2. Capsule globose-ovoid, dehiscent by 4 teeth; seeds disciform. Summer. — Native of Europe; rarely escaped from cultivation and established mostly as a roadside weed.

Petrorhagia saxifraga

Saponaria
SOAPWORT, BOUNCING-BET
Saponaria officinalis L.
BOUNCING-BET   *introduced*
Perennial herb from a horizontal rhizome and forming colonies. Stems coarse, erect, 4-8 dm tall, simple or branched, glabrous. Leaves 7-10 cm long, 2-4 cm wide, elliptic, elliptic-ovate, glabrous, rarely puberulent. Inflorescence congested and subcapitate to open and oblong-pyramidal, to 15 cm long; primary bracts foliaceous, the ultimate ones scarious. Flowers fragrant, frequently double; calyx 1.5-2.5 cm long, 20-nerved, the lobes long triangular, the tube often deeply bilobed; petals white or pinkish, appendages conspicuous, awl-shaped; stamens exsert; styles 2 (rarely 3). Capsule elliptic-oblong, dehiscent by 4 (rarely 6) teeth; seeds uniformly reticulate. Summer. — Native of the Old World; formerly in cultivation and now commonly weedy on roadsides and waste places.

Name from the Latin, *sapo*, soap, alluding to the mucilaginous juice which forms a lather with water.

Saponaria officinalis

Silene
CATCHFLY, CAMPION
Annual or perennial herbs. Leaves opposite, entire. Inflorescence simple or branched, sometimes reduced to a few-flowered or 1-flowered cyme. Flowers perfect or sometimes unisexual. Calyx sometimes inflated. Petals 5, the claw narrow, expanded distally into more or less prominent auricles, provided ventrally with a pair of appendages; blade usually exsert, entire, lobed or dissected. Stamens normally 10. Styles normally 3, sometimes 4, or occasionally 5. Ovary usually stipitate. Capsule 3-celled (or 1-celled), dehiscent normally by 6 teeth. Seeds kidney-shaped to globose, sometimes covered with small bumps.

1 Calyx not glandular or inflated . . . . *S. coronaria*
1 Calyx glandular, often inflated 2
2 Styles 5 . . . . *S. latifolia*
2 Styles 3 3
3 Calyx hairy or glandular . . . . *S. noctiflora*

3 Calyx glabrous 4
4 Corolla pink . . . . *S. armeria*
4 Corolla white 5
5 Annual; calyx tight around capsule . . . . *S. antirrhina*
5 Perennial; calyx somewhat inflated . . . . *S. vulgaris*

Silene antirrhina L.

SLEEPY CATCHFLY   *native / cons* **2**

Annual, glabrous or more or less puberulent. Stems 2-8 dm tall, simple or branched, erect or sometimes decumbent, usually with glutinous zones below the upper nodes. Basal leaves oblong lance-shaped; stem leaves oblong lance-shaped to linear, usually 3-6 cm long, 2-12 mm wide, glabrous or puberulent, the margins ciliate near the base. Inflorescence open, strict or with divaricate branches. Flowers numerous, rarely few or solitary; calyx 4-10 mm long, 10-nerved; petals white or pink, equaling or exceeding the calyx, frequently obsolete, 2-lobed; appendages minute or lacking. Capsule ovoid, 4-10 mm long, 3-celled; seeds with 3 or 4 rows of dorsal papillae. Summer. — In waste places or sandy soil.

Silene antirrhina

Silene armeria L.

NONE-SO-PRETTY   *introduced*

Atocion armeria (L.) Raf.

Annual with glabrous or rarely sparsely puberulent stems about 3 dm tall. Leaves ovate lance-shaped; basal leaves 2-5 cm long, 5-15 mm wide, sessile; upper leaves clasping. Inflorescence simple or open and compound, the ultimate cymes congested; calyx tubular, 13-17 mm long, 10-nerved; corolla pink or lavender; auricles lacking; appendages linear, 2-3 mm long; blades 4-7 mm long, obovate. Capsule nearly completely 3-celled; seeds 0.6 mm wide, rugose; carpophore (the stalk supporting the fruit) 7-8 mm long. June-July. — Native of Europe; once popular in cultivation; escaped as a weed in waste places.

Silene coronaria (Desr.) Clairv. ex Rchb.

ROSE CAMPION   *introduced*

Lychnis coronaria (L.) Desr.

Perennial, gray-tomentose, the stout stems rarely branched, 4-8 dm tall. Basal leaves 5-10 cm long, 1-3 cm wide; stem leaves 5-10 pairs, lance-shaped to oblong lance-shaped, usually smaller than the basal. Inflorescence few-flowered, the pedicels 5-10 mm long; calyx 12-15 mm long, 10-ribbed, the lobes narrowly lance-shaped, 4-7 mm long, connivent, twisted; petals crimson, 2-3 cm long, without auricles; appendages narrowly lance-shaped, 1.5-2.5 mm long, the blades broadly obovate, 10-15 mm long. Capsules 12-16 mm long, ovoid, dehiscent by 5 teeth. June-Aug. — Native of Europe; escaped from cultivation and established in many places.

Silene coronaria

Silene latifolia Poir.

WHITE CAMPION   *introduced*

Lychnis alba Mill.

Dioecious annual or perennial from a stout root. Stems 4-12 dm tall, coarsely pubescent, glandular above. Stem leaves as many as 10 pairs, lance-shaped to broadly elliptic, 3-10 cm long, 1-4 cm wide, 3-5-nerved, puberulent to hirsute, the lower petiolate, the upper sessile. Inflorescence usually much branched, the primary bracts foliar; flowers white, characteristically unisexual, opening in the evening; calyx 15-20 mm long, tubular in anthesis, becoming distended at maturity, 10-nerved in the staminate flower, 20-nerved in the pistillate, the lobes lance-shaped, 3-5 mm long; petals 2-4 cm long, the claw exsert, auriculate, the appendages 1-1.5 mm long, erose, the blade deeply bilobed. Capsule ovoid, 10-15 mm

Silene latifolia

long, dehiscent by 10 erect or spreading teeth. Summer. — Native of Europe; a common weed.

Silene noctiflora L.

NIGHT-FLOWERING CATCHFLY *introduced*

Annual, 2–8 dm tall, with simple or branched, coarsely hirsute stems and leaves. Leaves ovate lance-shaped, 5–12 cm long, 2–4 cm wide; basal leaves somewhat narrowed to a petiole; stem leaves narrower, sessile. Inflorescence loosely branched; flowers often unisexual; calyx about 15 mm long in anthesis, in fruit 2.5–5 cm long, the 10 nerves glandular, the lobes linear lance-shaped, 5–9 mm long; corolla white or pink; auricles 1–1.5 mm long; appendages broad, 0.5–1.5 mm long, entire or erose; blades 7–10 mm long, deeply 2-lobed. Capsule 3-celled; seeds 0.8–1 mm wide, uniformly rugose-papillate; carpophore 1–3 mm long. July–Sept. — Native of Europe; disturbed places such as roadsides, railways, fields. Plants superficially resemble *S. latifolia* which normally has 5 styles.

Silene vulgaris (Moench) Garcke

BLADDER-CAMPION *introduced*

Robust perennial. Stems 2–8 dm tall, from a creeping rhizome, usually glabrous and glaucous, often decumbent. Leaves ovate lance-shaped, 3–8 cm long, 1–3 cm wide, abruptly acuminate, sometimes ciliolate; stem leaves often clasping. Inflorescence open, 5–30-flowered; calyx campanulate, 1 cm long, papery in texture with reticulate veins, in fruit becoming much inflated; corolla white; auricles lacking; appendages inconspicuous or lacking; blades 3.5–6 mm long, deeply bilobed. Capsule 3-celled; seeds 1–1.5 mm, warty; carpophore 2–3 mm long. Summer. — Native of Europe, weedy in waste places.

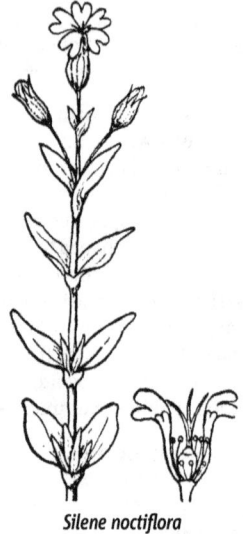

Silene noctiflora

Spergula

SPURRY

Spergula arvensis L.

CORN SPURRY *introduced*

Fleshy annual herb. Stems simple or much-branched, to 40 cm long, sparingly glandular-puberulent. Leaves whorled, 2–5 cm long, narrowly linear, clustered at the nodes in two opposite sets of 6–8; stipules small, connate. Inflorescence terminal, dichotomously branched, the pedicels reflexed; sepals 5, ovate, 2–3 mm long, obtuse, glandular-puberulent; petals 5, white, shorter than or somewhat surpassing the sepals; stamens 10 or sometimes 5; styles normally 5. Capsule broadly ovoid, somewhat longer than the sepals; seeds 1–1.5 mm wide, blackish, minutely roughened. May–Aug. — Native of Europe; a weed of cultivated ground and waste places.

Spergula arvensis

Stellaria

CHICKWEED

Low, spreading or erect perennials (ours), mostly without hairs. Stems slender, 4-angled. Flowers single in forks of stems or in few-flowered clusters at ends of stems; sepals green with translucent margins; petals white, lobed or deeply cleft (sometimes absent in *S. borealis*); stamens 10 or less; styles 3. Fruit an ovate or oblong capsule.

1 Plants large, the stems to 8 dm long; styles 5 .... *S. aquatica*
1 Plants smaller; styles 3–4 2
2 Leaves wider, not linear, mostly more than 1 cm wide .... *S. media*
2 Leaves narrow, linear or lance-shaped, less than 1 cm wide 3
3 Flowers single in forks of stems .... *S. borealis*
3 Flowers in branched cymes 4
4 Petals usually absent, or shorter than the sepals .... *S. borealis*

4   Petals much longer than sepals .... *S. graminea*

Stellaria aquatica (L.) Scop.
GIANT CHICKWEED   *introduced*
Myosoton aquaticum (L.) Moench
Perennial herb, spreading by rhizomes. Stems sprawling and matted, to
8 dm long, rooting at nodes, covered with gland-tipped hairs. Leaves
ovate to lance-shaped, 2-8 cm long and 1-4 cm wide, petioles short or absent. Flowers in open, leafy clusters at ends of stems; sepals 5-9 mm long;
petals white, much longer than sepals. Fruit a capsule; seeds 0.8 mm
long, covered with small bumps. June-Oct. — Streambanks, ponds, wet
or moist disturbed areas, often in partial shade.

Stellaria aquatica

Stellaria borealis Bigelow
NORTHERN STITCHWORT   *native / cons* 10
Perennial herb, spreading by rhizomes. Stems sprawling, to 5 dm long,
branched, angled. Leaves lance-shaped, narrowed at base, 1-5 cm long
and 2-8 mm wide, margins hairy. Flowers in clusters at ends of stems;
sepals 2-4 mm long; petals usually absent. Fruit a dark capsule, longer
than sepals; seeds 0.8 mm long, with very small bumps. June-Aug. —
Openings and hollows in conifer forests, margins of ponds and marshes.

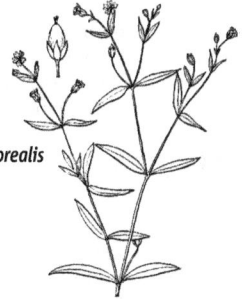

Stellaria borealis

Stellaria graminea L.
GRASS-LEAF STITCHWORT   *introduced*
Perennial herb. Stems 3-5 dm long, weak, 4-angled, the angles prominent
and sometimes scabrous. Leaves 1.5-5 cm long, 1.5-7 mm wide, linear to
linear lance-shaped, the base often obtuse and ciliate. Inflorescence terminal, many-flowered, diffuse, frequently extending to the middle of the
stem; bracts scarious, ciliolate; pedicels slender, spreading or reflexed;
sepals lance-shaped, in fruit 4.5-5.5 mm long, strongly 3-nerved, the margins scarious, commonly ciliolate, at least at the base; petals exceeding
the sepals. Capsule straw-colored, more or less equaling or somewhat
surpassing the sepals; seeds 0.8-1.2 mm long, kidney-shaped, dark reddish
brown, covered with small bumps. May-July. — Native of Europe; introduced in grassy places, fields, roadsides, and waste land.

Stellaria longifolia Muhl.
LONG-LEAVED STITCHWORT   *native / cons* 5
Perennial herb. Stems sprawling, prominently 4-angled, usually freely
branched, 1-5 dm long. Leaves spreading to ascending, linear to lance-
shaped, 2-5 cm long and 1-6 mm wide, widest at or above middle, tapered
at both ends. Flowers in branched clusters at ends of stems; sepals 3-5
mm long; petals longer than sepals. Fruit a green-yellow to brown capsule,
usually longer than the sepals; seeds light brown, about 1 mm long. May-
July. — Wet meadows and marshes, shrub thickets, swamps, streambanks,
pond margins.

Stellaria longifolia

Stellaria media (L.) Vill.
COMMON CHICKWEED   *introduced*
Weakly tufted annual. Stems to 4 dm long, with ascending branches, puberulent in lines. Leaves usually 1-3 cm long, ovate or obovate, glabrous,
frequently pustulate; upper leaves sessile; lower leaves with petioles that
may exceed the length of the blade, often ciliate toward the base or on
the petioles. Flowers solitary or in few-flowered, terminal, leafy cymes;
pedicels ascending, reflexed, frequently pubescent; sepals oblong, 3.5-6
mm long, pubescent and pustulate; petals shorter than the sepals; stamens 3-5. Capsule ovoid, somewhat surpassing the sepals; seeds 1-1.2
mm long, suborbicular, reddish brown, conspicuously covered with small

Stellaria media

bumps. — Introduced from the Old World but often appearing to be native; now a cosmopolitan weed of waste places, cultivated areas, meadows, and woodlands.

Celastraceae

BITTERSWEET FAMILY

Shrubs (*Euonymus*), vines (*Celastrus*), or glabrous perennial herbs (*Parnassia*) with simple, evergreen or deciduous, opposite or alternate leaves, and small, axillary or terminal, solitary or clustered flowers. Flowers perfect or unisexual, regular, polypetalous, usually 4-5-merous. Stamens as many as the petals and alternate with them; in Parnassia, staminodes (infertile stamens) attached to base of petals and divided into threadlike segments tipped with glandular knobs. Pistil 1, inserted on or surrounded by the disk; ovary 2-5-celled. Fruit a capsule.

Celastraceae now includes members of genus *Parnassia*.

1 Perennial herbs; leaves basal with a single stem leaf . . . . *Parnassia*
1 Shrubs or twining vines; leaves cauline, alternate or opposite. 2
2 Twining vines; leaves alternate . . . . *Celastrus*
2 Shrubs; leaves opposite . . . . *Euonymus*

Celastrus

BITTERSWEET

Celastrus scandens L.

AMERICAN BITTERSWEET   *native / cons* 3

Woody twiner climbing several meters high. Leaves deciduous, alternate, elliptic or oblong to ovate, acuminate, serrulate, 5-10 cm long. Panicles terminal, 3-8 cm long. Flowers dioecious or polygamo-dioecious, small, whitish or greenish, 5-merous. Staminate flowers with 5 stamens about as long as the petals, inserted on the margin of the cup-shaped disk. Pistillate flowers with rudimentary stamens and a well developed ovary, stout columnar style, and 3-lobed stigma. Fruit 3-valved, each valve covering 1 or 2 seeds enclosed in a fleshy orange aril, several in a cluster, subglobose, nearly 1 cm long. May-June. — Roadsides and thickets, usually in rich soil; occasionally cultivated and now established in open woods and thickets.

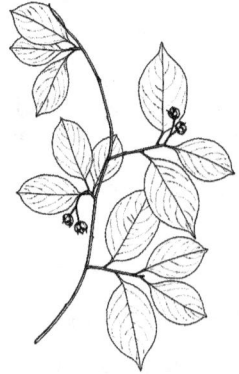
Celastrus scandens

Euonymus

SPINDLETREE

Euonymus alatus (Thunb.) Sieb.

BURNING BUSH   *introduced (invasive)*

Shrub to 2.5 m; twigs with 2-4 conspicuous corky wings. Leaves opposite, subsessile, elliptic to obovate, to 6 cm long, sharply serrulate, turning bright red in autumn, then deciduous. Flowers perfect, 4-merous, green, 6-8 mm wide, solitary or cymose in the leaf-axils. Fruit smooth, purplish; aril orange, completely covering the seed. May-June. — Native of e Asia, cultivated and locally escaped.

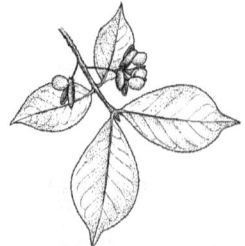
Euonymus alatus

Parnassia

GRASS-OF-PARNASSUS

Glabrous perennial herbs. Leaves all from base of plant but often with 1 stalkless leaf near middle of stalk, margins entire; petioles present. Flowers large, white, single at ends of stalks; calyx 5-lobed; petals white, veined, spreading; fertile stamens 5, alternating with petals; staminodes (infertile stamens) attached to base of petals and divided into threadlike segments tipped with glandular knobs; stigmas 4. Fruit a 4-chambered capsule with numerous seeds.

1 Sepals with narrow translucent margins; staminodes (sterile stamens) 3-parted, not widened at base; petals 12–16 mm long; leaves leathery and somewhat succulent . . . . *P. glauca*
1 Sepal margins green; staminodes 5 to many-parted; petals 5–13 mm long; leaves thin and membranous . . . . *P. parviflora*

Parnassia glauca Raf.

FEN GRASS-OF-PARNASSUS   *native / cons* **8**

Smooth perennial herb. Leaves from base of plant and usually with 1, more or less sessile stem leaf; broadly ovate to nearly round, 2-7 cm long and 1-5 cm wide; margins entire; petioles long. Flowers single atop a stalk 1-4 dm long; sepals ovate, 2-5 mm long, with a narrow, translucent margin; petals white with green veins, 1-2 cm long; staminodes 3-parted from near base, shorter than to equal to stamens. Fruit a capsule about 1 cm long. Aug-Sept. — Calcareous fens and wet meadows.

Parnassia parviflora DC.

SMALL-FLOWERED GRASS-OF-PARNASSUS   *endangered / native / cons* **10**

Smooth perennial herb. Leaves from base of plant, with a sessile, non-clasping leaf near middle of stalk; oval to ovate, 1-3 cm long, narrowed to base; margins entire; petioles present. Flowers single atop slender stalks 0.5-3 dm long; sepals ovate, 3-7 mm long, nearly as long as petals, green throughout; petals white with green veins, oval, 5-10 mm long; staminodes 5-7-parted near the slightly widened tip, shorter than the stamens. Fruit a capsule. July-Aug. — Lake Michigan shoreline in cracks in wet limestone pavement or on open, moist, sandy beaches and dunes.

Parnassia glauca

Ceratophyllaceae

HORNWORT FAMILY

Ceratophyllum

COON'S-TAIL, HORNWORT

Ceratophyllum demersum L.

COON'S-TAIL   *native / cons* **3**

Aquatic perennial herb, often forming large patches; roots absent, but plants usually anchored to substrate by pale, modified leaves. . Stems long, branched. Leaves in whorls of 5-12 at each node, stiff, 1-3 cm long, 1-2-forked; leaf segments linear, 0.5-1 mm wide, coarsely toothed. Flowers small, inconspicuous in leaf axils, staminate and pistillate flowers separate on same plant, staminate usually above pistillate on stems. Fruit an oval achene, 4-6 mm long, with 2 spines at base. — Shallow to deep water of lakes, ponds, backwater areas, ditches; water typically neutral or alkaline.

Our only genus of aquatic vascular plants with whorled, forked leaves.

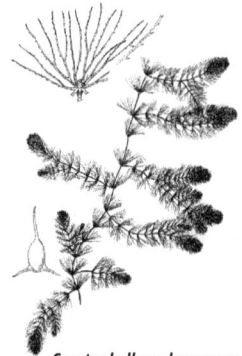

Ceratophyllum demersum

Cistaceae

ROCK-ROSE FAMILY

Crocanthemum

FROSTWEED

Crocanthemum bicknellii (Fernald) Janch.

HOARY FROSTWEED   *native / cons* **6**

Helianthemum bicknellii Fern.

Perennial herb. Stems solitary or few together, erect or nearly so, simple or sparsely branched at first anthesis, later producing numerous floriferous branches from the upper axils. Leaves linear-oblong to oblong lance-shaped, 2-3 cm long on the main axis, much smaller on the branches. Petaliferous flowers 5-12 in a loose terminal raceme, 15-25 mm wide; sepals densely stellate, the outer nearly or quite as long as the inner. Petals yellow, 5 in the first flowers of the season, soon deciduous; petals mostly absent in the later flowers; stamens numerous. Capsules of the petaliferous flowers 4-5 mm wide. Apetalous flowers numerous and crowded on short axillary branches, their capsules smaller, about 2 mm wide, strongly triquetrous; seeds minutely net-veined. June–July. — Dry, usually sandy soil.

Crocanthemum bicknellii

Convolvulaceae

MORNING-GLORY FAMILY

Herbs (ours), often twining, with alternate simple leaves and small to very large flowers. Flowers regular, perfect, mostly 5-merous. Sepals usually distinct to the base, imbricate, often of unequal size. Corolla rotate, funnelform, salverform, or tubular, entire or deeply to shallowly lobed. Stamens as many as the corolla lobes and alternate with them, inserted near the base of the corolla. Ovary usually 2-3-celled; styles 1 or 2; stigmas linear to capitate. Fruit in most genera a capsule.

1 Stigma 1, expanded into a head-like tip . . . . *Ipomoea*
1 Stigmas 2, linear and not expanded at tip 2
2 Bracts leaf-like, attached just below the calyx and nearly concealing it . . . . *Calystegia*
2 Bracts small, attached much below the calyx . . . . *Convolvulus*

Calystegia

BINDWEED

Much like *Convolvulus*, but the bracts usually large, inserted just beneath the calyx, and more or less concealing it; ovary more or less 1-chambered, the partition incomplete; stigmas oblong, cylindric, blunt; flowers usually solitary; ours rhizomatous perennials.

1 Petioles of leaves subtending flowers with petiole more than half length of blade midvein *C. sepium*
1 Petioles of leaves subtending flowers with petiole much less than half length of midvein . . . .
. . . . *C. spithamaea*

Calystegia sepium (L.) R. Br.

HEDGE-BINDWEED   *native / cons 2*

Stems twining or occasionally trailing, to 3 m long. Leaves long-petioled, triangular to oblong in outline, hastate or sagittate, 5-10 cm long, 1/4 to 3/4 as wide. Peduncles 5-15 cm long; bracts ovate or oblong, 1-2 cm long, nearly cordate at the base; corolla pink or white, 4-7 cm long. — Waste places, roadsides, fencerows, especially where disturbed.

Calystegia spithamaea (L.) Pursh

LOW BINDWEED   *native / cons 4*

Stems erect, at least to and including the flowering portion, the remainder often elongating and eventually declined. Leaves obovate-oblong, 3-8 cm long, acute to obtuse or rounded, at base rounded, truncate, or cordate, always more or less pubescent. Peduncles few, 2-8 cm long, produced 5-20 cm above the base of the plant; bracts oblong or ovate, very rarely cordate at base; corolla white or pink, 4-7 cm long. May–July. — Dry rocky or sandy soil, fields and open woods.

Calystegia sepium

Convolvulus
BINDWEED

Convolvulus arvensis L.
FIELD BINDWEED   *introduced (invasive)*
Perennial, deeply rooted herbs. Stems trailing or climbing, to 1 m long, often forming dense tangled mats. Leaves variable, triangular to oblong in outline, 2-5 cm long, cordate-ovate or hastate, glabrous or finely pubescent, the basal lobes spreading or descending. Flowers borne mostly 1-2 together on axillary peduncles exceeding the subtending leaves; bracts borne 5-20 mm below the flower; sepals elliptic, 3-5 mm long; corolla funnelform, usually white, sometimes pink, 15-20 mm long; stamens included, inserted near the base of the corolla. Capsule globose, 2-4-celled. May-Sept. — Native of Europe; naturalized in fields, roadsides, and waste places; often a troublesome weed.

Convolvulus arvensis

Ipomoea
MORNING-GLORY

Ipomoea purpurea (L.) Roth
COMMON MORNING-GLORY   *introduced*
Annual. Stems pubescent, twining to 5 m high. Leaves rounded-cordate, entire (very rarely 3-lobed), abruptly short-acuminate or merely apiculate, glabrous to sparsely pubescent. Peduncles nearly or quite equaling the subtending leaves, 1-5-flowered; sepals imbricate, ovate lance-shaped to oblong, acute or acuminate, hirsute toward the base, commonly 10-15 mm long; corolla funnelform or campanulate, the margin entire or shallowly lobed, blue, purple, white, or variegated, 4-6 cm long; ovary 3-celled; style 1, stigmas 3. Fruit a capsule. July-Sept. — Native of tropical America; formerly commonly cultivated; now escaped into fields, roadsides, and waste places; often a pernicious weed.

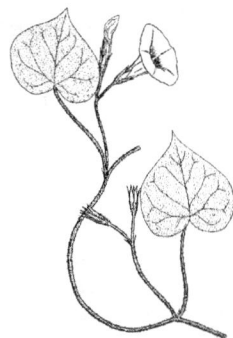

Ipomoea purpurea

Cornaceae
DOGWOOD FAMILY
Shrubs (ours). Flowers 4- or 5-parted. Fruit a drupe.

Cornus
DOGWOOD
Shrubs, or herbaceous shoots from a woody rhizome in bunchberry (*Cornus canadensis*). Leaves mostly opposite, sometimes alternate or whorled, simple, entire. Flowers in a rounded or flat-topped cluster, 4-parted, sepals and petals small. Fruit a berrylike drupe with 1-2 hard seeds.

1 Plants herbaceous from a woody rhizome, less than 3 dm tall; leaves whorled .... *C. canadensis*
1 Taller shrubs, 5 dm or more tall; leaves opposite or alternate 2
2 Leaves alternate on stems .... *C. alternifolia*
2 Leaves opposite 3
3 Twigs yellow or yellow-green with purple spots; leaves round in outline or nearly so ..... *C. rugosa*
3 Twigs not yellow, or if yellow not spotted; leaves longer than wide 4
4 Fruit white; young twigs densely short-hairy .... *C. obliqua*
4 Fruit blue; young twigs smooth or nearly so .... *C. alba*

Cornus alba L.
RED OSIER-DOGWOOD   *native / cons* **3**
   Cornus sericea L.
   Cornus stolonifera Michx.
Many-stemmed shrub, 1-3 m tall, forming thickets; branches upright or prostrate and rooting; twigs and young branches red; pith white. Leaves

opposite, green, ovate to oval, mostly 5-15 cm long and 2-7 cm wide, tapered to a tip, soft hairy on underside; margins entire; petioles to 2.5 cm long. Flowers small, white, many in flat-topped or slightly rounded clusters. Fruit a round, white or blue-tinged, berrylike drupe, 6-9 mm wide. May-Aug. — Swamps, marshes, shores, streambanks, floodplain forests, shrub thickets, calcareous fens; also on sand dunes.

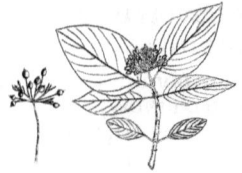

Cornus alba

Cornus alternifolia L. f.
PAGODA DOGWOOD, ALTERNATE-LEAF DOGWOOD   *native / cons* 7
  Swida alternifolia (L. f.) Small
Shrub, to 5 m tall; twigs red-green or brown, somewhat shiny, alternate on stems, pith white. Leaves alternate, sometimes crowded and appearing whorled near ends of stems, oval to ovate, 5-12 cm long and 3-7 cm wide, tapered to a sharp tip, underside finely hairy; lateral veins 4-5 pairs, these curving toward tip of blade; margins entire; petioles to 5 cm long. Flowers small, creamy-white, in crowded, flat-topped or rounded clusters at ends of stems. Fruit a round, blue, berrylike drupe, 6 mm wide, atop a red stalk. May-July. — Swamps, thickets, streambanks and springs; also in drier deciduous and mixed forests.

Cornus alternifolia

Cornus canadensis L.
BUNCHBERRY, DWARF CORNEL   *native / cons* 7
  Chamaepericlymenum canadense (L.) Graebn.
Perennial from horizontal, woody rhizomes, often forming large colonies. Stems erect, green, 1-2 dm tall, with a pair of small bracts on lower stem, topped with a whorl-like cluster of 4-6 leaves. Leaves oval to obovate, 4-7 cm long, tapered at both ends; lateral veins 2-3 pairs, arising from midvein below middle of blade; margins entire; petioles short or absent. Flowers small, yellow-green or creamy-white in a single cluster at end of a stalk 1-3 cm long; flowers surrounded by 4 white or pinkish, petal-like showy bracts, 1-2 cm long, these soon deciduous. Fruit a cluster of round, bright red berrylike drupes, the drupes 6-8 mm wide. June-July. — Cedar swamps, thickets and moist conifer forests, often on hummocks or rotting logs; also in drier, mixed conifer-deciduous forests.

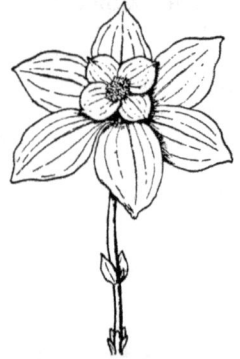

Cornus canadensis

Cornus obliqua Raf.
SILKY DOGWOOD   *native / cons* 4
  Cornus amomum subsp. *obliqua* (Raf.) J.S. Wilson
Shrub, 1-3 m tall; older branches red and gray-streaked, young twigs gray, finely hairy; pith brown. Leaves opposite, oval to ovate, 5-12 cm long and 2-5 cm wide, usually less than half as long, tapered to a sharp tip, lateral veins 4-6 on each side, underside finely hairy; margins entire; petioles 1-2 cm long, often curved and causing the leaves to droop. Flowers small, creamy-white, in flat-topped or slightly rounded, hairy clusters. Fruit a round, blue or blue-white, berrylike drupe, 8 mm wide, atop a long stalk. June-July (our latest flowering dogwood). — Conifer swamps, marshes, open bogs, calcareous fens, lakeshores, streambanks, wet dunes.

Cornus obliqua

Cornus rugosa Lam.
ROUND-LEAF DOGWOOD   *native / cons* 7
Shrub 1-3 m tall, the younger branches yellowish green, often shaded or mottled with red; pith white. Leaves ovate to rotund, usually 7-12 cm long, abruptly acuminate, broadly cuneate or usually rounded at base, minutely scaberulous above, softly white-pubescent beneath with erect, curled or curved hairs usually 0.5-1 mm long; lateral veins 7 or 8 on each side. Inflorescence flat or slightly convex. Drupes light blue, about 6 mm wide. May-July. — Moist or dry, sandy or rocky soil.

Cornus rugosa

Crassulaceae

STONECROP FAMILY

Sedum
STONECROP

Succulent perennial herbs. Leaves thick or terete, alternate, opposite, or whorled. and small or medium-sized, white to flowers. Flowers yellow or red-purple, 4-5-merous. Sepals united at base. Petals separate, often lance-shaped. Stamens 8 or 10, the epipetalous ones usually adnate to the petals at base. Nectarial scales short, at the base of the ovaries. Pistils distinct or nearly so, tapering into a stout style. Fruit a group of follicles; seeds numerous, small.

1   Petals deep pink .... *S. purpureum*
1   Petals yellow .... *S. acre*

Sedum acre L.
MOSSY STONECROP   *introduced*

Perennial from creeping stems, forming dense mats. Flowering stems 5-10 cm long. Leaves crowded, imbricate, terete, ovoid, 2-5 mm long, blunt. Inflorescence of a few short branched cymes. Flowers yellow, 8-10 mm wide; petals lance-shaped, spreading, about twice as long as the sepals. June-July. — Native of Eurasia; cultivated and escaped in dry, sandy soil.

Sedum purpureum (L.) J.A. Schultes
LIVE-FOREVER   *introduced*
     Hylotelephium telephium (L.) H. Ohba.
     Sedum telephium subsp. *purpureum* (L.) Schinz & Keller

Perennial from a thick caudex. Stems stout, erect, 3-6 dm tall. Leaves fleshy, green or bluish green, not glaucous, oblong to obovate, 3-6 cm long, commonly with several to many conspicuous irregular teeth, varying to entire. Inflorescence repeatedly branched, convex to hemispheric, the branchlets narrowly winged. Flowers densely crowded, red-purple, 6-8 mm wide; sepals triangular, about a third as long as the ovate lance-shaped petals. Late summer. — A highly variable Eurasian species, long cultivated for ornament and occasionally escaped.

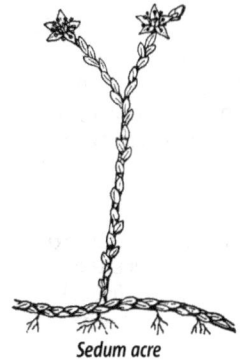

Sedum acre

Cucurbitaceae

CUCUMBER FAMILY

Annual or perennial vines, trailing or climbing by tendrils, with small to large, mostly white or yellow or greenish flowers, and simple, alternate, often lobed leaves. Flowers monoecious or dioecious, regular. Calyx 4-6 (usually 5)-lobed, sometimes to the very base of the tube. Stamens 1-5, distinct or wholly or partly united, usually 3. Ovary 1- or 3-celled; styles united, with a thick stigma. Fruit a dry or fleshy pepo, few-many-seeded.

ADDITIONAL SPECIES

     Several species of the Old World have been cultivated since antiquity, of which *Citrullus lanatus* (Thunb.) Matsum. & Nakai (watermelon) and *Cucumis melo* L. (muskmelon or canteloupe) occur in Wisconsin. Native of America and of aboriginal cultivation here are *Cucurbita pepo* L., (pumpkin), and *Cucurbita maxima* Duchesne (squashes). These species may grow in waste ground, especially where their seeds have been discarded, but none are long-persistent.

Echinocystis
WILD CUCUMBER

Echinocystis lobata (Michx.) Torr. & Gray
WILD CUCUMBER   *native / cons* **2**
Annual vining herb, to 5 m or more long. Leaves round in outline, with 3-

7 (usually 5) sharp, triangular lobes; petioles 3-8 cm long. Flowers white; staminate flowers 8-10 mm wide, with lance-shaped lobes, in long, upright racemes; pistillate flowers 1 to several on short stalks from leaf axils. Fruit green, ovate, inflated, 3-5 cm long, with soft prickles. Aug-Sept. — Floodplain forests, wet deciduous forests, streambanks, thickets, and waste ground.

Echinocystis lobata

Dipsacaceae
TEASEL FAMILY

Dipsacus
TEASEL

Coarse, tall, biennial or perennial herbs, little branched, with prickly stems. Leaves large, sessile or connate. Flowers small, in dense ovoid to cylindric heads subtended by a many-leaved involucre. Calyx short, 4-angled or 4-lobed. Corollas 4-lobed, the marginal ones not enlarged. Bracts of the involucre linear, often elongate. Receptacular bracts ovate or lance-shaped, acuminate into an awn surpassing the flowers. Involucel 4-angled, truncate or 4-toothed at the summit. True calyx very short, hairy, without appendages. Family included in Caprifoliaceae in the 2009 Angiosperm Phylogeny Group III system.

Dipsacus laciniatus L.
CUT-LEAF TEASEL   *introduced*
Stem leaves irregularly pinnately lobed or divided, more or less bristly. July-Aug. — Native of Europe; reported for Door County on a sunny roadside.

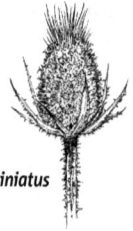

Dipsacus laciniatus

Droseraceae
SUNDEW FAMILY

Drosera
SUNDEW

Perennial herbs. Leaves all from base of plant, covered with stalked, sticky glands that trap and digest insects. Flowers white, several, on 1 side of erect, leafless stalks, the stalks nodding at tip; with 5 petals and 5 sepals; stamens mostly 5, styles 3. Fruit a dry, many-seeded capsule.

1   Leaves widely spreading, the blades round, wider than long .... *D. rotundifolia*
1   Leaves upright, blades linear, longer than wide .... *D. linearis*

Drosera linearis Goldie
SLENDER-LEAF SUNDEW   *threatened / native / cons* 10
Perennial insectivorous herb. Leaf blades linear, 2-5 cm long and 2 mm wide; petioles smooth, 3-7 cm long. Flowers 1-4 atop stalks 6-15 cm tall; flowers 6-8 mm wide; sepals 4-5 mm long; petals obovate, 6 mm long, white. Seeds black, less than 1 mm long, with small craterlike pits on surface. June-Aug. — Calcareous fens, wet areas between dunes near Lake Michigan; rarely in sphagnum moss.

Drosera rotundifolia L.
ROUND-LEAF SUNDEW   *native / cons* 7
Small, perennial insectivorous herb. Leaf blades more or less round, wider than long, 2-10 mm long and as wide or wider, covered with long, red, gland-tipped hairs; abruptly tapered to a petiole longer than blade; peti-

Drosera linearis

oles 2-5 cm long covered with gland-tipped hairs. Flowers 2-15 in a more or less 1-sided, racemelike cluster, on a leafless stalk 10-30 cm tall; flowers 4-7 mm wide, sepals 5, 4-5 mm long; petals white to pink, longer than sepals; stamens 5, shorter than petals. Seeds light brown, shiny and with fine lines, 1-1.5 mm long. July-Aug. — Swamps and open bogs, usually in sphagnum; wet sandy shores and openings.

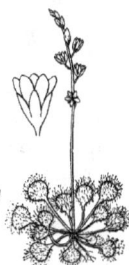

Drosera rotundifolia

Elaeagnaceae

OLEASTER FAMILY

Shrubs or trees. Leaves opposite or alternate, covered with small scales (lepidote). Flowers small, solitary or clustered, perfect or unisexual, regular, usually 4-merous. Hypanthium in the staminate flowers saucer-shaped to cup-shaped, in perfect or pistillate flowers prolonged into a short or elongate tube and persistently enclosing the ovary. Sepals present, often colored. Petals none. Stamens inserted at or just below the summit of the hypanthium, as many or twice as many as the sepals. Ovary 1-celled. Fruit drupe-like, the dry mature ovary enclosed by the base of the hypanthium.

1 Leaves alternate; stamens 4 . . . . *Elaeagnus*
1 Leaves opposite; stamens 8 . . . . *Shepherdia*

Elaeagnus
RUSSIAN-OLIVE, SILVER-BERRY

Elaeagnus umbellata Thunb.
AUTUMN OLIVE　　　　　　　　　　　　　　　*introduced (invasive)*
Shrubby tree to 5 m tall. Leaves alternate, soon green and glabrescent above; hypanthium tube about twice as long as the sepals. Fruit red, finely dotted with pale scales, juicy and edible, 6-8 mm wide, on pedicels about 1 cm long. May-June. — Native of e Asia, originally introduced as an ornamental shrub and for wildlife habitat, now spreading to many dry and wet habitats.

Shepherdia
BUFFALO-BERRY
Elaeagnus umbellata

Shepherdia canadensis (L.) Nutt.
RUSSET BUFFALO-BERRY　　　　　　　　　　*native / cons 7*
Unarmed shrub 1-3 m tall. Leaves opposite, ovate lance-shaped to ovate, varying to narrowly lance-shaped or elliptic, 3-5 cm long, obtuse, obtuse to rounded or subcordate at base, green and nearly glabrous above, densely scaly (lepidote) beneath. Flowers dioecious, in small clusters on twigs of the previous season; staminate flowers 4-6 mm wide; sepals ovate, spreading, much exceedmg the erect stamens; pistillate flowers similar, the mouth of the hypanthium closed by a dense tomentum. Fruit a yellowish red, inedible berry, 5-7 mm long. April-May. — Dry, sandy or stony, calcareous soil.

Shepherdia canadensis

Ericaceae

HEATH FAMILY

Ericaceae now includes former members of Monotropaceae and Pyrolaceae. The traditional Ericaceae are shrubs or scarcely woody shrubs. Leaves evergreen or deciduous, mostly alternate, simple, with entire or toothed margins. Flowers usually perfect (with both staminate and pistillate parts), urn- or vase-shaped, mostly white, pink, or cream-colored; stamens as many (or twice as many) as petals. Fruit a berry or dry capsule.

Former members of Monotropaceae (*Monotropa, Pterospora*) are mycotropic perennial herbs without chlorophyll, variously white to pink, red, purple, yellow or brown in color. Leaves much-reduced, scale-like, alternate. Flowers solitary or in a bracteate raceme, regular, perfect, mostly 4-5-merous; petals distinct or connate into a lobed tube, commonly about the same color as the stem; stamens mostly 6-10, mostly twice as many as the sepals or petals, distinct or shortly connate at base. Fruit a capsule or berry; seeds numerous and tiny.

Former Pyrolaceae (*Moneses, Orthilia, Pyrola*) are perennial herbs or half-shrubs, most dependent on wood-rotting fungi (mycotrophic). Leaves alternate to sometimes opposite or nearly whorled, often shiny, evergreen or deciduous. Flowers perfect, 5-parted, waxy and nodding. Fruit a capsule.

1 Leaves reduced to non-green scales; plants entirely white, yellow, reddish, orange, or maroon 2
1 Leaves green; plants normal green color 3
2 Stems to 2 dm tall; petals free; fruit erect .... ***Monotropa***
2 Stems 3–10 dm tall; petals joined and urn-like; fruit nodding on curved pedicels; rare ... ***Pterospora***
3 Leaves in a basal rosette; plants herbaceous 4
3 Leaves opposite, alternate, or whorled; plants woody (sometimes small subshrubs woody only at the base or prostrate creepers) 6
4 Style ± strongly bent downward, at least 4 mm long; inflorescence a symmetrical raceme .... ***Pyrola***
4 Style straight, short or long; inflorescence various (usually 1-flowered or a 1-sided raceme) 5
5 Inflorescence 1-flowered, the corolla 15–20 mm broad, flat (petals widely spreading); anthers prolonged into a short cylindrical tube below the pore; valves of capsule glabrous; style (not including prominent stigma lobes) 3–5 mm long .... ***Moneses***
5 Inflorescence racemose, the corolla 3–7 mm broad, ± bell-shaped (petals close about reproductive parts); anthers not prolonged into tubes; valves of capsule with cobwebby fibers on the margins when dehiscing; style various (but stigma only very shallowly lobed) .... ***Orthilia***
6 Leaves opposite or whorled; flowers 8–20 mm broad 9
6 Leaves all alternate; flowers in most species less than 8 mm wide 8
7 Leaves coarsely few-toothed; woody only at base, forming colonies by rhizomes .... ***Chimaphila***
7 Leaves entire; true woody, clump-forming shrub .... ***Kalmia***
8 Leaves narrow, linear to linear lance-shaped, more than 7 times longer than wide; margins revolute .... ***Andromeda***
8 Leaves ovate to oblong; margins various 9
9 Leaves less than 7 times longer than wide, dark green and leathery on upper surface, densely covered with woolly rust-red hairs on underside; margins revolute .... ***Rhododendron***
9 Leaves never both revolute and with rust colored hairs on underside 10
10 Leaves scurfy, densely scale-covered (especially on upper surface) .... ***Chamaedaphne***
10 Leaves various but not scurfy 11
11 Fruit fleshy; leaves evergreen or deciduous 12
11 Fruit dry or mealy; leaves evergreen 16
12 Plants trailing; leaves evergreen 13
12 Upright shrubs; leaves deciduous 15
13 Leaves with small bristles; fruit white .... ***Gaultheria***
13 Leaves glabrous; fruit red when ripe 14
14 Wet habitats; leaves small, to 15 mm long .... ***Vaccinium*** (cranberries)
14 Drier habitats; leaves 2 cm long or more .... ***Gaultheria***
15 Leaves with shiny, orange-yellow resinous glands (especially on underside) .... ***Gaylussacia***
15 Leaves without glands .... ***Vaccinium*** (blueberries)
16 Plants with stiff hairs; leaves ovate to broadly elliptic .... ***Epigaea***
16 Plants nearly glabrous or only finely hairy; leaves spatula-shaped, mostly widest above middle .... ***Arctostaphylos***

Andromeda
BOG-ROSEMARY

Andromeda glaucophylla Link
BOG-ROSEMARY native / cons 10
Andromeda polifolia L. var. glaucophylla (Link) DC.
Low upright or trailing shrub, 3-6 dm tall. Stems gray to blackish; twigs brown, with hairs in lines running down stems, or sometimes smooth.

Leaves evergreen and leathery, often blue-green, linear or narrowly oval, 2-5 cm long and 3-10 mm wide, the tip sharp-pointed and tipped with a small spine, the base tapered to the stem or a short petiole, dark green above and whitened below by short stiff hairs,; margins entire and distinctly rolled under. Flowers in drooping clusters at ends of branches, white or often pink, urn-shaped, 5-parted, 5-6 mm long, on curved stalks to 8 mm long. Fruit a rounded capsule to 5 mm wide, the style persistent from indented top of capsule; fruit drooping at first, but erect when mature. May-June. — Sphagnum bogs, black spruce and tamarack swamps.

Andromeda glaucophylla

Arctostaphylos
BEARBERRY

Arctostaphylos uva-ursi (L.) Spreng.
RED BEARBERRY　　　　　　　　　　　　*native / cons* **7**
Shrub, forming low mats to 1 m wide. Stems prostrate, freely branched. Leaves leathery, evergreen, alternate, entire, obovate, 1-3 cm long, obtuse or rounded at tip, tapering to the base. Flowers 5-merous, in short terminal racemes; calyx saucer-shaped, the sepals imbricate, distinct to the base, about 1.5 mm long; corolla white or tinged with pink, 4-6 mm long, the 5 rounded lobes spreading or recurved; stamens 10, much shorter than the corolla; ovary 5-celled, subtended by a 10-lobed disk. Fruit a bright red drupe, dry or mealy, 6-10 mm wide, with 5 bony nutlets. May-June. — Sandy or rocky soil.

Bearberry has a long history of medicinal uses, especially amongst American Indians; it is also the main component of a smoking mix known as kinnikinnick.

Arctostaphylos uva-ursi

Chamaedaphne
LEATHERLEAF

Chamaedaphne calyculata (L.) Moench
LEATHERLEAF　　　　　　　　　　　　*native / cons* **9**
Upright shrub to 1 m tall. Older stems gray, the outer bark shredding to expose the smooth, red inner bark; twigs brown, with fine hairs and covered with small, round scales. Leaves evergreen and leathery, becoming smaller toward ends of flowering branches, oval to oval, 1-5 cm long and 3-15 mm wide, the tip rounded or pointed, brown-green and smooth above, pale brown with a covering of small, round scales below; margins entire or with small rounded teeth; petioles short. Flowers white, urn-shaped or cylindric, in 1-sided, leafy racemes, hanging from axils of reduced leaves near ends of branches; 5-parted, 5-7 mm long, on stalks 2-5 mm long. Fruit a brown, rounded capsule to 6 mm wide, the hairlike style persistent from indented top of capsule; capsules persisting on branches for several years. May-June. — Open bogs, lakeshores and streambanks, often forming low, dense thickets.

Chamaedaphne calyculata

Chimaphila
PRINCE'S PINE

Chimaphila umbellata (L.) W. Bart.
PRINCE'S PINE, PIPSISSEWA　　　　　　*native / cons* **8**
Low, perennial, evergreen half-shrubs, from a creeping rhizome. Stems spreading, the flowering branches erect, 1-3 dm tall. Leaves thick, oblong lance-shaped, 3-6 cm long, acute or mucronate, sharply dentate especially toward the tip, nearly entire below the middle, tapering to a short petiole. Flowers 4-8, 10-15 mm wide, white or pink, corymbose on long peduncles; petals 5, distinct, widely spreading; stamens 10; ovary 5-celled, depressed-

Chimaphila umbellata

globose. Capsule erect, globose, opening from the top downward. June–Aug. — Dry woods, especially in sandy soil.

Epigaea
TRAILING ARBUTUS

Epigaea repens L.
TRAILING ARBUTUS   *native / cons* 7
Prostrate, creeping, evergreen shrub, often dioecious. Stems branched, 2-4 dm long, hirsute. Leaves leathery, alternate, entire, ovate or oblong, 2-10 cm long, apiculate, rounded to cordate at base, more or less pilose, especially when young; on pubescent petioles about half as long as the blade. Flowers pink to white, fragrant, perfect or unisexual, 5-merous, each closely subtended by 2 ovate bracts nearly or quite as long as the calyx, in short, crowded, terminal and axillary spikes 2-5 cm long; sepals distinct to the base, strongly imbricate; corolla salverform, the tube 8-15 mm long, densely pubescent within, the lobes 6-8 mm long; stamens included; ovary 5-celled. Capsule depressed-globose, hirsute, subtended by the persistent calyx and bracts, white-pulpy within. April–May. — Sandy or rocky acid soil.

Epigaea repens

Gaultheria
TEABERRY
Shrubs of diverse aspect, with alternate persistent leaves and usually white flowers in racemes or particles or (in our species) solitary in or just above the axils and closely subtended by 2 bracteoles. Flowers 4-5-merous. Calyx campanulate to saucer-shaped, deeply divided. Corolla tubular to campanulate, shallowly lobed. Stamens included. Ovary 4-5-celled. Capsule thin-walled, completely enclosed in the fleshy, white or colored, expanded calyx, forming a dry or mealy berry.

1 Leafy stems upright; flowers 5-parted; berries red .... *G. procumbens*
1 Leafy stems prostrate; flowers 4-parted; berries white .... *G. hispidula*

Gaultheria hispidula (L.) Muhl.
CREEPING SNOWBERRY   *native / cons* 8
Low, creeping, matted shrub. Stems 2-4 dm long, covered with brown hairs. Leaves crowded, evergreen, oval to nearly round, 4-10 mm long and to 5 mm wide, abruptly tapered to tip, green above, underside paler, with brown, bristly hairs; margins rolled under; petioles short. Flowers few, single in leaf axils, white, bell-shaped, 4-parted, 2-4 mm long, on curved stalks 1 mm long. Fruit a translucent, juicy, white berry 5-10 mm wide, slightly wintergreen-flavored. May–June. — Open bogs, swamps, wet conifer woods, often in moss on hummocks or downed logs.

Gaultheria procumbens L.
WINTERGREEN   *native / cons* 6
Low, creeping shrub. Leafy stems erect from a horizontal rhizome, 1-2 dm tall, bearing a few leaves crowded near the summit. Leaf blades elliptic or oblong, 2-5 cm long, entire or crenulate, glabrous; petioles 2-5 mm long. Flowers 5-merous, on nodding pedicels 5-10 mm long; calyx saucer-shaped; corolla barrel-shaped, 7-10 mm long, the rounded lobes about 1 mm long. Berry bright red, 7-10 mm wide, wintergreen-flavored. July–Aug. — Dry or moist woods in acid soil.

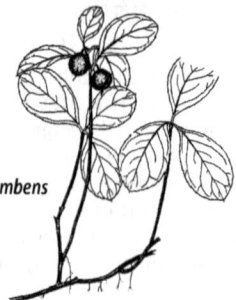

Gaultheria hispidula

Gaultheria procumbens

Gaylussacia
HUCKLEBERRY

Gaylussacia baccata (Wangenh.) K. Koch
BLACK HUCKLEBERRY   native / cons **6**
Medium shrub. Stems upright, much-branched, 3-10 dm long; branches
brown, finely hairy when young, dark and smooth with age. Leaves alternate, deciduous, leathery, oval, 2-5 cm long and 1-2.5 cm wide; dark
green above, paler below, both sides with shiny, orange-yellow resinous
dots; margins entire, often fringed with small hairs; petioles 2-4 mm
long. Flowers yellow-orange or red-tinged, cylindric, 5-lobed, 4-6 mm
long, in more or less 1-sided racemes from lateral branches, the flowers
on short, gland-dotted stalks 4-5 mm long. Fruit a red-purple to black,
berrylike drupe, 6-8 mm long, with 10 nutlets; edible but seedy. May-
June. — Open bogs, usually with tamarack and leatherleaf (*Chamaedaphne
calyculata*); more common in dry, acid, sandy or rocky habitats.

Gaylussacia baccata

Kalmia
LAUREL

Kalmia polifolia Wangenh.
BOG-LAUREL   native / cons **10**
Low evergreen shrub to 6 dm tall. Older stems dark; twigs swollen at
nodes, flattened and 2-edged in section, smooth, pale brown when young.
Leaves opposite, evergreen and leathery, linear to narrowly oval, 1-4 cm
long and 6-12 mm wide, tip blunt or narrowed to an abrupt point; dark
green and smooth above, white below with a covering of short, white
hairs, midrib on underside with large purple, stalked glands; margins
entire and rolled under; petioles absent. Flowers showy, pale to rose-pink,
in terminal clusters at ends of current year's branches, saucer-shaped, 5-
parted, 8-11 mm wide, on stalks to 3 cm long. Fruit a rounded capsule to
6 mm wide, tipped by the persistent style, the capsules in upright clusters.
May-June. — Sphagnum peatlands, black spruce and tamarack swamps.

Kalmia polifolia

Moneses
SINGLE-DELIGHT

Moneses uniflora (L.) Gray
ONE-FLOWERED SHINLEAF   native / cons **9**
    Pyrola uniflora L.
Low perennial herb from a very slender creeping rhizome. Stems to 10
cm long. Leaves deciduous, mostly at base of plant, opposite or in whorls
of 3, nearly round, margins entire or finely toothed. Flowers white, single
at end of long stalk, nodding, 1-2 cm wide; petals 5 distinct, widely spreading; stamens 10; ovary subglobose, concave at the summit, 5-celled. Capsule subglobose, opening from top downward. July-Aug. — Cedar swamps,
wet conifer or mixed conifer and deciduous forests.

Moneses uniflora

Monotropa
INDIAN-PIPE

White, yellow, pink, or red plants, turning black in drying, parasitic on soil-fungi. Stems erect.
Leaves small, scale-like. Flowers nodding, of the same color as the stem. Corolla urn-shaped or
broadly tubular; petals 4 or 5, distinct, all or some saccate at base. Sepals none, or 2-5. Stamens 8
or 10; filaments slender, pubescent. Ovary 4-5-celled; style short, thick. Capsules erect, ovoid to
subglobose.

1 Flowers single . . . . *M. uniflora*
1 Flowers few to many in a raceme . . . . *M. hypopithys*

Monotropa hypopithys L.

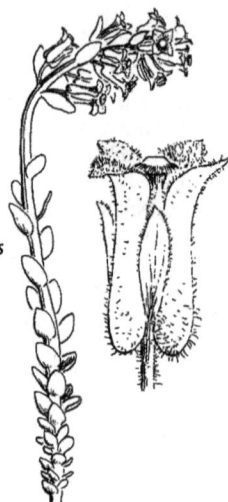

PINESAP   *native / cons 7*
Hypopitys americana (DC.) Small
Stems 1-3 dm tall, often gregarious, yellow, tawny, pink, or red, never pure
white, more or less pubescent. Raceme dense, at first nodding, erect at
anthesis. Flowers 8-12 mm long, the lower usually 4-merous, the terminal
often 5-merous and larger; sepals lance-shaped, erect; style shorter than
the ovary; stigma more or less villous at the margin. — Moist or dry
woods, usually in acid soil.   *Monotropa hypopithys*

Monotropa uniflora L.

ONE-FLOWER INDIAN-PIPE   *native / cons 5*
Stems 1-2 dm tall, usually solitary, commonly pure waxy white, rarely
pink or red. Flower solitary, nodding, odorless, 10-17 mm long; sepals
often absent; petals broadly oblong, slightly widened distally; style longer
than the ovary; stigma glabrous. June-Aug. — Rich woods in leaf-mold.

Orthilia
SIDEBELLS

Orthilia secunda (L.) House

ONE-SIDED SHINLEAF   *native / cons 7*
Pyrola secunda L.
Perennial herb. Leaves elliptic to subrotund, 1.5-4 cm long, entire to
crenulate, often separated by conspicuous internodes. Scape 8-20 cm tall,
bearing a crowded secund raceme; sepals semicircular to ovate, 0.5-1
mm long; petals white or greenish, about 5 mm long; style elongate,
exsert at anthesis, 5-lobed. June-July. — Moist woods and mossy bogs.

Separated from other members of family by its ovary subtended by a
10-lobed hypogynous disk, and petals with 2 rounded projections at base.

Orthilia secunda

Pterospora
PINEDROPS

Pterospora andromedea Nutt.

WOODLAND PINEDROPS   *endangered / native / cons 10*
Parasitic on soil fungi and forming a rounded subterranean mass of rhi-
zoids. Stems erect, simple, brown or purplish, 3-10 dm tall, glandular-pubescent, bearing numerous scale-like leaves especially toward the base.
Raceme many-flowered, 1-3 dm long; sepals densely glandular-pubescent;
corolla nodding, urn-shaped, white, 5-lobed, persistent on the fruit 6-7
mm long, its spreading-recurved lobes 1-2 mm long; stamens 10; ovary
5-celled. Capsule depressed- globose, nearly 1 cm wide; seeds 0.3 mm
long, with a terminal round wing, 0.7-1 mm wide. June-Aug. — Under
conifer trees (especially white pine as saprophytic on pines), usually in
drier, clayey soils.

Pterospora andromedea

Pyrola
WINTERGREEN, SHINLEAF

Perennial herbs from creeping rhizomes. Leaves few, broad, petiolate, nearly basal. Flowers in
an erect, terminal, long-peduncled raceme; regular, 5-merous. Stamens 10. Ovary 5-celled; style
short or elongate; stigma 5-lobed. Capsule dehiscent from the base upward.

1 Sepal lobes longer than wide 2
1 Sepal lobes shorter than wide 4
2 Leaves to 3 cm long; sepals ovate . . . . *P. chlorantha*
2 Leaves 3–7 cm long; sepals triangular . . . . *P. elliptica*
3 Petals white; sepals oblong . . . . *P. americana*
3 Petals pink; sepals triangular . . . . *P. asarifolia*

Pyrola americana Sweet
AMERICAN WINTERGREEN   *native / cons* **8**
  Pyrola rotundifolia L.
Perennial herb. Leaves firm in texture, broadly elliptic to subrotund,
rarely somewhat ovate or obovate, 2.5-7 cm long, broadly rounded above,
rounded, truncate, or short-cuneate at base; always somewhat decurrent
on the petioles. Scapes 1.5-3 dm tall, usually with 1 or 2 scale-leaves below
the raceme; sepals oblong or ovate-oblong, nearly twice as long as wide,
erose or undulate, not overlapping at base; petals white. July-Aug. — Dry
or moist woods, rarely bogs.   *Pyrola americana*

Pyrola asarifolia Michx.
PINK SHINLEAF   *native / cons* **8**
Perennial herb. Stems to 3 dm long. Leaves persisting over winter, all
near base of plant, kidney-shaped, 3-4 cm long and 3-5 cm wide, margins
shallowly rounded-toothed; flower stalk with 1-3 small, scale-like leaves.
Flowers nodding in a raceme; sepals triangular, 2-3 mm long; petals 5, 5-
7 mm long, pink to pale purple. Fruit a capsule opening from base upward.
June-Aug. — Cedar swamps, peatlands, marly wetlands, and interdunal
wetlands.

Pyrola chlorantha Sw.
GREEN-FLOWER WINTERGREEN   *native / cons* **7**
Perennial herb. Leaves obovate or broadly elliptic to subrotund, 1-3 cm
long, often shorter than the petiole, rounded to truncate at the summit,
rounded to broadly cuneate at base but scarcely decurrent. Scapes 1-2.5
dm tall; petals white, more or less veined with green; sepals broadly
ovate-triangular, broader than long, obtuse or subacute. June-Aug. — Dry
woods.

Pyrola elliptica Nutt.
ELLIPTIC SHINLEAF   *native / cons* **6**
Perennial herb. Leaves broadly elliptic, broadly oblong, or somewhat obovate, 3-7 cm long, commonly longer than the petiole, subacute to rounded
at the summit, acute to rounded at base but always decurrent partway
down the petiole. Scapes 1.5-3 dm tall; petals white, more or less veined
with green; sepals triangular, about as broad as long, very shortly acuminate. June-Aug.
— Dry upland woods.

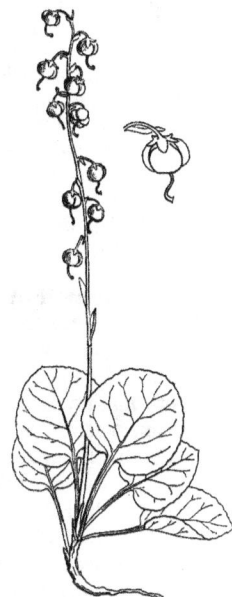

Pyrola asarifolia

Rhododendron
RHODODENDRON
Shrubs (ours), with alternate, deciduous or evergreen leaves and conspicuous flowers in terminal racemes or umbel-like clusters from scaly buds.
Calyx 5-lobed, the lobes much shorter than the corolla. Corolla campanulate to funnelform, conspicuous, regularly or commonly irregularly 5-
lobed. Stamens usually 5 or 10. Ovary 5-celled. Capsule 5-celled.
  Many species have been introduced and cultivated for their attractive
flowers.

Rhododendron groenlandicum (Oeder) Kron & Judd
RUSTY LABRADOR-TEA   *native / cons* **8**
  Ledum groenlandicum Oeder
Medium shrub, to 1 m tall. Older stems gray or red-brown; twigs covered
with woolly, curly brown hairs. Leaves alternate, evergreen and leathery,
fragrant when rubbed, narrowly oval to oblong, 2.5-5 cm long and 5-20
mm wide, rounded at tip; dark green and smooth above, the midvein
sunken; underside covered with tan to rust-colored curly hairs; margins

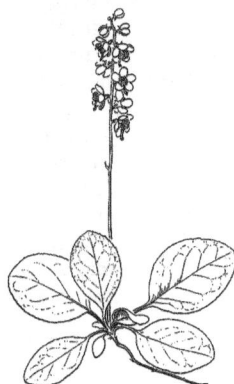

Pyrola elliptica

entire and rolled under; petioles short. Flowers creamy-white, in rounded clusters at ends of branches, 5-parted, to 1 cm wide, on finely hairy stalks 1-2 cm long. Fruit a lance-shaped capsule 5-6 mm long, the style persistent and hairlike; capsules splitting at base to release numerous small seeds, the empty capsules persistent on stems for several years. May-June. — Sphagnum bogs, swamps and wet conifer forests.

Rhododendron groenlandicum

Vaccinium
BLUEBERRY

Deciduous or evergreen shrubs. Leaves alternate, simple. Flowers 4- or 5-parted, single in leaf axils or in clusters in axils or at ends of branches; ovary inferior. Fruit a many-seeded, red, blue, or black berry.

Our members of the genus may be divided into two subgroups: blueberries (*V. angustifolium*, *V. myrtilloides*), and cranberries (*V. macrocarpon*, *V. oxycoccos*).

1 Leaves deciduous; berries blue to blue-black 2
1 Leaves evergreen; berries red 3
2 Leaf margins with small bristle-tipped teeth . . . . *V. angustifolium*
2 Leaves entire; margins sometimes finely hairy . . . . *V. myrtilloides*
3 Leaves blunt or rounded at tip (and sometimes notched), pale below; bracts on flower stalk green and leaflike (more than 1 mm wide) . . . . *V. macrocarpon*
3 Leaves tapered to pointed tip, white below; bracts on flower stalk red and narrow (less than 1 mm wide) . . . . *V. oxycoccos*

Vaccinium angustifolium Ait.
LOWBUSH BLUEBERRY *native / cons* 4

Low shrub 1-6 dm tall, forming colonies from surface runners. Older stems red-brown to black; twigs green-brown, with hairs in lines down stems, or sometimes smooth. Leaves deciduous, bright green oval, 2-5 cm long and 5-15 mm wide, smooth on both sides or sparsely hairy on veins; margins finely toothed with bristle-tipped teeth; petioles very short. Flowers in clusters, opening before or with leaves, white or pale pink, narrowly bell-shaped, 5-parted, 4-6 mm long. Fruit blue and wax-covered, 5-12 mm wide, edible and sweet. Flowering April-June, fruit ripening July-Aug. — Sphagnum peatlands and wetland margins; also in dry, sandy openings and forests.

Vaccinium angustifolium

Vaccinium macrocarpon Ait.
LARGE CRANBERRY *native / cons* 9

Evergreen trailing shrub. Stems slender, to 1 m or more long, with branches to 2 dm tall. Leaves leathery, oblong-oval, 5-15 mm long and 2-5 mm wide, rounded or blunt at tip, pale on underside; margins flat or slightly rolled under; petioles absent or very short. Flowers white to pink, 1 cm wide, 4-lobed, the lobes turned back at tips, single or in clusters of 2-6, on stalks 1-3 cm long, the stalks with 2 bracts above middle of stalk, the bracts green, 2-4 mm long and 1-2 mm wide. Fruit red, 1-1.5 cm wide, edible but tart, often over-wintering. Flowering June-July, fruit ripening Aug-Sept. — Sphagnum bogs, swamps and peaty pond margins. *V. macrocarpon* is the cultivated cranberry.

Vaccinium macrocarpon

Vaccinium myrtilloides Michx.
VELVET-LEAF BLUEBERRY *native / cons* 6

Low shrub, often forming colonies. Stems 3-6 dm long, red-brown to black with numerous wartlike lenticels; young twigs green-brown, densely velvety white-hairy. Leaves deciduous, thin and soft, oval, 2-5 cm long and 1-2.5 cm wide, dark green above, paler and soft hairy below, not waxy; margins entire and finely hairy; petioles very short. Flowers in clusters at ends of short, leafy branches, opening with leaves, creamy or green-white, tinged with pink, bell-shaped or short-cylindric, 5-parted, 4-5 mm long. Fruit blue, wax-covered, 6-9 mm wide; edible but tart. Flowering May-July, fruit ripening July-Sept. — Sphagnum bogs and swamps; also in dry to moist woods and clearings.

Vaccinium myrtilloides

Vaccinium oxycoccos L.
SMALL CRANBERRY   *native / cons* **9**

Evergreen trailing shrub. Stems slender, 0.5 m or more long, with upright branches 1-2 dm tall. Leaves leathery, ovate to oval or narrowly triangular, 2-10 mm long and 1-3 mm wide, pointed or rounded at tip, strongly whitened on underside; margins flat or strongly rolled under; petioles absent or very short. Flowers pale pink, 1 cm wide, 4-lobed, the lobes turned back at tips, single or in clusters of 2-4, on stalks 1-3 cm long, the stalks with 2 bracts at or below middle of stalk, the bracts red, scale-like, to 2 mm long and less than 1 mm wide. Fruit pale and red-speckled when young, becoming red, 6-12 mm wide, edible but tart. Flowering June-July, fruit ripening Aug-Sept. — Wet, acid, sphagnum bogs.

Vaccinium oxycoccos

Euphorbiaceae
SPURGE FAMILY

Euphorbia
SPURGE

Annual or perennial herbs of diverse form; with milky, often highly acrid juice. Leaves usually alternate and simple. Flowers greatly reduced, the staminate flowers consisting of a single stamen only, the pistillate flowers of a single pistil only. Several staminate flowers surround one pistillate flower inserted at the base of a cup-shaped involucre to form an inflorescence termed a cyathium. Around the margin of the cyathium are 4-5 glands, and in some species these have petal-like appendages, so that the whole cyathium mimics a single flower. Ovary 3-celled; styles 3, each bifid. Capsule 3-lobed, 3-seeded.

   The milky juice is poisonous and for some people produces a dermatitis similar to that caused by poison-ivy.

1 Annual herbs, the stems prostrate or ascending; leaves all opposite; leaf bases typically unequal . . 2
2 Stems with long soft hairs; capsules with stiff appressed hairs . . . . *E. maculata*
2 Stems glabrous or pubescent; capsules glabrous 3
3 Leaves entire; seeds round, smooth, with a white seed coat . . . . *E. polygonifolia*
3 Leaves finely toothed, at least along upper portion of blade; seeds angular, smooth, bumpy, or ridged, the seed coat brown or blackish . . . . *E. glyptosperma*
1 Annual or perennial herbs, the stems erect, leaves mostly alternate; leaf bases equal or nearly so . 4
4 Main stem leaves slender and linear, 1–3 cm long and 1–3 mm wide, crowded . . . . *E. cyparissias*
4 Main stem leaves 3–7 cm long, mostly 3–10 mm wide, less crowded . . . . *E. virgata*

Euphorbia cyparissias L.
CYPRESS SPURGE   *introduced (invasive)*

Perennial by horizontal rhizomes, gregarious. Stems 2-4 dm tall. Stem leaves very numerous, crowded, linear, 1-3 cm long, 1-nerved; leaves subtending the umbel similar; leaves of the umbel broadly cordate. Rays of the umbel usually 10 or more. Capsule about 3 mm long, slightly granular-roughened; seeds plump, smooth, 1.5-2 mm long. April-July. — Native of Eurasia; established on roadsides and waste ground. Fruit seldom produced.

Euphorbia cyparissias

Euphorbia glyptosperma Engelm.
RIB-SEED SANDMAT   *native / cons* 2
  Chamaesyce glyptosperma (Engelm.) Small
Plant glabrous throughout. Stems mostly prostrate, freely branched, 1-3
dm long, often forming mats. Leaves oblong or ovate, 4-15 mm long,
strongly inequilateral, minutely serrulate, especially on the rounded
summit. Appendages rather conspicuous. Capsule depressed-ovoid,
sharply 3-angled, about 1.5 mm long; seeds about 1 mm long, sharply 4-
angled, marked with 3 or 4 conspicuous transverse ridges. Dry sandy soil.
The teeth of the leaf are often visible only under a lens.

Euphorbia glyptosperma

Euphorbia maculata L.
SPOTTED SANDMAT   *native / cons* 0
  Chamaesyce maculata (L.) Small
Stems prostrate or nearly so, to 4 dm long, often forming circular mats,
sparsely to densely villous. Leaves dark green, often with a red spot,
oblong or ovate-oblong, varying to linear-oblong, usually 5-15 mm long,
almost always widest below the middle. Involucre cleft on one side; ovary
and capsule strigose. Capsule about 1.5 mm long; seeds 4-angled, about 1
mm long, with a few inconspicuous transverse ridges. — Common as a
weed in lawns, gardens, and waste places, also in meadows and open
woods.

Euphorbia maculata

Euphorbia polygonifolia L.
SEASIDE SANDMAT   *native / cons* 10
  Chamaesyce polygonifolia (L.) Small
Glabrous throughout. Stems usually prostrate, divergently branched or
forming mats. Leaves linear-oblong to oblong lance-shaped, usually 8-15
mm long, entire, slightly inequilateral at base. Petal-like appendages of
the involucre very small or none. Capsule 3-3.5 mm long; seeds smooth,
compressed-ovoid, gray, 2-2.6 mm long. — Sand dunes and sandy beaches
along Lake Michigan.

Euphorbia polygonifolia

Euphorbia virgata Waldst. & Kit.
LEAFY SPURGE   *introduced (invasive)*
  Euphorbia esula L.
Perennial from a deep root. Stems erect, 3-7 dm tall, glabrous, usually
with numerous alternate flowering branches below the umbel. Stem
leaves linear to narrowly oblong, 3-7 cm long, obtuse to mucronate; leaves
subtending the umbel shorter and broader, lance-shaped to ovate; leaves
of the umbel broadly cordate or kidney-shaped. Rays of the primary umbel
7 or more. Capsule 2.5-3 mm long, finely granular; seeds brown, globose-
ovoid, about 2 mm long. Summer. — Native of Eurasia; widely established
in North America; a troublesome noxious weed, sometimes infesting
large areas.

Euphorbia virgata

Fabaceae

PEA FAMILY

Perennial herbs, shrubs and trees. Leaves alternate, pinnately divided, the terminal leaflet sometimes modified as a tendril (*Lathyrus, Vicia*). Flowers in simple or branched racemes, perfect
(with both staminate and pistillate parts), irregular, 5-lobed, the upper lobe (banner) larger than
the other lobes, with 2 outer, lateral petals (wings), and 2 inner petals which are partly joined
(the keel), and enclosing the 10 stamens and style; pistil 1, ovary 1-chambered, maturing into a
pod.

ADDITIONAL SPECIES
  Genista tinctoria L. (Dyer's greenweed), introduced; reported from gravelly roadsides.

1 Trees or shrubs . . . . *Robinia*
1 Herbs 2
2 Leaves even-pinnate, ending in a tendril 3
2 Leaves not even-pinnate, tendril absent 4
3 Style round in cross-section, pubescent near the tip . . . . *Vicia*
3 Style flattened, pubescent along the inner side . . . . *Lathyrus*
4 Leaves divided into 3 or rarely 5 leaflets 5
4 Leaves with 5 or more leaflets 10
5 Leaflets toothed 6
5 Leaflets entire 8
6 Flowers reflexed in long slender racemes, white or yellow; pods small, straight, reflexed . . *Melilotus*
6 Flowers in rounded clusters 7
7 Pods straight; stamens joined to the corolla . . . . *Trifolium*
7 Pods curved or coiled; stamens free from the corolla . . . . *Medicago*
8 Fruit a loment, breaking into 1-seeded segments, or consisting of a single segment only . . . .
   . . . . *Desmodium*
8 Fruit a legume (like a pea pod), not breaking into 1-seeded segments 9
9 Leaflets without stipule-like appendages at their base; plants not twining or vining . . . . . . *Trifolium*
9 Leaflets with stipule-like appendages at their base; plants twining or vining . . . . *Amphicarpaea*
10 Leaflets 5–11 . . . . *Lotus*
10 Leaflets 11–31 11
11 Flowers in umbel-like clusters; pods linear, 4-angled, jointed . . . . *Securigera*
11 Flowers in racemes; pods neither 4-angled nor jointed . . . . *Astragalus*

Amphicarpaea
HOG-PEANUT

Amphicarpaea bracteata (L.) Fern.
AMERICAN HOG-PEANUT   *native / cons* 5
Twining, annual to short-lived perennial herb. Stems to 1 m long. Leaves
pinnately compound into 3 petioled leaflets. Flowers in racemes or panicles peduncled from many of the axils, each pedicel subtended at base
by a striate-veined bractlet, bearing several to many pale purple to whitish
flowers 12-18 mm long; calyx slightly irregular, the tube short-cylindric,
the lobes apparently 4 through the fusion of the upper two; standard
obovate, narrowed to the base, sometimes auricled below the middle;
wings and keel slightly shorter, with elongate slender claws exceeding
the blades. Pods (of petaliferous flowers) flat, oblong, pointed at both
ends, usually 3-seeded, coiled after dehiscence. Besides the pod-producing
petaliferous flowers, the plants bear nearly or completely apetalous flowers near the base of the stem, producing, often under the ground, small
1-seeded pods. Aug-Oct. — Woods and thickets.

Amphicarpaea bracteata

Astragalus
MILK-VETCH

Ours perennial herbs from a stout taproot, caudex, or rhizome. Leaflets numerous. Flowers
white, yellowish white, or purple, in long or short axillary racemes. Calyx tube campanulate to
cylindric; calyx lobes short, triangular or subulate. Standard obovate to rotund, usually exceeding
the wings. Stamens 10. Fruit a pod of various forms. Seeds few to many.

   Astragalus are most common in arid or semiarid regions of Asia and w North America. Many
species of the genus produce an alkaloid toxic to livestock. From their erratic movements,
animals so affected were termed loco (mad) and the name locoweed is commonly used for these
plants in the West. Other species are able to accumulate the element selenium, when growing
on seleniferous soils in the western states; such plants, if eaten by animals, cause sickness or
death.

1 Pod one-chambered . . . . *A. neglectus*
1 Pod divided to form 2 chambers . . . . *A. canadensis*

Astragalus canadensis L.
CANADIAN MILK-VETCH   *native / cons* 8
Stems erect, to 15 dm tall, glabrous to thinly strigose. Stipules connate, lance-shaped to deltoid, 3-6 mm long. Leaflets 13-29, oblong or elliptic, 1-3 cm long, 5-15 mm wide, glabrous or rarely strigose above, more or less strigose beneath with T-shaped hairs. Racemes long-peduncled, 5-12 cm long. Flowers white or yellowish white, spreading or somewhat reflexed, 12-15 mm long. Pods numerous in a crowded raceme, ovoid or oblong, 10-18 mm long, 2-celled, glabrous, nearly terete in cross-section. May-Aug. — Open woodlands, river banks and shores, usually in moist soil.

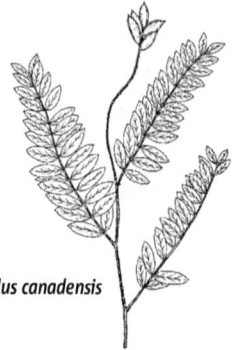

Astragalus canadensis

Astragalus neglectus (Torr. & Gray) Sheldon
COOPER'S MILK-VETCH   *end / native / cons* 9
Stems erect, 4-7 dm tall, nearly or quite glabrous. Leaflets 11-17, oblong to elliptic or obovate, 1-3 cm long, a fourth to a third as wide, glabrous above, strigose beneath. Racemes several, scarcely surpassing the subtending leaf, many-flowered; calyx tube cylindric, 4-6 mm long, strigose, the lobes about a third as long; corolla commonly white, 11-14 mm long. Pods erect, sessile, straight, ovoid, inflated, 1-2 cm long, glabrous. June. — Riverbanks and lakeshores, especially on limestone; disturbed forests and fields.

Astragalus neglectus

Desmodium
TICK-TREFOIL
Ours perennial herbs. Leaves petioled, 3-parted. Flowers small, white to purple or violet flowers, sometimes marked with yellow, often greenish in age, in elongate, simple or panicled racemes. Calyx tube more or less 2-lipped, slightly oblique; upper 2 calyx lobes connate for all or most of their length, the lower 3 separate. Standard oblong to nearly orbicular, narrowed at base; wings oblong; keel nearly straight. Fruit an indehiscent pod, elevated on a stalk above the persistent calyx and stamens, more or less beset with hooked hairs.

1 Pods conspicuously long-stalked, the stalk 2–3 times the length of the calyx . . . . *D. glutinosum*
1 Pods sessile or short-stalked . . . . *D. canadense*

Desmodium canadense (L.) DC.
SHOWY TICK-TREFOIL   *native / cons* 4
Stems erect, branched above, to 2 m tall, pubescent. Stipules linear-subulate, to 8 mm long, ciliate, otherwise glabrous. Petioles 2-20 mm, usually about 10 mm long, the petiole and leaf-rachis together up to 1/2 as long as the terminal leaflet. Leaflets oblong or lance-oblong, appressed-pubescent beneath, the terminal 5-9 cm long. Racemes densely flowered, with conspicuous ovate lance-shaped bracts. Flowers 10-13 mm long; calyx 5-7 mm long.; petiole of the fruit 2-4 mm long; joints commonly 3-5. July-Aug. — Moist soil, thickets, and riverbanks.

Desmodium glutinosum (Muhl.) Wood
POINTED-LEAF TICK-TREFOIL   *native / cons* 6
Hylodesmum glutinosum (Muhl. ex Willd.) H. Ohashi & R. R. Mill
Stems erect, 1-4 dm tall, bearing near the summit several long-petioled leaves and prolonged into a terminal panicle 3-8 dm long. Lateral leaflets asymmetrically ovate, acuminate; terminal leaflet round-ovate, 7-15 cm long and nearly as wide, long-acuminate. Flowers 6-8 mm long; calyx 2.5-3 mm long; petiole of the fruit 6-12 mm long, glabrous; joints seldom more than 3. July. — Rich woods.

Desmodium canadense

Lathyrus
VETCHLING, WILD PEA

Perennial herbs (ours). Leaves terminated by a tendril. Flowers few to many in a raceme; corolla small or medium-sized, red-purple to white or yellow; standard broadly obovate; wings obovate; keel upwardly curved. Pods flat to terete, 2-many-seeded. Most of our species superficially resemble *Vicia*.

ADDITIONAL SPECIES
   Lathyrus sylvestris L. (Narrow-leaf vetchling), introduced; in Door County, reported from a grassy open field. See key.

1 Leaflets 1-pair; introduced species 2
1 Leaflets 2 or more pairs; native species 4
2 Stem not winged .... *L. tuberosus*
2 Stem winged 3
3 Lobe of stipules more than half as wide as the winged stem; flowers 16–26 mm long . . . . *L. latifolius*
3 Lobe of stipules up to half as wide as the winged stem; flowers ca. 12–18 mm long. . . . . . *L. sylvestris**
4 Flowers yellow-white .... *L. ochroleucus*
4 Flowers purple, rarely white 5
5 Stipules leafy, nearly as large as the adjacent leaflets .... *L. japonicus*
5 Stipules much smaller than leaflets 6
6 Stems usually winged; racemes with usually 10–20 flowers; moist habitats .... *L. palustris*
6 Stems not winged; racemes with 2–6 flowers; dry woods .... *L. venosus*

Lathyrus japonicus Willd.
BEACH-PEA   *native / cons* 9
   Lathyrus maritimus Bigelow
Perennial, plants typically glabrous. Stems stout, decumbent to nearly erect, up to 1 m long. Stipules foliaceous, broadly ovate, 1.5-4 cm long, 1-2.5 cm wide, essentially symmetrical and therefore attached at the middle of the broadly truncate base. Leaflets 3-6 pairs, oblong to obovate, 3-5 cm long, about half as wide, the lowest pair near the base of the petiole. Peduncles equaling or shorter than the subtending leaves, bearing usually 5-10 purple flowers about 2 cm long; calyx irregular, the lowest lobe linear lance-shaped, nearly 2x as long as the triangular upper ones. June-Aug. — Beaches and lakeshores.

Lathyrus japonicus

Lathyrus latifolius L.
EVERLASTING-PEA   *introduced*
Perennial. Stems climbing or trailing, to 2 m long, broadly winged, 5-10 mm wide. Stipules lance-shaped, with a basal lobe, foliaceous, 1.5-4 cm long, usually wider than the stem. Petiole broadly winged, about as wide as the stem. Leaflets 2, lance-shaped to elliptic, 4-8 cm long, 1-3 cm wide. Peduncles 10-20 cm long, bearing a raceme of 4-10 handsome flowers 1.5-2.5 cm long; corolla purple, varying to pink or white; calyx lobes very unequal. June-Aug. — Native of s Europe; cultivated and escaping to roadsides and vacant land.

Lathyrus ochroleucus Hook.
CREAM VETCHLING   *native / cons* 7
Glabrous perennial. Stems to 8 dm long. Stipules semi-ovate, 1.5-3 cm long, the larger ones toothed, usually shorter than the petiole. Leaflets 3-5 pairs, thin, elliptic to ovate lance-shaped, 2.5-5 cm long. Racemes shorter than the subtending leaf, bearing 5-10 yellowish white flowers 12-18 mm long; calyx irregular, the upper lobes triangular, to half as long as the lower. May-July. — Dry upland woods and thickets.

Lathyrus ochroleucus

Lathyrus palustris L.

MARSH VETCHLING   *native / cons 5*

Perennial vining herb, spreading by rhizomes. Stems to 1 m long, strongly winged, climbing and clinging to surrounding plants by tendrils. Leaves pinnately divided, with 4-8 leaflets and a terminal leaflet modified into a tendril; leaflets linear to lance-shaped, 2-7 cm long and 3-20 mm wide; stipules prominent, more or less arrowhead-shaped, 1-3 cm long; margins entire; petioles absent. Flowers in racemes from leaf axils, 2-6 flowers per raceme, red-purple, drying blue to blue-violet; sepals irregular, 7-10 mm long; petals 12-20 mm long. Fruit a flat, many-seeded pod, 3-5 cm long. June-Aug. — Conifer swamps, wet meadows, marshes, streambanks, calcareous fens, low prairie.

Lathyrus tuberosus L.

EARTH-NUT VETCHLING   *introduced*

Perennial, with tuberously thickened roots. Stems wingless, 4-angled, to 8 dm long, branched above. Stipules lance-shaped, 5-12 mm long, with a single short basal lobe. Leaflets 2, oblong lance-shaped to elliptic, 1.5-3 cm long. Peduncles 5-8 cm long, bearing a crowded raceme of 2-5 red-purple flowers about 15 mm long, the standard broader than long; calyx lobes broadly triangular, shorter than the tube. June-Aug. — Native of Europe and w Asia.

Lathyrus palustris

Lotus

TREFOIL

Herbaceous or suffrutescent (woody at base) herbs. Leaves pinnately compound. Flowers solitary or umbellate. Calyx campanulate or obconic, the elongate teeth nearly equal. Petals clawed; standard obovate, not auriculate; keel petals usually distinctly beaked. Pod several-seeded, oblong to linear, terete (ours).

Lotus corniculatus L.

BIRD'S-FOOT-TREFOIL   *introduced*

Lotus corniculatus var. *arvensis* (Schkuhr) Ser.

Perennial herb. Stems prostrate, ascending, or erect, to 6 dm long. Leaves nearly sessile, without stipules; leaflets 5, elliptic to oblong lance-shaped, 5-15 mm long, the terminal one sessile. Flowers in long-peduncled head-like umbels from the upper axils; pedicels 1-3 mm long; calyx lobes linear-triangular, about equaling the tube; corolla yellow to brick-red, about 14 mm long. Pods 2-4 cm long. June-Aug. — Native of Europe; established in fields, meadows, and roadsides.

Lotus corniculatus

Medicago

ALFALFA, MEDICK

Herbs with 3-foliolate serrulate leaves, the terminal leaflet stalked. Flowers in axillary heads or short head-like racemes of small yellow or blue flowers. Calyx tube campanulate, the 5 lobes nearly equal. Standard obovate or oblong, longer than the oblong erect wings; keel blunt, shorter than the wings. Stamens all free from the corolla. Pod straight or coiled, glabrous or spiny, usually indehiscent, 1-several-seeded.

1 Plants perennial from a long taproot; flowers blue-violet or sometimes yellow .... *M. sativa*
1 Plants annual or biennial; flowers yellow .... *M. lupulina*

Medicago lupulina L.

BLACK MEDICK   *introduced*

Annual herb. Stems prostrate, widely spreading, or ascending, to 8 dm long; 4-angled. Leaflets elliptic to obovate, 1-2 cm long. Stipules lance-shaped, entire or toothed. Peduncles slender, much exceeding the subtending leaves, bearing a globose to short-cylindric head up to 1 cm long; peduncles and calyx glandular. Flowers yellow, 2-4 mm long. Pods nearly black, 2-3 mm long, kidney-shaped, 1-seeded, the conspicuous veins tending to be longitudinal. May-Sept. — Native of Europe and w Asia; common as a troublesome weed of roadsides, lawns, fields, railroads, and disturbed places.

Medicago sativa L.
ALFALFA   *introduced*
Perennial herb. Stems erect or decumbent, to 1 m tall. Leaflets oblong lance-shaped, 1.5-3 cm long, toothed at the tip. Stipules ovate lance-shaped, toothed. Peduncles erect, about equaling the subtending leaves, with a subglobose to short-cylindric head 1-3 cm long. Flowers blue, nearly 1 cm long, on pedicels 2-3 mm long; sepals linear lance-shaped, 2-3 mm long, about equaling the tube. Pod coiled into a loose spiral of 1-3 complete turns, finely pubescent. June-Sept. — Native probably of c and w Asia; long in cultivation and valued for hay and forage; commonly escaped or introduced on roadsides, fields, and disturbed places.

Medicago lupulina

Melilotus
SWEET-CLOVER
Annual or biennial herbs. Leaves 3-foliolate, serrulate, the terminal leaflet stalked. Flowers white or yellow, in elongate peduncled racemes from the upper axils; stipules partially adnate to the petiole. Calyx eventually deciduous, the tube campanulate, the lobes nearly equal. Petals separate; standard oblong to obovate, usually longer than the others; wings and keel coherent. Ovary short, sessile or somewhat stipitate. Pod ovate to rotund, slightly compressed to nearly globose, 1-4-seeded.

Melilotus officinalis (L.) Lam.
YELLOW SWEET-CLOVER   *introduced (invasive)*
Stems erect or ascending, 5-15 dm tall. Leaflets oblong lance-shaped to obovate, 1-2.5 cm long. Racemes 5-15 cm long including the peduncle. Pedicels 1.5-2 mm long, decurved. Flowers yellow, 5-7 mm long. Summer. — Native of Eurasia; established as a weed of waste places.
   Melilotus albus Medik (White sweet-clover) now often merged with *M. officinalis*.

Melilotus officinalis

Robinia
LOCUST
Trees or shrubs. Leaves odd-pinnate, stipules setaceous or modified into spines. Flowers white, pink, or purple flowers in axillary racemes. Calyx tube broadly campanulate, bilabiate; lower 3 calyx lobes about equal; upper 2 lobes connate for a third or more of their length. Corolla large; standard more or less reflexed; wings and keel long-clawed with a rounded lobe at base of the blade; the keel petals strongly upwardly curved. Pods elongate, flat, many-seeded.

1   Tree to 25 m tall; twigs and petioles glabrous; flowers white . . . . *R. pseudoacacia*
1   Shrubs to 3 m tall; stems gummy or bristly; flowers pink or rose-colored . . . . *R. hispida*

Robinia hispida L.
BRISTLY LOCUST   *introduced*
Stoloniferous shrub 1-2 m tall. Stems, peduncles, and calyx densely or sparsely hispid with glandular hairs up to 5 mm long, those of the stem persisting and becoming indurate. Leaflets 7-13, ovate-oblong to nearly rotund, commonly 3-6 cm long. Racemes usually 3-6-flowered. Flowers rose or pink-purple, 2.5-3 cm long; ovary densely glandular. Pods densely hispid, rarely developed. June-July. — Sometimes planted and occasionally escaped to roadsides and open woods.

Robinia hispida

Robinia pseudoacacia L.

BLACK LOCUST   *introduced (invasive)*

Tree up to 25 m tall, the younger stems and peduncles finely pubescent. Stipules frequently modified into stout woody thorns. Leaflets 7-19, oval or elliptic, 2-4 cm long, mucronate at the truncate or rounded apex. Racemes drooping, many-flowered, 1-2 dm long. Flowers white, very fragrant, 2-2.5 cm long; calyx finely pubescent, the upper lip truncate or broadly notched; ovary glabrous. Pods very flat, smooth, 5-10 cm long. June. — Commonly planted and escaped to roadsides, open woods, and waste land.

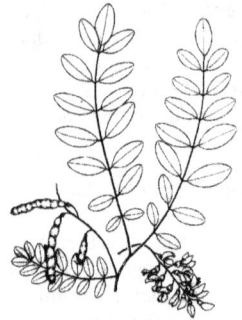

Robinia pseudoacacia

Securigera

CROWN-VETCH

Securigera varia (L.) Lassen

PURPLE CROWN-VETCH   *introduced*

  Coronilla varia L.

Perennial herb. Stems ascending, 3-5 dm long. Leaves sessile, 6-15 cm long; leaflets 11-21, oblong to obovate, 1-2 cm long, acute to rounded or retuse at the summit. Peduncles stout, equaling or surpassing the subtending leaves. Flowers in long-peduncled axillary umbels of 10-15 flowers, pink, the keel tipped with purple; calyx tube campanulate, bilabiate, the broad lower lip with 3 short triangular teeth, the upper lip narrow, triangular, shallowly cleft; petals about equal in length, clawed; standard orbicular; wings ovate-oblong; keel-petals upwardly curved. Pods linear, 4-angled, 2-4 cm long, with 3-7 joints. May-Sept. — Native of Eurasia and n Africa; introduced or escaped.

Securigera varia

Trifolium

CLOVER

Annual, biennial, or perennial herbs. Leaves 3-foliolate, serrulate. Flowers in heads, spikes, or head-like racemes or umbels. Calyx tube campanulate to tubular. Petals all separate or more or less united into a tube, usually withering and persistent after anthesis; standard ovate to obovate, often folded about the wings or with only the tip outwardly curved. Fruit short, straight, often included in the persistent calyx, 1-6-seeded. Valuable for forage and several species extensively cultivated. In the absence of fruit, our yellow-flowered species may be distinguished from *Medicago* by their strongly bilabiate calyx.

1 Flowers white, pink, or purple 2
1 Flowers yellow, turning brown with age 4
2 Flowers sessile or nearly so .... *T. pratense*
2 Flowers on short pedicels, these becoming reflexed with age 3
3 Flowers white; very common .... *T. repens*
3 Flowers pink or purple-tinged .... *T. hybridum*
4 Leaflets sessile; stipules linear .... *T. aureum*
4 Terminal leaflet on a short petiole; stipules ovate to lance-shaped .... *T. campestre*

Trifolium aureum Pollich

GREATER HOP CLOVER   *introduced*

Annual or biennial. Stems much-branched, mostly erect, 2-5 dm tall, appressed-hairy. Leaflets all sessile or nearly so, oblong lance-shaped, 1-2 cm long; petioles 5-12 mm long, about equaling the lance-oblong stipules. Heads short-cylindric, 1-2 cm long, on peduncles 1-4 cm from the upper axils; pedicels 0.5 mm. Flowers 5-7 mm long; calyx strongly 2-lipped, glabrous, the tube 5-nerved, 1 mm, the lobes lance-linear; corolla yellow; the standard obovate, conspicuously striate in age, usually serrulate; the wings somewhat spreading at the tip. May-Sept. — Native of Eurasia, weedy on roadsides and in waste places.

Trifolium aureum

Trifolium campestre Schreb.

LESSER HOP CLOVER   *introduced*

Annual. Stems much-branched, 1-4 dm tall, pubescent. Leaflets obovate, 8-15 mm long, the terminal one on a stalk 1-3 mm long; heads globose to short-cylindric, 8-15 mm long, compact, with usually 20-30 flowers; petioles 8-12 mm long, 2x as long as the stipules. Flowers 3.5-5 mm long; calyx as in *T. aureum;* corolla yellow, the standard obovate, with 5 conspicuous diagonal veins on each side, much exceeding the spoon-shaped, slightly divergent wings. May-Sept. — Native of Eurasia and n Africa; weedy on roadsides and in waste places.

Trifolium hybridum L.

ALSIKE CLOVER   *introduced*

Perennial. Stems erect or ascending, 3-8 dm tall. Stipules ovate lance-shaped, tapering to a long slender point. Leaflets oval to elliptic, broadly rounded to retuse at the summit. Heads numerous, not involucrate, globose, on peduncles 2-8 cm long. Flowers distinctly pediceled, 7-10 mm long; calyx glabrous, the linear lobes somewhat unequal, 1.7-2.5 mm long, slightly exceeding the tube; corolla white and pink, turning brown after anthesis; the standard obovate, about 2 mm longer than the obtuse wings. Summer. — Native of Eurasia, commonly escaped.

Trifolium hybridum

Trifolium pratense L.

RED CLOVER   *introduced*

Perennial. Stems erect, decumbent, or ascending, to 8 dm tall, sparsely to densely appressed-pubescent. Stipules oblong, the free portion abruptly narrowed to a short awn. Lower leaves long-petioled; upper leaves short-petioled to sessile. Heads sessile or on peduncles up to 2 cm long, globose or round-ovoid. Flowers 13-20 mm long; calyx glabrous to sparsely pilose, the tube 3-4 mm long, the lobes setaceous, one 4-7 mm long, four 2-5 mm long; corolla magenta, varying to nearly white; standard obovate, equaling or slightly exceeding the oblong obtuse wings. May-Aug. — Native of Europe, commonly planted for forage and escaped to fields and on roadsides.

Trifolium pratense

Trifolium repens L.

WHITE CLOVER   *introduced*

Perennial. Stems creeping, sending up long-petioled leaves and long-peduncled heads without involucres. Leaflets broadly elliptic to obovate, rounded or notched at the tip, 1-2 cm long. Flowers distinctly pediceled, 7-11 mm long; calyx glabrous, the tube to 3 mm long, its lobes narrowly triangular, unequal, the longest about equaling the tube; corolla white or tinged with pink; the standard elliptic-obovate, rounded at the summit, exceeding the obtuse wings. Summer. — Native of Eurasia, commonly planted and escaped to lawns and roadsides.

Trifolium repens

Vicia

VETCH

Annual or perennial herbs. Leaves 1-pinnate, with small stipules, the terminal leaflets in most species metamorphosed into tendrils. Flowers in short or elongate racemes from the axils, or in sessile or subsessile, few-flowered, axillary clusters. Calyx regular or irregular, often swollen at base. Standard with a broad claw overlapping the wings, its blade obovate to subrotund; wings oblong or narrowly obovate, adherent to and usually exceeding the keel. Ovary sessile or short-stipitate. Pods flat to terete. Seeds 2 to many.

1 Flowers single or in pairs, sessile from leaf axils . . . . *V. sativa*
1 Flowers in racemes on stalks from leaf axils 2
2 Flowers white or white tinged with purple, 3–7 mm long 3

2 Flowers blue or white, 8 mm long or more 4
3 Lobes of calyx nearly equal in length; pods hairy, mostly with 2 seeds . . . . *V. hirsuta*
3 Calyx lobes unequal; pods glabrous, mostly with 4 seeds . . . . *V. tetrasperma*
4 Calyx with a large swollen bump on one side of base . . . . *V. villosa*
4 Calyx only slightly swollen on one side of base 5
5 Margins of stipules sharply toothed; flowers 15–30 mm long . . . . *V. americana*
5 Margins of stipules entire; flowers to 13 mm long 6
6 Flowers white, in loose racemes . . . . *V. caroliniana*
6 Flowers blue, in dense racemes . . . . *V. cracca*

Vicia americana Muhl.
AMERICAN VETCH   *native / cons 4*
Perennial. Stems trailing or climbing, to 1 m long. Leaflets usually 4-7 pairs, elliptic to oblong, 1.5-3 cm long, 5-12 mm wide, obtuse to broadly rounded at the mucronate tip. Stipules all or mostly sharply serrate. Racemes shorter than the subtending leaves, loose, bearing 2-9 blue-purple flowers 17-27 mm long; calyx tube 3.5-5.5 mm long. May-July. — Moist woods.

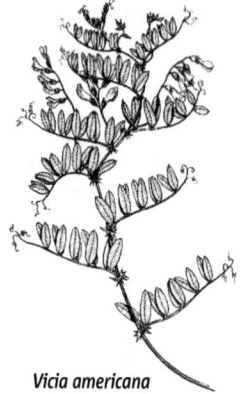

Vicia caroliniana Walt.
CAROLINA VETCH   *native / cons 7*
Perennial. Stems slender, trailing or climbing, to 1 m long. Leaflets usually 5-9 pairs, elliptic or oblong lance-shaped, 1-2 cm long, rounded or obscurely notched at the tip, mucronate, with usually 5-7 lateral veins on each side. Stipules lance-shaped, entire. Racemes, including the peduncle, 6-10 cm long, bearing 7-20 white flowers 7.5-12 mm long; calyx nearly regular, the tube 2-2.7 mm long, the lobes equal, broadly triangular, less than 1 mm long. Pod 1.5-3 cm long. May-June. — Moist woods and thickets.

Vicia americana

Vicia cracca L.
BIRD-VETCH   *introduced*
Perennial. Stems climbing or trailing, to 1 m long. Leaflets usually 5-10 pairs, linear, 1.5-3 cm long, mucronate. Stipules entire. Racemes long-peduncled, dense, secund, equaling or exceeding the subtending leaf, bearing numerous crowded blue flowers 9-13 mm long; calyx tube swollen at base, 2-3 mm long. Blade of the standard about as long as the claw. June-Aug. — Fields, roadsides, meadows.

Vicia hirsuta (L.) S.F. Gray
TINY VETCH   *introduced*
Annual. Stems slender, branched, decumbent or climbing, 3-7 dm long. Leaflets usually 6-8 pairs, linear or narrowly elliptic, 5-15 mm long, commonly mucronate and truncate, occasionally acute. Peduncles 1-3 cm long, bearing 3-8 whitish flowers 3-4 mm long; calyx lobes subulate, nearly equal. Pod flattened, about 1 cm long, usually pubescent, normally 2-seeded. May-Aug. — Native of Europe; introduced in fields, roadsides, and waste places.

Vicia cracca

Vicia sativa L.
COMMON VETCH   *introduced*
Annual. Stems slender, to 1 m long, ascending, erect, or tending to climb. Leaflets commonly 4-8 pairs, oblong to elliptic or obovate, 3-5 cm long, mucronate. Stipules often sharply serrate, bearing a glandular spot beneath. Flowers commonly paired in the axils, nearly sessile, violet or purple, rarely white, 2-3 cm long; calyx tube campanulate, 5-7 mm long; lobes nearly equal, the upper 3-7 mm, the lower 4-9 mm long. Pod flattened, brown; seeds flattened. — Native of Europe; long in cultivation; may persist after cultivation or escape into fields and roadsides.

Vicia sativa

Vicia tetrasperma (L.) Schreb.
LENTIL VETCH   *introduced*
Annual. Stems slender, branched, decumbent or climbing, 3-6 dm long.
Leaflets commonly 3 or 4 pairs, occasionally more, linear-oblong, 1-2 cm
long. Peduncles very slender, 1-3 cm long, bearing usually 1 or 2, sometimes
3 or 4, light purple to white flowers 4-6.5 mm long. Lowest calyx-lobe
linear, about equaling the tube, the upper much shorter and triangular.
Pod flat, glabrous, 1-1.5 cm long, usually 4-seeded. May-Aug. — Eurasian
introduction in Door County.

Vicia villosa Roth
HAIRY VETCH   *introduced*
Annual. Stems to 1 m long, more or less villous throughout, always softly
villous above and in the racemes with hairs 1-2 mm long. Leaflets usually
5-10 pairs, narrowly oblong to linear lance-shaped, obtuse and mucronate
to acute, 1-2.5 cm long. Racemes long-peduncled, dense, secund, bearing
usually 10-30 flowers; calyx irregular, villous, the tube 2-4 mm long,
swollen at base; upper lobes linear-triangular, 0.8-1.5 mm long; lateral
and lower lobes linear above a triangular base, the lowest 3-5 mm long,
long-villous; corolla slender, 12-20 mm long, the spreading blade of the
standard less than 1/2 as long as the claw. June-Aug. — Native of Europe;
introduced in fields, roadsides, and waste places.

Vicia villosa

Fagaceae

BEECH FAMILY

Trees or shrubs. Leaves alternate, simple, entire to lobed; the stipules early deciduous. Plants
monoecious. Staminate flowers in catkins or heads; corolla none; calyx small, deeply 4-8-parted;
stamens 3-20. Pistillate flowers solitary, or in small clusters or short spikes, more or less enclosed
by an involucre of numerous bracts; ovary commonly 3-celled or 6-celled, with 2 ovules in each
cell, of which only one matures. Fruit a 1-seeded nut, wholly or partly surrounded by the expanded
involucre.

1 Leaves entire, or toothed or lobed with fewer than 9 pairs of lateral veins .... *Quercus*
1 Leaf margins various; lateral veins more than 9 pairs 2
2 Bark smooth, silvery-gray; terminal buds single, narrow and more than 1 cm long; fruit bristly, the nut
   3-angled .... *Fagus*
2 Bark becoming furrowed or scaly; terminal buds 2 or more, less than 1 cm long; fruit not bristly, the
   nuts rounded .... *Quercus*

Fagus
BEECH

Fagus grandifolia Ehrh.
AMERICAN BEECH   *native / cons* 8
Tree, to 30 m tall, with smooth gray bark. Leaves simple, alternate,
straight-veined, a vein running to each tooth, short-petioled, ovate to
obovate, serrate to denticulate, densely silky when young, at maturity
glabrous above, beneath usually silky on the midvein and sometimes
more or less pubescent on the surface. Flowers appearing with the leaves,
the staminate flowers from the lower axils in small heads on drooping
peduncles, subtended by deciduous bracts, stamens 8-16; pistillate flowers
from the upper axils, usually in pairs at the end of a short peduncle, subtended and largely concealed by numerous subulate bracts; calyx adnate
to the ovary, with 6 short acuminate lobes; ovary 3-celled; styles 3. Fruit
a sharply 3-angled, 1-seeded nut, borne in pairs enclosed within the expanded 4-valved involucre. — Beech-maple forests mostly near Lake

Fagus grandifolia

Michigan; Wisconsin at western edge of species' range.

Quercus
OAK

Deciduous trees (or rarely shrubby). Leaves alternate, simple, lobed, pinnately veined. Plants monoecious (staminate and pistillate flowers separate but on same tree). Staminate flowers in slender naked catkins, catkins appearing with the leaves. Calyx divided to the base into 3-7 (usually 6) segments. Stamens 3-12. Pistillate flowers solitary or in small spikes, each subtended by a bract and surrounded by an involucre of many scales. Ovary 3-celled. Fruit a nut (acorn) partially enclosed by a cuplike structure (cupule).

Acorns are important food for many mammals and birds but are usually avoided by humans because of their high tannin content.

1 Lobes rounded, not tipped by bristles . . . . *Q. macrocarpa*
1 Lobes acute, bristle-tipped 2
2 Acorn cup covering lower one-third to lower one-half of acorn . . . . *Q. ellipsoidalis*
2 Acorn cup saucer-shaped; covering only base of acorn. . . . . *Q. rubra*

Quercus ellipsoidalis E.J. Hill
NORTHERN PIN OAK   *native / cons* 5

Medium-sized tree, the twigs soon glabrescent. Leaves smooth on both sides except for small tufts of stellate hairs in the vein-axils beneath; lateral lobes 2-3 pairs, separated by rounded sinuses, usually extending more than half-way to the midvein, sometimes broadest at the base, but usually widened and several-toothed distally, the sinuses then elliptic. Acorn cup turbinate, 9-14 mm wide, with closely appressed puberulent scales, covering about a third of the nut; nut ovoid to ellipsoid, 12-20 mm long. — Dry upland soil.

Quercus ellipsoidalis

Quercus macrocarpa Michx.
BUR-OAK   *native / cons* 5

Low shrub to tall tree, the latter with rough, deeply furrowed bark. Leaves obovate to obovate, cuneate at base, pale beneath with a close, fine, stellate pubescence, with 4-7 pairs of blunt or acute lateral lobes, a pair of sinuses near the middle usually deeper than the others. Acorns sessile or on a stout peduncle; cup deeply saucer-shaped to sub-globose, covering 1/3 to nearly all the nut, pubescent within, the marginal scales acuminate into slender awns forming a terminal fringe; nut depressed-ovoid, broadly rounded to almost retuse at the summit, to narrowly ovoid, 1-4 cm wide. — Moist woods and alluvial floodplains.

Quercus macrocarpa

Quercus rubra L.
NORTHERN RED OAK   *native / cons* 5

Medium to large tree. Young twigs glabrous, dark reddish brown. Leaves dull green, 10-20 cm long, soon glabrous throughout or with small tufts of pubescence persistent in the leaf-axils, 7-11-lobed, lobes roughly triangular in outline, broadest at the base, bristle-tipped and usually with a few lateral teeth, little if any longer than the central body of the blade. Acorn cup saucer-shaped or almost turbinate, 1.5-2 cm wide, enclosing about 1/4 to 1/3 of the nut; nut ovoid, 2-2.5 cm long. — Rich mesic forests, ridges, sandy plains with jack pine.

Quercus rubra

Gentianaceae

GENTIAN FAMILY

Annual, biennial or perennial herbs; plants usually glabrous. Leaves simple, entire, opposite or whorled, stem leaves without petioles. Flowers often showy, perfect (with both staminate and pistillate parts), regular, single at end of stems or in clusters; petals 4-5, blue, purple, white or green, joined for at least part of their length; stamens 4 or 5. Fruit a 2-chambered, many-seeded capsule enclosed by the withered, persistent petals.

1 Flowers pink . . . . *Centaurium*
1 Flowers blue, green tinged with purple, or white 2
2 Petals 4, spurred at base; flowers green, tinged with purple . . . . *Halenia*
2 Petals 4, with fringed lobes; or petals 5 and not spurred; blue, purple or white 3
3 Petals 4, fringed; flowers on stalks longer than the flowers; seeds covered with small bumps . . . . . .
   . . . . *Gentianopsis*
3 Petals 5, not fringed; flower stalks short or absent; seeds smooth . . . . *Gentiana*

Centaurium
CENTAURY

Centaurium pulchellum (Sw.) Druce
BRANCHED CENTAURY   *introduced*
Annual herb. Stems much branched, often from the base, 1-2 dm tall. Leaves sessile, lance-shaped or ovate lance-shaped, 1-2 cm long. Inflorescence a many-flowered terminal cyme; bracteal leaves linear. Flowers 4-merous; calyx tubular, about 9 mm long, deeply cleft into narrow segments; corolla tube slightly exceeding the calyx; corolla lobes pink, about 4 mm long; stamens inserted in the throut of the corolla; filaments slender, exsert; ovary elongate, 1-celled. Capsule oblong, thin-walled, invested by the persistent calyx and withered corolla. June-Sept. — Native of Europe; local in fields and waste places, often where salted in winter.

Centaurium pulchellum

Gentiana
GENTIAN
Perennial herbs, with thick, fibrous roots. Leaves opposite or whorled, simple, margins entire, petioles absent. Flowers large, blue, green-white or yellow, 5-parted, in clusters near ends of stems; petals forming a tubelike, shallowly lobed flower, the lobes alternating with a folded membrane as long or longer than petal lobes; stamens 5. Fruit a 2-chambered capsule.

1 Flowers blue (rarely white), remaining closed, the corolla lobes absent or reduced to small points . .
   . . . . *G. andrewsii*
1 Flowers blue, white or yellowish, opening, the corolla lobes prominent 2
2 Flowers yellowish or white, with greenish veins; leaves 5-veined . . . . *G. alba*
2 Flowers blue to purple; leaves 1–5-veined . . . . *G. rubricaulis*

Gentiana alba Muhl.
YELLOW GENTIAN   *native / cons 7*
   Gentiana flavida Gray
Perennial herb. Stems stout, 5-9 dm tall, almost always unbranched. Leaves divaricately spreading, lance-shaped or ovate lance-shaped, 5-13 cm long, 1.5-5 cm wide, long-acuminate, widest near the broadly rounded or subcordate sessile base, often distinctly 3-nerved. Inflorescence usually compact and many-flowered; calyx lobes ovate lance-shaped or ovate, 4-10 mm long, 2.5-5 mm wide, often spreading; corolla greenish white or yellowish white, 3-4.5 cm long, the broadly ovate lobes commonly 4-6 mm longer than the erose plaits. Seeds winged. Aug-Sept. — Clay soils in wooded ravines, open woodlands and woodland edges, bluffs, wet sandy prairies, and along railroads and in roadside ditches.

Gentiana andrewsii Griseb.
BOTTLE-GENTIAN   *native / cons 6*
Perennial herb. Stems erect, single or few together, 2-8 dm long, unbranched, smooth. Leaves opposite, lance-shaped, 4-12 cm long and 1-3

Gentiana andrewsii

cm wide, margins entire. Flowers 1 to many, stalkless in upper leaf axils, 3-5 cm long; sepals forming a tube around petals, the sepal lobes unequal, fringed with hairs; petals forming a tubelike flower, usually remaining closed, the folds between petal lobes finely fringed (use hand lens to see this) and longer than the petal lobes. Fruit a capsule; seeds winged. Aug–Sept. — Wet meadows, swamps and wet woods, thickets, low prairie, shores, ditches.

Gentiana rubricaulis Schwein.
GREAT LAKES GENTIAN   *native / cons* 7
Gentiana linearis var. *lanceolata* A. Gray
Perennial herb. Stems smooth, 3-7 dm long. Leaves pale green, lance-shaped, 4-8 cm long and 2-3 cm wide, margins entire. Flowers 3-5 cm long, green-blue below, blue above, narrowly open, in a cluster at end of stem; sepal lobes oblong, 4-12 mm long, chaffy and translucent near base. Fruit a capsule; seeds winged. — Wet meadows, peatlands, streambanks, thickets, conifer swamps; soils usually calcium-rich.

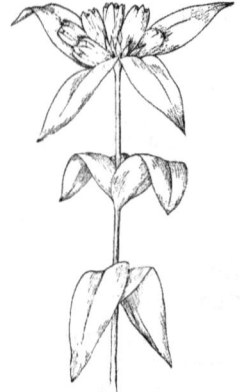

Gentiana rubricaulis

Gentianopsis
FRINGED-GENTIAN
Smooth, taprooted, annual or biennial herbs. Leaves opposite, stalkless, margins entire. Flowers 1 to several, showy, blue, sometimes tinged with white on outside, long-stalked, at ends of stems and branches, 4-parted; sepals oblong cone-shaped; petals deeply lobed, forming a tubular or bell-shaped flower, the lobes ragged or fringed at tips and sometimes on sides, without a folded membrane between the lobes (present in *Gentiana*); stamens 4. Fruit a capsule; seeds covered with bumps.

1   Upper leaves lance-shaped to ovate; petal lobes long-fringed across tip and sides, the fringes 2–5 mm
    long . . . . *G. crinita*
1   Upper leaves linear; tips of petal lobes ragged with short, fine teeth, and often fringed on sides . . .
    . . . . *G. virgata*

Gentianopsis crinita (Froel.) Ma
GREATER FRINGED-GENTIAN   *native / cons* 6
Annual or biennial herb. Stems erect, 2-7 dm long, usually branched above. Basal leaves spatula-shaped, smaller than stem leaves; stem leaves ovate, 2-6 cm long and 1-2.5 cm wide, the base usually clasping stem; margins entire. Flowers bright blue, 3-6 cm long, 4-parted, single at ends of main stems and branches, on stalks 5-20 cm long; sepals forming a tube, 1-2 cm long; petals joined to form a funnel-like to bell-shaped flower, the petal lobes fringed across tip and part way down sides with linear fringes 2-6 mm long. Fruit a capsule, broadest at middle. Aug-Oct. — Wet meadows, streambanks, ditches, wet woods; soils usually calcium-rich and sandy or gravelly.

Gentianopsis crinita

Gentianopsis virgata (Raf.) Holub
LESSER FRINGED-GENTIAN   *native / cons* 8
Gentianopsis procera (Holm) Ma
Annual herb, similar to *G. crinita* but smaller. Stems simple or few-branched, 1-5 dm long. Basal leaves spatula-shaped; stem leaves linear to linear lance-shaped, 2-5 cm long and 2-7 mm wide, tapered to a blunt tip, the base not clasping stem; margins entire. Flowers bright blue, 2-5 cm long, mostly 4-parted, single on stalks at ends of stems; sepal tube 6-15 mm long; petals forming a tubelike flower, flared toward tip, petal lobes ragged toothed across tips, often fringed on sides. Fruit a capsule. Sept-Oct. — Sandy and gravelly shores, wet meadows, fens, intradunal wetlands near Lake Michigan; soils usually calcium-rich.

Gentianopsis virgata

Halenia
SPURRED GENTIAN

Halenia deflexa (Sm.) Griseb.

SPURRED GENTIAN   *native / cons 7*

Annual herb. Stems erect, simple or few-branched, rounded 4-angled, 15–40 cm long. Leaves opposite, lower leaves spatula-shaped, narrowed to a petiole; stem leaves lance-shaped to ovate, 2-5 cm long and 1-2.5 cm wide, sessile; margins entire. Flowers green, tinged with purple, 10-12 mm long, 4-parted, on stalks to 4 cm long, in loose clusters of 5-9 flowers at ends of stems; petals lance-shaped, usually with downward-pointing spurs at base, the spurs to 5 mm long. Fruit an oblong capsule. July-Aug. — Cedar swamps, moist conifer woods (especially along shores), old logging roads.

Halenia deflexa

Geraniaceae
GERANIUM FAMILY

Geranium
WILD GERANIUM, CRANE'S-BILL

Annual or perennial herbs. Leaves usually opposite, palmately lobed, cleft, or divided, the stem leaves chiefly opposite. Flowers small or medium-sized, regular or nearly so, usually pedicellate in pairs at the ends of axillary peduncles. Sepals 5, imbricate. Petals 5, pink to purple, imbricate, alternating at base with 5 glands. Stamens 10. Fruit a carpel prolonged at maturity into beaks and eventually separating; seed 1 in each carpel.

1   Perennial, spreading by rhizomes; petals more than 11 mm long; leaves few, large . . . *G. maculatum*
1   Annuals or short-lived perennials; petals less than 11 mm long; stem leaves several to many 2
2   Leaves divided to their base. . . . . *G. robertianum*
2   Leaves deeply divided but not to their base . . . . *G. bicknellii*

Geranium bicknellii Britt.

NORTHERN CRANE'S-BILL   *native / cons 4*

Annual. Stems erect, usually with many ascending branches, eventually to 5 dm long. Leaves pentagonal in outline, the principal ones cleft nearly to the base with usually 5 segments, these deeply incised with several parrowly oblong lobes. Peduncles 2-flowered, the elongate pedicels glandular-villous; sepals at anthesis 7-9 mm long, including the conspicuous subulate tips; petals pink-purple, about equaling the sepals. Mature fruit, including the calyx, 20-25 mm long, the beak 4-5 mm long, the body 3 mm long, sparsely hirsute. May-Sept. — Open woods and fields, usually where sandy or gravelly.

Geranium bicknellii

Geranium maculatum L.

SPOTTED CRANE'S-BILL   *native / cons 4*

Perennial from a thick rhizome. Stems erect, 3-7 dm long. Basal leaves long-petioled, pedately 5-7-cleft into wedge-like segments; stem leaves a single pair, resembling the basal but short-petioled. Flowers few to several, rose-purple, 2.5-4 cm wide; calyx and pedicels pubescent but not glandular. Fruit erect, the beak 2-3 cm long; seed minutely reticulate. April-June. — Dry or moist woods.

Geranium robertianum L.

HERB-ROBERT   *native / cons 2*

Weak annual or winter-annual. Stems to 6 dm long, branched, spreading, villous. Leaves triangular in outline, 3-divided, the lateral segments often again divided and all segments pinnately lobed or cleft. Peduncles from

Geranium maculatum

most of the upper nodes, usually 2-flowered. Flowers pink to red-purple, 10-15 mm wide. Carpels detached from the cauline beak, each terminating in 2 slender filaments. May-Sept. — Damp rich woods.

Grossulariaceae

CURRANT FAMILY

Ribes
CURRANT, GOOSEBERRY
Medium shrubs. Stems upright to spreading, smooth, or with spines at nodes and sometimes also with bristles between nodes. Leaves alternate, palmately veined and palmately 3-5-lobed, margins toothed. Flowers 1 to several in short clusters, or few to many in racemes; green to white or yellow, perfect, regular, ovary inferior; sepals 5; petals 5, shorter than sepals; stamens 5, alternate with petals, styles 2. Fruit a many-seeded berry, usually topped by persistent, dry flower parts. *Ribes* are of two types: currants and gooseberries. Currants lack spines and bristles (except in *R. lacustre*) and the stalk of berry is jointed at its tip so that berries detach from stalks. Gooseberries have spines and bristles and the berry stalk is not jointed so that stalks remain attached to berries when picked.

1 Stems with spines or bristles, at least at the nodes; flowers single or in corymb-like clusters of 2–3 (gooseberries) 2
1 Flowers in racemes of 5 or more; stems without spines or bristles (except in *R. lacustre*; currants) . . 4
2 Ovary and fruit usually bristly; calyx lobes shorter than corolla tube .... *R. cynosbati*
2 Ovaries and fruit smooth (or bristly in *R. oxyacanthoides*); calyx lobes longer than corolla tube 3
3 Leaves with glands, at least on underside veins; fruit bristly or with gland-tipped hairs to sometimes smooth .... *R. oxyacanthoides*
3 Leaves without glands; fruit smooth .... *R. hirtellum*
4 Ovary and fruit bristly with gland-tipped hairs 5
4 Ovary and fruit neither bristly nor with gland-tipped hairs 6
5 Stems densely bristly .... *R. lacustre*
5 Stems unarmed .... *R. glandulosum*
6 Leaf underside dotted with shiny resinous glands; fruit black 7
6 Leaf underside without resinous glands 8
7 Flowers yellow to greenish; calyx glabrous or sparsely hairy; inflorescence bracts longer than pedicels .... *R. americanum*
7 Flowers white to greenish-white; calyx hairy; inflorescence bracts much shorter than pedicels .... *R. hudsonianum*
8 Flowers golden-yellow; fruit black .... *R. odoratum*
8 Flowers yellow-green; fruit red 9
9 Pedicels with scattered hairs and short-stalked glands .... *R. triste*
9 Pedicels glabrous .... *R. rubrum*

Ribes americanum P. Mill.
WILD BLACK CURRANT   *native / cons 4*
Shrub, 1-1.2 m tall. Stems without spines or bristles, young stems finely hairy; branches upright to spreading; twigs gray-brown and smooth, black with age. Leaves 3-8 cm long and 3-10 cm wide, 3-lobed and usually with 2 additional shallow lobes at base, dotted with shiny, yellow to brown resinous glands, especially on underside, smooth or short-hairy above, hairy below; margins coarsely toothed; petioles hairy and resin-dotted, 3-6 cm long. Flowers creamy-white to yellow, bell-shaped, 8-12 mm long; 6-15 in drooping racemes 3-8 cm long; each flower with a linear bract longer than the flower stalk, the stalks 2-3 mm long; sepals 4-5 mm long, rounded; petals blunt, 2-3 mm long; stamens about equaling petals. Fruit an edible, smooth, black berry. April-June. — Moist to wet forests, swamps, marsh and lake borders, streambanks.

Ribes americanum

Ribes cynosbati L.

EASTERN PRICKLY GOOSEBERRY   *native / cons* 3

Shrub to 6-9 dm tall, branches upright to spreading. Stems and branches with 1-3 spines at nodes, outer bark peeling off, inner bark brown-purple to black; young stems brown-gray, finely hairy. Leaves 3-8 cm long and 3-7 cm wide, 3-5-lobed, the lobes rounded at tips; upper surface dark green, sparsely hairy, underside paler, finely hairy and with gland-tipped hairs along veins; margins with coarse, round teeth; petioles 2.5-4 cm long, finely hairy and with scattered gland-tipped hairs. Flowers green-yellow, bell-shaped, 6-9 mm long, in clusters of 2-3 from spurs on old wood, on stalks with gland-tipped hairs. Fruit a red-purple berry, covered with stiff, brown spines. May-June. — Occasional in wet woods, swamps, thickets and streambanks; more typical in moist hardwood forests.

Ribes cynosbati

Ribes glandulosum Grauer

SKUNK CURRANT   *native / cons* 7

Ribes ruizii Rehder

Shrub to 8 dm tall. Stems sprawling, spines and bristles absent. Stems and leaves with skunklike odor when crushed; older stems smooth and dark as outer bark peels off; young stems smooth to finely hairy, brown-gray. Leaves 2-8 cm long and 4-8 cm wide, 3-5-lobed, smooth above, paler and finely glandular hairy below (at least along veins); margins toothed or double-toothed; petioles 3-6 cm long, finely hairy. Flowers yellow-green to purple, saucer-shaped, in loose upright clusters 3-6 cm long, on slender stalks; bracts very small, the stalks and bracts with gland-tipped hairs; sepals 2 mm long; petals 1-2 mm long. Fruit a dark red berry with bristles and gland-tipped hairs, 6 mm wide. June. — Cedar and tamarack swamps, cool wet woods, thickets and streambanks.

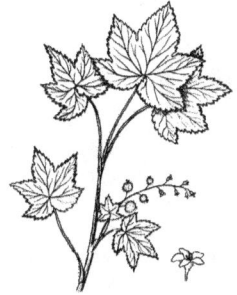

Ribes glandulosum

Ribes hirtellum Michx.

HAIRY-STEM GOOSEBERRY   *native / cons* 6

Shrub to 9 dm tall. Stems upright, outer bark pale, soon peeling to expose dark inner layer; young stems gray and smooth, or with 1-3 slender spines at nodes and scattered bristles between nodes. Leaves 2.5-5 cm long and 2-5 cm wide, with 3 or 5 pointed lobes, upper surface dark green, smooth to sparsely hairy, lower surface paler, hairy at least along veins, without glands; margins coarsely toothed and fringed with hairs; petioles 1-3 cm long, hairy, some of which are gland-tipped. Flowers green-yellow to purple, bell-shaped, 6-9 mm long, in clusters of 2-3 on short, smooth stalks; stamens as long or longer than sepals, the bracts fringed with long hairs. Fruit an edible, smooth, dark blue-black berry. June. — Cedar and tamarack swamps, thickets, shores, rocky openings.

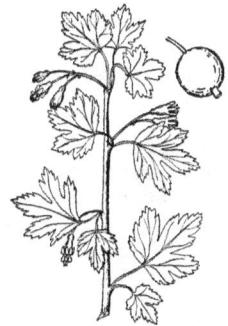

Ribes hirtellum

Ribes hudsonianum Richards.

HUDSON BAY CURRANT   *native / cons* 10

Shrub, 6-9 dm tall. Stems upright, spines and bristles absent; bark gray, with scattered yellow resin dots, peeling to expose inner purple-black bark. Leaves 5-9 cm long and 6-13 cm wide, 3-5-lobed, with unpleasant odor when rubbed, upper surface dark green and mostly hairless, underside paler, smooth to hairy and with yellow resin dots; margins coarsely toothed, the teeth with a hard tip; petioles 2.5-8 cm long, with fine hairs and resin dots. Flowers white, bell-shaped, 4-5 mm long, in small clusters on threadlike stalks. Fruit a smooth, blue-black berry, barely edible. June. — Cedar swamps, wet conifer woods and streambanks.

Ribes hudsonianum

Ribes lacustre (Pers.) Poir.

BRISTLY BLACK GOOSEBERRY   *native / cons* 9

Shrub to 1 m tall. Stems upright or spreading, densely bristly, long-spiny at nodes; older bark gray, peeling to expose dark inner bark. Leaves 4-8 cm long and 4-7 cm wide, with 3-5 deeply parted, pointed lobes, upper surface dark green and mostly smooth, underside paler with scattered gland-tipped hairs; margins cleft into rounded teeth; petioles 2.5-4 cm long, with gland-tipped hairs. Flowers yellow-green to pinkish, saucer-shaped, 4-5 mm wide, on stalks with dark, gland-tipped hairs, in arching or drooping clusters. Fruit palatable but insipid, red, becoming black or dark purple, covered with gland-tipped hairs. May-June. — Moist conifer woods, swamps, thickets, and rock outcrops.

Ribes odoratum H. Wendl.
BUFFALO-CURRANT   *introduced*
   Ribes aureum Pursh var. *villosum* DC.
Erect shrub. Leaves orbicular to cuneate-obovate in outline, broadly cuneate to truncate at base, deeply 3-lobed or rarely 5-lobed, ciliate, finely puberulent to glabrate beneath, the lobes entire in their lower half, entire or few-toothed across the summit. Flowers golden yellow; hypanthium above the ovary tubular, 11-15 mm long; sepals obovate, 5-6.5 mm long, broadly rounded above; petals 2.5-3.5 mm long, erose at the summit. Fruit yellow or black, edible. Apr-June. — Rocky hillsides; widely cultivated and sometimes escaped.

Ribes lacustre

Ribes oxyacanthoides L.
NORTHERN GOOSEBERRY   *threatened / native / cons* **9**
Shrub to 1 m tall. Stems upright with 1-3 spines to 1 cm long at nodes and smaller spines scattered between nodes; young stems gray-brown and finely hairy. Leaves 2.5-5 cm long and 2-5 cm wide, with 3-5 blunt or rounded lobes, upper surface sparsely hairy, some hairs tipped with glands, underside resin-dotted, hairy, some gland-tipped, especially along veins; margins coarsely toothed and hairy; petioles 0.5-3 cm long, with short hairs and scattered glands. Flowers green-yellow, bell-shaped, 6-9 mm long, in clusters of 2-3 on short stalks; stamens shorter than petals. Fruit a smooth, edible, blue-black berry. June. — Rocky and sandy shores, rocky openings, cold moist woods.

Ribes rubrum L.
GARDEN RED CURRANT   *introduced*
   Ribes sativum Syme
Erect shrub. Stems erect, nearly glabrous, crisped-puberulent; spines at nodes absent; prickles on internodes absent. Leaves commonly 5-lobed, the lateral lobes spreading. Flowers cream to pinkish. Fruit a bright red berry, glabrous, sour. — Native of the Old World; long in cultivation and occasionally escaped.

Ribes oxyacanthoides

Ribes triste Pallas
SWAMP RED CURRANT   *native / cons* **8**
Low shrub, 0.4-1 m tall. Stems spreading or lying on ground and rooting at nodes, spines and bristles absent; older stems smooth, purple-black, young stems short-hairy. Leaves 4-10 cm long and 4-10 cm wide, with 3-5 broad lobes, dark green and mostly smooth above, paler and usually finely hairy below; margins with both rounded and sharp teeth, the teeth with a hard tip; petioles 2.5-6 cm long, with scattered gland-tipped hairs. Flowers green-purple, 4-5 mm wide, on stalks 1-4 mm long, in drooping clusters of 5-12. Fruit a red berry, glabrous, sour-tasting. May-June. — Wet woods swamps, alder thickets, seeps.

Ribes triste

Haloragaceae

WATER-MILFOIL FAMILY

Perennial aquatic herbs. Leaves alternate or whorled, finely dissected. Flowers small, stalkless in axils of leaves or bracts, 3- or 4-parted, regular, perfect (with both staminate and pistillate parts), or imperfect, petals small or absent. Fruit small and nutlike, dividing into 3 or 4 segments (mericarps).

1   Flowers 4-parted; leaves mostly whorled, emersed leaves reduced to small bracts . . . . *Myriophyllum*
1   Flowers 3-parted; leaves alternate, emersed leaves not bract-like . . . . *Proserpinaca*

Myriophyllum

WATER-MILFOIL

Perennial aquatic herbs. Stems submerged, sparsely branched, freely rooting at lower nodes. Leaves mostly whorled, pinnately divided into threadlike segments, upper leaves often reduced to bracts. Flowers small, mostly imperfect, stalkless in axils of upper emersed leaves (the floral bracts) or axils of underwater leaves; staminate flowers above pistillate flowers; perfect flowers (if present) in middle portion of spike; sepals inconspicuous; petals 4 or absent; stamens 4 or 8; pistil 4-chambered. Fruit nutlike, 4-lobed, each lobe (mericarp) with 1 seed, rounded on back or with a ridge or row of small bumps.

1   Leaves simple, reduced to small, blunt-tipped scales; stems erect and crowded from creeping rhi-
     zomes . . . . *M. tenellum*
1   Leaves dissected into narrow segments 2
2   Flowers and bracts below flowers alternate on stem . . . . *M. alterniflorum*
2   Flowers and bracts below flowers whorled 3
3   Bracts surrounding staminate flowers deeply cleft . . . . *M. verticillatum*
3   Bracts surrounding staminate flowers sharply toothed or entire 4
4   Bracts sharply toothed and much longer than flowers . . . . *M. heterophyllum*
4   Bracts surrounding staminate flowers entire and not longer than flowers 5
5   Leaf segments mostly 5–12 on each side of midrib; small bulbs (turions) produced at ends of stems
     and in upper leaf axils . . . . *M. sibiricum*
5   Leaf segments many, 12–20 on each side of midrib; turions absent . . . . *M. spicatum*

Myriophyllum alterniflorum DC.

ALTERNATE-FLOWER WATER-MILFOIL   *native / cons* 10

Perennial herb. Stems very slender. Leaves in whorls of 3-5, usually less than 1 cm long and shorter than the stem internodes, pinnately divided. Flower spikes raised above water surface, 2-5 cm long; bracts mostly alternate, linear, shorter than the flowers; staminate flowers with 4 pink petals; stamens 8. Fruit segments 1-2 mm long, rounded on back and base. — Acidic lakes.

Myriophyllum alterniflorum

Myriophyllum heterophyllum Michx.

TWO-LEAF WATER-MILFOIL   *native / cons* 7

Perennial herb. Stems stout, to 3 mm wide, often red-tinged, to 1 m or more long. Leaves appearing whorled due to the very short internodes, 1.5-4 cm long, divided into threadlike segments. Flowers in spikes raised above water surface, 5-30 cm long; floral bracts whorled, smaller than foliage leaves, ovate, sharply toothed, spreading or curved downward. Flowers both perfect and imperfect; petals of staminate and perfect flowers 1-3 mm long; stamens 4. Fruit olive, more or less round, 2 mm long; fruit segments rounded or with 2 small ridges, beaked by the curved stigma. June-Aug. — Lakes, ponds and pools in streams; sometimes where calcium-rich.

Myriophyllum heterophyllum

Myriophyllum sibiricum Komarov
COMMON WATER-MILFOIL   *native / cons* 6
Myriophyllum exalbescens Fern.

Perennial herb; plants often whitish when dried. Stems to 1 m or more long. Leaves in whorls of 3-4, 1-4 cm long, with mostly 5-10 threadlike segments on each side of midrib; internodes between whorls about 1 cm long. Flowers in spikes with whorled flowers and bracts, raised above water surface, red, clearly different than underwater stems, 4-10 cm long; flowers imperfect, the upper staminate, the lower pistillate; floral bracts much smaller than the leaves, oblong to obovate; staminate flowers with pinkish petals (absent in pistillate flowers), 2-3 mm long; stamens 8, the yellow-green anthers conspicuous when flowering. Fruit more or less round, 2-3 mm long, the segments rounded on back. June-Sept. — Shallow to deep water of lakes, ponds, marshes, ditches and slow-moving streams; sometimes where calcium-rich. When flowering, the numerous red spikes of this species are conspicuous on water surface.

M. spicatum, introduced from Eurasia, is similar but has more finely divided leaves (12-24 threadlike segments on each side of midrib) and larger floral bracts.

Myriophyllum spicatum

Myriophyllum spicatum L.
EURASIAN WATER-MILFOIL   *introduced (invasive)*

Perennial herb, similar to *M. sibiricum.* Stems widening below head and curved to a horizontal position, usually many-branched near water surface, internodes between leaves mostly 1-3 cm long, turions absent. Leaves with more leaf segments per side (mostly 12-20) than in *M. sibiricum;* lower flower bracts often divided into comblike segments and often longer than the flowers. Fruit segments 2-3 mm long. Aug-Sept. — Lakes and ponds. Introduced from Europe and invasive in lakes throughout e USA.

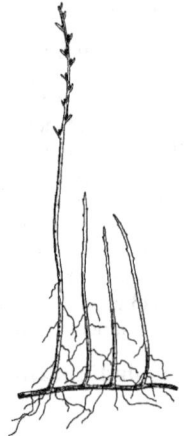

Myriophyllum tenellum Bigelow
SLENDER WATER-MILFOIL   *native / cons* 10

Perennial herb. Stems slender, 10-30 cm long, mostly upright and un-branched. Leaves absent or reduced to a few spaced scales. Flowers in spikes raised above water surface, 2-5 cm long; flower bracts mostly alternate, oblong to obovate, entire, shorter to slightly longer than the flowers. Fruit segments rounded on back and at base, 1 mm long. — Acidic lakes; often forming large colonies, especially in deep water.

Myriophyllum tenellum

Myriophyllum verticillatum L.
WHORLED WATER-MILFOIL   *native / cons* 8

Perennial herb, similar to *M. sibiricum,* but plants often larger. Stems 5-25 dm long. Leaves in whorls of 4-5, with 9-17 threadlike segments along each side of midrib, 1-5 cm long; lower and middle internodes between whorls mostly less than 1 cm long. Flowers perfect, or the lower pistillate and upper staminate; in spikes 4-12 cm long, the floral bracts much smaller than the leaves, with comblike segments, mostly longer than the flowers; petals blunt-tipped, 2-3 mm long, smaller in pistillate flowers; stamens 8. Fruit more or less round, 2-3 mm long, the segments rounded on back. July-Sept. — Lakes, ponds, quiet rivers.

Proserpinaca
MERMAID-WEED

Myriophyllum verticillatum

Proserpinaca palustris L.
COMMON MERMAID-WEED   *native / cons* 8

Perennial aquatic herb, often forming large colonies. Stems horizontal at base and often rooting; the flower-bearing branches erect, 1-4 dm tall. Leaves alternate; underwater leaves, if present, ovate in outline, 2-4 cm long, deeply divided into linear segments; emersed leaves narrowly lance-shaped, 2-6 cm long, margins with sharp, forward-pointing teeth. Flowers small, perfect, green or purple-tinged, 1-3 in axils of emersed leaves, stalk-less; sepals triangle-shaped, persistent; petals absent, stamens 3, stigmas 3. Fruit nutlike, 3-angled, 2-5 mm long and as wide, with 3 seeds. June-Aug. — Shallow water of ponds, streambanks and ditches, muddy shores, sedge meadows; usually where seasonally flooded.

Proserpinaca palustris

Hamamelidaceae
WITCH-HAZEL FAMILY

Hamamelis
WITCH-HAZEL

Hamamelis virginiana L.
AMERICAN WITCH-HAZEL   *native / cons 7*
Tall shrub to 5 m tall, with scurfy or glabrous twigs. Leaves broadly obo-vate, with several to many rounded teeth, base broadly rounded or sub-cordate, green on both sides, glabrous or stellate-pubescent beneath. Flowers in short-pediceled axillary clusters; 4-merous; sepals small, tri-angular, dull yellowish brown within; petals bright yellow or suffused with red, spreading, 1.5-2 cm long; stamens 4, opposite the sepals and much shorter than them, alternating with 4 small scale-like staminodia; styles 2. Fruit ovoid before dehiscence, 1-1.5 cm long, the hypanthium of-ten bearing the persistent sepals; seeds black, eventually discharged ex-plosively from the capsule. Oct-Nov; fruit ripe a year later. — Moist woods.
    Witch hazel extract is derived from the bark.

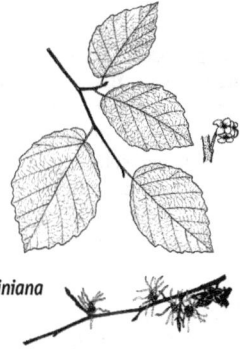

Hamamelis virginiana

Hypericaceae
ST. JOHN'S-WORT FAMILY

Glabrous annual or perennial herbs (shrubby in *Hypericum kalmianum*). Stems usually unbranched below, branched in head. Leaves simple, opposite, dotted with dark or translucent glands (visible when held to light), especially on underside; margins entire; petioles absent. Flowers few to many in clusters at ends of stems or from upper leaf axils, perfect, regular, sepals 5, petals 5, yellow or pink to green or purple; stamens 9-35, separate or joined near base into 3 or more groups; styles 3, ovary superior. Fruit a 3-chambered, many-seeded capsule.

1 Petals yellow; stamens 15–many .... *Hypericum*
1 Petals pink or purple; stamens 9 .... *Triadenum*

Hypericum
ST. JOHN'S-WORT

Shrubs or herbs. Leaves opposite, sometimes dotted with black and/or small transparent glands; margins entire. Flowers in clusters at ends of stems and upper leaf axils, yellow, perfect, regular, sepals 5, petals 5, stamens 5-many, separate or joined into 3 or 5 bundles. Fruit a capsule.

1 Styles joined at base, persisting on capsule as a straight beak; stamens many, distinct *H. kalmianum*
1 Styles free to base, the capsules not beaked; stamens few to many, joined at base into 3 or 5 bundles
  2
2 Plants 1–2 m tall; leaves 5 cm long or more; flowers 4 cm or more wide; styles 5 .... *H. ascyron*
2 Plants usually less than 1 m tall; leaves less than 5 cm long; flowers to 3 cm wide; styles 3 3
3 Petals spotted with black dots; stamens in 3 weak groups .... *H. perforatum*
3 Petals not spotted with black dots; stamens in 5 weak groups .... *H. canadense*

Hypericum ascyron L.
GREAT ST. JOHN'S-WORT   *native / cons* 6
Hypericum pyramidatum Ait.
Perennial herb. Stems upright, branched, 6-20 dm long. Leaves lance-shaped to oval, 4-10 cm long and 1-4 cm wide, base often clasping stem; petioles absent. Flowers few, 4-6 cm wide, mostly single on stalks from upper leaf axils; stamens numerous, joined at base into 5 bundles; petals bright yellow; styles 5, not persisting. Fruit an ovate, 5-chambered capsule, 15-30 mm long. July-Aug. — Streambanks, ditches, fen and marsh margins.

Hypericum ascyron

Hypericum canadense L.
LESSER CANADIAN ST. JOHN'S-WORT   *native / cons* 7
Annual or perennial herb, with short leafy stolons from base of plant. Stems upright, branched, 1-6 dm long. Leaves linear, 1-4 cm long and 1-4 mm wide, blunt-tipped, mostly 1-nerved, bracts much smaller; petioles absent. Flowers in open clusters at ends of stems and from upper leaf axils; sepals lance-shaped, 3-5 mm long; petals yellow, 2-3 mm long; stamens 12-22; styles 3 (sometimes 4), less than 1 mm long. Fruit a purple capsule 4-6 mm long. July-Sept. — Sandy shores, wetland margins, ditches.

Hypericum kalmianum L.
KALM'S ST. JOHN'S-WORT   *native / cons* 9
Branched shrub to 1 m tall; branches 4-angled, twigs flattened. Leaves linear, 2-4 cm long and 3-8 mm wide, often waxy on underside; margins sometimes rolled under; petioles absent. Flowers in clusters of 3-7 at ends of stems, yellow, 2-3.5 cm wide; stamens many, not joined; styles 5. Fruit a 5-chambered, ovate capsule, 7-10 mm long, beaked by the persistent style base. June-Sept. — Dunes (especially wet areas between dunes) and rocky lakeshores, often on limestone or where calcium-rich.

Hypericum kalmianum

Hypericum perforatum L.
COMMON ST. JOHN'S-WORT   *introduced (invasive)*
Perennial herb. Stems 4-6 dm tall, with numerous very leafy decussate branches. Leaves sessile, linear-oblong, commonly 2-4 cm long, on the main axis, about half as large on the branches. Flowers numerous, forming a large rounded or flattened compound cyme; sepals narrowly lance-shaped, acuminate, 4-6 mm long, with few or no black glands; petals oblong, 8-10 mm long, black-dotted near the margin. Seeds 1-1.3 mm long. June-Sept. — Native of Europe; common as a weed in fields, meadows, and roadsides.

Hypericum perforatum

Triadenum
MARSH ST. JOHN'S-WORT
Glabrous perennial herbs. Leaves opposite, entire and oval-shaped, ours dotted with small dark and transparent glands. Flowers pink to green-purple, in clusters at ends of stems and from leaf axils; stamens 9, in 3 groups of 3; sepals 5; petals 5; styles 3. Fruit a cylindric capsule.

1 Sepals 3-4 mm long, oval and rounded at tip; styles mostly less than 1 mm long . . . . *T. fraseri*
1 Sepals 5-8 mm long, lance-shaped and tapered to a tip; styles 2-3 mm long . . . . *T. virginicum*

Triadenum fraseri (Spach) Gleason
FRASER'S MARSH-ST. JOHN'S-WORT *native / cons* 8
Hypericum virginicum L. var. *fraseri* (Spach) Fern.
Perennial herb, with creeping rhizomes. Stems upright, mostly unbranched, red, smooth, 3-6 dm long. Leaves oval or ovate, 3-6 cm long and 1-3 cm wide, pinnately veined, rounded at tip, rounded or heart-shaped and clasping at the base, with dark dots and transparent glands

on underside. Flowers in clusters at ends of stems and from leaf axils; sepals 3-5 mm long, rounded at tip; petals pink to green-purple, 5-8 mm long; stamens 9, joined at base into 3 bundles, the bundles alternating with orange glands; styles 1-2 mm long. Fruit a purple, cylindric capsule, 7-12 mm long, abruptly narrowed to the 1 mm long persistent style beak. July-Aug. — Marshes, sedge meadows, open bogs, fens, sandy and calcium-rich shores.

Triadenum virginicum (L.) Raf.
VIRGINIA MARSH-ST. JOHN'S-WORT   *native / cons* **9**
  Hypericum virginicum L.
Perennial herb spreading by rhizomes, similar to T. fraseri but with larger flowers. Stems upright, mostly unbranched, red, smooth, 3-6 dm long. Leaves oblong, oval or ovate, 3-6 cm long and 1-3 cm wide, pinnately veined, rounded at tip, more or less heart-shaped and clasping at base, with dark dots and transparent glands on leaf underside. Flowers in clusters at ends of stems and from leaf axils; sepals 5-8 mm long, lance-shaped; petals pink to green-purple, 8-10 mm long; stamens 9, joined at base into 3 bundles; styles 1-2 mm long. Fruit a red-purple cylindric capsule, 8-12 mm long, gradually tapered to the 2-3 mm long persistent style beak. July-Aug. — Sphagnum bogs, wet meadows, shores; disjunct from Atlantic coast.

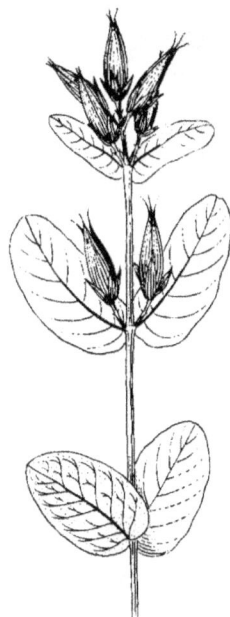

Triadenum virginicum

Juglandaceae
WALNUT FAMILY
Trees. Leaves alternate, odd-pinnate. Flowers monoecious. Staminate flowers in elongate catkins, each composed of a 2-6-lobed calyx, subtended by a narrow bract, and bearing few to many stamens on its upper side. Pistillate flowers terminating the young branches, each subtended by a perianth-like, cup-shaped involucre formed of connate bracts. Ovary 1-celled, tipped with 2 plumose stigmas. Fruit large, consisting of a fleshy or woody exocarp enclosing a nut; embryo large and oily, without endosperm.

1 Leaflets mostly 5–9, the terminal leaflet largest . . . . *Carya*
1 Leaflets 11–23, the lateral leaflets largest . . . . *Juglans*

Carya
HICKORY
Trees with hard heavy wood. All hickories are more or less stellate-pubescent, at least when young, and leaves, buds, and fruit also copiously covered with resin when young. Leaves odd-pinnate, the 3 terminal leaflets the largest. Flowers appear in spring as the leaves open. Staminate catkins slender, elongate, borne in peduncled groups of 3 at the summit of the previous year's growth or the base of that of the current year. Pistillate flowers solitary or in spikes of 2-10, terminating the branches, each subtended by a cup-shaped, 4-lobed, perianth-like involucre. Fruit a hard-shelled nut, enclosed within the expanded, 4-valved involucre.

Carya cordiformis (Wangenh.) K. Koch
BITTERNUT HICKORY   *native / cons* **6**
Tree; bark scaly. Winter-buds bright orange-yellow. Leaflets commonly 7 or 9, occasionally 5, rarely 11, the lateral lance-shaped to ovate lance-shaped, the terminal commonly long-cuneate at base and nearly or quite sessile. Involucre obovoid to subglobose, often somewhat flattened, 2.5-3.5 cm long, winged chiefly above the middle, splitting about to the middle; nut subglobose to obovoid, 1.5-3 cm long, at least two-thirds as thick, obscurely angled, otherwise smooth, tipped with a slender persistent point. Kernel bitter. — Dry or moist forests.

Carya cordiformis

Juglans
WALNUT

Trees. Leaves glandular-pubescent, odd-pinnate, the median lateral leaflets the largest. Staminate catkins protruding from the buds in autumn, elongating in spring, densely flowered, pendulous. Calyx spreading, 3-6-lobed, with 8-40 stamens on its upper side. Pistillate flowers in short spikes terminating the branches, composed of a 3-lobed, cup-shaped involucre. Fruit clammy-glandular. Nut indehiscent but 2-valved.

Juglans cinerea L.
BUTTERNUT, WHITE WALNUT   *native / cons* 6

Tree, to 30 m tall; bark grayish brown, with smooth ridges; pith dark brown. Leaflets commonly 11-17, oblong lance-shaped, somewhat pointed at tip. Nut ovoid, ovoid-oblong, or short-cylindric, very rough, marked with 2 or 4 obscure longitudinal ridges. — Rich moist soil.

Juglans cinerea

Lamiaceae

MINT FAMILY

Perennial, often aromatic, herbs. Stems usually 4-angled. Leaves simple, opposite, sharply toothed or deeply lobed. Flowers in leaf axils or in heads or spikes at ends of stems, perfect (with both staminate and pistillate parts), nearly regular to irregular; sepals 5-toothed or sometimes 2-lipped; petals white, pink, blue or purple, often 2-lipped; stamens 2 or 4; ovary 4-lobed, splitting into 4, 1-seeded nutlets when mature.

ADDITIONAL SPECIES
Origanum vulgare L. (Wild marjoram), roadsides.
Salvia azurea Michx. ex Lam. (Blue sage), native of s USA, grassy lawns (Chambers Island).

1 Calyx with a distinct cap or protuberance on the upper side of the tube . . . . *Scutellaria*
1 Calyx without a cap or protuberance on the tube 2
2 Upper lip of the corolla very short, or its lobes adjacent to the margins of the lower lip, the corolla
  thus appearing to be 1-lipped . . . . *Teucrium*
2 Upper lip of the corolla well developed, entire or 2-lobed, or the corolla regular or nearly so 3
3 Stamens included and hidden within the corolla tube . . . . *Glechoma*
3 Stamens exserted beyond the throat of the corolla 4
4 Stamens 2 . . . . *Group A*
4 Stamens 4 5
5 Inflorescence appearing axillary, the verticils (whorls of flowers around the stem) several to many,
  subtended by normal leaves and separated from one another by normal interodes, or the uppermost
  subtending leaves smaller and internodes shorter (not including plants with axillary spikes or racemes)
  . . . . *Group B*
5 Inflorescence appearing terminal, the verticils 1 to many, all or mostly subtended by bract-like leaves
  different from the main leaves, or separated by much shorter internodes (and including plants with
  lateral or axillary spikes) 6
6 Flowers single in the axils of each bract-like leaf, the verticils with 1 or 2 flowers . . . . *Physostegia*
6 Flowers 2–many in the axil of each bract-like leaf, the verticils with 4 or more flowers . . . . . *Group C*

GROUP A

1 Calyx distinctly 2-lipped . . . . *Hedeoma*
1 Calyx regular or nearly so, the lobes alike in size and shape 2
2 Corolla very irregular, 15–50 mm long . . . . *Monarda*
2 Corolla regular or nearly so, to 5 mm long . . . . *Lycopus*

GROUP B

1 Calyx regular or nearly so, the lobes of the upper and lower lips similar in shape and size 2
1 Calyx distinctly 2-lipped, the lobes of the upper and lower lips of different size and shape 5
2 Corolla about equally 4- or 5-lobed . . . . *Mentha*
2 Corolla strongly 2-lipped, the upper lip concave and arched over the stamens 3
3 Flowers distinctly pediceled, forming loosely flowered cymules (the clusters making up a cyme) . . .
    . . . . *Glechoma*
3 Flowers sessile in the cymules 4
4 Lower corolla lip with 2 yellow or white protuberances at its base . . . . *Galeopsis*
4 Lower corolla lip without protuberances . . . . *Leonurus*
5 Stamens projecting beyond the corolla . . . . *Mentha*
5 Stamens ascending under the upper lip of the corolla but not longer than the lip . . . . . *Clinopodium*

GROUP C

1 Stamens ascending under the upper corolla lip but not longer than the lip 2
1 At least some of the stamens protruding from the corolla 7
2 Calyx distinctly 2-lipped and irregular 3
2 Calyx regular or nearly so, the lobes all alike or differing in size only 4
3 Bracts broadly rounded, abruptly tapered at the tip to a short sharp point . . . . *Prunella*
3 Bracts awl-shaped, coarsely hairy . . . . *Clinopodium*
4 Leaves linear, entire, sessile 5
4 Leaves wider than linear, or the margins toothed, or petioled 6
5 Stems finely pubescent . . . . *Clinopodium*
5 Stems glabrous, or with small hairs on the angles only . . . . *Stachys*
6 Calyx 15-nerved; lower verticils often with distinct peduncles . . . . *Nepeta*
6 Calyx 5-10-nerved; lower verticils sessile . . . . *Stachys*
7 Inflorescence a dense or loose raceme in which the component verticils are plainly visible; flowers on
    distinctly short pedicels . . . . *Mentha*
7 Inflorescence otherwise . . . . *Agastache*

Agastache
GIANT-HYSSOP
Agastache scrophulariifolia (Willd.) Kuntze
PURPLE GIANT-HYSSOP   *native / cons 4*
Perennial herbs. Leaves varying from glabrous to villous beneath. Flowers small, numerous in dense, cylindric to somewhat tapering spikes to 15 cm long, 1.5-2 cm wide (including the corollas), usually continuous, occasionally with 1 or 2 separate verticils at base; bracteal leaves often projecting, round-ovate; calyx glabrous at anthesis, 7-9 mm long; calyx lobes 2-2.5 mm long. Stamens 4, exsert beyond the corolla lobes, the 2 lower stamens curved upward under the upper corolla-lip, the 2 upper stamens curved downward. Fruit a nutlet minutely pubescent at tip. Aug- Sept. — Upland woods.

Agastache scrophulariifolia

Clinopodium
WILD BASIL
Annual or perennial herbs. Calyx tubular to campanulate, conspicuously 10-13-nerved, often hairy in the throat, 2-lipped or regular, the lobes subulate to triangular. Corolla tube widened toward the summit; upper lip flat or slightly concave, straight to somewhat spreading, entire; lower lip deflexed, 3-lobed. Stamens 4, ascending under the upper lip of the corolla, the upper pair distinctly shorter than the lower. Nutlets smooth.

1 Stem leaves broadly elliptic-ovate, mostly 1 cm or more wide; flowers subtended by narrow, awl-shaped bracts, their margins fringed with long hairs . . . . *C. vulgaris*
1 Stem leaves linear to lance-shaped, usually less than 1 cm wide; flowers not with narrow bracts . . . 2
2 Plants glabrous . . . . *C. arkansanum*
2 Plants pubescent . . . . *C. acinos*

Clinopodium acinos (L.) Kuntze
BASIL-THYME   *introduced*
   Acinos arvensis (Lam.) Dandy
   Satureja acinos (L.) Scheele
Annual. Stems erect, 1-2 dm tall, usually branched from the base and occasionally above, finely pubescent. Leaves obovate to elliptic, 6-12 mm
long, scabrous, entire or with a few low teeth, on petioles 1-2 mm long.
Flowers 1-3 in each of the upper axils; pedicels 2-4 mm long, apparently
arising directly from the axil, not from the end of a peduncle; calyx 5-6
mm long, swollen on the lower side, constricted at the throat, the lips
equal and about 1/2 as long as the tube; lower lip cleft to the base into 2
narrow lobes; upper lip widened to the tip and almost truncate, the lobes
3 small teeth about 0.5 mm long; corolla pale purple, 7-10 mm long. June-
Sept. — Native of Europe; weedy on roadsides and waste places.

Clinopodium acinos

Clinopodium arkansanum (Nutt.) House
LIMESTONE WILD BASIL   *native / cons* –
   Clinopodium glabrum (Nutt.) Kuntze
   Satureja arkansana (Nutt.) Briq.
Satureja glabella var. angustifolia (Torr.) Svenson
Perennial. Stems glabrous except with a minute pubescent area at each
node, from short stolons; stolons for the following year usually developed
while the plant is in bloom and bearing ovate petioled leaves 3-10 mm
long. Leaves linear, entire, 1-2 cm long. Flowers 2-8 at each node in the
upper half of the plant, subtended by progressively reduced leaves and
each flower by 2 linear bracts; pedicels 3-10 mm long; calyx glabrous, 4-
6 mm long, the lips about 1/2 as long as the tube; corolla pale purple, 8-15
mm long. May-Aug. Sandy beaches, calcareous soil.

Clinopodium arkansanum

Clinopodium vulgare L.
WILD BASIL   *native / cons* 3
   Satureja vulgaris (L.) Fritsch
Perennial from short stolons. Stems erect, simple or occasionally branched
above, 2-6 dm tall. Leaves ovate, 2-4 cm long, entire or with a few low
teeth, on petioles to 1 cm long or the upper nearly sessile. Flowers numerous in a dense, subglobose, terminal, head-like glomerule, or in vigorous plants with 1 or 2 similar glomerules in the uppermost axils, mingled with numerous hirsute bracts about as long as the calyx; calyx
tubular, hirsute throughout, 9-10 mm long, the lips nearly as long as the
tube; upper lip cleft about half its length, its lobes subulate above a triangular base; lower lip cleft to its base into subulate lobes; corolla pale
purple, rosepurple, or pink, varying to white, 12-15 mm long. — Dry or
moist upland woods.

Clinopodium vulgare

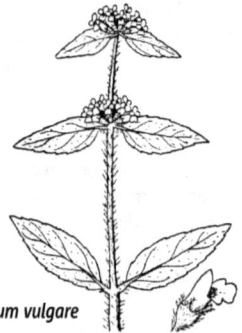

Galeopsis
HEMP-NETTLE
Annual herbs, mostly pubescent. Flowers small, crowded in dense verticils in the axils of the
upper foliage leaves. Calyx tube broadly tubular to campanulate, with 10 conspicuous ribs and
usually 10 intermediate ones; calyx lobes all equal, narrowly triangular, the strong midnerve
excurrent as a prominent spine. Corolla strongly 2-lipped, the tube exceeding the calyx, the
upper lip entire, erect, concave, the lower lip 3-lobed, bearing 2 protuberances at its base. Stamens
4, ascending under the upper corolla lip, the lower pair slightly the longer. Nutlets broadly
obovate, smooth.

ADDITIONAL SPECIES
   Galeopsis ladanum L. (Red Hemp-Nettle), introduced, moist, gravelly, disturbed places; see key.

1 Stems pubescent with tiny, recurved, eglandular hairs (longer glandular hairs may also be present as well); stem nodes not swollen . . . . *G. ladanum**
1 Stems pubescent with long, rigid, spreading to slightly recurved, eglandular hairs (shorter glandular hairs may also be present as well); stem nodes swollen 2
2 Stems stiffly hairy, the nodes swollen; leaves ovate, the margins coarsely toothed . . . . *G. bifida*
2 Stems with fine recurved hairs, nodes not swollen; leaves linear to lance-shaped, the margins shallowly toothed to entire . . . . *G. angustifolia*

Galeopsis angustifolia Ehrh.

NARROW-LEAF HEMP-NETTLE   *introduced*

Stems freely branched, 1-4 dm tall, never bristly, finely pubescent with minute recurved hairs. Leaves linear, entire, 1.5-3 cm long and rarely more than 5 mm wide. Verticils usually 2-4, few-flowered; calyx lobes about half as long as the canescent tube, to 3 mm long; corolla red, 2-2.5 cm long. June-Sept. — Native of Eurasia; introduced in waste places.

Galeopsis bifida Boenn.

BRITTLE-STEM HEMP-NETTLE   *introduced*

Galeopsis tetrahit var. *bifida* (Boenn.) Lej. & Courtois

Stems simple or branched, 3-8 dm tall, swollen at the nodes, hispid, often densely so, with long, straight, somewhat reflexed hairs. Leaves lance-shaped to ovate, 5-10 cm long, acuminate, crenate-serrate, pubescent on both sides, on petioles 1-3 cm long. Flowers white or pink or variegated, commonly with two yellow spots, borne in 2-6 dense verticils; calyx enlarged in fruit, its lobes eventually 5-10 mm long. Nutlets 3-4 mm long. June-Sept. — Native of Eurasia; introduced as a weed of gardens, roadsides, waste places, and forests.

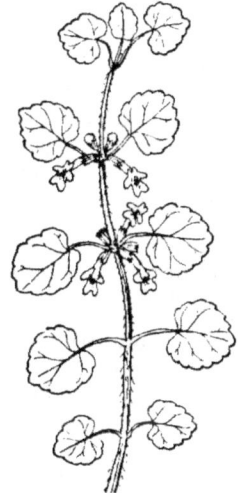

Galeopsis bifida

   Variable in density of pubescence, presence of glandular hairs, shape of leaf, and size of calyx and corolla.

Glechoma
GROUND-IVY

Glechoma hederacea L.

GROUND-IVY   *introduced (invasive)*

Perennial herb. Stems slender, creeping, eventually to 1 m long, villous to nearly glabrous. Leaves rotund to kidney-shaped, 1.5-4 cm wide, conspicuously crenate, long-petioled. Flowers blue, usually 3 in each axil; bractlets subulate, shorter than the calyx; calyx tubular, 5.5-9 mm long, 15-nerved, the 5 lobes triangular, about equal, about a 1/3 as long as the tube, with 3 nerves subtending each lobe, the middle one excurrent into a short awn; corolla much-surpassing the calyx, 2-lipped, upper lip shallowly 2-lobed; lower lip much larger, the lateral lobes short and rounded, the median lobe dilated; stamens 4, ascending under the upper corolla lip, and about equaling it. April-June. — Native of Eurasia; widely naturalized in yards, roadsides, cemeteries, and moist woods.

Glechoma hederacea

Hedeoma
FALSE PENNYROYAL

Hedeoma hispida

Hedeoma hispida Pursh

ROUGH FALSE PENNYROYAL   *native / cons 1*

Small, strongly scented, annual herb. Stems simple or branched from the base, occasionally branched above, 5-20 cm tall, pubescent with recurved hairs. Leaves linear, 1-2 cm long, sessile, entire. Flowers small, blue, pediceled, in axillary few-flowered verticils. Calyx tubular in anthesis, flask-shaped in fruit, gibbous on the lower side at base, strongly ribbed, 2-lipped, villous in the throat; upper lip of the calyx which is cleft to or below the middle into narrow ciliate teeth. Corolla tubular, weakly 2-

lipped, the upper lip erect, the lower spreading, 3-lobed. Stamens 2, ascending under the upper corolla-lip and about equaling it. Nutlets ovoid,
smooth. May-Aug. — Dry soil, sand dunes and barrens.

Leonurus
MOTHERWORT
Leonurus cardiaca L.
MOTHERWORT   *introduced*
Perennial, strongly scented herb. Stems stout, erect, to 1.5 m tall, finely
pubescent on the angles and nodes. Leaves long-petioled, the larger
broadly ovate, palmately lobed and sharply toothed, the upper progressively smaller and narrower, those subtending verticils commonly oblong
and merely 3-toothed. Flowers crowded in dense verticils subtended by
bracteal leaves and by linear bracts, forming long, interrupted, terminal
spikes; calyx tube 5-angled, 5-ribbed, nearly glabrous, 3-4 mm long; calyx
lobes nearly as long as the tube, the lower two somewhat the larger and
strongly deflexed. Corolla strongly bilabiate; upper lip erect, entire, pubescent; lower lip spreading or deflexed, 3-lobed, pale pink, the upper lip
white-villous. Stamens 4, about equal, ascending under the upper lip of
the corolla. Nutlets obpyramidal, 3-4-angled, truncate and pubescent at
their tips. June-Aug. — Native of c Asia; formerly cultivated as a home
remedy and now established in waste places, roadsides, and gardens.

Leonurus cardiaca

Lycopus
WATER-HOREHOUND
Perennial, unscented herbs. Stems erect, 4-angled. Leaves opposite, coarsely toothed or deeply
lobed, smaller on upper stems; petioles short or absent. Flowers small, in clusters in middle and
upper leaf axils, often appearing whorled; white to pink, the sepals and petals often dotted on
outer surface, 4-lobed, stamens 2. Fruit a nutlet.

1 Sepal lobes broad, triangular to ovate, to 1 mm long, shorter than to about as long as nutlets, the
   midvein not prominent . . . . *L. uniflorus*
1 Sepal lobes slender, 1–3 mm long, longer than nutlets, midvein prominent . . . . *L. americanus*

Lycopus americanus Muhl.
CUT-LEAF WATER-HOREHOUND   *native / cons 4*
Perennial herb, spreading by rhizomes. Stems erect, often branched, 2-8
dm long, upper stems smooth or short-hairy. Leaves opposite, lance-
shaped, 3-8 cm long and 1-4 cm wide, with glandular dots, smooth or
rough on upper surface, underside veins short-hairy; margins coarsely
and irregularly deeply toothed or lobed, the lowest teeth largest; nearly
stalkless or on short petioles. Flowers in dense, whorled clusters in leaf
axils; sepal lobes narrow, sharp-tipped, 1-3 mm long, longer than fruit;
petals white, sometimes pink to purple-dotted, 4-lobed, the upper lobe
wider and notched. Fruit a nutlet, 1-2 mm long. July-Sept. — Marshes,
wet meadows, shores, streambanks, ditches, calcareous fens, wetland
margins.

Lycopus americanus

Lycopus uniflorus Michx.
NORTHERN WATER-HOREHOUND   *native / cons 4*
Perennial herb. Stems smooth or short-hairy, 1-5 dm long. Leaves opposite,
lance-shaped to oblong, 3-6 cm long and 1-3 cm wide, margins with a
few outward-pointing teeth, petioles short or nearly absent. Flowers in
dense, whorled clusters in leaf axils; sepal lobes broad, triangular to
ovate, soft, rounded at tip, to 1 mm long, shorter to as long as nutlets;
petals white or pink, 2-3 mm long, 5-lobed, longer than sepals. Fruit a
nutlet 1-1.5 mm long. Aug-Sept. — Swamps, streambanks, thickets, wet
meadows, open bogs, calcareous fens, ditches; often with *L. americanus*.

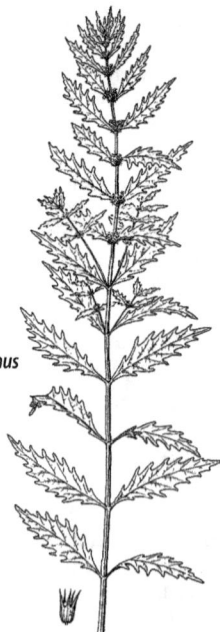

Mentha
MINT

Perennial herbs, spreading by rhizomes or stolons, with erect stems, serrate leaves, and small, blue to lavender flowers borne in the axils of the leaves or in terminal spikes or heads. Calyx regular or weakly 2-lipped, tubular to campanulate, 10-13-nerved, the lobes broadly triangular to subulate. Corolla tube slightly widened to the summit; corolla limb nearly regular and apparently 4-lobed; upper lobe, corresponding to the upper lip, usually somewhat wider than the others. Stamens 4, essentially uniform in length, straight, somewhat divergent, exsert from the corolla. Nutlets ovoid, smooth or roughened. All species bloom in summer.

HYBRIDS
Mentha x villosa Huds. (pro sp.) [*M. spicata* x *M. suaveolens*], Hairy mint, introduced, reported from a shaded roadside in Door county; see key.

1   Flowers in terminal spikes or heads, the internodes short . . . . *M. x piperita**
1   Flowers in axillary whorls spearated by internodes of normal length 2
2   Uppermost bracteal leaves small but still leaf like, with a short petiole, and exceeding the flowers . .
    . . . . *M. arvensis*
2   Uppermost bracteal leaves reduced to inconspicuous bracts usually not extending beyond the flowers . . . . *M. x villosa*

Mentha arvensis L.
AMERICAN WILD MINT   *native / cons* 3
Mentha canadensis L.

Perennial herb, strongly mint-scented, spreading by rhizomes and often also by stolons. Stems 2-8 dm long, 4-angled, hairy at least on stem angles. Leaves opposite, ovate to lance-shaped, 2-7 cm long and 0.5-3 cm wide, smooth or hairy; margins with sharp, forward-pointing teeth; petioles short. Flowers small, white or light pink to lavender, hairy, crowded in whorled clusters in middle and upper leaf axils; sepals 2-3 mm long, hairy and glandular; petals more or less regular to slightly 2-lipped, 4-6 mm long, glandular on outside, 4- or 5-lobed; stamens and style longer than petals. Fruit a smooth nutlet to 1 mm long, enclosed by the persistent sepals. July-Sept. — Wet meadows, marshes, swamps, thickets, streambanks, ditches, springs and other wet places.

Mentha arvensis

Mentha × piperita L.
PEPPERMINT   *introduced*

Perennial herb. Stems erect, glabrous or very nearly so, to 1 m. tall. Leaves lance-shaped or oblong lance-shaped, 4-8 cm long, sharply serrate, obtuse or rounded at base, glabrous; petioles of the principal leaves 4-15 mm long, those on the branches much shorter. Spikes 1 to several, terminating the stem and the upper lateral branches, 2-8 cm long, continuous, about 1 cm wide (excluding the corollas); calyx tubular, 3-4 mm long, the tube glabrous, the lance-subulate lobes glabrous or sparsely pilose. Considered to have originated by hybridization between *M. aquatica* and *M. spicata;* resembling the former in its large tubular calyx and blunt spikes, the latter in its elongate spikes and narrow leaves.

Of European origin; cultivated as an herb and commercially for its oil; escaped in wet soil.

Mentha × piperita

Monarda
BEEBALM

Erect perennial herbs (ours). Leaves lance-shaped to ovate, sessile or petiolate. Flowers conspicuous, densely aggregated into head-like clusters terminating the branches or also borne in the upper axils, subtended by foliaceous bracts and with linear bractlets. Calyx tubular, 13-15-nerved, regular, the 5 lobes alike or nearly so, much shorter than the tube. Corolla strongly bilabiate, the upper lip narrow, entire, straight or curved, the lower somewhat broader, spreading or deflexed,

3-lobed or with a central projecting tooth. Stamens 2, ascending under the upper corolla lip. Nutlets oblong, smooth.

1 Flowers yellowish, dotted with purple; stamens and style not exserted . . . . ***M. punctata***
1 Flowers lavender, rarely white; stamens and style strongly exserted beyond corolla; heads 2 or more and forming an interrupted spike . . . . ***M. fistulosa***

Monarda fistulosa L.

WILD BERGAMOT   *native / cons* **3**

Perennial. Stems erect, often branched, 5-12 dm tall, usually pubescent, at least above, rarely glabrous. Leaves 6-10 cm long, commonly deltoid-lance-shaped, varying to lance-shaped or rarely ovate, more or less serrate; rounded, truncate, or broadly acute at base, pubescent or essentially glabrous above or canescent beneath, on petioles 1-1.5 cm long. Heads 1.5-3 cm wide (excluding the corollas); bracteal leaves lance-shaped or ovate; calyx 7-10 mm long, puberulent, its throat densely hirsute within; calyx lobes subulate, 1-2 mm long; corolla pale lavender, 2-3 cm long, the upper lip densely villous at the summit. June-Sept. — Upland woods, thickets, and prairies.

Monarda fistulosa

Monarda punctata L.

HORSE-MINT   *native / cons* **3**

Perennial. Stems simple or branched, 3-10 dm tall, thinly canescent. Leaves lance-shaped or narrowly oblong, 2-8 cm long, more or less pubescent. Glomerules 2-5, or solitary on depauperate plants, the bracteal leaves lance-shaped to ovate, much exceeding the calyx, spreading or reflexed, often pale green to nearly white, or tinged with purple; calyx 5-9 mm long, densely villous in the throat, more or less villous externally at the summit, its lobes 1-1.5 mm long; corolla pale yellow, spotted with purple, the upper lip arched. June-Sept. — Sandy fields, sand dunes, open oak and pine woods, roadsides and disturbed areas.

Monarda punctata

Nepeta
CATNIP

Nepeta cataria L.

CATNIP   *introduced*

Perennial herb; stems, undersides of leaves, and inflorescences covered with grayish hairs. Stems erect, much branched, to 1 m tall. Leaves narrowly to broadly deltoid, 3-8 crn long, truncate or subcordate at base, coarsely crenate-dentate, on petioles about 1/2 as long as the blade. Flower clusters continuous or interrupted, 2-6 cm long, rather loosely many-flowered; calyx tubular, weakly 2-lipped, at anthesis about 7 mm long, its lobes about 1/2 as long as the tube; upper 3 lobes each 2-3-nerved, somewhat longer and wider than the 1-nerved lower lobes; corolla 10-12 mm long, dull white, the lower lobe dotted with pink or purple; stamens 4, ascending under the upper corolla lip and nearly equaling it, the upper pair slightly longer. July-Oct. — Native of se Europe and sw Asia, formerly cultivated for reputed medicinal properties, and now established in waste places, fencerows, and roadsides.

Physostegia
FALSE DRAGONHEAD

Physostegia virginiana (L.) Benth.

Nepeta cataria

OBEDIENCE   *native / cons* **4**

Perennial herb, spreading by rhizomes. Stems erect, 5-15 dm long, often branched near top, 4-angled. Leaves opposite, oval to oblong lance-shaped, 2-15 cm long and 1-4 cm wide, sometimes smaller upward; margins with sharp teeth; sessile, not clasping. Flowers in several racemes 5-20 cm long, the stalks short-hairy; sepals 4-8 mm long, often with some gland-tipped hairs; petals pink-purple or white with purple spots, 1.5-3 cm long, short-hairy to smooth. Fruit a nutlet, 2-3 mm long. July-Sept. — Sedge meadows, low prairie, shores, swamps, floodplain forests, thickets and ditches. Sometimes cultivated for its attractive flowers.

Physostegia virginiana

Prunella
SELF-HEAL
Prunella vulgaris L.
SELF-HEAL   *native-introduced / cons 1*

Perennial herb. Stems upright or sometimes spreading, 1-5 dm long, 4-angled. Leaves opposite, lance-shaped to oval or ovate, 2-8 cm long and 1-4 cm wide; lower leaves wider than upper; margins entire or with a few small teeth; petioles present. Flowers in dense spikes 2-5 cm long and 1-2 cm wide, with obvious bracts; sepals to 1 cm long, green or purple, with spine-tipped teeth; corolla 2-lipped, the upper lip hoodlike and entire, lower lip shorter and 3-lobed; petals blue-violet (rarely pink or white), 1-2 cm long; stamens 4, about as long as petals. Fruit a smooth nutlet. June-Oct. — Common in many types of wetlands (especially where disturbed): swamps, wet forest depressions, wet trails, streambanks; also in drier forests, fields and lawns.

Subsp. *vulgaris*, introduced from Europe and found in mostly disturbed places, has broad leaves half as wide as long. The native subsp. *lanceolata* has narrower leaves, 1/3 as wide as long.

Prunella vulgaris

Scutellaria
SKULLCAP
Perennial herbs, spreading by rhizomes. Stems erect or spreading, 4-angled. Leaves opposite, ovate to lance-shaped, margins toothed, petioled or nearly sessile. Flowers blue or blue with white markings, single on short stalks in axils of middle and upper leaves, or in racemes from leaf axils; calyx 2-lipped, with a rounded bump on upper side; corolla 2-lipped, pubescent on outer surface, upper lip hoodlike, lower lip more or less flat, 3-lobed; stamens 4, ascending into the upper corolla lip. Fruit a 4-parted nutlet.

1 Flowers single in leaf axils . . . . *S. galericulata*
1 Flowers in racemes from axils of stem leaves . . . . *S. lateriflora*

Scutellaria galericulata L.
HOODED SKULLCAP   *native / cons 5*
Scutellaria epilobiifolia A. Hamilton

Perennial. Stems erect or spreading, 2-8 dm long, unbranched or branched, 4-angled, short-hairy at least on angles of upper stem. Leaves opposite, lance-shaped to narrowly ovate, 2-6 cm long and 0.5-2.5 cm wide, upper surface smooth, underside short-hairy; margins with low, rounded, forward-pointing teeth; petioles very short. Flowers 2-lipped, single in leaf axils (and paired at nodes), on stalks 1-3 mm long; sepals 3-6 mm long; petals blue, marked with white, 15-25 mm long. Fruit a nutlet. June-Sept. — Shores, streambanks, marshes, wet meadows, swamps, thickets, bogs, ditches.

Scutellaria galericulata

Scutellaria lateriflora L.
BLUE SKULLCAP   *native / cons 5*

Perennial. Stems 2-6 dm long, usually branched, 4-angled, short-hairy on upper stem angles or smooth. Leaves opposite, ovate to lance-shaped, 3-8 cm long and 1.5-5 cm wide, smooth; margins coarsely toothed; petioles 0.5-2 cm long. Flowers 2-lipped, in elongate racemes from leaf axils; sepals

2-4 mm long; petals blue (rarely pink or white), 5-8 mm long. Fruit a nutlet. July-Sept. — Shores, streambanks, wet meadows, marshes, swamps, shaded wet areas.

Stachys
HEDGE-NETTLE
Erect perennial herbs, spreading by rhizomes; plants usually hairy. Stems 4-angled. Leaves opposite, margins entire or toothed, stalkless or with short petioles. Flowers in interrupted spikes at ends of stems, appearing whorled in more or less evenly spaced clusters; sepals more or less regular, with 5 equal teeth; corolla 2-lipped, petals pink, often with purple spots or mottles, upper lip concave, entire, lower lip spreading, 3-lobed; stamens 4, ascending under the upper lip. Fruit a dark brown, 4-lobed nutlet, loosely enclosed by the persistent sepals.

1 Plants glabrous; leaf petioles 8–25 mm long . . . . *S. tenuifolia*
1 Plants pubescent, at least on the stem angles; leaves with petioles to 10 mm long . . . . *S. pilosa*

Stachys pilosa Nutt.
HEDGE-NETTLE   *native / cons –*
Perennial herb Stems erect, rarely branched, 5-10 dm tall, villous on the sides and angles; hairs of the stem widely spreading. Leaves lance-shaped to ovate, 5-10 cm long, 2-4 cm wide, softly pubescent on both sides, usually with short fine hairs and longer bristles mingled, sharply serrate, sessile or nearly so. Verticils usually 6-flowered, the subtending leaves narrowly lance-shaped; calyx tube densely glandular-pubescent and also hirsute, the tube 3.5-5 mm long; calyx lobes glandular-hirsute, narrowly triangular, nearly as long as the tube, tapering to a stiff subulate tip. July-Aug. — Damp ground, ditch banks, beaches, and wet prairies.

Stachys tenuifolia Willd.
SMOOTH HEDGE-NETTLE   *native / cons 6*
Perennial herb. Stems 4-10 dm long, 4-angled, smooth, or with downward-pointing, bristly hairs on stem angles. Leaves opposite, lance-shaped to ovate, 6-14 cm long and 2-6 cm wide, more or less smooth; margins with sharp, forward-pointing teeth; petioles slender, 1-2 cm long or absent. Flowers in interrupted spikes at ends of stems or also in upper leaf axils; sepals 5-7 mm long, glabrous; petals pale red to purple, 1.5-2.5 cm long. Fruit a nutlet. July-Sept. — Floodplain forests, shores, streambanks, thickets, wet meadows.

Stachys tenuifolia

Teucrium
GERMANDER

Teucrium canadense L.
AMERICAN GERMANDER   *native / cons 4*
Perennial herb, spreading by rhizomes. Stems 3-10 dm long, mostly unbranched, 4-angled, long-hairy. Leaves opposite, lance-shaped or oblong, 4-12 cm long and 1.5-5 cm wide, upper surface smooth or sparsely hairy, underside with dense, matted hairs, margins irregularly finely toothed, petioles 5-15 mm long. Flowers in a dense spike-like raceme, 5-20 cm long; bracts present and narrowly lance-shaped; pedicels 1-3 mm long; sepals more or less regular, purple or green, 4-7 mm long, covered with long silky hairs and very short glandular hairs; corolla irregular, 10-16 mm long, with short gland-tipped hairs, upper lip absent, lower lip large; petals pink to purple; stamens 4, arched over the corolla. Fruit a golden nutlet. July-Sept. — Marshes, wet meadows, shores, streambanks, thickets, floodplain forests, ditches.

Teucrium canadense

Lentibulariaceae

BLADDERWORT FAMILY

Insectivorous herbs. Leaves floating, or in peat, muck, or wet soil. Flowers perfect (with both staminate and pistillate parts), irregular, 2-lipped, sometimes with a spur, 1 to several on an erect stem; stamens 2. Fruit a capsule.

Utricularia
BLADDERWORT

Mostly aquatic, annual or perennial herbs. Leaves underwater, alternate, entire or dissected into many linear segments, some with bladders which trap tiny aquatic invertebrates; or leaves in wet soil and rootlike or absent. Flowers perfect, irregular, 1 to several in a raceme atop stalks raised above water or soil surface, each flower subtended by a small bract; corolla yellow (ours), similar to a snapdragon flower, 2-lipped, the upper lip erect, entire or slightly 2-lobed, lower lip entire or 3-lobed, the corolla tube extended backward into a sac or spur, stamens 2. Fruit a many-seeded capsule.

1 Scapes appearing leafless; leaves simple or absent; plants of peat or moist sand or marl . *U. cornuta*
1 Scapes with leaves at base, the leaves dissected and with bladderlike traps; plants mostly floating in water 2
2 Leaf divisions flat in cross-section 3
2 Leaf divisions round in cross-section or threadlike 4
3 Bladders borne on leaves; smallest leaf divisions entire (visible with a 10x hand lens); flower with a sac or spur much shorter than lower lip . . . . *U. minor*
3 Bladders on branches separate from leaves; smallest leaf divisions finely toothed, the teeth spine-tipped; flower with a spur as long as lower lip . . . . *U. intermedia*
4 Plants large; leaves floating; scapes 1 mm or more wide; flowers 13 mm or more long, 5 or more per head; larger bladders more than 2 mm wide . . . . *U. macrorhiza*
4 Plants smaller; leaves floating or creeping; scapes threadlike; flowers to 12 mm long, 1–3 per head; larger bladders mostly less than 2 mm wide . . . . *U. geminiscapa*

Utricularia cornuta Michx.
HORNED BLADDERWORT   *native / cons* 10
Annual or perennial herb. Stems and leaves underground, roots with tiny bladders. Flowers yellow, with a downward-pointing spur 6-15 mm long, on stalks 1-2 mm long, 1-6 atop an erect stalk 10-25 cm long; bracts ovate, 1-2 mm long. Fruit a rounded capsule. June-Sept. — Acid lakes, shores, peatlands, calcareous pools between dunes, borrow pits.

Utricularia geminiscapa Benj.
HIDDEN-FRUIT BLADDERWORT   *native / cons* 9
Annual or perennial herb, similar to *U. macrorhiza* but smaller. Stems floating below water surface, sparsely branched. Leaves alternate, 1-2 cm long, branched into 4-7 segments and without bladders, or unbranched with bladders. Flowers yellow, 2-5 atop a slender stalk, 5-15 cm long, bracts below flowers 2-3 mm long; individual flower stalks 4-8 mm long, these arched when plants fruiting; cleistogamous flowers without petals more commonly produced, these single on leafless stalks 5-15 mm long along stems and often 1 at base of scape. July-Aug. — Acid lakes, pools in open bogs.

Utricularia intermedia Hayne
FLAT-LEAF BLADDERWORT   *native / cons* 9
Annual herb. Stems very slender, creeping along bottom in shallow water. Leaves alternate, 0.5-2 cm long, mostly 3-parted near base, then again divided 1-3x, the segments linear and flat, margins with small, bristly teeth; bladders 2-4 mm wide, borne on branches separate from leaves. Flowers yellow, 2-4 atop an emergent stalk 5-20 cm long; individual flower stalks

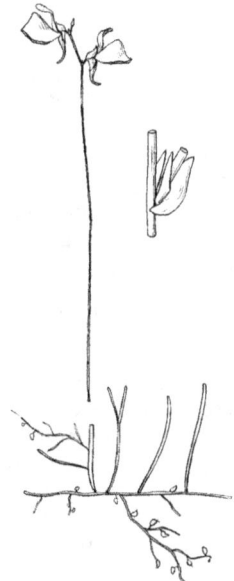

Utricularia cornuta

to 15 mm long, remaining erect in fruit; spur nearly as long as lower lip. Fruit a capsule. June–Aug. — Shallow water (usually alkaline), marly pools between dunes, calcareous fens, marshes, ponds and rivers. bogs and swamps.

Utricularia macrorhiza Le Conte
GREATER BLADDERWORT   *native / cons 7*
  Utricularia vulgaris L.
Perennial herb. Stems floating below water surface, sparsely branched, often forming large mats. Leaves alternate, 1–5 cm long, 2-forked at base and repeatedly 2-forked into segments of unequal length, the segments more or less round in section, becoming smaller with each branching, the final segments threadlike; bladders 1–4 mm wide, borne on leaf segments. Flowers yellow, 6–20 atop a stout stalk 6–25 cm long; lower flower lip 1–2 cm long, sometimes much smaller on late-season flowers, upper lip more or less equal to lower lip; spur about 2/3 as long as lower lip; stalks bearing individual flowers curved downward in fruit. Fruit a capsule. June–Aug. — Shallow water of lakes, ponds, peatlands, marshes and rivers.

Utricularia macrorhiza

Utricularia minor L.
LESSER BLADDERWORT   *native / cons 10*
Perennial herb. Stems few-branched, 10–30 cm long, creeping on bottom in shallow water or on wet soil. Leaves alternate, to 1 cm long, with few divisions, the segments slender, flat, the smallest segments strongly tapered to tip, margins entire; bladders 1–2 mm wide, 1–5 on leaves. Flowers pale yellow, 2–8 atop a threadlike stalk 4–15 cm long; individual flower stalks to 1 cm long, curved downward in fruit; lower lip of flower 4–8 mm long, 2x longer than upper lip; spur small, to half length of lower lip. Fruit a capsule. June–Aug. — Fens, open bogs, sedge meadows and marshes; often in shallow water and where calcium-rich.

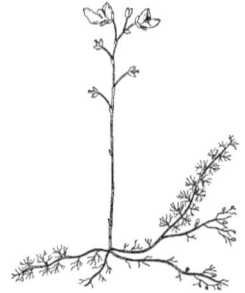

Utricularia minor

Linaceae

FLAX FAMILY

Linum

FLAX
Annual or perennial herbs. Leaves simple, alternate or opposite, narrow, margins entire, petioles absent. Flowers regular, perfect, 5-parted. Sepals separate, imbricate. Petals blue (ours). Stamens as many as the petals. Fruit a 10-chambered capsule.

1 Margins of inner sepals fringed with short hairs, the sepal tips long-tapered ...... *L. usitatissimum*
1 Margins of inner sepals entire, the sepal tips rounded or only short-pointed . . . . *L. perenne*

Linum perenne L.
WILD BLUE FLAX   *introduced*
Perennial. Stems mostly several from one root, 3–7 dm tall. Leaves erect or appressed, very numerous, linear, leaves 1–2 cm long; sepals ovate, 5–7 mm long, obtuse, minutely mucronate; petals blue, 15–20 mm long. Capsules subglobose, 6–9 mm long; carpels acuminate; false septa incomplete, strongly ciliate. May–July. — Prairies and meadows.

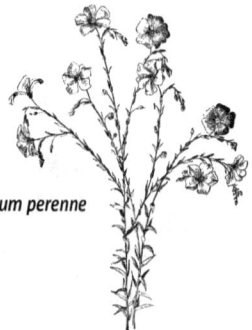

Linum perenne

Linum usitatissimum L.
CULTIVATED FLAX   *introduced*
Annual. Stems erect, usually solitary from a slender root, to 1 m tall. Leaves linear lance-shaped, 3-nerved. Sepals 7–9 mm long at maturity, the inner ciliate on the scarious margin; petals blue, 10–15 mm long. Cap-

sule globose, 6-10 mm wide; false septa very incomplete. Summer. — Of unknown origin; cultivated since prehistoric times for its fiber (linen) and more recently for its oil (linseed); sometimes escaped or adventive in fields and roadsides.

Lythraceae
LOOSESTRIFE FAMILY

Lythrum
LOOSESTRIFE

Lythrum salicaria L.
PURPLE LOOSESTRIFE   *introduced (invasive)*
Perennial herb, spreading and forming colonies by thick, fleshy roots which send up new shoots. Stems erect, 6-15 dm long, 4-angled, with many ascending branches, sometimes rather woody at base. Leaves opposite or sometimes in whorls of 3, becoming alternate and reduced to bracts in the head; lance-shaped, 3-10 cm long and 0.5-2 cm wide, mostly heart-shaped and clasping at base; margins entire; petioles absent. Flowers large and showy, 2 or more in axils of reduced upper leaves (bracts), in spikes 1-4 dm long at ends of branches; sepals joined, the calyx tube 4-6 mm long, hairy; petals 6, purple-magenta, 7-10 mm long; stamens usually 12, the stamens and styles of 3 different lengths. Fruit a capsule enclosed by the sepals. June-Sept. — Introduced from Europe and formerly planted as an ornamental, escaping to marshes, wet ditches, streambanks, cranberry bogs and shores, where a serious threat to our native flora and of little value to wildlife. In addition to spreading vegetatively, a single plant may produce thousands of seeds each year. To limit the spread of this species, plants should be pulled (including roots), bagged, and removed from infested sites.

Lythrum salicaria

Malvaceae
MALLOW FAMILY
Annual or perennial herbs with upright stems; trees in *Tilia*. Leaves alternate, entire to lobed or dissected, often round or kidney-shaped, palmately veined. Flowers single or in small, narrow clusters from leaf axils, with 5 united sepals (separate in *Tilia*) and 5 petals; stamens many and joined near base, forming a tube around the style. Fruit a capsule.

1 Trees; with inflorescence apparently borne at the middle of a tongue-shaped bract . . . . *Tilia*
1 Herbs or shrubs; inflorescences various, but never with a large, tongue-shaped bract 2
2 Calyx subtended by a series of 2 or more bracts . . . . *Malva*
2 Calyx without involucral bracts . . . . *Abutilon*

Abutilon
VELVETLEAF

Abutilon theophrasti Medik.
VELVETLEAF   *introduced*
Annual herb, softly pubescent throughout with stellate hairs. Stems stout, branched, 1-1.5 m tall. Leaves cordate, 10-15 cm long and as wide, toothed, on petioles of about the same length. Peduncles jointed above the middle, at first short, at maturity 2-3 cm long. Flowers yellow, 15-25 mm wide. Head of fruit 2-3 cm wide; carpels commonly 10-15, densely pubescent, with conspicuous, horizontally spreading beaks. July-Oct. — Native of

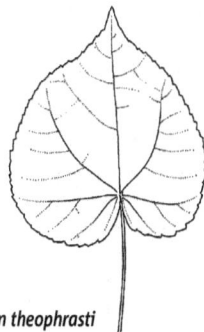

Abutilon theophrasti

Asia; established as a weed in cultivated fields and waste places but more abundant south of Wisconsin.

Malva
MALLOW

Annual, biennial, or perennial herbs. Leaves broad, serrate, crenate, lobed, or parted. Flowers solitary or fascicled in the axils. Bractlets of the involucel 3, linear to obovate. Calyx 5-lobed. Petals truncate, notched, or obcordate at tip. Carpels 10-20, beakless, 1-seeded, glabrous, pubescent, or rugose.

1 Leaves deeply divided into 3–7 segments .... *M. moschata*
1 Leaf margins entire or only shallowly lobed .... *M. neglecta*

Malva moschata L.
MUSK MALLOW   *introduced*

Perennial herb. Stems erect, 4–10 dm tall, roughly pubescent. Leaves orbicular in outline, 5-7-parted, the segments of the upper leaves again deeply pinnatifid. Flowers partly solitary on long pedicels from the upper axils but chiefly crowded in terminal clusters; bractlets linear to narrowly lance-shaped or oblong lance-shaped, ciliate, glabrous or nearly so on the back; petals white to pale purple, triangular, 2.4-3 cm long. Mature carpels rounded on the back, not rugose, densely pubescent. June–Sept. — Native of Europe; escaped from cultivation along roadsides and in waste places.

Malva moschata

Malva neglecta Wallr.
COMMON MALLOW   *introduced*

Biennial herb. Stems prostrate, procumbent, or ascending, to 1 m long, usually branched from the base. Leaves long-petioled, orbicular or kidney-shaped, 3-6 cm wide, shallowly 5-9-lobed, crenate, cordate at base. Flowers fascicled in the axils, on pedicels to 3 cm long; bractlets narrow; petals obcordate, 6-12 mm long, white or slightly tinged with pink or purple. Mature carpels usually 12-15, rounded on the back, not rugose or reticulate, usually finely pubescent, the whole ring of carpels presenting a crenate outline, the depressed central portion of the head about a 1/3 as wide as the head. May-Oct. — Native of Eurasia and n Africa; common as a weed in gardens and waste places.

Tilia
BASSWOOD, LINDEN

Tilia americana L.
BASSWOOD, LINDEN   *native / cons 5*

Tree to 35 m tall; bark gray to light brown, with narrow, well-defined fissures; twigs smooth, reddish-green, becoming light to dark gray, marked with dark wart-like bumps. Leaves broadly ovate to subrotund, palmately veined, cordate or truncate at the oblique base, sharply serrate, green beneath and glabrous to sparsely stellate-pubescent on the surface, with conspicuous tufts of hairs in the vein axils. Flowers fragrant, perfect, 5-merous, white or cream-colored in axillary cyme-like clusters, the long peduncle adnate about to the middle of a narrow, elongate, short-petioled, foliaceous bract; bracts glabrous or nearly so on both sides; sepals separate to the base petals narrowly oblong, 7-12 mm long, tapering to the base; stamens numerous, either all distinct or united into 5 bundles, one in front of each petal; ovary tomentose, 5-celled. Fruit nutlike, tomentose, 1-2-seeded. July. — Moist fertile soil.

The dried flowers have a long history of medicinal uses, and bees produce a fragrant honey from its blossoms.

Malva neglecta

Tilia americana

Menispermaceae
MOONSEED FAMILY

Menispermum
MOONSEED

Menispermum canadense L.
CANADIAN MOONSEED   *native / cons* 5
Dioecious woody twiners climbing 2-4 m high. Leaves simple, alternate, broadly ovate to nearly orbicular, 10-15 cm wide and long, palmately veined, shallowly 3-7-lobed to entire; slender-petioled. Flowers small, unisexual, usually 3-merous, regular, in racemes or panicles that arise just above the leaf-axils; perianth segments scarcely differentiated into calyx and corolla, normally in 4 alternating whorls, the 2 outer (calyx) exceeding the 2 inner (corolla); sepals 4-8, longer than the 4-8 petals; stamens 12-24. Drupe bluish-black, 6-10 mm long; stone flattened, thickened into 3 rough ridges over most of its margin. June-July. — Moist woods and thickets.

The drupes, suspected to be toxic, resemble wild grapes.

Menispermum canadense

Menyanthaceae
BUCKBEAN FAMILY

Menyanthes
BUCKBEAN

Menyanthes trifoliata L.
BUCKBEAN   *native / cons* 10
Perennial glabrous herb, with thick rhizomes covered with old leaf bases. Leaves alternate along rhizomes, palmately divided into 3 leaflets, the leaflets oval to ovate, 3-10 cm long and 1-5 cm wide, entire or sometimes wavy-margined; petioles 5-30 cm long, the base of petiole expanded and sheathing stem. Flowers in racemes on leafless stalks 2-4 dm long and longer than the leaves; bracts mostly 3-5 mm long; individual flowers on stalks 5-20 mm long; flowers perfect, regular, 5-parted, often of 2 types, some with flowers with long stamens and a shorter style, others with a long style and shorter stamens; sepal lobes 2-3 mm long; corolla funnel-shaped, 8-12 mm long, petals white, often purple-tinged, bearded with white hairs on inner surface; stamens 5. Fruit a rounded capsule, 6-10 mm wide; seeds shiny, yellow-brown. May-July. — Open bogs and fens (especially in pools and outer moat), cedar swamps, wet thickets.

Menyanthes trifoliata

Molluginaceae
CARPETWEED FAMILY

Mollugo
CARPETWEED

Mollugo verticillata

Mollugo verticillata L.
GREEN CARPETWEED   *introduced*
Annual herb. Stems prostrate or ascending, repeatedly forked, forming mats to 4 dm wide. Leaves in whorls of 3-8, narrowly to broadly oblong lance-shaped, 1-3 cm long, long-tapering to a short, scarcely differentiated petiole. Flowers perfect, 2-5 from each node, on pedicels 5-15 mm long,

sepals 5; petals 5, pale green to white, 4-5 mm wide; stamens 3 or 4. Capsules ovoid, 3 mm long, many-seeded. June-Sept. — Apparently native of tropical America; now a common weed in moist soil.

Montiaceae

MONTIA FAMILY

Claytonia
SPRINGBEAUTY
Glabrous perennial herbs from rounded tubers (ours). Leaves one or few from the base and a single opposite pair on the stem below the loose terminal raceme; the raceme with 5-15 long-pediceled flowers and sometimes a small bract below the lowest flower. Sepals ovate, persistent in fruit. Petals 5, white or pale pink with pink veins, oval or elliptic, spreading. Stamens 5, opposite the petals. Capsule ovoid, opening by inrolling valves. Flowers open from March to May, with stems and leaves withered by early summer. The family Montiaceae now includes *Claytonia*, formerly within Portulacaceae.

1 Leaves with distinct petiole; blades less than 5 times longer than wide . . . . *C. caroliniana*
1 Petiole not distinct; blades 5 times or more longer than wide . . . . *C. virginica*

Claytonia caroliniana Michx.
CAROLINA SPRINGBEAUTY   *native / cons* **8**
Perennial herb. Stem leaves, including the petiole, commonly 3-6 cm long, rarely to 9 cm long, the blade usually 10-15 mm wide, acute at base, clearly distinguished from the petiole. — Cool woods.

Claytonia virginica L.
VIRGINIA SPRINGBEAUTY   *native / cons* **6**
Perennial herb. Stem leaves long-tapering to base, the blade sessile or merging gradually into the short, poorly differentiated petiole, commonly 4-10 mm wide, rarely less than 7 cm long, including the petiole. — Damp woods and fields.

Myricaceae

Claytonia virginica

BAYBERRY FAMILY
Monoecious or dioecious shrubs, with alternate simple leaves; leaves resinous-dotted and fragrant. Flowers unisexual, without perianth, solitary in the axils of small bracts, aggregated into globose to cylindric catkins. Stamens 2-many, usually 4-8, the short filaments free or connate. Ovary 1-celled, subtended by 2-8 usually minute bractlets; ovule 1, basal; style very short; stigmas 2, linear, elongate.

1 Leaves entire or nearly so; wet habitats . . . . *Myrica*
1 Leaves pinnately lobed; dry sandy habitats . . . . *Comptonia*

Comptonia
SWEET-FERN   *Comptonia peregrina*

Comptonia peregrina (L.) Coult.
SWEET-FERN   *native / cons* **4**
Shrub. Stems much branched, to 1 m tall. Leaves linear-oblong, 6-12 cm long, about 1 cm wide, deeply pinnately lobed, resinous-dotted, more or less pubescent. Staminate catkins clustered, cylindric, 1-3 cm long, nodding; bracts quadrangular, resinous, villous, acuminate; pistillate catkins subglobose; bracts similar but concealed by the linear-subulate bractlets. Nutlets ellipsoid, blunt, 3-5 mm long, subtended by the elongate bractlets,

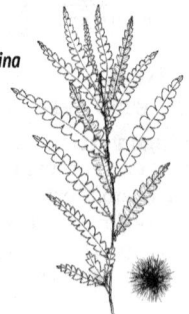

the whole fruit bur-like, 1-2 cm wide. April-May. — Dry, especially sandy soil.

Myrica
BAYBERRY

Myrica gale L.
SWEET GALE   *native / cons* **9**
Gale palustris Chev.
Much-branched shrub, 6-15 dm tall; bark dark gray to red-brown with small pale lenticels; twigs hairy, dotted with glands. Leaves alternate, deciduous, wedge-shaped, tapered to base, broadest above middle, 3-6 cm long and 1-2 cm wide, tip rounded and toothed, dark green on upper surface, paler below, dotted with shiny yellow glands, fragrant when rubbed; petioles short, 1-3 mm long. Staminate and pistillate flowers separate and on different plants, appearing before or with unfolding leaves; staminate flowers in catkins 1-2 cm long, with dark brown, shiny triangular scales; pistillate flowers in conelike, brown clusters 10-12 mm long. Fruit a flattened, ovate achene, resin-dotted, 2-3 mm long. April-May. — Lakeshores, marshes, swamps and bogs.

Myrica gale

Nymphaeaceae
WATER-LILY FAMILY

Aquatic, perennial herbs. Stems long and fleshy, from horizontal rhizomes rooted in bottom mud. Leaves large, leathery, mostly floating or emergent above water surface, heart-shaped to shield-shaped, notched at base, margins entire. Flowers showy, single on long stalks and borne at or above water surface, perfect, white or yellow, sepals 4-6, green or yellow; petals numerous, small to large and showy. Fruit a many-seeded, berrylike capsule, opening underwater when mature.

1   Flowers yellow, often red-tinged, sepals petal-like, true petals small; leaf blades oblong to oval or
    heart-shaped . . . . *Nuphar*
1   Flowers white (rarely pink), sepals green, true petals large and showy; leaf blades nearly round . . . .
    . . . . *Nymphaea*

Nuphar
YELLOW WATER-LILY

Nuphar variegata Dur.
YELLOW POND-LILY   *native / cons* **6**
Perennial aquatic herb, Leaves mostly floating, 10-25 cm wide, notch usually less than half as long as midvein, petioles flattened on upper side and narrowly winged; underwater leaves absent or few. Flowers 2.5-5 cm wide; sepals usually 6, yellow, red-tinged on inner surface, petal-like and forming a saucer-shaped flower; petals small and numerous; anthers 4-7 mm long, longer than filaments; disk at base of stigma green, 1 cm wide, with 10-15 rays. Fruit ovate, 2-4 cm long. June-Aug. — Ponds, lakes, quiet streams.

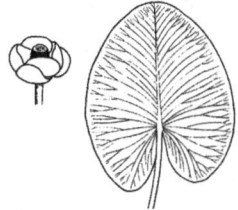

Nuphar variegata

Nymphaea
WATER-LILY

Nymphaea odorata Ait.
WHITE WATER-LILY   *native / cons* **6**
Nymphaea tuberosa Paine
Aquatic perennial herb, rhizomes sometimes with knotty tubers. Leaves floating, round, 1-3 dm wide, with a narrow notch, green and shiny on

upper surface, usually purple or red below. Flowers large and showy, white (rarely pink), usually fragrant, 7-20 cm wide, often opening in morning and closing in late afternoon (or remaining open on cool, cloudy days); sepals 4, green, 3-10 cm long; petals 17-25, about as long as sepals, oval, tapered to a rounded tip; stamens 40-100. Fruit round, mostly covered by sepals; seeds 2-4 mm long. June-Aug. — Shallow water of ponds and lakes, quiet water of rivers.

Oleaceae

Nymphaea odorata

OLIVE FAMILY

Trees or shrubs with opposite, simple or compound leaves. Flowers perfect or unisexual, regular, usually 4-merous. Calyx small or in some genera lacking. Corolla in our genera partially or wholly fused, or lacking (*Fraxinus*). Stamens 2-4, usually 2, inserted on the corolla tube if the corolla is present. Ovary 2-celled; seeds 2-several or rarely 1 in each cell. Fruit a drupe, capsule, or samara. Fsmily ncludes the widely cultivated and economically important *Olea europaea* L., the olive, native of the e Mediterranean region.

1 Trees; leaves pinnately compound . . . . *Fraxinus*
1 Shrubs; leaves simple . . . . *Syringa*

Fraxinus

ASH

Medium trees. Leaves deciduous, opposite, pinnately divided into leaflets. Flowers in clusters from axils of previous year's twigs, mostly single-sexed, staminate and pistillate flowers on different trees, rarely perfect, petals absent. Fruit a 1-seeded, winged samara.

1 Twigs densely hairy 2
2 Lateral leaflets tapered at base to a short winged petiole, or the leaflets sessile . . . . *F. pennsylvanica*
2 Lateral leaflets rounded at base, short petioles present . . . . *F. americana*
1 Twigs glabrous 3
3 Leaflets pale or waxy on underside, margins often entire; lateral leaflets on short petioles 5 mm or more long . . . . *F. americana*
3 Leaflets not waxy on underside, margins usually finely toothed; lateral leaflets sessile or on short petioles to 3 mm long 4
4 Lateral leaflets usually number 8, sessile; body of fruit flat in cross-section . . . . *F. nigra*
4 Lateral leaflets usually 4–6, on short petioles; body of fruit round in section . . . . *F. pennsylvanica*

Fraxinus americana L.

WHITE ASH   *native / cons 5*

Tree to 40 m. tall. Leaflets 5-9, usually 7, oblong to ovate or obovate, usually abruptly acuminate, entire or serrulate toward the summit, rounded to broadly acute at base, paler beneath, on wingless leaflet petioles. Samaras linear or oblong, 3-5 cm long, obtuse to retuse at the tip, the wing extending about 1/3 of the length of the terete body; the free portion, above the apex of the body, longer than the body itself; subtending calyx 1-1.5 or rarely 2 mm long, seldom cleft on one side only. — Rich moist woods; a valuable timber tree.

Fraxinus nigra Marsh.

BLACK ASH   *native / cons 8*

Fraxinus americana

Tree to 15 m tall, crown open and narrow; bark gray, thin, flaky; twigs smooth, round in section, dark green, becoming gray. Leaves opposite, pinnately divided into 7-11 stalkless (except for terminal) leaflets; leaflets lance-shaped to oblong, 7-13 cm long and 2.5-5 cm wide, long-tapered to a tip; margins with sharp, forward-pointing teeth. Flowers appear in spring before leaves, in open clusters on twigs of previous year; some perfect, some single-sexed, staminate and pistillate flowers on different trees. Fruit a 1-seeded samara, 2.5-4 cm long and 6-10 mm wide, the wing broad and rounded at tip, deciduous or persisting until following spring. April-May. — Floodplain forests, cedar swamps, wet depressions in forests.

Fraxinus pennsylvanica Marsh.
GREEN ASH   *native / cons* **2**
Tree to 15 m tall; bark dark gray or brown, thick, with shallow furrows and netlike ridges; twigs usually hairy for 1-3 years, becoming light gray or red-brown. Leaves opposite, pinnately divided into 7-9 leaflets; leaflets oblong lance-shaped to ovate, 7-13 cm long and 2.5-4 cm wide, upper surface smooth, underside smooth or hairy; margins entire or with few forward-pointing teeth; leaflet petioles short, smooth or hairy. Flowers appear in spring before or with leaves, in compact, hairy clusters on twigs of previous year; single-sexed, staminate and pistillate flowers on different trees. Fruit a 1-seeded, slender samara, 2.5-5 cm long, in open clusters persisting until following spring. April-May. — Floodplain forests, swamps, shores, streambanks.

Fraxinus nigra

Syringa
LILAC

Syringa vulgaris L.
COMMON LILAC   *introduced*
Much-branched deciduous shrub to 6 m tall, spreading and forming thickets. Leaves opposite, simple, entire, ovate, 5-10 cm long, short-acuminate, truncate to cordate at base. Flowers fragrant, in dense panicles 1-2 dm long; 4-merous; calyx small, campanulate, truncate to 4-toothed; corolla salverform, usually lilac, about 1 cm wide; stamens 2, included in the corolla tube. Fruit a 2-celled capsule with 2 seeds in each cell. May. — Native of se Europe, found near abandoned farms.

Lilac apparently does not spread by seed but is very long-lived and persists indefinitely after planting.

Fraxinus pennsylvanica

Onagraceae

EVENING-PRIMROSE FAMILY
Annual or perennial herbs. Leaves opposite to alternate, simple to pinnately divided, stalkless or short-petioled. Flowers usually large and showy, perfect (with both staminate and pistillate parts), regular, borne in leaf axils or in heads at ends of stems; sepals 8 or 4; petals 4, white, yellow or pink to rose-purple. Fruit a 4-chambered capsule; seeds many, with or without a tuft of hairs (coma).

1 Petals 2, small, white; leaves opposite; fruit with bristly hairs .... *Circaea*
1 Petals 4 (rarely absent), white, pink, or yellow; leaves alternate or opposite; fruit without bristly hairs
   2
2 Hypanthium prolonged beyond ovary into a tube below the petals; leaves alternate ..... *Oenothera*
2 Hypanthium scarcely if at all prolonged beyond ovary, not tube-like; leaves alternate or opposite ..
   .... *Epilobium*

Circaea
ENCHANTER'S NIGHTSHADE
Perennial herbs with opposite petioled leaves and small white flowers in one to few terminal racemes. Flowers 2-merous. Hypanthium shortly prolonged above the ovary, tubular. Petals obcordate or deeply notched. Stamens 2. Ovary 1-2-celled. Fruit reflexed, obovoid or pear-shaped, usually slightly compressed, beset with soft or stiff hooked bristles.

1 Plants to 6 dm tall; flowers and fruit well-spaced on stalk; calyx lobes more than 1.5 mm long; leaves rounded at base, the margins very shallowly toothed .... *C. canadensis*
1 Plants smaller, to 3 dm tall; open flowers clustered near top of stem; calyx lobes less than 1.5 mm long; leaves usually heart-shaped at base, margins sharply toothed .... *C. alpina*

Circaea alpina L.
ALPINE ENCHANTER'S NIGHTSHADE   *native / cons 7*
Perennial herb, spreading from rhizomes thickened and tuberlike at ends. Stems weak, 1-3 dm long, mostly smooth. Leaves opposite, ovate, 2-5 cm long and 1-3 cm wide; margins coarsely toothed; petioles flat on upper side, underside thin-winged along center. Flowers white, in short racemes of 10-15 flowers, becoming 1 dm long in fruit; sepals 1-2 mm long; petals to 2 mm long. Fruit a 1-seeded capsule, 2-3 mm long, covered with soft hooked bristles. June–Aug. — Cedar swamps (where often on rotting logs), low spots in forests.

Circaea canadensis (L.) Hill
COMMON ENCHANTER'S NIGHTSHADE   *native / cons 2*
    Circaea lutetiana L.
Stems erect, to 1 m tall, glabrous below, becoming minutely villosulous in the inflorescence. Leaves oblong-ovate, commonly 6–12 cm long, acuminate, very shallowly sinuate-denticulate, rounded or barely subcordate at base; petioles rounded or angled on the lower side. Racemes commonly many-flowered, to 2 dm long; petals 2.5-3.5 mm long. Fruit 3.5-5 mm long, beset with stiff bristles, equally 2-celled, 2-seeded, each half bearing normally 3 large and 2 small rounded ridges separated by narrow furrows. June–Aug. — Moist woods.

Circaea alpina

Epilobium
WILLOW-HERB, FIREWEED
Perennial herbs, often producing leafy rosettes or bulblike offsets (turions) at base of stem late in growing season. Leaves simple, opposite, alternate, or opposite below and becoming alternate above; stalkless or short-petioled. Flowers white to pink, single in axils of upper reduced leaves, or in spike or racemes at ends of stems; sepals 4; petals 4; stamens 8, the inner 4 stamens shorter than outer 4; ovary 4-chambered, maturing into a linear, 4-parted capsule, splitting from tip to release numerous brown seeds which are tipped with a tuft of fine hairs (coma).

1 Flowers showy in a terminal raceme; petals 8–16 mm long; stigma 4-parted .... *E. angustifolium*
1 Flowers single from upper leaf axils; petals 3–8 mm long 2
2 Stigma 4-parted .... *E. hirsutum*
2 Stigma entire, not 4-parted 3
3 Leaves entire or nearly so, the margins often revolute; stems round in cross-section, without lines of hairs on stem below base of each leaf 4
3 Leaf margins conspicuously toothed; stems 4-angled in section, with lines of hairs on stems below leaf bases 6
4 Stems with soft, straight hairs .... *E. strictum*
4 Stems finely hairy, the hairs appressed to stem 5
5 Upperside of leaves finely hairy .... *E. leptophyllum*
5 Upper surface of leaves glabrous or nearly so .... *E. palustre*
6 Tuft of hairs attached to tip of seeds (coma) white or nearly so, seeds with a broad, short beak; margins of stem leaves with mostly 10–30 teeth on a side .... *E. ciliatum*
6 Coma brown, seeds beakless; leaf margins with more than 30 teeth on a side .... *E. coloratum*

Epilobium angustifolium L.
FIREWEED   *native / cons 3*
    Chamaenerion angustifolium (L.) Scop.
Perennial herb. Stems erect, to 2 m or more tall. Leaves lance-shaped or linear lance-shaped, to 2 dm long, narrowed to a sessile or obscurely peti-

oled base, glabrous beneath; veinlets conspicuous. Racemes elongate, many-flowered, the lower flowers often exceeded by the subtending leaves, the upper with short bracts or none; petals purple, or rarely white, 10-15 mm long, clawed at base; style pubescent at base; stigmas soon revolute. Capsules 2.5-7 cm long. June-Sept. — In a variety of habitats, preferring moist soils rich in humus; often abundant after fires.

Epilobium ciliatum Raf.
AMERICAN WILLOWHERB   native / cons 3
   Epilobium glandulosum Lehm.
Perennial herb, with over-wintering leafy rosettes. Stems often branched, 3-10 dm long, smooth below, short-hairy above, especially in the head (where often with gland-tipped hairs). Leaves opposite, usually alternate near top; lance-shaped to ovate, 3-10 cm long and to 3 cm wide; margins with few, small, forward-pointing teeth; sessile or with short, winged petioles to 6 mm long. Flowers usually nodding when young, on stalks 3-10 mm long, on branches from upper leaf axils; sepals ovate, 2-5 mm long; petals white (or pink), notched at tip, 2-8 mm long. Fruit a linear capsule, 4-8 cm long, with gland-tipped hairs; seeds 1 mm long, the coma white. July-Sept. — Shores, streambanks, marshes, wet meadows, seeps, ditches and other wet places.

Epilobium angustifolium

Epilobium coloratum Biehler
PURPLE-LEAF WILLOWHERB   native / cons 3
Perennial herb, producing basal, leafy rosettes in fall; similar to American willowherb (*E. ciliatum*) but larger. Stems 5-10 dm long, much-branched in the head, smooth below, short-hairy above with hairs often in lines; stems and leaves often purple-tinged. Leaves mostly opposite, becoming alternate and smaller above, lance-shaped, 5-15 cm long and 0.5-3 cm wide, long-tapered to a pointed tip; margins finely toothed, with irregular sharp teeth; short-petioled to sessile. Flowers many on branches from upper leaf axils; sepals lance-shaped, 2-3 mm long; petals pink or white, 3-5 mm long, notched at tip; individual flowers on stalks to 10 mm long. Fruit a linear capsule, 3-5 cm long; seeds 1.5 mm long, the coma brown when mature. July-Sept. — Shores, seeps, swamps and wet woods, wet meadows, fens, ditches.

Epilobium coloratum

Epilobium hirsutum L.
HAIRY WILLOWHERB   introduced (invasive)
Perennial herb, spreading by rhizomes. Stems much-branched, 5-15 dm long, upper stems with a dense covering of soft, straight hairs. Leaves opposite (but bracts alternate), lance-shaped or oblong, 5-10 cm long and 1-3 cm wide, hairy on both sides, somewhat clasping at base; margins with sharp, forward-pointing teeth; petioles absent. Flowers upright on stalks from upper leaf axils, petals red-purple, 10-15 mm long, shallowly notched; stigma 4-lobed. Fruit a hairy, linear capsule, 5-8 cm long; coma nearly white. June-Sept. — Introduced from Eurasia, established in e Wisconsin in marshes, shores, wet meadows, and ditches.

Epilobium hirsutum

Epilobium leptophyllum Raf.
BOG WILLOWHERB   native / cons 8
Perennial herb, similar to marsh willowherb (*E. palustre*) but somewhat larger and more hairy. Stems simple or branched, 2-10 dm long, with short, incurved hairs. Leaves opposite or alternate, linear or linear lance-shaped, 2-7 cm long and 1-6 mm wide, upper surface hairy, underside hairy, at least on midvein, lateral veins indistinct; margins entire and rolled under; petioles short or more or less absent. Flowers erect in upper leaf axils on short, slender stalks to 1 cm long; petals light pink, 3-5 mm

long, entire or slightly notched at tip. Fruit a linear, finely hairy capsule, 4-5 cm long; the coma yellow-white. July-Sept. — Swamps, marshes, open bogs, sedge meadows, shores, streambanks and springs.

Epilobium palustre L.
MARSH WILLOWHERB   *native / cons* **8**
Perennial herb, from slender rhizomes or stolons. Stems simple or with a few branches above, 1-6 dm long, upper stem hairy with small incurved hairs. Leaves mostly opposite, lance-shaped, erect or ascending, 2-6 cm long and 3-15 mm wide, tapered to a rounded tip, upper surface smooth or with sparse hairs along midvein, underside smooth or finely hairy along midvein, lateral veins distinct; margins entire and often rolled under; sessile. Flowers few in upper leaf axils, on short stalks; petals white to pink, 3-5 mm long, notched at tip. Fruit a linear, finely hairy capsule; coma pale. July-Aug. — Open bogs and swamps.

Epilobium strictum Muhl.
DOWNY WILLOWHERB   *native / cons* **10**
Perennial herb, spreading by slender rhizomes; plants densely soft white-hairy. Stems erect, simple or branched above, 3-6 dm long. Lower leaves opposite, upper leaves alternate; lance-shaped, ascending, 2-4 cm long and 3-8 mm wide, tapered to a rounded tip; margins mostly entire, rolled under; sessile. Flowers on slender stalks from upper leaf axils; petals pink, 5-8 mm long, notched at tip. Fruit a linear, densely hairy capsule; coma pale brown. July-Aug. — Conifer swamps, sedge meadows, calcareous fens, marshes.

Epilobium palustre

Oenothera
EVENING-PRIMROSE
Annual, biennial, or perennial herbs with alternate, mostly narrow leaves and small to large, yellow, white, or pink flowers solitary in the axils or forming a terminal raceme. Flowers 4-merous. Hypanthium slenderly tubular (except one species), much prolonged above the ovary, deciduous from the fruit. Sepals at first connate, often with terminal or subterminal appendages, splitting at anthesis and reflexed. Petals commonly yellow, or in a few species white to pink, usually large and conspicuous but ephemeral. Stamens 8, equal or alternately unequal. Ovary 4-celled; stigma entire, 4-lobed, or deeply 4-cleft into elongate segments. Fruit a capsule; seeds numerous.

   Several of our species can be considered part of an *Oenothera biennis* "complex," including closely related (and sometimes difficult to distinguish) *O. biennis*, *O. oakesiana*, and *O. parviflora*.

1 Flowers white, becoming pink . . . . *O. nuttallii*
1 Flowers yellow 2
2 Bases of sepals contiguous . . . . *O. biennis*
2 Bases of awl-shaped sepals separate 3
3 Calyx, ovary, capsule, and upper leaves or bracts densely pubescent with appressed whitish non-glandular hairs; largest leaves typically less than 15 mm wide, finely toothed. . . . . *O. oakesiana*
3 Calyx, ovary, capsule, and other parts glabrate to sparsely pubescent, often with some long spreading hairs as well as shorter glandular hairs; largest leaves various, usually at least 15 mm wide, nearly entire. . . . . *O. parviflora*

Oenothera biennis L.
COMMON EVENING-PRIMROSE   *native / cons* **1**
Biennial herb. Stems 1-2 m tall, often suffused with red, terete or becoming somewhat angled in the inflorescence. Leaves lance-shaped to oblong, 1-2 dm long, acute or acuminate, entire to repand-dentate, often crisped on the margin, sessile or short-petioled, glabrous to sparsely pubescent but always green. Flowers several to many in a terminal raceme, the bracts resembling the leaves but much smaller; hypanthium tube 3-6

cm long; petals yellow, 15-25 mm long. July-Oct. — Fields, roadsides, and waste places.

Oenothera oakesiana (Gray) J.W. Robbins ex. S. Wats. & Coult.
OAKES' EVENING-PRIMROSE   *native / cons* **4**
Oenothera biennis var. oakesiana A. Gray
Oenothera parviflora var. oakesiana (Robbins) Fern.
Biennial herb with a taproot. Stems erect to procumbent, 1-6 dm tall. Leaves narrowly oblanceolate, 4-20 cm long, 0.5-3 cm wide. Flower petals yellow. July-Sept. Very similar to *O. parviflora* and *O. biennis;* differs from *O. parviflora* by its inflorescence lacking gland-tipped hairs; differs from *O. biennis* in having sepals separate at base (in *O. biennis* sepals close together at base). — Sandy or rocky shores, dunes, and clearings along the Lake Michigan; occasionally inland along railroads, sandy shores, or disturbed places; the common evening-primrose of Lake Michigan sand dunes and beaches.

Oenothera parviflora L.
SMALL-FLOWER EVENING-PRIMROSE   *native / cons* **2**
Biennial herb. Leaves commonly narrowly oblong lance-shaped to almost linear, green, glabrous or sparsely pubescent. Raceme crowded, the bracteal leaves commonly as long as the hypanthium; calyx in anthesis densely strigose to nearly glabrous, the actual end of the sepal represented by a prominent lobe or transverse ridge; petals yellow, 10-15 mm long. June-Oct.

Oenothera biennis

Orobanchaceae
BROOM-RAPE FAMILY
Annual, biennial, or perennial herbs; some genera fleshy, without green color, parasitic on the roots of other plants. Leaves opposite, alternate, or the leaves reduced to scales. Flowers mostly perfect, single or few from leaf axils, or numerous in clusters at ends of stems or leaf axils, usually with a distinct upper and lower lip; calyx 2-5-lobed or toothed, persistent in fruit; petals 4-5 (petals sometimes absent); stamens usually 4, inserted on the corolla tube. Fruit a several- to many-seeded 2-valved capsule. Now includes many former members of Scrophulariaceae.

1 Plants non-green, leaves converted to small, non-photosynthetic bracts 2
1 Plants green, leaves present and photosynthetic, green or sometimes strongly tinged with purple (often blackening upon drying). 4
2 Stems well-branched; the flowers nearly sessile in spike-like racemes . . . . ***Epifagus***
2 Stems unbranched or with only a few branches, the flowers single or few on long pedicels, or crowded into dense, spike-like racemes 3
3 Stems thickened and pinecone-like; calyx deeply parted on both upper and lower sides . ***Conopholis***
3 Stems not pinecone-like; calyx deeply parted on lower side . . . . ***Orobanche***
4 Stem leaves of fertile stems alternate . . . . ***Castilleja***
4 Stem leaves of fertile stems all opposite or nearly so, sometimes alternate below the flowers 5
5 Most leaves deeply pinnately divided; corolla cream-colored or yellow, 1.5–5 cm long . . . ***Pedicularis***
5 Stem leaves toothed or entire, not deeply pinnately lobed; corolla various colors, in most species less than 1.5 cm long 6
6 Calyx somewhat inflated in flower; conspicuously inflated and laterally compressed in fruit; flowers in a one-sided, leafy spike; leaf margins with large, forward-pointing teeth . . . . ***Rhinanthus***
6 Calyx not inflated 7
7 Flowers in the axils of alternate bracts, forming a terminal inflorescence . . . . ***Odontites***
7 Flowers solitary in the axils of opposite or whorled leaves or bracts 8
8 Corolla nearly regular, the 5 lobes similar . . . . ***Agalinis***
8 Corolla irregular, 2-lipped, the upper lip 2-lobed, the lower lip 3-lobed . . . . ***Melampyrum***

Agalinis
FALSE FOXGLOVE
Annual hemiparasitic herbs. Stems slender, erect, branched, usually 4-angled. Leaves opposite, linear, stalkless, smooth, or rough-to-touch on upper surface. Flowers showy, in clusters at ends of branches; sepals joined, the calyx 5-lobed, bell-shaped; petals united, corolla 5-lobed, bell-shaped, and nearly regular, only slightly 2-lipped, pink to purple; stamens 4, of 2 different lengths. Fruit a nearly round, many-seeded capsule.

1 Pedicels less than 6 mm long, shorter than or equaling the calyx . . . . **A. purpurea**
1 Longer pedicels more than 6 mm long, equaling or longer than the calyx . . . . **A. tenuifolia**

Agalinis purpurea (L.) Pennell
PURPLE FALSE FOXGLOVE   *native / cons 7*
Gerardia purpurea L.
Annual herb. Stems slender, 2-8 dm long, 4-angled, smooth to slightly rough, branched and spreading above. Leaves opposite, spreading, linear, 1-5 cm long and 1-3 mm wide; margins entire; petioles absent. Flowers on spreading stalks 2-5 mm long, in racemes on the branches; calyx 4-6 mm long, the lobes less than half the length of the tubular base; corolla purple, 2-3 cm long, the lobes spreading, 5-10 mm long. Fruit a round capsule, 4-6 mm wide. Aug-Sept. — Wet meadows, fens, shores of Lake Michigan and along inland lakes and ponds; moist areas between dunes, ditches; usually where sandy, often where calcium-rich.

Agalinis tenuifolia (Vahl) Raf.
COMMON FALSE FOXGLOVE   *native / cons 6*
Agalinis besseyana (Britt.) Britt.
Gerardia tenuifolia var. *parviflora* Nutt.
Annual herb. Stems slender, erect, 2-6 dm tall, smooth, usually with many branches. Leaves opposite, spreading, linear, 1-5 cm long and 1-3 mm wide, upper surface slightly rough; margins entire; petioles absent. Flowers on slender, ascending stalks, 1-2 cm long; calyx 3-5 mm long, with short teeth; corolla purple (rarely white), often spotted, 10-15 mm long, the lobes 3-5 mm long. Fruit a round capsule, 4-6 mm wide. Aug-Sept. — Wet meadows, low prairie, fens, shores, streambanks and ditches, usually where sandy.

Agalinis purpurea

Agalinis tenuifolia

Castilleja
INDIAN-PAINTBRUSH

Castilleja coccinea (L.) Spreng.
SCARLET INDIAN-PAINTBRUSH   *native / cons 6*
Annual herb, more or less pubescent. Stems usually simple, 2-6 dm tall. Leaves alternate; principal stem leaves diverse, varying from rarely entire to commonly 3-5-cleft, the segments linear, the lateral segments almost always shorter or narrower than the terminal. Flowers in dense terminal spikes, each subtended by a large bracteal leaf, these wholly or mostly scarlet (rarely pale), commonly deeply 3-lobed, occasionally 5-lobed. Spike at first dense, 4-6 cm long, elongating to as much as 2 dm in fruit; calyx 2-3 cm long, thin and membranous, often more or less scarlet, deeply divided into two lateral halves; each half gradually widened distally and at the summit broadly rounded or truncate; corolla greenish yellow, little surpassing the calyx, the minute lower lip less than 1/3 as long as the upper. Capsule ovoid or oblong, many-seeded. May-Aug. — Meadows, moist prairies, calcareous sandy or gravelly shores, swamps.

Castilleja coccinea

Conopholis
SQUAWROOT
Conopholis americana (L.) Wallr.
AMERICAN SQUAWROOT   *native / cons* **8**
Unbranched perennial herb; parasitic on the roots of oaks (especially *Quercus rubra*). Stems stout, 5-20 cm tall, entirely or mostly concealed by the numerous, fleshy, overlapping leaf-scales, pale brown or yellowish throughout. Leaf-scales ovate, to 2 cm long. Spike usually constituting half of the plant or more, 1.5-2 cm thick; calyx 8-13 mm long, subtended at base by 1 or 2 minute bractlets; corolla tubular, 10-15 mm long; curved downward, very irregular, the upper lip straight, entire or nearly so; the lower lip decurved, lobed. Capsule ovoid, tipped with the persistent style and capitate stigma. July-Sept. — Deciduous or mixed forests with oak trees.

Conopholis americana

Epifagus
BEECHDROPS
Epifagus virginiana (L.) W. Bart.
BEECHDROPS   *native / cons* **9**
Freely branched annual herb; parasitic on the roots of beech trees, with small, scattered, alternate leaf scales and numerous, nearly sessile, solitary, axillary flowers, forming a large panicle. Stems 1-5 dm tall, pale brown, usually marked with fine brown-purple lines, with numerous, elongate, ascending branches, the dead dry stems of the previous season persist through the winter and into the next summer. Leaf-scales triangular-ovate, 2-4 mm long. Flowers dimorphic, the lower small, pistillate, fertile, the upper perfect but sterile. Lower flowers about 5 mm long, calyx cup-shaped, strongly 5-ribbed, the 5 short lobes triangular-subulate; corolla not opening, promptly forced off by the developing ovary and persisting for a time on its tip; stamens none; style short. Upper flowers about 1 cm long, white, commonly with two stripes of brown-purple; calyx similar to lower flowers but larger; corolla tubular, shortly 4-lobed, about equaled by the stamens and style. Capsule about 5 mm long, dehiscent across the top. July-Sept. — In forests with beech and usually other trees, such as sugar maple, hemlock, oak.

Epifagus virginiana

Melampyrum
COW-WHEAT
Melampyrum lineare Desr.
AMERICAN COW-WHEAT   *native / cons* **7**
Annual herb, partially parasitic on other plants, often red-tinged when in open habitats. Stems usually branched, 1-4 dm long. Leaves opposite, lower leaves oblong lance-shaped, upper leaves linear or lance-shaped, often toothed near base; petioles short or absent. Flowers from upper leaf axils; calyx tube cup-shaped, calyx lobes 4 or 5, longer than the tube; corolla about 1 cm long, 2-lipped, the upper lip white, the lower pale yellow; stamens 4, ascending under the upper lip and not exsert. Fruit a capsule to 1 cm long. June-Aug. — In a variety of habitats, ranging from wet to dry forests and openings; in wetlands occasional in swamps and on hummocks in open fens.

Melampyrum lineare

Odontites
EYEBRIGHT

Odontites vulgaris Moench
EYEBRIGHT   *introduced*
Odontites serotinus Dumort.

Annual herb, usually parasitic on the roots of other plants. Stems 1-4 dm
tall, simple or more commonly branched, finely pubescent. Leaves opposite, lance-shaped, sessile with a broad base, 1-3 cm long, roughly pubescent, with 2 or 3 blunt teeth on each margin. Flowers nearly sessile from
the upper axils, forming a terminal, often secund spike or raceme; calyx
regular, campanulate, about 5 mm long, 4- lobed, shorter than the tube;
corolla tubular, about 1 cm long, pubescent, light red; upper lip nearly
straight, entire; lower lip spreading, shallowly 3-lobed; stamens 4, slightly
unequal, ascending under the upper corolla lip and included by it. Capsule
elliptic, pubescent, about 7 mm long, few-seeded, included in the mature
calyx. July-Sept. — Native of Europe; established as a weed in fields and
waste places.

Odontites vulgaris

Orobanche
BROOM-RAPE

Orobanche uniflora L.
NAKED BROOM-RAPE   *native / cons* 6
Aphyllon uniflorum (L.) Torr. & Gray

Parasitic on numerous plants. Stem proper only 1-3, or rarely 5 cm long,
all or mostly underground, bearing a few overlapping, obovate, glabrous
scales. Pedicels 1-4, usually 2, 6-20 cm long, erect, finely glandular-pubescent, without bractlets, each bearing a single white to violet flower
about 2 cm long. Flowers perfect; calyx lobes 5, about equal, triangular-
acuminate, slightly longer than the calyx tube. Stamens 4, about as long
as the corolla tube and inserted below its middle. Capsule 2-valved, many-
seeded, enclosed by the withering corolla. July-Sept. — Moist woods and
streambanks.

Orobanche uniflora

Pedicularis
LOUSEWORT

Perennial herbs (ours). Leaves either opposite, alternate, or scattered, sharply toothed to 2-pin-
natifid. Flowers yellow or purple in terminal spikes or racemes, each subtended by a bracteal
leaf. Calyx campanulate to tubular, entire or variously lobed, but usually longer on the upper
side. Corolla tube gradually enlarged distally; upper lip as long as or longer than the lower, very
concave or arched, often laterally compressed; lower lip more or less expanded, with two longitudinal folds below the sinuses. Stamens 4, ascending under the upper corolla-lip and not exsert.
Capsule compressed, ovate to oblong, pointed. Seeds several, not winged.

1 Stems glabrous or nealy so; leaves opposite; flowering in late summer . . . . *P. lanceolata*
1 Stems usually long-hairy; leaves alternate; flowering in early summer . . . . *P. canadensis*

Pedicularis canadensis L.
WOOD BETONY   *native / cons* 8
Perennial herb. Stems several, erect, 1.5-4 dm tall, sparsely villous. Leaves
chiefly basal, lance-shaped to oblong lance-shaped, pinnately lobed usually more than halfway to the midvein; lower leaves on petioles often
longer than the blade; stem leaves progressively reduced, short-petioled
to nearly sessile; bracteal leaves usually toothed only at the tip. Spikes
commonly solitary, at anthesis 3-5 cm, in fruit to 20 cm long; corolla yellow, maroon, or yellow on the lower lip and maroon on the upper; the

Pedicularis canadensis

upper lip bearing 2 slender teeth just below the rounded tip; lower lip shorter than the upper. Capsule oblong, about 15 mm long, 2x as long as the mature calyx, opening along the upper side. May-June. — Dry forests and savannas, moist hardwood forests, especially in openings; less often in conifer swamps, meadows, and grasslands.

Pedicularis lanceolata Michx.
SWAMP LOUSEWORT   *native / cons* **8**

Perennial herb; plants at least partially parasitic on other plants. Stems 3-8 dm long, more or less smooth, unbranched or few-branched. Leaves opposite, or in part alternate, mostly lance-shaped, 4-9 cm long and 1-2 cm wide, pinnately lobed; margins with small rounded teeth; lower leaves short-petioled, upper leaves sessile. Flowers more or less sessile, in spikes at ends of stems and from upper leaf axils; the spikes 2-10 cm long; calyx 2-lobed; corolla yellow, about 2 cm long, the upper lip entire and arched, lower lip upright. Capsule unequally ovate, mostly shorter than the calyx. July-Sept. — Wet meadows, calcareous fens, wetland margins, springs, streambanks.

Pedicularis lanceolata

Rhinanthus
YELLOW RATTLE

Rhinanthus minor L.
LITTLE YELLOW RATTLE   *introduced*
  Rhinanthus crista-galli L.

Annual herb. Stems 2-6 dm tall, simple or branched, retrorsely pubescent on two sides, glabrous on the other two. Leaves narrowly oblong to lanceolate, 1.5-5 cm long, conspicuously serrate; bracteal leaves equaling or exceeding the calyx, sharply laciniate; calyx about 1 cm long at anthesis; corolla 1-2 cm long, mostly yellow, often with pale or colored teeth on the upper lip or with dark markings on the lower. Capsule 10-15 mm long and wide; seeds pale yellowish brown, 4-6 mm long, the wing to 1 mm wide. Roadsides and waste places.

Rhinanthus minor

Oxalidaceae
WOOD-SORREL FAMILY

Oxalis
WOOD-SORREL

Perennial herbs (ours). Leaves basal or alternate on the stem, 3-foliolate, leaflets obcordate. Flowers solitary on axillary peduncles or in cymose or umbel-like clusters; perfect, regular, 5-merous, white, yellow, pink, or purple. Sepals usually imbricate. Stamens 10, alternately long and short. Ovary 5-celled; styles 5. Fruit a capsule.

1 Plants without stems, the leaves and scapes all from plant base; flowers white to pink . . *O. montana*
1 Plants with leaves from stems; flowers yellow 2
2 Hairs of the stem and petioles (or many of them) septate, blunt; plants rhizomatous; stipules absent
   . . . . *O. stricta*
2 Hairs of the stem and petioles pointed, nonseptate; plants without rhizomes, though often with
   stolons; stipules present, often conspicuous . . . . *O. dillenii*

Oxalis dillenii Jacq.
SLENDER YELLOW WOOD-SORREL   *native / cons* **0**

Tufted perennial herb; stems erect to decumbent (but scarcely creeping), light-green, with pointed hairs, to 40 cm tall; rhizomes absent. Leaflets 1-2 cm wide; stipules to 3 mm long or smaller. Petals yellow, 4-10 mm

long. Capsules 1.5-2.5 cm long, usually grayish hairy. Seeds brownish,
conspicuously rugose, the ridges whitish. Summer. — Mostly in disturbed
places such as gardens and lawns, along roadsides, in fields and gravel
pits; sometimes invading forests.

Plants frequently more branched than those of *O. stricta,* which often
has only a single erect stem.

Oxalis montana Raf.
NORTHERN WOOD-SORREL   *native / cons* 9
  Oxalis acetosella L.
Perennial herb, from slender, scaly rhizomes. Leaves single or 3-6 together,
all from base of plant, on stalks 4-15 cm long, these joined at base;
palmately divided into 3 leaflets, the leaflets notched at tips, sparsely
hairy. Flowers perfect, broadly bell-shaped, single atop stalks 6-15 cm
long (usually slightly taller than leaves), with a pair of small bracts above
middle of stalk; sepals 5, much shorter than petals; petals 5, white or
pink and with pink veins, 10-15 mm long. Fruit a smooth, nearly round
capsule. May-July. — Hummocks in swamps, wet depressions in forests,
moist wetland margins.

Oxalis dillenii

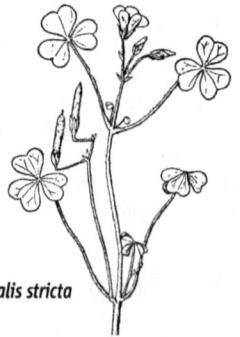

Oxalis montana

Oxalis stricta L.
COMMON YELLOW WOOD-SORREL   *native / cons* 0
  Oxalis fontana Bunge
Perennial herb. Stems erect or eventually decumbent, to 5 dm tall, more
or less pubescent with ascending or incurved hairs. Leaflets 1-2 cm wide.
Pedicels similarly pubescent, at maturity commonly abruptly divaricate
or deflexed. Inflorescence umbel-like. Flowers yellow, 5-10 mm long, rarely
more than 3 to a peduncle. Capsules 1-3 cm long, densely to thinly gray-
pubescent with mostly retrorse hairs 0.1-0.3 mm long, or with longer
ones intermingled, or wholly or partly glabrous. Summer. — In many different habitats: mostly a common weed of roadsides, railroads, gardens,
lawns, fields, and disturbed places; also in forests, especially along trails.

Oxalis stricta

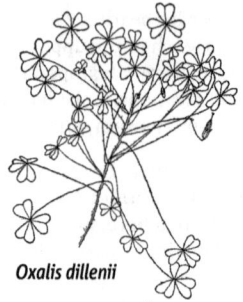

Papaveraceae
POPPY FAMILY
Herbs with watery, milky, or colored juice. Leaves alternate or rarely opposite. Flowers regular,
perfect. Sepals 2 or 3, early deciduous. Petals 4 or more (rarely absent), separate, conspicuous.
Stamens 6, or 12 or more. Ovary 1-celled or falsely 2-celled, of 2 or more carpels. Fruit a capsule,
dehiscent by terminal valves or longitudinally (rarely otherwise). The Fumariaceae (now included
here) were previously recognized as a separate family, differing in bilateral symmetry of the
flowers and watery juice. All of our members of Papaveraceae in the strict sense have colored
juice (yellow to red-orange or milky).

1 Corolla bilaterally symmetrical; juice watery, clear 2
2 Delicate vine . . . . Adlumia fungosa
2 Upright herbaceous plants 3
1 Corolla regular, juice colored whitish to yellow or red-orange. 5
3 Corolla white, with 2 spurs; leaves basal . . . . ***Dicentra***
3 Corolla pink, yellow or red-purple, 1-spurred; leaves alternate 4
4 Flowers pink, yellow-tipped; seeds 1–1.5 mm wide; plants erect, the terminal inflorescences definitely
   surpassing the leaves . . . . ***Capnoides***
4 Flowers yellow or purple; seeds 1.8–2.2 mm wide; plants ± spreading or sprawling, the terminal inflo-
   rescences barely if at all surpassing the leaves . . . . ***Corydalis***
5 Leaf 1 from base of plant; petals 8 or more . . . . ***Sanguinaria***
5 Plants with leafy stems; petals 4 . . . . ***Chelidonium***

Adlumia
ALLEGHENY-VINE
Adlumia fungosa (Ait.) Greene
ALLEGHENY-VINE   *native / cons 7*
Biennial vine, climbing by the upper part of the rachis of the 3-parted
leaves, During the first year acaulescent, with several ascending, non-
prehensile leaves; climbing to 3 m high the second year, with elongate
stems and large, delicate, prehensile leaves, their rachis elongate and the
uppermost leaflets greatly reduced. Flowers pink, in drooping axillary
panicles; sepals 2, scale-like, quickly deciduous; corolla bilateral, narrowly
flattened-ovoid, the outer petals constricted near the tip to form an ovate
appendage, the inner narrow, dilated at the summit into an oval appendage; corolla after anthesis persistent, enclosing the slender 2-valved
capsule. June-Sept. — Woods, rocky shores, often where calcareous.

Adlumia fungosa

Capnoides
ROCK-HARLEQUIN
Capnoides sempervirens (L.) Borkh.
ROCK-HARLEQUIN   *native / cons 7*
  Corydalis sempervirens (L.) Pers.
Biennial herb, glaucous. Stems slender, erect, 3-6 dm tall, divaricately
branched above; principal internodes 4-8 cm long. Lower leaves petioled,
upper leaves nearly sessile. Flowers in small panicles at the end of the
branches; bracts minute, lance-shaped; corolla pink, tipped with yellow,
12-17 mm long, the tube 8.5-12 mm, the spur 2.5-5 mm long; sepals broadly
ovate, 2-4 mm long. Capsule erect or nearly so, 2-4 cm long; seed 1.3-1.5
mm long. — Dry or rocky woods, gravelly shores, especially where disturbed.

Capnoides sempervirens

Chelidonium
CELANDINE
Chelidonium majus L.
CELANDINE   *introduced (invasive)*
Biennial herb with saffron-colored juice. Stems branched, 3-8 dm tall.
Stem leaves several, alternate, deeply pinnately parted (usually to the
midrib) into 5-9 segments; lateral segments often alternate, variously
toothed or lobed, the lowest often with a secondary basal division; terminal segments broadly obovate, 3-lobed. Umbel peduncled, several-flowered; sepals 2, deciduous, glabrous; petals 4, yellow, about 1 cm long; stamens numerous, with long slender filaments. Capsule cylindric, glabrous,
3-5 cm long. April-Sept. — Native of Eurasia; established in moist soil.

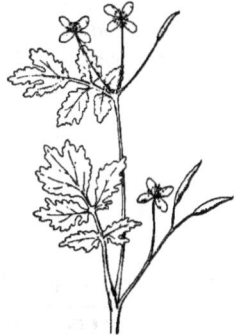

Chelidonium majus

Corydalis
FUMEWORT
Corydalis aurea Willd.
GOLDEN CORYDALIS   *native / cons 4*
Biennial herb. Stems 2-6 dm tall, erect or ascending, branched above.
Leaves cauline, alternate, 2-pinnately dissected. Racemes dense, 1-3 cm
long, often surpassed by the upper leaves; bracts lance-shaped; sepals
broadly ovate, 1.5-2 mm long, erose; corolla 12-15 mm long, of which the
spur constitutes less than 1/3; outer petals folded distally along the median
line into a conspicuous but wingless keel. Capsules slender, 2-valved, terminated by the slender persistent style, spreading or drooping, 1-2 cm
long; seeds black, shining, 2-2.5 mm wide. — Rocky banks, sandy soil.

Corydalis aurea

Dicentra
BLEEDINGHEARTS
Perennial herbs from rhizomes or a cluster of small tubers. Leaves basal or alternate, compound.
Flowers white to red-purple, scapose or in axillary racemes or panicles. Sepals 2, minute. Corolla
ovate or cordate, bilaterally symmetrical; petals weakly united, the outer two large, saccate or
spurred at base, spreading or ascending at the summit, the inner much narrower and more or
less dilated at the summit. Ovary slender, gradually tapering into the long style. Fruit a capsule.

1 Corolla sac-like, with small rounded spurs about as long as wide . leaf underside waxy . **D. canadensis**
1 Corolla with widely spreading spurs, these longer than wide; leaf green or only slightly waxy **D. cucullaria**

Dicentra canadensis (Goldie) Walp.
SQUIRREL-CORN   *native / cons* **8**
Very similar to *D. cucullaria* in foliage, size, and habit. Tubers fewer and
about 2x larger; corolla narrowly ovate, the spurs short, broadly rounded,
scarcely divergent. May. — Rich deciduous forests.

Dicentra cucullaria (L.) Bernh.
DUTCHMAN'S-BREECHES   *native / cons* **7**
Scapes and leaves from a dense cluster of small, white, grain-like tubers.
Leaves broadly triangular in outline, compound, the ultimate segments
linear, long-petioled. Scapes 1-3 dm tall, bearing a terminal raceme of 3-
12 nodding white flowers, suffused with yellow at the summit; corolla
15-20 mm long; spurs subacute, divergent. May. — Rich deciduous forests,
occasionally in swampy or relatively dry woods.

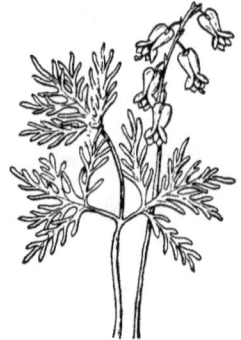

Dicentra canadensis

Sanguinaria
BLOODROOT
Sanguinaria canadensis L.
BLOODROOT   *native / cons* **6**
Perennial herb with red juice, from a stout rhizome which sends up a
single lobed leaf and a large white scapose flower. Leaves orbicular in
outline, 3-9-lobed, the lobes undulate to coarsely toothed. Scape 5-15 cm
tall at anthesis. Flowers white, varying rarely to pink, 2-5 cm wide; sepals
2; petals typically 8, often more and as many as 16; 4 petals usually longer
than the others and the flower quadrangular in outline; stamens numerous. Capsule 3-5 cm long, crowned by the persistent style, dehiscent
longitudinally. March-April. — Rich deciduous and floodplain forests. The
leaves continue to expand after anthesis and may grow to 2 dm wide.

Sanguinaria canadensis

Penthoraceae
PENTHORUM FAMILY
Penthorum
DITCH-STONECROP
Penthorum sedoides L.
DITCH-STONECROP   *native / cons* **3**
Perennial herb, spreading by rhizomes; plants often red-tinged. Stems 1-
6 dm long, smooth, upper stem often angled and with gland-tipped hairs.
Leaves alternate, lance-shaped, 2-10 cm long, 0.5-3 cm wide, tapered to
tip and base; margins with small, forward-pointing teeth; sessile or on
petioles to 1 cm long. Flowers star-shaped, perfect, 3-6 mm wide, on short
stalks, in branched racemes at ends of stems; sepals 5, green, triangular,
1-2 mm long; petals usually absent; stamens 10; pistils 5, joined at base
and sides to form a ring. Fruit a many-seeded capsule, the seeds about
0.5 mm long. July-Sept. — Streambanks, muddy shores and ditches.

Penthorum sedoides

Phrymaceae

LOPSEED FAMILY

Perennial herbs. Calyx tubular, 5-lobed. Fruit a dehiscent capsule. Previously, this family was monotypic with only genus *Phryma;* now includes *Mimulus.*

1 Flowers nearly sessile in pairs in terminal spike-like racemes, subtended by tiny bracts; upper calyx teeth bristle-like; fruit an achene, strongly reflexed . . . . *Phryma*
1 Flowers peduncled, borne singly in the axils of opposite leaves or bracts with expanded blades; calyx teeth not bristle-like; fruit a capsule, not strongly reflexed . . . . *Mimulus*

Mimulus
MONKEY-FLOWER
Mimulus ringens L.
ALLEGHENY MONKEY-FLOWER   *native / cons* 6
Smooth perennial herb, from stout rhizomes. Stems usually erect, 3-8 dm long, 4-angled and sometimes winged. Leaves opposite, oblong to lance-shaped, 4-12 cm long and 1-3.5 cm wide, upper leaves smaller; margins with forward-pointing teeth; petioles absent, the base of leaf clasping stem. Flowers single from upper leaf axils, on slender stalks 1-5 cm long and longer than the sepals; calyx regular, angled, 1-2 cm long, the lobes awl-shaped, 3-5 mm long; corolla blue-violet, 2-lipped, 2-3 cm long, the throat nearly closed, the upper lip erect and bent upward, lower lip longer and bent backward. Fruit a cylindric capsule, about as long as calyx tube. July-Aug. — Streambanks, oxbow marshes, swamp openings, floodplain forests, muddy shores, ditches; sometimes where disturbed.

Mimulus ringens

Phryma
LOPSEED
Phryma leptostachya L.
AMERICAN LOPSEED   *native / cons* 5
Perennial herb. Stems erect, 5-10 dm tall, simple or with a few divergent branches. Leaves opposite, ovate, 6-15 cm long; lower petioles to 5 cm long, the upper shorter or the uppermost sessile. Flowers pale purple to white, in elongate, long-peduncled, interrupted spike-like racemes terminating the stem and also from a few upper axils, opposite and horizontal; calyx 2-lipped, the upper 3 lobes bristle-like, with hooked tips when mature, about equaling the tube, the lower 2 very short, broadly triangular; corolla tube scarcely widened upward, the upper lip straight, the lower much longer, spreading, 3-lobed; stamens 4, included. Fruit an achene, contained in the persistent calyx. June-Aug. — Rich deciduous forests, especially moist areas in beech-maple woods, but also in drier forests with oak and sometimes with conifers.

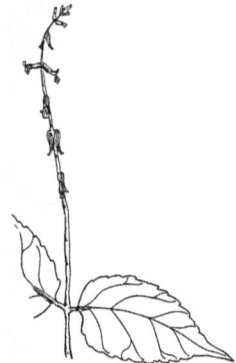

Phryma leptostachya

Plantaginaceae

PLANTAIN FAMILY

Annual or perennial herbs. Leaves simple, entire, all from base of plant. Flowers perfect, in terminal or axillary clusters or racemes, each flower subtended by bracts; flower parts mostly in 4s. Fruit a capsule opening at tip.

Plantaginaceae is the correct name for the family that encompasses not only the plantains with their reduced flowers, but also the related larger-flowered genera formerly placed in the Scrophulariaceae, as well as highly reduced aquatics, such as *Hippuris* (Hippuridaceae) and *Callitriche* (Callitrichaceae).

1   Flowers tiny, lacking a corolla, or corolla regular and scarious 2
1   Flowers usually conspicuous, with both calyx and corolla present, the corolla petal-like, usually con-
    spicuously bilaterally symmetrical. 4
2   Leaves in a basal rosette . . . . *Plantago*
2   Leaves opposite or whorled on an elongate stem 3
3   Leaves in whorls of 6–12 (usually 9) . . . . *Hippuris*
3   Leaves opposite . . . . *Callitriche*
4   Stem leaves all or mostly alternate on fertile stems (lowermost leaves sometimes opposite and rosette
    of larger basal leaves sometimes present) 5
4   Stem leaves all or mostly opposite (rarely whorled) on fertile stems (may be alternate beneath flow-
    ers) 8
5   Corolla nearly regular, the lobes equaling or exceeding the tube . . . . *Veronica*
5   Corolla bilaterally symmetrical, 2-lipped, the lobes distinctly shorter than the tube (including spur, if
    any) 6
6   Flowers all solitary in axils of leaves (nearly to base of plant); corolla pale purple and white; leaves,
    calyx, and stem with dense gland-tipped hairs . . . . *Chaenorrhinum*
6   Flowers in compact or elongate terminal inflorescences (half or less the height of the plant); corolla
    yellow, red, or blue; leaves, calyx, and usually stem glabrous and eglandular or nearly so. 7
7   Corolla 1.3–4 cm long (including spur), yellow (red-pink in a rare weedy annual); seeds strongly wrin-
    kled, tuberculate, ridged, or winged . . . . *Linaria*
7   Corolla 0.6–1.1 cm long, blue; seeds smooth or weakly pebbled . . . . *Nuttallanthus*
8   Inflorescence terminal and branched (paniculate); stamens 4 fertile plus 1 staminodium . *Penstemon*
8   Inflorescence a spike or raceme (no branched stalks), or flowers all axillary; stamens 2 or 4 fertile, in
    most genera with no staminodium (or only a very rudimentary one) 9
9   Corolla 2.3–3.5 cm long; sepals broadly ovate-orbicular, overlapping; stamens 4 fertile plus a filamen-
    tous elongate staminodium . . . . *Chelone*
9   Corolla less than 1.5 cm long; sepals linear-lanceolate to somewhat ovate, not conspicuously overlap-
    ping; stamens 2 or 4 (including any staminodia). . . . . *Veronica*

Callitriche
WATER-STARWORT
Callitriche palustris L.
VERNAL WATER-STARWORT   *native / cons* 6
Small, perennial aquatic herb; roots fibrous. Stems weak, slender, 10-20
cm long. Leaves opposite, of 2 types; underwater leaves mostly linear, 1-
2 cm long and to 1 mm wide, shallowly notched at tip, the leaf pairs connected at base by a narrow wing; floating leaves in clusters at ends of
stems or opposite along upper stems, 3-5-nerved, obovate to spatula-
shaped, rounded at tip, 5-15 mm long and 2-5 mm wide; leaves intermediate between underwater and floating leaves usually present. Flowers
tiny, either staminate or pistillate; usually 1 staminate and 1 pistillate
flower together in leaf axils, subtended by a pair of translucent bracts,
these soon deciduous. Fruit 4-chambered, separating when mature into
4 nutlets 1-1.5 mm long, broadest above middle, narrowly winged near
tip, pitted in vertical rows. June-Sept. — Shallow water of lakes, ponds,
streams; exposed mudflats.

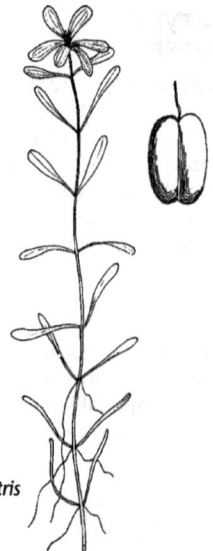

Callitriche palustris

Chelone
TURTLEHEAD
Chelone glabra L.
WHITE TURTLEHEAD   *native / cons* 7
Perennial herb. Stems erect, 5-10 dm long, rounded 4-angled, unbranched
or sometimes branched above. Leaves opposite, lance-shaped, to 15 cm
long and 1-3 cm wide, tapered to a sharp tip; margins with sharp, forward-pointing teeth; petioles very short or absent. Flowers in dense
spikes at ends of stems, 3-8 cm long; sepals 5; corolla white or light pink,
2.5-3.5 cm long. Fruit an ovate capsule. Aug-Sept. — Swamp openings,
thickets, streambanks, shores, wet meadows, marshes, calcareous fens.

Chelone glabra

ADDITIONAL SPOECIES
   Chelone obliqua L. (Red turtlehead), introduced, rarely escaped from cultivation, flowers deep pink; reported in Door County from moist soil in
mixed forest.

Hippuris
MARE'S-TAIL

Hippuris vulgaris L.
COMMON MARE'S-TAIL   *native / cons* 10
Perennial herb, from large, spongy rhizomes. Stems 2-6 dm long, unbranched, underwater and lax, or emersed and upright, densely covered
by the closely spaced whorls of leaves. Leaves numerous, in whorls of 6-
12, linear, 1-2.5 cm long and 1-3 mm wide, stalkless. Flowers very small,
perfect, stalkless and single in upper leaf axils, or often absent; sepals
and petals lacking; stamen 1, style 1, ovary 1-chambered. Fruit nutlike,
oval, 2 mm long. June-Aug. — Shallow water or mud of marshes, lakes,
streams and ditches.

Hippuris vulgaris

Linaria
TOADFLAX

Linaria vulgaris P. Mill.
BUTTER-AND-EGGS   *introduced (invasive)*
Glabrous perennial herb; spreading by rhizomes and forming colonies.
Stems erect, 3-8 dm tall. Leaves very numerous, pale green, narrow, 2-5
cm long, 2-4 mm wide, narrowed below to a petiolelike base. Flowers numerous in a compact spike, yellow with orange palate, 2-3 cm long, including the spur. Calyx deeply 5-parted. Corolla irregular, strongly bilabiate, spurred at base, the upper lip erect, 2-lobed, the lower 3-lobed;
stamens 4. Capsule round-ovoid, 8-12 mm long, the seeds winged. May-
Sept. — Native of Europe; fields, roadsides, and waste places.

Linaria vulgaris

Nuttallanthus
OLDFIELD-TOADFLAX

Nuttallanthus canadensis (L.) D.A. Sutton
OLDFIELD-TOADFLAX   *native / cons* 5
   Linaria canadensis (L.) Chaz.
Annual herb. Stems erect, 2-6 dm tall, glabrous, with several procumbent
or widely spreading sterile shoots from the base. Leaves narrowly linear,
1-3 cm long, those of the erect stems widely scattered, alternate, those of
the sterile shoots smaller, crowded, often opposite or whorled. Racemes
congested at anthesis, later elongate; pedicels 2-4 mm long; corolla blue,
the lips much longer than the tube; lower lip with 2 short white ridges.
Capsule 3-4 mm wide. May-Aug. Dry, open, sandy or rocky sterile ground;
oak savanna, jack pine plains, dried lake beds. Sometimes misidentified
as the less common *Lobelia kalmii,* with which it shares narrow leaves
and blue bilaterally symmetrical flowers, but *L. kalmii* is found in wetlands
and has milky juice.

Nuttallanthus canadensis

Penstemon
BEARDTONGUE
Perennial herbs (ours), the erect stems rising from a rosette of petioled basal leaves, the stem
leaves sessile and often clasping, the upper progressively reduced in size (the leaves of all our
species are about alike and are of little value in distinguishing species). Flowers white to blue-
violet or red-violet, in terminal clusters. Calyx herbaceous, deeply 5-parted, the lobes usually
unequal. Corolla tubular or trumpet-shaped, the tube much longer than the lobes, bilabiate, the

upper lip erect, 2-lobed, the lower equaling or longer than the upper, 3-lobed. Fertile stamens 4; sterile stamen present, about as long as the fertile stamens. Capsule ovoid or conic, many-seeded.

1 Throat of corolla inflated and broader than the tube; larger leaves often more than 1.5 cm wide; corolla white . . . . *P. digitalis*
1 Throat of corolla not broader than tube; leaves mostly less than 1.5 cm wide; corolla pale violet . . .
. . . . *P. gracilis*

Penstemon digitalis Nutt.
FOXGLOVE BEARDTONGUE   *introduced*
Stems to 1.5 m tall, typically glabrous and shining, often suffused with purple or somewhat glaucous. Stem leaves narrowly oblong to narrowly triangular, strictly glabrous beneath, the larger commonly 10-15 cm long. Panicle 1-3 dm long, with erect or strongly ascending branches; calyx at anthesis 6-7 mm long; corolla white or very faintly suffused with violet, usually marked with purple lines within, 23-30 mm long, the tube abruptly dilated near the middle into a wide throat. May-July. — Moist open woods and prairies.

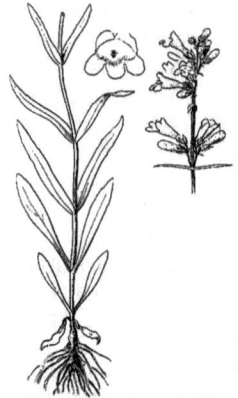

Penstemon gracilis Nutt.
SLENDER BEARDTONGUE   *native / cons 5*
Stems 3-5 dm tall, finely puberulent with minute reflexed hairs, often in two longitudinal strips, varying to glabrous. Stem leaves lance-shaped, 5-10 cm long, glabrous or finely pubescent. Inflorescence slender, the short lateral branches erect; calyx 6-9 mm long at anthesis; corolla 15-20 mm long, pale violet. — Prairies and open woods.

Penstemon gracilis

Plantago
PLANTAIN
Perennial or annual herbs. Leaves all from base of plant, simple. Flowers small, perfect or single-sexed, green, more or less stalkless in axils of small bracts, grouped into crowded spikes; sepals and petals 4. Fruit a capsule.

1 Leaves linear to lance-shaped, more than 5 times longer than wide 2
2 Leaves more than 5 mm wide . . . . *P. lanceolata*
2 Leaves less than 5 mm wide . . . . *P. patagonica*
1 Leaves narrowly ovate to ovate, less than 5 times longer than wide 3
3 Petioles green; bracts broadly ovate . . . . *P. major*
3 Petioles red-tinged at base; bracts narrowly lance-shaped . . . . *P. rugelii*

Plantago lanceolata L.
ENGLISH PLANTAIN   *introduced*
Perennial herb. Scapes to 6 dm tall, strigose above. Leaves narrowly lance-shaped to oblong lance-shaped, to 3 dm long, including the petiole, gradually tapering to both ends. Spikes very dense, at maturity 1-10 cm long; bracts broadly ovate, with narrow herbaceous center and broad scarious margins, often caudate-tipped. Outer two sepals united into one, broadly obovate, truncate, with two midveins; inner sepals ovate; corolla lobes 2-3 mm long. Capsule ellipsoid, 3-4 mm long, circumscissile near the base; seeds 2, black, 2-3 mm long, deeply concave on the inner face. Variable. — Native of the Old World; a common weed of lawns, roadsides, and waste places.

Plantago major L.
COMMON PLANTAIN   *introduced*
Perennial herb; closely resembling *P. rugelii* in form and often confused with it. Petioles commonly green and pubescent at base. Flowers sessile; bracts shorter than the sepals, broadly ovate or elliptic, with prominent

Plantago lanceolata

scarious margin and rounded elevated keel; sepals ovate, obtuse, the rounded keel about as wide as the scarious margins. Capsule ellipsoid, 2-4 mm long, circumscissile near the middle; seeds several, commonly 10 or more, about 1 mm long. — Native of Eurasia; naturalized in lawns, roadsides, and waste places.

Plantago major

Plantago patagonica Jacq.
WOOLLY PLANTAIN *introduced*
Annual herb, gray-villous throughout. Leaves linear, to 15 cm long, 3-8 mm wide. Scapes 5-15 or rarely 20 cm long. Spikes very dense cylindric, obtuse, 3-10 cm long, 4-6 mm wide; bracts linear, even the lowest scarcely exceeding the flowers; sepals, petals, and seeds. — Dry prairies.

Plantago rugelii Dcne.
AMERICAN PLANTAIN *native / cons* O
Perennial herb. Leaves broadly elliptic to oval, 5-20 cm long, many-nerved, narrowed at base; petiole margined, at base usually glabrous and tinged with purple. Spikes to 3 dm long, about 5 mm wide, comparatively loose, the axis frequently exposed; pedicels about 0.5 mm long. Bracts lance-shaped, 1/2 to 3/4 as long as the calyx; sepals ovate or oblong, the sharp keel much wider than the scarious margin; corolla lobes less than 1 mm long, reflexed after anthesis. Capsule narrowly ovoid, 4-6 mm long, dehiscent well below the middle; seeds 4-10, black, angular, about 2 mm long. — Lawns, gardens, roadsides, and waste places.

Plantago rugelii

Veronica
SPEEDWELL
Annual or perennial herbs. Leaves opposite, or becoming alternate in the head. Flowers single or in racemes from leaf axils or at ends of stems; sepals deeply 4-parted, enlarging after flowering; corolla blue or white, 4-lobed, somewhat 2-lipped, the tube shorter than the lobes; stamens 2. Fruit a flattened capsule, lobed or notched at tip; styles usually persistent on fruit.

ADDITIONAL SPECIES
Veronica verna L. (Spring speedwell), introduced; see key.

1 Flowers in racemes from leaf axils, or leaves more than 4 cm long, or both; plants perennial 2
1 Flowers single in axils of leafy bracts, or in terminal spikes; leaves less than 3 cm long; plants annual or perennial 8
2 Stems glabrous or nearly so; leaves toothed or entire 3
2 At least the upper stem pubescent; leaves toothed 6
3 Leaves with short petioles 4
3 Leaves sessile 5
4 Leaves mostly widest near leaf base; styles 2.5–3.5 mm long . . . . *V. americana*
4 Leaves widest above leaf middle; styles to 2.2 mm long . . . . *V. beccabunga*
5 Upper leaves lance-shaped, with wide, clasping bases; rachis of raceme stout and straight; capsules swollen . . . . *V. anagallis-aquatica*
5 Leaves mostly linear, narrowed to a sessile base; rachis of raceme slender and zigzagged; capsules strongly flattened . . . . *V. scutellata*
6 Flowers on long pedicels, the pedicels longer than the bracts . . . . *V. chamaedrys*
6 Flowers sessile or on short pedicels shorter than the bracts 7
7 Leaves more than 5 cm long, widest near base, on petioles 1 cm or more long; flowers in dense terminal spikes; styles persistent on capsule . . . . *V. longifolia*
7 Leaves less than 5 cm long, widest near middle, sessile or nearly so; spikes from leaf axils, loosely flowered; styles deciduous . . . . *V. officinalis*
8 Middle stem leaves pinnately divided . . . . *V. verna**
8 Leaves entire or toothed, not divided 9
9 Plants matted perennial herbs; stems finely hairy; leaves glabrous or nearly so (except when young) . . . . *V. serpyllifolia*
9 Plants erect annuals; stems and leaves glabrous or pubescent with mostly spreading hairs 10

10 Flowers blue; the capsules only shallowly notched at tip; stems fleshy; plants glabrous . *V. peregrina*
10 Flowers white; the capsules deeply notched at tip; stems not fleshy; plants pubescent . . . *V. arvensis*

Veronica americana Schwein.
AMERICAN-BROOKLIME   *native / cons* 9
Perennial, spreading by rhizomes; plants glabrous and succulent. Stems
erect to creeping, 1-6 dm long. Leaves opposite, ovate to lance-shaped (or
lower leaves oval), 2-8 cm long and 0.5-3 cm wide, upper leaves tapered
to a tip, lower leaves often rounded; margins with forward-pointing
teeth; petioles short. Flowers in stalked racemes from leaf axils; the
racemes with 10-25 flowers and to 15 cm long; corolla 4-lobed, blue (sometimes white), often with purple stripes. Fruit a more or less round, compressed capsule, 3-4 mm long, slightly notched at tip, the styles persistent,
2-4 mm long. July-Sept. — Streambanks and wet shores, hummocks in
swamps, springs.

Veronica americana

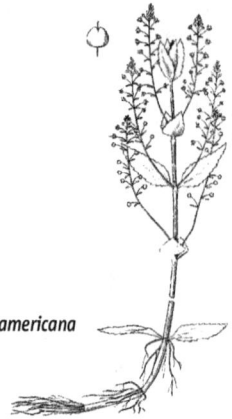

Veronica anagallis-aquatica L.
BLUE WATER SPEEDWELL   *native / cons* 4
Biennial or short-lived perennial, spreading by stolons or leafy shoots
produced in fall; plants more or less glabrous. Stems erect to spreading,
1-6 dm long, often rooting at lower nodes. Leaves opposite, lance-shaped
to ovate, 2-10 cm long and 0.5-5 cm wide, tapered to a blunt or rounded
tip; margins entire or with fine, forward-pointing teeth; petioles absent,
the leaves often clasping. Flowers in many-flowered racemes from leaf
axils, the racemes 5-12 cm long; corolla 4-lobed, blue or striped with purple, about 5 mm wide. Fruit a round, compressed capsule, 2-4 mm long,
notched at tip, the styles persistent, 1-2 mm long. June-Sept. — Wet,
sandy or muddy streambanks and ditches; often in shallow water.

Veronica arvensis L.
CORN SPEEDWELL   *introduced*
Annual, villous throughout. Stems erect or nearly so, simple or branched,
1-2 dm tall. Foliage leaves ovate, 6-12 mm long, obtuse, with 2-4 blunt
teeth on each side, palmately veined, the lower short-petioled, the upper
sessile. Inflorescence often constituting 2/3 of the plant, the bracteal
leaves progressively reduced in length and width and mostly entire.
Pedicels to 1.5 mm long; calyx lobes oblong, the lower pair 4-5 mm long,
the upper pair 3/4 as long; corolla blue, about 2 mm wide. Capsule 3-4
mm wide, nearly as long, deeply notched; style extending about as far as
the summit of the capsule lobes. April-June. — Native of Eurasia; established as an inconspicuous weed in gardens, lawns, and fields, and occasionally in open woods.

Veronica anagallis-aquatica

Veronica beccabunga L.
EUROPEAN SPEEDWELL, BROOKLIME   *introduced*
Perennial, closely related to V. americana and resembling it in habit.
Leaves elliptic or obovate, 2-6 cm long, broadest near or above the middle,
usually broadly rounded at the summit, obtuse to rounded at base, crenate.
Flowers blue, 5-7 mm wide. Capsules turgid, 3-4 mm wide, nearly as long,
not notched, the persistent style about 2 mm long. Summer. — Native of
Eurasia, muddy shores and streambanks; Door County.

Veronica chamaedrys L.
GERMANDER SPEEDWELL   *introduced*
Perennial, sparsely pubescent throughout, or in 2 strips only along the
stem. Stems prostrate, or ascending at the tip, 2-4 dm long. Leaves sessile
or nearly so, ovate, 2-3 cm long, obtuse, serrate, broadly rounded or truncate at base. Racemes few, erect from the upper axils, eventually 8-15 cm

Veronica chamaedrys

long including the peduncle, loosely 10-20-flowered; bracts lance-shaped, half to nearly as long as the slender pedicels; calyx lobes 3-5 mm long; corolla blue with dark blue lines and white orifice, about 1 cm wide; style (after corolla has fallen) 4-5 mm long. Capsules (rarely produced) 4-5 mm wide, very broadly and shallowly notched. May-June. — Native of Europe; introduced in moist gardens, roadsides, and fields.

Veronica longifolia L.
LONG-LEAF SPEEDWELL   *introduced*
Perennial. Stems erect, to 1 m tall. Leaves opposite or in whorls of 3, lance-shaped, 4-10 cm long, very sharply serrate; short-petioled. Racemes 1 or few, erect, spike-like, the axis pubescent but not glandular; corolla blue, pubescent in the throat, its lobes 4-5 mm long. Capsule little flattened, 3 mm long, smooth or puberulent, about half as long as the persistent style. June-Aug. — Native of Europe; introduced in fields, roadsides, and waste places.

Veronica officinalis L.
COMMON SPEEDWELL   *introduced*
Perennial, pubescent throughout. Stems prostrate or reclining, the flowering branches erect or ascending, 2-3 dm long. Leaves elliptic, 2.5-5 cm long, obtuse or rounded, uniformly serrate except toward the base, narrowed below to a short petiole. Racemes few, solitary or opposite, spike-like, commonly 3-6 cm long, often interrupted below, on a peduncle of about their own length; bracts linear lance-shaped, about equaling the flowers; calyx lobes oblong, 2-3 mm long; corolla pale violet with darker lines, 5-7 mm wide. Capsule reverse-triangular, glandular-puberulent, 4.5-5 mm wide, truncate or with a broad and very shallow notch; style at maturity to 3 mm long. May-July. — Dry fields and upland woods. At anthesis the main axis is often developed only slightly beyond the racemes, so that the inflorescence may appear terminal.

Veronica officinalis

Veronica peregrina L.
PURSLANE SPEEDWELL   *native / cons* 0
Small annual. Stems upright, 0.5-3 dm long, unbranched or with spreading branches, usually glandular-hairy. Lower leaves opposite, becoming alternate and smaller in the head, oval to linear, 5-25 mm long and 1-5 mm wide, rounded at tip; margins of lower leaves sparsely toothed, upper leaves entire; petioles short or absent. Flowers small, on short stalks from upper leaf axils; corolla 4-lobed, more or less white, about 2 mm wide. Fruit an oblong heart-shaped capsule, 2-4 mm long, notched at tip, the styles not persistent. May-July. — Mudflats, shores, ditches, temporary ponds, swales; also weedy in cultivated fields, lawns and moist disturbed areas.

Veronica peregrina

Veronica scutellata L.
GRASS-LEAF SPEEDWELL   *native / cons* 5
Perennial, spreading by rhizomes or leafy shoots produced in fall; plants smooth (or sometimes with sparse hairs). Stems slender, erect to reclining, 1-4 dm long, often rooting at lower nodes. Leaves opposite, linear to narrowly lance-shaped, 3-8 cm long and 2-10 mm wide, tapered to a sharp tip; margins entire or with small, irregularly spaced teeth; petioles absent. Flowers in racemes from leaf axils, the racemes with 5-20 flowers, as long or longer than the leaves; corolla 4-lobed, blue, 6-10 mm wide. Fruit a strongly flattened capsule, 3-4 mm long, notched at tip, the style persistent, 3-5 mm long. June-Sept. — Marshes, pond margins, hardwood swamps, thickets, springs, streambanks, wet depressions.

Veronica scutellata

Veronica serpyllifolia L.

THYME-LEAF SPEEDWELL   *native-introduced / cons* O

Perennial. Stems creeping and forming mats, the flowering ones upwardly curved and erect, 1-2 dm tall. Leaves ovate, elliptic, or nearly rotund, mostly 1-1.5 cm, occasionally 2 cm long, obtuse, entire or obscurely crenate, the lower often somewhat narrowed to a short petiole, the upper sessile. Racemes at first short and compact, soon elongating, in fruit to 10 cm long; bracteal leaves elliptic, to 3 cm long; pedicels 2-4 mm long; calyx lobes obtuse. Capsule notched at the apex for about 1/4 of its length, 3-5 mm wide, distinctly wider than long, but much longer than the persistent style. Variable. May-July. — Established in fields, meadows, and lawns, extending into moist open woods.

Veronica serpyllifolia

Polemoniaceae

PHLOX FAMILY

Perennial herbs (ours). Flowers perfect (with both staminate and pistillate parts), single or in clusters at ends of stems and from leaf axils; sepals and petals 5-parted and joined for part of length.

ADDITIONAL SPECIES

Ipomopsis rubra (L.) Wherry (Standing-Cypress), uncommon escape, reported for Door County from a partially shade roadside, in mixed hardwood forest.

Phlox

PHLOX

Erect perennial herbs. Leaves opposite, margins entire. Flowers pink, purple or rarely white, in stalked clusters at ends of stems and from upper leaf axils; sepals joined and tubelike; corolla 5-lobed, tubelike but flared outward at tip; stamens 5. Fruit a 3-chambered capsule.

1 Stems somewhat woody, trailing on ground . . . . *P. subulata*
1 Stems herbaceous, upright . . . . *P. paniculata*

Phlox paniculata L.

FALL PHLOX   *introduced*

Perennial herb. Stems erect, to 2 m tall. Leaves narrowly oblong or lance-shaped, 8-15 cm long, minutely ciliate, narrowed to an acute or obtuse base, usually glabrous above, the conspicuous lateral veins confluent to form a submarginal connecting vein. Inflorescence often large, of several panicled cymes, densely but minutely pubescent; calyx tube minutely puberulent or more commonly glabrous; calyx lobes glabrous; corolla red to purple, varying to white, 1.5-2 cm wide, its tube usually sparsely pubescent; one or more anthers at least partly exsert from the corolla tube. July-Sept. — Rich moist soil; cultivated in numerous horticultural varieties and escaped into roadsides and waste places.

Phlox subulata L.

MOSS-PINK   *introduced*

Perennial by a prostrate suffruticose stem, freely branched and producing numerous flowering branches 5-20 cm long. Leaves numerous and crowded, subulate, 5-20 mm long, usually ciliate, often with fascicles of smaller leaves in their axils. Cymes few-flowered; corolla rose-purple to pink or white, 12-20 mm wide, its lobes notched for 1/8 to 1/4 of their length. April-May. — Sandy or gravelly soil and rock-ledges; frequently cultivated and sometimes escaped.

Phlox subulata

Polygalaceae

MILKWORT FAMILY

Polygala

MILKWORT

Perennial herbs (ours). Leaves alternate or verticillate. Flowers perfect, in racemes. Sepals 5, the three outer small, the two inner (termed wings) much larger and often colored like the petals. Petals 3, all more or less united with each other and with the stamen-tube, the two upper ones similar, the lower one keel-shaped or boat-shaped with a fringe-like crest (in our species). Stamens 8 (or 6). Ovary 2-celled. Fruit a small capsule.

1   Flowers large, 13 mm long or more; stamens 6; leaves few, mostly near top of stem . . . . *P. paucifolia*
1   Flowers smaller, mostly less than 10 mm long; stamens 8; leaves distributed along stem or mostly
     near base . . . . *P. senega*

Polygala paucifolia Willd.

FRINGED POLYGALA   *native / cons* 7
Polygaloides paucifolia (Willd.) J. R. Abbott
Triclisperma paucifolia (Willd.) Nieuwl.
Perennial from a slender rhizome. Stems 8-15 cm tall, bearing below several scattered scale-like leaves 2-8 mm long and near the summit 3-6 elliptic to oval; leaves 1.5-4 cm long, usually pubescent only on the midrib and margin. Flowers 1-4, rose-purple varying to white, the obovate wings about 15 mm long; corolla about equaling the wings; stamens 6. Capsule suborbicular, about 6 mm long and wide, notched at the summit. May-June. — Moist rich woods.

Polygala senega L.

SENECA-SNAKEROOT   *native / cons* 8
Perennial. Stems commonly several from one base, 1-5 dm tall, usually unbranched, minutely puberulent. Leaves alternate, the lowest reduced or scale-like; stem leaves linear lance-shaped or wider. Racemes dense, 1.5-4 cm long, 6-8 mm thick, on a peduncle 1-3 cm long. Flowers white; wings broadly elliptic, 3-3.5 mm long, exceeding the corolla. Capsule suborbicular; seeds pubescent, 2-3 mm long; aril nearly or quite as long. May-June. — Dry or moist woods and prairies.

Polygala paucifolia

Polygonaceae

BUCKWHEAT FAMILY

Annual or perennial herbs, plants sometimes vining. Leaves alternate, simple, sometimes wavy-margined, otherwise entire; the nodes usually enlarged. Stipules joined to form a membranous or papery sheath (ocrea) around stem at each node. Flowers in spike-like racemes or small clusters from leaf axils (*Persicaria, Polygonum*), or in crowded panicles at ends of stems (*Rumex*). Flowers small, perfect (with both staminate and pistillate parts), regular, petals absent. In *Rumex* the sepals herbaceous, green to brown, in inner and outer groups, each group with 3 sepals, the 3 inner enlarging after flowering, becoming broadly winged, persisting to enclose the achene; stamens 4-8; ovary 1-chambered, styles 2-3; in other genera of family, sepals more or less petal-like, white to pink or yellow, mostly 5 (sometimes 4). Fruit a 3-angled or lens-shaped achene.

Polygonaceae recognized by presence of a stipular sheath (ocrea), which surrounds the stem above the attachment of each leaf. The similar reduced structure in the inflorescence is called an ocreola.

1   Tepals 6, greenish or reddish, scarcely petaloid, the 3 inner (but not the outer) ones enlarging in fruit
     and concealing the achene; stigmas a feathery tuft; plants in some species dioecious or polygamous
     and hence some flowers entirely staminate . . . . *Rumex*

1 Tepals 4–5, white to red and ± petaloid at least along the margins, uniform in size or the outer ones larger; stigmas usually not feathery and plants mostly with bisexual flowers 2
2 Stem and petioles with retrorse prickles; leaves hastate or sagittate (with acute basal lobes) ....
.... *Persicaria*
2 Stem and petioles without prickles; leaves various 3
3 Outer tepals winged or keeled in fruit, or plant somewhat twining or vine-like, or both; leaves ovate-cordate to broadly sagittate .... *Fallopia*
3 Outer tepals not winged or keeled; plant not twining; leaves various 4
4 Flowers 1–4 at a node, sessile or pediceled in the axils of foliage leaves or bracts; leaf blades jointed at the base, less than 2 (–2.4) cm broad; summit of ocrea silvery white, becoming lacerate-shredded; annuals .... *Polygonum*
4 Flowers numerous in peduncled terminal or axillary spikes, racemes, or panicles, often densely crowded; leaves not jointed at base of blade, in some species over 2.5 cm broad; summit of ocrea tinged with brown, shattering at maturity but not shredding; annuals or perennials ...... *Persicaria*

Fallopia
BLACK-BINDWEED

Annual or perennial, twining or stout and erect and forming large colonies. In the past, our species typically included in genus *Polygonum*.

1 Stems erect and stout, 1–3 m tall .... *F. japonica*
1 Stems twining and slender 2
2 Base of sheathing stipules with stiff, downward-pointing hairs .... *F. cilinodis*
2 Base of stipules not with stiff, downward-pointing hairs 3
3 Plants annual; styles united; achenes dull and finely roughened .... *F. convolvulus*
3 Plants perennial; styles divergent; achenes shiny and smooth .... *F. scandens*

Fallopia cilinodis (Michx.) Holub
FRINGED BLACK BINDWEED *native / cons* **1**
Polygonum cilinode Michx.

Perennial, pubescent, varying to nearly glabrous. Stems twining, trailing, or occasionally erect, to 2 m long, nearly terete. Leaves ovate, deeply cordate at base; ocreae very oblique, reflexed-bristly at base. Racemes long-peduncled, mostly branched, 4-10 cm long, the small flower clusters remote; perianth white, 1.5-2 mm long; styles separate, divergent. Achenes very glossy, black, scarcely surpassed by the calyx. July-Aug. — Dry woods and thickets.

Plants in open sun are often erect, with stouter red stems, the red color extending into the leaf veins.

Fallopia convolvulus (L.) Á. Löve
BLACK BINDWEED *introduced*
Polygonum convolvulus L.

Annual. Stems trailing or twining, to 1 m long, angled, minutely scabrous in lines, as are also the petioles and often the leaf veins. Leaves hastate to triangular-cordate, broadly V-shaped to cordate at base; ocreae smooth. Racemes interrupted, 2-6 cm long, naked or with a few small leaves at base; flowers in clusters of 3-6; pedicels 1-2 mm long, jointed near the summit; perianth 1.5-2 mm long, green without, white within; outer 3 sepals often narrowly winged on the midrib; styles united. Achenes dull black, 3-4 mm long, not exceeded by the calyx. June-Sept. — Native of Europe; roadsides, railway tracks, and waste ground.

Fallopia convolvulus

Fallopia japonica (Houtt.) Ronse Decr.
JAPANESE KNOTWEED *introduced (invasive)*
Polygonum cuspidatum Sieb. & Zucc.
Reynoutria japonica Houtt.

Perennial and spreading by long rhizomes. Stems stout, 1-3 m tall. Leaves broadly ovate, 8-15 cm long, 5-12 cm wide, abruptly acuminate, broadly

Fallopia japonica

truncate at base, the basal angles prominent. Racemes numerous from most of the upper axils, often branched, forming a series of panicles 8-15 cm long; perianth white or greenish white; outer sepals narrowly winged along the midrib; styles 3; stigmas minute. Achenes triangular, about 3 mm long, enclosed by the enlarged calyx. Aug-Sept. — Native of Japan; sometimes planted but often escaping to form large colonies.

Fallopia scandens (L.) Holub
FALSE BUCKWHEAT   native / cons 3
Polygonum scandens L.

Perennial. Stems twining, to 5 m long, sharply angled, often minutely scabrous on the angles, as are also the petioles, peduncles, and often the leaf veins, varying to glabrous. Leaves ovate or broadly cordate; ocreae oblique, smooth. Racemes from most of the upper axils, 5-11 cm long, usually unbranched, interrupted, leafless or with 1-2 small leaves; pedicels winged above the joint, even at anthesis; perianth 1.5-2.5 mm long, white; outer 3 sepals at maturity with broadly winged midrib. Achenes very glossy, black, 3-5 mm long, much exceeded by the calyx. Aug-Sept. — Moist woods, thickets, and roadsides. Variable in the shape of the wings and the length of the fruit.

Fallopia scandens

Persicaria
LADY'S-THUMB, SMARTWEED
Annual and perennial herbs. Flowers pink or sometimes white, in terminal spikes. The genus was formerly included in *Polygonum*.

1 Tepals 4; styles elongate, persistent and becoming hard and stiff .... *P. virginiana*
1 Tepals usually 5; styles short, not persistent or becoming hard and stiff 2
2 Stems with downward-pointing prickles on the stem angles .... *P. sagittata*
2 Stems smooth to hairy, but not prickly 3
3 Perennial herbs from rhizomes or stolons 4
3 Taprooted annual herbs 6
4 Flowers in 1 or 2 terminal racemes .... *P. amphibia*
4 Flowers in several to many terminal and axillary racemes 5
5 Perianth dotted with glands .... *P. punctata*
5 Perianth not dotted with glands .... *P. hydropiperoides*
6 Sheathing stipules (ocreae) fringed with bristles at tip 7
6 Ocreae entire or irregularly cut, not fringed with bristles 10
7 Perianth dotted with glands 8
7 Perianth not dotted with glands 9
8 Tepals usually 4; achenes dull .... *P. hydropiper*
8 Tepals 5; achenes shiny .... *P. punctata*
9 Small stipules at base of each inflorescence (ocreolae) fringed with long hairs 2–3 mm long
  .... *P. longiseta*
9 Small stipules at base of each inflorescence entire, or with a few short hairs to 1 mm long ....
  .... *P. maculosa*
10 Outer sepals strongly 3-nerved, each nerve ending in an anchor shaped fork; racemes nodding to
   erect .... *P. lapathifolia*
10 Outer sepals with faint, irregularly forked nerves; racemes erect .... *P. pensylvanica*

Persicaria amphibia (L.) Delarbre
WATER SMARTWEED   native / cons 5
Polygonum amphibium L.

Perennial floating or emergent herb, from spreading rhizomes. Stems to 1 m or more long, leaves and habit variable. Submerged plants smooth, usually branched, the branches floating, branch tips often upright and raised above water surface; leaves floating, leathery, oval, 4-20 cm long and 1-4 cm wide, rounded at tip; stipules (ocreae) membranous; petioles 1-8 cm long. Exposed plants hairy; leaves stalkless or with short petioles.

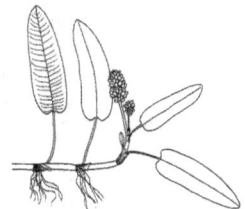

Persicaria amphibia

Flowers pink to red, in 1-2 spike-like racemes from branch tips, the racemes 2-15 cm long and 1-2 cm wide; sepals 5-lobed to below middle, 4-5 mm long; stamens 5. Achenes lens-shaped, 2-4 mm long, shiny dark brown. June-Sept. — Ponds, lakes, marshes, bog pools, backwater areas, quiet streams.

Persicaria hydropiper (L.) Delarbre
MILD WATER-PEPPER   *introduced*
Polygonum hydropiper L.

Annual herb. Stems red, erect to sprawling, 2-6 dm long, sometimes rooting at lower nodes, branched or unbranched, peppery-tasting. Leaves lance-shaped, 3-8 cm long and to 2 cm wide, hairless except for short hairs on veins and margins, nearly stalkless or with a short petiole; stipules (ocreae) membranous, 5-15 mm long, swollen and fringed with bristles. Flowers green and usually white-margined, continuous in slender racemes, often nodding at tip; sepals 5, 3-4 mm long, with glandular dots; stamens 4 or 6. Achenes dull, dark brown, 3-angled or lens-shaped, 2-3 mm long. July-Oct. — Muddy shores, streambanks, floodplains, marshes, ditches and roadsides.

Persicaria amphibia

Persicaria hydropiperoides (Michx.) Small
SWAMP SMARTWEED   *native / cons 6*
Polygonum hydropiperoides Michx.

Perennial herb, spreading by rhizomes. Stems erect to sprawling with upright tips, to 1 m long, usually branched, nearly smooth or with short hairs. Leaves linear to lance-shaped, 4-12 cm long and to 2.5 cm wide, petioles short; stipules (ocreae) membranous, 5-15 mm long, with stiff hairs and fringed with bristles. Flowers green, white or pink, in 2 to several slender racemes, 1-6 cm long, often interrupted near base; sepals 2-3 mm long, 5-lobed to just below middle, without glandular dots or only the inner sepals slightly glandular; stamens 8. Achenes black, shiny, 3-angled with concave sides, 2-3 mm long. July-Sept. — Shallow water or wet soil; ponds, marshes, swamps, bogs and fens, streambanks, lakeshores and ditches.

Persicaria hydropiper

Persicaria lapathifolia (L.) Delarbre
DOCK-LEAF SMARTWEED   *native / cons 2*
Polygonum lapathifolium L.

Annual herb. Stems erect to sprawling, unbranched or few-branched, 2-15 dm long. Leaves lance-shaped, 4-20 cm long and 0.5-5 cm wide, smooth above, often densely short-hairy on leaf undersides; petioles to 2 cm long, smooth to glandular; stipules (ocreae) 5-20 mm long, entire or with irregular, jagged margins. Flowers deep pink, white or green, crowded in erect or nodding racemes 1-5 cm long; sepals 3-4 mm long, 4- or 5-lobed to below middle, the outer 2 sepals strongly 3-nerved; stamens usually 6. Achenes brown, lens-shaped, 2-3 mm long. July-Sept. — Marshes, wet meadows, shores, streambanks, ditches and cultivated fields. Common and weedy.

Persicaria longiseta (Bruijn) Kitag.
BRISTLY LADY'S-THUMB   *introduced*
Polygonum caespitosum Blume

Annual. Stems glabrous or nearly so, freely branched, soon decumbent, to 1 m long. Leaves thin, dark green, lance-shaped to elliptic or oblong lance-shaped; ocreae minutely strigose or glabrous, ciliate with bristles 5-10 mm long. Racemes dense 2-4 cm long, about 5 mm thick; ocreolae overlapping, their cilia 2-3.5 mm long, often equaling or surpassing the flowers. Achenes black, smooth and shining, trigonous, about 2 mm long. — Native of e Asia; waste places, preferably in moist soil.

Persicaria hydropiperoides

Persicaria maculosa Gray
LADY'S-THUMB   *introduced*
   Polygonum persicaria L.
Annual herb. Stems upright to spreading, 2-8 dm long, unbranched to branched, often red. Leaves lance-shaped, 3-15 cm long and 0.5-3 cm wide, smooth or with few hairs, underside usually dotted with small glands, leaves stalkless or on petioles to 1 cm long; ocreae 5-15 mm long, fringed with bristles, with short hairs. Flowers pink to rose, crowded in straight, cylindric racemes 1-4 cm long and 0.5-1 cm wide; sepals 2-4 mm long, 5-lobed to near middle; stamens 6. Achenes black, shiny achene, lens-shaped or sometimes 3-angled, 2-3 mm long. July-Sept. — Muddy shores, streambanks, ditches and cultivated fields, often weedy.

Persicaria pensylvanica (L.) M. Gómez
PINKWEED   *native / cons* **1**
   Polygonum pensylvanicum L.
Annual herb. Stems erect, 3-20 dm long, unbranched to widely branching. Leaves lance-shaped, 3-15 cm long and 1-4 cm wide, smooth except for short hairs on margins; petioles to 2.5 cm long; stipules (ocreae) 0.5-1.5 cm long, entire or with an irregular, jagged margin, hairless, not fringed with bristles. Flowers pink to white, in dense racemes 2-3 cm long, the flower stalks with gland-tipped hairs; sepals 3-5 mm long, 5-parted to below middle, the outer sepals faintly nerved; stamens 8 or less. Achenes dark brown to black, shiny, lens-shaped, to 3 mm long. June-Sept. — Streambanks, exposed shores, marshes, fens, ditches and cultivated fields.

Persicaria pensylvanica

Persicaria punctata (Elliott) Small
DOTTED SMARTWEED   *native / cons* **5**
   Polygonum punctatum Ell.
Annual or perennial herb. Stems erect to spreading, 4-10 dm long, un-branched to branched. Leaves narrowly lance-shaped or oval, 4-15 cm long and 1-2 cm wide, smooth except for small short hairs on margins, underside usually dotted with small glands; petioles short; stipules (ocreae) 5-15 mm long, smooth or with stiff hairs and fringed with bristles. Flowers green-white; in numerous slender, loosely flowered racemes, in-terrupted in lower portion, to 10 cm long; sepals 3-4 mm long, with glan-dular dots, 5-parted to about middle; stamens 6-8. Achenes dark, shiny, lens-shaped or 3-angled, 2-3 mm long. Aug-Sept. — Floodplain forests, marshes, shores, streambanks and cultivated fields.

Persicaria sagittata (L.) H.Gross
ARROW-LEAF TEARTHUMB   *native / cons* **6**
   Polygonum sagittatum L.
Slender annual herb. Stems 4-angled, weak, usually supported by other plants, 1-2 m long, with downward pointing prickles on stem angles, petioles, leaf midribs and flower stalks. Leaves lance-shaped to oval, ar-rowhead-shaped at base, 3-10 cm long and to 2.5 cm wide, the basal lobes pointing downward; petioles long on lower leaves, shorter above; stipules (ocreae) 5-10 mm long, with a few hairs on margins. Flowers white or pink; in round racemes to 1 cm long, on long slender stalks at ends of stems or from leaf axils; sepals 3 mm long, 5-parted to below middle. Achenes brown to black, shiny, 3-angled, 2-3 mm long. July-Sept. — Swamps, marshes, wet meadows and burned wetlands.

Persicaria punctata

Persicaria virginiana (L.) Gaertn.
JUMPSEED   *native / cons 7*
   Antenoron virginianum (L.) Roberty & Vautier
   Polygonum virginianum L.
Perennial herb. Stems erect from a rhizome, 5-10 dm tall. Leaves lance-shaped to ovate, to 15 cm long, acute to rounded at base, varying from roughly pubescent to glabrous on either or both sides; petioles to 2 cm long; ocreae pubescent and long-ciliate. Racemes very slender, terminal, 1-4 dm long, the ocreolae much separate toward the base, becoming contiguous or overlapping toward the summit, 1-3-flowered; pedicels divergent, jointed at the tip; sepals 4, greenish white, or suffused with pink, about 2.5 mm long, the 2 lateral exterior and somewhat smaller than the median, scarcely changed in fruit. Achenes lens-shaped, ovate, about 4 mm long; styles persistent, hooked at the tip. Aug, Sept. — Moist woods.

Persicaria virginiana

Polygonum
SMARTWEED, KNOTWEED, TEARTHUMB
Annual herbs (ours). Stems erect to sprawling, often swollen at nodes. Leaves arrowhead-shaped to lance-shaped or oval; stipules joined to form a tubular sheath (ocrea) around the stem above each node; the ocreae (plural) membranous or papery, entire or with an irregular, jagged margin or fringed with bristles. Flowers small, green, white or pink, from leaf axils; sepals usually 5, petal-like, green-white to pink; stamens 8 or less; styles 2-3. Fruit a brown to black achene, lens-shaped or 3-angled.

1 Perianth abruptly narrowed above achene ("bottle-shaped") 2
1 Perianth not narrowed above achene 3
2 Leaves yellow-green; fruiting perianth divided for about three-fourths of its length . . . . . . *P. erectum*
2 Leaves blue-green; fruiting perianth divided for about one-third its length . . . . *P. achoreum*
3 Outer 3 tepals flat, shorter than or equaling inner 2 tepals . . . . *P. aviculare*
3 Outer 3 tepals hood-like, longer than inner 2 tepals 4
4 Plants prostrate; leaves 2–4 times longer than wide . . . . *P. aviculare*
4 Plants upright; leaves 4–12 times longer than wide . . . . *P. ramosissimum*

Polygonum achoreum Blake
LEATHERY KNOTWEED   *native / cons 0*
Annual, closely resembling *P. erectum*. Stems erect or ascending, freely branched, 1-5 dm tall. Leaves elliptic to obovate, thin, bright- or bluish green, 1-3 cm long, broadly rounded at the tip; ocreae to 1 cm long, mostly scarious, 3-nerved. Pedicels about equaling the calyx; calyx about 3 mm long, green, unequally 5-lobed to above the middle; outer lobes at maturity distinctly exceeding the inner, cucullate at the tip, exceeding the achene. Achenes dull yellow-brown, finely and uniformly granular. — Sandy and gravelly roadsides, barnyards, gardens, railroads.

Polygonum achoreum

Polygonum aviculare L.
YARD KNOTWEED   *native / cons 0*
Annual. Stems erect to prostrate, much branched, the branches commonly equaling the central axis, or the latter suppressed. Leaves linear to elliptic or oblong, 1-3 cm long, 1-8 mm wide, narrowed to the base, veinless or faintly veined. Flowers short-pediceled, included in the ocrea or barely exsert; calyx 2-3 mm long, lobed to below the middle; sepals oblong or ovate, green with white or pink margins, appressed at maturity. Achenes ovoid, dark brown, 2-2.5 mm long, finely puncticulate. — Common weed of waste ground, streets, and lawns; also common on beaches and around salt marshes.

Polygonum aviculare

Polygonum erectum L.

ERECT KNOTWEED   *native / cons* **2**

Annual. Stems erect or ascending, 1-5 dm tall, with numerous branches. Leaves oval to obovate, 1-4 cm long, thin, bright- or bluish green, broadly rounded, acute at base; ocreae to 1 cm long, 3-5-nerved. Pedicels shorter than to equaling the calyx; calyx about 3 mm long, green, unequally 5-lobed to below the middle; outer lobes at maturity distinctly exceeding the inner, narrowly keeled. Achenes dimorphic; either shining, punctate, dark brown, broadly ovoid, included, about 2.5 mm long; or dull brown, ovoid, exsert, 3-3.5 mm long. — Common weed in waste ground.

Polygonum ramosissimum Michx.

YELLOW-FLOWER KNOTWEED   *native / cons* **6**

Annual. Stems erect, 3-10 dm tall, freely branched; lower internodes to 5 cm long, the upper progressively shorter. Leaves linear, mostly 1-6 cm long, 2-5 mm wide, flat; lateral veins inconspicuous or obsolete. Flowers from the upper ocreae only, 1-3 together, exsert on pedicels longer than the calyx, forming racemes to 15 cm long; calyx about 3 mm long, 5-parted nearly to the base, the outer 3 sepals notably exceeding the inner, especially in fruit. Achenes black, ovoid, 3 mm long, smooth and shining. — Sandy fields and meadows, sandy or gravelly shores of Lake Michigan.

Polygonum erectum

Rumex

DOCK, SORREL

Perennial, sometimes weedy, herbs (annual in *R. fueginus*). Leaves large and clustered at base of plants, or leafy-stemmed; mostly oblong to lance-shaped, flat to wavy-crisped along margins, usually with petioles. Membranous sheaths around stems present at nodes (ocreae). Flowers in crowded whorls in panicles at ends of stems; flowers small and numerous, green but turning brown; sepals in 2 series of 3, the inner 3 sepals (valves) enlarging, becoming winged and loosely enclosing the achene, giving the appearance of a 3-winged fruit, the midvein of the valve often swollen to produce a grainlike tubercle on the back; stamens 6; styles 3. Fruit a brown, 3-angled achene, tipped with a short slender beak.

1   At least some of the leaves arrowhead-shaped, the basal lobes pointing outward ..... *R. acetosella*
1   Leaves not arrowhead-shaped with basal lobes 2
2   Margins of mature valves entire or shallowly lobed, not toothed 3
2   Margins of mature valves with coarse or spine-tipped teeth 5
3   Flower pedicels without a large swollen joint; base of grain distinctly above base of valve ....
     .... *R. britannica*
3   Flower stalks with a large swollen joint below the middle or near base; base of grain even with base
     of valve 4
4   Leaves crisp-margined (crinkled); grains two-thirds as wide as long .... *R. crispus*
4   Leaf margins flat; grains narrower, up to half as wide as long .... *R. triangulivalvis*
5   Plants annual from fibrous roots; grains 3 .... *R. fueginus*
5   Plants perennial from a stout taproot; grain 1 .... *R. obtusifolius*

Rumex acetosella L.

COMMON SHEEP SORREL   *introduced (invasive)*

Perennial herb. Stems erect, simple or branched, 1-4 dm tall. Leaves variable, usually 3-lobed, the terminal lobe narrowly elliptic to oblong, the lateral much smaller, triangular, divergent; leaf base below the lobes truncate to long-cuneate. Inflorescence sometimes half as long as the plant; outer sepals lance-shaped; inner sepals in the staminate flower 1.5-2 mm long, obovate, in the pistillate flower broadly ovate. Achenes about 1.5 mm long, shining golden brown. — Naturalized from Eurasia in fields, lawns, and waste places, soils acidic; often a troublesome weed.

Rumex acetosella

Rumex brittannica L.

GREAT WATER-DOCK   *native / cons* 8
   Rumex orbiculatus Gray
Perennial herb. Stems stout, unbranched, 2-2.5 m long. Leaves lance-shaped or oblong lance-shaped, lower leaves 30-60 cm long, upper leaves 5-15 cm long; margins flat. Flowers in panicles to 5 dm long; valves rounded, flat at base, 5-8 mm long and as wide, smooth or with small teeth; grains 3, narrowly lance-shaped, the base distinctly above base of valve. June-Aug. — Marshes, fens, streambanks and ditches, often in shallow water.

Rumex brittannica

Rumex crispus L.

CURLY DOCK   *introduced*
Perennial herb, from a thick taproot. Stems stout, upright, usually single, 5-15 dm long. Basal leaves large, 10-30 cm long and 1-5 cm wide, on long petioles, often drying early in season; stem leaves smaller and with shorter petioles, oval to lance-shaped, margins strongly wavy-crisped (crinkled). Flowers in large branched panicles, the panicle branches more or less upright; flower stalks drooping at tips, 5-10 mm long, swollen-jointed near base; valves heart-shaped to broadly ovate, 4-5 mm long and as wide, margins more or less smooth; grains 3, swollen, often of unequal size, rounded at ends. Fruit a brown achene, 2-3 mm long. July-Sept. — Wet meadows, ditches, fields, and other wet and disturbed areas; weedy.

Rumex fueginus Phil.

GOLDEN DOCK   *native / cons* 2
Annual herb. Stems hollow, to 8 dm long, much-branched. Leaves mostly on stems, smaller upward, lance-shaped to linear, 5-20 cm long and 0.5-4 cm wide, wedge-shaped or heart-shaped at base, margins flat to wavy-crisped. Flowers in large open panicles, the panicle branches more or less upright, leafy, the flower stalks jointed near base; valves triangular-ovate, 2-3 mm long, the margins lobed into 2-3 spine-tipped teeth on each side; grains 3. Fruit a light brown achene, 1-2 mm long. July-Aug. — Marshes, shores, streambanks and ditches, sometimes where brackish.

Rumex obtusifolius L.

BITTER DOCK   *introduced*
Perennial herb. Stems stout, to 12 dm long, usually unbranched. Lower leaves oblong or ovate, to 30 cm long and 15 cm wide, heart-shaped or rounded at base; upper leaves smaller. Flowers in much-branched panicles, flower stalks longer than fruit, jointed near base; valves triangular-ovate, 4-5 mm long, with 2-4 spine-tipped teeth on each side; grains large and with tiny wrinkles. Fruit a shiny, red-brown achene. June-Aug. — Floodplain forests and openings, cultivated fields and disturbed areas.

Rumex crispus

Rumex triangulivalvis (Danser) Rech.f.

WILLOW-LEAF DOCK   *native / cons* 1
   Rumex salicifolius Weinm.
Perennial taprooted herb. Stems smooth, 3-10 dm long, usually branched from base and with short branches on stem. Leaves mostly on stems, not much smaller upward, narrowly lance-shaped, tapered at both ends, pale waxy green, 5-16 cm long and 1-3 cm wide, margins mostly flat. Flowers in panicles 1-3 dm long, panicle branches few and more or less upright, with small linear leaves at base; flower stalks 2-4 mm long, swollen and jointed near base; valves thick, triangular, 3-6 mm long and wide, margins smooth or shallowly toothed; grains usually 3. Fruit a brown achene, 2 mm long. June-Aug. — Wet meadows, marshes, shores, streambanks, ditches and other low areas, sometimes where brackish.

Rumex triangulivalvis

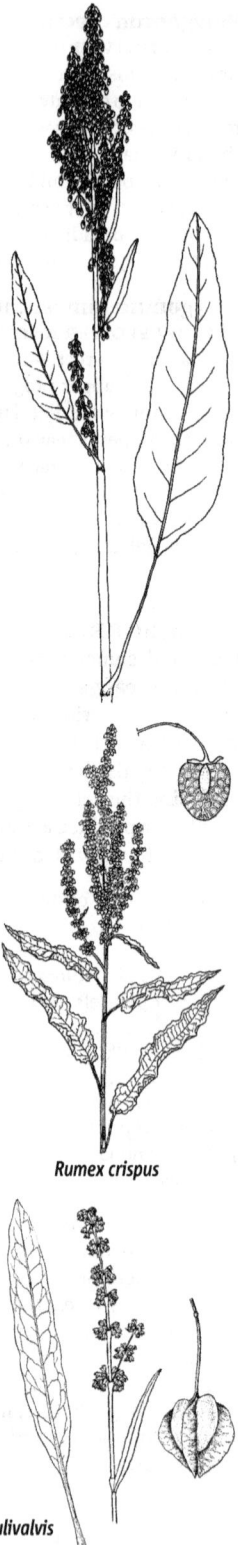

Portulacaceae
PURSLANE FAMILY

Portulaca
PURSLANE

Portulaca oleracea L.

COMMON PURSLANE   *introduced*

Stems prostrate, fleshy, usually purplish red, glabrous, repeatedly branched, forming large mats. Leaves succulent, flattened, spatulate to obovate-cuneate, 1-3 cm long, commonly rounded at the tip; stem leaves usually alternate, occasionally opposite. Flowers ephemeral, opening only in the sunshine, yellow, sessile, 5-10 mm wide, solitary or in small terminal glomerules; sepals 2; petals 4-6, commonly 5; stamens 6-10. Capsule opening near the middle, many-seeded; seeds with low blunt tubercles. All summer. — Reputedly native of w Asia, but now widely distributed as a familiar weed; sometimes cooked for greens.

Portulaca oleracea

Primulaceae
PRIMROSE FAMILY

Annual or perennial herbs. Leaves simple, opposite (sometimes whorled in *Lysimachia*), or leaves all basal. Flowers perfect (with both staminate and pistillate parts), regular, single from leaf axils, or in clusters at ends of stems; sepals 4-5, petals mostly 5 (varying from 4-9), joined, tube-shaped below and flared above, deeply cleft to shallowly lobed at tip; ovary superior, style 1; stamens 5. Fruit a 5-chambered capsule.

1 Leaves all from base of plant, inflorescence an umbel at end of naked stalk .... *Primula*
1 Leaves from stem; inflorescence various 2
2 Leaves in a single whorl near end of stem; flowers 7-merous .... *Trientalis*
2 Leaves opposite or in several whorls; flowers 5–6-merous .... *Lysimachia*

Lysimachia
LOOSESTRIFE

Perennial herbs, spreading by rhizomes. Stems erect. Leaves mostly opposite (sometimes appearing whorled), ovate or lance-shaped. Flowers 5-parted, yellow, single on stalks from leaf axils or in racemes or panicles; sepals green; petals bright to pale yellow. Fruit a capsule.

Purple loosestrife (*Lythrum salicaria*), an introduced weed of wetlands, is a member of the Lythraceae.

1 Plants creeping; leaves opposite, nearly round .... *L. nummularia*
1 Plants upright; leaves opposite or whorled, longer than wide 2
2 Flowers in terminal racemes or panicles .... *L. terrestris*
2 Flowers solitary or in clusters or spikes from the leaf axils 3
3 Leaves rounded or heart-shaped at base; petioles 1–3 cm long, fringed with hairs .... *L. ciliata*
3 Leaves tapered to their base; petioles absent or short, smooth or fringed with hairs 4
4 Flowers many in racemes from leaf axils; flowers mostly 6-merous .... *L. thyrsiflora*
4 Flowers 1 to several from the leaf axils, 5-merous 5
5 Leaves narrowly linear, to 5 mm wide .... *L. quadriflora*
5 Leaves lance-shaped to ovate, usually more than 8 mm wide 6
6 Flowers in clusters of several from leaf axils .... *L. punctata*
6 Flowers usually single from each leaf axil 7
7 Main leaves in whorls of 3 or more leaves; corolla lobes entire .... *L. quadrifolia*
7 Leaves opposite or whorled; corolla lobes ragged-toothed at tip .... *L. lanceolata*

Lysimachia ciliata L.
FRINGED YELLOW-LOOSESTRIFE   *native / cons* 5
Perennial herb, spreading by rhizomes. Stems upright, 3-12 dm long, un-branched or with few branches above. Leaves ovate to lance-shaped, 4-15 cm long and 2-6 cm wide, rounded to heart-shaped at base, green above, slightly paler below; margins fringed with short hairs; petioles 0.5-5 cm long, fringed with hairs. Flowers yellow, single from upper leaf axils, on stalks 2-7 cm long; sepal lobes lance-shaped, often with 3-5 parallel red-brown veins; petal lobes rounded and finely ragged at tip, 4-10 mm long and 3-9 mm wide, with a short slender tip. Fruit a capsule, 4-7 mm wide. June-Aug. — Usually shaded wet areas, such as shores, streambanks, wet meadows, ditches, floodplains, wet woods and thickets.

Lysimachia lanceolata Walt.
Lysimachia ciliata

LANCE-LEAF YELLOW-LOOSESTRIFE   *native / cons* 6
Perennial herb. Stems erect or nearly so, producing long slender stolons from the base, 3-6 dm tall, the lateral branches barely longer than the subtending leaves. Lower stem leaves ovate to obovate, petioled; principal leaves linear, to 15 cm long, paler beneath, tending to be folded along the midvein, scabrellate on the margin or ciliate near the base, gradually tapering below, with no distinction of petiole and blade. Sepals firm or thick, 5-7 mm long; petals 7-10 mm long. June-July. — Moist or wet woods or prairies.

Lysimachia nummularia L.
CREEPING-JENNY   *introduced*
Perennial herb, often forming mats. Stems creeping, to 5-6 dm long. Leaves opposite, dotted with black glands, round or broadly oval, 1-2.5 cm long; petioles short. Flowers single in leaf axils, on stalks to 2.5 cm long; sepals leaflike, triangular; petals yellow, dotted with dark red, 10-15 mm long. Fruit a capsule, shorter than sepals. June-Aug. — Swamps, floodplain forests, streambanks, shores, meadows and ditches.

Lysimachia punctata L.
LARGE YELLOW-LOOSESTRIFE   *introduced*
Lysimachia nummularia

Perennial herb, spreading by shallow stolons. Stems erect, rarely branched, to 1 m tall, pubescent. Leaves chiefly in whorls of 3 or 4, occasionally only opposite, lance-shaped, 5-10 cm long, pubescent. Flowers in axillary whorls, usually more numerous than the subtending leaves; uppermost whorls with smaller leaves and shorter internodes, simulating a raceme; pedicels 1-2 cm long, pubescent; sepals linear lance-shaped, pubescent; corolla yellow, 2-3 cm wide. June-July. — Native of se Europe and sw Asia.

Lysimachia quadriflora Sims
FOUR-FLOWER YELLOW-LOOSESTRIFE   *native / cons* 9
Perennial herb, spreading by rhizomes which form clusters of basal rosettes. Stems upright, 3-10 dm long. Leaves opposite, sometimes appearing whorled; stem leaves stalkless, often ascending, linear, 3-8 cm long and 2-7 mm wide, margins smooth or rolled under, sometimes fringed with a few hairs near base. Flowers yellow, single in clusters at ends of stems and branches, on stalks 1-4 cm long; sepal lobes lance-shaped; petal lobes oval, 7-12 mm long, entire or finely ragged at tip. Fruit a capsule, 3-5 mm wide. July-Aug. — Wet meadows, pond and marsh margins, low prairie, calcareous fens; often where sandy and calcium-rich.

Lysimachia quadrifolia L.
WHORLED YELLOW-LOOSESTRIFE   *native / cons* 6
Lysimachia quadriflora

Perennial herb. Stems erect, 3-9 dm tall, glabrous or sparsely pubescent, rarely branched. Leaves chiefly in whorls of 4 (3-6), lance-shaped, 5-10 cm long, widely spreading. Flowers from many of the median nodes, usually one from the axil of each leaf, on spreading pedicels 2-5 cm long; sepals oblong lance-shaped; petals yellow with dark lines, oblong or elliptic, 6-8 mm long. June-July. — Moist or dry upland soil, chiefly in open woods.

Lysimachia terrestris (L.) B.S.P.
SWAMPCANDLES　　　　　　　　　　　　　　　　　*native / cons* 7
Perennial herb, spreading by shallow rhizomes. Stems smooth, 4-8 dm long, usually branched. Leaves opposite, dotted with glands, narrowly lance-shaped, 5-10 cm long and 2-4 cm wide, with small bulblike structures produced in leaf axils late in season; bracts awl-like, 3-8 mm long. Flowers yellow, in a single, crowded, upright raceme, 1-3 dm long; sepals lance-shaped; petal lobes oval, 5-7 mm long, with dark lines, on stalks 8-15 mm long. Fruit a capsule, 2-3 mm wide. June-Aug. — Marshes, fens, thickets, muddy shores, and ditches.

Lysimachia thyrsiflora L.
SWAMP LOOSESTRIFE　　　　　　　　　　　　　　　*native / cons* 7
Perennial upright herb, spreading by rhizomes; plants conspicuously dotted with dark glands. Stems smooth or with patches of brown hairs, 3-7 dm long, unbranched, or branched on lower stem. Leaves opposite, linear to lance-shaped, 4-12 cm long and 0.5-4 cm wide, smooth above, smooth or sparsely hairy below; petioles absent. Flowers yellow, crowded in dense racemes from leaf axils, on spreading stalks 2-5 cm long; mostly 6-parted; sepal lobes awl-shaped; petal lobes linear, 3 mm long; stamens 2x longer than petals. Fruit a capsule, 2-4 mm wide. June-Aug. — Many types of wetlands: thickets, shores, fens and bogs, marshes, low places in conifer and deciduous swamps, often in shallow water.

Lysimachia thyrsiflora

Primula
PRIMROSE

Primula mistassinica Michx.
MISTASSINI PRIMROSE　　　　　　　　　　　　　　*native / cons* 10
Perennial herb. Stems to 25 cm long. Leaves all at base of plant, oblong lance-shaped, 2-7 cm long, long tapered to base, smooth on upper surface, smooth or often white-yellow powdery below; margins with outward pointing teeth; bracts below flowers awl-shaped, 3-6 mm long. Flowers 1-2 cm wide, 2-10 in a cluster atop a leafless stalk; sepals joined, shorter than petals; petals joined, tubelike and flared at ends, pink and sometimes with a yellow center. Fruit an oblong, upright capsule to 1 cm long. May-June. — Moist rock ledges near Lake Michigan.

Trientalis
STARFLOWER

Trientalis borealis Raf.
STARFLOWER　　　　　　　　　　　　　　　　　*native / cons* 7
　　Lysimachia borealis (Raf.) U. Manns & A. Anderb.
Low perennial herb, with slender rhizomes. Stems 1-2 dm tall, bearing usually a small scale-leaf near the middle and at the summit a whorl of lance-shaped acuminate leaves 4-10 cm long, from the axils of which appear 1 or several white flowers on slender pedicels 2-5 cm long. Flowers ordinarily 7-merous; calyx deeply divided into nearly separate lance-shaped sepals; corolla rotate, 8-14 mm wide, with very short tube and lance-shaped to ovate lobes; stamens inserted at base of the corolla. Fruit a 5-valved, many-seeded capsule. May-June. — Rich woods, hummocks in swamps and bogs.

Primula mistassinica

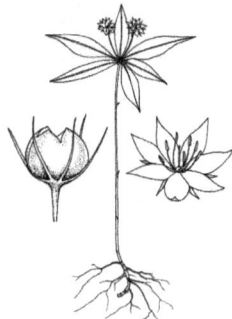

Trientalis borealis

Ranunculaceae

BUTTERCUP FAMILY

Annual or perennial, aquatic or terrestrial herbs (or vines in *Clematis*). Leaves simple to compound, usually alternate, sometimes opposite or whorled, or all at base of plant. Flowers mostly white or yellow, usually with 5 (occasionally more) separate petals and sepals, or petals absent and then with petal-like sepals; sepals leafy and green or petal-like and colored; flowers perfect (with both staminate and pistillate parts), stamens usually numerous; pistils several to many, ripening into beaked achenes or dry capsules (follicles).

1 Vines; leaves opposite; fruit with a long, feathery style .... *Clematis*
1 Herbs; leaves alternate or from base of plant; fruit not with a long, feathery style 2
2 Flowers spurred or strongly irregular .... *Aquilegia*
2 Flowers regular and unspurred 3
3 Stem leaves whorled. .... *Anemone*
3 Stem leaves alternate, or all leaves from base of plant 4
4 Flowers yellow, or leaves simple and not lobed, or plants aquatic 5
4 Flowers not yellow; leaves compound or 3-lobed; plants not aquatic 6
5 Leaves all alike, unlobed; sepals yellow, large and petal-like; petals absent .... *Caltha*
5 Leaves usually of 2 types (stem leaves different from basal leaves), or leaves deeply lobed or divided; sepals green; petals yellow or white .... *Ranunculus*
6 Plants with a naked scape and solitary flowers, the leaves basal 7
6 Plants with leafy stems; flowers 1 to several 8
7 Leaves glabrous, parted into 3 leaflets; sepals white .... *Coptis*
7 Leaves hairy, 3 lobed but not divided into leaflets; sepals white, pink, or blue .... *Hepatica*
8 Flowers few, not in racemes or panicles .... *Enemion*
8 Flowers several to many in terminal racemes or panicles 9
9 Fruit an achene .... *Thalictrum*
9 Fruit berry-like, white or red .... *Actaea*

Actaea

BANEBERRY

Perennial herbs. Leaves 2-3x 3-partedly compound, the leaflets sharply toothed. Flowers small, white, in a dense, long-peduncled, terminal raceme; the raceme at anthesis short and congested with short-pediceled flowers; the axis and pedicels elongating later, the pedicels becoming widely divergent. Sepals 3-5, obovate, petal-like. Petals 4-10, deciduous, obovate, clawed at base. Stamens numerous; filaments usually distinctly widened toward the summit. Pistil 1; stigma broad, sessile, 2-lobed. Fruit a several-seeded berry.

1 Fruit red, on slender pedicels .... *A. rubra*
1 Fruit white, on thicker pedicels .... *A. pachypoda*

Actaea pachypoda Ell.

WHITE BANEBERRY, DOLL'S EYES   *native / cons 6*

Stems 4-8 dm tall. Leaflets usually completely glabrous beneath. Petals 2.5-4 mm long, usually spatulate; stigma wider than the diameter of the ovary. Fruiting pedicels red, 1-2 mm thick; berries globose, normally white, with persistent stigma. May-June. — Rich woods.

Actaea rubra (Ait.) Willd.

RED BANEBERRY   *native / cons 7*

Stems 4-8 dm tall. Leaflets commonly pubescent on the veins beneath. Petals 2.5-4 mm long, spatulate to obovate; stigma not so wide as the diameter of the ovary. Fruiting pedicels slender, 0.4-0.7 mm thick; berries usually red, ellipsoid. May-June. — Rich woods.

Actaea rubra

Anemone
THIMBLEWEED
Perennial herbs from a rhizome or caudex. Basal leaves few to several, deeply palmately divided; stem erect with a whorl of 3 or more involucral leaves subtending one or more elongate peduncles. Flowers white to blue or red or greenish. Petals none. Sepals 4-20, petal-like. Stamens numerous. Pistils numerous, in a subglobose to cylindric head, pubescent. Achenes flattened, clavate, or fusiform, tipped with the persistent style.

1 Achenes hidden by long, cottony hairs 2
1 Achenes nearly glabrous or only short-hirsute 3
2 Involucral leaves 2 or 3; main leaflets thin and shallowly divided; achenes in an ovoid cluster ......
   .... *A. virginiana*
2 Involucral leaves 4 or more; main leaflets thickened and deeply divided; achenes borne in an cylindrical
   cluster .... *A. cylindrica*
3 Involucral leaves with petioles; basal leaves absent or few .... *A. quinquefolia*
3 Involucral leaves with petioles; basal leaves usually many .... *A. canadensis*

Anemone canadensis L.
ROUND-LEAF THIMBLEWEED   *native / cons* 4
Anemonastrum canadense (L.) Mosyakin
Perennial herb, from slender rhizomes, often forming large patches. Stems erect, 1-6 dm long, unbranched below the head. Leaves all from base of plant and with long petioles except for 2-3 stalkless leafy bracts below the head; 4-15 cm wide, deeply 3-5-lobed, round to kidney-shaped in outline, underside with long silky hairs, margins sharp-toothed. Flowers mostly single at ends of stalks, white and showy, 2-5 cm wide; sepals 5, petal-like, 1-2 cm long; petals absent; stamens and pistils many. Achenes clustered in a round, short-hairy head; achene body flat, 3-5 mm long and wide, beak 2-4 mm long. May-Aug. — Wet openings, streambanks, thickets, ditches and roadsides.

Anemone canadensis

Anemone cylindrica Gray
LONG-HEAD THIMBLEWEED   *native / cons* 6
Perennial herb. Stems stiffly erect, 3-10 dm tall. Basal and involucral leaves similar, the basal few to several, the involucral 3-10 and commonly 2x as many as the peduncles, both types petioled, broadly rounded in outline, deeply 5-parted into segments which are incised or sharply toothed only above the middle. Inflorescence usually of 2-6 erect peduncles 1-3 dm long, some of them often bearing a secondary involucre. Flowers greenish white, about 2 cm wide. Fruit a dense cylindric spike 20-35 mm long, about 8 mm thick. Achenes and style densely woolly, the style about 0.5 mm long, outwardly curved. June-Aug. — Dry open woods.

Anemone quinquefolia L.
WOOD-ANEMONE   *native / cons* 6
Delicate perennial herb from a slender horizontal rhizome. Stems 1-2 dm tall. Basal leaf solitary, long-petioled; leaflets 3 or apparently 5, coarsely and unevenly toothed or incised, chiefly above the middle; involucral leaves similar but smaller, the lateral leaflets commonly incised on the outer margin. Peduncle villous; sepals white or suffused with red beneath, usually 5, 10-22 mm long. Achenes fusiform, 3-4 mm long. April-June. — Moist woods.

Anemone quinquefolia

Anemone virginiana L.
TALL THIMBLEWEED   *native / cons* 5
Anemone riparia Fern.
Perennial herb. Leaf segments cuneate at base, the margins straight or nearly so. Flowers white, greenish white, or even red, 2-3 cm wide. Head

of fruit slenderly ovoid or nearly cylindric, usually about 8 mm thick. Achenes densely woolly, the styles 1-1.5 mm long, strongly ascending, the stigma often incurved. June-Aug. — Rocky banks and open woods.

Aquilegia
COLUMBINE

Aquilegia canadensis L.
RED COLUMBINE   *native / cons* 5
Perennial herb from a stout caudex-like rhizome. Stems at anthesis 3-10 dm tall, with few to several large basal leaves. Leaves compound, stem leaves gradually reduced upward, with fewer leaflets, the uppermost 3-foliolate or simple; leaflets broadly obovate, crenately toothed or lobed. Flowers nodding, 3-4 cm long; sepals 5, red; petals 5, the blade yellow, prolonged backward from the base into an elongate red spur; stamens numerous, projecting in a column; pistils usually 5, erect, each prolonged into a slender style. Fruit a several-seeded follicle. April-June. — Dry woods, rocky cliffs and ledges.

Aquilegia canadensis

Caltha
MARSH-MARIGOLD

Caltha palustris L.
COMMON MARSH-MARIGOLD   *native / cons* 6
Loosely clumped, succulent perennial herb. Stems smooth, 2-6 dm long, hollow. Leaves mostly from base of plant, becoming smaller upward, heart-shaped to kidney-shaped, 4-10 cm wide, usually with 2 lobes at base; margins smooth or shallowly toothed; lower leaves with long petioles, stem leaves with shorter petioles. Flowers bright yellow, showy at ends of stems or in leaf axils, 2-4 cm wide; sepals 4-9, petal-like, 12-20 mm long; petals absent; stamens many; pistils 4-15, with short styles. Fruit a follicle, 10-15 mm long. March-June. — Shallow water, swamps, wet woods, thickets, streambanks, calcareous fens, marshes, springs.

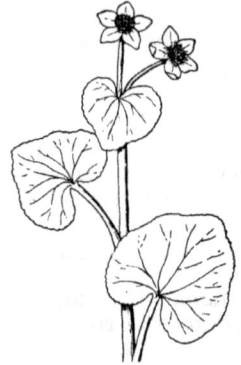
Caltha palustris

Clematis
VIRGIN'S BOWER

Clematis virginiana L.
VIRGIN'S BOWER   *native / cons* 4
Perennial, woody vine. Stems slender, to 5 m long or more, trailing on ground or over shrubs, smooth, brown to red-purple. Leaves opposite, divided into 3 leaflets, the leaflets ovate, 4-8 cm long and 2.5-5 cm wide; margins sharp-toothed or lobed; petioles 5-9 cm long. Staminate and pistillate flowers separate and on separate plants, in many-flowered, open clusters from leaf axils, on stalks 1-8 cm long, usually shorter than leaf petioles; sepals 4, creamy-white, 6-10 mm long; petals absent; stamens numerous; pistils numerous; style elongate. Fruit a rounded head of hairy brown achenes tipped with feathery, persistent styles 2.5-4 cm long. July-Sept. — Thickets, streambanks, moist to wet woods, rocky slopes.

Clematis virginiana

Coptis
GOLDTHREAD

Coptis trifolia (L.) Salisb.
THREE-LEAF GOLDTHREAD   *native / cons* 8
    Coptis groenlandica (Oeder) Fern.
Perennial herb, with slender, bright yellow rhizomes. Leaves from base of plant on long petioles, evergreen, divided into 3-leaflets, the leaflets

Coptis trifolia

shallowly lobed, with rounded teeth tipped by an abrupt point. Flowers single, white, 10-15 mm wide, on a stalk 5-15 cm long from base of plant; sepals 4-7, petal-like; petals absent; pistils 3-7, narrowed to a short, slender style. Fruit a beaked follicle 8-13 mm long. May-June. — Wet conifer woods and swamps, often on mossy hummocks.

Enemion
FALSE RUE-ANEMONE
Enemion biternatum Raf.
EASTERN FALSE RUE-ANEMONE   *native / cons 7*
   Isopyrum biternatum (Raf.) Torr. & Gray
Perennial herb; roots bearing numerous small tuber-like thickenings. Stems slender, erect or ascending, 1-4 dm long. Basal leaves long-petioled, 2-3x 3-parted; stem leaves short-petioled or sessile, 1-2x 3-parted or the uppermost trifoliolate; leaflets broadly obovate, 3-lobed. Flowers white, 1.5-2 cm wide. Follicles commonly 4, ovoid, compressed, divergent. April-May. — Moist woods.

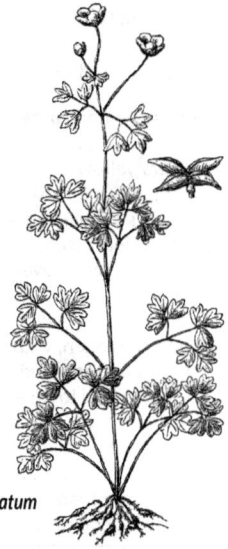

Enemion biternatum

Hepatica
LIVERWORT
Perennial herbs. Leaves basal, simple, lobed, and several 1-flowered scapes bearing a calyx-like involucre of 3 entire bracts immediately below the flower. Sepals 5-12, petal-like. Petals none. Stamens numerous. Pistils numerous; ovary tapering into a short style. Achenes conic to fusiform, pubescent. The genus is very close to *Anemone* and often merged with it, differing only in the simple leaves and the position of the involucre (see species descriptions for synonyms). Leaves persist during the winter, the new leaves appearing after the very early blooming flowers.

1   Leaves lobed nearly to middle of blade, the lobes rounded . . . . *H. americana*
1   Leaves lobed to more than middle of blade, the lobes acute . . . . *H. acutiloba*

Hepatica acutiloba DC.
SHARP-LOBE HEPATICA   *native / cons 7*
   Anemone acutiloba (DC.) G. Lawson
   Hepatica acutiloba DC.
   Hepatica nobilis var. *acuta* (Pursh) Steyermark
Leaves 3-lobed or occasionally 5-7-lobed, deeply cordate at base, the lobes acute. Scapes 5-15 cm long, villous, as are also the petioles. Bracts acute, about equaling the sepals. Flowers 12-25 mm wide. March-April. — Dry or moist woods.

Hepatica acutiloba

Hepatica americana (DC.) Ker-Gawl.
ROUND-LOBE HEPATICA   *native / cons 7*
   Anemone americana (DC.) Hara
   Hepatica nobilis var. *obtusa* (Pursh) Steyermark
   Hepatica triloba Chaix
Very similar to *H. acutiloba* except in leaves and bracts. Leaves averaging smaller, 3-lobed, the lobes broadly obtuse or rounded, the terminal one often wider than long. Bracts obtuse. March-April. — Rich beech-maple forests, as for *H. acutiloba*, but more often found on drier sites with aspen, oak, hickory, pine; sometimes with spruce or cedar.

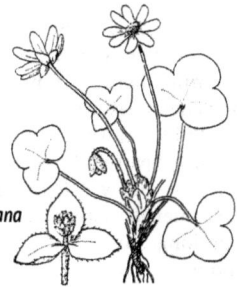

Hepatica americana

Ranunculus
BUTTERCUP, CROWFOOT, SPEARWORT
Aquatic, semi-aquatic, or terrestrial annual and perennial herbs. Stems erect to sprawling, sometimes floating in water. Leaves simple, or compound and finely dissected, often variable on same plant; alternate on stem or all from base of plant; petioles short to long. Flowers borne above water surface in aquatic species; sepals usually 5, green; petals usually 5, yellow or white,

often fading to white, usually with a small nectary pit covered by a scale near base of petal; stamens and pistils numerous. Achenes many in a round or cylindric head; achene body thick or flattened, tipped with a straight or curved beak.

1 Flowers white; leaves divided into linear or threadlike segments; plants typically aquatic (submersed) .... *R. aquatilis*
1 Flowers yellow; leaves simple to deeply lobed or divided into narrow segments; plants submersed, emergent, or terrestrial 2
2 All leaves simple and entire, or shallowly lobed with rounded teeth .... *R. flammula*
2 All, or at least stem leaves, deeply lobed, divided, or compound 3
3 Basal and stem leaves distinctly different in shape, the basal leaves mostly with rounded teeth, the stem leaves deeply divided .... *R. abortivus*
3 Basal and stem leaves similar, all deeply lobed, divided, or compound 4
4 Achenes swollen, without a sharp-edged margin 5
4 Achenes flattened, with a sharp or winglike margin 7
5 Petals 2–4 mm long; achenes to 1.2 mm long, nearly beakless; plants terrestrial or in water only part of season .... *R. sceleratus*
5 Petals 4–14 mm long; achenes 1.2–2.5 mm long, beaked; plants underwater or exposed later in season 6
6 Petals more than 7 mm long; achene body more than 1.6 mm long, achene margin thickened and white-corky below the middle .... *R. flabellaris*
6 Petals less than 7 mm long; achene body less than 1.6 mm long, achene margin rounded but not thickened .... *R. gmelinii*
7 Petals 2–5 mm long 8
7 Petals 7–15 mm long 9
8 Beak of achene strongly hooked .... *R. recurvatus*
8 Beak of achene straight or only slightly curved .... *R. pennsylvanicus*
9 Style short and outcurved; introduced and weedy species 10
9 Style elongate and nearly straight; native, non-weedy species .... *R. hispidus*
10 Stems creeping; terminal segment of the main leaves stalked .... *R. repens*
10 Stems upright; terminal segment of the main leaves not stalked, usually with green tissue extending to the lateral segments .... *R. acris*

Ranunculus abortivus L.
KIDNEY-LEAF BUTTERCUP *native / cons* **1**
Biennial or perennial herb. Stems upright, 2-5 dm long, branched above, smooth or with fine hairs. Leaves at base of plant round to kidney-shaped, margins with rounded teeth, some leaves lobed; petioles long; stem leaves 3-5-divided into linear segments, margins entire or broadly toothed, petioles absent. Flowers yellow, petals 2-3 mm long, shorter than sepals. Achenes in a short, round head; achene body swollen, 1-2 mm long, with a very short, curved beak. April-June. — Wet to moist woods, floodplains, wet meadows, thickets, ditches; especially where soils disturbed or compacted.

Ranunculus abortivus

Ranunculus acris L.
MEADOW-BUTTERCUP *introduced*
Perennial herb, with fibrous roots. Stems hairy, to 1 m long, with few branches, most leaves on lower part of stem. Leaves kidney-shaped, deeply 3-7-divided, the segments again lobed or dissected; branch leaves much smaller, 3-parted. Flowers numerous; sepals 5, half length of petals; petals 5, bright yellow, 6-15 mm long, obovate, often with a rounded notch at tip. Achenes in a round head; achene body flat, 2-3 mm long, beak 0.5 mm long. June-Aug. — Common weed of fields, thickets, ditches and shores.

Ranunculus acris

Ranunculus aquatilis L.
LONG-BEAK WATER-CROWFOOT *native / cons* **8**
Perennial aquatic herb; plants mostly smooth. Stems underwater or floating, 3-8 dm long, with a few branches, rooting from lower nodes. Leaves

round to kidney-shaped in outline, 2-3x divided into narrow threadlike segments 1-2 cm long, stiff and not collapsing when removed from water; leaf segments tipped with tiny transparent spine; petioles absent or to 4 mm long. Flowers at or below water surface, single from upper leaf axils, 1-1.5 cm wide; sepals 5, purple-green, spreading, 2-4 mm long; petals 5, white, yellow at base, 4-9 mm long. Achenes 15-25 in a round head; achene body obovate, ridged, the beak thin and straight, 1-1.5 mm long. May-Aug. — Ponds, lakes, streams, rivers and ditches.

Our plants sometimes treated as 2 species as follows:

1   Styles (at least the longest) and achene beaks 0.6–1.1 mm long, more
    than 1/3 the length of the achene body . . . . **R.longirostri**Godr.
1   Styles and achene beaks very short, less than 0.6 mm long, less than
    about 1/3 the length of the body. . . . . **R. trichophyllus** Chaix

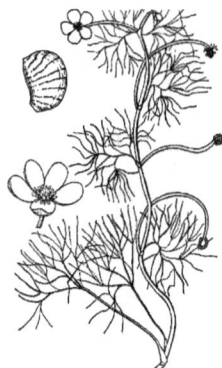

Ranunculus flabellaris Raf.

Ranunculus aquatilis

GREATER YELLOW WATER BUTTERCUP   *native / cons* **8**

Perennial herb; plants smooth or sometimes hairy when growing out-of-water. Stems floating, or upright from a sprawling base when exposed, branched, rooting at lower nodes, 3-7 dm long. Underwater leaves 3-parted into linear segments 1-2 mm wide, exposed leaves (when present) round to kidney-shaped in outline, 2-10 cm long and 2-12 cm wide, divided into 3 segments, the segments again 3-divided. Flowers 1 to several at ends of stems; sepals 5, green-yellow, 4-8 mm long; petals 5-8, bright yellow, 6-15 mm long. Achenes 50-75 in a round to ovate head; achene body obovate, to 2 mm long, the margin thickened and corky below middle, beak broad, flat, 1-1.5 mm long. May-July. — Shallow water or muddy shores of ponds, quiet streams, swamps, woodland pools, marshes and ditches.

Ranunculus flabellaris

Ranunculus flammula L.

CREEPING SPEARWORT   *native / cons* **9**

Perennial herb, spreading by stolons; plants often covered with appressed hairs. Stems sprawling, rooting at nodes, unbranched or few-branched, with upright shoots 4-15 cm long. Leaves in small clusters at nodes, simple, linear or threadlike, 1-5 cm long and 1.5 mm wide, margins more or less entire; upper leaves smaller and with shorter petioles than lower. Flowers single at ends of stems; sepals 5, yellow-green, 2-4 mm long, with stiff hairs; petals 5, yellow, obovate, 3-5 mm long. Achenes 10-25 in a round head; achene body swollen, obovate, 1-1.5 mm long, smooth, the beak short, to 0.5 mm long. June-Aug. — Sandy, gravelly, or muddy shores; shallow to deep water, water usually acid.

Ranunculus gmelinii DC.

LESSER YELLOW WATER BUTTERCUP   *endangered / native / cons* **10**

Perennial herb, similar to yellow water-crowfoot (*R. flabellaris*) but plants aquatic or at least partly underwater; smooth or sometimes with coarse hairs. Stems usually sprawling and rooting at nodes, 1-5 dm long, sparsely branched. Leaves all on stem or with a few basal leaves on long petioles, deeply 3-lobed or dissected, the segments again forked 2-3 times; underwater leaf segments 2-4 mm wide; exposed leaves to 2 cm long and 1.5-2.5 cm wide. Flowers usually 1 to several at ends of stems; sepals 5, green-yellow, 3-6 mm long; petals 5-8, yellow, 4-8 mm long. Achenes 50-70 in a round to ovate head; achene body obovate, 1-1.5 mm long, the margin rounded, not corky-thickened, the beak broad and thin, 0.4-0.7 mm long, somewhat curved. July-Aug. — Muddy streambanks and lakeshores, cold springs, pools in swamps and bogs.

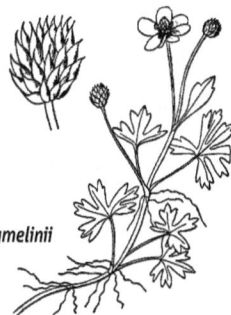

Ranunculus flammula

Ranunculus gmelinii

dicots

Ranunculus hispidus Michx.

NORTHERN SWAMP BUTTERCUP   *native / cons* 6
Perennial herb; stems and leaves variable. Stems upright, 2-9 dm long,
smooth to coarsely hairy. Leaves from base of plant and on stems, the
basal leaves larger and with longer petioles than stem leaves; 3-lobed,
heart-shaped in outline, 3-14 cm long and 4-20 cm wide, with appressed
hairs on veins, upper leaves usually strongly toothed. Flowers 1 to several;
sepals 5, yellow-green, 5-11 mm long, hairy; petals 5-8, yellow, fading to
white, 7-15 mm long and 3-10 mm wide. Achenes 15-30 or more in a round
head; achene body obovate, 2-4 mm long, smooth, winged on margin,
the beak straight, 2-3 mm long. May-July. — Wet woods, floodplains and
swamps, thickets, lakeshores, wet meadows and fens.

Ranunculus pensylvanicus L.f.

BRISTLY CROWFOOT   *native / cons* 5
Annual or short-lived perennial herb. Stems erect, hollow, 3-8 dm long,
branched or unbranched. Leaves at base of plant withering early, larger
and with longer petioles than the few stem leaves; 4-12 cm long and 4-15
cm wide, with appressed hairs, 3- lobed and coarsely toothed, the terminal
leaflet stalked. Flowers few, on short stalks; sepals 5, yellow, 4-5 mm long;
petals 5, pale yellow, fading to white, shorter than the sepals, 2-4 mm
long; stamens 15-20. Achenes many, in a rounded cylindric head 10-15
mm long; achene body flattened, 2-3 mm long, smooth, the beak stout,
0.5-1.5 mm long. July-Aug. — Marshes, wet meadows, ditches and streambanks, often in muck.

Ranunculus hispidus

Ranunculus recurvatus Poir.

HOOKED CROWFOOT   *native / cons* 5
Perennial herb. Stems 2-7 dm long, usually hairy, branches few. Leaves
broadly kidney-shaped or round in outline, 3-parted to below middle,
covered with long, soft hairs; petioles present on all but uppermost leaves.
Flowers on stalks at ends of stems; sepals curved downward, to 6 mm
long; petals pale yellow, 4-6 mm long; styles strongly hooked. Achenes
in a short-cylindric head; achene body flat, round, sharp-margined, to 2
mm long; beak 1 mm long, hooked or coiled. May-June. — Moist deciduous
forests (especially in openings), swamps; also in drier woods; southward
also in partial shade in calcareous fens.

Ranunculus recurvatus

Ranunculus repens L.

CREEPING BUTTERCUP   *introduced*
Perennial herb. Stems normally creeping and rooting at the nodes, rarely
ascending or erect; hirsute, strigose, or rarely glabrous. Leaves 3-parted,
petioled, the segments broadly obovate in outline, cleft or lobed, sharply
toothed. Petals 8-15 mm long, about two-thirds as wide. Achenes obovate,
2.5-3.5 mm long, sharply but narrowly margined; beak triangular, usually
somewhat curved, 0.8-1.5 mm long. May-July. — Native of Europe; introduced in fields, lawns, roadsides, and wet meadows.

Ranunculus sceleratus L.

CURSED CROWFOOT   *native / cons* 3
Weedy annual herb; plants smooth, sometimes partly submersed in shallow water. Stems upright, hollow, 1-6 dm long, branched above and with
many flowers. Leaves from base of plant less deeply parted and with
longer petioles than stem leaves; upper stem leaves small; leaves deeply
3-parted, the main lobes again lobed, heart-shaped at base, rounded at
tip, 1-6 cm long and 3-8 cm wide. Flowers numerous at ends of stalks
from upper leaf axils and branches; sepals 5, 2-3 mm long, yellow-green,
tips curved downward; petals 5, light yellow, fading to white, 3-5 mm

Ranunculus sceleratus

long.Achenes numerous in a short-cylindric head 4-11 mm long; achene body obovate,1 mm long,slightly corky-thickened on margins; beak tiny, blunt. May-Sept. — Muddy shores, streambanks, wet meadows, ditches, marshes and other wet places.

Thalictrum
MEADOW-RUE
Perennial herbs.Leaves alternate,compound.staminate and pistillate flowers separate,in panicles on separate plants; sepals 4-5, green or petal-like but soon deciduous; petals absent; stamens numerous, the stalks (filaments) long and slender; pistils several to many. Fruit a ribbed or nerved achene.

1   Upper stem leaves with long petioles; leaflets glabrous and not glandular; flowering in April or May before leaves fully expanded; plants less than 1m tall . . . . *T. dioicum*
1   Upper stem leaves sessile or nearly so (the 3 stalked leaflets appearing to arise together from the node); leaflets glabrous, or hairy, or with small, short-stalked glands; flowering in summer after leaves expanded; plants often more than 1 m tall 2
2   Underside of leaflets with very short hairs (rarely smooth), not glandular; leaves odorless . . . .
. . . . *T. dasycarpum*
2   Underside of leaflets with small beads and hairs tipped with gray or amber exudate; leaves with strong odor when crushed . . . . *T. revolutum*

Thalictrum dasycarpum Fisch.& Avé-Lall.
PURPLE MEADOW-RUE   *native / cons* 4
Perennial herb, from a short rootstock. Stems purple-tinged, 1-2 m long, branched above.Leaves divided into 3-4 groups of leaflets; leaflets 15 mm or more long, mostly tipped with 3 pointed lobes, dark green above, underside sparsely short-hairy, not waxy and without gland-tipped hairs; margins usually slightly turned under; stem leaves mostly without petioles.Flowers in panicles at ends of stems; staminate and pistillate flowers separate and on different plants (sometimes with some perfect flowers); sepals 3-5 mm long, lance-shaped; anthers linear and sharp-tipped, 2-3 mm long, filaments white; stigmas straight, 2-4 mm long. Achenes 4-6 mm long, ribbed, in a round cluster.June-July. — Wet to moist meadows, swamps, thickets, streambanks.

Thalictrum dioicum L.
EARLY MEADOW-RUE   *native / cons* 7
Perennial herb, dioecious. Stems 3-7 dm tall at anthesis. Leaves all with long petioles, the uppermost 3-6 cm long and subtending the inflorescence; stipules of the upper leaves broadly ovate, mostly much wider than long.Filaments and anthers yellow or greenish yellow.Mature achenes sessile or subsessile, about 4 mm long,strongly ribbed,straight and essentially symmetrical. — Moist woods. Flowering with or before the expansion of leaves on deciduous trees.

Thalictrum dasycarpum

Thalictrum revolutum DC.
WAXY-LEAF MEADOW-RUE   *native / cons* 6
Thalictrum amphibolum Greene
Perennial herb, from short rootstocks, with strong odor when crushed. Stems more or less smooth, often purple-tinged, 0.5-1.5 m long. Lowest leaves with petioles, middle and upper leaves stalkless; leaves divided into 3-4 groups of leaflets; leaflets variable in shape and size, usually 3-lobed,some 1-2 lobed,upper surface smooth,underside leathery and conspicuously net-veined, finely hairy with gland-tipped hairs, margins turned under.Flowers in panicles at ends of stems; staminate and pistillate flowers separate and on different plants (sometimes with some perfect flowers); anthers linear, 2-3 mm long, filaments threadlike, 2-5 mm

Thalictrum revolutum

long; pistils 6-12, stigmas 2-3 mm long. Fruit an oval or lance-shaped achene, 4-5 mm long, ridged, with tiny gland-tipped hairs. June-July. — Streambanks, thickets, moist meadows.

Rhamnaceae

BUCKTHORN FAMILY

Shrubs, trees, or woody vines with simple, opposite or alternate leaves and small flowers. Flowers perfect or unisexual, regular, 4-5-merous. Petals present or lacking, small, separate. Stamens as many as and alternate with the sepals, opposite and often enfolded by the petals. Ovary 1, sessile on the disk or immersed in it; styles 2-5, united for all or part of their length. Fruit a capsule or drupe.

1 Leaves 3-veined from base of leaf; flowers white in many-flowered, stalked clusters; fruit a capsule. . . . . *Ceanothus*
1 Leaves not 3-veined from base; flowers greenish, single or few to a cluster; fruit a fleshy drupe . . . 2
2 Leaf margins entire or nearly so . . . . *Frangula*
2 Leaf margins toothed . . . . *Rhamnus*

Ceanothus

BUCKBRUSH

Low shrubs. Leaves alternate, 3-nerved, glandular-serrate. Flowers small, white, in sessile or short-peduncled umbels aggregated into terminal or axillary panicles. Sepals inflexed, at length deciduous above the hypanthium. Petals long-clawed. Stamens at anthesis free and exsert. Ovary 3-angled and 3-celled, immersed in the disk; style 3-lobed. Fruit a 3-lobed capsule-like drupe subtended by the persistent hypanthium.

Many species highly ornamental. Characteristic are the leaves with 3 pairs of prominent parallel veins extending from the leaf base to the outer margins of the leaf tips.

1 Leaves elliptic, less than 2 cm wide; inflorescences at ends of current year's shoots . . . . *C. herbaceus*
1 Leaves ovate, mostly more than 2 cm wide; inflorescences from leaf axils . . . . *C. americanus*

Ceanothus americanus L.

NEW JERSEY-TEA   *native / cons* 9

Shrub to 1 m tall, often freely branched. Leaves narrowly to broadly ovate, 3-8 cm long, usually more than half as wide, broadly cuneate to rounded or subcordate at base, the lateral nerves commonly naked for 1-3 mm at base. Inflorescences on axillary peduncles, the lower peduncles progressively longer and to 2 dm long; panicle short-cylindric to ovoid, occasionally branched, often subtended by 1-3 reduced leaves, the umbels usually separated by distinct internodes. Fruit depressed-obovoid, 5-6 mm long. June-July. — Upland woods, barrens.

Ceanothus americanus

Ceanothus herbaceus Raf.

PRAIRIE REDROOT   *native / cons* 8

Bushy shrub to 1 m tall. Leaves oblong to elliptic, 2-6 cm long, 1-2 cm wide, the lateral nerves never naked and often arising unevenly 1-3 mm above the base of the leaf; leaf underside usually pubescent. Panicles several to many, terminating the leafy branches of the season, on peduncles rarely to 5 cm long, hemispheric to short-ovoid, the component umbels very close together. Fruit 4-5 mm long. May-June. — Sandy or rocky soil.

Ceanothus herbaceus

Frangula
FALSE BUCKTHORN

Frangula alnus P. Mill.
GLOSSY FALSE BUCKTHORN   *introduced (invasive)*
   Rhamnus frangula L.
Shrub to 7 m tall. Leaves usually obovate-oblong, 5–8 cm long, commonly
more than half as wide, acute to abruptly short-acuminate, entire or with
a few marginal glands near the tip. Umbels sessile, 2–8-flowered; pedicels
usually unequal, 3–10 mm long. Flowers perfect, 5-merous; petals broadly
obovate, scarcely clawed, cleft at the tip, 1–1.4 mm long; styles connate to
the tip. Fruit red ripening nearly black, 2–3-stoned. May–June. — Native
of Eurasia; escaped from cultivation, especially in wet soil.

Frangula alnus

Rhamnus
BUCKTHORN
Shrubs or small trees. Leaves simple, alternate or opposite, pinnately veined, usually with
stipules. Flowers perfect, or staminate or pistillate, regular, single or few from leaf axils; sepals
joined, 4- or 5-parted; petals 4 or 5. Fruit a purple-black, berrylike drupe with 2–4, 1-seeded
stones.

1 Leaves less than 8 cm long; petals 4 . . . . *R. lanceolata*
1 Leaves more than 8 cm long; petals absent . . . . *R. alnifolia*

Rhamnus alnifolia L'Hér.
ALDER-LEAF BUCKTHORN   *native / cons* 8
   Endotropis alnifolia (L'Hér.) Hauenschild
Shrub to 1 m tall, forming low thickets. Leaves alternate, oval to ovate, 6–
10 cm long and 3–5 cm wide, green above, paler green below; margins
with low, rounded teeth; petioles grooved, 5–12 mm long; stipules linear,
to 1 cm long, deciduous before fruit mature. Flowers appearing with
leaves in spring, in clusters of 1–3 flowers from leaf axils; yellow-green,
usually 5-parted, 3 mm wide, on short stalks, with both stamens and
pistils but one or other is nonfunctional, sepals 1–2 mm long, petals
absent. Fruit a purple-black, berrylike drupe, 6–8 mm wide, with 1–3 nut-
let-like stones. May–June. — Conifer swamps, thickets, sedge meadows,
wet depressions in deciduous forests; usually where calcium-rich.

Rhamnus alnifolia

Rhamnus lanceolata Pursh
LANCE-LEAF BUCKTHORN   *native / cons* 8
Shrub to 1–2 m tall. Leaves alternate, lance-shaped to oval, 3–7 cm long
and 1–3 cm wide, tapered to a short tip; margins with small, incurved,
forward-pointing teeth; petioles to 1 cm long. Flowers appearing with
leaves in spring, staminate and pistillate flowers separate, staminate
flowers usually 2–3 in leaf axils, pistillate flowers usually single; sepals
green-yellow, 4-lobed, petals to 1 mm long. Fruit a black, berrylike drupe,
with 2 nutlike stones. — Calcareous fens.

Rhamnus lanceolata

Rosaceae
ROSE FAMILY
Shrubs and perennial, biennial, or annual herbs. Leaves evergreen or deciduous, mostly alternate
and simple or compound. Flowers perfect (with both staminate and pistillate parts), regular,
with 5 sepals and petals; stamens numerous. Fruit an achene, capsule, or fleshy fruit with numerous embedded seeds (drupe), or a fleshy fruit with seeds within (pome).

ADDITIONAL SPECIES

Alchemilla monticola Opiz (Hairy Lady's-mantle), introduced, rarely escaping; reported in Door County from moist soil in mixed hardwood forest.

Cotoneaster divaricatus Rehder & E.H.Wilson (Spreading cotoneaster), introduced, reported from edge of mixed hardwood forest.

1 Plants trees, shrubs, or erect to trailing, thorny to bristly brambles 2
1 Plants herbs (sometimes woody at base), not thorny or bristly 13
2 Leaves mostly compound; branches or stems often thorny or bristly 3
2 Leaves simple; branches and stems smooth or only with long stout spines 6
3 Stems biennial, prickly or bristly; leaves 3-parted or palmately compound; fruit a tight cluster of juicy drupelets; flowers usually white . . . . *Rubus*
3 Stems perennial, smooth or thorny; leaves pinnately compound; fruit various but not a cluster of dru-pelets; flowers white, pink, or yellow 4
4 Flowers pink (rarely white or yellow), 2 cm or more wide; stems thorny; fruit fleshy, red to orange . .
. . . . *Rosa*
4 Flowers white or yellow, mostly less than 2 cm wide; stems smooth; fruit various 5
5 Flowers solitary or few in an inflorescence, the petals yellow; leaflets entire . . . . *Potentilla fruticosa*
5 Flowers many in a crowded inflorescence, the petals white; leaflets toothed . . . . *Sorbus*
6 Style and ovary 1; fruit a drupe; leaves unlobed . . . . *Prunus*
6 Styles 2 or more; fruit a pome, or a cluster of drupelets or dry fruits 7
7 Ovary superior 8
7 Ovary inferior 9
8 Leaves mostly 3–5 lobed; bark shredding into long strips . . . . *Physocarpus*
8 Leaves not lobed; bark not shredding into long strips . . . . *Spiraea*
9 Leaves with red or black appressed glands along midrib of leaf upper surface . . . . *Aronia*
9 Leaves without glands on midrib 10
10 Branches never thorny; flower petals white, lance-shaped and usually more than 2 times longer than wide . . . . *Amelanchier*
10 Branches sometimes with stout spines; petals less than 2 times longer than wide 11
11 Branches normally with spines; leaves toothed and often slightly lobed 12
11 Branches without spines; leaves toothed but not lobed . . . . *Malus*
12 Spines shiny; bud scales glabrous; petals white; seeds within hard nutlets . . . . *Crataegus*
12 Spines dull; bud scales hairy; petals pinkish; seeds within papery carpels . . . . *Malus*
13 Leaves 3-parted or palmately compound 14
13 Leaves pinnately compound or divided 19
14 Styles long, jointed near middle, the lower portion persistent on the achene as a long beak . . *Geum*
14 Styles short, neither jointed nor persistent on the fruit 15
15 Calyx with bractlets about as large as sepals, the calyx appearing 10-lobed 16
15 Calyx without bractlets between the sepals, the calyx 5-lobed 18
16 Petals white; fruit fleshy and red; leaflets 3 . . . . *Fragaria*
16 Petals yellow or white; fruit dry; leaflets 3, 5, or 7 17
17 Flowers yellow, leaflets 3, 5, or 7, regularly toothed, deciduous . . . . *Potentilla*
17 Flowers white; leaflets 3, entire except for a 3 (–5)-toothed apex, evergreen . . . . *Sibbaldia*
18 Petals yellow; fruit an achene . . . . *Geum fragarioides*
18 Petals white or pink; fruit fleshy drupelets . . . . *Potentilla*
19 Calyx 5-lobed, small bractlets absent; receptacle flat or concave 20
19 Calyx 10-lobed, small bractlets alternating with sepals; receptacle hemispherical or conical 21
20 Petals yellow; floral tube with hooked bristles at tip; inflorescence an elongate raceme . . *Agrimonia*
20 Petals absent; floral tube not bristly; inflorescence short . . . . *Poterium*
21 Styles elongating and becoming longer than achene, persistent as a beak atop the achene . . . *Geum*
21 Styles short, deciduous 22
22 Petals deep maroon to purple; sepals red tinged; stem usually decumbent, the lower portion in water or wet ground, rooting at nodes. . . . . *Comarum*
22 Petals yellow or white; sepals green; stem usually erect, or with slender stolons; mostly upland . . 23
23 Pubescence not glandular; petals deep yellow . . . . *Potentilla*
23 Pubescence glandular-viscid; petals white to pale yellow . . . . *Drymocallis*

Agrimonia
AGRIMONY, GROOVEBURR

Perennial herbs from stout rhizomes. Stems erect, simple or branched above. Leaves pinnately compound, mostly below middle of stem; stipules foliaceous, usually deeply toothed or laciniate. Flowers in long, interrupted, spike-like racemes, the short peduncle subtended by a laciniate bract, the very short pedicels by a pair of 3-lobed bractlets. Hypanthium obconic to hemispheric, with hooked bristles and small resinous glands. Sepals spreading at anthesis, later incurved and forming a beak on the fruit. Petals 5, yellow, 5-8 mm wide. Stamens 5-15. Pistils 2. Fruit an achene.

1 Inflorescence rachis covered with small glands, the pubesence sparse or absent .... *A. gryposepala*
1 Inflorescence rachis without glands or nearly so, but rachis covered with appressed to spreading hairs
    2
2 Leaflet underside velvety to touch, the hairs spreading .... *A. pubescens*
2 Leaflet underside smooth or rouch-to-touch, the hairs usually appressed .... *A. striata*

Agrimonia gryposepala Wallr.

TALL HAIRY AGRIMONY   *native / cons* **2**

Perennial herb, roots fibrous. Stems stout, to 15 dm tall, glandular and sparsely or densely long-hirsute throughout. Principal leaflets of the larger leaves 5-9, ovate lance-shaped to elliptic or obovate, coarsely and often bluntly serrate, glabrous or or nearly so on the surface, sparsely hirsute on the veins; stipules large and leaflike, usually 1-2 cm wide. Axis glandular, hirsute with long spreading hairs; pedicel hirsute. Hypanthium glandular only, or also with a few short stiff hairs near the base, 3-5 mm long at maturity, expanded at the summit. July-Aug. — Moist or dry, open woods.

Agrimonia pubescens Wallr.

SOFT AGRIMONY   *native / cons* **5**

Perennial herb, roots tuberously thickened. Stems stout, to 1 m tall or more, densely pubescent. Principal leaflets of the larger leaves 5-13, lance-shaped to elliptic or narrowly obovate, coarsely serrate, glabrous, scabrel-late, or sparsely pubescent above, velvety-pubescent beneath; stipules lance-shaped to semi-ovate. Axis eglandular, densely pubescent. Mature hypanthium campanulate, 2.5-3 mm long, nearly as wide, more or less covered with short stiff ascending hairs, especially toward the base. July-Aug. Dry, open woods.

Agrimonia gryposepala

Agrimonia striata Michx.

WOODLAND AGRIMONY   *native / cons* **3**

Perennial herb, roots fibrous. Stems stout and coarse, to 1 m tall or more, hirsute below, pubescent and glandular above. Principal leaflets of the larger leaves 7-11, the upper 5 commonly directed forwards, ovate lance-shaped, coarsely serrate, glabrous or nearly so above, sparsely pubescent beneath, especially on the veins; stipules lance-shaped, 1-2 cm long. Axis eglandular, densely pubescent with ascending hairs, commonly also with some long flexuous hairs. Flowers densely crowded; peduncle and pedicel short, the 3-cleft bractlet commonly surpassing the hypanthium; mature hypanthium reflexed, turbinate, 4-5 mm long, deeply furrowed. July-Aug. — Dry or moist woods.

Agrimonia striata

Amelanchier
SERVICEBERRY

Trees or shrubs, without thorns. Leaves simple, alternate, serrate. Flowers in short leafy racemes terminating the branches of the season and opening with or before the leaves. Hypanthium ob-conic, campanulate, or saucer-shaped. Sepals 5, spreading to recurved, persistent. Petals 5, white, oblong to oval or obovate. Stamens usually 20, shorter than the petals. Ovary 5-celled; styles 5. A confusing genus, similar to Crataegus in that hybridization, polyploidy, and asexual vegetative reproduction have reulted in a wide variety of forms.

1 Tip of ovary glabrous; leaf blades short-acuminate, finely and closely serrate with 22–45 teeth per side 2
1 Tip of ovary tomentose; leaf blades variously shaped and toothed 3
2 Leaves just beginning to unfold at flowering time, densely white-tomentose beneath, otherwise green, retaining some of the pubescence on petioles and along midrib beneath into maturity. . . . . . .
   . . . . *A. arborea*
2 Leaves mostly half-grown at flowering time, usually bronze-red, glabrous or nearly so, completely glabrous at maturity. . . . . *A. laevis*
3 Larger leaves with 25–50 fine teeth on a side (more than twice as many teeth as lateral veins), acute, at flowering time open though not fully grown and often glabrous or soon becoming so. . *A. interior*
3 Larger leaves with fewer than 20 (–25) teeth on a side (no more than 2x as many teeth as lateral veins), the blades at flowering time ± folded and white-tomentose beneath, when mature the tip acute to rounded 4
4 Most leaves coarsely toothed (2–5 teeth per cm toward tip when mature), the veins prominent and running to tips of the teeth (or a principal fork into the teeth) at least toward tip of blade; petals 10–20 mm long; plants typically solitary or in tall clumps with many stems . . . . *A. sanguinea*
4 Most leaves finely toothed at least toward apex (5–8 teeth per cm when mature), the veins anastomosing and becoming indistinct near the margin, at most with weak veinlets ending in the teeth; petals 5–9 mm long; plants typically spreading underground and forming colonies of low shrubs . . .
   . . . . *A. spicata*

Amelanchier arborea (Michx. f.) Fern.
DOWNY SERVICEBERRY   *native / cons* 6
Tall shrub or small tree, rarely to 10 m tall. Leaves typically obovate, sharply and finely serrate nearly to the rounded or cordate base, densely pubescent beneath when young, at maturity nearly glabrous. Racemes drooping, many-flowered; pedicels 5-20 mm long, often covered with silky hairs; hypanthium glabrous; sepals 2-3 mm long, reflexed; petals narrowly oblong, 10-15 mm long. Pomes maroon-purple, 6-10 mm wide, insipid. April-May. — Usually in dry sandy open forests with red maple, aspen, oaks, or jack pine; sometimes in moist or swampy forests and along forest borders.

Amelanchier arborea

Amelanchier interior Nielsen
INLAND SERVICEBERRY   *native / cons* 7
Shrub or small tree. Stems 1-10 m tall, often straggling or arching; twigs glabrous at flowering. Leaves broadly ovate, 3-7 cm long and 2-5 cm wide, acute to short-acuminate, base rounded to subcordate, upper surface green, sparsely pubescent or glabrous by flowering time; margins serrate nearly to the base; petioles 1-3 cm long. Inflorescence 4-12-flowered, drooping or nodding; pedicels glabrous or nearly so; hypanthium campanulate, 3-6 mm wide; sepals recurving after flowering, 2-5 mm; petals white, obovate, 6-15 mm long; stamens 20; styles 5. Pomes purple-black, globose, 6-8 mm wide, sweet. May-June. — Sandy open savannas and dunes, shallow soil on rock outcrops and shores; sometimes at borders of hardwood forests and conifer swamps.

Amelanchier laevis Wieg.
SMOOTH SERVICEBERRY   *native / cons* 6
Tall erect shrub or tree, to 10 m tall. Leaves elliptic to ovate, to 8 cm long at maturity, abruptly acute to short-acuminate, finely and sharply serrate nearly to the rounded or subcordate base, at anthesis about half grown and glabrous beneath or rarely with a few scattered hairs. Racemes many-flowered; pedicels glabrous, 1-3 cm long; hypanthium glabrous externally; sepals 3-4 mm long, reflexed; petals oblong, 10-18 mm long or rarely

Amelanchier laevis

more; ovary glabrous at the tip. Pomes dark purple, mostly 10-15 mm wide, sweet. May. — Most often in dry sandy open forests and savannas, rocky sites, sandy bluffs and shores; also on river banks and forest and bog margins.

Amelanchier sanguinea (Pursh) DC.
NEW ENGLAND SERVICEBERRY   *native / cons 7*
Amelanchier humilis Wieg.
Erect or straggling shrub or small tree, to 3 m tall, usually growing in clumps of several stems. Leaves about half grown at anthesis and then tomentose beneath, eventually glabrous, oblong to subrotund or sub-quadrate, to 7 cm long, finely or coarsely toothed, often only above the middle; veins often prominent and running to the teeth, especially in upper portion of blade. Inflorescences 4-10-flowered, soon arching or drooping; pedicels hairy; hypanthium saucer-shaped, 3.5-7.5 mm wide; sepals recurving or spreading after flowering, 3.5-5 mm; petals white, linear to narrowly spatulate, 11-18 mm long; stamens 20; styles 5; ovary summit rounded, densely hairy. Pomes dark purple or almost black, 5-8 mm wide, sweet. May-June. — Dry, open, sandy savannas and clearings; sandy thickets, borders of forests, gravelly shores, and low dunes.

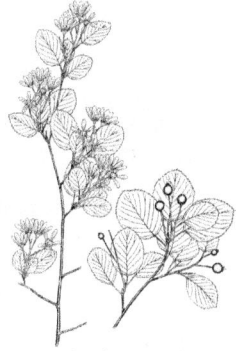

Amelanchier sanguinea

Amelanchier spicata (Lam.) K. Koch
RUNNING SERVICEBERRY   *native / cons 5*
Amelanchier stolonifera Wieg.
Stoloniferous shrub 3-10 or rarely 15 dm tall, forming colonies. Leaves a quarter to half grown at anthesis and then densely tomentose beneath, at maturity glabrous and much paler beneath, ovate to oblong, or obo-vate-oblong, usually 2-5 cm long, finely and sharply toothed; lateral veins curved forward, branched and anastomosing near the margin; teeth al-most always more than twice as many as the veins. Racemes dense; pedicels thinly pubescent, soon glabrescent, the lowest 7-15 mm long. Pomes purple-black, glaucous, 7-12 mm wide, sweet. May. — Dry, sandy plains, dunes, and savannas, usually with jack pine or oaks, often little taller than the associated shrubby species of *Comptonia* and *Vaccinium.*

Amelanchier spicata

Aronia
CHOKEBERRY
Aronia prunifolia (Marsh.) Rehder
PURPLE CHOKEBERRY   *native / cons 7*
Shrub, 1-2.5 m tall; twigs gray to purple, smooth or hairy. Leaves alternate, oval or obovate, 3-8 cm long and 1-4 cm wide, upper surface dark green and smooth (except for dark, hairlike glands along midveins), underside paler, smooth or hairy; margins with small, rounded, forward-pointing teeth, the teeth gland-tipped; petioles to 1 cm long. Flowers 5-10 mm wide, in clusters of 5-15 at ends of stems and short, leafy branches; sepals usually glandular; petals white, 4-6 mm long. Fruit a dark purple to nearly black, berrylike pome, 8-11 mm wide, not persisting into winter. May-June. — Tamarack swamps, open bogs, thickets, marshes and shores.

ADDITIONAL SPECIES
Completely glabrous plants are sometimes recognized as *Aronia melanocarpa* (Michx.) Elliott (black chokeberry).

Aronia prunifolia

Comarum
MARSHLOCKS
Comarum palustre L.

MARSH CINQUEFOIL   *native / cons* **8**
Potentilla palustris (L.) Scop.

Perennial herb, from long, stout rhizomes. Stems 3-8 dm long, ascending to sprawling or floating in shallow water, often rooting at nodes, more or less woody at base; lower stems smooth, upper stems sparsely hairy. Leaves all from stem, pinnately divided or nearly palmate, with 3-7 leaflets; leaflets oblong to oval, 3-10 cm long and 1-3 cm wide, mostly rounded at tip, underside waxy; margins with sharp, forward-pointing teeth; lower leaves long-petioled, upper leaves nearly sessile; stipules forming wings around petioles of lower leaves, becoming shorter upward. Flowers single or paired from leaf axils, or in open clusters; sepals dark red or purple (at least on inner surface), ovate to lance-shaped, 6-20 mm long; petals 5 (sometimes 10), very dark red, 3-5 mm long, with a short slender tip; stamens about 25, dark red. Achenes red to brown, smooth, 1 mm long. June-Aug. — Open bogs (especially in pools and wet margins), conifer swamps, shores.

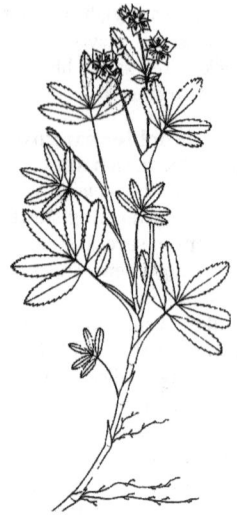

Comarum palustre

Crataegus
HAWTHORN
Small trees or shrubs with usually spiny branches. Leaves simple, deciduous, alternate, serrate or dentate and otherwise entire or variously lobed. Flowers perfect, regular, in corymbs or rarely single or 2 or 3 together. Petals 5, white or rarely pink. Sepals 5. Stamens 5-20(the number of stamens can be counted on the flowers and also on the fruit using remnants of the filaments). Ovary of 1-5 carpels; styles 1-5, persistent. Fruit a globose pome, red or rarely yellow, blue, or black at maturity, with 1-5 bony nutlets. The leaves of sterile shoots or of the ends of branches (vegetative leaves) are often differently shaped and more deeply incised than those of the flowering branchlets.

Because of the apparent instability of many species and their tendency to hybridize, there is no generally agreed upon consensus regarding Wisconsin's hawthorns. Included in the key and descriptions are the most common, well-defined species reported for Door County.

1 Nutlets with deep to shallow pits or depressions on their lateral surfaces; flowering in late May or June 2
1 Nutlets plane, not pitted laterally; flowering in April–early June 3
2 Mature leaf blades thin, the veins (except sometimes for midrib) scarcely if at all impressed above, strigose above and usually pubescent beneath; inflorescences, branchlets of current year, and petioles all usually villous or lightly tomentose; thorns 2.5–5 cm long, often sparse or even absent; stamens ca. 20; late flowering (usually early June) . . . . *C. calpodendron*
2 Mature leaf blades ± leathery, thickened at margins, the veins usually deeply impressed above, glabrous to pubescent on both surfaces; inflorescences, new branchlets, and petioles sparsely villous to glabrous (if inflorescence somewhat villous, at least the young branchlets nearly always glabrous, the veins deeply impressed, and/or the stamens ca. 10); thorns 2.5–9.5 cm long, usually numerous; stamens ca. 20 or 10; mid-season flowering (about late-May in Door County) . . . . *C. succulenta*
3 Blades of at least the floral leaves (in many species also the vegetative leaves) acute to broadly or (more commonly) narrowly tapered or cuneate at the base 4
3 Blades of both floral and vegetative leaves mostly broadly rounded, truncate, or subcordate at the base. . . . . *C. macrosperma*
4 Blades (especially of floral leaves) mostly obovate to oblong-elliptic, broadest above or rarely at the middle, unlobed or very obscurely lobed near the apex, mostly 1.5–3 or more times as long as broad, usually thick or even stiff and leathery . . . . *C. punctata*
4 Blades (at least of floral leaves) mostly elliptic to ovate, broadest at or below the middle, often lobed, usually 1–1.5 times as long as broad, often thin . . . . *C. chrysocarpa*

Crataegus calpodendron (Ehrh.) Medik.

PEAR HAWTHORN   *native / cons* 5

Arborescent shrub or small tree; branchlets straight, slender, villous or tomentose while young. Leaves ovate or rhombic, 5-9 cm long, 4-8 cm wide, coarsely serrate except near the base and usually more or less divided above the middle into 3-5 pairs of shallow, often irregular, lateral lobes, short-villous above and pubescent or rarely glabrous beneath, dull yellow-green, thin but firm and with the veins slightly impressed above at maturity; petioles stout, 1-1.5 cm long, wing-margined sometimes nearly to base. Flowers 1.3-1.5 cm wide, in many-flowered tomentose corymbs; sepals coarsely glandular-serrate; stamens about 20, anthers pink or rarely white. Fruit oblong or obovoid or rarely subglobose, 7-9 mm thick, scarlet, glabrous when ripe, with a narrow elevated calyx, and 2-3 nutlets. May-June; fruit ripe Oct. — Open woods and thickets, usually along small rocky streams.

Crataegus calpodendron

Crataegus chrysocarpa Ashe

FIREBERRY HAWTHORN   *native / cons* 4

Stout, intricately branched shrub or rarely a small tree to 5-6 m tall; branchlets very thorny. Leaves elliptic, oval, or suborbicular, 4-6 cm long and 2-4 cm wide, lobed, serrate except near the base with gland-tipped teeth, roughened above with short appressed hairs while young; firm, dark yellow-green, the veins impressed above at maturity; petioles slender, sometimes slightly glandular, mostly 1/4 to 1/2 as long as the blades. Flowers 1.3-1.6 cm wide; stamens about 10, anthers white or pale yellow; sepals nearly entire or finely glandular-serrate. Fruit dull or dark red or rarely dull yellow, with thin flesh, remaining hard or dry or becoming mellow late in the season; nutlets 3-4. May; fruit ripe Sept.-Oct. Thickets and rocky ground along streams. — Sandy hillsides, stream and river banks, forest borders, roadsides, fields, pastures; sometimes in wet places.

Crataegus chrysocarpa

Crataegus macrosperma Ashe

BIG-FRUIT HAWTHORN   *native / cons* –

Much-branched shrub to 5-6 m high. Older stems stout and armed with straight or slightly curved thorns 3- 10 cm long. Leaves ovate, acuminate at tip and cuneate to truncate at base, short-pilose above; margins indented with 4-6 small lobes on each side and serrate nearly to the base; petioles 1-3 cm long, somewhat winged and grooved near the blade; leaves of the vegetative shoots more deeply lobed. Flowers 15-18 mm wide, in loose, pubescent corymbs; stamens usually 10 or less; anthers pink. Fruit oblong to nearly globose, sometimes slightly angular, 8-12 mm thick, bright red; nutlets 3-5. May-June. — Open woods, thickets, fields, and along river banks and rocky ridges.

Crataegus macrosperma

Crataegus punctata Jacq.

DOTTED HAWTHORN   *native / cons* 2

Tree to 8-10 m tall with an open top of stiff spreading branches, armed with slender gray thorns 4-6 cm long, and often with compound thorns on trunk. Leaves obovate, or sometimes oblong-elliptic on shoots, dull yellowish green, mostly 2.5-6 cm long, 1.5-3 cm wide, cuneate or attenuate at base, serrate sometimes only above the middle, usually slightly lobed at least toward the apex, often deeply lobed on vegetative shoots, firm, the veins distinctly impressed on the upper surface at maturity; leaves covered with short appressed hairs above while young. Flowers in many-flowered corymbs; stamens about 20, anthers pink or pale yellow; corymbs and calyx tube gray-pubescent. Fruit subglobose or short-oblong, appearing pyriform while immature, usually 1.2-1.5 cm wide, dull red or or-

Crataegus punctata

236 ROSACEAE (Rose Family)   *dicots*

ange-red, pale-dotted, with thick flesh becoming mellow or slightly succulent; nutlets usually 5, rounded and ridged on the back. May-June;
fruit ripe Sept-Oct. — Thickets and borders of woods, often in rocky
ground.

Crataegus succulenta Schrad.
FLESHY HAWTHORN   *native / cons* 5
Tree to 7 or 8 m tall or sometimes an arborescent shrub, with slender
branchlets, glabrous or rarely slightly hairy while young, often armed
with long chestnut-brown thorns becoming gray and compound on the
larger branches. Leaves elliptic, rhombic or rarely ovate, finely serrate
and usually indented with 4-6 pairs of shallow lobes above the middle,
glabrous or slightly villous along the veins beneath, firm to nearly leathery, roughened with short appressed hairs on the upper surface while
young; petioles 1-2 cm long, wing-margined above, usually eglandular.
Flowers 1.3-1.7 cm wide, in many-flowered, slightly villous or glabrous
corymbs; stamens about 10-20, anthers white or pink; sepals glandular-
serrate, reflexed after anthesis, usually deciduous from the mature fruit.
Fruit subglobose, 0.7-1.2 cm wide, bright red, glabrous, lustrous; nutlets
2-3. May-June; fruit ripe Sept. — Thickets, pastures, and borders of woods,   *Crataegus succulenta*
usually in dry or rocky ground.

Drymocallis
WOODBEAUTY
Drymocallis arguta (Pursh) Rydb.
TALL WOODBEAUTY   *native / cons* 7
   Potentilla arguta Pursh
Perennial from a stout rhizome, more or less viscid-pubescent throughout.
Stems erect, 3-10 dm tall, simple to the inflorescence. Leaves pinnately
compound, the basal leaves long-petioled; leaflets 7-11, or only 5 in the
uppermost leaves. Flowers white, cream or pale yellow, 12-18 mm wide,
crowded in a slender, elongate inflorescence; sepals ovate, much longer
than the lance-shaped bractlets, nearly as long as the petals. Achenes
obovoid, pale brown, 1 mm long, finely striate. June-July. — Dry woods.

Drymocallis arguta

Fragaria
STRAWBERRY
Perennial herbs, usually spreading freely by runners and forming colonies. Leaves basal, 3-foli-
olate, serrate. Flowers several on peduncles. Hypanthium saucer-shaped. Sepals alternating with
foliaceous bracts of nearly equal size. Petals white, obovate to subrotund. Stamens numerous,
sometimes abortive. Pistils numerous. Fruit consisting of numerous minute achenes on the
greatly enlarged, red, juicy receptacle, subtended by the persistent calyx and bracts.

1 Terminal center tooth of leaflets smaller than the tooth on either side of it; calyx lobes appressed to
   fruit .... *F. virginiana*
1 Terminal center tooth of leaflets as large or larger than the tooth on either side of it; calyx lobes
   spreading away from fruit .... *F. vesca*

Fragaria vesca L.
THIN-LEAVED WILD STRAWBERRY   *native / cons* 3
Leaflets sessile or nearly so, ovate to obovate, more or less silky beneath;
principal lateral veins diverging from the midvein at an angle of about
45 degrees; teeth sharp and divergent. Peduncles at anthesis usually
shorter than the leaves, exceeding them at maturity; pedicels of unequal
length, eventually forming a panicle-like inflorescence; petals commonly
5-7 mm long. April-June. — Hardwood and mixed forests, cedar and tamarack swamps, shores and forest edges.

In nearly all leaves, a line connecting the apices of the 2 uppermost lateral teeth passes across the projecting terminal tooth, which is usually more than half as wide as the lateral ones, measured from sinus to sinus.

Fragaria virginiana Duchesne
THICK-LEAVED WILD STRAWBERRY   *native / cons* **1**
Leaflets petioled, glabrate to sericeous beneath; principal lateral veins commonly divergent from the midrib at an angle of about 30 degrees; teeth blunter and less divergent than in *F. vesca.* Inflorescence with as many as 12 flowers on pedicels of about uniform length, forming a corymbiform cluster usually shorter than the leaves; petals usually 7-10 mm long. April-June. — In a diversity of deciduous, mixed, and coniferous forests, clearings, dry sandy forests, roadsides, and fields; more often in dry open sunny places than *F. vesca.*

Fragaria vesca

In nearly all leaves, a line connecting the tips of the 2 uppermost lateral teeth passes above the tip of the small terminal tooth, which is usually less than half as wide as the adjacent lateral ones, measured from sinus to sinus.

Fragaria virginiana

Geum
AVENS
Perennial herbs. Lower leaves pinnately lobed or divided, upper leaves smaller, less divided or entire. Flowers yellow, white or purple; 1 to many in clusters at ends of stems; petals 5; stamens 10 to many. Fruit an achene.

1 Leaves all 3-foliolate and basal . . . . *G. fragarioides*
1 Leaves mostly pinnately compound or divided, if 3-foliolate, then cauline 2
2 Calyx bell-shaped; reddish; flowers nodding; petals yellow, tinged with purple . . . . *G. rivale*
2 Calyx lobes spreading, green; flowers upright; petals white or yellow 3
3 Plants flowering 4
3 Plants fruiting 6
4 Petals white to pale yellow . . . . *G. canadense*
4 Petals bright yellow 5
5 Terminal leaflet of basal leaves much larger than lateral segments; lower portion of style with short-stalked glands . . . . *G. macrophyllum*
5 Terminal leaflet various; lower portion of style without glands . . . . *G. allepicum*
6 Receptacle glabrous or only sparsely hairy (remove a few achenes to check); the achene beak with short-stalked glands. . . . . *G. macrophyllum*
6 Receptacle densely hairy; achene beaks neither with glands nor the pedicels with dense long hairs 7
7 Stem leaves pinnately compound; achenes many (150 or more) in each head, the achene beak with long hairs at base . . . . *G. allepicum*
7 Stem leaves mostly 3-parted; achenes less than 100 in each head, the achene beak glabrous . . . .
. . . . *G. canadense*

Geum aleppicum Jacq.
YELLOW AVENS   *native / cons* **3**
Perennial herb. Stems erect or ascending, to 1 m long, branched above, covered with coarse hairs. Leaves variable, basal leaves pinnately divided into 5-7 oblong leaflets, wedge-shaped at base, petioles long-hairy; stem leaves divided into 3-5 segments, stalkless or short-petioled; margins coarsely toothed. Flowers 1 to several, short-stalked, on branches at ends of stems; sepals lance-shaped; petals 5, yellow; style jointed. Achenes usually long-hairy. June-July. — Swamps, wet forests, wet meadows, marshes, calcareous fens, ditches and roadsides.

Geum aleppicum

Geum canadense Jacq.
WHITE AVENS   *native / cons* 2

Perennial herb. Stems slender, 4–10 dm tall, glabrous or sparsely pubescent below, above and on the pedicels becoming densely velvety-puberulent, often with a few scattered longer hairs. Basal leaves long-petioled, commonly 3-foliolate with obovate leaflets; upper leaves short-petioled, 3-foliolate with oblong lance-shaped, sharply serrate leaflets; uppermost leaves mostly simple, lance-shaped, nearly sessile. Pedicels finely velvety hairy, with or without long scattered hairs; petals white, obovate, about as long as the sepals or distinctly exceeding them. Head of fruit obovoid, 10–15 mm long; receptacle densely bristly, the hairs protruding among the ovaries at anthesis but shorter than the mature achenes. Achenes 2.5–3.5 mm long, excluding the style. May–June. — Dry or moist woods.

Geum canadense

Geum fragarioides (Michx.) Smedmark
BARREN STRAWBERRY   *native / cons* 6
  Waldsteinia fragarioides (Michx.) Tratt.

Perennial rhizomatous herb with the aspect of a strawberry. Leaves basal, 3-foliolate, 1–2 dm long including the petioles, ± winter-green; leaflets broadly obovate, rounded at tip, serrate with numerous broad teeth and commonly also shallowly and irregularly lobed, the lateral leaflets unsymmetrical. Flowers in a cyme on a naked or bracted peduncle, the peduncles about equaling the leaves; hypanthium obconic; sepals triangular; petals yellow, obovate, 5–10 mm long, obtuse or rounded, much exceeding the sepals; stamens numerous, the slender filaments erect and persistent after anthesis. Fruit an achene. April–May. — Moist or dry woods, thickets, thin soil over rock outcrops.

Geum fragarioides

Geum macrophyllum Willd.
BIG-LEAF AVENS   *native / cons* 6

Perennial herb. Stems to 1 m long, unbranched, or branched above, bristly-hairy. Leaves pinnately divided, basal leaves stalked, the terminal segment large, 3–7-lobed, with much smaller segments intermixed; stem leaves smaller, deeply 3-lobed or divided into 3 leaflets, short-stalked or stalkless; margins sharply toothed. Flowers 1 to several on branches at ends of stems; sepals triangular, bent backward; petals yellow, obovate, 4–7 mm long; style jointed. Achenes finely hairy. May–July. — Moist to wet forest openings, streambanks, wet meadows.

Geum macrophyllum

Geum rivale L.
PURPLE AVENS   *native / cons* 8

Perennial herb. Stems erect, 3–8 dm long, mostly unbranched, hairy. Basal leaves large, 1–4 dm long, pinnately divided, the terminal 1–3 leaflets much larger than other segments; stem leaves smaller, 2–5 on stem, pinnately divided or 3-lobed; margins shallowly lobed and coarsely toothed. Flowers mostly nodding, few on pedicels at ends of stems, the pedicels with short gland-tipped hairs and longer coarse hairs; sepals 5, purple, triangular, 6–10 mm long, ascending; petals 5, yellow to pink with purple veins, tapered to a clawlike base; stamens many; styles jointed above middle, the portion above joint deciduous, lower portion persistent and curved in fruit. Fruit a long-beaked, hairy achene, 3–4 mm long, grouped into round heads. May–July. — Conifer swamps, wet forests, bogs, fens, wet meadows; often where calcium-rich.

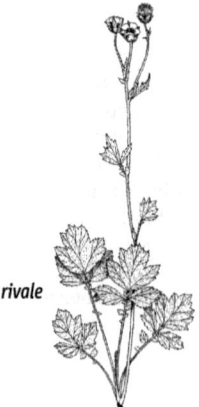

Geum rivale

Malus
APPLE

Trees (ours), sometimes thorny, with simple, alternate, toothed or lobed leaves and large flowers in simple umbels or umbel-like clusters on dwarf lateral branches (fruit-spurs). All bloom in April or May. Hypanthium globose to obovoid. Sepals 5, spreading or ascending or recurved. Petals 5, elliptic to obovate, short-clawed. Stamens 15-50, shorter than the petals. Ovary inferior, 3-5-celled (5-celled in our species); styles as many as the cells, separate or connate at base. Fruit a fleshy pome, each cell normally with 2 seeds.

Malus pumila P. Mill.
CULTIVATED APPLE   *introduced*
  Pyrus malus L.

Widely spreading tree to 15 m tall. Leaves elliptic to ovate, finely serrate, permanently pubescent beneath. Flowers white, tinged with pink, about 3 cm wide; hypanthium densely tomentose, open at the mouth; anthers yellow; calyx persistent on the fruit. — Native probably of w Asia; long in cultivation and occasionally escaped. Persistent from planted trees; also growing from seed in old fields and along fences and roads.

Malus pumila

Physocarpus
NINEBARK

Physocarpus opulifolius (L.) Maxim.
NINEBARK   *native / cons* 6

Much-branched shrub, 2-3 m long; twigs greenish, slightly angled or ridged, smooth or finely hairy; bark of older stems shredding in long thin strips. Leaves alternate, ovate in outline, mostly 3-lobed, dark green above, paler and often sparsely hairy below; margins irregularly toothed; petioles 1-2 cm long, with a pair of small, deciduous stipules at base. Flowers 5-parted, white, 5-10 mm wide; many in stalked, rounded clusters at ends of branches. Fruit a red-brown pod, 5-10 mm long, in round clusters; seeds 1-2 mm long, shiny, 3-4 in each pod. June-July. — Streambanks, lakeshores, swamps, rocky shores.

Physocarpus opulifolius

Potentilla
CINQUEFOIL

Annual or perennial herbs, or woody in shrubby cinquefoil (*P. fruticosa*); stolons present in some species. Leaves pinnately or palmately divided, alternate or mostly from base of plant. Flowers perfect, regular; sepals 5, alternating with small bracts, the sepals and bractlets joined at base to form a saucer-shaped hypanthium; petals 5, yellow; stamens many; pistils numerous. Fruit a group of many small achenes, surrounded by the persistent hypanthium.

1 Shrub; leaflets 5–7, 1–2 cm long	*P. fruticosa*
1 Herbs (or woody only at base)	2
2 Flowers solitary on naked pedicels from nodes of creeping stems	3
2 Flowers few to many in cymes	4
3 Leaves pinnately compound; leaf underside densely white-hairy	*P. anserina*
3 Leaves palmately compound; leaf underside coarsely hairy	*P. simplex*
4 Leaf undersides with long straight hairs or glabrous, but not woolly hairy	5
4 Leaf undersides woolly hairy	8
5 Main leaves below inflorescence 3-parted	6
5 Main leaves below inflorescence usually 5-parted or more	7
6 Petals and sepals about same length; stamens usually 20; achenes ridged	*P. norvegica*
6 Petals much shorter than the sepals; stamens 5–10; achenes smooth	*P. rivalis*
7 Plants erect, unbranched up to the inflorescence	*P. recta*
7 Plants often not erect, often much branched	*P. intermedia*
8 Leaves white-woolly on underside, the pubescence concealing the leaf surface	*P. argentea*
8 Leaves only thinly hairy on underside, the hairs mostly straight	*P. intermedia*

Potentilla anserina L.
SILVERWEED   *native / cons –*
  Argentina anserina (L.) Rydb.
Perennial herb, with a stout rootstock and spreading by stolons to 1 m
long. Leaves all at base of plant except for a few clustered leaves on
stolons, pinnately divided into 7-25 leaflets; leaflets oblong or obovate,
1.5-5 cm long and 0.5-2 cm wide, lower leaflets much smaller; upper surface green and smooth to gray-green and silky-hairy, underside densely
white-hairy; margins with sharp, forward-pointing teeth; stipules brown,
membranous, at base of petiole. Flowers single from leafy axils of stolons,
on stalks 5-15 cm long; sepals white silky-hairy; petals yellow, oval to
obovate, 5-10 mm long; stamens 20-25. Fruit a light brown achene. May-
Sept. — Wet meadows, marshes, sandy and gravelly shores and streambanks, Lake Michigan shoreline; soils often calcium-rich.

Potentilla anserina

Potentilla argentea L.
SILVERY CINQUEFOIL   *introduced*
Perennial herb, at first acaulescent, soon producing one or more long
stolons which root and have small clusters of leaves at the nodes. Leaves
erect, oblong lance-shaped in outline, to 3 dm long, pinnately compound
with numerous leaflets often alternating with others much smaller; axis
and peduncles villous; leaflets narrowly elliptic, to 4 cm long, sharply
toothed, tomentose beneath and also with long appressed hairs. Flowers
yellow, 15-25 mm wide, on naked peduncles, about as long as the leaves.
Achenes about 2.5 mm long, deeply furrowed on the summit and back.
May-Sept. — Wet sandy beaches.

Potentilla fruticosa L.
SHRUBBY CINQUEFOIL   *native / cons* 9
  Dasiphora fruticosa (L.) Rydb.
  Potentilla floribunda Pursh
Much-branched shrub, 0.5-1 m tall; twigs brown to red, covered with
long, silky- white hairs; bark of older branches shredding. Leaves alternate,
pinnately divided; leaflets 3-7 (mostly 5), the terminal 3 leaflets often
joined at base, oval to oblong, 1-2 cm long and 3-7 mm wide, tapered at
each end, upper surface dark green, underside paler, with silky hairs on
both sides or at least on underside; margins entire, often rolled under;
short-stalked. Flowers 5-parted, bright yellow, 1-2.5 cm wide, 1 to few in
clusters at ends of branches; bracts much narrower than the ovate sepals;
stamens 15-20. Fruit a small head of hairy achenes surrounded by the 10-
parted calyx. June-Sept. — Calcareous fens, lakeshores, open bogs, conifer
swamps, wet meadows.

Potentilla argentea

Potentilla intermedia L.
DOWNY CINQUEFOIL   *introduced*
Perennial from a stout root. Stems erect or decumbent, leafy, 3-7 dm tall.
Leaves digitate; leaflets of the principal leaves 5, oblong lance-shaped in
outline, commonly 3-5 cm long, deeply and often irregularly serrate above
the cuneate base, villous or sericeous beneath. Inflorescence much-
branched, many-flowered; flowers 8-10 mm wide; petals yellow, about
equaling the sepals. Achenes with prominent longitudinal ridges. — European; roadsides, waste places.

Potentilla fruticosa

Potentilla norvegica L.
STRAWBERRY-WEED   *native / cons* 0
Annual herb. Stems stout and leafy, commonly branched and many-flowered, hirsute below. Leaves 3-foliolate; leaflets elliptic to broadly obovate,
to 8 cm long, crenately toothed. Flowers yellow, nearly 1 cm wide; bractlets

and sepals ovate lance-shaped, about equal at anthesis, the sepals expanding in fruit to 16 mm long; petals nearly as long as the sepals; stamens usually 20. Achenes pale brown, flattened, about 1 mm long, with curved longitudinal ridges. June–Aug. — In a wide variety of moist or dry habitats, usually where somewhat disturbed: roadsides, railroads, fields, shores, meadows, rock outcrops, gardens.

Potentilla norvegica

Potentilla recta L.
SULPHUR CINQUEFOIL   *introduced*

Perennial herb. Stems erect, simple to the inflorescence, 4–8 dm tall, pubescent. Leaves digitately compound, the basal and lower long-petioled with 5–7 leaflets, the upper short-petioled to sessile, smaller, with only 3 leaflets; leaflets radially divergent, narrowly oblong lance-shaped, deeply toothed. Inflorescence many-flowered, flattened; sepals and bractlets ovate lance-shaped, about equal; petals yellow, about 1 cm long. Achenes striate with low curved ridges. June–Aug. — Native of Europe; weedy in dry soil, roadsides, fields, railroads, gravel pits; invading dry open forests.

Sessile or short-stalked glands are usually present on the leaflet underside.

Potentilla simplex Michx.
OLDFIELD CINQUEFOIL   *native / cons* 2

Perennial herb. Stems and basal leaves from a short rhizome to 8 cm long. Stems at first erect or ascending, soon widely spreading or arching and rooting at the tips, to 1 m long, very slender, with long internodes, villous to glabrate. Leaflets 5, oblong lance-shaped to elliptic, to 7 cm long, usually less than half as wide, with numerous teeth in the upper 2/3. Peduncles slender, the lowest arising at the end of the second well developed internode. Flowers yellow, 10–15 mm wide. April–June. — Dry open sandy forests, fields, roadsides, and sandy barrens; also in moist thickets and deciduous forests, and on rocky ledges.

Potentilla recta

Poterium
BURNET

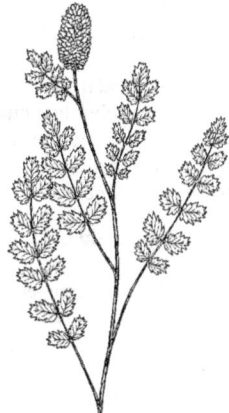

Poterium sanguisorba L.
SALAD-BURNET   *introduced*
   Sanguisorba minor Scop.

Perennial from a caudex-like rhizome. Stems 2–7 dm tall. Basal and lower leaves numerous, 1-pinnate with usually 7–17 leaflets; upper leaves progressively reduced; leaflets ovate, 5–20 mm long, with 3–7 deep sharp teeth on each side. Heads densely flowered, several on elongate peduncles, short-ovoid to globose, 8–20 mm long; flowers 4-merous, subtended by ciliate bracts, the lower staminate, the upper pistillate or perfect; hypanthium contracted at the mouth, more or less 4-angled, not prickly; sepals green or brown, 2.5–5 mm long; petals none; stamens numerous, the long filaments drooping; pistils 2. Fruit a pair of achenes, enclosed by the indurate hypanthium. May–June. — Native of Eurasia, established on roadsides, and waste places.

Poterium sanguisorba

Prunus
PLUM, CHERRY

Trees or shrubs. Leaves alternate, simple, serrate, often with petiolar glands. Flowers umbellate or solitary from axillary buds or short lateral branches, or racemose and terminal. Hypanthium cup-shaped, obconic, or urn-shaped. Sepals spreading or reflexed, usually soon deciduous. Petals 5, white to pink or red, elliptic to obovate, spreading. Stamens about 20, with slender exserted filaments. Pistil 1. Fruit a 1-seeded drupe, the exocarp fleshy or juicy, the endocarp hard.

Many members of the genus are important for their edible fruits or attractive flowers.

1 Flowers 20 or more in elongate racemes; inflorescence bracts absent 2
1 Flowers 1 to several, in umbel-like clusters 3
2 Tree; leaves 2 times longer than wide, the margins with incurved teeth; fruit black ...... *P. serotina*
2 Shrub; leaves less than 2 times longer than wide, the margins with sharp, outward pointing teeth; fruit
   dark red to purple .... *P. virginiana*
3 Plants in flower 4
3 Plants in fruit and with fully developed leaves 7
4 Sepals glabrous 5
4 Sepals hairy, at least on upper surface near base 6
5 Leaf margins mostly entire below middle; leaves widest above middle; flower pedicels mostly less
   than 1 cm long .... *P. pumila*
5 Leaf margins finely toothed for their entire length; leaves widest at or below middle; pedicels mostly
   more than 1 cm long .... *P. pensylvanica*
6 Sepals entire or with a few small teeth at tip .... *P. americana*
6 Sepal margins with gland-tipped teeth .... *P. nigra*
7 Leaves glabrous, obovate, distinctly widest above the middle; the margins entire or nearly so below
   the middle of the leaf .... *P. pumila*
7 Leaves not as above 8
8 Leaves finely toothed, the teeth much less than 1 mm long; leaves more than 2 times longer than
   wide, widest below middle .... *P. pensylvanica*
8 Leaves coarsely toothed, the teeth 1 mm or more long; leaves less than 2 times longer than wide,
   widest at or above the middle .... *P. americana*

Prunus americana Marsh.

WILD PLUM   native / *cons* 3

Shrub or small tree to 8 m tall, spreading from the roots and forming thickets. Leaves obovate, 6-10 cm long at maturity, sharply acuminate, coarsely serrate, glandless, 1-2 mm long. Flowers 2-4 in an umbel; petals white, 10-12 mm long; sepals pubescent on the upper (inner) side, often toothed toward the end, nearly glandless. Fruit red, glaucous, about 2 cm wide; stone flattened. April-May. — Moist woods, roadsides, and fencerows.

Prunus americana

Prunus pensylvanica L.f.

PIN-CHERRY   native / *cons* 4

Slender shrub or small tree to 10 m tall. Leaves mostly lance-shaped, 6-12 cm long, long-acuminate, finely and irregularly serrate with rounded teeth, the gland near the sinus; petioles usually glandular at the summit. Flowers in umbel-like clusters of 2-5, on pedicels 1-1.5 cm long; sepals glabrous; petals white, about 6 mm long, villous on the back near the base. Fruit red, juicy, about 6 mm wide; stone subglobose. April-May. — Dry or moist woods and forest clearings, often abundant after fires.

Prunus pumila L.

SAND-CHERRY   native / *cons* 9

Low, diffusely branched, decumbent or prostrate shrub, seldom more than 1 m tall, rarely to 3 m. Leaves oblong lance-shaped, 4-10 cm long, long-tapering at base, the margin firm or cartilaginous, finely and re-motely serrate with glandular teeth, glabrous, often glaucous beneath; petioles 5-12 mm long. Flowers in clusters of 2-4, on pedicels 4-12 mm long; sepals glandular-serrulate; petals white, elliptic, 4-8 mm long. Fruit nearly black, subglobose, 10-15 mm wide, edible. May. — Sand dunes and sandy soil, Lake Michigan shores, dry or rocky woods.

Prunus pensylvanica

Prunus serotina

Prunus serotina Ehrh.

WILD BLACK CHERRY   native / *cons* 3

Tree to 25 m tall, but often blooming when less than 5 m tall. Leaves firm, lance-shaped to oblong, 6-12 cm long, finely serrate with slender or blunt incurved teeth. Racemes terminating leafy twigs of the current season,

8-15 cm long; pedicels 3-6 mm long; sepals oblong or triangular, 1-1.5 mm long, entire or sparsely glandular-erose, persistent under the fruit; petals white, subrotund, about 4 mm long. Fruit dark purple or black, 8-10 mm wide, edible when fully ripe. May. — Formerly a forest tree, now more common as a weedy tree of roadsides, waste land, and forest margins.

Prunus virginiana L.
CHOKE-CHERRY   *native / cons* 3
Usually a shrub, sometimes a tree to 10 m tall. Leaves thin, oblong to obovate, 5-12 cm long, sharply serrate with slender ascending teeth. Racemes terminating leafy twigs of the season, 6-15 cm long; pedicels usually 5-8 mm long; sepals broadly triangular, 1-1.5 mm long, conspicuously glandular-erose, deciduous soon after anthesis; petals white, subrotund, about 4 mm long. Fruit dark red or crimson, 8-10 mm wide, astringent, scarcely edible. May. — In a wide variety of habitats, from rocky hills and dunes to borders of swamps.

Prunus virginiana

Rosa
ROSE
Shrubs or woody vines, usually thorny. Leaves pinnately compound with 3-11 serrate leaflets, the stipules commonly large and adnate to the petiole. Hypanthium globose to urceolate with a constricted orifice. Sepals usually long-attenuate, often persistent in fruit. Petals large, spreading at anthesis, white to yellow or red. Stamens very numerous, inserted near the orifice of the hypanthium. Ovaries numerous, inserted on the bottom or also on the sides of the hypanthium; styles usually barely exsert, distinct or united. Fruit a bony achene; mature hypanthium commonly colored, pulpy or fleshy.

1 Styles joined to form a column, protruding from the hypanthium opening . . . . *R. multiflora*
1 Styles distinct, styles not exserted from hypanthium opening, only the stigmas protruding 2
2 Sepals not all same shape and size, the outer sepals pinnately divided into lance-shaped segments; hypanthium opening small, 1 mm wide; styles short-exserted . . . . *R. rubiginosa*
2 Sepals entire; hypanthium opening 2–4 mm wide; styles not exserted 3
3 Young twigs densely hairy; petals 3–5 cm long; introduced species . . . . *R. rugosa*
3 Young twigs glabrous or nearly so; petals 2–3 cm long; native species 4
4 Pedicel and hypanthium with stalked glands; sepals spreading and then deciduous 5
4 Pedicel and hypanthium glabrous; sepals persistent on fruit and typically upright 6
5 Leaf margins with fine teeth; internodal prickles absent . . . . *R. palustris*
5 Leaf margins coarsely toothed; internodal prickles many . . . . *R. carolina*
6 Stems usually not prickly or bristly, or with slender prickles only on lower internodes . . . . . . *R. blanda*
6 Stems densely prickly on most internodes . . . . *R. acicularis*

Rosa acicularis Lindl.
BRISTLY ROSE   *native / cons* 6
   Rosa sayi Schwein.
Stems to 1 m tall, usually densely beset with straight slender thorns, even on the flowering lateral branches. Stipules pubescent, glandular on the margin, when young densely covered with short-stipitate glands, as are also the bracts; rachis usually pubescent and glandular; leaflets 3-7, usually 5, oblong-elliptic, coarsely serrate. Flowers usually solitary; hypanthium and pedicel glabrous; petals pink, 2-3 cm long Fruit ellipsoid to pyriform or globose. — Upland woods, hills, and rocky banks.

Rosa blanda Ait.
SMOOTH ROSE   *native / cons* 4
Stems to 1.5 m tall, unarmed or with few to many slender prickles toward the base, not extending upon the flowering branches. Stipules entire to glandular-dentate; leaflets commonly 5 or 7, narrowly oblong to oval,

Rosa acicularis

glabrous or nearly so, coarsely toothed, especially above the middle. Flowers solitary or corymbose; pedicel and hypanthium glabrous; petals pink, 2-3 cm long. — Dry woods, hills, dunes.

Thorns may be entirely absent, or may extend a variable distance up the stem.

Rosa carolina L.
PASTURE ROSE   *native / cons* **4**

Stems rarely more than 1 m tall, simple or little branched, usually armed with copious internodal thorns; thorns straight, slender and terete to the base, the infrastipular thorns scarcely differentiated. Stipules glandular-dentate to entire; leaflets 3, 5, or 7, oblong to oval or nearly rotund, often more than half as wide as long, coarsely toothed, in some forms glandular on the margin. Flowers usually solitary; pedicels and hypanthium stipitate-glandular; sepals attenuate into linear tips, or rarely with narrow foliaceous appendages; petals pink, 2-3 cm long. — Upland woods, dunes.

Rosa multiflora Thunb.
MULTIFLORA ROSE   *introduced (invasive)*

Stems climbing or scrambling. Stipules conspicuously serrate and glandular-ciliate; leaflets commonly 7-9, elliptic to obovate, obtuse or merely acute. Inflorescence many-flowered; sepals often prolonged into lance-shaped appendages; petals commonly white, 1-2 cm long. — Native of Asia; previously cultivated as a hedgerow plant; sometimes escaping as the seeds are readily spread by birds eating the fruit.

Rosa carolina

Rosa palustris Marsh.
SWAMP ROSE   *native / cons* **7**

Stems much-branched, prickly, to 2 m tall; twigs red-brown, smooth, with a pair of broad-based, downward-curved prickles at nodes; bristles between nodes absent. Leaves alternate, pinnately divided into usually 7 leaflets; leaflets oval or obovate, 2-6 cm long and 1-2 cm wide; underside midrib often soft-hairy; margins finely toothed; stipules narrow; petioles present. Flowers single at ends of leafy branches, or in small clusters of 2-5; petals pink, 2-3 cm long; flower stalks, sepals and hypanthium with stalked glands. Fruit more or less round, red-orange, 6-10 mm wide. July-Aug. — Open bogs, conifer swamps, thickets, shores and streambanks; increasing in disturbed wetlands.

Rosa palustris

Rosa rubiginosa L.
SWEETBRIER   *introduced*
   Rosa eglanteria L.

Stems 1-2 m tall, armed with stout decurved thorns. Leaflets 5-9, commonly 7, broadly elliptic, ovate, or subrotund, 1-3 cm long, mostly rounded at the base, beneath more or less pubescent and densely stipitate-glandular. Flowers pink, 3-5 cm wide; hypanthium smooth or sparsely glandular-bristly, not thickened around the very narrow orifice; some or all sepals pinnately lobed, glandular-ciliate; styles very pubescent, somewhat exsert. — Native of Europe; cultivated and escaped.

Rosa rugosa Thunb.
RUGOSA ROSE   *introduced*

Stems 1-2 m tall, densely thorny, the infrastipular thorns larger, decurved; younger parts of the stem, young thorns, and thorn bases densely pubescent. Leaflets usually 7-9, rugose above. Pedicel bristly and pubescent; hypanthium smooth.

— Native of e Asia; commonly cultivated and occasionally escaped; notable for its large "rose-hips" (fruit), rich in vitamin C.

Rosa rubiginosa

Rubus
BLACKBERRY, RASPBERRY, DEWBERRY

Perennials, woody at least at base, usually with bristly stems. Stems biennial in some species, the first year's canes called primocanes, the second year's growth termed floricanes. Leaves alternate, palmately lobed or divided. Flowers 5-parted, usually perfect, white to pink or rose-purple; stamens many. Fruit a group of small, 1-seeded drupes forming a berry.

1 Stems without bristles or prickles 2
1 Stems with bristles or prickles 3
2 Leaves simple .... *R. parviflorus*
2 Leaves with 3–5 leaflets .... *R. pubescens*
3 Leaves whitish or gray-hairy on underside; fruit separating easily from receptacle when ripe (raspberries) 4
3 Leaves green on both sides, underside veins hairy; fruit falling with receptacle when ripe (dewberries and blackberries) 5
4 Stems erect or spreading, with stiff straight bristles; fruit red .... *R. idaeus*
4 Stems arching, often rooting at tip, with broad-based, recurved prickles; fruit black .. *R. occidentalis*
5 Plants low and trailing (less than 0.5 m tall), often rooting at nodes; flowers 1 to several in a cluster; fruit red to red-purple (dewberries) .... *R. hispidus*
5 Plants tall, to 2 m; stems erect, neither rooting at nodes nor arching and rooting at tips; flowers numerous in elongate clusters; fruit black (blackberries) 6
6 Stems smooth or with scattered prickles; leaves glabrous .... *R. canadensis*
6 Stems glandular-hairy, with broad-based prickles or covered with spreading bristles; leaves hairy on underside veins .... *R. allegheniensis*

Rubus allegheniensis Porter
COMMON BLACKBERRY   *native / cons* **2**

Stems 0.5-3 m tall, mostly erect, the young primocanes often sparsely glandular. Primocane leaves usually 5-foliolate with the intermediate pair long-petiolulate, softly pubescent beneath; terminal leaflet usually 1-2 dm long, widest near the middle, sharply serrate; lateral leaflets smaller; armature of the stem of nearly straight spines spreading at right angles, much flattened at their base; armature of the petioles, pedicels, and lower side of the midveins commonly present and consisting of spines similarly flattened but prominently hooked. Inflorescence racemose, many-flowered, the lower 1, 2, or rarely 3 flowers subtended by leaves, the others by stipules only; pedicels glandular; flowers about 2 cm wide. — Forests and forest edges, clearings, old fields, roadsides; usually on dry uplands, occasional in marshy or swampy ground.

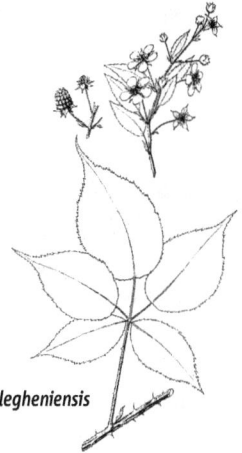

Rubus allegheniensis

Rubus elegantulus Blanch.
SMOOTH BLACKBERRY, DEWBERRY   *native / cons* **4**
   Rubus canadensis L.

Stems erect or nearly so, to 2 m tall, spineless, or occasionally beset with stout straight spines from expanded bases. Primocane leaves commonly 5-foliolate, typically glabrous beneath, rarely softly pubescent, the 3 central leaflets long-petiolulate, the lower pair subsessile to short-petiolulate; lateral leaflets lance-shaped to broadly oblong; terminal leaflet commonly 1-2 dm long, ovate, always conspicuously acuminate, sharply serrate. Floricane leaflets much smaller, the terminal one widest above the middle, always acute to short-acuminate, sometimes entire in the basal half. Inflorescence racemose, many-flowered, the flowers on glandless pedicels 2-4 cm long, subtended by stipules 1-2 cm long and serrate. — Woods, clearings, fields, roadsides; occasionally in moist soil.

Rubus elegantulus

Rubus hispidus L.
BRISTLY DEWBERRY *native / cons* **4**
Stems trailing or low-arching, often rooting at tip, with slender bristles or spines 2-5 mm long, these sometimes gland-tipped, not much widened at base. Leaflets 3 (rarely 5), ovate to obovate, 2-5 cm long and 1-3 cm wide, upper surface dark green and slightly glossy, slightly paler and more or less glabrous below, some leaves persisting through winter; margins with rounded teeth; petioles finely hairy and bristly; stipules linear, persistent. Flowers single in upper leaf axils or in open clusters of 2-8 at ends of short branches; sepals joined, the lobes ovate, tipped with a small dark gland; petals 5, white, 5-10 mm long. Fruit red-purple, less than 1 cm wide, sour, not easily separated from receptacle. June-Aug. — Conifer swamps, wet hardwood forests, thickets, wetland margins; usually where shaded.

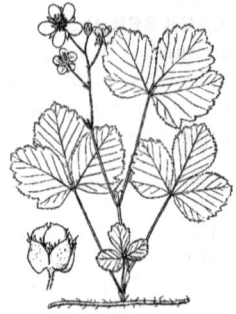

Rubus hispidus

Rubus idaeus L.
WILD RED RASPBERRY *native / cons* **3**
  Rubus strigosus Michx.
Stems erect or spreading, to 1.5 m long, biennial; young stems bristly with slender, often gland-tipped hairs; older stems brown, smooth. Primocane leaves divided into 3 or 5 leaflets, floricane leaflets usually 3; leaflets ovate to lance-shaped, upper surface dark green and smooth or sparsely hairy, underside gray-hairy; margins with sharp, forward-pointing teeth; petioles with bristly hairs; stipules slender, soon deciduous. Flowers in clusters of 2-5 at ends of stems and 1-2 from upper leaf axils; sepals with gland-tipped hairs; petals 5, white, shorter than the sepals. Fruit red, about 1 cm wide, edible, separating from receptacle when ripe. May-Aug. — Thickets, moist to wet openings, streambanks; often where disturbed. Our native plants sometimes considered a variety (*R. idaeus* var. *strigosus*) of the cultivated red raspberry (*R. idaeus* L.) from Europe.

Rubus idaeus

Rubus occidentalis L.
BLACK RASPBERRY *native / cons* **2**
Stems erect or ascending, or sometimes arching and rooting at the tip, not glandular, glaucous the first year, becoming glabrous the second, sparsely beset with stout, straight or hooked spines with expanded bases, as are also the petioles and especially the pedicels. Leaflets commonly 3, occasionally 5 on the primocanes ; uppermost leaves of the floricane often simple; terminal leaflet broadly ovate, rounded or subcordate at base, sharply and irregularly serrate; lower leaflets similar but smaller and narrower; all thinly gray-tomentose beneath. Flowers 3-7 in a dense umbel-like cluster; often 1 or 2 flowers also from the upper axils; petals white, shorter than the sepals, narrowly obovate, at first erect, soon deciduous. Fruit commonly black, rarely yellowish, about 1 cm wide. May-June. — Dry or moist woods, fields, and thickets. Often cultivated in many horticultural varieties.

Rubus occidentalis

Rubus parviflorus Nutt.
THIMBLEBERRY *native / cons* **7**
Stems unarmed, 1-2 in. tall, with shredding bark, the younger parts, petioles, and pedicels stipitate-glandular. Leaves rotund to kidney-shaped in outline, 1-2 dm wide, lobed to 1/3 of their width, the lobes serrate. Flowers few, white, in a long-peduncled cluster; sepals with a long caudate tip; petals elliptic-obovate, 1.5-2 cm long. Fruit edible, red, 1.5-2 cm wide, with pubescent coherent drupelets. May-July. — Open woods and thickets.
  Where abundant, a favorite for making jam.

Rubus parviflorus

Rubus pubescens Raf.
DWARF RASPBERRY   *native / cons 7*
Low perennial. Stems long-creeping at or near soil surface, with upright,
hairy branches 1-3 dm tall; the branches herbaceous but woody at base,
bristles absent; sterile branches arching to trailing, often rooting at
nodes; flowering branches erect, with few leaves. Leaves alternate, divided
into 3 leaflets; leaflets oval, 2-6 cm long and 1-4 cm wide, tapered to a
sharp point; margins with coarse, forward-pointing teeth, often entire
near base; petioles hairy. Flowers on glandular-hairy stalks, 1-3 in loose
clusters at ends of erect branches, sometimes with 1-2 flowers from leaf
axils; petals 5, white or pale pink, to 1 cm long. Fruit bright red, round, 5-
15 mm wide, the drupelets large, juicy, edible, not separating easily from
receptacle. May–July. — Conifer swamps, wet deciduous woods, rocky
shores.

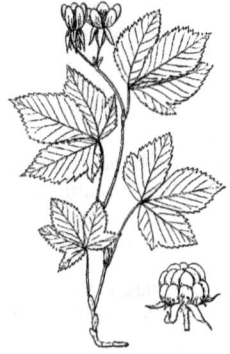

Rubus pubescens

Sibbaldia
FIVEFINGERS

Sibbaldia tridentata (Aiton) Paule & Soják
SHRUBBY-FIVEFINGERS   *native / cons 7*
  Potentilla tridentata Ait.
  Sibbaldiopsis tridenta (Aiton) Rydb.
Stems woody at base, from a caudex, often 2 or 3 together, 1-3 dm tall,
sparsely strigose. Leaves mostly near the base, digitately compound;
leaflets 3, firm in texture, oblong lance-shaped, 15-25 mm long, entire
near base, 3-toothed at the truncate tip, glabrous above, obscurely strigose
beneath. Flowers several in a flattened cyme, white, about 10 mm wide;
bractlets lance-shaped, somewhat shorter than the ovate-triangular
sepals; ovary and achenes villous. June–Aug. — Open sandy places, dry
savannas of jack pine and oak; rocky and gravelly shores, rock outcrops.

Sibbaldia tridentata

Sorbus
MOUNTAIN-ASH
Trees or shrubs. Leaves odd-pinnate with normally 11-17 serrate leaflets; flowers white, numerous,
in repeatedly branched, round or flattened clusters. Hypanthium obconic. Sepals triangular, as-
cending. Petals 5, obovate to orbicular, rounded or cuneate at base, spreading. Stamens 15-20.
Fruit a small pome, each cell with 1 or 2 elongate flattened seeds.

1 Leaflets tapered to a tip, 3–5 times longer than wide; petals obovate, to 4 mm long; fruit 5–6 mm
   wide . . . . *S. americana*
1 Leaflets rounded at tip or abruptly tapered to a tip, 2–3 times longer than wide; petals orbicular, 4–5
   mm long; fruit 8–10 mm wide 2
2 Leaflets glabrous, pale on underside; inflorescence branches and pedicels glabrous or nearly so . . . .
   . . . . *S. decora*
2 Leaflets soft-hairy on underside; inflorescence branches and pedicels with soft hairs . . . *S. aucuparia*

Sorbus americana Marsh.
AMERICAN MOUNTAIN-ASH   *native / cons 7*
  Pyrus americana (Marsh.) DC.
Shrub or tree to 10 m tall, the young twigs glabrous or nearly so. Winter
buds glutinous, with glabrous or sparsely ciliate scales. Leaflets lance-
shaped to narrowly oblong, long-acuminate, 5-9 cm long, sharply serrate,
paler and usually glabrous beneath. Inflorescence 6-15 cm wide; hypanthium and sepals glabrous; petals obovate, 3-4 mm long, conspicuously
longer than the stamens. Fruit bright red, 4-6 mm wide. May–June; fruit
in late summer. — In moist or wet soil; swamps (both cedar and deciduous), stream banks, forest borders.

Sorbus americana

Sorbus aucuparia L.
EUROPEAN MOUNTAIN-ASH, ROWAN   *introduced*
Pyrus aucuparia (L.) Gaertn.
Tree to 10 m tall, the young twigs more or less villous. Winter buds white-
villous, not glutinous. Leaflets oblong, 3-5 cm long, serrate, paler and
usually long-villous beneath, at least when young. Inflorescence 10-20
cm wide; hypanthium densely white-villous; petals orbicular, 4-5 mm
long, about equaling the stamens. Fruit bright red, about 10 mm wide.
May-June. — Native of Europe; planted for ornament and escaped into
moist woods.

Sorbus decora (Sarg.) Schneid.
NORTHERN MOUNTAIN-ASH   *native / cons* 8
Pyrus decora (Sarg.) Hyl.
Shrub or tree to 10 m tall, the young twigs glabrous or nearly so. Winter
buds glutinous, the principal scales glabrous on the back, the inner
usually conspicuously brown-ciliate. Leaflets oblong, 4-7 cm long, acute
or very shortly acuminate, sharply serrate, paler and glabrous or sparsely
pilose beneath. Inflorescence 6-15 cm wide; hypanthium glabrous or
sparsely pilose; petals orbicular, 4-5 mm long, about equaling the stamens.
Fruit bright red, 8-10 mm wide. May-June. — Moist or dry, often rocky
soil; wooded dunes and bluffs, forest margins.

Sorbus decora

Spiraea
MEADOWSWEET
Shrubs with simple leaves and terminal or lateral clusters of white, pink, or purple flowers.
Flowers 5-merous. Hypanthium cup-shaped or turbinate. Petals small, widely spreading. Stamens
15 to many. Pistils commonly 5, alternate with the sepals; styles terminal; ovules 2-several. Follicles firm in texture, dehiscent along the ventral suture. Most species are attractive flowering
shrubs.

1   Leaves glabrous on both sides; flowers white to pinkish .... *S. alba*
1   Leaf underside densely covered with light brown, woolly hairs; flowers rose-pink ..... *S. tomentosa*

Spiraea alba Du Roi
MEADOWSWEET   *native / cons* 4
Much-branched shrub, often forming colonies. Stems somewhat angled
or ridged, 0.5-1.5 m long, smooth or short-hairy when young, becoming
red-brown and smooth. Leaves alternate, often crowded on stems, oval
to oblong lance-shaped, 3-7 cm long and 1-2 cm wide, smooth on both
sides; margins with sharp, forward-pointing teeth; petioles 2-8 mm long;
stipules absent. Flowers small, 6-8 mm wide, many in a narrow, pyramid-shaped panicle 5-25 cm long at ends of branches; sepals 5; petals 5,
white. Fruit a group of 5-8 small follicles, each with several seeds; the
fruiting branches often persistent over winter. June-Aug. — Wet meadows,
streambanks, lakeshores, conifer swamps; soils often sandy.

Spiraea tomentosa L.
HARDHACK   *native / cons* 6
Sparsely branched shrub to 1 m tall. Young stems covered with brown
woolly hairs, becoming smooth and red-brown. Leaves alternate, lance-
shaped to ovate, 2-5 cm long and 0.5-2 cm wide; more or less smooth
above, underside gray-green to tan, densely covered with feltlike hairs,
the veins prominent; margins with coarse, forward-pointing teeth; petioles 1-4 mm long or absent. Flowers small, 3-4 mm wide, in spirelike panicles 5-15 cm long at ends of stems, the panicle branches covered with
reddish woolly hairs; petals 5, pink or rose (rarely white). Fruit a cluster
of small, hairy follicles, often persisting over winter. July-Sept. — Open

Spiraea tomentosa

bogs, conifer swamps, thickets, lakeshores, wet meadows; soils often sandy.

Rubiaceae

MADDER FAMILY

Herbs (ours). Leaves simple, opposite or whorled. Flowers small, perfect (with both staminate and pistillate parts), white to green, single or in loose or round clusters; petals joined, 3-4-lobed; stamens 3-4; ovary 2-chambered. Fruit a capsule (*Galium*), or berry (*Mitchella*).

1 Leaves whorled .... *Galium*
1 Leaves opposite .... *Mitchella*

Galium

BEDSTRAW

Annual or perennial herbs, from slender rhizomes. Stems 4-angled, ascending to reclining, smooth or bristly. Leaves entire, in whorls of 4-6. Flowers small, perfect, regular, 1 to several from leaf axils or in clusters at ends of stems; sepals absent; petals joined, 3-4-lobed, white; stamens 3-4; styles 2, ovary 2-chambered and 2-lobed, maturing as 2 dry, round fruit segments which separate when mature.

1 Fruit with bristly hairs 2
1 Fruit smooth or nearly so 5
2 Main leaves in whorls of 5 or more 3
2 Leaves in whorls of 4 or less 4
3 Annual herb; leaves in whorls of 7 or more; flowers white, blooming completed by early summer; stems very rough-to-touch .... *G. aparine*
3 Perennial herb; leaves in whorls of up to 6; flowers greenish, blooming beginning in early summer; stems rough or smooth .... *G. triflorum*
4 Leaves linear to linear lance-shaped, usually less than 5 mm wide; flowers white in a large panicle .. .. *G. boreale*
4 Leaves broader, often more than 5 mm wide; flowers greenish to purple, in few-flowered clusters .. .. *G. lanceolatum*
5 Leaves tipped with a short spine or at least sharp-pointed 6
5 Leaves rounded or blunt at tip 7
6 Leaves and stems with rough, downward-pointing hairs .... *G. asprellum*
6 Leaves and stems smooth or with short, upward-pointing hairs .... *G. mollugo*
7 Lobes of corolla 4, mostly longer than wide .... *G. labradoricum*
7 Lobes of corolla 3, mostly wider than long 8
8 Leaves in whorls of 4; flowers and fruit on long, curved, rough-hairy pedicels .... *G. trifidum*
8 Leaves usually in whorls of 5 or more; flowers and fruit on straight glabrous pedicels 9
9 Pedicels 0.5-4 mm long and often curved at maturity, solitary or in pairs in leaf axils or at ends of branches but not on a common peduncle; corolla less than 1 mm wide; mature fruit to 1 mm long; leaves mostly 2.5-7 mm long .... *G. brevipes*
9 Pedicels (at least the longest) 3-8 mm long and nearly always straight at maturity, often on a peduncle; corolla 1-1.8 mm wide; mature fruit 1-2 mm long; leaves mostly 5.5-14 (-22) mm long *G. tinctorium*

Galium aparine L.

STICKY-WILLY, CLEAVERS   *native / cons* **2**
Annual herb. Stems weak, prostrate or reclining on bushes, 3-10 dm long, with stiff, downward-pointing hairs. Leaves on the principal stems in whorls of 8, typically oblong lance-shaped, mostly 3-8 cm long, rounded to an apiculate tip, retrorsely hispid on the margins and midvein. Peduncles axillary, exceeding the subtending leaves, divaricately branched, few-flowered. Fruit with hooked bristly hairs, 2-4 mm long. May-June. — — Damp ground, usually in shade.

Galium aparine

Galium asprellum Michx.
ROUGH BEDSTRAW   *native / cons* 7
Perennial herb. Stems spreading or reclining on other plants, much-branched, to 2 m long, 4-angled, with rough, downward-pointing hairs on stem angles (which cling tightly to clothing). Leaves 6 in a whorl or 5-whorled on branches, narrowly oval, usually widest above middle, 1-2 cm long and 4-6 mm wide, tapered to a sharp tip; underside midvein and margins with rough hairs; petioles absent. Flowers in loose, few-flowered clusters at ends of stems and from upper leaf axils; corolla 4-lobed, white, 3 mm wide. Fruit smooth. July-Sept. — Swamps, streambanks, thickets, marshes, wet meadows, calcareous fens.

Galium asprellum

Galium boreale L.
NORTHERN BEDSTRAW   *native / cons* 5
Perennial herb. Stems erect, 2-8 dm long, 4-angled, smooth or with short hairs at leaf nodes, sometimes slightly rough-to-touch. Leaves in whorls of 4, linear to lance-shaped, 1.5-4 cm long and 3-8 mm wide, 3-nerved, tapered to a small rounded tip; margins sometimes fringed with hairs; petioles absent. Flowers many, 3-6 mm wide, in branched clusters at ends of stems; corolla lobes 4, white. Fruit with short, bristly hairs, or smooth when mature. June-Aug. — Streambanks, shores, thickets, swamps, moist meadows; also in drier woods and fields.

Galium brevipes Fern. & Weig.
LIMESTONE SWAMP BEDSTRAW   *native / cons* 10
   Galium trifidum subsp. *brevipes* (Fernald & Wiegand) A. Löve & D. Löve
Perennial herb. Stems scabrous, forming sprawling, tangled mats. Leaves whorled, 4 at each node. Flowers 1 per peduncle, the peduncles very short, to only 4 mm long. Fruit smooth, lacking bristles. The very small pedicels (usually ± recurved), fruits, corollas, and leaves, if all are present, are distinctive. July-Aug. — Marshes, thickets; exposed calcareous shores, interdunal hollows, ditches.

Galium boreale

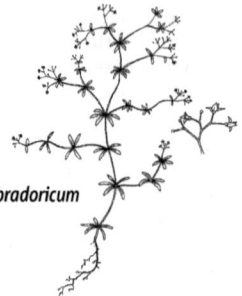

Galium labradoricum (Wieg.) Wieg.
NORTHERN BOG BEDSTRAW   *native / cons* 10
Perennial herb. Stems simple or branched, 1-3 dm long, 4-angled, hairy at leaf nodes, smooth on stem angles. Leaves in whorls of 4, soon curved downward, oblong lance-shaped, 1-1.5 cm long and 1-2 mm wide, blunt-tipped; underside midvein and margins with short, bristly hairs; petioles absent. Flowers single or in small groups on stalks from leaf axils; corolla lobes 4, white. Fruit smooth, dark. June-July. — Conifer swamps, sphagnum bogs, fens, sedge meadows.

Galium labradoricum

Galium lanceolatum Torr.
LANCE-LEAF WILD LICORICE   *native / cons* 7
   Galium circaezans Michx. var. *lanceolatum* (Torr.) Torr. & Gray
Perennial herb. Stems slender, branched from the base, erect or ascending, 3-7 dm tall, the stems glabrous or nearly so. Leaves in whorls of 4, thin, the lower elliptic, the upper lance-shaped, 3-8 cm long, 1-2.5 cm wide, long-tapering to an acute or acuminate apex, 3-5-nerved, minutely ciliate, smooth above, finely pubescent on the midvein and sometimes on the other veins beneath. Inflorescence widely divaricate, 1-3-forked; corolla glabrous, turning purple with age, its lobes acuminate. Fruit deflexed, uncinate-hispid, 3 rnm. long. June-July. — Dry woods and thickets.

Galium lanceolatum

Galium mollugo L.
FALSE BABY'S-BREATH   *introduced*
Erect perennial from a decumbent base. Stems 3-10 dm tall, smooth to finely pubescent. Leaves in whorls of 6 or 8, narrow, oblong lance-shaped,

10-20 (rarely 25) mm long, acute or apiculate, scabrous on the margin. Inflorescences several from the upper axils, forming a loose, open, elongate, divaricately branched panicle 1-3 dm long; corolla lobes white, acuminate. Fruit smooth, 1.5 mm long. May-July. — Meadows, fields, roadsides, and lawns.

Galium tinctorium (L.) Scop.
STIFF MARSH BEDSTRAW   *native / cons* 5
Galium trifidum subsp. *tinctorium* (L.) Hara
Perennial herb. Stems slender, weak, 4-angled, with rough hairs on angles. Leaves in whorls of 4 or sometimes 5-6, linear to oblong lance-shaped, 1-2.5 cm long, tapered to a narrow base, dark green and dull; underside midvein and margins with rough hairs; petioles absent. Flowers in clusters of 2-3, on slender, smooth, straight stalks at ends of stems; corolla lobes 3, white. Fruit smooth. July-Sept. — Conifer swamps, open bogs, fens, thickets, wet shores and marshes.

   Plants are similar to *G. trifidum* and are sometimes considered a variety of that species.

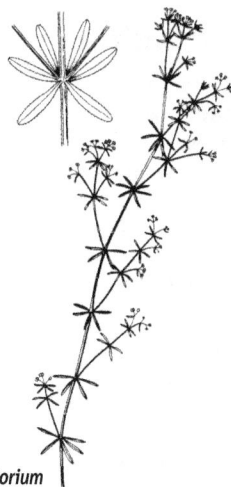

Galium tinctorium

Galium trifidum L.
NORTHERN THREE-LOBED BEDSTRAW   *native / cons* 6
Perennial herb. Stems slender, weak, 2-6 dm long, much-branched, sharply 4-angled, with rough, downward-pointing hairs on stem angles. Leaves in whorls of 4, linear to oblong lance-shaped, 5-20 mm long and 1-3 mm wide, blunt-tipped, dark green and dull on both sides; underside midvein and margins often rough-hairy; petioles absent. Flowers small, on 2-3 slender stalks from leaf axils or at ends of stems, the stalks much longer than the leaves; corolla lobes 3, white. Fruit dark, smooth. June-Sept. — Lakeshores, streambanks, swamps, marshes, bogs, springs.

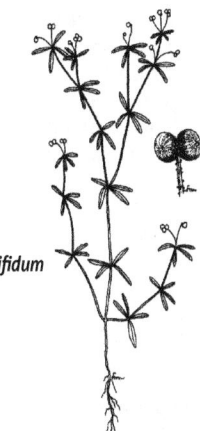

Galium trifidum

Galium triflorum Michx.
SWEET-SCENTED BEDSTRAW   *native / cons* 5
Perennial herb. Stems prostrate or scrambling, 2-8 dm long, 4-angled, smooth or with rough, downward-pointing hairs on stem angles. Leaves shiny, in whorls of 6 (or 4 on smaller branches), narrowly oval to oblong lance-shaped, 2-5 cm long and to 1 cm wide, 1-nerved, tipped with a short, sharp point, slightly vanilla-scented, underside midvein with rough hairs, margins with rough, forward-pointing hairs; petioles absent. Flowers 2-3 mm wide, on slender stalks from leaf axils and at ends of stems, the stalks with 3 flowers or branched into 3 short stalks, each with 1-3 flowers; corolla lobes 4, green-white. Fruit 2-lobed, covered with hooked bristles. June-Aug. — Moist to wet woods, hummocks in cedar swamps, wetland margins and shores, clearings.

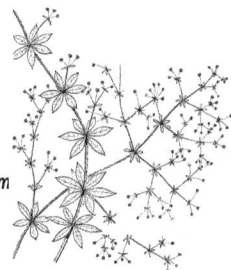

Galium triflorum

Mitchella
PARTRIDGE-BERRY

Mitchella repens L.
PARTRIDGE-BERRY   *native / cons* 6
Creeping perennial herb. Stems rooting at the nodes, 10-30 cm long, forming mats. Leaves evergreen, petioled, round-ovate, 1-2 cm long. Flowers 4-merous, dimorphic, in pairs, their hypanthia united, mostly terminal, the common peduncle shorter than the subtending leaves; corolla white, funnelform, 10-14 mm long, with elongate tube and 4 short, spreading or recurved lobes villous on the inner face; ovary 4-celled; stigmas 4. Fruit a scarlet berry, composed of the ripened hypanthia and ovaries of the 2 flowers, 5-8 mm wide, crowned with the short sepals, edible but insipid, persistent through the winter; seeds 8. May-July. — Dry or moist woods.

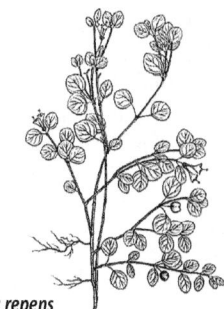

Mitchella repens

Rutaceae

RUE FAMILY

Mostly trees or shrubs with alternate, simple or compound leaves and small flowers. Flowers perfect or unisexual, usually regular. Stamens usually as many or 2x as many as the petals. Carpels commonly as many as the petals, in some genera fewer, separate, or weakly united (often by the styles only), or completely connate into a compound ovary. Fruit commonly separating into segments, in some genera a capsule, drupe, or berry. Most parts of the plant contain oil-glands; those of the leaves appear as translucent dots. The most important economic genus is *Citrus* L., including cultivated varieties of orange, grapefruit, lemon, lime, citron, and tangerine; the two Wisconsin species are the northernmost members of the family.

1 Leaflets 3 .... *Ptelea*
1 Leaflets 5–11 .... *Zanthoxylum*

Ptelea

HOP-TREE

Ptelea trifoliata L.

COMMON HOP-TREE   *native / cons* **6**

Deciduous shrub or small tree, without spines. Leaves alternate, 3-foliolate, long-petioled; leaflets sessile, ovate, elliptic, or ovate-oblong, entire or serrulate. Flowers small, greenish white or yellowish white, with staminate, pistillate, and perfect flowers on the same plant, and produced together in terminal cymes 4-8 cm wide; sepals, petals, and stamens 4 or 5, the latter imperfect or abortive in the pistillate flowers; petals oblong, pubescent, 4-7 mm long. Fruit a thin, flat, circular samara, 15-25 mm wide, the broad wing completely surrounding the indehiscent 2-celled body, reticulately veined, with the odor of hops. May-June. — Moist or rich woods and thickets.

Ptelea trifoliata

Zanthoxylum

PRICKLY ASH

Zanthoxylum americanum P. Mill.

PRICKLY ASH   *native / cons* **3**

Tall dioecious shrub or rarely a small tree to 8 m tall, foliage strongly aromatic. Stems thorny. Leaves alternate, odd-pinnately compound; leaflets 5-11, oblong to elliptic or ovate, crenate or entire, pubescent beneath, at least when young. Flowers greenish or whitish, in short-peduncled, sessile, axillary clusters on branches of the previous year; sepals none; petals 4 or 5, fringed at the tip; stamens 4 or 5, alternate with the petals; ovaries 3-5. Fruit (from each ovary) a firm-walled or somewhat fleshy follicle, about 5 mm long, the surface pitted, dehiscent across the top, with 1 or 2 seeds. April-May. — Moist woods and thickets.

Zanthoxylum americanum

Salicaceae

WILLOW FAMILY

Deciduous trees or shrubs. Leaves alternate, margins entire or toothed; stipules often present at base of leaf petiole, these usually soon falling. Flowers borne in catkins near ends of branches. Flowers imperfect, the staminate and pistillate flowers on separate plants, usually appearing before leaves open, or in a few species after leaves open; flowers without petals or sepals, each flower with either 1 or 2 enlarged basal glands (*Salix*) or a cup-shaped disk (*Populus*). Fruit a dry, many-seeded capsule; seeds small, covered with long, silky hairs.

1 Large trees; leaves heart-shaped to ovate, mostly less than 2 times longer than wide; buds often sticky and covered by 2 or more overlapping scales; catkins drooping, flowers subtended at base by a cup-shaped disk; stamens many, 12–80 . . . . ***Populus***
1 Shrubs and trees; leaves ovate, lance-shaped or linear, 2 or more times longer than wide; buds covered by 1 scale; catkins upright or drooping, flowers subtended by 1 or 2 enlarged glands; stamens 2–8 . . . . ***Salix***

Populus
ASPEN, POPLAR, COTTONWOOD

Trees with deciduous, ovate to triangular leaves. Flowers in drooping catkins that develop and mature before and with leaves in spring; staminate and pistillate flowers on separate trees; base of flower with a cup-shaped disk; stamens 10-80. Fruit a 2-4 chambered capsule with many small seeds, these covered with long, white hairs which aid in dispersal by the wind.

1 Leaf petioles round in section, leaf underside often stained brown from resin . . . . *P. balsamifera*
1 Leaf petioles strongly flattened, leaf underside not stained brown 2
2 Leaf underside and petioles densely woolly hairy . . . . *P. alba*
2 Leaf underside and petioles glabrous 3
3 Leaves strongly triangular in shape . . . . *P. deltoides*
3 Leaves ovate to nearly round 4
4 Leaf margins coarsely wavy-toothed; leaves 7–13 cm long . . . . *P. grandidentata*
4 Leaf margins finely sharp-toothed; leaves less than 7 cm long . . . . *P. tremuloides*

Populus alba L.
WHITE POPLAR   *introduced (invasive)*

Tree with widely spreading branches and whitish gray bark; terminal bud and young twigs tomentose. Leaves white-tomentose beneath, palmately 3-7-lobed on the elongate shoots at the end of the branches, on the short lateral shoots ovate, irregularly dentate. Pistillate catkins 4-6 cm long. Capsules narrowly ovoid. — Native of Eurasia, sometimes planted as a fast-growing tree but short -lived and spreads by root sprouts.

Populus alba

Populus balsamifera L.
BALSAM-POPLAR   *native / cons* **4**

Medium to large tree to 20 m or more tall; trunk 30-60 cm wide; crown open, somewhat narrow; bark smooth when young, becoming dark gray and furrowed; twigs red-brown when young, becoming gray; leaf buds fragrant, very resinous and sticky. Leaves resinous, ovate, 8-13 cm long and 4-7 cm wide, tapered to a long tip, rounded or somewhat heart-shaped at base, dark green and somewhat shiny above, white-green or silvery and often stained with rusty brown resin below; margins with small, rounded teeth; petioles round in section, 3-4 cm long. Catkins densely flowered, drooping, appearing before leaves; scales fringed with long hairs, early deciduous; pistillate catkins 10-13 cm long; pistillate flowers with 2 spreading stigmas; stamens 20-30. Capsules ovate, 6-8 mm long, crowded on short pedicels. April-May. — Swamps, floodplain forests, shores, streambanks, forest depressions, moist dunes.

Populus balsamifera

Populus deltoides Bartr.
PLAINS COTTONWOOD   *native / cons* **2**

Large tree to 30 m or more tall, with a large trunk (often 1 m or more wide) and a broad, rounded crown; bark gray to nearly black, deeply furrowed; twigs olive-brown to yellow, turning gray with age; leaf buds very resinous and sticky, shiny, covered by several tan bud scales. Leaves smooth, broadly triangular, 8-14 cm long and 6-12 cm wide, short-tapered to tip, heart-shaped or truncate at base; margins with forward-pointing, incurved teeth, 2-5 large glands usually present at base of blade near petiole; petioles strongly flattened, 3-10 cm long; stipules tiny, early de-

Populus deltoides

ciduous. Catkins loosely flowered, drooping, appearing before leaves; scales fringed, soon falling; flowers subtended by a cup-shaped disk 2-4 mm wide; pistillate catkins green, 7-12 cm long in flower, to 20 cm long in fruit; pistillate flowers with 3-4 spreading stigmas; staminate catkins dark red, soon deciduous; stamens 30-80. Capsules ovate, 6-12 mm long, on pedicels 3-10 mm long. April-May. — Floodplains, streambanks and bars, shores, wet meadows, ditches.

Populus grandidentata Michx.
BIG-TOOTH ASPEN   *native / cons* 3
Small or large tree; bark light greenish gray when young, becoming dark brown in age. Terminal buds dull brown, finely pubescent. Leaf blades broadly ovate in outline, 8-12 cm long, with 5-10 large, projecting, round-pointed teeth on each side, the lowest veins strongly ascending; petioles strongly flattened. Scales of the catkins shallowly cleft into 5-7 lance-shaped lobes; stamens 5-12; stigmas 4. Capsules slenderly conic, 3-5 mm long, on pedicels 1-2 mm long. — Dry or moist soil; common northward where it usually grows in drier soil than *P. tremuloides.*

Populus grandidentata

Populus tremuloides Michx.
QUAKING ASPEN   *native / cons* 2
Slender tree with light grayish green bark, becoming dark and furrowed in age; terminal buds brown, shining, glabrous or nearly so. Leaves broadly ovate to orbicular, 3-10 cm long, abruptly pointed, broadly cuneate (rarely) to truncate or subcordate at base, finely and regularly serrate or crenate to nearly entire; lowest lateral veins strongly ascending, the venation hence apparently palmate; petioles strongly flattened. Scales of the catkins cleft to below the middle into 3-5 lance-shaped lobes; stamens, stigmas, and capsules as in *P. grandidentata.* — Dry or moist soil, especially in cut-over land.

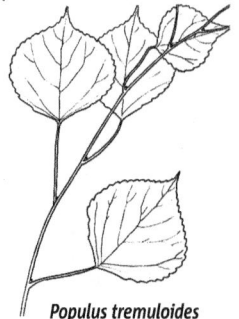

Populus tremuloides

Salix
WILLOW
Shrubs and trees. Leaves variable in shape, petioles glandular in some species; stipules early deciduous or persistent, sometimes absent. Catkins (aments) stalkless or on leafy branchlets, usually shed early in season. Staminate and pistillate flowers on separate plants; staminate flowers with mostly 2-3 stamens (to 8 in some species). Fruit a 2-chambered, stalked or stalkless capsule.

1 Leaves opposite or nearly so; young branches often dark purple .... *S. purpurea*
1 Leaves alternate; branches various colors 2
2 Leaf petioles with glands at or near base of blade 3
2 Petioles without glands 7
3 Trees, usually with a single trunk; leaves narrow 4
3 Small trees or shrubs, usually with several to many stems; leaves broader 5
4 Leaves often curved sideways (scythe-shaped), tapered to a long, slender tip; vigorous shoots with large stipules; native species .... *S. nigra*
4 Leaves not curved sideways, tapered to a short tip; stipules small, early deciduous; introduced .....
    .... *S. fragilis*
5 Leaves not waxy on underside .... *S. lucida*
5 Leaves waxy-coated on underside 6
6 Young leaves sparsely hairy; margins with small forward-pointing teeth; flowering in early summer
    .... *S. amygdaloides*
6 Young leaves without hairs; margins with small, gland-tipped, forward-pointing teeth; flowering summer or fall .... *S. serissima*
7 Mature leaves hairy, at least on underside 8
7 Mature leaves without hairs (sometimes hairy on petiole and midvein) 15
8 Leaves linear or narrowly lance-shaped 9

8 Leaves broadly lance-shaped, oblong, or ovate 11
9 Underside of leaves with felt-like covering of white tangled hairs; young twigs white-hairy; plant of peatlands, often where calcium-rich . . . . *S. candida*
9 Leaves not with felt-like hairs; twigs smooth or sparsely hairy 10
10 Leaf margins with widely spaced sharp teeth; petioles 1–5 mm long; colony-forming shrub of sandy banks . . . . *S. interior*
10 Leaf margins with small teeth at least above middle of blade; petioles 3–10 mm long; stems clustered but not forming large colonies . . . . *S. petiolaris*
11 Leaves rounded or heart-shaped at base; margins toothed; stipules present and persistent 12
11 Leaves tapered to base; margins entire or toothed; stipules usually falling early 13
12 Leaves oblong lance-shaped, tapered to a long tip; young leaves reddish . . . . *S. eriocephala*
12 Leaves obovate to oblong, tapered to a short tip; young leaves not reddish . . . . *S. cordata*
13 Small branches widely spreading; young leaves with white hairs; catkins appearing with leaves in spring; catkin bracts yellow or straw-colored; capsules on pedicels 2–5 mm long . . . . *S. bebbiana*
13 Small branches not widely spreading; young leaves with some red or copper-colored hairs; catkins appearing before leaves in spring; catkin bracts dark brown to black; capsules on pedicels 1–3 mm long 14
14 Leaf upperside smooth or the veins slightly raised, the underside sparsely hairy; twigs often shiny . . . . *S. discolor*
14 Leaf upperside somewhat wrinkled, the veins sunken, the leaf underside densely woolly hairy; twigs dull . . . . *S. humilis*
15 Leaves green on both sides or slightly paler on underside, not glaucous or white-hairy below 16
15 Leaves glaucous or white-hairy on underside 18
16 Leaves wider, ovate to oblong-ovate . . . . *S. cordata*
16 Leaves linear to linear lance-shaped 17
17 Many-stemmed, colony-forming shrub . . . . *S. interior*
17 Single-stemmed tree . . . . *S. nigra*
18 Leaf margins entire to shallowly lobed or with irregular teeth, sometimes revolute 19
18 Leaf margins distinctly toothed 21
19 Leaf margins entire and somewhat revolute . . . . *S. pedicellaris*
19 Leaf margins irregularly toothed, the teeth sharp or rounded 20
20 Leaves dull green above, wrinkled below; catkins appearing with leaves; bracts of pistillate catkins green-yellow to straw-colored . . . . *S. bebbiana*
20 Leaves dark green and shiny above; catkins appearing before leaves; bracts of pistillate catkins dark brown to black . . . . *S. discolor*
21 Leaves broadly elliptic, ovate, or obovate, rounded or somewhat abruptly short-tapered to tip . . . . *S. myricoides*
21 Leaves narrowly to broadly lance-shaped, long- or short-tapered to tip 22
22 Leaves more or less equally tapered from middle of blade to tip and base . . . . *S. petiolaris*
22 Leaves unequally tapered, the tip tapered to a point; base usually rounded or heart-shaped 23
23 Young twigs glabrous; stipules small or absent; bracts of pistillate catkins pale yellow and soon deciduous . . . . *S. amygdaloides*
23 Young twigs gray-hairy; stipules large; bracts of pistillate catkins dark brown to black, persistent . . . . *S. eriocephala*

Salix amygdaloides Anderss.
PEACH-LEAF WILLOW					native / cons 4

Shrub or tree to 15 m tall, often with several trunks; twigs gray-brown to light yellow, shiny and flexible. Leaves smooth, lance-shaped, long-tapered to tip, 5-12 cm long and 1-3 cm wide, yellow-green above, waxy-white below, margins finely toothed; petioles 5-20 mm long and often twisted; stipules small and early deciduous. Catkins appearing with leaves and loosely flowered; pistillate catkins 3-12 cm long, on leafy branches 1-4 cm long; catkin bracts deciduous, pale yellow, long hairy especially on inner surface; stamens 3-7 (usually 5). Capsules smooth, ovate, 3-7 mm long, on pedicels 1-3 mm long. May-June. — Floodplains, streambanks, lake and pond borders.

Salix amygdaloides

Salix bebbiana Sarg.

BEBB'S WILLOW, BEAKED WILLOW   *native / cons 7*

Shrub or small tree to 8 m tall, stems 1 to several; twigs yellow-brown to dark brown, usually with short hairs. Leaves oval to ovate or obovate, 4–8 cm long and 1-3 cm wide, dull gray-green, hairy or sometimes smooth on upper surface, waxy-gray, hairy and wrinkled below, the veins distinctly raised on lower surface; margins entire to shallowly toothed; petioles 5-15 mm long; stipules deciduous or persistent on vigorous shoots. Catkins appearing before leaves in spring; pistillate catkins loose, 2-6 cm long, on short leafy branches to 2 cm long; catkin bracts persistent, red-tipped when young, turning brown, long hairy; stamens 2. Capsules ovate, 5-8 mm long, finely hairy, on pedicels 2-6 mm long. May–June. — Swamps, thickets, wet meadows, streambanks, marsh borders.

Salix bebbiana

Salix candida Flueggé

SAGE WILLOW   *native / cons 10*

Low shrub to 1.5 m tall; twigs much-branched, covered with dense, matted white hairs. Leaves linear-oblong, tapered at tip, 4-10 cm long and 0.5-2 cm wide, dull, dark green and sparsely hairy above, veins sunken, densely white-hairy below; margins entire and rolled under; petioles 3-10 mm long; stipules persistent, 2-10 mm long, white-hairy. Catkins appearing with leaves in spring; pistillate catkins 1-5 cm long, on leafy branches 0.5-2 cm long; catkin bracts persistent, brown, hairy; stamens 2. Capsules ovate, 4-8 mm long, white-hairy, on pedicels to 1 mm long. May–June. — Fens, bogs, open swamps, streambanks, usually where calcium-rich.

Salix candida

Salix cordata Michx.

HEART-LEAF WILLOW   *endangered / native / cons 10*

Shrub 2-3 m tall, the vegetative parts all more or less hairy; twigs stoutish, the seasonal ones and buds densely gray-tomentose, the older less so. Leaves ovate lance-shaped to broadly ovate, 4-6 or 8 cm long, 1.5-3 or 4 cm wide, abruptly acuminate, glandular-dentate-serrate, often with stout spinulose teeth (extremely variable), rounded or cordate at base, green on both sides, more or less lanate, strongly nerved beneath; petioles stout, 4-8 mm long, somewhat clasping; stipules cordate-ovate, 6-15 mm long, dentate. Catkins appearing with leaves in spring, 5-8 cm long, on 3-5-leaved peduncles 1-2.5 cm long; catkin bracts brown, densely long-villous; stamens 2; filaments glabrous. Capsules lance-shaped, 5-8 mm long, glabrous; pedicels 0.5-1 mm long, glabrous. — Open sand dunes and sandy shores.

Salix cordata

Salix discolor Muhl.

PUSSY-WILLOW   *native / cons 2*

Shrub or small tree to 5 m tall; twigs yellow-brown to red-brown, dull, smooth with age or with patches of fine hairs. Leaves oval and short-tapered to tip, 3-10 cm long and 1-4 cm wide, dark green and smooth above, underside red-hairy when young, becoming white-waxy, smooth and not wrinkled; margins entire or with few rounded teeth; petioles without glands; stipules deciduous, or often persistent on vigorous shoots. Catkins appearing and maturing before leaves in spring; pistillate catkins 4-8 cm long, stalkless, sometimes with 2 or 3 small, brown, bractlike leaves at the base; stamens 2. Capsules ovate with a long neck, 6-10 mm long, densely gray-hairy, on pedicels 2-3 mm long. April–May. — Swamps, fens, streambanks, floodplains, marsh borders.

Salix discolor

Salix eriocephala Michx.

MISSOURI WILLOW   *native / cons* 4
   Salix cordata Muhl.
   Salix rigida Muhl.

Shrub or small tree to 6 m tall; twigs red-brown to dark brown, hairy when young. Leaves lance-shaped or oblong lance-shaped, 5-12 cm long and 1-3 cm wide, red-purple and hairy when young, upper surface becoming smooth and dark green, underside becoming pale-waxy; margins finely toothed; petioles without glands, 3-15 mm long; stipules persistent (especially on vigorous shoots), ovate or kidney-shaped, to 12 mm long, hairless, toothed. Catkins appearing with or slightly before leaves in spring; pistillate catkins 2-6 cm long, on short leafy branches to 1 cm long; catkin bracts persistent, brown to black, hairy; stamens 2. Capsules ovate with a long neck, 4-6 mm long, without hairs, on pedicels 1-2 mm long. April-May. — Shores, streambanks, floodplains, ditches and wet meadows.

Salix eriocephala

Salix × fragilis L.

CRACK WILLOW   *introduced*

Large tree to 20 m tall and 1 m diameter; twigs greenish to dark red, glabrous, very brittle at base and deciduous in strong winds. Leaves large, lance-shaped, 7-12 or 15 cm long, 2-3.5 cm wide, with 5-6 glandular serrations per cm of margin, dark green above, glaucescent to glaucous beneath, glabrous at maturity; petioles 7-15 mm long, glandular above at the outer end; stipules wanting or small, semicordate, and early deciduous. Catkins appearing with leaves in spring, lax, 4-8 cm long, on leafy peduncles 1-3 or 5 cm long, bearing 2-5 small leaves; catkin bracts greenish yellow, crisp-villous, deciduous. Capsules narrowly conic, 4-5.5 mm long, glabrous; pedicels 0.5-1 mm long. April-May. — Introduced to North America from Europe in colonial times for ornament, shade, and gunpowder charcoal; common in farmyards and pastures and sometimes escaped. Considered of hybrid origin from *S. alba* and *S. euxina*.

Salix × fragilis

Salix humilis Marsh.

UPLAND WILLOW   *native / cons* 6

Shrub 1-3 m tall; twigs yellowish to brown, pubescent to glabrate. Leaves oblong lance-shaped to narrowly obovate, 3-10 or 15 cm long, 1-2 or 3 cm wide, acute to abruptly short-acuminate, somewhat revolute, entire or sparingly undulate-crenate, dark green and often puberulent above; underside glaucous, somewhat rugose, and more or less gray-pubescent, becoming glabrate; stipules lance-shaped, acute, dentate, often deciduous. Catkins precocious, sessile or nearly so, oval-obovoid, 1.5-3 cm long, 1.5-2 cm wide; scales oblong lance-shaped, 1.5-2 mm long, blackish, long-villous; stamens 2; filaments long, free, glabrous. Capsules narrowly lance-shaped, 7-9 mm long, gray-pubescent; pedicels 1-2 mm long, pubescent. March-April. — Open woodlands, dry barrens.

Salix humilis

Salix interior Rowlee

SANDBAR WILLOW   *native / cons* 6
   Salix exigua Nutt. subsp. *interior* (Rowlee) Cronq.

Shrub to 4 m tall, spreading by rhizomes and often forming dense thickets; twigs yellow-orange to brown, smooth. Leaves linear to lance-shaped, tapered at tip and base, 5-14 cm long and 5-15 mm wide, green on both sides but paler below, at first hairy but soon usually smooth; margins with widely spaced, large teeth; petioles without glands, 1-5 mm long; stipules tiny or absent. Catkins appearing with leaves in spring on short leafy branches (and plants sometimes again flowering in summer); pistillate catkins loosely flowered, 2-8 cm long; catkin bracts deciduous, yel-

Salix interior

low; stamens 2. Capsules narrowly ovate, 5-8 mm long, hairy when young, smooth when mature, on pedicels to 2 mm long. May-June. — Shores, streambanks, sand and mud bars, ditches and other wet places; often colonizing exposed banks.

Salix lucida Muhl.
SHINING WILLOW   *native / cons* **5**

Shrub or small tree to 5 m tall; twigs yellow-brown or dark brown, smooth and shiny. Leaves lance-shaped to ovate, long-tapered and asymmetric at tip, 4-12 cm long and 1-4 cm wide, shiny green above, pale below, red-hairy when young, but soon smooth; margins with small, gland-tipped teeth; petioles with glands near base of leaf; stipules often persistent, strongly glandular. Catkins appearing with leaves in spring; pistillate catkins 2-5 cm long, on leafy branches 1-3 cm long; catkin bracts deciduous, yellow, sparsely hairy; stamens 3-6. Capsules ovate with a long neck, 4-7 mm long, not hairy, on short pedicels to 1 mm long. May. — Swamps, shores, wet meadows, moist sandy areas.

Salix lucida

Salix myricoides Muhl.
BLUELEAF WILLOW   *native / cons* **8**
  Salix glaucophylloides Fern.

Shrub to 4 m tall; twigs yellow to dark brown, hairy when young. Leaves thickened, lance-shaped to ovate or oval, 4-12 cm long and 1.5-5 cm wide, dark green above, strongly waxy-white below; margins with gland-tipped teeth; petioles 5-12 mm long; stipules 5-10 mm long. Catkins appearing shortly before or with leaves; pistillate catkins 2-8 cm long, on leafy branches 5-15 mm long; catkin bracts deciduous, 1-2 mm long, brown-black and long hairy; stamens 2. Capsules lance-shaped, 5-8 mm long, not hairy, on pedicels 1-3 mm long. May. — Dune hollows and sandy shorelines, fens; wet, calcium-rich sites.

Salix myricoides

Salix nigra Marsh.
BLACK WILLOW   *native / cons* **4**

Medium tree to 15 m tall, trunks 1 or several, crown rounded and open; bark dark brown, furrowed, becoming shaggy; twigs bright red-brown, often hairy when young. Leaves commonly drooping, linear lance-shaped, 6-15 cm long and 0.5-2 cm wide, long-tapered to an often curved tip, green on both sides but satiny above and paler below, lateral veins up-turned at tip to form a more or less continuous vein near leaf margin; margins finely toothed; petioles 3-8 mm long, hairy, usually glandular near base of blade; stipules to 12 mm long, heart-shaped, usually deciduous. Catkins appearing with leaves in spring; pistillate catkins 3-8 cm long, on leafy branches 1-3 cm long; stamens usually 6 (varying from 3-7); catkin bracts yellow, hairy, deciduous. Capsules ovate, 3-5 mm long, without hairs, on a short pedicel to 2 mm long. May. — Streambanks, lakeshores and wet depressions; not tolerant of shade.

Salix nigra

Salix pedicellaris Pursh
BOG WILLOW   *native / cons* **8**

Short, sparsely branched shrub 4-15 dm tall; twigs dark brown and smooth. Leaves oblong lance-shaped to obovate, tapered to tip or blunt and often with a short point, 3-6 cm long and 0.5-2 cm wide, silky hairy when young, soon hairless, green on upper surface, white-waxy below, veins slightly raised on both sides; margins entire, often slightly rolled under; petioles without glands, 2-8 mm long; stipules absent. Catkins appearing with leaves in spring; pistillate catkins 2-4 cm long, on leafy branches 1-3 cm long; catkin bracts persistent, yellow-brown, hairy on inner surface near tip; stamens 2. Capsules lance-shaped, 4-7 mm long,

Salix pedicellaris

without hairs, on pedicels 2-3 mm long. May-June. — Bogs, fens, sedge meadows, interdunal wetlands.

Salix petiolaris

Salix petiolaris Sm.
MEADOW WILLOW   *native / cons* **6**
Shrub to 5 m tall; twigs red-brown to dark brown, sometimes with short, matted hairs when young, smooth with age. Leaves narrowly lance-shaped, 4-10 cm long and 1-2.5 cm wide, hairy when young, becoming smooth, dark green above, white-waxy below; margins entire or with small, gland-tipped teeth; petioles without glands, 3-10 mm long; stipules absent. Catkins appearing with leaves in spring; pistillate catkins 1-4 cm long, stalkless or on short branches to 2 cm long; catkin bracts persistent, brown, with a few long, soft hairs; stamens 2. Capsules narrowly lance-shaped, 4-8 mm long, finely hairy, on pedicels 2-4 mm long. May. — Wet meadows, fens, streambanks, shores, open bogs, floating sedge mats, ditches. Common.

Salix purpurea

Salix purpurea L.
PURPLE WILLOW, BASKET WILLOW   *introduced*
Shrub to 2.5 m tall; twigs smooth, green-yellow to purple. Leaves more or less opposite (unique among our willows), smooth, linear to oblong lance-shaped, 4-9 cm long and 7-16 mm wide, purple-tinged, somewhat waxy below, veins raised and netlike on both sides; margins entire near base, irregularly toothed near tip; petioles short; stipules absent. Catkins appearing with and maturing before leaves in spring; pistillate catkins 2-3.5 cm long, stalkless; catkin bracts black; stamens 2 but often joined. Capsules ovate, 3-4 mm long, short-hairy, stalkless. May-June. — Introduced from Europe, occasionally escaping to lakeshores and streambanks; mostly near Lake Michigan.

Salix serissima (Bailey) Fern.
AUTUMN WILLOW   *native / cons* **8**
Shrub to 4 m tall; twigs gray, yellow or dark brown, shiny and smooth. Leaves smooth, oval to lance-shaped, 4-10 cm long and 1-3 cm wide, red and hairless when young; green and shiny above, usually white-waxy below; margins with small gland-tipped teeth; petioles with glands near base of leaf; stipules usually absent. Catkins appearing with or after leaves in spring; pistillate catkins 2-4 cm long, on leafy branches 1-4 cm long; catkin bracts deciduous, light yellow, long hairy; stamens 3-7. Capsules narrowly cone-shaped, 7-10 mm long, smooth, on pedicels to 2 mm long. Late May-July (our latest blooming willow). — Fens, cedar and tamarack swamps, marshes, floating sedge mats, streambanks and shores, often where calcium-rich.

Salix serissima

Santalaceae

SANDALWOOD FAMILY

Herbs (woody in *Arceuthobium*), usually root-parasites. Leaves simple, alternate or opposite. Flowers perfect or unisexual (plants usually monoecious or dioecious), in terminal or axillary clusters, or solitary. Hypanthium wholly or partly enclosing the ovary, with 3-several sepals on its margin and a conspicuous disc in its center. Stamens as many as the sepals, opposite them, inserted on their base. Ovary 1-celled. Fruit a nut or drupe, indehiscent, 1-seeded. *Santalum album* furnishes the fragrant sandalwood of the East Indies. Both *Comandra* and *Geocaulon,* though bearing green leaves, are hemiparasitic, and are apparently always attached (by means of modified roots, or haustoria) to some other plant. Both species also serve as alternate hosts for the canker-producing Comandra blister rust fungus (*Cronartium comandrae*), which in Wisconsin infects trees of jack pine.

260 SANTALACEAE (Sandalwood Family)

1 Plant an essentially leafless, non-green parasite on the branches of coniferous trees . . *Arceuthobium*
1 Plant leafy, green, terrestrial 2
2 Flowers green-purple, 2–3 from leaf axils; fruit a juicy orange to red drupe . . . . *Geocaulon*
2 Flowers white, numerous in a terminal inflorescence; fruit a dry green or yellowish drupe *Comandra*

Arceuthobium
DWARF-MISTLETOE
Arceuthobium pusillum Peck
EASTERN DWARF-MISTLETOE   native / cons 10
Woody, parasitic plants on conifers, dioecious, attached to trees by haustoria, lacking ordinary roots, but with chlorophyll. Stems short, to 2 cm
long, usually only 5-10 mm long, simple or with a few short branches,
greenish brown. Leaves opposite, scale-like, about 1 mm wide. Flowers
perfect or unisexual, regular, resembling short lateral branches until expanded, solitary or few in the axils of the leaves; perianth simple. Fruit
united with the receptacle, berry-like or drupe-like, about 2 mm long, on
a short recurved pedicel about equaling the subtending leaf. June-July.
— Chiefly on trees of black spruce, rarely on tamarack or white spruce;
reported on white pine. On spruce it often produces witches' brooms,
and may become so abundant as to endanger the host tree. Usually a
single host tree supports only one sex of the mistletoe.

Arceuthobium pusillum

Comandra
BASTARD TOADFLAX
Comandra umbellata (L.) Nutt.
BASTARD TOADFLAX   native / cons 6
Perennial herb; stems 1-3 dm tall, from a rhizome near the surface of the
soil. Leaves narrowly oblong to oval, 2-4 cm long, blunt or subacute, green
on both sides, the lateral veins obscure and (except the basal) scarcely
differentiated from the veinlets. Flowers white (rarely pinkish), bright
green at their base; cymules terminal or subterminal, usually forming a
flat-topped cluster; sepals oblong, 2-3 mm long. Fruit a dry or slightly
fleshy green or yellowish drupe 4-6 mm long. May-July. — Shores, upland
woods, and rock bluffs.

Comandra umbellata

Geocaulon
FALSE TOADFLAX
Geocaulon lividum (Richards.) Fern.
FALSE TOADFLAX   endangered / native / cons 10
   Comandra livida Richards.
Perennial herb, from a slender rhizome; at least partially parasitic on
other plants. Stems smooth, 1-3 dm long. Leaves alternate, oval or ovate,
1-3 cm long and 1-1.5 cm wide, rounded at tip; margins entire; petioles
short. Flowers greenish, usually 3 on slender stalks from leaf axils, the
lateral 2 flowers typically staminate, the middle flower perfect; sepals 4-
5, triangular, 1-2 mm long; petals absent. Fruit a round, orange or red
drupe, about 6 mm wide. June-Aug. — Cedar swamps, open bogs; more
commonly in sandy conifer woods and forested dune edges.

Geocaulon lividum

Sapindaceae
SOAPBERRY FAMILY
Soapberry Family now includes former members of Aceraceae.

Acer

MAPLE

Trees or shrubs. Leaves opposite, simple or compound. Staminate and pistillate flowers borne on same or separate plants. Flowers with 5 sepals and 5 petals (sometimes absent), clustered into a raceme or umbel). Fruit a samara with 2 winged achenes joined at base.

1 Leaves pinnately compound . . . . *A. negundo*
1 Leaves simple 2
2 Leaf sinuses between main leaf lobes sharp at their base 3
3 Leaves deeply lobed to middle of blade or below, the lobes long and narrow . . . . . . *A. saccharinum*
3 Leaf lobes shorter and wider 4
4 Leaves with downy white hairs on underside, tips of twigs with appressed hairs; shrubs or small trees; fruit persistent on plants until autumn . . . . *A. spicatum*
4 Leaves not downy-hairy on underside; twigs glabrous; medium trees; fruit shed in early summer . . . . *A. rubrum*
2 Leaf sinuses rounded at their base 5
5 Leaf undersides and petioles covered with downy hairs; margins of the lobes undulate or entire; blades drooping at their edges; base of petioles with stipules . . . . *A. nigrum*
5 Leaf undersides and petioles glabrous or nearly so; margins of the lobes coarsely toothed; blades not drooping at their edges; base of petioles without stipules 6
6 Leaf blades large, the margins finely doubly toothed; bark with vertical white stripes; small trees or shrubs . . . . *A. pensylvanicum*
6 Leaf blades smaller, the margins only coarsely toothed; bark not vertically white-striped; medium to large trees 7
7 Leaf petioles exuding milky juice when broken; twigs stout; samara wings widely divergent; bark becoming closely fissured, not scaly . . . . *A. platanoides*
7 Leaf petioles not exuding milky juice when broken; twigs slender; samara wings less divergent; bark becoming deeply furrowed and plate-like . . . . *A. saccharum*

Acer negundo L.

BOXELDER   *native / cons* 0

Tree to 20 m tall, the trunk soon dividing into widely spreading branches; bark brown, ridged when young, becoming deeply furrowed; twigs smooth, green and often with waxy-coated. Leaves opposite, compound, leaflets 3-7, oval to ovate, coarsely toothed or shallowly lobed, upper surface light green and smooth, underside pale green and smooth or hairy. Flowers either staminate or pistillate and on separate trees, appearing with leaves in spring; petals absent; staminate flowers in drooping, umbel-like clusters, pistillate flowers in drooping racemes. Fruit a paired samara 3-4.5 cm long. — Floodplain forests, streambanks, shores; also fencerows, drier woods and disturbed areas.

Distinguished from the ashes (*Fraxinus*) by its paired fruit (vs. single in ash) and its green or waxy twigs.

Acer negundo

Acer nigrum Michx. f.

BLACK MAPLE   *native / cons* 5

Acer saccharum var. *nigrum* (Michx. f.) Britt.

Resembling A. saccharum in habit and size, but with darker, thicker, more deeply furrowed bark. Leaves commonly with somewhat decurved or drooping sides, similarly lobed but the lobes with fewer and shorter, often obtuse or rounded teeth, darker and duller above than those of *Acer saccharum*. Samaras similar in shape, 2.7-4.5 cm, averaging 3.5 cm long. — In moist soil, often associated with *Acer saccharum*.

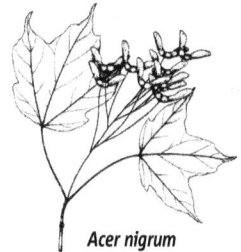

Acer nigrum

Acer pensylvanicum L.

STRIPED MAPLE   *native / cons* 6

Shrub or slender tree to 12 m tall. Leaves 3-lobed, glabrous on both sides at maturity, finely and sharply serrate, the teeth commonly 7-12 per centimeter. Flowers mostly polygamo-monoecious, borne singly on slender pedicels along a drooping axis, forming a slender, peduncled, terminal raceme 3-10 cm long, each raceme commonly either pistillate or staminate throughout; petals bright yellow, narrowly obovate, 5-8 mm long, scarcely surpassing the oblong-oblong lance-shaped sepals; stamens 6-8. Samaras 25-30 mm long, scarcely veined over the seed, the halves diverging at 90-120 degrees. May-June. — Moist woods.

Acer platanoides L.
NORWAY MAPLE   *introduced*

Tree with widely spreading crown. Leaves resembling those of Acer saccharum, with 5-7 sharply acuminate lobes and a few large teeth; juice milky (best seen at the base of a detached petiole). Flowers yellow, in erect rounded corymbs, the obovate petals 5-6 mm long, widely spreading. Samaras 35-45 mm long, scarcely distended over the seed, the halves divergent at an angle of about 180 degrees. Apr.-May. — Native of Europe; planted as a shade tree and established as a weedy tree in vacant lots.

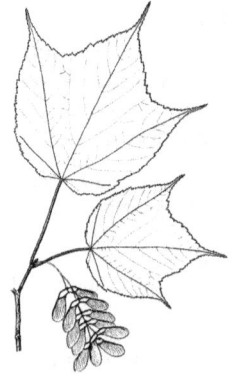

Acer pensylvanicum

Acer rubrum L.
RED MAPLE   *native / cons* 3

Tree to 25 m tall; bark gray and smooth when young, becoming darker and scaly; twigs smooth, reddish with pale lenticels. Leaves opposite, 3-5-lobed (but not lobed to middle of blade), coarsely doubly toothed or with a few small lobes, upper surface green and smooth, underside pale green to white, smooth or hairy. Flowers either staminate or pistillate, usually on different trees but sometimes on same tree, in dense clusters, opening before leaves in spring; sepals oblong, 1 mm long, petals narrower and slightly longer. Fruit a paired samara, 1-2.5 cm long. F— loodplain forests, swamps; also common in drier forests.

Distinguished from silver maple (*Acer saccharinum*) by its shallowly lobed leaves vs. the deeply lobed leaves of silver maple.

Acer rubrum

Acer saccharinum L.
SILVER MAPLE   *native / cons* 2

Tree to 30 m tall; bark gray or silvery when young, becoming scaly; twigs red-brown, smooth. Leaves opposite, deeply 5-lobed to below middle of blade, sharply toothed, upper surface pale green and smooth, underside silvery white; petioles usually red-tinged. Flowers either staminate or pistillate, usually on different trees but sometimes on same tree, in dense clusters, opening before leaves in spring. Fruit a paired samara, each fruit 3-5 cm long, falling in early to mid-summer. — Floodplain forests, swamps, streambanks, shores, low areas in moist forests.

Acer saccharum Marsh.
SUGAR MAPLE   *native / cons* 5

Tree to 40 m tall, with straight central trunk when growing in a forest and a widely spreading network of branches when in the open. Leaves about as wide as long, 3-5-lobed, the lobes usually bearing a few large sharp teeth. Flowers unisexual in umbels from the terminal or uppermost lateral buds, appearing as the leafbude open, drooping on slender pedicels up to 8 cm long; calyx gamosepalous, campanulate, 2.5-6 mm long, more or less hirsute; petals none. Samaras 2.5-4 cm, averaging 3 cm long, the seed-bearing portions diverging at right angles to the pedicel, the wings curved forward. Apr,-May. — Rich woods, especially in calcareous soils.

Acer saccharinum

Acer spicatum Lam.
MOUNTAIN MAPLE   *native / cons* 6

Shrub or small tree, occasionally 10 m tall. Leaves 3-lobed or obscurely 5-lobed, softly pubescent beneath, coarsely and irregularly serrate, the

Acer saccharum

teeth 2-3 per centimeter, each tipped with a minute sharp gland. Flowers mostly polygamo-monoecious, produced in fascicles of 2-4 along an erect axis, forming a slender, terminal, long-peduncled panicle, each flower long-pediceled, the terminal one of each fascicle usually perfect, the others sterile; petals greenish, very narrowly linear-oblong lance-shaped, about 3 mm long, much exceeding the sepals; stamens usually 8. Samaras 18-25 mm long, conspicuously reticulate-veined over the seed, the halves diverging at about a right angle. June. — Moist woods.

Acer spicatum

Sarraceniaceae

PITCHERPLANT FAMILY

Sarracenia
PITCHERPLANT

Sarracenia purpurea L.
PITCHERPLANT *native / cons* **10**

Perennial insectivorous herb. Flower stalks leafless, 3-6 dm long. Leaves clumped, hollow and vaselike, curved and upright from base of plant, 1-2 dm long and 1-5 cm wide, green or veined with red-purple, winged, smooth on outside, upper portion of inside with downward-pointing hairs, tapered to a short petiole at base. Flowers large and nodding, 5-6 cm wide, single at ends of stalks, perfect; sepals 5; petals 5, obovate, dark red-purple, curved inward over yellow style; ovary large and round. Fruit a 5-chambered capsule; seeds small and numerous. May-July. — Sphagnum bogs, floating bog mats, occasionally in calcium-rich wetlands.

Sarracenia purpurea

Saxifragaceae

SAXIFRAGE FAMILY

Perennial herbs. Leaves alternate, opposite or basal. Flowers perfect (with both staminate and pistillate parts), regular, single on stalks or in narrow heads. Sepals 5; petals 5; stamens 5 or 10, stigmas 2 or 4. Fruit mostly a 2-parted capsule.

1 Petals entire; flowers and fruit on slender pedicels mostly more than 5 mm long; scapes without any leaves .... *Tiarella*
1 Petals deeply fringed and comb-like; flowers and fruit on short thick pedicels less than 5 mm long; scape naked or with a single pair of leaves .... *Mitella*

Mitella
MITREWORT, BISHOP'S CAP

Perennial rhizomatous herbs. Leaves basal or alternate from the rhizome, the flowering stems leafless or few-leaved, bearing a terminal raceme of small white, greenish, or purple flowers. Flowers perfect, regular, 5-merous. Hypanthium turbinate to saucer-shaped, adnate to the base of the ovaries, bearing the sepals, petals, and stamens at its margin. Petals narrow, deeply pinnatifid or fimbriate to entire. Stamens 10 (in our species), shorter than the sepals. Pistils 2. Carpels short, dehiscent along the ventral suture.

1 Plants small, the scape naked; the basal leaves not lobed or only slightly so; flowers green-yellow .. .. *M. nuda*
1 Plants larger, with a pair of nearly sessile leaves on the scape below the inflorescence; the basal leaves clearly 3-lobed; flowers white .... *M. diphylla*

Mitella diphylla L.

TWO-LEAF MITREWORT *native / cons* **8**
Flowering stems 1-4 dm tall, sparsely pubescent below, glandular-puberulent above the stem leaves. Basal leaves long-petioled, ovate-rotund,
shallowly 3-5-lobed, crenate, cordate at base, pubescent. Stem leaves 2,
sessile, smaller, 3-lobed, the middle lobe elongate. Raceme 5-15 cm long;
pedicels 1-2 mm long. Flowers white, 5-6 mm wide; petals deeply fimbri-
ate-pinnatifid, about 2 mm long; seeds few, black, smooth, shining, 1-1.5
mm long. May-June. — Rich woods.

The follicles at dehiscence diverge and open widely, exposing the
shiny seeds.

Mitella diphylla

Mitella nuda L.

NAKED MITREWORT *native / cons* **9**
Small perennial herb, spreading by rhizomes or stolons. Leaves all from
base of plant, or with 1 small leaf on flower stalk, rounded heart-shaped,
1-3.5 cm wide, both sides with sparse coarse hairs; margins with rounded
teeth; petioles 2-8 cm long. Flowers small, green, on short stalks, in
racemes of 3-12 flowers, on a glandular-hairy stalk 10-25 cm tall; calyx
lobes 5, 1-2 mm long; petals green, pinnately divided into usually 4 pairs
of threadlike segments, the segments 2-4 mm long; stamens 10. Fruit a
capsule, splitting open to reveal the black, shiny, 1 mm long seeds. June-
July. — Hummocks in swamps and alder thickets, ravines, seeps, moist
mixed conifer and deciduous forests.

Mitella nuda

Tiarella
FOAM-FLOWER

Tiarella cordifolia L.

FOAM-FLOWER *endangered / native / cons* **10**
Perennial herbs with long stolons. Flowering stems 1-3.5 dm tall, glandular-puberulent. Leaves basal, broadly cordate-ovate, shallowly 3-5-lobed,
crenate, sparsely pubescent. Flowers in a raceme atop an erect, usually
leafless stem; the raceme at first short and crowded, elongating to 1 dm;
pedicels 5-10 mm long; hypanthium small, campanulate; sepals 2-3.5
mm long, blunt; petals white, elliptic, clawed, 3-5 mm long; stamens 10;
pistils 2, united at base, unequal in length. Follicles 2, thin-walled, the
larger about 10 mm long; seeds several, black, smooth, shining. May-
early June. — Rich, mesic hardwood forests.

Tiarella cordifolia

Scrophulariaceae

FIGWORT FAMILY

Annual, biennial or perennial herbs. Leaves mostly opposite or alternate (*Verbascum*). Flowers
single or few from leaf axils, or numerous in clusters at ends of stems or leaf axils, perfect (with
both staminate and pistillate parts), usually with a distinct upper and lower lip; sepals and
petals 4-5 (petals sometimes absent); stamens usually 4; pistil 2-chambered. Fruit a several- to
many-seeded capsule. Formerly a much larger family, now many of our genera segregated into
other families, especially Orobanchaceae and Plantaginaceae.

1 Stem leaves of fertile stems mostly alternate (a basal rosette may be present and the lower leaves
 may sometimes be opposite) . . . . *Verbascum*
1 Stem leaves of fertile stems all opposite or nearly so, sometimes alternate below the flowers . . . . . .
 . . . . *Scrophularia*

Scrophularia
FIGWORT
Perennial herbs. Stems 4-angled. Leaves opposite. Flowers 2-lipped, in open terminal clusters. Corolla brownish, bilaterally symmetrical and ± 2-lipped. Stamens 5, 4 plus a staminodium (sterile filament) under the upper corolla lobe.

1 Main leaves coarsely toothed or cleft, the teeth mostly long-tapered to their tip; leaf base tapered and extending downward along the petiole; sterile stamen green-yellow . . . . *S. lanceolata*
1 Main leaves evenly toothed, the teeth rounded at tip or abruptly tapered to a short point; leaf base rounded to heart-shaped, not extending downward along petiole; sterile stamen brown to purple . . . . *S. marilandica*

Scrophularia lanceolata Pursh
LANCE-LEAF FIGWORT   *native / cons* 4
Perennial herb, often forming dense colonies. Stems erect, to 2 m tall, glabrous, or minutely glandular in the inflorescence, the sides flat or shallowly grooved. Leaves ovate or ovate lance-shaped, 8-20 cm long, sharply serrate or incised or doubly serrate, truncate to broadly rounded at the base, glabrous beneath; petioles commonly 1.5-3 cm long, rarely as much as a third as long as the blade, narrowly margined to the base. Panicle 1-3 dm long, tending to be cylindric, rarely more than 8 cm wide; corolla 7-11 mm long, dull reddish brown except the yellowish green lower lobe. Capsule dull brown, 6-10 mm long. Late May-July. — Roadsides, railroads, old roads; forests, especially in clearings and edges; fields, fencerows, shores, swamp borders.

Scrophularia marilandica L.
CARPENTER'S-SQUARE   *native / cons* 4
Perennial herb, often forming colonies. Stems erect, to 3 m tall, glabrous below, becoming sparsely glandular in the inflorescence, the sides, especially in the older portions, roundly angled and prominently grooved. Leaves ovate or ovate lance-shaped, to 2.5 cm long, finely to coarsely serrate, broadly rounded to shallowly cordate at base; petioles slender, not margined, 1.5-5 cm long, those of the principal leaves commonly to half as long as the blades. Panicle loosely and irregular branched, tending to be pyramidal, often 10-15 cm wide; corolla 5-8 mm long, reddish brown. Capsule 4-7 mm long, somewhat shining. July-Aug, its flowering period rarely overlapping that of *S. lanceolata.* — Less common than *S. lanceolata* but in similar habitats: riverbank thickets and floodplains; open woods, especially in clearings and on margins; roadsides. Not always easily distinguished from *S. lanceolata,* though often taller and more branched.

Scrophularia lanceolata

Verbascum
MULLEIN
Biennial herbs (ours), producing a rosette of leaves the first year, from which the tall flowering stem rises the following season. Leaves alternate, entire, crenate, or rarely deeply toothed. Flowers yellow, white, or blue, in one to many spike-like racemes. Calyx regular, deeply 5-parted. Corolla rotate or saucer-shaped, nearly regular, the 3 lower lobes slightly larger than the 2 upper. Stamens 5, all fertile, more or less dimorphic. Capsule ovoid to globose, the 2 valves more or less cleft at the tip; seeds numerous, marked with longitudinal ridges.

1 Leaves not extending downward on stem (or only very shortly so) . . . . *V. phlomoides*
1 Leaves extending downward on stem to about next lower leaf . . . . *V. thapsus*

Verbascum phlomoides L.
ORANGE MULLEIN   *introduced*
Stems 5-12 dm tall. Leaves oblong to ovate lance-shaped, tomentose on both sides, sessile, only slightly decurrent on the stem. Raceme at first

dense and spike-like, at full maturity usually elongating and exposing the axis between the flower clusters; corolla as in *V. thapsus,* but commonly 25-35 mm wide; stamens as in *V. thapsus.* June-Sept.

Native of Europe; locally established as a weed, but never as common as *V. thapsus.*

Verbascum thapsus L.
GREAT MULLEIN   *introduced*
Plants usually densely gray-tomentose throughout. Stems stout and erect, 1-2 m tall. Lower leaves oblong or oblong lance-shaped, to 3 dm long, petioled; upper leaves progressively reduced, sessile, decurrent along the stem to the next leaf below. Raceme spike-like, very dense, 2-5 dm long, about 3 cm thick, usually solitary; corolla yellow, 12-22 mm wide; upper 3 filaments short, densely white-villous; lower 2 filaments much longer, glabrous or nearly so. June-Sept. — European weed of fields, roadsides, and disturbed places.

Verbascum thapsus

Simaroubaceae

QUASSIA-WOOD FAMILY

Ailanthus
TREE-OF-HEAVEN

Ailanthus altissima (P. Mill.) Swingle
TREE-OF-HEAVEN   *introduced (invasive)*
Rapidly growing weed tree; bark smooth and light gray, often becoming somewhat rougher with light tan fissures with age; branches few; twigs smooth to lightly pubescent, reddish or chestnut, with lenticels. Leaves odd-pinnate; leaflets 11-41, narrowly oblong, entire except with a few coarse glandular teeth at base. Dioecious, with staminate and pistillate flowers being borne on different individuals (sometimes flowers perfect); staminate trees produce 3-4x as many flowers as the pistillate trees, and are foul-smelling. Flowers greenish or greenish yellow, about 5 mm wide, in large terminal pyramidal panicles 1-3 dm long; 5-merous; stamens 10 in the staminate flowers, fewer in the perfect ones, usually absent in the pistillate. Samaras 3-5 cm long, 7-13 mm wide, twisted, with the seed near the center. — Native of e Asia; introduced as a shade tree and now established in vacant lots, roadsides, and occasionally in open woodlands.

Ailanthus altissima

Solanaceae

POTATO FAMILY

Herbs or shrubs, rarely climbing, or in the tropics small trees. Leaves alternate or appearing opposite. Flowers perfect, almost always 5-merous, regular (in most of our genera) or irregular. Calyx gamosepalous, persistent in fruit. Corolla rotate to funnelform or tubular. Stamens inserted on the corolla tube and alternate with its lobes, as many as the petals in most genera with regular flowers. Ovary commonly 2-celled (ours). Fruit a capsule or berry.

A large family, most numerous in tropical America, and with many plants, such as tomato, potato, eggplant, and peppers, of economic importance.

1 Plants woody near base, a sprawling or climbing vine; fruit a red berry . . . . *Solanum dulcamara*
1 Plants herbs; fruit various 2
2 Plants in flower 3
2 Plants in fruit 6

Leucophysalis
FALSE GROUND-CHERRY

Leucophysalis grandiflora (Hook.) Rydb.
LARGE FALSE GROUND-CHERRY   *native / cons* **6**

> *Chamaesaracha grandiflora* (Hook.) Fernald
> *Physalis grandiflora* Hook.

Annual herb; thinly villous and more or less viscid. Stems to 1 m tall. Leaves ovate or ovate lance-shaped, 5-12 cm long, acute or short-acuminate, entire, at base rounded or broadly cuneate and decurrent along the petiole. Flowers commonly 2-4 from the upper nodes, on pedicels 10-15 mm long; calyx lobes narrowly triangular, acuminate; corolla white with pale yellow center, rotate, 3-4 cm wide; filaments slender; anthers about 3 mm long; fruiting calyx round-ovoid, open at the end, about 15 mm long, nearly filled by the berry; berry 8-14 mm long, ovoid, beaked, puberulent. June-Aug. — Dry sandy soil.

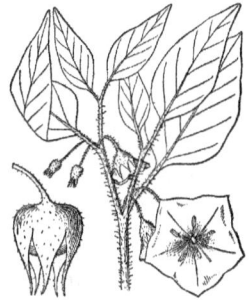

Leucophysalis grandiflora

Physalis
GROUND-CHERRY

Annual or perennial herbs, commonly widely branching. Leaves alternate or falsely opposite. Flowers solitary or few at the nodes, white, greenish yellow, or yellow, often with a darker center. Calyx at anthesis small, 5-lobed. Calyx tube enlarging promptly after anthesis, at maturity completely enclosing the berry or barely open at the summit, often greatly enlarged, commonly 5-angled, the calyx lobes scarcely enlarged. Corolla rotate to campanulate, shallowly lobed or entire. Stamens inserted near the base of the corolla. Berry many-seeded, pulpy.

1   Upper stems with soft, spreading hairs . . . . *P. heterophylla*
1   Upper stems with short, stiff, appressed hairs or glabrous . . . . *P. longifolia*

Physalis heterophylla Nees
CLAMMY GROUND-CHERRY   *native / cons* **3**

Perennial. Stems erect or spreading, often much branched; pubescence of the younger parts, pedicels, and calyx distinctly villous, composed of slender spreading hairs. Leaves ovate to rhombic, 3-8 cm long, shallowly and irregularly sinuate-dentate, varying to entire, at base broadly rounded or subcordate, often inequilateral, not decurrent, more or less pubescent on each side. Pedicels at anthesis about 1 cm long, to 3 cm in fruit; calyx lobes deltoid or ovate; corolla 15-20 mm long. Fruiting calyx ovoid, 3-4 cm long. June-Sept. — In dry or sandy soil.

Physalis heterophylla

Physalis longifolia Nutt.
LONGLEAF GROUND-CHERRY   *native / cons* **2**

Perennial. Stems erect, 4-8 dm tall, usually divergently branched, the younger parts nearly glabrous to densely puberulent, the hairs always ascending or appressed, rarely more than 0.5 mm long. Leaves thin, ovate to lance-shaped, entire to sinuate-dentate, glabrous or nearly so, long-petioled. Pedicels at anthesis 1-2 cm long; calyx minutely pubescent in longitudinal strips along the nerves; calyx lobes triangular or ovate, 3-4 mm long, densely ciliate; corolla about 15 mm long. Fruiting calyx ovoid or short cylindric, 3-4 cm long. July-Aug. — Moist or dry fields, open woods.

Physalis longifolia

Solanum
NIGHTSHADE
Herbs or vines. Corolla rotate or broadly campanulate, regular. Stamens 5. Fruit a many-seeded berry. Our species bloom in summer, often continuing into the fall.

ADDITIONAL SPECIES
Solanum physalifolium Rusby (Ground-cherry nightshade), introduced; see key.

1 Plants climbing or trailing vines, woody at base; flowers usually light blue or violet ... *S. dulcamara*
1 Plants upright herbs; flowers white 2
2 Plants sicky-hairy; calyx enlarging to cover lower half of berry .... *S. physalifolium* *
2 Plants glabrous to pubescent, the hairs appressed and not gland-tipped; calyx covering only bottom of berry .... *S. ptychanthum*

Solanum dulcamara L.
CLIMBING NIGHTSHADE   *introduced*
Perennial vine, climbing 2-4 m high, the stems somewhat woody at base. Leaves either simple or deeply lobed, the simple ones and the terminal segment of the lobed ones ovate, 4-10 cm long, entire, rounded or sub-cordate at base; lobes, when present, 1 or 2, basal, divergent, lance-shaped or ovate, much smaller than the terminal segment. Inflorescences arising from the internodes or opposite the leaves, peduncled, loosely branched, the pedicels jointed at base; corolla pale violet or blue, varying to white, about 1 cm wide. Berry red. — Native of Eurasia; moist thickets.

Solanum ptychanthum Dunal
WEST INDIAN NIGHTSHADE   *native / cons 1*
   Solanum nigrum L.
Annual. Stems erect, 3-6 dm tall, often widely branched. Leaves thin, long-petioled, ovate lance-shaped or ovate, cuneately narrowed to the base. Inflorescences lateral from the internodes, peduncled, umbel-like, 2-10-flowered, the pedicels not jointed at base; corolla white or very pale violet, 5-9 mm wide. Berries globose. — Both native and Eurasian forms present; a cosmopolitan weed.

Solanum ptychanthum

Staphyleaceae
BLADDERNUT FAMILY

Staphylea
BLADDERNUT

Staphylea trifolia L.
AMERICAN BLADDERNUT   *native / cons 7*
Erect shrub to 5 m tall, with striped bark. Leaves opposite, long-petioled, 3-foliolate; leaflets oblong to elliptic or ovate, 5-10 cm long at maturity, finely serrate, broadly acute to rounded at base, the terminal leaflet long-stalked, the lateral leaflets sessile or short-petioled and often oblique at base. Inflorescence a terminal drooping panicle 4-10 cm long including the peduncle. Flowers regular, perfect, 5-merous; sepals and petals distinct, imbricate, long-pediceled, greenish white, campanulate, 8-10 mm long;

Staphylea trifolia

sepals nearly as long as the erect petals; ovary 3-lobed. Fruit a 3-lobed, thin-walled, inflated capsule about 5 cm long. May. — Moist deciduous forests and thickets, especially on riverbanks and floodplains.

Thymelaeaceae

MEZEREUM FAMILY

Dirca
LEATHERWOOD

Dirca palustris L.

EASTERN LEATHERWOOD　　　　　　　　　　*native / cons* **9**

Freely branched shrub 1-2 m tall, bark very tough and pliable, twigs jointed. Leaves alternate, entire, obovate, 5-8 cm long, 1/2 to 2/3 as wide, usually rounded at base, glabrous at maturity, on petioles 2-5 mm long. Flowers perfect, regular, pale yellow, 7-10 mm long, subtended by hairy bud scales in early spring before the leaves appear, in lateral clusters of 2-4; hypanthium narrowly funnelform, the limb slightly spreading; stamens 8, protruding about 3 mm from the top of the tube; sepals minute; petals none. Fruit an ellipsoid drupe, about 8 mm long. Spring. — Rich moist deciduous woods.

Dirca palustris

Ulmaceae

ELM FAMILY

Ulmus
ELM

Trees. Leaves alternate, simple, short-petioled or subsessile, inequilateral; margins usually doubly serrate. Flowers perfect, in short racemes or, by abbreviation of the axis, in fascicles. Calyx campanulate, 4-9-lobed. Corolla none. Stamens as many as the calyx segments, exsert. Ovary compressed, 1-celled. Fruit a flat, 1-seeded samara, usually short-stipitate and often surmounted by the persistent or enlarged styles.

1　Leaf blades small, 3–7 cm long . . . . *U. pumila*
1　Leaf blades larger, 7–18 cm long . . . . *U. americana*

Ulmus americana L.

AMERICAN ELM　　　　　　　　　　*native / cons* **3**

Tree to 25 m tall, trunk to 1 m wide, crown broadly rounded or flat-topped, smaller branches usually drooping; bark gray, furrowed, breaking into thin plates with age; twigs brown, smooth or with sparse hairs, often zigzagged; buds red-brown. Leaves to 15 cm long and 7-8 cm wide, oval, pointed at tip, base strongly asymmetrical, upper surface dark green and smooth, lower surface pale and smooth or soft-hairy; margins coarsely double-toothed; petioles short, usually yellow. Flowers small, green-red, hairy, in drooping clusters of 3-4; appearing before leaves unfold in spring. Samaras 1-seeded, oval, 1 cm wide, with a winged, hairy margin, notched at tip. — Floodplain forests, streambanks and moist, rich woods; less common now than formerly due to losses from Dutch elm disease.

Ulmus americana

Ulmus pumila L.

SIBERIAN ELM　　　　　　　　　　*introduced (invasive)*

Tree 15 to 30 m; crowns open; bark gray to brown, deeply furrowed with interlacing ridges; branches not winged; twigs gray-brown, pubescent;

buds dark brown, ovoid, glabrous; scales light brown, shiny, glabrous to slightly pubescent. Leaves narrowly elliptic to lanceolate, to 6.5 cm long and 2-3.5 cm wide, base generally not oblique, margins singly serrate; upper surface with some pubescence in axils of veins, lower surface glabrous; petioles 2-4 mm long. Inflorescences tightly clustered fascicles of 6-15 sessile flowers; flowers and fruits not pendulous, sessile; calyx shallowly 4-5-lobed, glabrous; stamens 4-8; anthers brownish red; stigmas green. Samaras yellow-cream, orbiculate, 10-14 mm wide, broadly winged, glabrous, tip notched 1/3-1/2 its length. — Escaping from cultivation to waste places, roadsides, fencerows.

Distinguished from our other elms by its singly serrate leaf margins.

Ulmus pumila

Urticaceae

NETTLE FAMILY

Annual or perennial herbs with watery juice, sometimes with stinging hairs. Leaves alternate or opposite, simple, with petioles. Flowers small, green, in simple or branched clusters from leaf axils, staminate and pistillate flowers usually separate, on same or separate plants; sepals joined, 3-5-lobed; petals absent; ovary superior, 1-chambered. Fruit an achene, often enclosed by the sepals which enlarge after flowering.

1 Leaves alternate 2
1 Leaves opposite 3
2 Plants large and coarse, with stiff stinging hairs, the leaves sharply toothed . . . . *Laportea*
2 Plants smaller, without stinging hairs; leaves entire . . . . *Parietaria*
3 Plants with stinging hairs; leaves lance-shaped . . . . *Urtica*
3 Plants without stinging hairs; leaves ovate . . . . *Pilea*

Laportea
WOOD-NETTLE

Laportea canadensis (L.) Weddell
CANADIAN WOOD-NETTLE   native / cons 4
Perennial herb, spreading by rhizomes. Stems somewhat zigzagged, 5-10 dm long. Leaves alternate, 8-15 cm long, ovate and narrowed to a tip, with small stinging hairs, margins coarsely toothed. Flowers small, green, staminate and pistillate flowers separate but borne on same plant; staminate flowers in branched clusters from lower leaf axils, shorter than leaf petioles; pistillate flowers in open, spreading clusters from upper axils, usually much longer than petioles. Fruit a flattened achene, longer than the 2 persistent sepals. July-Sept. — Floodplain forests, rich moist woods, low places in hardwood forests, streambanks.

Differs from stinging nettle (*Urtica dioica*) by its shorter size, broader, alternate leaves, and the longer spike-like heads arising from upper leaf axils.

Laportea canadensis

Parietaria
PELLITORY

Parietaria pensylvanica Muhl.
PENNSYLVANIA PELLITORY   native / cons 2
Annual pubescent herb; monoecious or polygamous. Stems erect, 1-4 dm tall, simple or rarely branched. Leaves alternate, entire, thin, lance-shaped, 3-8 cm long, 3-nerved from above the cuneate base, slightly scabrellate above, with a slender petiole. Flowers green, from the middle and upper

axils, subtended and exceeded by narrow green bracts about 5 mm long; staminate flowers with deeply 4-parted calyx and 4 stamens; calyx of the pistillate flowers tubular at base, 4-lobed. Fruit a smooth, shining achene about 1 mm long, loosely enclosed by the expanded calyx. June-Sept. — Moist to dry forests, gravelly shores, disturbed sites.

Parietaria pensylvanica

Pilea
CLEARWEED

Annual herbs, sometimes forming colonies from seeds of previous year. Stems erect to sprawling, smooth, translucent and watery. Leaves opposite, stinging hairs absent, thin and translucent, ovate, with 3 major veins from base of leaf, margins toothed. Flowers green, staminate and pistillate flowers separate, borne on same or different plants, in clusters from leaf axils; staminate flowers with 4 sepals and 4 stamens; pistillate flowers with 3 sepals, ovary superior. Fruit a flattened, ovate achene.

1 Achenes 1–1.5 mm wide, olive-green to dark purple with a narrow pale margin, covered with low
   bumps . . . . *P. fontana*
1 Achenes to 1 mm wide, green to yellow, often marked with purple spots, smooth . . . . *P. pumila*

Pilea fontana (Lunell) Rydb.
LESSER CLEARWEED   *native / cons 7*
Annual herb. Stems 1-4 dm long, often sprawling. Leaves opposite, 2-6 cm long and 1-4 cm wide; petioles 0.5-5 cm long. Flowers in clusters, staminate flowers usually innermost when mixed with pistillate flowers. Fruit a dark olive-green to purple achene, 1-1.5 mm wide, with a narrow pale margin; sepals persistent, shorter to slightly longer than achene. Aug-Sept. — Lakeshores, riverbanks, swamps, marshes and springs.

Pilea pumila (L.) Gray
CANADIAN CLEARWEED   *native / cons 3*
Annual herb, similar to *P. fontana,* but sometimes taller (to 5 dm). Leaves opposite, usually larger (to 12 cm long and 8 cm wide), thinner and more translucent than in *P. fontana;* petioles to 8 cm long. July-Sept. — Swampy woods (often on logs), wooded streambanks, floodplain forests, wet depressions, rocky hollows; usually in partial shade.

Pilea pumila

Urtica
STINGING NETTLE

Urtica dioica L.
STINGING NETTLE   *native / cons 1*
Stout perennial herb, often forming dense patches from spreading rhizomes. Stems 8-20 dm tall, usually unbranched, with stinging hairs on stems and leaves, the hairs irritating to skin. Leaves opposite, ovate to lance-shaped, 5-15 cm long and 2-8 cm wide; margins coarsely toothed; petioles 1-6 cm long; stipules lance-shaped, to 15 mm long. Flowers small, green, staminate and pistillate flowers separate but mostly on same plants; flower clusters branched and spreading from leaf axils, the clusters usually longer than petioles, all of one sex or a mix of staminate and pistillate flowers, the pistillate clusters usually above the staminate clusters when both present on a plant. Fruit an ovate achene, 1-2 mm long, enclosed by the inner pair of sepals. July-Sept. — Moist woods, thickets, ditches, streambanks and disturbed areas.

Urtica dioica

Verbenaceae

VERBENA FAMILY

Verbena
VERVAIN

Annual or perennial herbs. Leaves usually opposite, simple, entire to somewhat lobed. Inflorescence of terminal spikes, usually densely many-flowered, often flat-topped, sometimes elongate with scattered flowers. Flowers solitary in the axil of a usually narrow bractlet. Calyx usually tubular, 5-angled, 5-ribbed, unequally 5-toothed, not at all or but slightly changed in fruit. Corolla funnelform, the limb weakly 2-lipped, 5-lobed. Stamens 4, inserted in the upper half of the corolla tube, included. Ovary mostly 4-lobed. Fruit mostly enclosed by the mature calyx, separating at maturity into 4 linear-oblong nutlets.

1 Plants densely gray-hairy; leaves sessile . . . . *V. stricta*
1 Plants not densely hairy; leaves with petioles . . . . *V. hastata*

Verbena hastata L.
COMMON VERVAIN   *native / cons* 3

Perennial herb; plants with short, rough hairs. Stems stout, erect, 4–12 dm tall, 4-angled, sometimes branched above. Leaves opposite, lance-shaped to oblong lance-shaped, 4–12 cm long and 1–5 cm wide; margins with coarse, forward-pointing teeth and sometimes lobed near base; petioles short. Flowers small, numerous, slightly irregular, in long, narrow spikes 5–15 cm long at ends of stems, the spikes elongating as flowers open upward from base; calyx unequally 5-toothed, 1–3 mm long; corolla dark blue to purple, 5-lobed, trumpet-shaped, slightly 2-lipped, 2–4 mm wide. Fruit 4-angled, splitting into 4 nutlets. July–Sept. — Marshes, wet meadows, shores, streambanks, openings in swamps, ditches.

Verbena hastata

Verbena stricta Vent.
HOARY VERVAIN   *native / cons* 3

Perennial. Stems 2–12 dm tall, subterete, simple or branched above, rather densely pale-pubescent or hirsute. Leaves ovate, elliptic, or suborbicular, 3–10 cm long, sessile or nearly so, thick-textured, sharply serrate, or incised, hirsute and rugose above, densely hirsute and prominently veined beneath. Flowers in 1 or several spikes, usually quite densely compact at anthesis and in fruit; bractlets lance-shaped, about as long as the calyx, hirsute, ciliate; calyx 4–5 mm long, densely hirsute; corolla deep blue or purple, or white, its tube protruding slightly beyond the calyx, pubescent outside, its limb 8–9 mm wide. Nutlets ellipsoid, about 2.5 mm long, raised-reticulate above, striate below. June–Sept. — Barrens, fields, and roadsides.

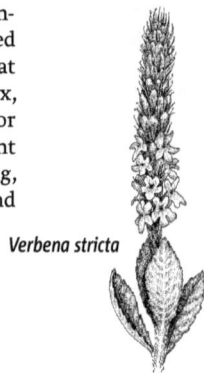

Verbena stricta

Violaceae

VIOLET FAMILY

Viola
VIOLET

Perennial herbs, with or without leafy stems. Leaves all at base of plant or alternate on stems; petioles with membranous stipules. Flowers perfect, nodding, and single at ends of stems, with 5 unequal sepals, 2 upper petals, 2 lateral, bearded petals, and 1 lower petal prolonged into a nectar-holding spur at its base. Fruit an ovate capsule which splits to eject the numerous seeds.

1 Plants with stems; leaves and flowers borne on the upright stems 2
1 Plants without stems; leaves and flowers borne directly from rootstock 7
2 Corolla solid yellow, or white with a yellow center; stipules entire or jagged-tooth on margins 3
2 Corolla creamy-white to yellow-orange, or lavender to blue, with or without a yellow center; stipules fringed or deeply lobed 4
3 Corolla yellow; stipules ovate, widened above base before tapering to tip .... *V. pubescens*
3 Corolla white with yellow center; stipules long-tapered from base to tip .... *V. canadensis*
4 Stipules deeply lobed near base into long oblong segments .... *V. arvensis*
4 Stipules fringed with short, slender segments 5
5 Leaves narrowly ovate to triangular, tapered to a rounded tip, often densely pubescent, the hairs tiny; margins entire or nearly so; corolla dark blue .... *V. adunca*
5 Leaves ovate to kidney-shaped, glabrous or only slightly hairy; the margins with rounded or sharp teeth; corolla creamy-white, lavender or light blue 6
6 Corolla solid light blue; lateral petals bearded on inner surface; stem leaves broadly ovate or kidney-shaped; margins with low, rounded teeth .... *V. labradorica*
6 Corolla light blue to lavender with a darker purple spot; lateral petals not bearded; stem leaves ovate to oblong-ovate; margins with scattered sharp teeth .... *V. rostrata*
7 Style tipped by a slender, recurved hook; stolons green and cord-like; introduced species mostly of lawns and parks .... *V. odorata*
7 Style slightly expanded into a spatula-like tip, not hooked; stolons slender and pale or absent; native species 8
8 Corolla blue; petals not bearded within; spur more than 2 times longer than wide .... *V. selkirkii*
8 Corolla white or purple; petals sometimes bearded within; spur less than 2 times longer than wide 9
9 Corolla white 10
9 Corolla purple; lateral petals bearded within; stolons absent 12
10 Leaves dull, upper and lower surface without hairs, lower surface not paler than upper; margins nearly entire or with low rounded teeth; petioles often with long, soft hairs .... *V. macloskeyi*
10 Leaves shiny and smooth on upper surface, or dull and hairy on either upper or lower surface; underside paler than upper surface; margins with sharp teeth 11
11 Plants with stolons and horizontal rhizomes; upper and lower surface of leaves sparsely to densely hairy with short hairs less than 1 mm long .... *V. blanda*
11 Plants without stolons, rhizomes turned upright; leaves often shiny and smooth on upper surface, or densely hairy on upperside with hairs about 1–2 mm long and smooth on underside .... *V. renifolia*
12 Leaf blades lobed or divided .... *V. sagittata*
12 Leaf blades toothed on margin, not lobed or divided 13
13 Leaf blades distinctly longer than wide 14
13 Leaf blades as wide as or wider than long 16
14 Plants of wet places; leaves glabrous or nearly so; sepal margins not fringed with hairs 15
14 Plants of dry or rocky habitats; leaves sparsely to densely hairy; sepal margins usually fringed with hairs .... *V. sagittata*
15 Lateral petals with long, threadlike hairs on inner surface; spurred petal densely hairy within ......
.... *V. affinis*
15 Lateral petals with short, knob-tipped hairs on inner surface; spurred petal without hairs *V. cucullata*
16 Sepals long-tapered to a sharp tip; lateral petals with short, knob-tipped hairs on inner surface; spurred petal without hairs .... *V. cucullata*
16 Sepals oblong to broadly lance-shaped, rounded at tip; lateral petals with long, threadlike hairs on inner surface 17
17 Flowers held above the leaves; leaves and stems without hairs, leaves rounded at tip, margins with rounded teeth; the spurred petal densely hairy within; plants of wetlands .... *V. nephrophylla*
17 Flowers overtopped by leaves; leaves and stems usually hairy, leaves tapered to a pointed tip, margins with sharp, forward-pointing teeth; spurred petal glabrous to only slightly hairy within; plants of moist forests .... *V. sororia*

Viola adunca Sm.
HOOK-SPURRED VIOLET   *native / cons* **6**
Perennial herb. Stems several to many, 2-8 cm tall at anthesis, at first erect, later becoming prostrate and spreading and to 15 cm long. Leaves ovate to suborbicular, obtuse, crenulate, subcordate at base; stipules linear lance-shaped with fimbriate teeth. Peduncles long and slender; sepals

narrowly lance-shaped; petals violet; spur 4-6 mm long, either straight
and blunt or tapering to a sharp incurved point. Capsules 4-5 mm long,
ellipsoid; seeds dark brown. — Dry sandy open places, often with jack
pine and oaks; crevices in rock outcrops.

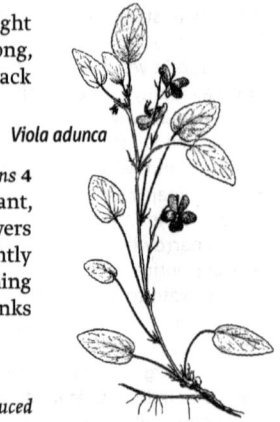

Viola adunca

Viola affinis Le Conte
SAND VIOLET   *native / cons* 4

Perennial herb, spreading by rhizomes. Leaves all from base of plant,
hairless, narrowly heart-shaped; margins with rounded teeth. Flowers
violet, bearded within with long, threadlike hairs, atop stalks slightly
longer than leaves. Fruit a purple-flecked capsule on horizontal or arching
stalks, seeds dark. April-May. — Swamps, floodplain forests, streambanks
and lakeshores.

Viola arvensis Murr.
EUROPEAN FIELD-PANSY   *introduced*

Pubescent to glabrate annual. Stems often branched from the base, to 30
cm tall, with reflexed hairs on the angles. Leaves variable, the lowest
ones with orbicular to ovate blades, the upper oblong to elliptic or narrowly elliptic, blunt or acutish, all distantly crenate. Flowers 1-1.5 cm
long, 0.8 cm wide; petals shorter than the broadly lance-shaped sepals,
all pale yellow or occasionally with purplish-tinged tips. Capsules globose,
5-7 mm long; seeds brown. — Native of Europe; mostly in cultivated or
abandoned fields.

Viola blanda Willd.
SWEET WHITE VIOLET   *native / cons* 5

Perennial herb, spreading by short rhizomes (and stolons later in season).
Stems smooth. Leaves all from base of plant, heart-shaped, dark green
and satiny, 2-5 cm wide, upper surface near base of blade usually with
short, stiff white hairs; petioles usually red. Flowers white, fragrant, on
stalks shorter than longer than leaves; lower 3 petals with purple veins
near base, all more or less beardless; upper 2 petals narrow, twisted backward, 2 side petals forward-pointing. Fruit a purple capsule 4-6 mm long,
seeds dark brown. April-May. — Hummocks in swamps and bogs, low
wet areas in deciduous and conifer forests.

Viola affinis

Viola blanda

Viola canadensis L.
TALL WHITE VIOLET   *native / cons* 7

Perennial herb, arising from a short, woody rhizome; plants glabrous or
minutely pubescent. Stems numerous, 2-4 dm tall, with several long-
petioled basal leaves. Stem leaves numerous, the lower widely spaced,
the upper crowded toward the apex, the blades cordate, 5-10 cm long;
upper leaves becoming shorter-petioled and with truncate or broadly
cuneate base. Stipules lance-shaped, slightly sscarious; peduncles slende.
Sepals lance-shaped, ciliate; petals white inside, with yellowish eye-spot
and brown-purple veins near the base, purplish-tinged outside. Capsules
globose-ellipsoid, 5-7 mm long; seeds brown. — Mesic deciduous woods.

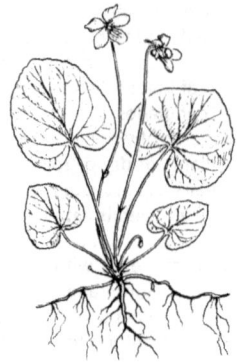

Viola cucullata Ait.
MARSH BLUE VIOLET   *native / cons* 7

Perennial herb, spreading by short, branched rhizomes; plants smooth.
Leaves all from base of plant, ovate to kidney-shaped, to 10 cm wide,
heart-shaped at base; margins coarsely toothed; blade angled from the
upright petioles. Flowers light purple or white, dark at center, on slender
stalks longer than leaves; the 2 side petals densely bearded with short
hairs, the hairs mostly knobbed or club-tipped. Fruit a cylinder-shaped
capsule, seeds dark. April-June. — Swamps, sedge meadows, shady seeps;
occasionally in bogs and low areas in forests.

Viola

Viola labradorica Schrank
ALPINE VIOLET   *native / cons* 4
Viola adunca Sm. var. *minor* (Hook.) Fern.
Perennial herb; plants smooth. Leaves in clumps from rhizomes, at first all from base of plants, later with leafy, horizontal stems to 15 cm long; light green, ovate to kidney-shaped, 1-2.5 cm wide; margins with rounded teeth; petioles 2-6 cm long. Flowers pale blue, side petals bearded on inner surface. Fruit 4-5 mm long, seeds dark brown. April-June. — Swamps, streambanks, moist hardwood forests.

Viola macloskeyi Lloyd
WILD WHITE VIOLET   *native / cons* 7
Viola pallens (Banks) Brainerd
Small perennial herb (our smallest violet), spreading by rhizomes and stolons. Leaves all from base of plant, heart-shaped to kidney-shaped, 1-3 cm wide at flowering, later to 8 cm wide, underside orange-tinged; margins with rounded teeth. Flowers white, on upright stalks equal or longer than leaves, 3 lower petals purple-veined near base, 2 side petals beardless or with sparse hairs. Fruit a green capsule 4-6 mm long, seeds olive-black. April-July. — Marshes, sedge meadows, open bogs and swamps, alder thickets; sometimes in shallow water.

Viola nephrophylla Greene
NORTHERN BOG VIOLET   *native / cons* 9
Low perennial herb, spreading by short rhizomes. Leaves all from base of plant, smooth, heart-shaped to kidney-shaped, 1-4 cm long and 2-6 cm wide, rounded at tip; margins with rounded teeth; petioles slender, 2-16 cm long. Flowers single, nodding on slender stalks, the stalks longer than leaves. Flowers violet, bearded near base on inside, or upper pair of petals not bearded. Fruit a capsule 5-10 mm long. May-sometimes again flowering in Aug or Sept. — Wet meadows, calcareous fens, low areas between dunes, streambanks, rocky shores.

Viola odorata L.
SWEET BLUE VIOLET   *introduced*
Perennial herb; plants finely pubescent, from a stout or wiry, long-creeping rhizome, and spreading by numerous leafy stolons rooting at the nodes. Leaves broadly ovate to orbicular, 2-6 cm long, rounded or obtuse at the apex, evenly crenate, cordate at base. Peduncles equaling the leaves or often shorter. Flowers very fragrant, typically deep violet, varying through many paler shades to white; sepals narrowly oblong, obtuse, ciliate; lateral petals usually bearded; style hook-shaped; cleistogamous flowers on recurved peduncles. Capsules broadly ovoid, pubescent, purplish; seeds large and cream-color. — Native of Europe; commonly cultivated in many forms (both single-flowered and double-flowered), and sometimes escaped from gardens to waste places or woods.

Viola pubescens Ait.
YELLOW FOREST VIOLET   *native / cons* 5
Perennial herb; plants softly pubescent, with 1 or 2 (rarely more) stout upright stems arising from a brown woody rhizome bearing coarse fibrous roots. Leaves 2-4 near the summit, occasionally accompanied by a long-petioled kidney-shaped-cordate root leaf; stem leaves orbicular-ovate, short-pointed, crenate-dentate, cordate or truncate-decurrent at base, 4-10 cm long, and usually about 1 cm wider than long; stipules broadly ovate, the apex blunt or shallowly toothed. Flowers on downy pubescent peduncles rising little above the leaves; sepals lance-shaped, ciliate; petals clear yellow with brown-purple veins near the base, the lateral

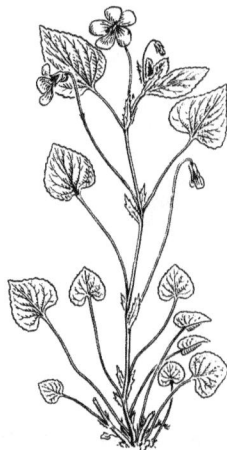
Viola labradorica

Viola macloskeyi

Viola nephrophylla

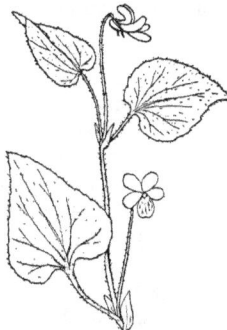
Viola pubescens

ones bearded. Capsules 10-12 mm long, woolly or glabrous; seeds pale brown. — Mesic woods.

Viola renifolia Gray
KIDNEY-LEAF WHITE VIOLET   *native / cons 7*
Perennial herb, spreading by long rhizomes. Leaves all from base of plant, mostly kidney-shaped, rounded at tip, varying from smooth and shiny above to hairy on lower surface only; margins with few rounded teeth. Flowers white, all bearded or beardless, 3 lower petals purple-veined at base. Capsules 4-5 mm long, seeds brown and dark-flecked. May-July. — Cedar swamps, sphagnum hummocks in peatlands.

Viola renifolia

Viola rostrata Pursh
LONG-SPUR VIOLET   *native / cons 8*
Perennial herb, from a branched woody rhizome. Stems glabrous, erect or spreading, 5-12 cm tall, becoming 15-25 cm tall after anthesis. Leaves ovate, 2-4 cm long, cordate at base, all but the lowermost acute or pointed at the apex; stipules lance-shaped, fimbriate-toothed to above the middle. Flowers on slender peduncles held well above the foliage; sepals narrowly lance-shaped, eciliate; petals light violet with darker veins forming a pronounced dark eye at the center of the flower; spur 10-16 mm long, slightly upcurved at the tip. Capsules ellipsoid, 5-6 mm long; seeds light yellow-brown. — Shady slopes and moist woods, usually in deep humus.

Viola sagittata Ait.
ARROWHEAD VIOLET   *endangered / native / cons 6*
Perennial herb. Leaves lance-shaped to oblong lance-shaped, hastately or sagittately incised, lobed, or toothed at the subcordate or truncate base, otherwise distantly toothed; leaves of late summer often nearly deltoid and merely crenate at the base; petioles usually longer than the blades Flowers 2-2.5 cm across, violet-purple, the lower petals usually with prominent dark veining; sepals glabrous; cleistogamous flowers on erect peduncles; seeds numerous, brown. April-June. — Open, dry pine and oak woods, usually where sandy.

Viola rostrata

Viola sagittata

Viola selkirkii Pursh
GREAT-SPUR VIOLET   *native / cons 8*
Perennial herb; plants rather delicate, from a slender, elongate, non-stoloniferous rhizome. Leaves all from base of plant, with minute spreading hairs on the upper surface, otherwise glabrous, at anthesis 1.5-3 cm long, later becoming larger, broadly ovate-cordate, the basal sinus narrow, the basal lobes converging or overlapping. Flowers numerous, about 1.5 cm across; sepals lance-shaped; petals pale violet, beardless, the spur large and blunt, 5-7 mm long. Capsules ellipsoid, 4-6 mm long; seeds buff-colored. — Deciduous woods and shady ravines; preferring calcareous soils.

Viola sororia Willd.
HOODED BLUE VIOLET   *native / cons 3*
   Viola septentrionalis Greene
Perennial herb, spreading by short rhizomes. Leaves all from base of plant, ovate to heart-shaped, sometimes expanding to 10 cm wide in summer, with long hairs; margins with rounded teeth; blades angled from the upright petioles. Flowers blue-violet, on stalks about as high as leaves, the 2 side petals densely bearded with hairs 1 mm long and not club-tipped. Fruit a purple-flecked capsule, seeds dark brown. April-June. — Moist hardwood forests; occasionally in swamps, floodplain forests and along rocky streambanks. Wisconsin state flower.

Viola sororia

Vitaceae

GRAPE FAMILY

Mostly woody vines, climbing by tendrils. Leaves alternate, simple or compound; tendrils and flower clusters produced opposite the leaves. Flowers regular, 4-5-merous, perfect or unisexual, with a hypanthium and usually with a cup-shaped disk. Calyx small or almost lacking. Petals small. Stamens as many as the petals and opposite them. Ovary 2-celled; style very short; stigma 1, slightly 2-lobed. Fruit a berry; seeds 4 or by abortion fewer.

1 Stems brown-pithy inside, the bark shredding into strips; leaves simple .... *Vitis*
1 Stems white-pithy, the bark tight, not shredding; leaves simple or palmately compound ....
   .... *Parthenocissus*

Parthenocissus

CREEPER

Woody vines, trailing or climbing by tendrils. Leaves palmately compound with typically 5 leaflets (ours). Flowers small in panicles borne opposite the leaves or aggregated into terminal clusters, perfect or unisexual, 5-merous. Disk none. Petals separate and spreading at anthesis. Stamens short, erect. Berries with thin flesh and 1-4 seeds.

Parthenocissus inserta (Kerner) Fritsch
THICKET-CREEPER   *native / cons* 4
Parthenocissus vitacea (Knerr) A.S. Hitchc.
Very similar to *P. quinquefolia* (shown in illustration, not known from Door County but common in Wisconsin) in habit and foliage; tendrils few-branched, almost always without adhesive disks. Leaflets glossy green above, glabrous to thinly pubescent beneath. Inflorescence forked at the summit of the peduncle, the two branches both divergent, producing a broad rounded cluster. June. — Moist soil.

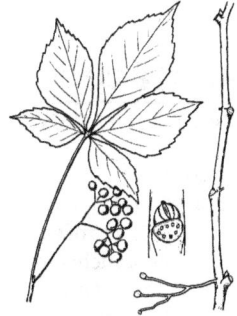

Parthenocissus quinquefolia

Vitis

GRAPE

Woody vines, climbing by tendrils. Flowers actually or functionally unisexual, with a hypogynous disk, 5-merous. Calyx essentially none. Petals cohering at the tip, separating at the base, falling early. Sterile flower with 5 erect stamens and a rudimentary pistil. Fertile flower with a well-developed pistil and 5 short, reflexed, functionless stamens. Fruit a juicy berry; seeds ovoid, 4, or fewer by abortion. Our species bloom in May or June.

1 Leaf underside with felt-like covering of white or rust-colored hairs .... *V. labrusca*
1 Leaf underside smooth and greenish, any hairs not felt-like and soon disappearing or only remaining
   in small patches as leaves mature .... *V. riparia*

Vitis labrusca L.
FOX GRAPE   *introduced*
High-climbing vine; pith interrupted at the nodes by a diaphragm; tendrils or flower clusters from 3 or more successive nodes. Leaves firm, round-cordate, 1-2 dm long and wide, usually shallowly 3-lobed, occasionally unlobed or lobed nearly to the middle, shallowly serrate, the lower surface persistently covered with a dense tomentum concealing the surface, reddish or rusty in color when young and usually also at maturity, occasionally ashy-gray at maturity. Peduncles and young branches eventually glabrous. Panicles ovoid, 4-8 cm long. Berries dark red to nearly black, 1-2 cm wide. — Woods, roadsides, and thickets.
    The size of the leaf serrations varies greatly. The species has contributed to the parentage of many cultivated grape varieties.

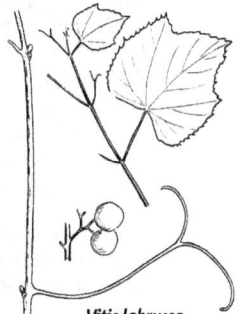

Vitis labrusca

Vitis riparia Michx.
RIVER-BANK GRAPE   *native / cons* **2**
   Vitis vulpina subsp. *riparia* (Michx.) R.T. Clausen
Woody, climbing vine to 5 m or more long; young branches green or red,
hairy, becoming smooth. Leaves alternate, heart-shaped in outline, 1-2
dm long and as wide, with a triangular tip and 2 smaller lateral lobes,
leaf base with a U-shaped indentation, upper surface smooth, bright
green, underside paler and sparsely hairy along veins; margins with
coarse, forward-pointing teeth; petioles shorter than blades. Flowers
small, sweet-scented, green-white to creamy, in stalked clusters 5-10 cm
long. Berries dark blue to black, 6-12 mm wide, with a waxy bloom, sour
when young, becoming sweeter when ripe in fall. May-July.— Floodplain
forests, moist sandy woods, streambanks, thickets, sand dunes.

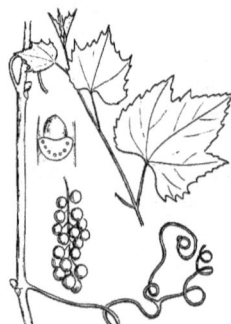

Vitis riparia

MONOCOTS

Acoraceae

CALAMUS FAMILY

Acorus
SWEETFLAG, CALAMUS
Perennial herbs of wetlands, rhizomes and leaves pleasantly scented.
Rhizomes branched, creeping at or near surface. Leaves sword-shaped,
equitant, bright green, with 1-6 prominent veins parallel along length
of leaf. Inflorescence a solitary spadix, borne from near midway of
leaf, nearly cylindric, tapering, apex obtuse; true spathe absent. Flowers bisexual; tepals 6, light brown; stamens 6; ovaries 1, usually 3-
locular. Fruit light brown to reddish berries with darker streaks. Seeds
1-6 (-14), embedded in a mucilagenous jelly.

Acorus americanus (Raf.) Raf.
AMERICAN SWEETFLAG   *native / cons* **7**
   Acorus calamus var. *americanus* (Raf.) H.D. Wulff.
Similar to the less common, introduced *A. calamus* and long considered a
variety of it. The introduced *A. calamus* is common throughout Wisconsin
but not yet reported from Door County.— Marshes, wet meadows, edges
of rivers.

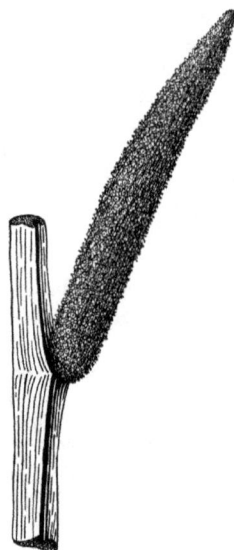

Acorus americanus

Alismataceae

WATER-PLANTAIN FAMILY
Perennial, aquatic or emergent herbs; plants swollen and tuberlike at base. Leaves all from base
of plant and clasping an erect stem; underwater leaves often ribbonlike; emergent leaves broader.
Flowers perfect (with both staminate and pistillate parts) or imperfect, in racemes or panicles
at ends of stems, with 3 sepals and 3 petals; stamens 6 or more. Fruit a compressed achene,
usually tipped by the persistent style.

1   Leaves often arrowhead-shaped; pistils or achenes in several series around a large, round receptacle,
    and forming a dense, round head. . . . . *Sagittaria*
1   Leaves never arrowhead-shaped; achenes in a single whorl on a small, flat receptacle or few in a
    small loose head . . . . *Alisma*

Alisma
WATER-PLANTAIN
Perennial herbs, from cormlike rootstocks. Leaves emersed or floating, ovate to lance-shaped, never arrowhead-shaped. Flowers perfect, in whorled panicles, sepals 3, green; petals 3, white or light pink; stamens 6. Fruit a flattened achene in a single whorl on a flat receptacle, style beak small or absent.

1 Flowers larger; petals about 4 mm long and to 4 mm wide .... *A. triviale*
1 Flowers smaller; petals to 2.5 mm long and 2 mm wide .... *A. subcordatum*

Alisma subcordatum Raf.
AMERICAN WATER-PLANTAIN   *native / cons* **3**
Alisma plantago-aquatica L. var. *parviflorum* (Pursh) Torr.
Leaves ovate to oval, 3-15 cm long and 2-12 cm wide, rounded to nearly heart-shaped at base; petioles long. Flowers clustered on slender stalks 1-10 dm long, in whorls of 3-10; sepals 3; petals white, 3-5 mm long. Fruit an achene, 2-3 mm long, with a central groove. July-Sept. — Shallow water marshes, shores, ditches.

Alisma triviale Pursh
NORTHERN WATER-PLANTAIN   *native / cons* **4**
Alisma brevipes Greene
Alisma plantago-aquatica L. var. *americanum* J.A. Schultes
Leaves usually long-petioled, the blade elliptic to broadly ovate, rounded to subcordate at base, 3-18 cm long. Inflorescence on a scape 1-10 dm long. Flower pedicels in whorls of 3-10; sepals obtuse, 2-3 mm long; petals white, about 4 mm long. Fruit an achene 2-3 mm long, usually with a median dorsal groove. June-Sept. — Marshes, ponds, and streams.

Alisma subcordatum

Sagittaria
ARROWHEAD
Perennial or annual herbs, with fleshy or tuberous rootstocks. Leaves sheathing, all from base of plant, variable in shape and size. Emersed and floating leaves usually arrowhead-shaped with large lobes at base, or sometimes ovate to oval and without lobes; underwater leaves often linear in a basal rosette, normally absent by flowering time. Flowers in a raceme of mostly 3-flowered whorls; upper flowers usually staminate, lower flowers usually pistillate or sometimes perfect; sepals 3, green, persistent; petals 3, white, deciduous; stamens 7 to many; pistils crowded on a rounded receptacle. Fruit a crowded cluster of achenes in more or less round heads, the achenes flattened and winged, beaked with a persistent style.

1 Emersed leaves not arrowhead-shaped, basal lobes absent .... *S. graminea*
1 Emersed leaves all or mostly arrowhead-shaped, with large basal lobes .... *S. latifolia*

Sagittaria graminea Michx.
GRASS-LEAF ARROWHEAD   *native / cons* **9**
Perennial herb, with rhizomes. Underwater plants sometimes only a rosette of bladeless, ribbonlike petioles (phyllodes) to 1 cm wide; emergent leaves lance-shaped to oval, never arrowhead-shaped, 3-20 cm long and 0.5-3 cm wide, tapered to a blunt tip. Flowers imperfect, the staminate flowers usually above the pistillate, clustered in more or less round heads, 5-12 mm wide, the heads on spreading stalks 1-4 cm long; with 2-10 whorls of flowers along an unbranched stalk mostly shorter than leaves; bracts broadly ovate, joined in their lower portion, 2-8 mm long; sepals ovate, bent backward in fruit; petals white, equal or longer than sepals. Fruit a winged achene, 1-2 mm long, beak small or absent. June-Sept. — Shallow water and shores.

    S. graminea includes *Sagittaria graminea* Michx. var. *cristata* (Engelm.) Bogin, sometimes treated as a separate species *Sagittaria cristata* Engelm.,

Sagittaria graminea

and distinguished as follows:

1 Achenes with beak 0.4–0.6 mm long; anthers clearly shorter than the filaments . . . . *S. cristata*
1 Achenes with beak minute, scarcely discernable, ca. 0.2 mm long; anthers as long as or longer than filaments . . . . *S. graminea*

Sagittaria latifolia Willd.

DUCK-POTATO   *native / cons* 3

Perennial herb, with rhizomes and edible tubers in fall. Leaves variable; emersed leaves arrowhead-shaped, mostly 8-40 cm long and 1-15 cm wide, lobes typically narrow on plants in deep water to broad on emersed plants; plants sometimes with bladeless, expanded petioles (phyllodes). Flowers staminate above and pistillate below, clustered in more or less round heads 1-2.5 cm wide, at ends of slender, spreading stalks 0.5-3 cm long, in whorls of 2-15 along a stalk 2-10 dm tall; bracts tapered to a tip or blunt, 0.5-1 cm long; sepals ovate, bent backward by fruiting time; petals white, 7-20 mm long. Fruit a winged achene, 2-4 mm long, the beak projecting horizontally, 1-2 mm long. July-Sept. — Shallow water, shores, marshes and pools in bogs.

Araceae

Sagittaria latifolia

ARUM FAMILY

Perennial herbs with alternate, simple or compound, often fleshy leaves. Flowers small and numerous, mostly single-sexed, staminate flowers usually above pistillate, crowded in a cylindric or rounded spadix subtended by a leaflike spathe; sepals 4-6 or absent; petals absent; stamens mostly 2-6; pistils 1-3-chambered. Fruit a usually fleshy berry, containing 1 to few seeds, or the entire spadix ripening as a fruit.

Now included in the Araceae are the tiny duckweeds (*Lemna*), aquatic genera formerly treated as their own family (Lemnaceae). These are small perennial herbs, floating at or near water surface, single or forming colonies. Plants thallus-like (not differentiated into stems and leaves), the thallus (or frond) flat or thickened; the roots, if present, unbranched, 1 or several from near center of leaf underside; reproducing vegetatively by buds from 1-2 pouches on the sides, the parent and budded plants often joined in small groups. Flowers rare, either staminate or pistillate, in tiny reproductive pouches on margins of the leaves, subtended by a small spathe within the pouch; sepals and petals absent; staminate flowers 1-2, consisting of 1 anther on a short filament; pistillate flower 1 (a single ovary), in same pouch as staminate flowers. Fruit a utricle with 1 to several seeds.

1 Plants tiny floating or submerged aquatic species less than 1 mm long, without differentiation into leaves or stems. . . . . *Lemna*
1 Plants large, with clearly differentiated normal leaves and with rhizomes or tubers 2
2 Leaves compound . . . . *Arisaema*
2 Leaves simple, not divided 3
3 Leaves broadly heart-shaped, abruptly tapered to a tip; spathe white, long-stalked; flowering in spring
    . . . . *Calla*
3 Leaves rounded ovate, tapered to a rounded tip; spathe green-yellow to purple-brown, short-stalked or stalkless; flowering in late winter to early spring . . . . *Symplocarpus*

Arisaema

JACK-IN-THE-PULPIT

Arisaema triphyllum (L.) Schott

JACK-IN-THE-PULPIT   *native / cons* 5

Perennial herb, from bitter-tasting corms. Stems 3-12 dm long. Leaves usually longer than the flower stalk, mostly 2, divided into 3 leaflets, the terminal leaflet oval to ovate, the lateral leaflets often asymmetrical at base. Flowers staminate or pistillate and usually on separate plants; the pistillate flowers on a fleshy spadix, the spadix subtended by a green, purple-striped spathe, rolled inward below, expanded and arched over the spadix above, abruptly tapered to a tip. Fruit a cluster of shiny red berries, the fruit about 1 cm wide. April-July. — Moist forests, cedar swamps.

Calla
WATER-ARUM

Arisaema triphyllum

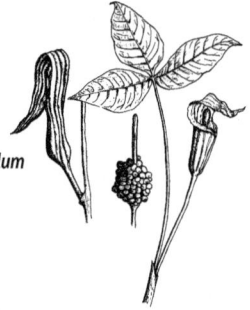

Calla palustris L.
WATER-ARUM　　　　　　　　　　　　　　　　　*native / cons* **9**
Perennial herb, from thick rhizomes, the rhizomes creeping in mud or floating in water. Leaves broadly heart-shaped, abruptly tapered to a tip, 5-15 cm long and about as wide; petioles stout, 1-2 dm long (or longer when underwater). Flowers perfect or the uppermost staminate, on a short-cylindric spadix, 1.5-3 cm long, shorter than the spathe; the spathe white, ovate, tipped with a short, sharp point to 1 cm long; sepals and petals absent; stamens 6. Fruit a fleshy, few seeded berry, turning red when ripe, 8-12 mm long. May-July. — Bog pools, swamps, shores and wet ditches.

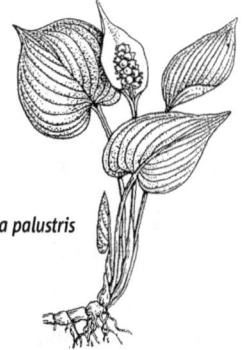

Calla palustris

Lemna
DUCKWEED
Small perennial floating herbs, with 1 root per frond (or roots sometimes absent on oldest and youngest leaves). Blades single or 2 to several and joined in small colonies, floating on water surface or underwater (ivy-leaf duckweed, *L. trisulca*), varying from round, ovate, to obovate or oblong, tapered to a long point (petiole) in star-duckweed; green or often red-tinged; upper surface flat to slightly convex, underside flat or convex. Reproductive pouches 2, on margins of frond. Flowers uncommon, consisting of 2 stamens (staminate flowers) and a single pistil (pistillate flower) in each pouch. Fruit an utricle with 1 to several seeds. Reproduction mostly by budding of new leaves from the reproductive pouches.

1　Fronds denticulate toward the tip, tapered to a slender stipitate base, the stipe often as long as the main body and commonly attached to the parent frond; colonies star-shaped, usually submersed . . . . *L. trisulca*
1　Fronds entire on the margin, nearly rounded and not obviously stipitate at the base, solitary or in tight colonies, these not star-shaped, floating on the water surface or stranded on mud 2
2　Fronds with several about equal sized, small papillae on the upper surface from the midline to the tip (often obscure), often red-tinged on the lower surface . . . . *L. turionifera*
2　Fronds lacking papillae; fronds never red beneath . . . . *L. minor*

Lemna minor L.
COMMON DUCKWEED　　　　　　　　　　　　*native / cons* **4**
Fronds nearly orbicular to elliptic-obovate, broadest near the middle, 2-4 mm long, symmetric or nearly so, green to yellowish green, never red-tinged or mottled on either surface (as in L. turionifera), obscurely 3 (5)-nerved; both surfaces flat to weakly convex, the upper surface with a low papilla at the apex and often one above the node, usually with a low median ridge or row of smaller papillae between them, the lower surface never inflated. Turions not produced. Root sheath not winged, root tip rounded. Fruit ovoid to ellipsoid, wingless, 1-seeded. July-Sept. — Quiet or stagnant water of ponds, oxbows, shores, slow-moving rivers, ditches.

Lemna minor

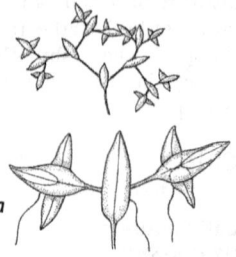

Lemna trisulca L.

IVY-LEAF DUCKWEED   *native / cons* 6

Perennial floating herb, forming tangled colonies just below water surface, floating at surface only when flowering; roots single from underside of frond or absent. Fronds several to many, joined to form star-shaped colonies; oblong lance-shaped, 5-20 mm long, tapered to a slender base (petiole), flat on both sides. Fruit a 1-seeded utricle. — Ponds, streams, ditches.

Lemna trisulca

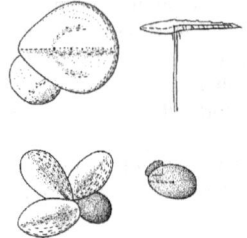

Lemna turionifera Landolt

TURION DUCKWEED   *native / cons* 2

Fronds single or in groups of several, obovate, usually flat and not humped, 1-4 mm long and 1-1.5 times longer than wide, veins 3, small white dots (papillae) present on midline of upper surface (visible with naked eye but clearer with 10x hand lens); underside of frond usually red or purple and redder than upper side point of root, upper surface (especially near tip) sometimes red-spotted. Turions sometimes present, dark-green to brown, 1-1.6 mm wide, without roots, sinking to bottom and forming new plants. — Quiet water of ponds and lakes.

Lemna turionifera

Symplocarpus
SKUNK-CABBAGE

Symplocarpus foetidus (L.) Salisb.

SKUNK-CABBAGE   *native / cons* 8

Perennial, foul-smelling herb, from thick rootstocks. Leaves all from base of plant, ovate to heart-shaped, 3-8 dm long and to 3 dm wide, strongly nerved; petioles short, channeled. Flowers appearing before leaves in late winter or early spring, perfect; the spathe ovate, curved over spadix, 8-15 cm long, green-purple and often mottled; sepals 4. Fruit round, 8-12 cm wide; seeds 1 cm thick. Feb-May (our earliest flowering native plant). — Floodplain forests, swamps, streambanks, calcareous fens, moist wooded slopes.

Symplocarpus foetidus

Commelinaceae

SPIDERWORT FAMILY

Commelina
DAY-FLOWER

Commelina communis L.

COMMON DAY-FLOWER   *introduced*

Annual herb. Stems succulent, at first erect, later diffuse and rooting from the lower nodes, to 8 dm long. Leaves alternate, ovate lance-shaped, 5-10 cm long, 1-3 cm wide. Inflorescence a small cyme, closely subtended by a folded, heart-shaped spathe from which the pedicels protrude; spathe (folded) cordate, 15-25 mm long, about half as wide, glabrous or minutely pubescent, its margins free to the base. Sepals 3, herbaceous, somewhat unequal, two usually somewhat united at base. Petals 3, the two upper blue, ovate to kidney-shaped, the lower one much smaller, usually white, sometimes absent. Fertile stamens 3; sterile stamens 3, smaller than the fertile. Fruit a 2-celled capsule, the upper cell abortive, 4-seeded. July-Sept. — Moist or shaded ground, sometimes a weed in gardens.

Commelina communis

Cyperaceae

SEDGE FAMILY

Mostly perennial, grasslike, rushlike or reedlike plants. Stems 3-angled or more or less round in section, solid or pithy. Leaves 3-ranked or reduced to sheaths at base of stem; leaf blades, when present, grasslike, parallel-veined, often keeled; sheaths mostly closed around the stem. Flowers small, perfect (with both staminate and pistillate parts), or single-sexed, each flower subtended by a bract (scale); perianth of 1 to many (often 6) small bristles, or a single perianth scale, or absent; stamens usually 3; ovary 2-3-chambered, contained in a saclike covering (perigynium) in *Carex*, maturing into an achene, stigmas 3 or 2. Flowers arranged in spikelets (termed spikes in *Carex*), the spikelets single as a terminal or lateral spike, or several to many in various types of heads, the head often subtended by 1 to several bracts.

1 Achenes enclosed in a closed sac (perigynium) subtended by a scale, the style protruding through the apex; flowers strictly unisexual (sedges with exclusively staminate flowers should be keyed here) . . .
. . . . *Carex*
1 Achenes not enclosed in a closed sac, naked beside the subtending scale; at least some flowers bisexual 2
2 Scales of spikelets 2-ranked; spikelets flattened in cross-section and always more than one per inflorescence. 3
2 Scales of spikelets spirally arranged (or if 2-ranked, the spikelet solitary); spikelets round or several-angled in cross-section, solitary or several to many per inflorescence. 4
3 Stems usually angled, solid; inflorescences terminal; achenes without subtending bristles. . . *Cyperus*
3 Stems round, hollow; inflorescences in the axils of stem leaves; achenes with subtending bristles . . .
. . . . *Dulichium*
4 Spikelet or cluster of spikelets borne on one side of the stem at the base of a single erect to somewhat angled or curved involucral bract that appears to be a continuation of the stem . . . . . *Schoenoplectus*
4 Spikelet or spikelets terminating the stem or borne both terminally and laterally; if more than one spikelet, the inflorescence with (1–) 2 to several spreading to reflexed, leaflike involucral bracts . . . 5
5 Spikelet solitary and terminal on the stem (very rarely a few smaller accessory spikelets occur at the base of the terminal spikelet in the bladeless genus Eleocharis) 6
5 Spikelets several to many on the stem, terminal or lateral 9
6 Sheaths totally bladeless or at most with an apical tooth up to 1 mm long; achenes usually with an apical tubercle formed by the expanded and persistent base of the style . . . . *Eleocharis*
6 Upper sheaths with short green blades 0.3–12 cm long; achenes blunt at apex, tubercle absent. . . . 7
7 Achenes subtended by 1–8 bristles less than twice as long as the achenes, or bristles absent . . . .
. . . . *Trichophorum*
7 Achenes subtended by conspicuous silky, white or tawny, hair-like bristles many times as long as the achenes 8
8 Bristles numerous, 12–50 or more; rhizomes erect, very short. . . . . *Eriophorum*
8 Bristles 6; rhizomes horizontal and short-creeping. . . . . *Trichophorum alpinum*
9 Achenes subtended by (12–) 15–50 conspicuous, silky, white or tawny, hair-like bristles many times as long as the achenes . . . . *Eriophorum*
9 Achenes subtended by 1–8 bristles, or bristles absent 10
10 Leaves flat or folded; with a definite, keeled midrib 11
10 Leaves inrolled and wiry; rounded on the back and without a definite midrib. 12
11 Achenes with a conspicuous tubercle formed by the expanded, persistent style base . . *Rhynchospora*
11 Achenes blunt at apex, without a tubercle; style base, if expanded, not persistent to maturity *Scirpus*
12 Styles 2-cleft; achenes subtended by slender bristles . . . . *Rhynchospora*
12 Styles 3-cleft; achenes lacking bristles. . . . . *Cladium*

Carex

SEDGE

Perennial grasslike plants. Stems mostly 3-angled. Leaves 3-ranked, margins often finely toothed. Flowers either staminate or pistillate, with both sexes in same spike, or in separate spikes on same plant, or the staminate and pistillate flowers on different plants. Staminate flowers with 3 or rarely 2 stamens; pistillate flowers with style divided into 2 or 3 stigmas. Achenes lens-shaped or flat on 1 side and convex on other (in species with 2 stigmas), or achenes 3-angled or nearly round (in species with 3 stigmas), enclosed in a sac called the perigynium (singular) or perigynia (plural).

Carex is the largest genus of plants in Door County (and in Wisconsin). To aid in identification, *Carex* are placed in sections [groups of closely related species; section divisions largely follow those of Reznicek et al. (2011) and Hipp (2008)]. As evident in the keys, identification of *Carex* is often based on characteristics of the mature perigynium (the sac around the achene), and a hand lens or dissecting microscope are often useful.

QUICK ENTRY KEY TO CAREX SECTIONS

Spike 1 per stem, all flowers attached to main stem in terminal spike . . . . go to couplet 2
Spikes 2 or more per stem, all flowers staminate . . . . go to couplet 7
Spikes 2 or more per stem, at least some flowers pistillate, stigmas 2, achenes flat to biconvex in cross-section . . . . go to couplet 9
Spikes 2 or more per stem, at least some flowers pistillate, stigmas usually 3, achenes usually more or less 3-angled in cross-section, body of perigynium pubescent . . . . go to couplet 29
Spikes 2 or more per stem, at least some flowers pistillate, stigmas usually 3, achenes usually more or less 3-angled in cross-section, body of perigynium glabrous . . . . go to couplet 33

KEY TO SECTIONS OF CAREX

1 Spike solitary, terminal (entirely staminate, entirely pistillate, or mixed) 2
1 Spikes 2 or more, sometimes crowded but distinguishable by the lobed appearance of inflorescence or protruding bracts or visible short segments of rachis between spikes 7
2 Styles 2-cleft; achenes 2-sided (lenticular); basal sheaths brown 3
2 Styles 3-cleft; achenes 3-sided (or nearly terete); basal sheaths brown or purple-red 4
3 Plants with slender rhizomes; perigynia obscurely or not at all serrate, plump (usually at least as convex on upper face as on the lower), the lowermost tending to be remote (as much as 1 mm apart at points of attachment); spikes without empty basal scales; anthers to 3 mm long . . . .
   . . . . *Carex sect. Physoglochin*
3 Plants densely tufted, not rhizomatous; perigynia minutely but strongly and regularly serrate on upper portion and beak, flattened, crowded; spikes usually with 1–2 empty basal scales; anthers 2–3.5 mm long . . . . *Carex sect. Stellulatae (C. exilis)*
4 Spikes unisexual (either staminate or pistillate); perigynia pubescent . . . . *Carex sect. Acrocystis*
4 Spikes containing both staminate and pistillate flowers; perigynia usually glabrous. 5
5 Perigynia minutely pubescent . . . . *Carex sect. Acrocystis*
5 Perigynia glabrous 6
6 Lower pistillate scale leaflike, at least on most spikes, much exceeding the perigynium; perigynia distinctly beaked, the body plump and filled by the mature achene. . . . . *Carex sect. Phyllostachyae*
6 Lower pistillate scale not leaflike, scarcely if at all exceeding perigynium; perigynia essentially beakless or linear-lanceolate (tapering into an indistinct beak) . . . . *Carex sect. Leptocephalae*
7 All spikes staminate 8
7 At least some spikes bisexual or pistillate 9
8 Leaves flat, lax and spreading; usually in swamps and marshes *Carex sect. Deweyanae (C. bromoides)*
8 Leaves channeled, stiff and erect; fens and other calcareous open wetlands . . . *Carex sect. Stellulatae*
9 Styles 2-cleft; achenes 2-sided 10
9 Styles 3-cleft; achenes 3-sided (or nearly terete) 28
10 Lateral spikes peduncled, or if sessile, then elongate; terminal spike often entirely staminate 11
10 Lateral spikes sessile, short, often crowded; terminal spike at least partly pistillate (rarely staminate)
   13
11 Plants slender, the stems to 3 dm tall and less than 1 mm thick (excluding leaf bases) even near the base; terminal (staminate or sometimes mixed) spike solitary, ca. 1 cm long; lowermost bract usually with a short sheath 2–7 mm long; perigynia white-pulverulent or golden-yellow at maturity 12
11 Plants coarse, the stems over (3–) 5 dm tall and usually over 1 mm thick, at least toward base; staminate spikes often 2 or more, mostly 2.5–7 cm long; lowermost bract essentially sheathless (rarely with very short sheath); perigynia neither white-pulverulent nor golden-yellow . . . *Carex sect. Phacocystis*
12 Lowermost pistillate spike sessile or nearly so (except rarely one arising from near base of plant); terminal spike staminate; perigynia green or slightly glaucous, crowded . . . . *Carex sect. Phacocystis*

12 Lowermost pistillate spike nearly always peduncled; terminal spike often pistillate near apex, or the pistillate spikes loosely flowered; fresh perigynia white-pulverulent or golden-yellow. . . . . *Carex sect. Bicolores*
13 Stems arising mostly singly from rhizome or stolon 14
13 Stems tufted, the tufts with or without connecting rhizomes 17
14 Perigynia plumply plano-convex to nearly terete in cross-section, not winged or sharply margined; plants of sphagnum bogs, cedar swamps, etc 15
14 Perigynia strongly flattened, with distinctly winged or sharply edged margins; plants mostly of wet or dry open habitats 16
15 Scales pale-hyaline with green midrib; perigynia apiculate or with very small beak; at least the lower few-flowered spikes ± separated; plants clumped from short, slender rhizomes *Carex sect. Glareosae*
15 Scales rich brown; perigynia with distinct beak ca. 0.5 mm long; spikes crowded as if in a single head; stems arising from axils of old decumbent stems (stolons) . . . . *Carex sect. Chordorrhizae*
16 Perigynia mostly over 2 mm wide; staminate flowers only at the base of some or all spikes . . . . *Carex sect. Ovales*
16 Perigynia mostly not over 2 mm wide; staminate flowers not restricted to base of spikes . . . . *Carex sect. Holarrhenae*
17 Staminate flowers at the base of some or all spikes, not at the apex (note especially the terminal spike) 18
17 Staminate flowers at the apex of some or all spikes (even when anthers have fallen, protruding filaments are usually visible). 23
18 Perigynia with thin-winged margins, at least narrowly so along apical part of body and basal part of beak, strongly flattened and scale-like (in some species elongate), appressed and overlapping (or in some species spreading at the tips) 19
18 Perigynia at most with a ridge along the margin, not winged, the achene plumply filling at least the apical part of the body all the way to the margins 20
19 Bracts not resembling the leaves, narrower than 2 mm most or all their length and not over twice as long as the inflorescence; perigynia various . . . . *Carex sect. Ovales*
19 Bracts leaflike, the broadest 2–4 mm wide, many times exceeding the spikes (which are crowded in a dense head); perigynia narrowly lanceolate, not over 1 mm wide. . . . . . . *Carex sect. Cyperoideae*
20 Body of perigynium elliptic or nearly so (except in *C. arcta*) with at most a very short beak, and with rounded or slightly margined edges, nearly or entirely filled by the achene. . . . *Carex sect. Glareosae*
20 Body of perigynium ovate or lanceolate or prominently beaked, sharp-edged, only 1/2 to 2/3 filled by achene (very spongy around and below base of achene) 21
21 Mature perigynia loosely to strongly appressed-ascending, 4–5.7 mm long; anthers 1.3–2.6 mm long . . . . *Carex sect. Deweyanae*
21 Mature perigynia strongly spreading to reflexed, 2–3.6 mm long; anthers 0.8–2 mm long. 22
22 Spikes 7–15, usually crowded, except sometimes the lowest, the inflorescence axis mostly concealed; beaks not bidentate . . . . *Carex sect. Glareosae* (*C. arcta*)
22 Spikes 3–8, not usually crowded, inflorescence axis clearly visible; beaks clearly bidentate with teeth 0.1–0.4 mm long . . . . *Carex sect. Stellulatae*
23 Stems stout (often 1.5 mm thick at ca. 3 cm below inflorescence) and very sharply angled (or even narrowly winged), soft and easily compressed (flattened in pressing); wider leaves 5–10 mm broad, with rather loose sheaths; perigynia spongy-thickened basally, on short slender stalks; anthers 1.3–2.6 mm long. . . . . *Carex sect. Vulpinae*
23 Stems slender (not over 1.5 mm thick at ca. 3 cm below inflorescence, or rarely so in some species), firm, not wing-angled nor easily compressed (hence, not flattened in pressing); leaves, perigynia, and anthers various 24
24 Spikes 10 or fewer, usually greenish at maturity, crowded or remote in a simple inflorescence (one spike, no branches, at each node of it) 25
24 Spikes numerous (10–many), yellowish or brownish at maturity; inflorescence tending to be compound, at least its lower nodes with 2 or more spikes crowded on a lateral branch 27
25 Perigynia elliptic, essentially beakless, very plump (nearly terete) and filled by the achene; at least the lower spikes well separated, containing 1–5 perigynia . . . . *Carex sect. Dispermae*
25 Perigynia ± ovate, beaked, plano-convex or lenticular; spikes various. 26
26 Mature perigynia brownish; some spikes (especially terminal) entirely or mostly staminate or staminate at their bases only . . . . *Carex sect. Stellulatae*
26 Mature (not over-ripe) perigynia generally greenish; no spikes entirely or mostly staminate (a few may have stamens at their base in addition to their apex). . . . . *Carex sect. Phaestoglochin*

27 Pistillate scales terminating in a distinct rough awn; bracts, at least lower ones, very slender and exceeding spikes or branches; ventral surface of leaf sheaths usually transversely wrinkled or puckered (very rarely smooth) . . . . *Carex* sect. *Multiflorae*

27 Pistillate scales acute or minutely cuspidate; bracts mostly short, inconspicuous, or absent; leaf sheaths smooth ventrally . . . . *Carex* sect. *Heleoglochin*

28 Perigynia at least sparsely puberulent, pubescent, hispidulous, or scabrous 29

28 Perigynia glabrous (in some species, papillose or granular, but not even sparsely puberulent or scabrous) 33

29 Perigynia 12–18 mm long, in 1–2 short-oblong to spherical spikes 2–3.5 cm wide . . . .
. . . . *Carex* sect. *Lupulinae*

29 Perigynia 2–11 mm long, in 2–5 elongate, cylindrical spikes less than 2 cm wide 30

30 Perigynia with distinct and definite slender beak and/or the apex with 2 firm teeth 31

30 Perigynia beakless or merely apiculate ("beak" not over 0.4 mm long) and the apex not toothed . . . .
. . . . *Carex* sect. *Digitatae*

31 Leaves hairy . . . . *Carex* sect. *Hirtifoliae*

31 Leaves glabrous (often rough or scabrous, but not hairy) 32

32 Pistillate spikes not over 10 mm long (occasionally 12 mm in *C. communis*); achenes mostly with very convex or rounded sides (the angles thus obscured), at least apically, very tightly enveloped by the perigynium, especially on the apical half; anthers 1.5–3.7 mm long; plants of dryish habitats . . . .
. . . . *Carex* sect. *Acrocystis*

32 Pistillate spikes mostly over 10 mm long; achenes with flattish to slightly concave sides (the angles thus evident), the summit (especially around base of style) loosely enveloped by the perigynium; anthers 2.5–4.7 mm long; plants of dry to wet habitats . . . . *Carex* sect. *Paludosae*

33 Leaf sheaths (at least at apex) finely pubescent; blades often also pubescent or at least strongly hispidulous, especially toward base of plant 34

33 Leaf sheaths and blades completely glabrous (though sometimes scabrous) 35

34 Basal sheaths pale brown, leaf blades and stems glabrous or scabrous, perigynia with ca. 50 fine, impressed nerves . . . . *Carex* sect. *Griseae* (*C. hitchcockiana*)

34 Basal sheaths reddish purple tinged, leaf blades and stems pubescent, perigynia 5–12 nerved . . . . . .
. . . . *Carex* sect. *Hymenochlaenae*

35 Perigynia rounded to broadly tapered at summit, beakless or essentially so (the tiny beak or apiculus less than 0.5 mm long if distinct, or up to 0.8 mm long if vaguely defined, often strongly bent or curved); beak or apiculus (if present) never toothed (or teeth scarcely 0.1 mm long) 36

35 Perigynium abruptly contracted or more gradually tapering to a definite slender beak 0.5 mm or more long, or to an indistinct tapering beak 1 mm or more long; beak in some species with short apical teeth 45

36 Leaf blades not over 0.5 mm broad, linear-filiform; perigynia dark brown or nearly black at maturity, 2 mm or less long, in few-flowered spikes, of which at least the upper ones are on peduncles usually surpassing the sessile staminate spike . . . . *Carex* sect. *Albae*

36 Leaf blades 0.5 mm or more broad; perigynia and spikes various (but not as above) 37

37 Bract of lowest pistillate spike sheathless (at most with a thin scarious sheath 1–3 mm long) 38

37 Bract of lowest pistillate spike with a sheath 4 mm or more long 40

38 Terminal spike partly pistillate; pistillate spikes nearly or quite sessile and erect or ascending; roots glabrous or nearly so . . . . *Carex* sect. *Racemosae*

38 Terminal spike normally entirely staminate; spikes and roots various 39

39 Pistillate spikes mostly drooping at maturity on slender peduncles; species of wet peatlands with roots with dense felt-like pubescence . . . . *Carex* sect. *Limosae*

39 Pistillate spikes erect or ascending, sessile or peduncled; roots glabrous . . . . go to couplet 43

40 Terminal spike bearing some perigynia (very rarely a few individuals with one entirely staminate); plants very strongly reddish tinged at base 41

40 Terminal spike entirely staminate; plants reddish or not at base 42

41 Staminate flowers at apex of terminal spike, pistillate flowers at base; cauline sheaths bladeless or with rudimentary blades to 2 (rarely 4) cm long; pistillate spikes short-cylindric, bearing fewer than 10 perigynia, very long-peduncled, some elongate peduncles usually arising from base of plant . . . .
. . . . *Carex* sect. *Digitatae* (*C. pedunculata*)

41 Staminate flowers at base of terminal spike, pistillate flowers at apex; cauline sheaths with well-developed blades; pistillate spikes linear-cylindric, bearing more than 10 perigynia, on peduncles about as long as the spike or shorter, all arising from the upper part of the stem . . . .
. . . . *Carex* sect. *Hymenochlaenae*

42 Perigynia concave- or at least cuneate-tapering toward the base, 3-angled and often somewhat broadly spindle-shaped 43
42 Perigynia convex-rounded toward the base, nearly or quite circular in cross-section (or very obscurely triangular), ellipsoid-cylindric to nearly spherical 44
43 Plants with elongate deep or shallow rhizomes and very slender, firm stems; leaf blades 1–4 mm wide . . . . *Carex* sect. *Paniceae*
43 Plants without elongate rhizomes, the stems sharply triangular, sometimes nearly wing-margined, rather weak and easily compressed, soon shriveling after maturity of the fruit; leaf blades usually more than 4 (and up to 35) mm wide . . . . *Carex* sect. *Laxiflorae*
44 Larger perigynia 4–5 mm long, the nerves not raised above the surface at maturity. . . . . *Carex* sect. *Griseae*
44 Larger perigynia 2–3.5 mm long; nerves various . . . . *Carex* sect. *Granulares*
45 Lower pistillate scales leaflike or bract-like, much exceeding the perigynia; achenes abruptly constricted to a short thick base; body of perigynium nearly terete, essentially nerveless except for 2 ribs; anthers 0.5–1.6 mm long . . . . *Carex* sect. *Phyllostachyae*
45 Lower pistillate scales scarcely if at all exceeding the perigynia; achenes not abruptly constricted at the base; perigynia and anthers various. 46
46 Perigynia in densely crowded spherical to very short-cylindric spikes, spreading and with the lowermost usually reflexed, usually strongly few-ribbed; at least the uppermost pistillate spikes ± sessile and often crowded; the terminal spike (staminate or partly pistillate) often sessile or short-peduncled 47
46 Perigynia in elongate or long-peduncled spikes or both, all ascending, 2-ribbed or variously many-nerved; inflorescences various, but the upper spikes often not crowded and the terminal spike often long peduncled 48
47 Perigynia 2–6.2 mm long; basal sheaths brown . . . . *Carex* sect. *Ceratocystis*
47 Perigynia 11–18 mm long; basal sheaths reddish purple tinged . . . . *Carex* sect. *Lupulinae*
48 Bract of lowest pistillate spike sheathless (or pistillate spikes all crowded at base of plant in *Carex tonsa*); check several stems; rarely, a pistillate spike will be borne abnormally low on the stem and this spike may then have a sheath, which should be disregarded in keying 49
48 Bract of lowest pistillate spike consistently with sheath 4 mm or more long 54
49 Pistillate scales subtending at least some of the perigynia terminated by a distinct slender scabrous awn; perigynia 3–9 mm long. 50
49 Pistillate scales smooth-margined and awnless or very short-awned, or at most with a scabrous margin toward an acuminate (sometimes inrolled) apex (occasionally a long rough awn in species with perigynia more than 9 mm long); perigynia 4–18 mm long 51
50 Scales toward apex of pistillate spikes merely acuminate or with awns shorter than their bodies (the latter easily visible, about half as long as perigynia or longer); staminate spikes 2 or more; body of perigynium rather gradually tapered to a beak 1.5 mm long, including the short (not over 0.8 mm) teeth. . . . . *Carex* sect. *Paludosae*
50 Scales toward apex of pistillate spikes ordinarily with awns (as on the other pistillate scales) nearly or fully as long as their bodies (the latter small and mostly hidden among the bases of the densely crowded perigynia); staminate spike solitary (or very rarely a second smaller one present); body of perigynium tapered or strongly contracted into a beak 1.2–3.5 mm long, including teeth up to 2.2 mm long . . . . *Carex* sect. *Vesicariae*
51 Perigynia strongly inflated, not tight around the achene, 2–8 mm wide 52
51 Perigynia not inflated, ± tightly enclosing achene, 1–1.6 mm wide. 53
52 Perigynia 4–12 mm long, 6–12 (–15)-nerved . . . . *Carex* sect. *Vesicariae*
52 Perigynia 12–17 (–18) mm long, 15–20-nerved. . . . . *Carex* sect. *Lupulinae*
53 Pistillate spikes linear-cylindric, drooping or curving on slender peduncles; perigynia (somewhat twisted) and achenes strongly angled, the latter with concave sides; tall plants (stems over 3 dm high) with scattered thin leaves. . . . . *Carex* sect. *Hymenochlaenae* (*C. prasina*)
53 Pistillate spikes short, thick, and few-flowered, often crowded at base of plant; perigynia and achenes very convex-sided; low plants (stems less than 1 dm high) with crowded, very stiff leaves. . . . . *Carex* sect. *Acrocystis* (*C. tonsa*)
54 Perigynia (6–) 9–17 (–18) mm long; beak teeth usually conspicuous and stiff. . . . . . . go to couplet 50
54 Perigynia 2–6.5 (–9) mm long; beak teeth absent or weak and inconspicuous. 55
55 Perigynia with several to many conspicuous fine nerves on each side 57
55 Perigynia with 2 (–3) main ribs, the sides otherwise nerveless or with much less prominent nerves 59
56 Nerves of perigynia very numerous (20–65) and impressed, giving a longitudinally corrugated appearance; awns of pistillate scales rough or even ciliate . . . . *Carex* sect. *Griseae*

56 Nerves of perigynia several to many (5–40) and slightly raised; awns of pistillate scales absent, smooth, or rough 57
57 Awns rough and/or summit of pistillate scales minutely ciliate; lower spikes drooping on long very thin peduncles; beak slightly bidentate at maturity; plants strongly reddish at base . . . . *Carex* sect. *Hymenochlaenae*
57 Awns of pistillate scales usually smooth or absent; lower spikes mostly not drooping; beak not bidentate; plants pale, brown, or reddish at base 58
58 Perigynia sharply triangular with flattish sides, short-tapering at the base; stems bluntly trigonous, firm and not easily compressed; anthers mostly 3–4.5 mm long or lower pistillate spikes on elongate filiform spreading or drooping peduncles . . . . *Carex* sect. *Careyanae*
58 Perigynia rounded-triangular with swollen sides, long-tapering to a stalk-like base; stems sharply triangular to nearly wing-margined, easily compressed; anthers mostly 1.5–3 mm long and lower pistillate spikes usually on erect or ascending peduncles . . . . *Carex* sect. *Laxiflorae*
59 Lowermost pistillate spikes erect or ascending at maturity 60
59 Lowermost pistillate spikes drooping on long slender peduncles at maturity 61
60 Staminate spike well-peduncled; perigynia convex-sided toward the base; bracts with poorly developed blades; plants mat-forming from long-creeping rhizomes . . . . *Carex* sect. *Paniceae* (*C. vaginata*)
60 Staminate spike sessile or nearly so; perigynia tapered-cuneate toward the base; bracts with well-developed blades; plants tufted . . . . *Carex* sect. *Laxiflorae* (*C. leptonervia*)
61 Pistillate spikes not over 15 mm long . . . . *Carex* sect. *Chlorostachyae*
61 Pistillate spikes mostly 20 mm or more long . . . . *Carex* sect. *Hymenochlaenae*

Carex Section Acrocystis

First sedges to flower each year, fruits maturing in April or May and soon shed. Basal leaf sheaths in most species becoming fibrous with age. Perigynium beaks bidentate, less than 0.5 mm long. — Most common in dry woods, prairies, and open sandy places; less common in mesic woods or wetlands.

Similar to section Digitatae but basal sheaths not becoming fibrous.

1 Pistillate spikes on stems of varying length, at least some of the stems short (up to 5 cm long) and partly hidden among the tufted leaf bases; anthers 1.5–2 mm long 2
1 Pistillate spikes all on elongate stems (none borne on short basal peduncles); anthers various 4
2 Bract of the lowest non-basal pistillate spike leaflike, equaling or exceeding the tip of the staminate spike; remnants of old leaves only slightly breaking into fibrous shreds at the base . . . . *C. deflexa*
2 Bract of the lowest non-basal pistillate spike scale-like or bristle-like, not exceeding the staminate spike (or all spikes often on short basal stems, but foliage and stems stiffer and much more scabrous than in *C. deflexa*, which nearly always have some elongate stems); remnants of old leaves breaking into copious fibrous shreds at the base 3
3 Perigynia 3.2–4 mm long, the beak 1.2–1.6 mm, about half as long as the body or longer . . . *C. tonsa*
3 Perigynia 2.5–3 mm long, the beak 0.4–0.9 mm, about 1/4–1/3 as long as the body . . . . *C. umbellata*
4 Main body of perigynium, not including spongy-tapered base or beak, orbicular to short-obovoid, about the same diameter as length; anthers 2.1–3.7 mm long; plants either with the widest leaves 3–8 mm broad or with elongate shallow rhizomes 5
4 Main body of perigynium elliptic (to slightly obovoid or oblong), definitely longer than thick; anthers 1.3–2.5 mm long; plants with mostly narrow leaves and lacking stout elongate rhizomes (and otherwise not fitting either lead of couplet 5) 6
5 Widest leaves (at least the oldest dry ones) 3–5 mm broad; cauline leaves above base of plant (when present on stem) usually with the ligule longer than the width of the leaf; bract subtending the middle (and sometimes the lowest) pistillate spike(s) scarious-lobed at base, blade awn-like to leaflike, usually green, arising from between the lobes; staminate spike ca. 1–2 (–2.5) mm thick; plants without elongate rhizomes . . . . *C. communis*
5 Widest leaves 1.5–3 mm broad; cauline leaves with ligule no longer than the width; bracts subtending middle pistillate spikes tapered to apex, without an elongate awn-like or leaflike blade (the lowermost bract often green but seldom lobed); staminate spike 2–3.5 (–5) mm thick; plants with stout, shallow elongate rhizomes with fibrous sheaths . . . . *C. pensylvanica*
6 Widest leaves (at least the oldest dry ones) 3–5 mm broad; bract subtending the middle (and sometimes also the lowest) pistillate spike(s) scarious-lobed at base, the blade awn-like or leaflike, usually green, arising from between the lobes . . . . *C. communis*
6 Widest leaves not over 3 mm broad; bracts either scale-like or leaflike and lacking a scarious-lobed base 7

7 Perigynia 2–3 mm long, minutely puberulent to short-hairy; stems very slender (seldom over 0.4 mm thick) and mostly surpassed by the leaves . . . . *C. deflexa*
7 Perigynia 3–4 mm long, definitely short-hairy; stems usually 1 mm or more in thickness and surpassing the leaves. . . . . *C. peckii*

Carex communis Bailey
FIBROUS-ROOT SEDGE   *native / cons* 6

Plants tufted; rootstocks short, ascending, scaly, reddish purple. Stems 1.5-5 dm long, rough on angles above, purplish red at base, the old leaves conspicuous; sterile shoots numerous. Well-developed leaves several to a fertile culm, near base; blades 2-5 cm long and 2-4 mm wide, flat, flaccid, light-green, rough especially towards the tip and on margins. Terminal spike staminate; lateral spikes 2-3, pistillate; lowest bract scale-like, hyaline-margined and purplish tinged at base, the upper reduced or scale-like; scales reddish purple or -brown with hyaline margins and 3-nerved green or straw-colored center. Perigynia 3-10 to a spike, 3-3.5 mm long, ascending, light-green, puberulent, 2-keeled, the spongy base 0.75 mm long; beak 0.5 mm, flattish, bidentate. Achenes triangular with convex sides, light-brown with lighter angles, minutely pitted, truncate and bent-apiculate; stigmas 3, reddish brown. — Mesic forests.

Carex communis

Carex deflexa Hornem.
NORTHERN SEDGE   *native / cons* 7

Plants loosely tufted. Stems 1-2 dm long, purple-tinged at base, shorter than the leaves. Leaves soft, 1-3 mm wide. Spikes either staminate or pistillate; staminate spike short, to 5 mm long; pistillate spikes on long, slender stalks near base of plant and also 2-4 spikes on stem near staminate spike; bract leaflike, to 2 cm long; pistillate scales ovate, shorter than perigynia. Perigynia green, oblong-ovate, 2-3 mm long, covered with short hairs, abruptly tapered to a small beak about 0.5 mm long. Achenes 3-angled; stigmas 3. June-Aug. — Moist woods and swamps, wetland margins, often where sandy or in sphagnum moss.

Carex peckii Howe
PECK'S SEDGE   *native / cons* 7
Carex nigromarginata var. *elliptica* (Boott) Gleason

Plants tufted and stoloniferous; rootstocks slender, scaly. Stems to 65 cm long, roughened beneath spikes, reddish purple at base; sterile stems long, the well developed leaves towards the top. Well-developed leaves several to a fertile culm, on lower 1/4; blades 1.5-4 cm long and 1-1.5 mm wide (larger on sterile stems), flat, green, roughened on margins and towards apex. Terminal spike staminate; lateral spikes pistillate, in an inflorescence 8-20 mm long; scales reddish brown with white-hyaline margins. Perigynia 3-12 to a spike, 3.5 mm long, ascending, grayish or yellowish green, hirsute-pubescent, 2-ridged, the base spongy, 0.5 mm long; beak 0.5 mm long, obliquely cut, bidentate, hyaline at mouth. Achenes yellowish brown, triangular with convex sides and blunt green angles, minutely apiculate; stigmas 3, dark-reddish brown. — Open woods.

Carex deflexa

Carex pensylvanica Lam.
PENNSYLVANIA SEDGE   *native / cons* 3

Plants tufted and stoloniferous; stolons horizontal, slender, scaly, fibrillose, reddish. Stems 5-40 cm long, smooth or roughened on angles above, reddish purple at base, clothed with old leaves, often fibrillose; sterile shoots reddish purple at base, the sheaths puberulent, becoming filamentose. Well-developed leaves 2-several to a fertile culm; blades to 3 cm long and 1.5-3 mm wide, flat above, canaliculate towards base, often roughened especially towards tip. Terminal spike staminate; lateral spikes 1-4, pistillate; bracts scale-like, enlarged at base, hyaline-margined and reddish

Carex peckii

brown; pistillate scales reddish purple, with white-hyaline margins and lighter center. Perigynia 4–20 to a spike, 2.5–3 mm long, ascending, dull- or yellowish green, puberulent, 2-keeled, the spongy base to 0.75 mm long; beak 0.75 mm long, bidentate, hyaline and often purplish tinged at orifice. Achenes triangular with convex sides and narrow angles, brown, minutely pitted, minutely apiculate; stigmas 3, reddish brown. — Common in a wide range of dry to mesic woods.

Carex pensylvanica

Carex tonsa (Fern.) Bickn.
SHAVED SEDGE   *native / cons* **4**
   Carex rugosperma Mackenzie

Plants tufted; rootstocks stoutish, branching; stolons short-ascending. Stems to 15 cm long, stiff, roughened, reddish brown-tinged and fibrillose at base. Leaves numerous; blades 5–25 cm long and 2–4.5 mm wide, channeled with revolute margins, thick, stiff, deep-green, rough towards the tip. Terminal spike staminate; pistillate spike occasionally present near terminal spike; basal pistillate spikes long-peduncled; bract of upper spike setaceous, reddish at base; pistillate scales conspicuous, whitish or straw-colored, with 3-nerved greenish or straw-colored center. Perigynia 3–20 to a spike, 3.5–4.5 mm long, appressed-ascending, compressed-orbicular, somewhat leathery, light-green, sparsely pubescent, 2-keeled, the base 0.75 mm long; beak to 2.5 mm long, 2-edged, serrulate, bidentate. Achenes triangular with convex sides and sharp angles, brownish, shiny, truncate and minutely apiculate; stigmas 3. — Dry sandy fields and open woods.

Carex umbellata Schkuhr
PARASOL SEDGE   *native / cons* **6**

Plants densely tufted; rootstocks short, stout. Stems to 15 cm long, stiff, rough on angles, reddish brown-tinged and fibrillose at base. Leaves numerous; blades to 3 dm long and 1.5–2.5 mm wide, channeled towards base, flat and rough above, with revolute margins, firm, light-green. Terminal spike staminate; lateral spikes 3–4, pistillate or androgynous; bract of upper spike scale-like, reddish tinged at base; pistillate scales hyaline with several-nerved green center, the upper reddish brown-tinged. Perigynia 4–20 to a spike, 2–3 mm long, ascending, triangular-orbicular, dull-green, pubescent above, 2-keeled, the base 0.5 mm long; beak to 1 mm long, 2-edged, bidentate, hyaline-tipped. Achenes triangular with convex sides and sharp angles, filling perigynia, brownish black, shining, minutely pitted, minutely apiculate; stigmas 3. — Dry, often calcareous fields.

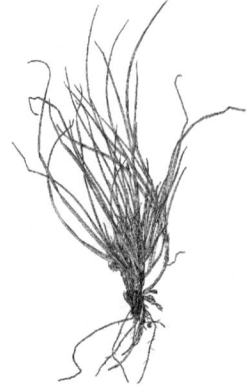

Carex tonsa

Carex umbellata

Carex Section Albae

One member of the section in Door County. Rhizomes elongate, the plants forming mats. Leaf blades involute, wiry. Perigynia becoming dark in age, beaks short, white-tipped.

Carex eburnea Boott
BRISTLE-LEAF SEDGE   *native / cons* **8**

Plants tufted; rootstocks long, slender, brownish. Stems 1–3.5 dm, obtusely triangular, brownish tinged at base. Well-developed leaves 3–6 to a fertile culm, near base; blades 5–25 cm long and 0.5 mm wide, often recurved-spreading, involute, firm, green, roughened. Terminal spike staminate; lateral spikes 2–4, pistillate, on peduncles 1–2.5 cm long; bracts bladeless, tubular, greenish or greenish yellow with white margins; pistillate scales whitish with green midrib, often yellowish brown-tinged. Perigynia 2–6 to a spike, 2 mm long, triangular, light-green or brownish, shining, puncticulate, 2-ribbed, finely nerved; beak short, cylindric, obliquely cut,

Carex eburnea

hyaline at orifice. Achenes triangular with concave sides and thickened angles, closely enveloped, brown, granular, apiculate; stigmas 3, brownish. — Dry sand prairies, and rarely in fens.

Carex Section Bicolores

Plants short, colonial, loosely tufted, shoots arising singly or few in a clump; rhizomes elongate; bases brown. Terminal spike staminate or gynecandrous, hidden by the crowded lateral spikes. Perigynia plump, golden to whitish, weakly veined; margins and apex rounded, beakless to short-beaked. Stigmas 2. — Calcium-rich sites where somewhat disturbed.

1 Mature perigynia golden-orange when fresh (drying dark brown or, especially if immature, white); terminal spikes mostly all staminate (occasionally with a very few perigynia); pistillate scales loosely spreading, distinctly shorter than the mature perigynia (usually averaging 3/4 or less as long), most of them acute to cuspidate. . . . . *C. aurea*
1 Mature perigynia white-pulverulent when fresh; terminal spikes usually staminate at base only, with several to numerous perigynia apically; pistillate scales appressed, nearly (averaging about 3/4) to quite as long as the perigynia, most of them blunt to acute . . . . *C. garberi*

Carex aurea Nutt.
GOLDEN-FRUIT SEDGE   *native / cons* 5
Plants small, loosely tufted. Stems upright, 3-angled, 5-30 cm long. Leaves 1-4 mm wide. Spikes 2-5 per stem, the lower spikes stalked; spikes at ends of stems staminate, 3-18 mm long; lateral spikes pistillate, 8-20 mm long, the spikes clustered to widely spaced; bract of lowest spike longer than the head; pistillate scales white-tinged to yellow-brown, with a green midvein, tipped with a short, sharp point, shorter than the perigynia. Perigynia with short white hairs when young, becoming a distinctive gold-orange when mature (drying paler), round to obovate, beakless or with a very short beak, several-ribbed, 2-3 mm long. Achenes dark brown to black, lens-shaped; stigmas 2. May-July. — Moist to wet meadows, low swales, wet woods and along sandy or gravelly shores; often where calcium-rich.

Carex garberi Fern.
ELK SEDGE   *threatened / native / cons* 10
Similar to *Carex aurea;* one distinction between the 2 species is terminal spike of *C. garberi* is tipped with pistillate flowers (with staminate flowers below); in *Carex aurea,* terminal spike is of staminate flowers only. Also, in *C. garberi,* the perigynia are more granular, more crowded, and more overlapping than in *C. aurea.* — Wet sandy, gravelly, or marly shores, limestone pavements, interdunal flats, and edges of cedar thickets, typically near Lake Michigan.

Carex aurea

Carex Section Careyanae

Resembling section Laxiflorae in appearance, but stems generally firm. Perigynia acutely angled, tightly enclosing the achene; veins more than 40, impressed in fresh plants, raised when dried.

1 Bases of plants and staminate scales pale or brownish, not reddish; perigynia to 4 mm long . . . .
. . . . *C. platyphylla*
1 Bases of plants and staminate scales strongly reddish-tinged; perigynia 4–6.5 mm long . . . .
. . . . *C. plantaginea*

Carex plantaginea Lam.
PLANTAIN-LEAF SEDGE   *native / cons* 10
Plants tufted; rootstocks short. Stems 2.5-6 dm long, purple-tinged at base, the lower bladeless sheaths conspicuous. Basal leaves and those of sterile stems 15-35 cm long and 10-25 mm wide, flat, the mid-nerve prominent below, 2 lateral nerves prominent above, roughened on margins;

fertile stem leaves bladeless or nearly so; sheaths purple-tinged. Terminal spike staminate; lateral spikes about 3, pistillate, the lower on slender peduncles; bracts bladeless, purple-tinged; pistillate scales white-hyaline with green midrib, purplish tinged. Perigynia 4-12 to a spike, 4-5 mm long, triangular; beak 1 mm, erect or curved, the orifice entire, hyaline. Achenes triangular with deeply concave sides, filling perigynia, brownish, apiculate; stigmas 3. — Mesic forests.

Carex platyphylla Carey
BROAD-LEAF SEDGE   *native / cons* **10**
Plants tufted; rootstocks short. Stems 15-30 cm long, yellowish brown at base, the basal sheaths cinnamon-brown-edged. Basal leaves 1-2 dm long and 9-25 mm wide, flat, glaucous, the mid-nerve prominent below, the mid-laterals above, minutely roughened on margins; blades of fertile stems 5 cm long and 5 mm wide or less; sheaths red-dotted. Terminal spike staminate; lateral spikes 2-4, pistillate, on rough peduncles, the lower nearly basal; bracts with long sheaths entire or undulate on margins; pistillate scales acuminate to awned, cinnamon-brown-tinged, white-hyaline with green midrib. Perigynia 4-10 to a spike, 2.5-3.5 mm long, appressed-ascending, early deciduous, triangular, olive-green; beak short, erect or curved, the orifice entire, hyaline. Achenes triangular with deeply concave sides, filling perigynia, apiculate; stigmas 3. — Forests of sugar maple and beech.
Leaves strongly glaucous.

Carex plantaginea

Carex Section Ceratocystis
Plants tufted; rhizomes short; bases brown. Terminal spike staminate, occasionally androgynous. Lateral spikes pistillate, densely flowered, globose to oblong. Perigynia strongly veined, abruptly beaked; beak toothed, generally reflexed. Stigmas 3. — Usually where wet and calcareous.

1 Larger perigynia ca. 2–3 mm long, horizontally spreading, the beak about 1/4 to nearly 1/2 as long as the body . . . . *C. viridula*
1 Larger perigynia (3–) 3.5–6.2 mm long, at least the beaks becoming conspicuously reflexed on lower half of spike, the beak nearly or fully half as long as the body 2
2 Pistillate scales at maturity strongly flushed with shiny brown or reddish color, hence conspicuous in the spike; widest leaves 3–5 mm wide . . . . *C. flava*
2 Pistillate scales greenish or yellowish, the same color as the perigynia and essentially invisible in the spikes; widest leaves 1.5–4 mm wide . . . . *C. cryptolepis*

Carex cryptolepis Mackenzie
NORTHEASTERN SEDGE   *native / cons* **8**
Plants tufted. Stems 2-6 dm long and longer than leaves. Leaves 2-4 mm wide. Spikes staminate or pistillate; staminate spikes short-stalked or stalkless, the stalk shorter than the pistillate spikes; pistillate spikes 3-4, the upper 2 spikes grouped, the third separate, the fourth spike lower on stem, short-cylindric, 1-2 cm long, stalkless; bracts leaflike and spreading; pistillate scales narrowly ovate, same color as perigynia and as long as perigynia body. Perigynia yellow-brown when mature, lower ones curved outward and downward, body obovate, 3-5 mm long, 2-ribbed and several nerved, contracted into a smooth beak 1-1.5 mm long. Achenes 3-angled; stigmas 3. June-Aug. — Wet meadows and marshy areas, peatlands, swamp margins; often where calcium-rich. Similar to *C. flava*.

Carex cryptolepis

Carex flava L.
YELLOW-GREEN SEDGE   *native / cons* **6**
Plants densely tufted, from short rootstocks. Stems stiff, 1-7 dm long, usually longer than the leaves. Leaves 4-8 to a stem, mostly near base, 3-5 mm wide. Terminal spike staminate (or rarely partly pistillate), stalkless

or short-stalked; pistillate spikes 2-5, sometimes with staminate flowers at tip, the uppermost spikes nearly stalkless, the lower stalked; bracts conspicuous, leaflike, spreading outward, much longer than the head; pistillate scales ovate, narrower and much shorter than the perigynia, red-tinged except for the pale, three-nerved middle and the narrow translucent margins. Perigynia 15-35, crowded in several to many rows, 4-6 mm long, obovate, yellow-green becoming yellow with age, conspicuously ribbed, tapered to a slender, finely toothed beak about as long as the body, the tip notched. Achenes obovate, 3-angled, yellow-brown; stigmas 3. May-Aug. — Wet, peaty meadows, often where calcium-rich.

Carex flava

Carex viridula Michx.
LITTLE GREEN SEDGE   *native / cons* **6**

Plants tufted. Stems stiff, slightly 3-angled, 0.5-4 dm long, longer than leaves. Leaves 1-3 mm wide; sheaths white-translucent. Spikes either staminate or pistillate (or sometimes mixed), the terminal spike staminate or with a few pistillate flowers at tip or middle, 3-15 mm long, short-stalked or stalkless, longer than the pistillate spikes or clustered with them; lateral spikes pistillate, 2-6, ovate to short-cylindric, 5-10 mm long, clustered and stalkless above, the lower spikes often separate and on short stalks; bracts leaflike, usually upright, much longer than the heads; pistillate scales brown on sides, rounded or with a short, sharp point, about equal to perigynia. Perigynia yellow-green to brown, rounded 3-angled, obovate, 2-4 mm long, 2-ribbed, tapered to a slightly notched beak 0.5-1 mm long. Achenes 3-angled; stigmas 3. May-Aug. — Wet meadows, sandy lake margins, fens and seeps; often where calcium-rich.

Carex viridula

Carex Section Chlorostachyae
One member of the section in Door County. Plants small, densely tufted, with fibrous basal leaf sheaths and small beadlike perigynia borne in slender spikes on threadlike stalks.

Carex capillaris L.
HAIR-LIKE SEDGE   *native / cons* **9**

Plants small, densely tufted. Stems slender, 3-angled, 1.5-4 dm long. Leaves mostly at base of plant and much shorter than stems, 1-3 mm wide; sheaths tight. Spikes either staminate or pistillate; terminal spike staminate, 4-8 mm long; lateral spikes 1-4, separated on stem, loosely flowered, short-cylindric, on threadlike, spreading to drooping stalks 5-15 mm long; pistillate scales white, translucent on outer edges, green or light brown in middle, blunt or acute at tip, shorter but usually wider than perigynia, deciduous. Perigynia shiny brown to olive-green, ovate, round in section, 2-4 mm long, 2-ribbed, otherwise without nerves, tapered to a translucent-tipped beak 0.5 mm or more long. Achenes 3-angled with concave sides; stigmas 3. June-July. — Alder thickets, wetland margins, usually in shade.

Carex capillaris

Carex Section Chordorrhizae
One member of the section in Door County. Plants stoloniferous, the stolons arching and rooting.

Carex chordorrhiza Ehrh.
ROPE-ROOT SEDGE   *native / cons* **10**

Plants from long, creeping stems. Flowering stems upright, rounded 3-angled in section, 1-3 dm tall, single or several together, arising from axils of dried leaves on older, reclining sterile stems. Leaves several on stem, the lower ones often bladeless, 1-2 mm wide; sheaths translucent. Spikes 3-8, with both staminate and pistillate flowers, staminate flowers

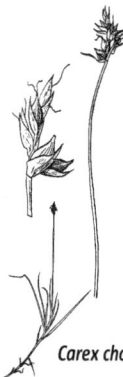

Carex chordorrhiza

borne above pistillate, crowded in an ovate head 5-15 mm long; bracts absent; pistillate scales dark brown, ovate, about equaling the perigynia. Perigynia brown, compressed, ovate, 2-3.5 mm long, leathery, with many nerves on both sides; beak short. Achenes lens-shaped; stigmas 2. May-Aug. — Open floating mats around lakes and ponds, fens, conifer swamps, interdunal hollows.

Carex Section Cyperoideae

One member of the section in Door County. Similar to the Ovales and sometimes placed within that section; distinguished by leafy bracts more than 3 times as long as the inflorescence and very long-tapering perigynia, the beak often twice as long as the body.

Carex sychnocephala Carey
Many-Head Sedge   native / cons 9

Plants tufted, from fibrous roots. Stems many and crowded, rounded 3-angled, 0.5-6 dm long. Leaves 1.5-4 mm wide; sheaths tight, white-translucent. Spikes with both staminate and pistillate flowers, pistillate flowers borne above staminate, densely clustered in ovate heads 1.5-3 cm long; bracts leaflike, 2-4 per head, the longest bracts much longer than the heads; pistillate scales thin and translucent with a green midvein, 2/3 length of perigynia, tapered to a tip or with a short sharp point. Perigynia green to straw-colored, flat, lance-shaped, 5-7 mm long and to 1 mm wide, narrowly wing-margined, spongy at base when mature, tapered to a finely toothed, notched beak 3-5 mm long. Achenes lens-shaped; stigmas 2. June-Aug. — Wet meadows, sandy lakeshores, marshes.

Carex sychnocephala

Carex Section Deweyanae

One member of the section in Door County. Plants tufted; rhizomes mostly short; bases brown. Inflorescence slender, open, at least the lowest spike(s) distinct; bracts setaceous. Spikes mostly gynecandrous, lateral spikes sometimes pistillate, mixed, or (rarely) staminate. Perigynia appressed to ascending, ovate to lanceolate, plano-convex, slender; base spongy; beak distinct, margins serrate, tip bidentate. Achenes mostly filling the perigynium body. Usually in moist to wet shaded places.

Carex deweyana Schwein.
DEWEY'S SEDGE   *native / cons 7*

Plants loosely tufted, from short rhizomes. Stems weak and spreading, 2-12 dm long, rough-to-touch below the head. Leaves shorter than stems, yellow-green to waxy blue-green, soft, flat, 2-5 mm wide; sheath tight. Spikes 2-6, the lower separate, the upper grouped, forming a head 2-6 cm long and often drooping near tip; terminal spike with staminate flowers at base, lateral spikes usually pistillate, the perigynia upright; pistillate scales ovate, blunt to short-awned at tip, thin and translucent with green center, slightly shorter than perigynia. Perigynia flat on 1 side and convex on other, 4-6 mm long, pale-green, very spongy at base, the beak 2-3 mm long, finely toothed and weakly notched. Achenes lens-shaped, nearly round, yellow-brown; stigmas 2. May-Aug. — Thickets, swamps, and moist to dry woods.

Carex deweyana

Carex Section Digitatae

Basal sheaths not fibrous. Bracts reduced to bladeless sheaths. Perigynium beaks untoothed, mostly less than 0.5 mm long. Similar to section Acrocystis but basal sheaths not fibrous.

1 Terminal spike pistillate at base; basal spikes usually present, on long very thin peduncles; pistillate scales abruptly truncate and awned; anthers 2–3 mm long . . . . *C. pedunculata*

1 Terminal spike usually entirely staminate; basal spikes not present; pistillate scales not awned; anthers various 2

2   Staminate spike 4–6 (–8) mm long; pistillate spikes less than 10 mm long; pistillate scales obtuse, minutely ciliate, distinctly shorter than the perigynia; anthers 1–1.5 mm long . . . . *C. concinna*
2   Staminate spike 10–22 mm long; pistillate spikes (often staminate at their tips) (8–) 10 mm long; pistillate scales mostly acute to acuminate, glabrous, and equaling or exceeding the perigynia; anthers 2–3.5 mm long . . . . *C. richardsonii*

Carex concinna R. Br.

LOW NORTHERN SEDGE   *threatened / native / cons* 10

Plants loosely tufted; rootstocks slender, often long, brownish black, scaly, ascending. Fertile stems 5-20 cm long, erect or incurved, roughened on angles above, dark-brownish tinged and fibrillose at base. Well-developed leaves 5-9 to a fertile culm, near base, the upper reduced; blades 5-10 cm long and 2-2.5 mm wide, flat with recurved margins, involute at base, thick, light-green. Terminal spike staminate; lateral spikes 2-3, pistillate, the lower rarely long-peduncled; bracts reduced to sheaths 7 mm long or less, light-reddish brown-tipped; pistillate scales dark-reddish brown with white-hyaline margins and obsolete or straw-colored midrib, hairy. Perigynia 5-12 to a spike, 3-3.5 mm long, ascending, obtusely triangular, whitish or greenish, hirsute, 2-ribbed and several-nerved; beak chestnut-brown, the orifice hyaline. Achenes triangular with sides convex above, closely enveloped; stigmas 3, short, blackish. — Open, moist, sandy places, usually where underlain by limestone.

Carex concinna

Carex pedunculata Muhl.

LONG-STALK SEDGE   *native / cons* 7

Rootstocks stout, woody, branching. Fertile stems 2-3 dm long, decumbent, roughened on angles, purple-tinged at base; sterile shoots purple-tinged. Upper leaves of fertile stems bladeless, the sheaths loose, reddened at base; basal leaves 3-5; blades 15-35 cm long and 2-3 mm wide, flat, thickish, pale-green, glaucous, roughened, especially on margins and nerves above. Terminal spike staminate or usually androgynous; lateral spikes 3-4, pistillate or usually androgynous, the lowest basal, long-peduncled; bracts sheathing, reddish purple-tinged at base; pistillate scales ciliate, purple with 3-nerved green center. Perigynia 1-8 to a spike, 3.5-4.5 mm long, appressed, triangular, deep-green, minutely puberulent, 2-ridged, spongy at base; beak minute, usually bent, the orifice entire. Achenes triangular with concave sides, closely enveloped in upper part of perigynia, short bent-apiculate; stigmas 3, deciduous. — Rich, mesic forests.

Carex pedunculata

Carex richardsonii R. Br.

RICHARDSON'S SEDGE   *native / cons* 7

Plants loosely tufted; rootstocks long, ascending, slender, brownish black, scaly. Stems 15-35 cm long, roughened on angles above, dark-brownish at base. Well-developed leaves 6-10 to a fertile culm, near base; blades 1-2.5 dm long and 2-2.5 mm wide, thick, light-green, rough on margins, especially towards the tip; upper leaves bladeless, the sheaths reddish purple with hyaline margins. Terminal spike staminate; lateral spikes usually 2, pistillate; bracts bladeless, purple-tinged and white-hyaline-margined; pistillate scales dark-purplish with hyaline margins and lighter midvein. Perigynia 10-25 to a spike, 2.5-3 mm long, ascending, obscurely triangular, straw-colored or light-brownish above, appressed-pubescent, 2-keeled; beak 0.5 mm, the orifice obliquely cut. Achenes triangular with sides convex above, closely enveloped, brownish, shining, conic-apiculate; stigmas 3, blackish. — Dry sandy meadows and barrens; rarely in fens.

Carex richardsonii

Carex Section Dispermae

One member of the section in Door County. Plants slender, shoots arising singly or in small bunches from pale, slender rhizomes; spikes few-flowered, androgynous; perigynia spreading, darkening at maturity, plump.

Carex disperma Dewey
SOFT-LEAF SEDGE   *native / cons* 10
Plants small, loosely tufted, from slender rhizomes. Stems slender, weak,
3-angled, 1-4 dm long, shorter to longer than leaves. Leaves soft and
spreading, 1-2 mm wide; sheaths tight, translucent. Spikes with both
staminate and pistillate flowers, staminate flowers borne above pistillate,
2-5, few flowered and small, with 1-6 perigynia and 1-2 staminate flowers,
to 5 mm long, stalkless, separate or upper spikes grouped in interrupted
heads 1.5-2.5 cm long; bracts sheathlike and resembling the pistillate
scales, or threadlike and to 2 cm long; pistillate scales white, translucent
except for the darker midrib, tapered to tip or short-awned, 1-2 mm long.
Perigynia convex on both sides to nearly round in section, oval, 2-3 mm
long, strongly nerved and rounded on the margins, beak tiny. Achenes
lens-shaped, oval; stigmas 2. May-July. — Hummocks in conifer swamps
and alder thickets, wetland margins; usually where shaded.

Carex disperma

Carex Section Glareosae

Tufted sedges of wetlands, soils often peaty. Spikes distinct, mostly nonoverlapping (except
Carex arcta which has spikes overlapping, the upper not separated), mostly or all gynecandrous,
lateral spikes sometimes pistillate. Perigynia ascending to spreading; margins rounded in most
species, smooth or finely serrate, often finely papillose.

1   Lowest bract bristle-like, several times as long as its spike; perigynia mostly 2.8–3.8 (–4) mm long, in-
    cluding very short smooth beak; spikes widely separated, containing 1–5 perigynia each. *C. trisperma*
1   Lowest bract absent or at most about twice as long as its spike (if rarely prolonged, the perigynia
    smaller and often with serrulate beak); perigynia and spikes various. 2
2   Perigynia broadest near the base of the body, with a conspicuous beak 0.7–1.1 mm long; spikes
    mostly 7–15, usually ± overlapping or crowded into an ovoid to narrowly pyramidal head 2–4.5 cm
    long. . . . . *C. arcta*
2   Perigynia broadest at or near the middle of the body; beak essentially absent or less than 0.6 mm
    long; spikes 2–8, at least the lower spikes well separated or, if crowded, the inflorescence only 0.6–3
    cm long 3
3   Perigynia 3–9 per spike (occasionally one or two spikes on a plant, especially terminal one, with as
    many as 15), loosely spreading, becoming rich brown in age; largest leaves 1–2 mm wide; foliage and
    perigynia green when fresh . . . . *C. brunnescens*
3   Perigynia mostly 10–many per spike, appressed-ascending, greenish or dull brown in age; largest
    leaves 2–2.7 (–3.7) mm wide; foliage and perigynia glaucous or gray-green at least when fresh. . . . .
    . . . . *C. canescens*

Carex arcta Boott
NORTHERN CLUSTER SEDGE   *native / cons* 8
Plants loosely to densely tufted, from very short thick rhizomes. Stems
2-8 dm long, soft, sharply triangular, very rough-to-touch above. Leaves
clustered near base, light-green, flat, 2-4 mm wide, very rough; sheaths
loose, purple-dotted. Spikes 5-15, each with both staminate and pistillate
flowers, the staminate small and below the pistillate; flowers crowded
in oblong heads, 1.5-3 cm long, upper spikes densely packed, lower spikes
slightly separate; pistillate scales ovate, acute, translucent with a brown-
tinged center, shorter than the perigynia. Perigynia flat on 1 side and
convex on other, ovate, 2-3 mm long, green to straw-colored or brown
when mature, covered with white dots, widest near the broad base, tapered to a sharp-toothed, notched beak 0.5-1.5 mm long. Achenes lens-
shaped, brown; stigmas 2. June-Aug. — Floodplain forests, old river channels, swamps and wetland margins.

Carex arcta

Carex brunnescens (Pers.) Poir.
BROWNISH SEDGE   *native / cons* **7**
Plants densely tufted, from a short fibrous rootstock. Stems sharply 3-angled, to 5 dm long, smooth or slightly rough-to-touch below the head. Leaves 1-3 mm wide; sheaths tight, thin and translucent. Spikes 5-10 in a head 2-5 cm long, all with pistillate flowers borne above staminate, each spike with 5-15 perigynia, lower spikes separated; lowermost bract bristlelike, shorter or longer than lowermost spike; pistillate scales ovate, rounded or acute at tip, shorter than the perigynia. Perigynia 3-angled, not winged or sharp-edged, 2-3 mm long, faintly nerved on both sides, not spongy-thickened at base, tapered at tip to a short, minutely notched beak, the beak and upper body finely toothed and white-dotted. Achenes lens-shaped; stigmas 2. June-Aug. — Wet forests and swamps, peatland margins.

Carex brunnescens

Carex canescens L.
HOARY SEDGE   *native / cons* **8**
Plants tufted. Stems 2-6 dm long. Leaves waxy blue- or gray-green, 2-4 mm wide, mostly near base of plant and shorter than stems. Spikes 4-8, silvery green or grayish, with both staminate and pistillate flowers, the staminate below the pistillate, ovate to cylindric, 5-10 mm long, the lower spikes more or less separate, each spike with 10-30 perigynia. Perigynia flat on one side and convex on other, 2-3 mm long and 1-2 mm wide, with a beak to 0.5 mm long, not noticeably finely toothed on the margins; pistillate scales shorter than perigynia. Achenes lens-shaped; stigmas 2. May-July. — Peatlands, tamarack swamps, floating mats, swamps, alder thickets, wet forest depressions.

Similar to *C. brunnescens* but leaves waxy blue-green rather than green and spikes somewhat larger and silver-green vs. brown.

Carex canescens

Carex trisperma Dewey
THREE-SEED SEDGE   *native / cons* **9**
Loosely tufted perennial, with short, slender rhizomes. Stems very slender and weak, 2-7 dm long. Leaves 1-2 mm wide. Spikes 1-3 (usually 2), stalkless, 1-4 cm apart in a slender, often zigzagged head, each spike with 2-5 perigynia and a few staminate flowers at the base; lowest spike subtended by a bristlelike bract 2-4 cm long; pistillate scales ovate, translucent with a green center, shorter or equal to the perigynia. Perigynia flat on 1 side and convex on other, oval, 3-4 mm long, finely many-nerved, tapered near tip to a short, smooth beak 0.5 mm long. Achenes oval-oblong, filling the perigynia; stigmas 2. May-Aug. — Forested wetlands and conifer swamps, alder thickets.

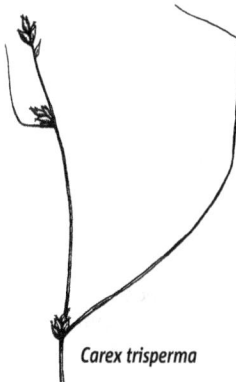

Carex trisperma

Carex Section Granulares
Plants tufted or shoots arising singly from elongate rhizomes. Pistillate spikes oblong to narrowly oblong, densely packed with perigynia. Pistillate scales and perigynia dotted or finely streaked with red. Perigynia more than 25 per pistillate spike; veins 25-40, raised.

1   Staminate spike long-peduncled, elevated above summit of uppermost pistillate spikes; lowest pistillate spike usually on a separate basal peduncle; stems mostly solitary from elongate rhizomes; widest leaves 1.5–4 mm broad . . . . ***C. crawei***
1   Staminate spike sessile or nearly so; lowest pistillate spike not on a basal peduncle; stems clumped, without elongate rhizomes; widest leaves 4.5–10 mm broad . . . . ***C. granularis***

Carex crawei Dewey
CRAWE'S SEDGE   *native / cons* **8**
Plants from long-creeping rhizomes. Stems single or several together, faintly 3-angled, 0.5-4 dm long. Leaves 1-4 mm wide. Spikes either stami-

nate or pistillate, cylindric, densely flowered, 1-3 cm long, terminal spike staminate; lateral spikes pistillate, 2-5, separate, the lowest spike near base of plant; bract leaflike, the blade shorter than the terminal spike; pistillate scales red-brown with a pale or green midrib, shorter and narrower than the perigynia. Perigynia green to brown, ovate, 2-3.5 mm long, many-nerved; beak absent or very short, entire to notched. Achenes 3-angled; stigmas 3. May-July. — Wet to moist meadows, marly lakeshores, ditches, especially where calcium-rich.

Carex crawei

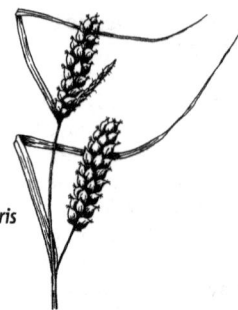

Carex granularis Muhl.
LIMESTONE-MEADOW SEDGE   native / cons 3

Plants tufted, from short rhizomes. Stems rounded 3-angled, 1-5 dm long. Leaves often longer than stems, 3-13 mm wide; sheaths membranous on front, divided-with small swollen joints on back. Spikes either all staminate or pistillate, the terminal spike staminate, stalkless; the lateral spikes pistillate, clustered around the staminate spike; bracts longer than the head; pistillate scales brown, tapered to tip or with a short, sharp point, half as long as perigynia. Perigynia crowded in several rows, green or olive to brown, oval to obovate, 2-3 mm long, 2-ribbed, strongly nerved; beak tiny or absent, entire to slightly notched. Achenes 3-angled; stigmas 3. May-July. — Wet to moist meadows and swales, streambanks and pond margins, especially where calcium-rich.

Carex granularis

Carex Section Griseae
Perigynia round or obtusely angled in cross-section, many-veined; veins impressed on both fresh and dried plants. Pistillate scales awned.

Carex hitchcockiana Dewey
HITCHCOCK'S SEDGE   native / cons 8

Plants tufted; rootstocks short. Stems 1.5-7 dm long, roughened above, brownish tinged at base. Leaves 3-4 to a culm, 1-2.5 dm long and 3-7 mm wide, flat, thin, light-green, the midvein conspicuous below, roughened on margins and towards tip on veins; sheaths conspicuously prolonged, cinnamon-brown-tinged; ligule ciliate. Terminal spike staminate, the peduncle rough; lateral spikes 3-4, pistillate, the rachis zigzag, the lowest separate; bracts leaflike, reduced upwards, the sheaths rough-hairy; pistillate scales rough-awned, serrulate, keeled, all white-hyaline with 3-nerved green center. Perigynia 1-9 to a spike, 4.5-5 mm long, ascending, obtusely triangular, somewhat leathery, yellowish or grayish green, puncticulate, spongy at base; beak 1 mm long, straight or bent, the orifice hyaline, entire. Achenes triangular, filling perigynia, yellowish brown, granular, bent-apiculate; stigmas 3, red-brown. — Mesic woods.

Carex hitchcockiana

Carex Section Heleoglochin
Plants densely tufted; bases brown. Stems narrowing toward the tip, typically arching at maturity. Inner band of the leaf sheaths smooth, pigmented toward the summit. Leaf blades less than 3 mm wide (ours). Spikes androgynous, the lower branched. Perigynia plano-convex to biconvex, darkening at maturity, mostly less than 3 mm long; beak short-triangular, scabrous on the margin, bidentate. — Wetlands, primarily in peaty soils.

1 Leaf sheaths whitish or pale ventrally except for purplish dots; inflorescence ± crowded, the lowermost spike (or branch) usually at least slightly overlapping the next above it (occasionally separated by a distance no more than its total length); perigynia tending to spread at maturity, therefore not concealed by the scales . . . . *C. diandra*
1 Leaf sheaths strongly tinged with copper color toward their summits ventrally; inflorescence interrupted, the lowermost spikes (or branches) often well separated or even peduncled; perigynia appressed at maturity, nearly or completely concealed by the large scales . . . . *C. prairea*

Carex diandra Schrank
LESSER TUSSOCK SEDGE   *native / cons* **9**
Plants densely tufted. Stems sharply 3-angled, 3-8 dm long, usually longer than leaves. Leaves 1-3 mm wide; sheaths white with fine pale lines, translucent on front or slightly copper-colored at mouth. Spikes with both staminate and pistillate flowers, staminate flowers borne above pistillate, clustered in ovate heads 1-4 cm long; bracts small and inconspicuous, shorter than the spikes; pistillate scales brown, tapered to tip or with a short sharp point, about equaling the perigynia. Perigynia brown, shiny, unequally convex on both sides, broadly ovate, 2-3 mm long, beak finely toothed, entire to notched, 1-2 mm long. Achenes lens-shaped; stigmas 2. May-July. — Wet meadows, ditches, peatlands (especially calcareous fens), floating mats.

Carex diandra

Carex prairea Dewey
PRAIRIE SEDGE   *native / cons* **10**
Plants densely tufted, from short rootstocks. Stems sharply 3-angled, 5-10 dm long. Leaves 2-3 mm wide; sheaths translucent, yellow-brown or bronze-colored. Spikes with both staminate and pistillate flowers, staminate flowers borne above pistillate, ovate, 4-7 mm long, lower spikes usually separate, in linear-oblong heads 3-8 cm long; bracts small; pistillate scales red-brown, tapered to tip, as long as and covering most of perigynia. Perigynia dull brown, flat on 1 side and convex on other, lance-shaped to ovate, 2-3 mm long, tapered to a finely toothed, unequally notched beak 1-2 mm long. Achenes lens-shaped; stigmas 2. May-July. — Wet meadows, calcareous fens, marshes, tamarack swamps and peaty lakeshores.

Carex prairea

Carex Section Hirtifoliae
One member of the section in Door County; recognized by the soft pubescence covering the entire plant, including the distinctly beaked, 2-ribbed perigynia.

Carex hirtifolia Mackenzie
PUBESCENT SEDGE   *native / cons* **5**
Plants loosely tufted; rootstocks slender, branched. Stems 3-6 dm long, pubescent, roughened above, brownish red at base. Well-developed leaves 3-4 to a fertile culm, more on sterile stems; blades to 35 cm long and 3-7 mm wide, flat, flaccid, hirsute, the nerves prominent above; sheaths cinnamon-brown ventrally. Terminal spike staminate; lateral spikes 2-4, pistillate; lowest bract 1.5-7 cm long, the upper shorter; pistillate scales whitish, ciliate, with green excurrent midrib. Perigynia 10-25 to a spike, 3.5-5 mm long, triangular, green, pubescent, nerveless; beak ca. 1 mm long, obliquely cut, 2-toothed. Achenes sharply triangular with concave sides, short-apiculate; stigmas 3, reddish brown. — Rich mesic woods.

Carex hirtifolia

Carex Section Holarrhenae
One member of the section in Door County; distinguished by the green-veined inner band of its leaf sheaths.

Carex sartwellii Dewey
SARTWELL'S SEDGE   *native / cons* **7**
Plants colony-forming, from long black rhizomes. Stems single or few together, stiff, sharply 3-angled, 3-8 dm long. Leaves 2-4 mm wide, few per stem, the lowest leaves small and without blades; sheaths with green lines on front, and a translucent ligule around stem. Spikes with both staminate and pistillate flowers, staminate flowers above pistillate, or upper spikes staminate; clustered or lower spikes separate, 5-10 mm

Carex sartwellii

long, in cone-shaped heads, 3-6 cm long; bracts small, the lower bracts sometimes bristlelike and longer than the spike; pistillate scales brown with a prominent green midvein, about equal to perigynia. Perigynia tan to brown, flat on 1 side and convex on other, ovate, 2.5-3.5 mm long, finely nerved on both sides, sharp-edged, tapered to a short, finely toothed beak. Achenes lens-shaped; stigmas 2. May-July. — Wet to moist meadows, marshes, fens and shores, often where calcium-rich.

Carex Section Hymenochlaenae

Includes nearly all of the forest understory sedges with long, nodding pistillate spikes. Superficially similar to section Gracillimae but plants more delicate. Terminal spike wholly staminate. Perigynia 8-45 per spike, narrow and long-tapering to the beak. — Woodlands and wetlands.

1 Terminal spike gynecandrous; sheaths ± softly pubescent or perigynia essentially beakless (except *C. prasina*). 2
1 Terminal spike staminate; sheaths glabrous (except *C. castanea*) and perigynia conspicuously beaked 4
2 Perigynia strongly angled, gradually tapering into a beak 1–1.5 mm long; bract of lowest pistillate spike sheathless or with sheath up to 1.2 cm long; terminal spike mostly staminate, with at most a few perigynia at apex . . . . **C. prasina**
2 Perigynia obscurely angled or nearly terete, essentially beakless or beak less than 0.5 mm long; bract of lowest spike with sheath 1.5–8 cm or more in length; terminal spike staminate at base, pistillate toward apex 3
3 Perigynia 1.3–1.6 mm wide, beakless; sheaths and blades glabrous. . . . . **C. gracillima**
3 Perigynia 1.7–2.5 mm wide, abruptly contracted to a short beak; sheaths and leaf blades ± softly pubescent, at least below (sometimes very sparsely so) . . . . **C. formosa**
4 Leaf sheaths and blades (at least toward the base) ± hairy; pistillate spikes 1–2.5 cm long **C. castanea**
4 Leaf sheaths and blades glabrous (at most the lowermost bladeless sheaths minutely hispidulous); pistillate spikes mostly (2–) 2.5–6.5 cm long 5
5 Basal sheaths reddish purple for at least several cm above the base; perigynia clearly nerved between the 2 ribs . . . . **C. arctata**
5 Basal sheaths brown, lacking any trace of reddish purple color (at most a small trace on the smaller sheaths in *C. prasina*); perigynia 2-ribbed, but otherwise nerveless or faintly nerved . . . . . . **C. prasina**

Carex arctata Boott.
DROOPING WOODLAND SEDGE   *native / cons* **5**
Plants tufted. Stems 3-9 dm long, the basal sheaths purple-tinged at base. Leaves 2-3 per fertile culm; blades 2-3 dm long and 5-10 mm wide long, ca. 3 mm wide, flat, thin, soft, deep-green, roughened on margins and towards the tip; sheaths minutely roughened, yellowish brown-tinged and red-dotted ventrally. Terminal spike staminate; lateral spikes 3-5, pistillate, nodding on slender rough peduncles; lowest bract leaflike, the upper reduced; pistillate scales awned, greenish white with green center, ciliate, thin. Perigynia 15-45 to a spike, 3-5 mm long, ascending, deep-green, puncticulate, 2-ribbed; beak 0.75 mm long, bidentate, hyaline above, ciliate between teeth. Achenes triangular with sides concave below, closely enveloped, yellowish brown, granular, apiculate; stigmas 3, short, blackish. Deciduous forests. Similar to *C. gracillima* but perigynia of *C. arctata* taper to the beak and are constricted at base to form a short stipe, and its terminal spikes are staminate.

Carex castanea Wahlenb.
CHESTNUT-COLOR SEDGE   *native / cons* **6**
Plants tufted. Stems 3-10 dm long, purple-tinged at base. Leaves 3-6 mm wide, softly hairy. Spikes either staminate or pistillate; the terminal spike staminate, upright atop a long stalk; lateral spikes pistillate, usually 3, on slender, drooping stalks, short cylindric; pistillate scales ovate, brown-tinged, about as long as perigynia. Perigynia lance-shaped, 4-6 mm long,

Carex castanea

somewhat 3-angled, strongly 2-ribbed with several faint nerves, tapered to a notched beak up to half length of body. Achenes 3-angled; stigmas 3. June–July. — Swamps, moist openings, wetland margins and ditches.

Carex formosa Dewey
HANDSOME SEDGE *threatened / native / cons* **9**
Plants densely tufted. Stems 3-8 dm long, dark maroon at base; flowering stems 0.5-1 mm thick, scabrous on angles within inflorescence. Leaves flat, 3-6 mm wide, glabrous on undersurface, pilose on uppersurface and margins; sheaths pubescent; lowest bracts shorter or equal to tip of the inflorescence. Spikes usually all gynecandrous, pubescent; lateral spikes 2-4, well separated, on slender peduncles to 10 cm long and usually much longer than spikes, drooping at maturity, pistillate except for 1 or 2 basal staminate flowers in each; pistillate scales hyaline tinged with chestnut brown, the broad green midrib red-dotted, shorter than the mature perigynia. Perigynia intermediate in size between those of *C. gracillima,* copiously red dotted, 2-ribbed, loosely enveloping achene, tip narrowed to abrupt beak; beak minutely bidentate, less than 0.5 mm long; stigmas 3. — Rich mesic forests, usually where soils calcareous.

Carex gracillima Schwein.
GRACEFUL SEDGE *native / cons* **5**
Plants tufted; rootstocks short, slender. Stems 2-9 dm long, purple-tinged at base. Leaves 3-4 to a culm, on lower half; blades 1-3 dm long, flat, flaccid, deep-green, roughened on margins and towards tip; sheaths yellowish brown-tinged and reddish dotted. Terminal spike gynaecandrous or staminate; lateral spikes 3-4, pistillate, separate on slender nodding roughish peduncles; lowest bract leaflike, sheathing, the upper shorter; scales whitish or yellowish brown with green midrib. Perigynia 10-45 to a spike, 2.5-3.5 mm long, ascending, puncticulate, few-nerved, beakless. Achenes triangular with concave sides and thick angles; stigmas 3, short, blackish. — Mesic to wet forests, sometimes in drier oak woods.

Carex gracillima

Carex prasina Wahlenb.
DROOPING SEDGE *native / cons* **10**
Plants tufted. Stems 3-8 dm long, brown or green at base. Leaves 3-5 mm wide. Terminal spike staminate or with a few pistillate flowers at tip; pistillate spikes 2-4, widely separated, cylindric, 2-5 cm long and 5 mm wide, curved or nodding, lower spikes on long stalks, the upper stalks much shorter; upper bract more or less sheathless; pistillate scales ovate to obovate, shorter than the perigynia, tipped with an awn or short point. Perigynia 3-4 mm long, ovate, 3-angled, tapered to beak. Achenes 3-angled; stigmas 3. May–June. — Springs, seeps and low areas in deciduous woods, shaded streambanks.

Carex prasina

Carex Section Laxiflorae
Plants tufted; bases pale to brown or occasionally reddish. Stems weak, ascending to decumbent, sharply triangular in cross-section, angles sometimes winged. Perigynia triangular in cross-section with rounded edges, 25-40-veined (except *Carex leptonervia*); beak (in our species) abrupt, short, often bent. — Woodland species.

1 Sides of perigynia with at most 1 main nerve, otherwise nerveless or each with up to 6 obscure nerves; perigynium with a straightish or slightly bent short beak . . . . *C. leptonervia*
1 Sides of perigynia each with 7 or more conspicuous nerves; perigynium with straightish or strongly bent beak 2
2 Angles of bract sheaths granular-papillose; beak of perigynium usually straight or slightly bent . . . . *C. ormostachya*
2 Angles of bract sheaths minutely ciliate-serrulate; beak of perigynium strongly bent 3

3   Widest leaves 8 mm or more broad; pistillate scales broadly obtuse or truncate, at most scarcely toothed at apex; staminate spike sessile or nearly so . . . . *C. albursina*
3   Widest leaves often less than 8 mm broad; pistillate scales acuminate, awned, or cuspidate; staminate spike sessile or short-peduncled . . . . *C. blanda*

Carex albursina Sheldon
WHITE BEAR SEDGE   *native / cons 7*

Plants loosely tufted; rootstocks short. Stems 1-6 dm, winged, roughened on angles, dark-brown at base. Well-developed leaves 2-5 to a fertile culm; blades 6-25 cm long and 7-15 mm wide, thin, flaccid, light-green, the midvein prominent below, the midlaterals above, roughened towards tip and on margins; sheaths loose. Terminal spike staminate; lateral spikes 3-4, pistillate, the lower separate; bracts sheathing, rough-edged, the lower leaflike; scales white-hyaline with 3-nerved green center. Perigynia 3-18 to a spike, 3-4 mm long, erect, obtusely triangular, yellowish green, the base spongy; beak 0.5 mm long, bent, the orifice entire, hyaline. Achenes triangular with concave sides, closely enveloped, yellowish brown; stigmas 3, reddish brown. — Mesic forests.

Carex albursina

Carex blanda Dewey
EASTERN WOODLAND SEDGE   *native / cons 3*

Plants tufted; rootstocks short. Stems 1-6 dm long, slightly winged, 2-edged and flattened in drying, minutely serrulate above, brownish at base; sterile stem leaf blades 1-3.5 dm long and 4-15 mm wide, flat, thin, flaccid, light-green, the midvein prominent below, roughened on margins; blades of fertile stems smaller. Terminal spike staminate or gynaecandrous; lateral spikes 2-5, pistillate, the lower separate on slender, 2-edged peduncles; bracts leaflike; pistillate scales awned, greenish white with 3-nerved green center. Perigynia 8-25 to a spike, 3-4 mm long, ascending, obtusely triangular, yellowish green, strongly nerved, the base spongy; beak 0.5 mm long, bent, the orifice entire, hyaline. Achenes triangular, yellowish brown, granular; stigmas 3, short, reddish brown. — Mesic to wet deciduous forests, sometimes in moist open places, tolerant of disturbance.

Carex blanda

Carex leptonervia (Fern.) Fern.
NERVELESS WOODLAND SEDGE   *native / cons 6*

Plants tufted; rootstocks slender. Stems 1.5-7 dm long, weakly erect or decumbent, the angles minutely serrulate, brownish at base. Leaves to 3.5 dm long and 3-10 mm wide, flat, flaccid, deep-green, the midvein prominent below, roughened on margins towards tip; sheaths enlarged upward; ligule long. Terminal spike staminate; lateral spikes 2-4, pistillate, the lower 1-2 separate, rough-peduncled; bracts with sheath-margins serrulate; pistillate scales white-hyaline, brownish tinged; anthers 1.5 mm. Perigynia 10-20 to a spike, 3.5-4.5 mm long, erect-ascending, obtusely triangular, light-green, glandular-puncticulate, 2-ribbed, the base spongy; beak 0.5 mm long, the orifice entire, oblique. Achenes triangular with concave sides, filling perigynia, brownish, granular; stigmas 3, reddish brown. — Woodlands.

Carex leptonervia

Carex ormostachya Wieg.
NECKLACE SPIKE SEDGE   *native / cons 6*

Plants tufted; rootstocks short. Stems 2-6 dm long, minutely granular, crenulate on angles, purplish at base. Leaves 5-20 cm long and 2-8 mm wide, flat, flaccid, light-green, the midvein prominent below, roughened on margins towards the sharp tip; sheaths tight. Terminal spike staminate; lateral spikes 3-5, pistillate, loosely flowered below, the lower separate, the peduncles minutely serrulate; pistillate scales mucronate to awned, hyaline with 3-nerved greenish center, sometimes reddish brown-tinged. Perigynia 10-20 to a spike, 2.5-3.5 mm long, appressed-ascending, obtusely triangular, dull-brownish, strongly nerved, the base spongy; beak minute, straight or bent, the orifice hyaline, truncate or oblique. Achenes triangular with concave sides and blunt angles, yellowish; stigmas 3, short, reddish brown. — Woodlands.

Carex Section Leptocephalae

One member of the section in Door County. Plants soft, very slender, rhizomatous; spike androgynous (with staminate flowers at tip, pistillate flowers below); perigynia few.

Carex leptalea Wahlenb.
BRISTLY STALK SEDGE   *native / cons* **9**

Densely tufted perennial. Stems slender, rounded 3-angled, 1-7 dm long, equal or longer than leaves. Leaves narrow, 0.5-1.5 mm wide; sheaths tight, white, translucent on front. Spikes single on the stems, few-flowered, 5-15 mm long, with both staminate and pistillate flowers, the staminate flowers borne above pistillate; bracts absent; pistillate scales rounded or with a short sharp point, shorter than the perigynia (or the tip of lowest scale sometimes longer than the perigynium). Perigynia yellow-green, nearly round in section to slightly flattened, oblong to oval, 3-5 mm long, finely many-nerved, beakless or with a short beak. Achenes 3-angled; stigmas 3. May-July. — Swamps, alder thickets, open bogs, calcareous fens; usually in partial shade.

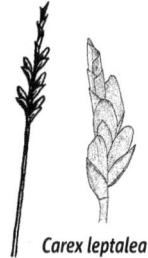

Carex leptalea

Carex Section Limosae

Plants loosely tufted or stems arising singly, strongly rhizomatous; bases reddish. Roots covered in a dense yellow felt-like tomentum. Vegetative shoots becoming decumbent, behaving like stolons, producing shoots at the nodes. Pistillate spikes pendulous on slender stalks. Perigynia pale, short-beaked, papillose. Stigmas 3. — Bogs and fens.

1 Pistillate scales nearly or quite as broad as the perigynia and often only slightly if at all longer; staminate spike (12–) 15–30 (–50) mm long; plants strongly stoloniferous . . . . *C. limosa*
1 Pistillate scales distinctly narrower than perigynia, generally with narrowly acuminate tips much exceeding them; staminate spike 5–12 (–15) mm long; plants loosely clumped . . . . *C. magellanica*

Carex limosa L.
MUD SEDGE   *native / cons* **10**

Plants loosely tufted, from long, scaly, yellow-felted rhizomes. Stems sharply 3-angled, 3-5 dm long, longer than leaves, usually rough-to-touch above. Leaves involute, 1-3 mm wide; sheaths translucent, shredding into threadlike fibers near base. Spikes either all staminate or pistillate, the terminal spike staminate; the lower 1-3 spikes pistillate, drooping on lax, threadlike stalks 1-3 cm long; pistillate scales brown, rounded or with a short, sharp point, about same size as perigynia. Perigynia waxy blue-green, ovate, flattened except where filled by achene, 2.5-4 mm long, strongly 2-ribbed with a few faint nerves on each side; beak tiny. Achenes 3-angled; stigmas 3. May-July. — Open bogs and floating mats, calcareous fens.

Poor sedge (*C. magellanica*) similar but has scales much narrower than perigynia; *C. buxbaumii* also similar but lacks yellow roots.

Carex magellanica Lam.
POOR SEDGE   *native / cons* –
   Carex paupercula Michx.

Plants loosely tufted, from slender, branching, yellow-felted rhizomes. Stems slender, 1-8 dm long, longer than the leaves, red-brown at base. Leaves 3-12 on lower half of stem, flat but with slightly rolled under mar-

Carex limosa

gins, 2-4 mm wide, the dried leaves of previous year conspicuous; sheaths red-dotted. Terminal spike staminate (or sometimes with a few pistillate flowers at tip), on a long stalk, usually upright; pistillate spikes 1-4 (rarely with several staminate flowers at base), clustered, usually drooping on slender stalks; lowest bract leaflike, equal or longer than the head; pistillate scales lance-shaped to ovate, narrower but usually longer than the perigynia, brown or green in center, margins brown. Perigynia broadly ovate or oval, 2-3 mm long, flattened and 2-ribbed, with several evident nerves, pale or somewhat waxy blue-green, covered with many small bumps, the tip rounded and barely beaked. Achenes 3-angled, obovate, 2 mm long; stigmas 3. July-Aug. — Open bogs, partly shaded peatlands, floating mats, cedar swamps and thickets, usually in sphagnum moss.

Carex magellanica

Carex Section Lupulinae
Distinctive sedges of wet forests; recognized by the strongly inflated, ribbed perigynia, 1-2 cm long.

1 Pistillate spikes cylindrical or short-oblong, usually definitely longer than broad; sheath of uppermost leaf usually 1.7 cm or longer; style strongly bent and contorted immediately above the body of the achene; beak of perigynium nearly or quite as long as the body . . . . *C. lupulina*
1 Pistillate spikes spherical or nearly so, scarcely if at all longer than wide; sheath of uppermost leaf absent or less than 1.5 (–2.5) cm; style straight or sinuous or contorted (especially in *C. intumescens*) just below or at the middle; beak of perigynium much shorter than the body 2
2 Perigynia (7–) 10–31 per spike, radiating in all directions, narrowed at the base to a broad cuneate stalk, sometimes hispidulous basally; pistillate spikes 1–2 (–3) . . . . *C. grayi*
2 Perigynia 2–8 (–12) per spike, mostly spreading-ascending, rounded at the base, glabrous (and often very shiny); pistillate spikes (1–) 2–5 . . . . *C. intumescens*

Carex grayi Carey
GRAY'S SEDGE   *native / cons 7*
Rhizomes absent. Stems single or forming small clumps, 3-9 dm long, rough on upper stem angles, sheaths at base of stem persistent, red-purple. Leaves 5-12 mm wide. Spikes either staminate or pistillate; terminal spike staminate, stalked; pistillate spikes 1-2, rounded, stalked; bracts leaflike; pistillate scales ovate, body shorter than perigynia but sometimes tipped with an awn to 7 mm long. Perigynia 10-30 per spike, spreading in all directions, not shiny, 10-20 mm long, strongly nerved, tapered from widest point to a notched beak 2-3 mm long. Achenes with a persistent, withered style; stigmas 3. June-Sept. — Wet forests.

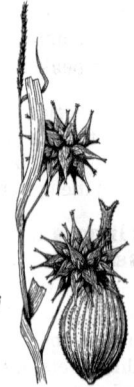

Carex grayi

Carex intumescens Rudge
GREATER BLADDER SEDGE   *native / cons 5*
Rhizomes absent. Stems single or in small clumps, 3-9 dm long, rough on upper stem angles; sheaths at base of stem persistent, red-purple. Leaves 4-12 mm wide, bracts leaflike. Spikes either staminate or pistillate, or sometimes staminate spikes with a few pistillate flowers; terminal spike staminate, stalked; pistillate spikes 1-4, rounded, on stalks to 1.5 cm long; pistillate scales narrowly ovate, shorter and narrower than perigynia. Perigynia 1-12 per spike, spreading in all directions, satiny (not dull), 10-17 mm long, tapered to a beak 2-4 mm long. Achenes flattened; stigmas 3. May-Aug. — Mixed and deciduous moist forests, kettle wetlands in woods, swamps and alder thickets.

Carex intumescens

Carex lupulina Muhl.
HOP SEDGE   *native / cons 6*
Plants loosely tufted, from rhizomes. Stems stout, 3-12 dm long. Leaves much longer than head, 4-15 mm wide; upper sheaths white and translucent, the lower sheaths brown. Spikes either all staminate or pistillate,

the upper spike staminate, short-stalked, 2-5 cm long; pistillate spikes 2-6, clustered or overlapping, the lowermost sometimes separate; bracts leaflike and spreading, much longer than head; pistillate scales narrowly ovate, tapered to tip or with a short awn, much shorter than the perigynia. Perigynia many, upright, dull green-brown, lance-shaped, inflated, 10-20 mm long, many-nerved, tapered to a finely toothed bidentate beak 5-10 mm long. Achenes 3-angled; stigmas 3. June-Aug. — Wet woods, swamps, wet meadows and marshes, ditches and shores.

Shining bur sedge (*C. intumescens*) is similar but differs from hop sedge by having fewer, uncrowded perigynia which are olive-green and glossy.

Carex lupulina

Carex Section Multiflorae

Plants tufted; bases fibrous, brown or pale. Inner band of the leaf sheaths hyaline, corrugated. Inflorescence compound, cylindrical, densely flowered, stiff. Bracts setaceous. Spikes androgynous, at least the lowest branched. Perigynia plano-convex, weakly or inconspicuously spongy at the base. Primarily in wetlands. Characterized by the corrugated inner band of the leaf sheaths; firm, narrow stems; and densely flowered, straight, compound inflorescence.

Carex vulpinoidea Michx.
COMMON FOX SEDGE   *native / cons* 2

Plants densely tufted, from short rootstocks. Stems stiff, sharply 3-angled, 3-9 dm long. Leaves 2-4 mm wide; sheaths tight, cross-wrinkled and translucent on front, mottled green and white on back. Spikes with both staminate and pistillate flowers, staminate flowers borne above pistillate; heads oblong to cylindric, 3-9 cm long, with several spikes per branch at lower nodes; bracts small and bristlelike, longer than the spikes; pistillate scales awn-tipped, the awns equal or longer than the perigynia. Perigynia yellow-green, becoming straw-colored or brown when mature, flat on 1 side and convex on other, ovate to nearly round, 2-3 mm long, abruptly contracted to a notched, finely toothed beak 1 mm long. Achenes lens-shaped, 1-2 mm long; stigmas 2. May-Aug. — Wet to moist meadows, marshes, lakeshores, streambanks, roadside ditches.

Carex vulpinoidea

Carex Section Ovales

In general, Ovales are characterized by a tufted habit, brownish basal sheaths, and sterile shoots with both nodes and internodes; this is in contrast to the sterile shoots of most species of *Carex*, where the stem-like portion is formed only of overlapping leaf sheaths, and nodes and internodes are absent. Mature perigynia (and often a dissecting microscope) are often needed for accurately identifying species in this large group. Considering the preferred moisture regime of Door County Ovales may help narrow the list of possible species:

Wetlands: *C. bebbii, C. cristatella, C. scoparia, C. tenera.*
Non-wetlands: *C. brevior, C. foenea.*

1 Pistillate scales about or fully as long as the perigynia and nearly the same width as the beaked portion (not necessarily the body), so that the tip of each perigynium is largely concealed; anthers 1.5–3 mm long. . . . . *C. foenea*
1 Pistillate scales (or most of them) both shorter and narrower than beaks of perigynia, so the mature perigynia are largely exposed at the tip; anthers various 2
2 Pistillate scales in the middle or lower portions of the spikes acuminate with a subulate tip or awned . . . . *C. scoparia*
2 Pistillate scales obtuse, acute or acuminate, sometimes inconspicuous in the spikes 3
3 Mature perigynia more than 2 mm broad at widest part 4
3 Mature perigynia not over 2 mm broad 5
4 Perigynium body obovate, widest above the middle; leaf sheaths green-nerved ventrally nearly to the summit with at most a narrow V-shaped hyaline area . . . . *C. albolutescens*
4 Perigynium body lanceolate, ovate, elliptic, or orbicular, widest at or below the middle; leaf sheaths various, some with prominent hyaline band near the apex ventrally. . . . . *C. brevior*

5 Perigynia thin, ± scale-like, often not winged to the base; leaf sheaths somewhat expanded towards apex and bearing narrow wings continuous with midrib and edges of leaf blade, blades 3–7 mm wide; vegetative shoots tall, conspicuous, and with numerous leaves spaced along upper 1/2 of stem . . . . *C. cristatella*

5 Perigynia thicker, plano-convex, winged to the base; leaf sheaths with ± rounded edges, not distinctly expanded towards apex, blades 1–4.5 mm wide; vegetative shoots usually inconspicuous, with leaves relatively few and clustered at apex 6

6 Perigynia 2.6–4 times longer than wide, the bodies lanceolate, the distance from beak tip to top of achene 2.2–5.0 mm. 7

6 Perigynia less than 2.5 times longer than wide, the bodies obovate, orbicular, or ovate, the distance from beak tip to top of achene 0.8–2.2 mm 8

7 Inflorescences dense or open, spikes usually overlapping; perigynia usually ascending . . . *C. scoparia*

7 Inflorescences nodding or flexuous, spikes separated; perigynia spreading . . . . *C. echinodes*

8 Inflorescences on tallest stems compact, 1.5–3 times as long as wide, erect, the spikes overlapping; lowest inflorescence internodes 1–6 (–7.5) mm long, 1/12–1/5 (–1/4) the total length of the inflorescence . . . . *C. bebbii*

8 Inflorescences on tallest stems elongate, open proximally, (2.5–) 3–5.1 times as long as wide, often arching or nodding; lowest inflorescence internodes (5–) 7–19 mm long, mostly 1/5–1/3 (–1/2) the total length of the inflorescence 9

9 At least some sheaths papillose near the collar (30–40x), not prominently whitish mottled; perigynium beaks appressed or ascending in the spikes, exceeding pistillate scales by 0–0.8 mm; beaks and shoulders of perigynia stramineous to reddish brown at maturity. . . . . *C. tenera*

9 Sheaths totally smooth, often whitish mottled; perigynium beaks spreading, mostly exceeding pistillate scales by 0.7–1.6 mm; beaks and shoulders of perigynia greenish to yellowish or greenish brown at maturity . . . . *C. echinodes*

Carex bebbii Olney
BEBB'S SEDGE   native / cons 4

Plants tufted. Stems sharply 3-angled, 2-8 dm long. Leaves 2-5 mm wide; sheaths white, thin and translucent. Spikes 5-10, with both staminate and pistillate flowers, pistillate flowers above staminate, 5-8 mm long, clustered in an ovate head 1.5-3 cm long; pistillate scales tapered to tip, narrower and slightly shorter than the perigynia. Perigynia green to brown, flat on 1 side and convex on other, ovate, 2.5-3.5 mm long, finely nerved on back, nerveless on front, wing-margined, with a finely toothed beak 1/3-1/2 the length of the body, shallowly notched at tip. Achenes lens-shaped; stigmas 2. June-Aug. — Wet to moist meadows, marshes, streambanks, ditches and other wet places; calcareous fens.

Carex brevior (Dewey) Mackenzie
SHORT-BEAK SEDGE   native / cons 3

Plants tufted; rootstocks short, somewhat woody, black. Stems 3-10 dm long, clothed with old leaves. Well-developed leaves 3-6 to a culm, on lower third; blades 1-2 dm long and 1.5-4 mm wide, thickish, light-green, roughened towards tip especially on margins; sheaths tight, white-hyaline ventrally; sterile shoots conspicuous, the leaves at top. Spikes 3-10, gynaecandrous, in a narrow head; lowest bracts lowest often 1-4 cm long, the upper acuminate or awned; scales yellowish brown with hyaline margins and 3-nerved green center. Perigynia 8-20 to a spike, 4-5.5 mm-long, ascending-spreading, thick, leathery, green above, greenish white beneath, strongly nerved dorsally, winged, serrulate; beak 1 mm long, flat, serrulate, obliquely cut, bidentate, reddish brown-tipped. Achenes yellowish brown; stigmas 2, long, reddish brown. — Open places.

Carex cristatella Britt.
CRESTED SEDGE   native / cons 4

Plants tufted, from short rhizomes. Stems sharply 3-angled, 3-10 dm long. Leaves 3-7 mm wide; sheaths loose, with fine green lines. Spikes with both staminate and pistillate flowers, pistillate flowers borne above sta-

Carex bebbii

Carex cristatella

minate; spikes 5–12, crowded in an ovate to oblong head; bracts much reduced; pistillate scales tapered to tip, shorter than the perigynia. Perigynia
widely spreading when mature, green to pale brown, flat on 1 side and
convex on other, 2.5–4 mm long, faintly nerved on both sides, strongly
winged above the middle, tapered to a finely toothed, notched beak 1–2
mm long. Achenes lens-shaped; stigmas 2. June–Aug. — Wet meadows,
ditches, floodplains, marshy shores and streambanks.

Carex foenea Willd.
BRONZE-HEAD OVAL SEDGE   *native / cons* **4**
Carex aenea Fernald

Plants densely tufted. Stems 2–12 dm long. Leaves 3–6 per fertile culm,
green, 8–30 cm long and 2–4 mm wide. Inflorescences open, usually with
widely spaced spikes, brown or greenish brown; bracts scalelike, sometimes bristlelike. Spikes usually 3–7, nodding, copper-colored; pistillate
scales usually reddish brown, with 3-veined green or brown midstripe,
equaling, and more or less covering the perigynia. Perigynia erect-ascending, green or brown, conspicuously 4–9-veined 3–5 mm long, margin
flat, including small wing 0.2–0.4 mm wide; beak white or brown, white
margined at tip, flat, serrulate, distance from beak tip to achene 1.7–2.5
mm. Achenes dark brown at maturity, ovoid-orbicular; stigmas 2. — Dry
open sandy places, roadsides, cut-over forests.

Carex foenea

Carex scoparia Schkuhr
POINTED BROOM SEDGE   *native / cons* **4**

Plants densely tufted, sometimes spreading by surface runners. Stems
2–10 dm long, sharply 3-angled. Leaves 1–3 mm wide; sheaths tight, white-
translucent. Spikes 4–10, with both staminate and pistillate flowers, pistillate flowers borne above staminate, ovate to broadest at middle, clustered or separate, in a narrowly ovate head; bracts small, the lowest often
bristlelike; pistillate scales slightly shorter than perigynia. Perigynia
greenish white, flat, 3–7 mm long, margins narrowly winged, tapered to a
finely toothed, slightly notched beak 1–2 mm long. Achenes lens-shaped;
stigmas 2. May–July. — Wet meadows and openings, swamps and sandy
lakeshores.

Carex scoparia

Carex tenera Dewey
QUILL SEDGE   *native / cons* **4**

Plants tufted, from short rhizomes. Stems slender, sharply 3-angled, 3–8
cm long, rough-to-touch above. Leaves 0.5–3 mm wide; sheaths white-
translucent on front, mottled green and white on back. Spikes 4–8, with
both staminate and pistillate flowers, pistillate flowers borne above staminate, ovate to round, loose in nodding heads; bracts small, sometimes
bristlelike, longer than the spike; pistillate scales slightly shorter than
perigynia. Perigynia ovate, flat on 1 side and convex on other, straw-colored when mature, 2.5–4 mm long, wing-margined, tapered to a notched,
finely toothed beak 1–2 mm long. Achenes lens-shaped; stigmas 2. June–
Aug. — Wet to moist meadows, streambanks, floodplains and moist woods.

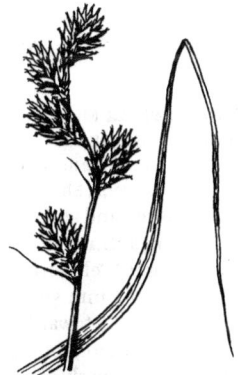

ADDITIONAL TAXON

Carex tenera var. *echinodes* sometimes treated as separate species *Carex
echinodes* (Fernald) P. Rothr., found in mesic to wet forests, in mostly s
Wisconsin. Inflorescences nodding, similar to those of *C. tenera* var. *tenera,*
but the perigynium tips arch outward and are often 1 mm or more longer
than the pistillate scales; see key.

Carex tenera

Carex Section Paludosae
Mostly slender, long-rhizomatous plants, with red basal leaf sheaths (and ladder-fibrillose), and
pubescent perigynia. *Carex lacustris* is somewhat different, having glabrous perigynia.

1 Perigynia glabrous . . . . *C. lacustris*
1 Perigynia pubescent 2
2 Leaf blades involute to triangular-channeled, 0.7–2 (–2.2) mm wide, those of vegetative shoots espe-
   cially long-prolonged into a curled, filiform tip; leaves and lowermost bracts with the midvein low,
   rounded, and forming an inconspicuous keel (at least proximally). . . . . *C. lasiocarpa*
2 Leaf blades flat or folded into an M-shape except at the base and near the tip, (2–) 2.2–4.5 (–6.5) mm
   wide, not prolonged into a long filiform tip; leaves and lowest bract with the midvein forming a
   prominent and sharply pointed keel for much of the length . . . . *C. pellita*

Carex lacustris Willd.

LAKEBANK SEDGE   *native / cons* 6

Plants large, tufted, from scaly rhizomes. Stems erect, 3-angled, 6-13 dm
long, rough-to-touch. Leaves 6-15 mm wide; sheaths often red-tinged,
the lower ones disintegrating into a network of fibers. Spikes either staminate or pistillate, the upper 2-4 staminate, stalkless; the lower 2-4
spikes pistillate, erect, usually separate, stalkless or short-stalked, cylindric; bracts leaflike, some or all longer than the head; pistillate scales
awned or tapered to tip, the body shorter than the perigynia, the sides
thin and translucent to pale brown. Perigynia olive, flattened to nearly
round in section, narrowly ovate, 5-7 mm long, with more than 10 raised
nerves, tapered to a smooth beak about 1 mm long. Achenes 3-angled;
stigmas 3. May-Aug. — Swamps, marshes, kettle wetlands, wetland margins, usually in shallow water; low areas in tamarack swamps.

Carex lasiocarpa Ehrh.

SLENDER SEDGE   *native / cons* 9

Carex lanuginosa Michx. var. *americana* (Fern.) Boivin

Colony-forming perennial, from long, scaly rhizomes. Stems loosely tufted,
3-angled, 3-10 dm long. Leaves elongate and inrolled, 1-2 mm wide; sheaths
tinged with yellow-brown. Spikes either all staminate or pistillate, usually
the upper 2 staminate; the staminate spikes slender, on a long stalk; the
lower 1-3 spikes pistillate, widely separate, more or less stalkless, cylindric;
bracts leaflike, the lowest usually longer than the stem; pistillate scales
purple-brown with a green center, narrowly ovate. Perigynia dull brown
green, obovate, nearly round in section, 3-5 mm long, densely soft hairy,
contracted to a beak about 1 mm long, the beak teeth erect. Achenes yellow-brown, 3-angled with concave sides; stigmas 3. June-Aug. Peatlands
and wet peaty soils, open bogs, pond margins (where a pioneer mat-for-
mer).

Carex lasiocarpa

Carex pellita Muhl. ex Willd.

WOOLLY SEDGE   *native / cons* 4

Carex lanuginosa auct. non Michx.
Carex lasiocarpa Ehrh. var. *latifolia* (Boeckl.) Gilly

Plants tufted and stoloniferous; stolons long, horizontal, scaly. Stems 3-
10 dm long, stiff, rough above, dark-purplish red at base, the lower sheaths
filamentose;. Well-developed leaves 2-5 to a fertile culm; blades 2-6 dm
long and 1.5-5 mm wide, flat with revolute margins, septate-nodulose,
rough especially towards the tip; sheaths purplish tinged. Usually 2 upper
spikes staminate, long-peduncled; lower 2-3 spikes pistillate; bracts
sheathless or nearly so; pistillate scales acuminate, mucronate or awned,
ciliate, reddish brown with hyaline margins and 3-nerved green center.
Perigynia 25-75 to a spike, 2.5-3.5 mm long, ascending, suborbicular, inflated, leathery, dull-brownish green, densely hairy, many-ribbed; beak 1
mm, bidentate. Achenes triangular with concave sides and blunt angles,
loosely enveloped, yellowish brown, punctate; stigmas 3, blackish. —
Swamps.

Carex pellita

Carex Section Paniceae

Plants colonial, shoots arising singly or few together; rhizomes elongate; bases brown to maroon. Leaf blades typically stiff. Terminal spike staminate, typically raised above the uppermost pistillate spike. Lateral spikes generally cylindrical, ascending (except *Carex vaginata*). Perigynia several-veined, mostly short-beaked, papillose (except *C. vaginata*). — Calciphiles, growing mostly in wet soils.

The section is fairly distinctive and easy to recognize, apart from *C. vaginata*, which is morphologically distinct.

1 Perigynium with a beak 1 mm long . . . . *C. vaginata*
1 Perigynium beakless, indistinctly beaked, or contracted to beak less than 0.5 mm 2
2 Perigynia strongly ascending, beakless or tapering to an erect, very short straight beak; leaves stiff, thick, channeled, strongly glaucous. . . . . *C. livida*
2 Perigynia ascending to spreading, tapering to a bent apex; leaves relatively thin and flexible, flat or folded, green to somewhat glaucous 3
3 Bladeless basal sheaths and proximal leaf sheaths strongly tinged with reddish purple; plants forming loose clumps to extensive closed colonies of vegetative shoots from superficial rhizomes; perigynia ± 2-ranked; plants of rich forests. . . . . *C. woodii*
3 Bladeless basal sheaths and proximal leaf sheaths brownish, green, or faintly, irregularly tinged with reddish purple; plants usually with vegetative shoots widely scattered and inconspicuous from deep rhizomes; perigynia 3–6-ranked; plants of moist, usually sunny habitats . . . . *C. tetanica*

Carex livida (Wahlenb.) Willd.

LIVID SEDGE   *native / cons* **10**

Plants forming small clumps, from long slender rhizomes. Stems erect, to 6 dm long, light brown at base. Leaves 6-12 on lower third of stem, strongly waxy blue-green, channeled, 0.5-4 mm wide, dried leaves of the previous year conspicuous; sheaths thin. Terminal spike staminate (or rarely with both staminate and pistillate flowers, the staminate below the pistillate), linear; pistillate spikes 1-3, the lowest more or less separate, sometimes long-stalked, the upper grouped, stalkless or short-stalked, with 5-15 upright perigynia; bracts leaflike, sometimes longer than the head; pistillate scales shorter than the perigynia, light purple with broad green center and white translucent margins. Perigynia slightly flattened and rounded 3-angled, 2-5 mm long, strongly waxy blue-green, with small dots, two-ribbed and with fine nerves, tapered to a beakless tip. Achenes 3-angled with prominent ribs, brown-black; stigmas 3. July-Aug. — Wet meadows and fens, especially where calcium-rich.

Carex livida

Carex tetanica Schkuhr

RIGID SEDGE   *native / cons* **9**

Tufted perennial from slender rhizomes. Stems 3-angled, 1-6 dm long, rough-to-touch above. Leaves 1-5 mm wide; sheaths tight, white or yellow and translucent. Spikes either all staminate or pistillate, terminal spike staminate; lateral spikes pistillate, usually widely separated, the lower spikes short-cylindric, stalked, loosely flowered with perigynia in 3 rows; bracts shorter than the head; pistillate scales purple-brown on margins, as wide as but shorter than the perigynia. Perigynia green, faintly 3-angled, obovate, 2-4 mm long, 2-ribbed; beak tiny, bent. Achenes 3-angled with concave sides; stigmas 3. May-July. — Wet meadows and openings, marshy areas.

Carex vaginata Tausch

SHEATHED SEDGE   *native / cons* **10**

Carex saltuensis Bailey

Perennial, from long rhizomes. Stems 2-6 dm long, several together. Leaves 2-5 mm wide, not scale-like at base of stem. Terminal spike staminate, 1-2 cm long; pistillate spikes 1-3, sometimes staminate at tip,

Carex tetanica

loosely spreading, widely separated, the lower stalks long, the upper shorter; bracts with loose sheaths and blades shorter than the spikes; pistillate scales purple-brown, sometimes with a narrow green center. Perigynia usually in 2 rows, the lower separate, the upper overlapping, 3-5 mm long, narrowly obovate, with a curved beak 1 mm long. Achenes 3-angled, nearly filling the perigynia; stigmas 3. June-Aug. — Swamps and thickets, especially where calcium-rich.

Carex woodii Dewey
PRETTY SEDGE   *native / cons* 9
   Carex tetanica Schkuhr var. *woodii* (Dewey) Wood
Plants loosely tufted and stoloniferous; stolons slender, purple, scaly. Stems 3-7 dm long, roughened above; sterile shoots numerous, long. Well-developed leaves 2-4 to a fertile culm, near base; blades 5-20 cm long and 2.5-4 mm wide, flat with revolute margins, flaccid, light-green, white-lined below; sheaths loose, overlapping, white or yellowish hyaline ventrally. Terminal spike staminate, the peduncle roughish; lateral spikes 2-3, pistillate, on slender roughish peduncles; bracts long-sheathing, the sheaths tight; pistillate scales purplish or reddish brown with hyaline margins and 3-nerved green center. Perigynia 6-15 to a spike, 3.5-4 mm long, ascending, yellowish green, puncticulate, 2-keeled, lightly nerved; beak 0.5 mm long, excurved, the orifice oblique, entire, hyaline. Achenes triangular with concave sides and blunt angles, closely enveloped, yellowish brown; stigmas 3, reddish brown. — Dry woodlands.

Carex vaginata

Carex Section Phacocystis

Plants often cespitose; rhizomes short or long. Lower leaf sheaths brown to red, fibrous in some species. Terminal spike typically staminate, ascending. Lateral spikes pistillate or androgynous, ascending to nodding or drooping, elongate. Perigynia biconvex with distinct marginal veins. Stigmas 2. — Habitats range from floodplains and wet forests (*Carex crinita*), to sedge meadows (*C. stricta*), bogs and marshes (*C. aquatilis*), and wet roadsides and ditches.

1 Pistillate spikes on lax peduncles, at length drooping, the scales prominently awned; body of achene with an irregular notch, constriction, or wrinkle on one side . . . . **C. crinita**
1 Pistillate spikes erect or strongly ascending, often sessile, the scales acute or acuminate, not awned; body of achene smooth and regular 2
2 Fertile stems of current year with conspicuous bladeless sheaths at base, not surrounded by dried-up bases of the previous year's leaves but arising laterally; lowest bract usually shorter than to approximately equaling the inflorescence . . . . **C. stricta**
2 Fertile stems of current year mostly lacking bladeless sheaths at base, arising centrally from tufts of dried-up bases of previous years leaves; lowest bract usually conspicuously longer than the inflorescence . . . . **C. aquatilis**

Carex aquatilis Wahlenb.
WATER SEDGE   *native / cons* 7
Plants large, tufted or forming turfs; spreading by many slender rhizomes. Stems 3-12 dm long, 3-angled, usually rough-to-touch below the spikes. Leaves waxy blue-green, 2-7 mm wide; sheaths white or purple-dotted. Spikes 3-5, the upper spikes staminate, the middle and lower spikes pistillate or often with staminate flowers borne above pistillate; pistillate scales tapered to tip. Perigynia pale green to yellow-brown or red-brown, broadest near tip, not inflated, 2-3 mm long; beak tiny. Achenes lens-shaped; stigmas 2. May-Aug. — Wet meadows, marshes, shores, streambanks, kettle lakes, ditches and fens.

Carex aquatilis

Carex crinita Lam.
FRINGED SEDGE   *native / cons* 6

Plants large, densely tufted. Stems 5-15 dm long. Leaves 7-13 mm wide, lowest stem leaves reduced to scales; sheaths smooth. Spikes staminate or pistillate, drooping on slender stalks; staminate spikes 1-3, above pistillate spikes; pistillate spikes 2-5, narrowly cylindric; bract leaflike, without a sheath; pistillate scales rounded and notched at tip with pale midvein prolonged into a toothed awn to 10 mm long, scale edges coppery-brown. Perigynia green, 2-ribbed, nerves faint or absent, round in cross-section, abruptly tapered to a tiny beak. Achenes lens-shaped; stigmas 2. May-July. — Swamps and alder thickets, wet openings, ditches and potholes.

Carex crinita

Carex stricta Lam.
TUSSOCK SEDGE   *native / cons* 7

Plants densely tufted, from long scaly rhizomes, forming large raised hummocks to 1 m tall. Stems 3-angled, 3-10 dm long, rough-to-touch. Leaves 2-6 mm wide, the lower leaves reduced to sheaths around the base of stem; sheaths white to red-brown on front, green on back, the lower sheaths breaking into ladderlike thin strands. Spikes mostly all staminate or pistillate (sometimes mixed), the upper 1-3 spikes staminate, the terminal spike 1.5-5 cm long, the lower 2-5 spikes pistillate or some with staminate flowers borne above pistillate; lowest bract leaflike; pistillate scales equal or longer than the perigynia but narrower. Perigynia green at tip and margins, golden to yellow-brown in middle, with white or brown bumps, convex on both sides to nearly flat, 2-3 mm long, 2-ribbed with a few faint nerves on both sides; beak short, to only 0.3 mm long. Achenes lens-shaped; stigmas 2. May-July. — Often dominant sedge of wet meadows, marshes, fens, shores, streambanks, ditches.

Carex stricta

Carex Section Phaestoglochin

Plants tufted; rhizomes short or inconspicuous; bases pale to brown, occasionally reddish. Inner band of the leaf sheaths hyaline, corrugated or smooth. Spikes all or mostly androgynous, simple in most taxa, the lower branched in some species. Perigynia mostly plano-convex, beaks typically bidentate. — Mostly upland species of forests and open, sometimes disturbed, habitats.

1 Leaf sheaths loose, white with green veins or mottled green and white on back; wider blades 5–10 mm broad . . . . ***C. sparganioides***
1 Leaf sheaths tight and slender and uniform green or whitish on back; wider blades 0.9–4.3 mm broad 2
2 Perigynia mostly ascending and not widely spreading, at most with thin spongy area at base not conspicuously puckered in drying (unless immature), the margin above flat or slightly incurved . . . . ***C. muehlenbergii***
2 Perigynia mostly widely spreading at maturity, conspicuously spongy-thickened at their bases and there puckered in drying, the wire-like margin above the base tending to turn inward 3
3 Wider leaf blades mostly 0.9–1.8 (very rarely 2.5) mm broad; stigmas reddish to dark brown, slender and elongate (when intact), often protruding 1–1.5 mm or more, often reflexed but otherwise straight or slightly sinuous . . . . ***C. radiata***
3 Wider leaf blades mostly (1.5–2.7 mm broad; stigmas very dark reddish brown, comparatively short and stout, strongly curled . . . . ***C. rosea***

Carex muehlenbergii Schkuhr ex Willd.
MUHLENBERG'S SEDGE   *native / cons* 4

Plants tufted; rootstocks short, somewhat woody, dark, fibrillose. Stems 2-9 dm long, stiff, rough above, light-brownish at base, the old leaves conspicuous. Well-developed leaves 5-10 to a stem, on lower fifth; blades 1-3 dm long and 2-4 mm wide, flat or channeled, thick, light-green, roughened on margins and towards the tip; sheaths tight, yellowish brown-tinged at mouth. Spikes 3-10, androgynous; bracts setiform, short; staminate flowers few; scales greenish hyaline with 3-nerved green center. Perigynia 8-20 to a spike, 3-3.5 mm long, ascending or spreading, plano-convex, somewhat leathery, pale-green, many-ribbed with sharp slightly raised margins, serrulate above; beak 1 mm long, bidentate, the teeth hyaline within. Achenes filling perigynia, lenticular; stigmas 2, long, reddish brown. — Sand soils and dry places.

Carex radiata (Wahlenb.) Small
EASTERN STAR SEDGE   native / cons 4
Plants tufted. Stems slender, 2.5-5 dm long, weak, roughened above, light-brown to blackish tinged and fibrillose at base. Well-developed leaves 4-6 to a fertile stem, on lower fourth; blades 1-2 mm wide, flat, light-green; sheaths tight. Spikes 4, androgynous; staminate flowers few; lowest bract setaceous, the upper smaller; scales thin, white-hyaline with green midvein. Perigynia 2-6 to a spike, 2-3 mm long, exceeding scales, deep-green, erect, nerveless or nearly so, serrulate above, spongy at base; beak to 1 mm long, bidentate, white-hyaline between teeth. Achenes lenticular, filling perigynia; stigmas 2, short, twisted, dark-brownish red. — Dry woods.

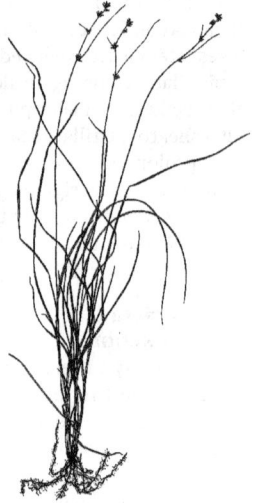

Carex radiata

Carex rosea Schkuhr
ROSY SEDGE   native / cons 4
  Carex convoluta Mackenzie
Plants tufted; rootstocks short, dark, fibrillose. Stems slender 2-5 dm long, smooth or serrulate above, light-brownish tinged and fibrillose at base. Well-developed leaves 3-6 to a stem, on lower third; blades 3 dm long and 1-2 mm wide, flat, light-green, serrulate on margins and on veins towards tip; sheaths tight. Spikes 4-8, androgynous; bracts to 10 cm long, the upper reduced; scales thin, greenish hyaline with green midrib; staminate flowers inconspicuous. Perigynia 4-12 to a spike, 3-3.5 mm long, exceeding scales, ascending or widely radiating, plano-convex, light-green, nerveless or nearly so, serrulate above, spongy at base; beak ca. 0.5 mm long, bidentate. Achenes lenticular, filling perigynia; stigmas 3, long, light-reddish brown. — Dry woodlands.

Carex rosea

Carex sparganioides Muhl.
BUR-REED SEDGE   native / cons 6
Plants tufted; rootstocks short, dark, somewhat woody. Stems 3-7.5 dm long, ascending or erect, narrowly margined, serrulate above, brownish yellow-tinged at base. Well-developed leaves 3-6 to a stem; blades 2-4 dm long and 5-10 mm wide, flat, weak, serrulate on margins, roughened on veins; sheaths loose, overlapping, green-and-white-mottled dorsally, the lower sheaths transversely rugulose. Spikes 6-12, androgynous; bracts short, often rudimentary; scales thin, greenish hyaline; staminate flowers inconspicuous. Perigynia 5-50 to a spike, 3 mm long, ascending or spreading, plano-convex, nerveless or nearly so, with sharp and somewhat raised margins, serrulate above; beak serrulate, bidentate, white-hyaline within. Achenes lenticular; stigmas 2, short, reddish brown. — Dry woods.

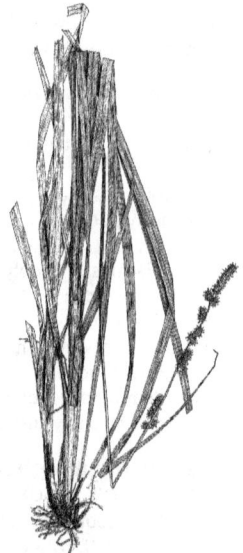

Carex Section Phyllostachyae

One member of the section in Door County. Plants tufted, bases brown. Bracts lacking. Lateral spikes absent or basal, pistillate or androgynous. Terminal spike androgynous. Lowest pistillate scale foliose, suggesting the lowest bract of the inflorescence in most other sections, exceeding the tip of the spike. Perigynia 2-ribbed, beak untoothed.

Carex sparganioides

Carex backii Boott
BACK'S SEDGE   *native / cons* **8**
Plants tufted; rootstocks short, dark-brown. Stems to 25 cm long, weak,
narrowly winged, serrulate on angles, enlarged upward. Well-developed
leaves 2-6 to a stem, near base; blades 1-3 dm long and 2.5-6 mm wide,
flat, erect or curved, thickish, deep-green papillate, roughened especially
on margins and towards tip; sheaths thin and hyaline ventrally, yellowish
brown-tinged, oblique at mouth. Spikes 1-3, androgynous, the lower long-
peduncled; pistillate scales bract-like, 3-4 cm long and 5 mm wide, nerved,
tapering; staminate flowers few. Perigynia 2-5 to a spike, 5-6 mm long,
erect on a zigzag winged rachis, concealed by scales, light-green, many-
nerved, 2-keeled, spongy at base; beak 2 mm, 2-edged the orifice entire,
truncate, hyaline, tawny-tinged below. Achenes triangular-globose with
convex sides, closely enveloped, yellowish green or blackish, triangular;
stigmas 3, short, dark. — Dry woods.

Carex backii

Carex Section Physoglochin
One member of the section in Door County. — Sphagnum moss peatlands.

Carex gynocrates Wormsk.
NORTHERN BOG SEDGE   *native / cons* **10**
  Carex dioica L.
Small perennial, from long, slender rhizomes. Stems single or few together, 0.3-3 dm long, smooth, brown at base. Leaves clustered near base
of plant, blades inrolled and threadlike, to 1 mm wide. Spikes only 1 per
stem, all staminate or all pistillate, or with both staminate and pistillate
flowers and with the staminate flowers borne above the pistillate; the
staminate spike or portion of spike narrowly cylindric, the pistillate spike
or portion short-cylindric; bract absent; pistillate scales brown or red-
brown, tapered to tip, shorter but wider than perigynia. Perigynia 4-10,
widely spreading, yellow to dark brown, shiny, plump, obovate, 2-4 mm
long, spongy at base, abruptly contracted to the beak; beak nearly entire
to unequally notched, 0.5 mm long. Achenes lens-shaped; stigmas 2.
June-July. — Conifer swamps and open peatlands, usually in sphagnum
and wet, peaty soils.

Carex gynocrates

Carex Section Racemosae
Plants loosely to densely tufted; rhizomes variable in length; bases dark red, generally fibrous;
roots not clothed with yellow felt. Terminal spike gynecandrous (in our species). Pistillate scales
dark, often black. Perigynia pale, often greenish, very short-beaked to beakless, smooth or papillose, 2-ribbed, inconspicuously veined (in our species). Stigmas 3.

Carex buxbaumii Wahlenb.
BROWN BOG SEDGE   *native / cons* **8**
Loosely tufted perennial, from long rhizomes. Stems single or few together, 3-angled, 3-10 dm long, rough-to-touch above, red-tinged near
base. Leaves 1-3 mm wide, the lowest leaves without blades; lower sheaths
shredding into thin strands, the upper sheaths membranous and purple-dotted. Spikes 2-5, terminal spike with pistillate flowers above staminate and larger than the lateral spikes, lateral spikes pistillate, short-
cylindric, stalkless or nearly so; bracts leaflike, the lowest shorter than
the head; pistillate scales dark brown, tapered to an awn at tip. Perigynia
light green, golden brown near base, oval, 2.5-3.5 mm long, 2-ribbed, with
6-8 faint nerves on each side; beak tiny, notched. Achenes 3-angled; stigmas 3. May-Aug. — Wet meadows and fens, shallow marshes, hollows in
patterned peatlands.

Carex buxbaumii

Carex Section Stellulatae

Plants tufted; rhizomes short; bases brown, not fibrous. Inflorescence mostly open, spikes readily distinguished from each other, the lowest in our more common species not overlapping; bracts inconspicuous or lacking. Spikes 2-10 (solitary in *Carex exilis*), gynecandrous (unisexual in *C. sterilis*). Perigynia spreading to reflexed, typically plano-convex, widest at the base, generally chestnut brown to dark brown or even blackish at maturity; margins acute; base spongy; beak generally bidentate, margins finely serrate. Achenes much smaller than the perigynia. — Wetlands.

The distinctions between species in this section are subtle; however, the species have habitat preferences that help with field identification. When examining perigynia, view the lowest 2-3 perigynia in the spike; the upper perigynia are very similar in all of our species.

1 Spikes solitary; leaves involute; anthers 2–3.6 mm long. . . . . *C. exilis*
1 Spikes 2–8; leaves flat or plicate; anthers 0.6–2.2 mm long 2
2 Terminal spikes entirely staminate. . . . . *C. sterilis*
2 Terminal spikes partly or wholly pistillate 3
3 Terminal spikes without a distinct clavate base of staminate scales, staminate portion less than 1 mm long. . . . . *C. sterilis*
3 Terminal spikes with a distinct clavate base of staminate scales mostly 1–8 mm long. 4
4 Lower perigynia mostly 2.9–3.6 mm long, 1.8–3.6 times as long as wide; beaks 0.9–2 mm long, mostly 0.5–0.8 times as long as the body. . . . . *C. echinata*
4 Lower perigynia mostly 1.9–3 mm long, 1–2 times as long as wide; beaks 0.4–0.9 mm long, mostly 0.2–0.5 times as long as body . . . . *C. interior*

Carex echinata Murr.

STAR SEDGE   *native / cons* 8
Carex angustior Mackenzie
Plants tufted. Stems 1-6 dm long, rough above. Leaves scale-like at base of stem; leaves with blades 3-6 on lower stem, 1-3 mm wide. Spikes 3-7, stalkless, few-flowered; terminal spike with a slender staminate portion near its base; lateral spikes usually all pistillate; bract small; pistillate scales shorter than perigynia, yellow-tinged with green midvein. Perigynia 5-15 and crowded in each spike, spreading or curved downward, green or light brown, flat on 1 side and convex on other, spongy-thickened at base, 3-4 mm long, tapered to a toothed, notched beak 1-2 mm long. Achenes lens-shaped; stigmas 2. July-Sept. — Swamp margins, wet sandy lakeshores, hummocks in peatlands.

Carex echinata

Carex exilis Dewey

COASTAL SEDGE   *thr / native / cons* 10
Plants densely tufted. Stems stiff, 2-7 dm long and longer than the leaves. Leaves narrow and rolled inward. Spike usually 1, either staminate or pistillate, or with both staminate and pistillate flowers, the staminate below the pistillate, 1-3 cm long; lateral spikes (if present) 1 or 2 and much smaller than terminal spike; lower 2 scales empty and upright; pistillate scales red-brown with translucent margins, about as long as perigynia. Perigynia spreading or drooping, flat on 1 side and convex on other, 3-5 mm long, spongy-thickened at base, tapered to a toothed beak to 2 mm long. Achenes lens-shaped; stigmas 2. June-Aug. — Sphagnum peatlands, interdunal wetlands near Lake Michigan; coastal disjunct.

Carex interior Bailey

INLAND SEDGE   *native / cons* 7
Plants densely tufted. Stems slender, sharply 3-angled, 1-6 dm long, equal or longer than the leaves. Leaves 1-2 mm wide; sheaths tight, thin and translucent. Spikes 2-4, the terminal spike with pistillate flowers borne above staminate (or rarely all staminate), the lateral spikes pistillate (or rarely with pistillate flowers borne above staminate), more or less over-

Carex exilis

lapping; bracts small or absent; pistillate scales much shorter than the perigynia. Perigynia green-brown to brown, filled to margins by the achenes, sharp-edged but not wing-margined, 2-3 mm long, the base spongy so that achene fills upper perigynium body, tapered to a finely toothed beak to 1 mm long; the beak teeth small, not longer than 0.3 mm. Achenes lens-shaped; stigmas 2. May-Aug. — Swamps, tamarack bogs, alder thickets, wet meadows and wetland margins.

Carex sterilis Willd.
DIOECIOUS SEDGE	*native / cons* **10**
Carex muricata L. var. *sterilis* (Willd.) Gleason

Plants tufted. Stems stiff, 1-7 dm long, longer than the leaves, rough-to-touch on the upper stem angles. Leaves 3-5 from lower part of stem, 1-4 mm wide, rough, lower stem leaves reduced to scales. Spikes 3-8, stalkless, clustered or the lower separate; staminate and pistillate flowers mostly on separate plants; pistillate scales red-brown with green midvein and translucent marginperigynium (l), pistillate scale (r)s, about as long as body of perigynia. Perigynia 5-25, the lower spreading, 2-4 mm long, red-brown, flat on 1 side and convex on other, spongy-thickened at base, tapered to a finely toothed, notched beak 0.5-1.5 mm long, the beak teeth sharp, to 0.5 mm long. Achenes lens-shaped; stigmas 2. April-June. — Spring-fed calcareous fens, calcium-rich wet meadows.

Similar to inland sedge (*C. interior*).

Carex sterilis

Carex Section Vesicariae

Includes typical bottlebrush sedges of former section Pseudocypereae, with pistillate spikes tightly packed with perigynia, and pistillate scales with scabrous awns conspicuous between the perigynia; and also former section Vesicariae, with pistillate spikes often narrower, longer, less densely packed with perigynia in some species, and pistillate scales mostly not awned, hidden by the perigynia.

1 Pistillate scales with a prominent, scabrous awn; often the body also ciliate 2
1 Pistillate scales smooth-margined, obtuse to acuminate, awnless (rarely the lowermost awned in *C. utriculata*) 4
2 Perigynia spreading to ascending, thin-textured, inflated, round in cross-section; many nerves separated by more than three times their width; longest beak teeth 0.3–0.7 mm long . . . . *C. hystericina*
2 Perigynia reflexed at maturity, hard-walled, uninflated, flattened-triangular in cross-section, strongly and closely nerved with most nerves separated by less than three times their width; longest beak teeth 0.7–2.2 mm long 3
3 Spikes 12–18 mm thick; beak teeth strongly outcurved, the longest 1.3–2.1 mm long . . . . . *C. comosa*
3 Spikes 9–12 mm thick; beak teeth straight or slightly outcurved, the longest 0.7–1.2 mm . . . .
   . . . . *C. pseudocyperus*
4 Perigynia 4–7 mm thick; achenes with a deep notch or constriction on one angle . . . . *C. tuckermanii*
4 Perigynia 2.5–3.5 mm thick; achenes symmetrical, not notched on one angle 5
5 Lowest pistillate bract 3–9 times as long as the entire inflorescence; mature perigynia 7–12 mm long, at least the lower reflexed or widely spreading; staminate spike often 1, its base (or base of lowest staminate spike if more than one) slightly if at all elevated above summit of the crowded pistillate spikes (rarely lower spike remote) . . . . *C. retrorsa*
5 Lowest pistillate bract less than 3 times as long as inflorescence; perigynia 4–7.5 mm long, ascending or spreading; staminate spikes mostly 2–4, generally well elevated above the pistillate spikes 6
6 Colonial from long-creeping rhizomes; widest leaves 5–12 mm wide; ligules about as long as wide; basal sheaths usually spongy-thickened with little or no red tingeing; perigynia (at least those on lower portion of fully mature spike) widely spreading; stems bluntly triangular and sparsely and irregularly scabrous below the inflorescence . . . . *C. utriculata*
6 Tufted; widest leaves 3–6) mm wide; ligules longer than wide; basal sheaths not spongy-thickened and often tinged with reddish purple; perigynia ascending; stems sharply triangular and scabrous-angled below the inflorescence . . . . *C. vesicaria*

Carex comosa Boott

BEARDED SEDGE   *native / cons* **5**

Plants large, often forming large clumps. Stems stout, sharply 3-angled, 5-15 dm long. Leaves 5-12 mm wide; sheaths translucent on front, with small swollen joints on back. Spikes either staminate or pistillate; terminal spike staminate; lateral spikes pistillate, 3-5, cylindric, the lower spikes longer stalked and drooping when mature; bracts leaflike, much longer than the head; pistillate scales with translucent margins, tapered into a long, rough awn. Perigynia numerous, spreading outward when ripe, flattened 3-angled, lance-shaped, 5-8 mm long, shiny, strongly nerved, gradually tapered to the 2-3 mm long beak, the beak with curved teeth 1-2 mm long. Achenes 3-angled; stigmas 3. June-Aug. — Marshes, wetland margins, floating mats, ditches.

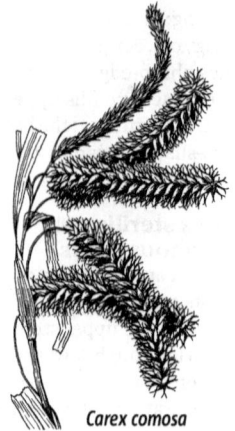

Carex comosa

Carex hystericina Muhl.

PORCUPINE SEDGE   *native / cons* **3**

Plants from short rhizomes, often forming large clumps. Stems upright or leaning, 3-angled, 2-10 dm long, usually longer than the leaves. Leaves yellow-green, 3-8 mm wide; sheaths white, thin and translucent on front, green to yellow or red on back, the lower sheaths breaking into threadlike fibers. Spikes either all staminate or pistillate, the terminal spike staminate, usually short-stalked and often with a bract; lateral spikes pistillate or occasionally with staminate flowers above pistillate, 1-4, short-cylindric, separate or clustered, the lower spikes usually nodding on slender stalks, the upper spikes short-stalked and upright; pistillate scales small, narrow and much shorter than the perigynia, tipped with a rough awn. Perigynia spreading or upright, green to straw-colored, ovate, round in section when mature, 5-8 mm long, strongly nerved, abruptly tapered to a slender, toothed beak 3-4 mm long; the beak teeth to 1 mm long. Achenes 3-angled with concave sides; stigmas 3. May-July. — Swamps, alder thickets, wet meadows and ditches, calcareous fens.

Carex hystericina

Carex pseudocyperus L.

CYPRESS-LIKE SEDGE   *native / cons* **8**

Plants large, tufted. Stems stout, 3-10 dm long, 3-angled, rough-to-touch. Leaves 5-15 mm wide; sheaths translucent, yellow-tinged on back. Spikes either all staminate or pistillate, the terminal spike staminate; lateral spikes pistillate, 2-6, cylindric, lower spikes drooping on slender stalks; bracts much longer than the head; pistillate scales tipped by an awn, the awn shorter or longer than the perigynia. Perigynia spreading, 3-angled, 4-6 mm long, shiny, strongly nerved, tapered to a toothed beak, the beak teeth 0.5-1 mm long. Achenes 3-angled; stigmas 3. June-Aug. — Marshy lake margins, swamps, fens, wet ditches.

Similar to *Carex comosa*.

Carex retrorsa Schwein.

RETRORSE SEDGE   *native / cons* **6**

Plants densely tufted. Stems 4-10 dm long. Leaves 3-4 dm long and 4-10 mm wide, flat and soft; sheaths dotted with small bumps. Spikes either all staminate or pistillate, or the terminal 1-2 spikes with both staminate and pistillate flowers, the staminate above the pistillate, stalkless or lowest spike on a slender stalk; lower spikes 3-8, pistillate; pistillate scales conspicuous, shorter and narrower than the perigynia. Perigynia crowded in rows, spreading or the lowest perigynia angled downward, smooth and shiny, 6-13-nerved, 7-10 mm long, somewhat inflated, tapered to a long, smooth beak 2-4 mm long, the beak teeth short, to 1 mm long. Achenes dark brown, 3-angled, loose in the lower part of the perigynium; stigmas 3. June-Aug. — Floodplain forests, swamps, thickets and marshes.

Carex pseudocyperus

Carex tuckermanii Dewey
TUCKERMAN'S SEDGE   *native / cons* **8**

Plants tufted, from short rhizomes. Stems 4-8 dm long. Leaves 2-4 dm long and 3-6 mm wide, soft and flat. Spikes either staminate or pistillate; staminate spikes usually 2, separated, raised above pistillate spikes; pistillate spikes 2-4, separated, cylindric. Perigynia overlapping and ascending in 6 rows, 7-10 mm long and 4-7 mm wide, inflated, tapered to a notched beak 2 mm long. Achenes 3-angled, obovate, with a deep indentation near the middle of 1 angle; stigmas 3. June-Aug. — Swamps, alder thickets, low areas in forests, pond margins.

Carex tuckermanii

Carex utriculata Boott
NORTHWEST TERRITORY SEDGE   *native / cons* **7**
  Carex rostrata Stokes var. *utriculata* (Boott) Bailey

Plants large, densely tufted, from short rootstocks, also forming turfs from long rhizomes. Stems bluntly 3-angled, 3-12 dm long, spongy at base. Leaves strongly divided with swollen joints 4-12 mm wide; sheaths white-translucent on front, divided with swollen joints on back. Spikes either staminate or pistillate, the upper 2-5 staminate, held well above the pistillate spikes; lower 2-5 spikes pistillate or sometimes 1 or 2 with staminate flowers above the pistillate, usually separate, cylindric, the upper spikes stalkless or short-stalked, lower spikes stalked, upright; bracts shorter to slightly longer than the head; pistillate scales acute to awn-tipped, body of scale shorter than perigynia. Perigynia upright at first to widely spreading when mature, in many rows, yellow-green to brown, shiny, nearly round in section, inflated, 3-8 mm long, strongly 7-9-nerved, contracted to a toothed beak 1-2 mm long, the teeth mostly straight, 0.5 mm long. Achenes 3-angled; stigmas 3. June-Aug. — Wet meadows, marshes, fens, swamps and lakeshores.

Carex utriculata

Carex vesicaria L.
LESSER BLADDER SEDGE   *native / cons* **7**

Plants tufted, from stout, short rhizomes. Stems 3-10 dm long, sharply 3-angled and rough-to-touch below the head, not spongy at base (as in *C. utriculata*). Leaves 2-7 mm wide; sheaths white-translucent on front, not conspicuously divided-with small swollen joints on back, the lowest sheaths often shredding into ladderlike fibers. Spikes either all staminate or pistillate, the upper 2-4 staminate, held well above the pistillate; lower 1-3 spikes pistillate, separate, cylindric, stalkless or short-stalked, erect; lowest bract usually longer than the head; pistillate scales acute to awn-tipped, shorter to as long as perigynia. Perigynia upright and overlapping in rows, dull yellow-green to brown, inflated, 3-8 mm long, strongly nerved, abruptly tapered to a toothed beak 1-2 mm long, the teeth 0.5-1 mm long. Achenes 3-angled; stigmas 3. June-Aug. — Wet meadows, marshes, forest depressions and shores.

Carex vesicaria

Carex Section Vulpinae
One member of the section in Door County. Plants tufted; bases generally pale. Stems thick, spongy, weak, the angles narrowly winged, scabrous. Inflorescence longer than wide (ours), ovate to cylindrical. Bracts setaceous. Spikes densely flowered, the lower branched, mostly or all androgynous (the terminal always androgynous). Perigynia plano-convex, bases spongy. — Wetlands. The thick, spongy stems, branched lower spikes, and spongy perigynium bases are characteristic.

Carex stipata Muhl.
STALK-GRAIN SEDGE   *native / cons* **2**
Plants densely tufted. Stems 3-angled and slightly winged, 2-12 dm long.
Leaves 4-8 mm wide; sheaths cross-wrinkled on front, divided with small
swollen joints on back. Spikes with both staminate and pistillate flowers,
staminate flowers borne above pistillate, clustered or the lowest spikes
often separate; bracts small and sometimes bristle-like, longer than the
spike; pistillate scales tapered to a tip or with a short, sharp point, half
to 3/4 as long as the perigynia. Perigynia yellow-green to dull brown, flat
on 1 side and convex on other, 3-5 mm long, strongly several-nerved on
both sides, tapered to a finely toothed, notched beak 1-3 mm long. Achenes
lens-shaped; stigmas 2. May-July. — Floodplain forests and swamps,
thickets, wet meadows, wetland margins and ditches; usually not in
sphagnum bogs.

Carex

Cladium
SAW-GRASS

Cladium mariscoides (Muhl.) Torr.
SMOOTH SAW-GRASS   *native / cons* **10**
Grasslike perennial, spreading by rhizomes and forming colonies. Stems
single or in small groups, stiff, slender, smooth, 0.3-1 m tall. Leaves 1-3
mm wide, upper portion round in section, middle portion flattened. Flowers in lance-shaped spikelets, 3-5 mm long, in branched clusters (umbels)
at end of stem and also with 1-2 clusters on slender stalks from leaf axils;
uppermost flower perfect, the style 3-parted; middle flowers staminate;
lowest scale of each spikelet empty; scales overlapping, ovate, brown;
bristles absent. Achenes dull brown, 2-3 mm long, pointed at tip; tubercle
absent. June-Aug. — Shallow water, sandy or mucky shores, floating bog
mats, calcium-rich wet meadows, seeps, fens.

Cladium mariscoides

Cyperus
FLAT SEDGE

Small to medium, annual or perennial, grasslike plants. Stems often clumped, unbranched,
sharply 3-angled. Leaves mostly from base of plants, with 1 or more leaflike bracts near top of
stems, the blades flat or folded along midvein. Flower heads in umbels at ends of stems; the
spikelets many, grouped in 1 to several rounded or cylindric spikes. Flowers perfect; bristlelike
sepals and petals absent; stamens 1-3; styles 2-3-parted. Achenes lens-shaped or 3-angled, beakless.

1 Achenes lens-shaped; stigmas 2 . . . . *C. bipartitus*
1 Achenes 3-angled; stigmas 3 2
2 Rachilla of spikelets continuous, scales gradually deciduous, falling from base of rachilla to apex . . .
   . . . . *C. esculentus*
2 Rachilla of spikelets articulated; scales persistent and then falling all at once from the rachilla. 3
3 Rachilla articulating at the base of each scale. . . . . *C. odoratus*
3 Rachilla not separating into joints (short segments) . . . . *C. schweinitzii*

Cyperus bipartitus Torr.
SHINING FLAT SEDGE   *native / cons* **3**
  Cyperus niger var. *rivularis* (Kunth) V. Grant
  Cyperus rivularis Kunth
  Pycreus rivularis (Kunth) Palla
Tufted grasslike annual. Stems 3-angled, 1-3 dm tall. Leaves usually shorter
than stems; leaves and bracts 0.5-2 mm wide, the bracts usually 3, longer
than the spikes. Spikelets linear, 10-15 mm long and 2-3 mm wide, in
clusters (spikes) of 3-10, the spikes stalkless or on stalks to 10 cm long;
scales overlapping, ovate, shiny, purple-brown on margins; stamens 2 or

Cyperus bipartitus

3; style 2-parted, the lower 1/3 not divided. Achenes lens-shaped, 1-2 mm long, hidden by the scales. July-Sept. — Wet, sandy, gravelly or muddy shores, streambanks, wet meadows, ditches.

Cyperus esculentus L.
CHUFA   *native / cons* **O**
Grasslike perennial, with rhizomes ending in small tubers. Stems single, 3-angled, erect, 2-7 dm long. Leaves light green, mostly from base of plant, about as long as stems, 3-10 mm wide, with a prominent midvein; the bracts 3-6, usually much longer than the spikes. Spikelets linear, 3-12 cm long and 1-2 mm wide; pinnately arranged on a stalk, forming loose cylindrical spikes, the spikes to 5 cm long and 1-2 mm wide; scales straw-colored, 2-3 mm long, overlapping; stamens 3; style 3-parted. Achenes pale brown, 3-angled, 1-2 mm long. July-Sept. — Sandy or muddy shores, streambanks, marshes, ditches and other wet places; weedy in wet or moist cultivated fields.

Cyperus esculentus

Cyperus odoratus L.
RUSTY FLAT SEDGE   *native / cons* **4**
   Cyperus engelmannii Steud.
   Cyperus ferruginescens Boeckl.
   Cyperus speciosus Vahl
Stout, grasslike, fibrous-rooted annual. Stems tufted or single, 3-angled, 2-7 dm long. Leaves mostly from base of plant, shorter to longer than flowering stems, the blades 2-8 mm wide; the involucral bracts much longer than the spikes. Spikelets linear, 1-2 cm long, pinnately arranged along a stalk, forming several to many cylindrical spikes, the spikes stalkless or stalked; scales red-brown, 2-3 mm long, overlapping; stamens 3; style 3-parted. Achenes brown, 3-angled, 1-2 mm long. July-Sept. — Sandy or muddy shores, floating mats, ditches, wet cultivated fields.

Cyperus odoratus

Cyperus schweinitzii Torr.
SAND FLAT SEDGE   *native / cons* **4**
   Cyperus × *mesochoreus* Geise
   Mariscus schweinitzii (Torr.) T. Koyama
Grasslike perennial. Stems rough, 1-8 dm tall. Leaves 2-4 mm wide; bracts 3-6, usually much longer than the inflorescence. Sessile spike obconic to oblong; rays 1-6, rarely more than 10 cm long; spikelets 5-15 in the sessile spike, fewer in the peduncled ones, all crowded, ascending, flattened, usually 8-12-flowered. Scales broadly ovate-elliptic to rotund, the body 3-3.7 mm long, many-nerved, the uppermost with a conspicuous mucro to 1 mm long. Achenes oblong, 2-3 mm long. — Dry or moist sandy soil.

Dulichium
THREE-WAY SEDGE

Dulichium arundinaceum (L.) Britt.
THREE-WAY SEDGE   *native / cons* **9**
Grasslike perennial, spreading by rhizomes and often forming large colonies. Stems stout, erect, 3-10 dm long, jointed, hollow, rounded in section. Leaves 3-ranked, flat, short, 4-15 cm long and 3-8 mm wide; lower leaves reduced to sheaths. Flower heads from leaf axils, in linear clusters of 5-10 spikelets, the clusters 1-2.5 cm long; scales lance-shaped, green to brown, 5-8 mm long. Flowers perfect; sepals and petals reduced to 6-9 downwardly barbed bristles; stamens 3; style 2-parted. Achenes light brown, oblong, 2-4 mm long, beaked by the persistent, slender style. July-Sept. — Shallow marshes, wet meadows, shores, bog margins.

Dulichium arundinaceum

Eleocharis
SPIKE-RUSH

Small to medium rushlike plants, perennial from rhizomes or annual, often forming large, matlike colonies. Stems round, flattened, or angled in section. Leaves reduced to sheaths at base of stems. Flower head a single spikelet at tip of stem; scales of the spikelets spirally arranged and overlapping. Flowers perfect; sepals and petals bristlelike or absent, the bristles usually 6 if present; stamens 3; styles 2-3-parted, the base of style swollen and persistent as a projection (tubercle) atop the achene, or sometimes joined with the achene body. Achenes rounded on both sides or 3-angled.

1 Tubercle a slender or tiny conical continuation of the body of the achene, slightly differentiated in texture or color, not separated by a constriction or shaped as a distinct apical cap; stigmas 3; tip of leaf sheath without a prominent tooth . . . . *E. quineflora*
1 Tubercle differentiated in shape as well as texture, and usually separated from body of achene by a narrow constriction to form a distinct apical cap; stigmas 2 or 3; leaf sheaths sometimes with a prominent tooth at tip 2
2 Achenes 3-sided (the angles sharp, or obscure and the achene plumply rounded); styles 3-cleft; surface of achene normally ridged, reticulate, roughened, or in a few species only minutely punctate 3
2 Achenes 2-sided (lenticular or biconvex); styles 2- or 3-cleft; surface of achene smooth, usually shiny
   5
3 Achenes white or pearly, with prominent longitudinal ridges connected by numerous tiny cross-bars; basal scales of spikelet fertile . . . . *E. acicularis*
3 Achenes greenish, yellow, golden, brown, black (rarely whitish), and reticulate, smooth, or roughened; basal scales of spikelets sterile 4
4 Stems very strongly flattened and often twisted, with obscure ridges; scales at middle of spikelet reddish brown with narrow, deeply bifid scarious whitish tips to 1 mm long . . . . *E. compressa*
4 Stems slightly or not at all flattened, prominently ridged; scales at middle of spikelet deep reddish brown to nearly black, with short, entire, lacerate, or bifid tips mostly less than 0.6 mm long. . . . . . .
   . . . . *E. elliptica*
5 Plants perennial, with stiff stems and rhizomes; scales acute to acuminate at tip (or somewhat obtuse); achenes 1.5–2.8 mm long, including tubercle; anthers ca. 1–3 mm long . . . . *E. palustris*
5 Plants annual, with soft, easily compressed, densely tufted stems; scales broadly rounded at tip; achenes 1–1.5 mm long, including the strongly flattened tubercle; anthers to 0.7 mm long. *E. obtusa*

Eleocharis acicularis (L.) Roemer & J.A. Schultes
NEEDLE SPIKE-RUSH   *native / cons* 5

Small, tufted, mat-forming perennial, from slender rhizomes. Stems threadlike, 3-15 cm long and to 0.5 mm wide, somewhat 4-angled and grooved; sheaths membranous, usually red at base. Spikelets narrowly ovate, 3-6 mm long and 1-1.5 mm wide; scales with a green midvein and chaffy margins; sepals and petals reduced to 3-4 bristles or absent; style 3-parted. Achenes gray, rounded 3-angled, ridged, to 1 mm long; tubercle cone-shaped, constricted at base. May-Sept. — Shallow water, exposed muddy or sandy shores, marshes and streambanks.

Eleocharis acicularis

Eleocharis compressa Sullivant
FLAT-STEM SPIKE-RUSH   *native / cons* 9
   Eleocharis elliptica var. *compressa* (Sullivant) Drapalik & Mohlenbrock
   Eleocharis tenuis var. *atrata* (Svens.) Boivin

Tufted perennial, from stout black rhizomes. Stems flattened and often twisted, 1.5-4 dm long and 0.5-1 mm wide, shallowly grooved; sheaths red or purple at base. Spikelets ovate, 4-10 mm long and 3-4 mm wide; lowest scale sterile and encircling the stem; fertile scales with a green midvein, purple-brown on sides, and white translucent margins; sepals and petals absent or reduced to 1-5 bristles; style 3-parted. Achenes yellow-brown, covered with small bumps, somewhat 3-angled, 1-1.5 mm long; tubercle small, constricted at base. May-Aug. — Wet calcareous meadows, swamps, ditches.

Eleocharis compressa

Eleocharis elliptica Kunth

ELLIPTIC SPIKE-RUSH   *native / cons* **7**

Mat-forming perennial, with long rhizomes. Stems subterete to sometimes compressed, often with 5-10 ridges, to 90 cm long and 0.3-0.8 mm
wide, spongy; sheaths persistent, not splitting, dark red at base, usually
red-brown at tip, tooth to 0.5 mm long (usually present on some stems).
Spikelets ovoid, 3-8 cm long and 2-3 mm wide, scales spreading in fruit,
10-30, brown, midrib region often paler, entire or shallowly notched; perianth bristles absent or rarely 1-3, pale brown, to 1/2 of achene length,
sparsely retrorsely spinulose; style 3-parted. Achenes yellow, orange, or
medium brown, obpyriform, angles evident to prominent, ca. 1 mm long,
rugulose (visible at 10x), with 12-20 horizontal ridges in a vertical series;
tubercle brown to whitish, depressed, apiculate. — Wet, stony or sandy
places, often where marly; shores, marshes, swamps.

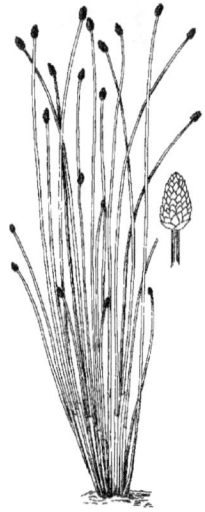

Eleocharis obtusa (Willd.) J.A. Schultes

OVOID SPIKE-RUSH   *native / cons* **3**

Tufted, fibrous-rooted annual. Stems slender, round in section, ribbed,
0.5-5 dm long and 0.5-2 mm wide; sheaths green, with a small tooth.
Spikelets ovate to cylindric, 4-15 mm long and 2-4 mm wide; scales orange-brown, with a green midvein and pale margins; sepals and petals
reduced to 6-7 brown bristles, or absent; styles 2- or 3-parted. Achenes
lens-shaped, light to dark brown or olive, shiny, ca. 1 mm long; tubercle
flattened-triangular, 2/3 to nearly as wide as the broad top of achene.
June-Sept. — May form large colonies, especially on exposed mud flats
and drying shores of receding lakes.

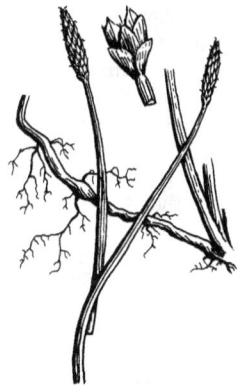

Eleocharis obtusa

Eleocharis palustris (L.) Roemer & J.A. Schultes

COMMON SPIKE-RUSH   *native / cons* **6**

   Eleocharis calva var. *australis* (Nees) St. John
   Eleocharis erythropoda Steud.
   Eleocharis mamillata (H.Lindb.) H.Lindb.
   Eleocharis smallii Britt.

Perennial, spreading by rhizomes. Stems single or in small clusters, slender to stout, round in section, 1-8 dm long and 1-3 mm wide; sheaths red
or purple at base. Spikelets long-ovate, 5-30 mm long and 2-4 mm wide,
wider than stems; lowest scale sterile, encircling the stem; fertile scales
lance-shaped to ovate, 2-5 mm long, brown or red-brown, with a green or
pale midvein; sepals and petals reduced to usually 4, pale brown, barbed
bristles; style 2-parted. Achenes lens-shaped, yellow to brown, 1-2 mm
long; tubercle flattened-triangular, constricted at base. May-Aug. — Shallow water of marshes, wet meadows, muddy shores, bogs, ditches, streambanks and swamps.

Eleocharis palustris

   A variable and common species known by a number of synonyms.

Eleocharis quinqueflora (F.X. Hartm.) Schwarz

FEW-FLOWER SPIKE-RUSH   *native / cons* **8**

   Eleocharis bernardina Munz & Johnston
   Scirpus pauciflorus Lightf.
   Scirpus quinqueflorus F.X. Hartmann

Small, tufted perennial, spreading by rhizomes. Stems threadlike, grooved,
1-3 dm long and less than 1 mm wide. Spikelets ovate, 4-8 mm long and
2-3 mm wide; scales ovate, brown, chaffy on margins, 2-5 mm long; sepals
and petals reduced to bristles or absent; style 3-parted. Achenes gray-
brown or brown, 3-angled, 1-3 mm long; tubercle slender, joined to the
achene and beaklike. June-Aug. — Wet, sandy or gravelly shores and flats
marshes and fens; often where calcium-rich.

Eleocharis quinqueflora

Eriophorum
COTTON-GRASS

Grasslike perennials. Stems clumped or single, round to rounded 3-angled in section. Leaves mostly at base of plant, the blades flat, folded or inrolled; upper leaves often reduced to bladeless sheaths. Flower heads at ends of stems, with 1 or several spikelets; spikelets resemble cottonballs when mature; scales many, spirally arranged, chaffy on margins; involucral bracts leaflike in species with several spikelets in the head, or reduced to scales in species with 1 spikelet at end of stems (*E. vaginatum*). Flowers perfect; sepals and petals numerous, reduced to long, cottony, persistent, white to tawny brown bristles; stamens 3; styles 3-parted. Achenes brown, more or less 3-angled, sometimes with a short beak formed by the persistent style.

1 Head a single spikelet at end of stem; leaflike bracts absent . . . . *E. vaginatum*
1 Head of 2 or more spikelets; leaflike bracts present 2
2 Scales 3–7-nerved, copper-brown on sides . . . . *E. virginicum*
2 Scales with 1 nerve, sides olive-green to nearly black. 3
3 Midvein of scale slender, fading before reaching tip of scale . . . . *E. angustifolium*
3 Midvein of scale widening toward tip of scale and reaching scale tip . . . . *E. viridicarinatum*

Eriophorum angustifolium Honckeny
THIN-SCALE COTTON-GRASS *native / cons* 9
Eriophorum polystachion L.

Grasslike perennial, spreading by rhizomes and forming colonies. Stems mostly single, 2-8 dm long and 2-3 mm wide, more or less round in section, becoming 3-angled below the head. Leaves few, flat or folded along midrib, 3-8 mm wide, often dying back from the tips; sheaths sometimes red, dark-banded at tip. Spikelets 3-10, clustered in heads 1-3 cm wide when mature, the heads drooping on weak stalks; involucral bracts leaflike, often black at base, the main bract upright and usually longer than the head; scales lance-shaped, brown or purple-green, 4-6 mm long, the midvein not extending to tip of scale; bristles bright white, 2-3 cm long. Achenes brown to nearly black, 2-3 mm long. May-July. — Bogs, calcareous fens, wet meadows.

Similar to *E. viridicarinatum*, which see.

Eriophorum vaginatum

Eriophorum vaginatum L.
TUSSOCK COTTON-GRASS *native / cons* 10
Eriophorum spissum Fern.

Densely tufted, grasslike perennial, forming large hummocks. Stems stiff, rounded 3-angled, 2-7 dm long. Leaves at base of stems, mostly shorter than stems, only 1 mm wide, with 1-3 inflated, bladeless sheaths on stem. Spikelets clustered in a single head at end of stems; involucral bracts absent; scales narrowly ovate, purple-brown to black, with white margins, spreading when mature; bristles usually white (rarely red-brown). Achenes obovate, 3-4 mm long. June. — Sphagnum bogs and tamarack swamps.

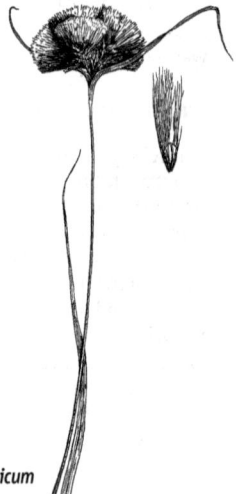

Eriophorum virginicum L.
TAWNY COTTON-GRASS *native / cons* 10

Large grasslike perennial, with slender rhizomes. Stems single or in small groups, stiff, erect, to 1 m long, leafy, mostly smooth. Leaves flat, 2-4 mm wide, the uppermost often longer than the head. Spikelets in dense clusters of several to many at ends of stems, on short stalks of more or less equal lengths, the clusters wider than long; involucral bracts 2-3, leaflike, spreading or bent downward, unequal, much longer than the head; scales ovate, thick, copper-brown with a green center; bristles tawny or copper-brown. Achenes light brown, 3-4 mm long. July-Aug. — Sphagnum moss peatlands.

Eriophorum virginicum

Eriophorum viridicarinatum (Engelm.) Fern.

DARK-SCALE COTTON-GRASS *native / cons* **10**

Grasslike perennial, forming colonies from spreading rhizomes. Stems mostly single, more or less round in section, 3-7 dm long. Leaves flat except at tip, the uppermost leaves 10-15 cm long; sheaths green. Spikelets usually 20-30, clustered in heads at ends of stems, on short to long, finely hairy stalks; involucral bracts 2-4, not black at base, longer or equal to head; scales narrowly ovate, black-green, the midvein pale, extending to tip of scale; bristles white. Achenes brown, 3-4 mm long. May-July. — Bogs and open conifer swamps.

Similar to *E. angustifolium,* but usually with more spikelets, the scale midvein extending to tip of scale, and leaf sheaths not dark-banded at tip.

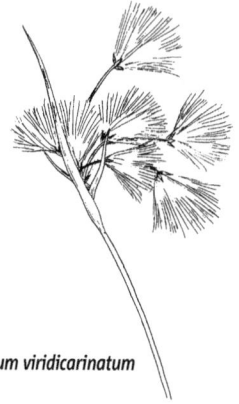

Eriophorum viridicarinatum

Rhynchospora

BEAK SEDGE

Grasslike perennials (ours), clumped or spreading by rhizomes. Stems erect, leafy, usually 3-angled or sometimes round. Leaves flat or rolled inward. Spikelets clustered in dense heads, the heads open to crowded; scales overlapping in a spiral. Flowers perfect, or sometimes upper flowers staminate only; sepals and petals reduced to usually 6 (1-20) bristles or sometimes absent; stamens usually 3; styles 2-parted, swollen at base and persistent on the achene as a tubercle. Achenes lens-shaped.

1 Spikelets white to tan; bristles 8 or more . . . . *R. alba*
1 Spikelets brown, dark olive-green or nearly black; bristles 5–6 2
2 Scales dark olive-green to black; bristles with upward-pointing barbs, at least some of the bristles longer than the tubercle . . . . *R. fusca*
2 Scales brown; bristles with downward pointing barbs (rarely smooth), the bristles shorter to as long as the tubercle . . . . *R. capillacea*

Rhynchospora alba (L.) Vahl

WHITE BEAK SEDGE *native / cons* **9**

Tufted, grasslike perennial. Stems slender, erect, 1-6 dm long. Leaves bristlelike, 0.5-3 mm wide, shorter than the stems. Spikelets in 1-3 rounded heads, 5-20 mm wide, at or near ends of stems, the lateral heads usually long-stalked; the spikelets oblong, narrowed at each end, 4-5 mm long, white, becoming pale brown; bristles 8-15, downwardly barbed, about equaling the tubercle. Achenes lens-shaped, brown-green, 1-2 mm long; tubercle triangular, about half as long as achene. June-Sept. — Bogs, open conifer swamps of black spruce and tamarack, fens.

Rhynchospora capillacea Torr.

NEEDLE BEAK SEDGE *native / cons* **10**

Small, tufted, grasslike perennial. Stems slender, 0.5-4 dm long. Leaves threadlike, rolled inward, to only 0.5 mm wide, much shorter than the stem. Spikelets in 1-2 small, separated clusters, each cluster subtended by 1 to several short, bristlelike bracts; the spikelets ovate, 3-7 mm long; scales overlapping, ovate, brown with a paler, sharp-tipped midvein; bristles 6, downwardly barbed, longer than the achenes; style 2-parted. Achenes lens-shaped, satiny yellow-brown, 2 mm long; tubercle dull brown, narrowly triangular, about 1 mm long. June-Aug. — Calcareous fens, interdunal flats, wet sandy or gravelly shores, seeps; usually where calcium-rich.

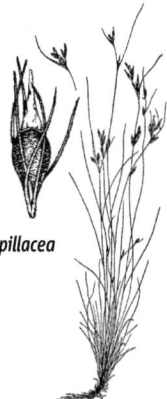

Rhynchospora alba

Rhynchospora capillacea

Rhynchospora fusca (L.) Ait. f.

BROWN BEAK SEDGE *native / cons* **10**

Tufted, grasslike perennial, spreading by short rhizomes and forming colonies. Stems slender, 3-angled, 1-3 dm long. Leaves very slender, rolled inward, mostly shorter than the stems. Spikelets spindle-shaped, dark brown, 4-6 mm long, in 1-4 loose clusters, the lower clusters on long stalks, each cluster subtended by an erect, leafy bract, the bract longer than the cluster; bristles 6, upwardly barbed; style 2-parted. Achenes light brown, 1-1.5 mm long; tubercle flattened-triangular, nearly as long as achene. — Wet sandy shores, interdunal wetlands, sedge meadows, bog mats.

Rhynchospora fusca

Schoenoplectus
CLUB-RUSH
Perennial or annual, tufted or rhizomatous herbs. Stems cylindric to strongly 3-angled, smooth, spongy with internal air cavities. Leaves basal, rarely 1(-2) on stem; sheaths tubular; ligules membranous; blades well-developed to rudimentary. Inflorescences terminal, head-like to openly paniculate; spikelets 1-100 or more; involucral bracts 1-5, leaflike, proximal bract erect to spreading. Spikelets terete; scales deciduous, spirally arranged, each subtending a flower, or proximal scale empty, midrib usually prolonged into short awn, margins ciliate. Flowers bisexual; perianth of 0-6(-8) spinulose bristles shorter than to somewhat longer than the achene; stamens 3. Achenes biconvex to trigonous, with apical beak, rugose or with transverse wavy ridges.

1 Spikelets (at least several of them) distinctly pediceled (sometimes congested in *S. acutus*); stems terete, often over 1 m tall 2
2 Stems firm and dark olive-green when fresh; spikelets ovoid to cylindrical (often 2.5 or more times as long as wide), usually in a stiffer, sometimes condensed, inflorescence; scales dull, pale or whitish brown, the midrib not strongly contrasting, the margins often more copiously ciliate than in *S. tabernaemontani*, and the backs copiously flecked with shiny red dots, often puberulent; mature achenes ca. 2.2–2.7 mm long, including apiculus, completely hidden by the scales. . . . . *S. acutus*
2 Stems rather soft and easily compressed, pale blue-green when fresh; spikelets ovoid (about twice as long as wide, or shorter), in an open, lax inflorescence; scales ± shiny, rich orange-brown, often with prominent greenish midrib, the margins ciliate but the backs essentially glabrous (puberulence and swollen red flecks, if any, limited to region of midrib); mature (dark gray or lead-colored) achenes ca. 1.6–2.1 (–2.4) mm long, including apiculus, barely covered by the scale . . . . *S. tabernaemontani*
1 Spikelets 1-few, crowded, sessile or nearly so (rarely one on a short pedicel); stems 3-angled or terete (if terete, then slender, soft, and not over 1 m tall) 3
3 Spikelet 1, strongly ascending, the involucral bract surpassing its tip by not more than 15 (–20) mm; leaves normally many, hair-like, submersed; stem seldom over 1 mm thick; anthers (2.1–) 2.5–3.5 mm long; achenes 3-sided, the body ca. 2.5–3 mm long . . . . *S. subterminalis*
3 Spikelets usually more than 1 and the involucral bract surpassing them by more than 15 mm; leaves stiff and stems thicker; anthers and achenes various . . . . *S. pungens*

Schoenoplectus acutus (Muhl.) A. & D. Löve
HARDSTEM CLUB-RUSH   native / cons 6
Scirpus acutus Muhl.
Perennial, from stout rhizomes and often forming large colonies. Stems round in section, 1-3 m long. Leaves reduced to 3-5 sheaths near base of stem, blades absent, or upper leaves with blades to 25 cm long; main bract erect, appearing as a continuation of stem, 2-10 cm long, eventually turning brown. Spikelets 5-15 mm long and 3-5 mm wide, in clusters of mostly 3-7, the clusters grouped into a branched head of up to 60 spikelets, the head appearing lateral from side of stem, the branches stiff and spreading; scales chaffy, mostly translucent, 3-4 mm long, often with red-brown spots, usually tipped with an awn to 1 mm long; bristles 6, unequal, usually shorter than achene; style 2-parted (rarely 3-parted). Achenes light green to dull brown, flat on 1 side and convex on other, 2-3 mm long, the style beak small, to 0.5 mm long. May-Aug. — Usually emergent in shallow to deep water (1-2 m deep) of marshes, ditches, ponds and lakes; sometimes where brackish.

Schoenoplectus acutus

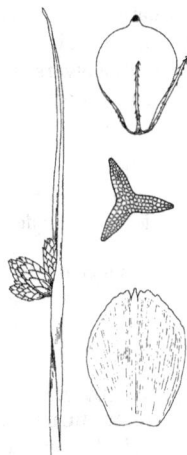

Schoenoplectus pungens (Vahl) Palla
COMMON THREESQUARE   *native / cons* 5
Scirpus americanus var. *pungens* (Vahl) Barros & Osten
Scirpus pungens Vahl

Perennial, from slender rhizomes and forming colonies. Stems erect to somewhat curved, 2-12 dm long, 3-angled, the sides concave to slightly convex. Leaves mostly 1-3 near base of stem, usually folded, or channeled near tip, reaching to about middle of stem and 1-3 mm wide; main bract erect, sharp-tipped, resembling a continuation of the stem, 2-15 cm long. Spikelets 5-20 mm long and 3-5 mm wide, clustered in heads of 1-6 stalkless spikelets, the head appearing lateral; scales brown and translucent, 3-5 mm long, notched at tip, with a midvein extended into a short awn 1-2 mm long; bristles 4-6, unequal, shorter than achene; style 2-3-parted. Achenes light green or tan to dark brown, 3-angled or flat on 1 side and convex on other, 2-3 mm long, the beak to 0.5 mm long. May-Sept. — Shallow water (to about 1 m deep), wet sandy, gravelly or mucky shores, streambanks, wet meadows, ditches, seeps and other wet places.

Schoenoplectus pungens

Schoenoplectus subterminalis (Torr.) Soják
SWAYING CLUB-RUSH   *native / cons* 9
Scirpus subterminalis Torr.

Aquatic perennial, spreading by rhizomes. Stems slender, weak, round in section, to 1 m or more long, floating or slightly emergent from water surface near tip. Leaves many, threadlike, channeled, from near base of stem and extending to just below water surface; bract 1-6 cm long, appearing to be a continuation of stem. Spikelets single at ends of stems, with several flowers, light brown, narrowly ovate, tapered at each end, 7-12 mm long; scales thin, 4-6 mm long, light brown with a green midvein; bristles shorter to about as long as achene, downwardly barbed; style 3-parted. Achenes 3-angled, brown, 2-4 mm long, tipped with a slender beak to 0.5 mm long. July-Aug. — In water to about 1 m deep of lakes, ponds and bog margins.

Schoenoplectus subterminalis

Schoenoplectus tabernaemontani (K.C. Gmel.) Palla
SOFT-STEM CLUB-RUSH   *native / cons* 4
Scirpus tabernaemontani K.C. Gmel.
Scirpus validus Vahl

Perennial, spreading by rhizomes and sometimes forming large colonies. Stems stout, smooth, erect, 1-3 m long, round in section. Leaves reduced to 4-5 sheaths at base of stem, or upper leaves with a blade to 7 cm long; main bract erect, 1-10 cm long, shorter than the head. Spikelets redbrown, 4-12 mm long and 3-4 mm wide, single or in clusters of 2-5 at ends of stalks, the stalks spreading or drooping, the clusters in paniclelike heads; scales ovate, light to dark brown, 2-3 mm long, the midvein usually extended into a short awn to 0.5 mm long; bristles 4-6, downwardly barbed, equal or longer than achene; style 2-parted. Achenes flat on 1 side and convex on other, brown to black, about 2 mm long, tapered to a very small beak to 0.2 mm long. June-Aug. — Shallow water and shores of lakes, ponds, marshes, streams, and ditches.

Schoenoplectus

Similar to hardstem club-rush (*S. acutus*) but the stems easily crushed between the fingers, plants generally smaller and more slender, and the head more open.

Scirpus
BULRUSH
Stout, rushlike perennials, mostly spreading by rhizomes. Stems unbranched, 3-angled or round in section, solid or pithy. Leaves broad and flat, to narrow and often folded near tip, or reduced to sheaths at base of stems; involucral bracts several and leaflike, or single and appearing like a

continuation of the stem. Spikelets single, or in panicle-like or umbel-like clusters at ends of stems, or appearing lateral from the stem; the spikelets stalked or stalkless; scales overlapping in a spiral. Flowers perfect; sepals and petals reduced to 1-6 smooth or downwardly barbed bristles, or sometimes absent; stamens 2 or 3; styles 2-3-parted. Achenes lens-shaped, flat on 1 side and convex on other, or 3-angled, usually tipped with a beak.

1 Lower sheaths red-tinged. . . . . *S. microcarpus*
1 Sheaths green or brown. 2
2 Spikelets many in dense, more or less round heads; bristles about as long as achene or shorter . . . .
   . . . . *S. atrovirens*
2 Spikelets few in open clusters; bristles much longer than achene 3
3 Mature bristles equal or only slightly longer than scales, spikelets not woolly . . . . *S. pendulus*
3 Mature bristles longer than scales, giving spikelets woolly appearance . . . .
   . . . . *S. cyperinus* complex (see desc.)

Scirpus atrovirens Willd.
DARK-GREEN BULRUSH   *native / cons 3*
Loosely tufted perennial, with short rhizomes. Stems 3-angled, leafy, 0.5-1.5 m long. Leaves mostly on lower half of stem, blades ascending, usually shorter than the head, 6-18 mm wide; bracts 3-4, leaflike, to 15 cm long, mostly longer than the head. Spikelets many, 2-8 mm long and 1-3 mm wide, crowded in rounded heads at end of stems, the heads on stalks to 12 cm long; scales brown-black, translucent except for the broad green midvein, 1-2 mm long, tipped by an awn to 0.5 mm long; bristles 6, white or tan, shorter or equal to the achene; style 3-parted. Achenes tan to nearly white, compressed 3-angled, about 1 mm long, with a short beak 0.2 mm long. June-Aug. — Wet meadows, shores, ditches, streambanks, swamps, springs and other wet places.

Scirpus atrovirens

Scirpus cyperinus (L.) Kunth
WOOL-GRASS   *native / cons 4*
Coarse, densely tufted perennial, rhizomes short. Stems leafy, to 2 m tall, rounded 3-angled to nearly round in section. Leaves flat, 3-10 mm wide, rough-to-touch on margins; sheaths brown; bracts 2-4, leaflike, spreading, usually drooping at tip, often red-brown at base. Spikelets numerous, ovate, 3-8 mm long and 2-3 mm wide, appearing woolly due to the long bristles, in clusters of 1 to several spikelets; the spikelet clusters grouped into large, spreading, branched heads at ends of stems; scales ovate, 1-2 mm long; bristles 6, smooth, brown, much longer than achene and scale; styles 3-parted. Achenes white to tan, flattened 3-angled, 0.5-1 mm long, with a short beak. July-Sept. — Wet meadows, marshes, swamps, ditches, bog margins, thickets; where wet or in very shallow standing water.

Scirpus cyperinus

The *Scirpus cyperinus* complex, including this species and *S. atrocinctus* Fern. (reported from Door County), is often regarded as one highly variable species. Alternately, the 2 taxa can be separated as follows:
1 Spikelets all or mostly all sessile in clusters of (2–) 3–7 or more. . . . . *S. cyperinus*
1 Spikelets mostly pediceled, the ultimate branches of the inflorescence typically bearing 1 central,
   sessile spikelet with 2–3 pediceled ones . . . . *S. atrocinctus*
Scirpus atrocinctus flowers and fruits earlier than *S. cyperinus*, often with inflorescences fully developed by late June, and achenes ripe by late July. *S. atrocinctus* readily hybridizes with *S. cyperinus* to form hybrid swarms. Scales of *S. atrocinctus* are usually distinctly blackened, at least near the tip.

Scirpus microcarpus J. & K. Presl
RED-TINGE BULRUSH   *native / cons 6*
Perennial, from stout rhizomes. Stems single or few together, 5-15 dm long, weakly 3-angled. Leaves several along stem, flat, ascending, 7-15

mm wide, the upper leaves longer than the head, margins rough-to-touch; sheaths often red-tinged; bracts 3-4, leaflike, to 2-3 dm long. Spikelets numerous, 3-6 mm long and 1-2 mm wide; in a loose, spreading, umbel-like head, the head formed of clusters of 4-20 or more spikelets on stalks to 15 cm long; scales 1-2 mm long, brown and translucent except for green midvein; bristles 4-6, white to tan, downwardly barbed, longer than achene; style 2-parted. Achenes lens-shaped, pale tan to nearly white, about 1 mm long, the beak tiny. June-July. — Streambanks, wet meadows, marshes, wet shores, thickets, swamps, springs; not in dense shade.

Scirpus microcarpus

Scirpus pendulus Muhl.
RUFOUS BULRUSH　　　　　　　　　　　　native / cons 4
Loosely tufted perennial, from short, thick rhizomes. Stems upright, rounded 3-angled, to 1.5 m long, lower stem covered by old leaf bases. Leaves several on stem, flat, 4-10 mm wide, shorter than head; bracts leaflike, 3 or more, shorter than the head, pale brown at base. Spikelets many, cylindric, 4-10 mm long and 2-4 mm wide; in an open, umbel-like head at end of stem, the spikelets drooping and clustered in groups of 1 stalkless and several stalked spikelets; scales about 2 mm long, red-brown with a green midvein; bristles 6, brown, smooth, longer than achene and about as long as scale; style 3-parted. Achenes compressed 3-angled, light brown, about 1 mm long, with a short, slender beak. June-Aug. — Marshes, wet meadows, streambanks, swamp openings and ditches.

Scirpus pendulus

Trichophorum
LEAFLESS-BULRUSH
Tufted perennials. Stems 3-angled or terete. Leaves basal or nearly so; sheaths bladeless or with very short blades less than 1 cm long and to 1 mm wide. Inflorescences terminal; spikelets 1; involucral bracts 1, suberect, scale-like, tip mucronate or awned. Spikelets with 3-9 spirally arranged scales, each subtending a flower. Flowers bisexual; perianth of 0-6 bristles, straight, shorter than to about 20 times as long as the achene, smooth or scabrous; stamens 3. Achenes 3-angled or plano-convex.

1　Stems more or less round in section, smooth . . . . *T. caespitosum*
1　Stems 3-angled, rough on angles. . . . . *T. alpinum*

Trichophorum alpinum (L.) Pers.
ALPINE LEAFLESS-BULRUSH　　　　　　　native / cons 10
Eriophorum alpinum L.
Scirpus hudsonianus (Michx.) Fern.
Perennial, from short rhizomes. Stems single to clustered, slender, 1-4 dm long, sharply 3-angled, rough-to-touch on the angles. Leaves reduced to scales at base of stem, with 1-2 leaves upward on stem, these with short narrow blades 5-15 mm long. Spikelets single at ends of stems, brown, 5-7 mm long, with 10-20 flowers, involucral bract awl-shaped, shorter than spikelet, sometimes absent; scales ovate, blunt-tipped, yellow-brown; bristles 6, white, flattened, longer than the scales, when mature forming a white tuft 1-2 cm longer than the spikelet. Achenes 3-angled, dull brown, 1-4 mm long. — Open bogs, conifer swamps, wet meadows, wet sandy shores; sometimes where calcium-rich.

Trichophorum caespitosum (L.) Hartman
TUFTED LEAFLESS-BULRUSH　　　　　　thr / native / cons 10
Scirpus caespitosus L.
Densely tufted perennial, rhizomes short. Stems slender, smooth, more or less round in section, 1-4 dm long. Leaves light brown and scalelike at base of stems, and also usually 1 leaf upward on stem, the blade narrow,

Trichophorum alpinum

short, to 6 mm long. Spikelets 1 at end of stems, brown, 4-6 mm long, several-flowered; scales yellow-brown, deciduous, the lowest scale about as long as spikelet; bristles 6, usually slightly longer than achene; style 3-parted. Achenes brown, 3-angled, 1.5 mm long. — Open bogs, cedar swamps, calcareous fens, wet swales between dunes.

Hydrocharitaceae

TAPE-GRASS FAMILY

Aquatic, perennial herbs. Stems leafy, the leaves opposite (*Najas*), whorled (*Elodea*), or plants stemless with clusters of long, linear, ribbonlike leaves (*Vallisneria*). Flowers usually either staminate or pistillate and borne on separate plants, small and stalkless, or in a spathe at end of a stalk; sepals 3; petals 3 or absent; staminate flowers with 3 or more stamens; stigmas 3. Fruit several-seeded, maturing underwater.

1 Leaves very long and ribbon-like (mostly 3–11 mm wide), in a basal rosette . . . . *Vallisneria*
1 Leaves to 6 (–12) cm long, opposite or whorled 2
2 Leaves whorled, entire . . . . *Elodea*
2 Leaves opposite, minutely denticulate to visibly toothed . . . . *Najas*

Elodea

WATERWEED

Aquatic perennial herbs, rooting from lower nodes or free-floating. Stems slender, leafy, branched. Leaves crowded near tip of stem, mostly in whorls of 3-4, or opposite, stalkless; margins finely sharp-toothed. Flowers either staminate or pistillate and on separate plants, tiny, single in upper leaf axils, subtended by a 2-parted spathe, usually extended to the water surface by a long, threadlike hypanthium, or stalkless and breaking free to float to water surface in staminate flowers of *E. nuttallii;* sepals 3; petals 3 or absent, white or purple; staminate flowers with 9 stamens; pistillate flowers with 3 stigmas, the stigmas entire or 2-parted. Fruit a capsule, ripening underwater. *Anacharis* or *Philotria* in older floras.

1 Leaves mostly 2 mm or more wide; staminate flowers long-stalked in a spathe, the spathe more than 7 mm long, extended to water surface by a long, threadlike hypanthium . . . . *E. canadensis*
1 Leaves to 1.5 mm wide; staminate flowers stalkless in a spathe, the spathe 2–4 mm long, breaking free to float to water surface at flowering time . . . . *E. nuttallii*

Elodea canadensis Michx.

CANADIAN WATERWEED   *native / cons 3*
  Anacharis canadensis (Michx.) Planch.

Submerged perennial herb. Stems round in section, usually branched, 2-10 dm long. Leaves bright green, firm; lower leaves opposite, reduced in size, ovate or lance-shaped; upper leaves in whorls of 3, the uppermost crowded and overlapping, lance-shaped, 5-15 mm long and about 2 mm wide, rounded at tip. Flowers at ends of threadlike stalks, 2-30 cm long; staminate flowers in spathes from upper leaf axils, the spathes about 10 mm long and to 4 mm wide; sepals green, 3-5 mm long; petals white, 5 mm long; stamens 9; pistillate flowers in spathes from upper leaf axils, the spathes 10-20 mm long, extended to water surface by a threadlike hypanthium; sepals 2-3 mm long; petals white, 2-3 mm long. Fruit a capsule, 5-6 mm long, tapered to a beak 4-5 mm long. June-Aug. — Shallow to deep water of lakes, streams and ditches.

Elodea canadensis

Elodea nuttallii (Planch.) St. John

WESTERN WATERWEED   *native / cons 7*
  Anacharis nuttallii Planch.

Submerged perennial herb. Stems slender, round in section, usually branched, 3-10 dm long. Lower leaves opposite, reduced in size, ovate to lance-shaped; upper leaves in whorls of 3 (or sometimes 4), not densely overlapping at tip, linear to lance-shaped, 6-13 mm long and 0.5-1.5 mm wide, tapered to a pointed tip. Staminate flowers in stalkless spathes from middle leaf axils, the spathes ovate, 2-3 mm long, the flowers single and stalkless in the spathe, breaking free and floating to water surface and then opening; sepals green or sometimes red, 2 mm long; petals absent or very short (to 0.5 mm long); stamens 9; pistillate flowers in cylindric spathes from upper leaf axils, the spathes 1-2.5 cm long, extended to water surface by a threadlike stalk to 10 cm long; sepals green, about 1 mm long; petals white, longer than sepals. Fruit a capsule, 5-7 mm long. June-Aug. — Shallow to deep water of lakes, streams and ditches.

Similar to *E. canadensis* but less common, smaller and more delicate overall, the leaves narrower, paler green, and not closely overlapping at stem tips, and the staminate flowers not elevated on a long slender stalk.

Elodea nuttallii

Najas
WATERNYMPH

Aquatic annual herbs, roots fibrous, rhizomes absent. Stems wavy, with slender branches. Leaves simple, opposite or in crowded whorls, stalkless, abruptly widened at base to sheath the stem; margins toothed to nearly entire, the teeth sometimes spine-tipped. Flowers either staminate or pistillate, separate on same plant or on different plants, tiny, single and stalkless in leaf axils, enclosed by the sheathing leaf bases; staminate flowers a single anther within a membranous envelope (spathe), this surrounded by perianth scales, the scales sometimes joined into a tube; pistillate flowers surrounded by 1-2 spathes, pistils 1, stigmas 2-4, style usually persistent. Fruit a 1-seeded achene.

Najas flexilis (Willd.) Rostk. & Schmidt
WAVY WATERNYMPH   *native / cons* 6
Stems branched, 5-40 cm long. Leaves densely clustered at tips of stems, linear, tapered to a long slender point, spreading or ascending, 1-4 cm long and to 0.5 mm wide; margins with tiny sharp teeth. Achenes oval, olive-green to red, the beak 1 mm or more long; seeds straw-colored, shiny, 2-4 mm long. July-Sept. — Ponds, lakes, streams.

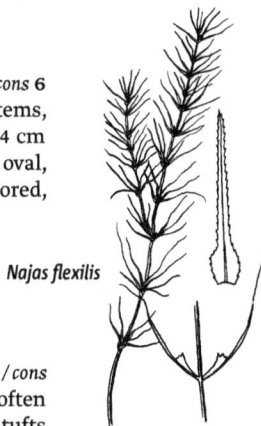

Vallisneria
EEL-GRASS

Najas flexilis

Vallisneria americana Michx.
AMERICAN EEL-GRASS   *native / cons*
Submerged perennial herb, fibrous rooted, spreading by stolons and often forming large colonies. Stems absent. Leaves long and ribbonlike, in tufts from a small crown, to 1 m or more long and 3-10 mm wide, rounded at tip, margins smooth. Flowers either staminate or pistillate and on separate plants; staminate flowers small, about 1 mm wide, in a many-flowered head, the head within a stalked spathe from base of plant, the stalk 3-15 cm long; sepals 3, petals 1, stamens 2; the staminate flowers released singly from the spathe and floating to water surface where they open; pistillate flowers single in a spathe, on long slender stalks that extend to water surface, the stalk contracting and coiling after flowering to draw the fruit underwater; sepals 3, petals small, 3; stigmas 3. Fruit a cylindric, curved capsule, 4-10 cm long. July-Sept. — Shallow to sometimes deep water of lakes and streams.

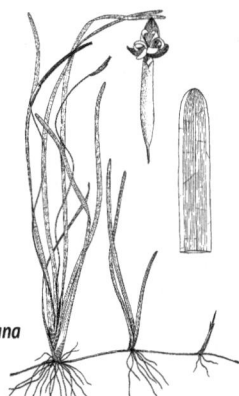

Vallisneria americana

Iridaceae

IRIS FAMILY

Perennial herbs with rhizomes, bulbs, or fibrous roots. Leaves parallel-veined, narrow, 2-ranked, the margins joined to form an edge facing the stem (equitant). Flowers perfect, with 6 petal-like segments, single or in clusters at ends of stem, stamens 3, style 3-parted. Fruit a 3-chambered capsule.

1 Flowers more than 2 cm wide; stems not winged; leaves more than 6 mm wide . . . . *Iris*
1 Flowers to 2 cm wide; stems winged; leaves to 6 mm wide . . . . *Sisyrinchium*

Iris

IRIS, FLAG

Perennial herbs, spreading by thick rhizomes. Stems erect. Leaves swordlike, erect or upright, the margins joined to form an edge facing the stem. Flowers 1 or several at ends of stems; yellow or blue-violet; sepals 3, spreading or bent downward, longer and wider than the petals; petals 3, erect or arching; stamens 3; styles 3-parted, the divisions petal-like and arching over the stamens. Fruit an oblong capsule.

1 Plants dwarf, the flowering stems less than 15 (usually less than 10) cm tall. . . . . *I. lacustris*
1 Plants more than 15 cm tall 2
2 Flowers yellow, mature capsules spreading or pendant . . . . *I. pseudacorus*
2 Flowers blue (white in albinos), capsules erect 3
3 Base of expanded portion of sepal with a bright yellow spot, finely pubescent with hairs as long as the thickness of the sepal; outer spathe bracts of uniform texture and color; seeds round to D-shaped, irregularly (but shallowly) pitted . . . . *I. virginica*
3 Base of expanded portion of sepal at most with a greenish yellow spot, with papillae shorter than thickness of the sepal; outer spathe bracts with the margins generally darker and more shiny than the rest of the dull surface; seeds D-shaped, with a regularly pebbled surface . . . . *I. versicolor*

Iris lacustris Nutt.

DWARF LAKE IRIS   *threatened / native / cons* 9
Perennial herb, mostly less than 15 cm tall. Leaves broadly linear, curved-arching, 4-6 cm long at anthesis, later to 18 cm long and 5-10 mm wide. Petals blue or rarely, white; perianth tube dull yellow, 1-2 cm long, dilated upwards, shorter than the sepals and petals; perianth limb 4-5 cm across. Fruit a capsule; seeds dark brown. May. — Sandy beach ridges and stabilized dunes near Lake Michigan.

Iris lacustris

Iris pseudacorus L.

PALE-YELLOW IRIS   *introduced*
Perennial herb, from thick rhizomes. Stems 0.5-1 m long, shorter or equal to the leaves. Leaves sword-shaped, stiff and erect, waxy, 1-2 cm wide. Flowers several at end of stems, yellow, 7-9 cm wide, sepals spreading, upper portion marked with brown; petals erect, narrowed in middle, 1-2.5 cm long. Capsules oblong, 6-angled, 5-9 cm long. May-June. — Lakeshores, streambanks, marshes, ditches.

Iris versicolor L.

NORTHERN BLUEFLAG   *native / cons* 5
Perennial herb, from thick, fleshy rhizomes and forming colonies. Stems more or less round in section, often branched above, 4-9 dm long. Leaves sword-shaped, erect or arching, somewhat waxy, 2-3 cm wide, usually shorter than stem. Flowers several on short stalks at ends of stems, blue-violet, 6-8 cm wide; sepals spreading, unspotted, or with a green-yellow spot near base, surrounded by white streaks and purple veins; petals erect, about half as long as sepals. Capsules oblong, 3-6 cm long. June-July. — Marshes, shores, wet meadows, open bogs, swamps, thickets, forest depressions; often in shallow water.

Iris versicolor

Iris virginica L.

SOUTHERN BLUEFLAG *native / cons* **5**

Perennial herb, from thick rhizomes, often forming large colonies. Stems more or less round in section, to 1 m long. Leaves sword-shaped, erect or arching, 2-3 cm wide, usually longer than stems. Flowers several on short-stalks at ends of stems, blue-violet, often with darker veins, 6-8 cm wide; sepals spreading, curved backward at tip, with a hairy, bright yellow spot near base; petals shorter than sepals. Capsules ovate to oval capsule, 4-7 cm long. May-July. — Swamps, shores, streambanks, marshes, ditches.

Iris virginica

Sisyrinchium

BLUE-EYED-GRASS

Tufted perennial herbs, from fibrous roots. Stems slender, leafless, flattened or winged. Flowers in an umbel at end of stem, above a pair of erect green bracts (spathe), blue-violet or rarely white, with 6 spreading segments, the segments joined only at base, the tips rounded but with an small bristle. Leaves narrow and linear, from base of plant, the margins joined and turned to form an edge facing the stem. Fruit a rounded capsule; seeds round, black.

1 Margins of outer spathe distinct nearly to the base, spathes greenish, slightly or not at all purple tinged; stems less than 2 mm wide . . . . *S. campestre*
1 Margins of outer spathe fused for 1–5 mm, spathes often strongly purple tinged or stems often more than 2 mm wide (or both) 2
2 Plant very slender, the stem usually 1 mm or less wide and barely margined or narrowly winged; largest leaves to 1.5 mm wide; capsule 2.5–4 mm long . . . . *S. mucronatum*
2 Plants stout, the stem usually 2–2.5 mm wide, winged; largest leaves 2–3 mm wide; mature capsule 5–7 mm long . . . . *S. montanum*

Sisyrinchium campestre Bickn.

PRAIRIE BLUE-EYED-GRASS *native / cons* **7**

Plants pale green, scarcely glaucescent, drying pale. Stems erect, 1-4 dm tall. Scapes narrowly winged, 1-2 mm wide. Spathes scabridulous, often purple-tinged, the inner bracts 1-2 cm long; perianth pale blue or white, 8-10 mm long. Capsules 2-3 mm long, pale or straw-color. — Meadows, sandy places, and open woodlands.

Sisyrinchium montanum Greene

STRICT BLUE-EYED-GRASS *native / cons* **5**

Plants pale-green and waxy. Stems stiff and erect, leafless, flattened and winged, 1-5 dm tall and 2-4 mm wide. Leaves mostly from base of plant, narrow and grasslike, about half as long as stem, 1-3 mm wide. Flowers in head of 1 to several flowers at end of stem, subtended by a spathe, the spathe of 2 bracts, the outer bract 3-7 cm long, the inner bract about half as long; the flower segments (tepals) blue-violet with a yellow center, 5-15 mm long, with a short, slender tip. Capsules more or less round, 4-7 mm wide, pale brown, on an erect stalk shorter than the inner bract. May-July. — Wet meadows, shores, thickets, ditches, swales; also in drier woods and fields.

Sisyrinchium montanum

Sisyrinchium mucronatum Michx.

NEEDLE-TIP BLUE-EYED-GRASS *native / cons* –

Plants dark green. Stems very slender, to 1 mm wide, leafless, margins not or barely winged. Leaves from near base of plant, narrow and linear, to 1.5 mm wide. Flowers in a single head at end of stem, subtended by a spathe, the spathe of 2 bracts, the bracts often purple-tinged, the outer bract 2-3 cm long, the inner bract shorter, 1-2 cm long; the segments (tepals) deep violet-blue, 8-10 mm long, tipped with a sharp point. Capsules more or less round, 2-4 mm long, pale brown, on spreading stalks. May-June. — Wet meadows, calcareous fens.

Sisyrinchium mucronatum

Juncaceae

RUSH FAMILY

Distinguished from grasses and sedges by the presence of a true perianth of 6 tepals and a 3-many-seeded capsule rather than a 1-seeded grain (grasses) or achene (sedges). No ligule (as in the grasses) is present at junction of leaf blade and sheath; however an auricle (an ear-like appendage) may occur at top of leaf sheath.

1 Foliage completely glabrous; capsules usually many-seeded. . . . . *Juncus*
1 Foliage hairy, at least toward summit of sheaths; capsules 3-seeded . . . . *Luzula*

Juncus

RUSH

Clumped or rhizomatous rushes, mostly perennial (annual in Juncus bufonius). Stems erect and unbranched. Leaves from base of plant or along stem, alternate, round in section, flat to involute, or reduced to sheaths at base of stem. Flowers perfect, regular, in compact to open clusters of few to many flowers, subtended by 1 or several leaflike involucral bracts; sepals and petals of 6 chaffy, scalelike, green to brown tepals (perianth); stamens 6 or 3; stigmas 3, ovary 1 or 3-chambered. Capsule many-seeded; seeds with a short slender tip or with a tail-like appendage at each end.

1 Head from side of stem, the involucral bract erect, round in section and appearing to be a continuation of stem; basal and stem leaves absent 2
1 Head at end of stem; basal leaves present, stem leaves present or absent 3
2 Stems densely clumped; stamens 3 . . . . *J. effusus* (J. pylaei)
2 Stems single from rhizomes, the stems often in rows; stamens 6 . . . . *J. balticus*
3 Leaves flat or somewhat channeled 4
3 Leaves round, inrolled, or narrowly channeled 7
4 Plants annual, to 10 cm tall . . . . *J. bufonius*
4 Plants perennial, more than 10 cm tall 5
5 Stems clumped; leaves with spreading blades extend from base to middle of stem. . . . *J. compressus*
5 Stems not clumped; leaves with spreading blades only near base of stem 6
6 Auricles flaplike, prolonged into a membranaceous or scarious projection 3-5 mm long . . . . *J. tenuis*
6 Auricles shorter, not flaplike, prolonged up to 2 mm beyond the sheath, cartilaginous . . . . *J. dudleyi*
7 Heads with 1–2 flowers; the head less than 1/4 total height of plant . . . . *J. pelocarpus*
7 Heads with 3–13 flowers; the head more than 1/4 total height of plant 8
8 Leaf sheath distinctly ribbed; seeds 1–2 mm long, with short to long white tails. 9
8 Leaf sheath smooth or veined; seeds less than 1 mm long, without white tails or with very short dark tails 10
9 Petals short-tapered to a rounded tip; seeds with a short tail (about 1/10 of body length). . . . .
. . . . *J. brachycephalus*
9 Petals short-tapered to a pointed tip; seeds with long or short tails . . . . *J. brevicaudatus*
10 Flowers in dense round heads, the flowers radiating in all directions 11
10 Heads various but the flowers not radiating in all directions 12
11 Heads to 10 mm wide; most capsules longer than sepals . . . . *J. nodosus*
11 Heads more than 10 mm wide; capsules shorter to nearly equaling the sepals . . . . *J. acuminatus*
12 Branches of head ascending; petals shorter than sepals . . . . *J. alpinoarticulatus*
12 Branches of head spreading; petals as long or slightly longer than sepals . . . . *J. articulatus*

Juncus acuminatus Michx.

KNOTTY-LEAF RUSH native / cons 6

Tufted perennial rush. Stems erect, slender, 2-8 dm tall, with 1-2 leaves. Leaves from stem and at base of plant, round to compressed in section, 5-40 cm long and 1-3 mm wide; auricles rounded, 1-2 mm long; bract erect, round, 1-4 cm long, shorter than the head. Flowers in an open, pyramid-shaped inflorescence, 5-12 cm long and less than half as wide, composed of 5-50 rounded heads 6-10 mm wide, each head with 5-30 flowers, the branches spreading, 1-10 cm long; tepals lance-shaped, green

Juncus acuminatus

or straw-colored, 3-4 mm long; stamens 3, shorter than the tepals. Capsule oval, straw-colored to light brown, 3-4 mm long, about as long as the tepals, tipped with a short, blunt point. June-Aug. — Wet sandy shores, streambanks and ditches; not in open bogs.

Juncus alpinoarticulatus Chaix
NORTHERN GREEN RUSH   *native / cons* **6**
   Juncus alpinus Vill.
Perennial rush, spreading by rhizomes. Stems in small clumps, 1.5-4 dm long. Leaves mostly from base of plant and with 1-2 stem leaves, round in section, hollow, with small swollen joints, 2-12 cm long and 0.5-1 mm wide; sheaths green to red, auricles rounded, 0.5-1 mm long; bract round in section, 2-6 cm long and shorter than the head. Flowers in an open panicle of 5-25 heads, 2-15 cm long and 1-5 cm wide, the heads oblong pyramid-shaped, 2-6 mm wide, mostly 2-5-flowered, the branches upright, 1-7 cm long; tepals green to brown, 2-3 mm long, the inner tepals shorter, the margins chaffy; stamens 6. Capsule oblong, 3-angled, straw-colored to chestnut brown, satiny, 2-3 mm long, slightly longer than the tepals, tapered to a rounded tip. June-Sept. — Sandy or gravelly shores, stream-banks, fens; often where calcium-rich.

Juncus alpinoarticulatus

Juncus articulatus L.
JOINT-LEAF RUSH   *native / cons* **–**
Perennial rush, with coarse white rhizomes. Stems usually tufted, 2-6 dm long. Leaves from stem and at base of plant, more or less round in section, hollow, with small swollen joints, 4-12 cm long and 1-3 mm wide; sheaths green or sometimes red, auricles rounded, about 1 mm long; bract erect, round in section, 1-4 cm long, shorter than the head. Flowers in open panicles, 4-10 cm long and 3-6 cm wide, composed of 3-30 heads, the heads rounded, 6-8 mm wide, 3-10-flowered, panicle branches erect to widely spreading, 1-4 cm long; tepals green to dark brown, 2-3 mm long; stamens 6. Capsule oval, dark brown, shiny, 3-4 mm long, longer than the tepals, tapered to a tip. July-Sept. — Sandy, gravelly or mucky shores, streambanks and springs.

Juncus articulatus

Juncus balticus Willd.
SMALL-HEAD RUSH   *native / cons* **5**
   Juncus arcticus Willd.
Perennial rush, spreading by stout, brown to black rhizomes. Stems slender and tough, dark green, 3-9 dm long, in rows from the rhizomes. Leaves reduced to red-brown sheaths at base of stem; bract erect, round in section, 1-2 dm long, longer than the head and resembling a continuation of stem. Flowers single on stalks, in dense to spreading heads, the heads appearing lateral, extending outward from stem 1-7 cm; tepals lance-shaped, dark brown, 3-5 mm long, margins chaffy; stamens 6. Capsule ovate, somewhat 3-angled, red-brown, 3-4 mm long, shorter to slightly longer than the tepals, tapered to a sharp point. May-Aug. — Wet sandy or gravelly shores, interdunal wetlands near Lake Michigan, meadows, ditches, marshes, seeps.

Juncus balticus

Juncus brachycephalus (Engelm.) Buch.
SMALL-HEAD RUSH   *native / cons* **9**
Densely tufted perennial rush. Stems erect, round in section, 3-7 dm long. Leaves from stem and base of plant, round in section, 2-20 cm long and 1-2 mm wide, often spreading; auricles rounded, to 1 mm long; bract erect, round in section, 1-5 cm long, shorter than the head. Flowers in an open raceme or panicle of 10-80 heads, 5-25 cm long and 2-12 cm wide, the heads oval, 2-5 mm wide, 2-6-flowered, branches upright to spreading, 1-5 cm long; tepals lance-shaped, green to light brown, 3-nerved, 2-3 mm

long, margins chaffy; stamens 3 or sometimes 6. Capsule ovate, more or
less 3-angled, light brown, 3-4 mm long, longer than the tepals, abruptly
narrowed to a short beak. June-Sept. — Sandy or gravelly shores, streambanks, open bogs, calcium-rich springs.

Juncus brachycephalus

Juncus brevicaudatus (Engelm.) Fern.
NARROW-PANICLE RUSH   *native / cons* **6**

Densely tufted perennial rush. Stems erect, round in section, 1.5-5 dm
long. Leaves from stem and base of plant, round in section, hollow, with
small swollen joints, 3-20 cm long and 1-2 mm wide; sheaths green or
sometimes red, auricles rounded, 1-2 mm long; bract erect, round in section, 2-7 cm long, shorter to longer than the head. Flowers in a raceme or
panicle of 3-35 heads, 3-12 cm long and 1-4 cm wide, the heads oval, 2-6
mm wide, 2-7-flowered, branches upright, 0.5-3.5 cm long; tepals green
to light brown, often red-tinged near tip, 3-nerved, 3-4 mm long, margins
chaffy; stamens 3. Capsule oval, 3-angled, dark brown, 3-5 mm long, longer
than the tepals, tapered to a sharp point. Aug-Sept. — Wet meadows,
marshes, fens, sandy lshores.

Juncus brevicaudatus

Juncus bufonius L.
TOAD RUSH   *native / cons* **3**

Small annual rush. Stems tufted, erect to spreading, 5-20 cm long. Leaves
from stem and at base of plant, flat or channeled, 1-7 cm long and to 1
mm wide, usually shorter than stem; sheaths green to red or brown, auricles absent; bract erect, 1-10 cm long, shorter than the head. Flowers
single, mostly stalkless, with 1-7 flowers along each branch of the inflorescence, the inflorescence comprising half or more of the entire length
of plant; tepals lance-shaped, green to straw-colored, 4-6 mm long, margins chaffy; stamens 6. Capsule ovate, brown or green, 3-4 mm long,
rounded at tip, shorter than the tepals. June-Aug. — Sandy or silty shores,
mud flats, streambanks, wet compacted soil of trails and wheel ruts.

Juncus compressus Jacq.
ROUND-FRUIT RUSH   *introduced*

Tufted, perennial rush. Stems erect, flattened, 2-7 dm long. Leaves from
plant base and 1 or 2 along stem, flat or channeled, 5-20 cm long and to
1.5 mm wide; auricles rounded, to 1 mm long; bract erect, somewhat
bent, flat or folded, 2-8 cm long, often longer than head. Flowers on short
stalks 1-5 mm long, with 1-2 flowers along each branch of the inflorescence, the inflorescence 3-7 cm long and 1-3 cm wide, branches upright;
tepals ovate, light to dark brown, 1-3 mm long, margins translucent; stamens 6. Capsule nearly round, light brown, 2-3 mm long, longer than
tepals. June-Aug. — Wet meadows, disturbed wet areas, ditches along
highways where forming dark green colonies, often where saline.

Juncus dudleyi Wieg.
DUDLEY'S RUSH   *native / cons* **4**

Juncus tenuis var. *dudleyi* (Wiegand) F.J. Herm.

Tufted perennial rush, from branching rhizomes. Stems 2-10 dm long.
Leaves basal, 2-3; auricles carilaginous, yellowish, to 0.4 mm long; blade
flat, 5-30 cm long; bract usually longer than the head. Inflorescences
compact to loose and lax; primary bract usually exceeding inflorescence.
Capsules tan, ca. 3 mm long. — Damp to drier open places, including
lakeshores, marsh margins, ditches.

Juncus bufonius

Juncus effusus L.
LAMP RUSH   *native / cons* **4**

Densely tufted perennial rush. Stems erect, round in section, to about 1
m long. Leaves reduced to bladeless sheaths at base of stem, the sheaths

to 2 dm long, mostly red-brown; bract round in section, 10-30 cm long, appearing like a continuation of stem, longer than the head. Flowers in a many-flowered inflorescence, with 2-4 flowers along each branch of the inflorescence, the inflorescence appearing lateral, the branches upright to spreading or bent downward; tepals lance-shaped, green to straw-colored, 2-3 mm long; stamens 3. Capsule broadly ovate, olive-green to brown, 2-3 mm long, about as long as tepals, sometimes tipped with a short point. June-July. — Marshes, shores, thickets, streambanks, bog margins, wet meadows.

ADDITIONAL SPECIES

Juncus pylaei Laharpe, sometimes treated as a species or as *J. effusus* var. *pylaei* (Laharpe) Fern. & Weig., and apparently the common taxon in Wisconsin. *J. effusus* has soft, broad stems rather smooth because the ridges are tiny and numerous and readily flattened by pressing; *J. pylaei* has fewer, prominent ridges; and stems that are narrower, sturdier, and generally not flattened by pressing.

Juncus effusus

Juncus nodosus L.

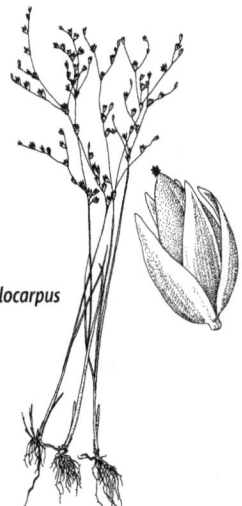

INLAND RUSH   native / cons 6

Perennial rush, spreading by rhizomes. Stems erect, slender, round in section, 1.5-6 dm long. Leaves on stem and one at base of plant, round in section, hollow, with small swollen joints, 3-30 cm long and 1-2 mm wide, upper leaves usually longer than the head; sheaths green, their margins green, becoming yellow and membranous toward tip, auricles rounded, yellow, 0.5-1 mm long; bract erect to spreading, round in section, 2-12 cm long, usually much longer than head. Flowers in a raceme or panicle of several heads, 1-6 cm long and 1-3 cm wide, the heads more or less round, 6-10 mm wide, 6-20-flowered, the branches erect to spreading, 0.5-3 cm long; tepals narrowly lance-shaped, green to light brown, 3-4 mm long; the margins narrowly translucent; stamens 6. Capsule awl-shaped, brown, 4-5 mm long, longer than the tepals, tapered to a sometimes curved beak. July-Sept. — Sandy, gravelly or clayey shores and streambanks, wet meadows, fens, ditches, springs; often where calcium-rich.

Juncus pelocarpus E. Mey.

BROWN-FRUIT RUSH   native / cons 8

Perennial rush, spreading by rhizomes and forming colonies. Stems erect, round in section, 1-4 dm long. Leaves from stem and at base of plant, round in section, very slender, 2-10 cm long and about 1 mm wide; auricles absent or short and straw-colored; bract erect, round in section, 2-4 cm long, shorter than the head. Flowers single or paired in a much-branched inflorescence, 5-15 cm long and 4-10 cm wide, the flowers on mostly 1 side of each branch, the branches upright to widely spreading, 1-4 cm long, with at least some of the flowers usually replaced by clusters of awl-shaped leaves; tepals ovate, dark brown, about 2 mm long, margins chaffy; stamens 6. Capsule narrowly ovate, dark brown, satiny, 2-3 mm long, equal or slightly longer than tepals, tapered to a slender beak. July-Aug. — Shallow water, sandy or mucky shores, bog margins.

Juncus nodosus

Juncus pelocarpus

Juncus tenuis Willd.

POVERTY RUSH   native / cons 1

Tufted, perennial rush. Stems erect, round in section to slightly flattened, 1-6 dm long. Leaves near base of stem, flat to broadly channeled, 10-15 cm long and to 1 mm wide; sheaths green, the margins yellow and glossy, auricles triangular, 1-3 mm long; bracts 1-3 (usually 2), the lowest erect, flat, 6-10 cm long, longer than the head. Flowers stalkless or on short stalks to 3 mm long, on branches with 1-7 flowers, in a crowded to spread-

ing head 2-5 cm long; tepals lance-shaped, green to straw-colored or light brown, 3-5 mm long, margins narrowly translucent; stamens 6. Capsule ovate, green to straw-colored, 2-5 mm long, shorter or equaling the tepals, rounded at tip. June-July. Wet meadows, shores, streambanks, springs, common in disturbed places (often where soils compacted) such as trails, roadsides, ditches; also in drier woods and meadows.

Luzula
WOOD-RUSH
Luzula acuminata Raf.
HAIRY WOOD-RUSH   *native / cons* 6
Perennial grasslike herb. Stems tufted, often stoloniferous 1-4 dm tall. Basal leaves elongate, to 3 dm long and 10 mm wide; stem leaves 2-4, shorter and somewhat narrower, all sparsely pilose, with a blunt callous tip. Inflorescence 3-6 cm high, the loosely spreading, almost filiform pedicels mostly simple with a single terminal flower, or a few with a lateral flower, or a few branched and bearing 3 or 4 flowers; perianth segments lance-shaped, usually chestnut-brown in the center, with scarious margins, 2.6-4.3 mm long, the petals slightly exceeding the sepals; stamens 6. Capsule ovoid, 3.2-4.5 mm long, mucronate; seeds purple-brown, the body about 1 mm long, tipped with a pale appendage (caruncle) nearly as long. May-June. — Moist or dry woods and forest openings, meadows, streambanks, hillsides.

Distinguished from *Juncus* by the presence of few to many long hairs on the leaves, especially toward the base of the blades.

Juncus tenuis

Luzula acuminata

Juncaginaceae
ARROW-GRASS FAMILY

Triglochin
ARROW-GRASS
Grasslike perennial herbs, clumped from creeping rhizomes, often in brackish habitats. Stems slender, leafless. Leaves all from base of plant, slender, linear, round or somewhat flattened in section, sheathing at base. Flowers perfect, regular, on short stalks in a spike-like raceme at end of stem; flower segments (tepals) 6; stigmas 3 or 6, styles short or absent; stamens 6, anthers stalkless, nearly as large as tepals. Fruit of 3 or 6 carpels, these splitting when mature into 1-seeded segments.

1 Plants generally small and slender; stigmas 3; fruit linear, clublike toward tip . . . . *T. palustris*
1 Plants larger, usually 3 dm or more tall; stigmas 6; fruit short-cylindric . . . . *T. maritima*

Triglochin maritima L.
SEASIDE ARROW-GRASS   *native / cons* 10
Tufted perennial herb, from a thick crown and spreading by rhizomes. Stems more or less round in section, leafless, 2-8 dm long. Leaves upright to spreading, somewhat flattened, to 5 dm long and 1-3 mm wide. Flowers 2-3 mm wide, in densely flowered, spike-like racemes 1-4 dm long; the flowers on upright stalks 4-6 mm long, the stalks extending downward on the stem as a wing; tepals 6, 1-2 mm long; stigmas 6; stamens 6. Fruit of 6 ovate carpels, 2-5 mm long and 1-3 mm wide, the carpel tips curved backward. June-Aug. — Sandy, gravelly, or marly lakeshores and streambanks; marshes, brackish wetlands.

Plants larger than marsh arrow-grass (*T. palustris*) and the fruit ovate rather than linear.

Triglochin maritima

Triglochin palustris L.

MARSH ARROW-GRASS   *native / cons* **10**

Small tufted perennial herb. Stems slender, leafless, 2-4 dm long. Leaves erect, round in section, to 3 dm long and 1-2 mm wide. Flowers small, 1-2 mm wide, in loosely flowered racemes, 10-25 cm long; the flowers on erect stalks, 2-5 mm long; tepals 6, 1-2 mm long; stigmas 3; stamens 6. Fruit of 3 narrow, clublike carpels, 5-8 mm long and 1 mm wide, splitting upward from base into 3 segments. June-Sept. — Sandy, gravelly, or marly lakeshores and streambanks, calcareous fens, marshes, interdunal swales; often where calcium-rich.

Triglochin palustris

Liliaceae

LILY FAMILY

Perennial herbs, from corms, bulbs or rhizomes. Stems leafy or leafless. Leaves linear to ovate, usually from base of plant, sometimes along stem, alternate to opposite or whorled. Flowers perfect (with both staminate and pistillate parts), regular; sepals and petals of 6 petal-like tepals in 2 series of 3; stamens 6; ovary superior or inferior, 3-chambered. Fruit a capsule or round berry.

Under the Angiosperm Phylogeny Group system (APG III, 2009; APG IV, 2016), genera in the Liliaceae have been placed into various new familes. However, the family designations are still in a state of flux, and may change in the future. As a convenience, our genera are retained within the traditional Lily Family grouping, with the proposed new family assignment noted in the following table and in parentheses after each genus name.

Key 1. Begin here for unknown specimens belonging to traditionally defined Liliaceae.

1 Sepals and petals of quite different color and/or texture, the former green or brownish . . . . *Trillium*
1 Sepals and petals both colored and petal-like, usually similar in shape (tepals) or the sepals of different size and shape 2
2 Ovary inferior (flowers bisexual) 3
2 Ovary superior (or flowers unisexual) 4
3 Ovary clearly inferior; uncommon garden escape . . . . *Narcissus*
3 Ovary half-inferior, part of it adnate to the perianth, glabrous (at most granular-roughened); native . . . . *Anticlea*
4 Flowers or inflorescences lateral, arising from the axils of alternate cauline leaves or scales 5
4 Flowers or inflorescences terminal on scapes or leafy (simple or branched) stems 6
5 Leaves scale-like, mostly brownish or yellowish, those on the much-branched upper portion of the plant subtending short green filiform branches (often mistaken for leaves) . . . . *Asparagus*
5 Leaves broad, flat, green (scale-like leaves or bracts may be present in addition to normal leaves) . . . . (go to Key 2)
6 Leaves all withering before plant flowers . . . . *Allium* (A. tricoccum)
6 Leaves present at flowering time. 7
7 Leaves all in one or two whorls on the stem . . . . *Medeola*
7 Leaves alternate or basal, or if in whorls these more than 2 or some alternate leaves also present 8
8 Flowers more than 3.5 cm long 9
8 Flowers less than 3.5 cm long 13
9 Leaves perfoliate . . . . *Uvularia*
9 Leaves not perfoliate 10
10 Principal leaves cauline, not crowded toward base of plant . . . . *Lilium*
10 Principal leaves basal or nearly so; stem leafless above or with very small bracts 11
11 Flowers 2–many . . . . *Hemerocallis*
11 Flowers solitary 12
12 Leaves basal . . . . *Erythronium*
12 Leaves cauline . . . . *Lilium*

13 Flowers in a many (7–60 or more) flowered umbel on an unbranched stem or scape; plants with odor of onion or garlic. . . . . *Allium*
13 Flowers solitary or in a raceme, panicle, or corymb, on a simple or branched stem (if in a few-flowered umbel, then either the stem branched or forked or flowers less than 7) 14
14 Plants with principal leaves clearly along the stem; basal leaves (at least of current season) absent or at most apparently one 15
14 Plants with the principal leaves all basal (at ground level) or nearly so; stem leaves absent, reduced to bracts, or much smaller or fewer than basal leaves. 18
15 Stem forked or branched; perianth over 12 mm long . . . . (go to Key 2)
15 Stem unbranched (above the ground and below the inflorescence); perianth usually less than 12 mm long 16
16 Ovary with 1 style; fruit a berry; inflorescence a raceme (tepals up to 7 mm long) or if a panicle, the tepals less than 3 mm long; leaves ovate . . . . (go to Key 2)
16 Ovary with 3 styles (one on each lobe); fruit a capsule; inflorescence a panicle; tepals 5–13 mm long; leaves very elongate 17
17 Plants with at least some principal leaves clearly along the stem; inflorescences paniculate (except in depauperate individuals) . . . . *Anticlea*
17 Plants with the principal leaves all basal (at ground level) or nearly so, stem leaves absent, reduced to bracts, or at most much smaller or fewer than basal leaves; inflorescences racemose . . . . . . *Triantha*
18 Flower solitary . . . . *Erythronium*
18 Flowers 2 or more in an inflorescence 19
19 Tepals united for half or more of their length 20
19 Tepals completely separate or united at base only 21
20 Perianth blue, less than 6 mm long; leaves linear; plants bulbous . . . . *Muscari*
20 Perianth white, 5–10 mm long when mature; leaves elliptic; plants not bulbous . . . . *Convallaria*
21 Ovary with a single style; flowers bisexual . . . . *Clintonia*
21 Ovary with 3 styles (1 on each lobe) in bisexual or pistillate flowers (or flowers all staminate) 22
22 Plants with at least some principal leaves clearly cauline; inflorescences paniculate . . . . *Anticlea*
22 Plants with the principal leaves all basal or nearly so, stem leaves absent, reduced to bracts, or at most much smaller or fewer than basal leaves; inflorescences racemose . . . . *Triantha*

Key 2. Fruit red, blue, or black berries, except in *Uvularia*, which has a capsule.

1 Leaves all in 1 or 2 whorls on the stem . . . . *Medeola*
1 Leaves alternate or basal 2
2 Plants with the leaves all basal or nearly so; stem leaves absent, reduced to bracts, or at most much smaller or fewer than the basal leaves 3
2 Plants with leaves clearly along the stem 4
3 Flowers yellow, in an umbel; fruit blue . . . . *Clintonia*
3 Flowers white, in a raceme; fruit (rarely produced) orange or red . . . . *Convallaria*
4 Plant unbranched 5
4 Plant branched (above the ground) 6
5 Flowers in a terminal raceme or panicle, white, tepals separate; ripe fruit red, with dark stripes in one species . . . . *Maianthemum*
5 Flowers in the axils of leaves, greenish, greenish white, or yellowish, tepals united most of their length; ripe fruit blue to black . . . . *Polygonatum*
6 Perianth pale to deep yellow; fruit a glabrous capsule; stem and pedicels glabrous. . . . . *Uvularia*
6 Perianth greenish, rose-purple, white, or creamy, fruit a pubescent to glabrate or tuberculate red berry; stem (at least when young) and pedicels often pubescent. *Streptopus*

Allium

ONION, LEEK, GARLIC (AMARYLLIDACEAE)

Biennial or perennial herbs from a coated bulb, with a strong odor of onion or garlic, the leaves usually narrow, basal or on the lower part of the stem, the scape-like stem erect, terminated by an umbel subtended by 1-3 bracts, the flowers white to pink or purple, in some species wholly or partly replaced by sessile bulblets. Flowers perfect, in umbels. Perianth segments 6, uniform in color, but the inner circle often somewhat different in shape or size, withering and persistent below the capsule. Stamens 6. Ovary 3-celled. Capsule short, ovoid, globose, or obovoid, 3-lobed, loculicidal; seeds black, 1 or 2 in each cell.

Allium includes the cultivated onion, garlic, chives, and leek.

1 Leaves usually over 2 cm wide, flat, petiolate, withering before the plant flowers . . . . . . **A. tricoccum**
1 Leaves linear, flat or terete, less than 2 cm wide (usually less than 1 cm), not petiolate, present at flowering time . . . . **A. cernuum**

Allium cernuum Roth
NODDING ONION — native / cons 7

Bulb slenderly conic, very gradually tapering into the stem. Leaves several, arising near together at the surface of the soil, shorter than the stem, commonly 2-4 mm, occasionally to 8 mm wide. Scape 3-6 dm tall, abruptly declined or decurved near the summit. Umbel nodding, many-flowered, without bulblets; pedicels 12-25 mm long, becoming rigid in fruit; perianth segments white to rose, ovate or elliptic, 4-6 mm long, obtuse or subacute; stamens exsert; filaments barely widened toward the base. Capsule obovoid, 3-lobed, about 4 mm long, each valve (and each lobe of the ovary) bearing 2 erect triangular processes near the summit. July-Aug. — Dry woods, rocky banks.

Allium tricoccum Ait.
WILD LEEK — native / cons 6

Bulb ovoid, 2-6 cm long, its coats finely fibrous-reticulate. Leaves 2-3 dm long, including the slender petiole, the blades flat, lance-elliptic, 1-2 dm long, 2-6 cm wide. Scape 1.5-6 dm tall. Umbel erect, subtended by 2 ovate deciduous bracts; pedicels 1-2 cm long; perianth segments ovate to oblong-obovate, white, 5-7 mm long obtuse, about equaling the stamens; filaments greatly widened toward the base. Capsule depressed, deeply 3-lobed, each valve often gibbous on the back below the middle. — Rich woods, often in large colonies.

The leaves develop in early spring and disappear before the flowers appear in June and July.

Anticlea
DEATH-CAMAS (MELANTHIACEAE)
Anticlea elegans (Pursh) Rydb.
MOUNTAIN DEATH-CAMAS — native / cons 9
Zigadenus elegans Pursh

Perennial herb, from an ovate bulb; plants waxy, especially when young. Stems erect, 2-6 dm long. Leaves mostly from base of plant, linear, 2-4 dm long and 4-12 mm wide; stem leaves much smaller. Flowers green-yellow or white, in a raceme or panicle, 1-3 dm long, the branches upright, subtended by large, lance-shaped, green or purplish bracts; tepals 6, obovate, 7-12 mm long, usually purple-tinged near base; stamens 6. Fruit an ovate capsule, 10-15 mm long; seeds 3 mm long. July-Aug. — Sandy or rocky shores of Lake Michigan, open bogs, calcareous fens.

Highly toxic.

Allium tricoccum

Asparagus
ASPARAGUS (ASPARAGACEAE)
Asparagus officinalis L.
ASPARAGUS — introduced

Perennial herb of various forms, the leaves reduced to small scales and replaced functionally by branches sometimes leaflike in appearance. Stems perennial from a rhizome, freely branched, to 2 m tall; ultimate branches filiform, 8-15 mm long; pedicels solitary or paired, lateral, 5-10 mm long, jointed in the middle. Flowers perfect or unisexual, greenish

Anticlea elegans

white, campanulate, 3-5 mm long; stamens 6; stigmas 3; ovary 3-celled. Fruit a red, spherical berry, about 8 mm wide, with a few large rounded seeds. May-June. — Native of Europe; commonly cultivated in gardens and escaped in waste places.

Clintonia
BLUEBEAD-LILY (LILIACEAE)

Clintonia borealis (Ait.) Raf.
YELLOW BLUEBEAD-LILY   *native / cons 7*
Perennial herbs from a rhizome, bearing 2-4 ample basal leaves, the bases sheathing a leafless erect scape bearing a few-flowered umbel of conspicuous flowers. Leaves 2-5, dark glossy green, oblong to elliptic or obovate, eventually to 3 dm long, abruptly acuminate, finely ciliate. Scape 1.5-4 dm tall, usually pubescent at the summit, or glabrous at maturity. Umbel 3-8-flowered; pedicels 1-3 cm, long, erect in fruit, softly pubescent. Flowers perfect, nodding; perianth segments distinct, narrow, greenish yellow, 15-18 mm long; stamens 6, inserted on the very base of the perianth; ovary 3-celled; stigma obscurely 3-lobed. Fruit a blue berry, rarely varying to white, spherical, about 8 mm wide, containing a few to several seeds. May-June. — Rich moist woods and swamps.

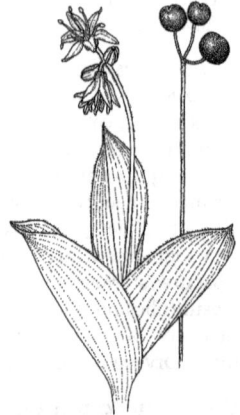
Clintonia borealis

Convallaria
LILY-OF-THE-VALLEY (ASPARAGACEAE)

Convallaria majalis L.
EUROPEAN LILY-OF-THE-VALLEY   *introduced*
Perennial herb from a rhizome, the short stem bearing a few leafless sheaths and 2 or 3 broad leaves, the scape terminating in a bracted raceme. Leaves narrowly elliptic, to 2 dm long, acuminate. Scape 1-2 dm tall. Raceme loosely flowered, one-sided; bracts small, lance-shaped; pedicels drooping. Flowers perfect, white, fragrant, 6-9 mm long; perianth globose-campanulate, with 6 short recurved lobes; stamens 6, inserted on the perianth near its base; style straight, included; ovary 3-celled. Fruit a spherical many-seeded red berry, about 1 cm wide. May. — Widely distributed in n Eurasia, commonly cultivated, and occasionally escaped near gardens.

Convallaria majalis

Erythronium
TROUT-LILY, FAWN-LILY (LILIACEAE)
Perennial from a deep solid corm, the slender stem about half underground; leaves borne near middle of the stem and therefore appearing basal, usually mottled with brown, lance-shaped to oblance-shaped or elliptic; scape bearing a single nodding flower. Our species grow in colonies, producing numerous 1-leaved sterile plants and a few 2-leaved fertile ones; reproduction often vegetatively by bulbs at the tip of slender lateral offshoots from the corm or stem. Perianth segments (tepals) separate to the base but connivent, lance-shaped, at anthesis spreading and usually eventually recurved. Stamens 6. Ovary 3-celled; style slender below, thickened above to the 3 short stigmas. Fruit an obovoid to oblong capsule.

1 Flowers white . . . . *E. albidum*
1 Flowers yellow . . . . *E. americanum*

Erythronium albidum Nutt.
SMALL WHITE FAWN-LILY   *native / cons 7*
Scape stout, 1-2 dm tall. Perianth segments (tepals) normally bluish white, varying to light pink, often suffused with green or blue externally, yellow at base within, without marginal glands; stigmas stout, separate, divergent from the linear-clavate style. April-May. — Moist woods.

Erythronium americanum Ker-Gawl.
TROUT-LILY   *native / cons* 7

Scape stout, 1-2 dm tall. Perianth segments (tepals) normally yellow, often spotted toward the base within or darker colored without, the petals bearing a glandular spot on each margin near the base; stigmas very short, scarcely separate, terminating the club-shaped style. April-May. — Moist woods.

Hemerocallis
DAY-LILY (ASPHODELACEAE)

Tall perennial herbs, with numerous, elongate, linear, basal leaves and leafless scapes bearing a terminal cluster of large flowers, each lasting a single day. Perianth funnelform, its segments spreading or recurved, cormate below into a short tube. Stamens inserted at the summit of the tube. Ovary 3-celled with numerous ovules; style slender, declined; stigma capitate.

Hemerocallis fulva (L.) L.
ORANGE DAY-LILY   *introduced (invasive)*

Scapes about 1 m tall. Flowers tawny-orange, about 12 cm wide. June-July. — Long in cultivation and freely escaped.

Erythronium albidum (left)
Erythronium americanum (right)

Lilium
LILY (LILIACEAE)
Lilium philadelphicum L.
WOOD-LILY   *native / cons* 9

Tall perennial herb from a scaly bulb. Stems erect, 3-8 dm long. Leaves all from stem, narrowly lance-shaped, 4-10 cm long and 3-9 mm wide, parallel-veined; lower leaves alternate, upper leaves opposite or whorled; petioles absent. Flowers 1-5, erect, large and showy, on stalks 1-8 cm long at ends of stem; perianth campanulate or funnelform, the segments orange-red, yellow and dark-spotted toward base, lance-shaped, 4-8 cm long and 0.8-2.8 cm wide; stamens 6; stamens and pistil about as long as tepals; stigma 3-parted; ovary superior. Fruit a more or less 3-angled capsule 2.5-4 cm long, with numerous closely packed, flat seeds. June-July. — Wet meadows, fens and open bogs, seeps, ditches; also in drier meadows and woods.

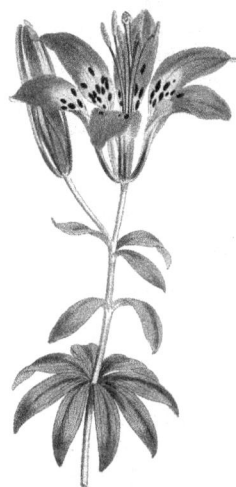

Maianthemum
FALSE SOLOMON'S-SEAL (ASPARAGACEAE)

Lilium philadelphicum

Perennial herbs from a slender creeping rhizome, the erect or ascending stems bearing few to many alternate, sessile or nearly sessile leaves, and a terminal raceme or panicle of small white flowers. Perianth regular, spreading, the segments equal and distinct, 6 (4 in *M. canadense*). Stamens 6 (4 in *M. canadense*). Ovary globose, 2- or 3-celled; style very short; stigma obscurely 2- or 3-lobed. Fruit a globose berry, usually with only 1 or 2 seeds.

Includes species formerly included in *Smilacina*.

1   Perianth of 4 parts; leaves 3 or fewer (very rarely 4, usually 2), sometimes pubescent beneath . . . . *M. canadense*
1   Perianth of 6 parts; leaves often more than 3 (1–4 in one species, where completely glabrous) 2
2   Inflorescence a panicle; perianth 1–2.5 mm long, the stamens up to 3 mm long . . . . . *M. racemosum*
2   Inflorescence a raceme; perianth 2.5–9 mm long, exceeding the stamens 3
3   Stem leaves more than 6, finely pubescent beneath (rarely almost glabrous); uppermost leaves surpassing the top of the inflorescence . . . . *M. stellatum*
3   Stem leaves 1–4, completely glabrous; inflorescence almost always overtopping leaves . *M. trifolium*

Maianthemum canadense Desf.
FALSE LILY-OF-THE-VALLEY   *native / cons* **5**
Stems erect, 5-20 cm long, spreading by rhizomes. Leaves usually 2 along stem, ovate, heart-shaped at base, 3-10 cm long; petioles short or absent. Flowers small, white, 4-6 mm wide, stalked, in a short raceme at end of stem, the raceme 3-6 cm long; tepals 4, spreading; stamens 4; style 2-lobed. Fruit a pale red berry, 3-4 mm wide; seeds 1-2. May-July. — Moist to dry woods; also on hummocks in swamps, open bogs and thickets.

Maianthemum racemosum (L.) Link
FALSE SOLOMON'S-SEAL   *native / cons* **5**
  Smilacina racemosa (L.) Desf.
Stems usually curved-ascending, 4-8 dm tall, finely pubescent. Leaves spreading horizontally in two ranks, elliptic, 7-15 cm long, 2-7 cm wide, obtuse or rounded at base, short-acuminate, finely pubescent beneath. Panicle peduncled or rarely sessile, ovoid to cylindric, 3-15 cm long. Flowers very numerous, short-pediceled, 3-5 mm wide. Berry red, dotted with purple. May-June. — Rich woods.

Maianthemum canadense

Maianthemum stellatum (L.) Link
STARRY FALSE SOLOMON'S-SEAL   *native / cons* **5**
  Smilacina stellata (L.) Desf.
Stems ascending or usually erect, 2-6 dm tall, finely pubescent or glabrous. Leaves spreading or oftener strongly ascending, usually folded along the midvein, sessile and somewhat clasping, mostly lance-shaped, 6-15 cm long and 2-5 cm wide, gradually tapering to the acute tip, finely pubescent beneath. Raceme short-peduncled or nearly sessile, 2-5 cm long, with few to several flowers 8-10 mm wide. Fruit black or green with black stripes, 6-10 mm wide. May-June. — Moist, especially sandy soil of woods and shores.

Maianthemum trifolium (L.) Sloboda
THREE-LEAF FALSE SOLOMON'S-SEAL   *native / cons* **10**
  Smilacina trifolia (L.) Desf.
Stems erect, 1-5 dm long at flowering time, from long rhizomes. Leaves alternate, smooth, usually 3 (2-4), oval or oblong lance-shaped, 6-12 cm long and 1-4 cm wide; petioles absent. Flowers small, white, 8 mm wide, stalked, 3-8 in a raceme; tepals 6, spreading; stamens 6. Fruit a dark red berry, 3-5 mm wide; seeds 1-2. May-June. — Open bogs, conifer swamps, thickets.

Maianthemum stellatum (left)
Maianthemum trifolium (right)

Medeola
CUCUMBER-ROOT (LILIACEAE)
Medeola virginiana L.
INDIAN CUCUMBER-ROOT   *native / cons* **9**
Perennial herb from a thick, tuber-like, horizontal rhizome, the slender stem bearing 2 whorls of leaves and a sessile, few-flowered, terminal umbel of greenish yellow flowers. Stems erect, 3-7 dm tall; when young, more or less covered with flocculent wool, which persists about the base of the leaves. Lower leaves 5-11 in a whorl, oblong-oblance-shaped, 6-12 cm long and a fourth to a third as wide, acuminate at both ends; upper leaves normally 3 in a whorl, ovate, 3-6 cm long, half to two-thirds as wide, rounded at base, acuminate. Umbel 3-9-flowered; pedicels spreading or deflexed, 15-25 mm long; perianth segments essentially similar, separate, recurved, about 8 mm long; ovary ovoid, 3-celled. Fruit a globose, few-seeded, dark purple berry. May-July. — Rich woods.

Medeola virginiana

Muscari
GRAPE-HYACINTH (ASPARAGACEAE)
Muscari botryoides (L.) P. Mill.
COMMON GRAPE-HYACINTH   *introduced*
Perennial herbs from a bulb, the linear leaves basal, the short erect scape
bearing a dense raceme of flowers. Leaves flat, oblong lance-shaped, at
maturity to 2.5 dm long and 1 cm wide. Scape 1-2 dm tall at anthesis, to 4
dm in fruit. Raceme compact, ovoid-cylindric, 2-4 cm long at anthesis,
elongating in fruit. Flowers blue, all fertile except a few at the tip, nodding,
exceeding their slender pedicels; perianth globular, 4-5 mm long; stamens
6, inserted on the perianth tube, included. Fruit a 3-celled capsule, distinctly 3-angled or almost winged, with 2 angular seeds in each cell.
Spring. — Commonly cultivated and sometimes escaped.

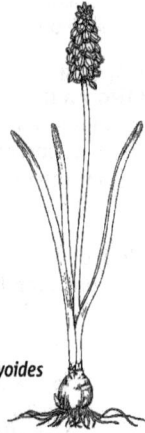

Muscari botryoides

Polygonatum
SOLOMON'S-SEAL (ASPARAGACEAE)
Perennial herbs from a horizontal knotty rhizome, the stem erect or arching, bearing in the
upper portion numerous alternate leaves in two ranks and short, axillary, 1-15-flowered peduncles
with pendent, white to greenish or yellow flowers. Perianth regular, tubular, shortly 6-lobed.
Stamens included, inserted on the perianth tube. Ovary ovoid to globose, 3-celled; style slender,
shorter than the perianth; stigma capitate, obscurely 3-lobed. Fruit a dark blue or black, several-
seeded berry.

1 Leaves completely glabrous . . . . *P. biflorum*
1 Leaves finely pubescent on the veins beneath. . . . . *P. pubescens*

Polygonatum biflorum (Walt.) Ell.
KING SOLOMON'S-SEAL   *native / cons* 4
Stems slender, 4-6 dm tall, erect or arching. Leaves sessile or somewhat
clasping, lance-elliptic, 5-10 cm long, 1-3 cm wide, glabrous, paler and
glaucous beneath, with 1-5 (rarely 9) nerves, only the midnerve prominent
to the apex Peduncles slender, 1-2-flowered. Flowers greenish white, 14-
20 mm long. May-July. — Moist woods and thickets.

Polygonatum pubescens (Willd.) Pursh
HAIRY SOLOMON'S-SEAL   *native / cons* 6
Stems slender, 5-9 dm tall, mostly erect. Leaves narrowly elliptic to broadly
oval, 4-12 cm long, 1-6 cm wide, narrowed below to a short petiole, glabrous
above, glaucous and pubescent on the veins beneath, with 3-9 prominent
nerves. Peduncles slender, usually 1-2 (rarely 3-4)-flowered; pedicels usually shorter than the peduncle. Flowers yellowish green, 10-13 mm long.
May-July. — Moist woods and thickets.

Streptopus
Polygonatum biflorum
TWISTED STALK (LILIACEAE)
Perennial herbs from a rhizome, often branched, with alternate sessile or clasping leaves and
small, greenish white to purple, solitary or paired, axillary flowers. Perianth campanulate to rotate, its segments separate to the base, essentially alike, the outer whorl usually slightly wider.
Stamens 6, adnate to the base of the perianth; filaments widened at base; anthers oblong to
linear, apiculate or aristate. Ovary 3-celled with several ovules; style slender (in our species), 3-
cleft, 3-lobed, or entire. Fruit a red, ellipsoid to subglobose, many-seeded berry.

1 Leaves entire or minutely denticulate, strongly clasping at the base, glaucous beneath; nodes and
   upper internodes glabrous (lower internodes sometimes hispid); tepals spreading or curving from
   near the middle; flowers whitish green . . . . *S. amplexifolius*
1 Leaves prominently ciliate on the margins, the cilia usually visible to the naked eye, sessile or slightly
   clasping (the larger ones subtending branches more strongly clasping), sometimes paler but not

glaucous beneath; nodes and upper internodes ± pubescent or sparsely hispidulous; tepals spreading or recurved only at the tips; flowers usually pinkish (or even maroon) . . . . *S. lanceolatus*

Streptopus amplexifolius (L.) DC.
CLASPING TWISTED STALK   *native / cons* 10

Stems 4-10 dm tall, glabrous. Leaves ovate, varying to ovate or ovate-lance-shaped, cordate and clasping at base, entire or very minutely toothed, the principal leaves 6-12 cm long and 2-5.5 cm wide. Free portion of the peduncle and the pedicel together 3-5 cm long, jointed at about 2/3 of its length, above the joint 1-flowered or sometimes 2-flowered and abruptly deflexed or twisted. Perianth segments greenish white, about 1 cm long, spreading from near the middle; anthers 1-pointed; stigma entire or barely 3-lobed. Berry red, usually ellipsoid, about 15 mm long. June,-July. — Rich moist woods.

Streptopus amplexifolius

Streptopus lanceolatus (Ait.) Reveal
LANCE-LEAF TWISTED STALK   *native / cons* 7

Streptopus roseus Michx.

Stems simple or in larger plants commonly branched, 3-8 dm tall, sparsely and finely pubescent, especially at the nodes. Leaves ovate lance-shaped, broadly rounded to a sessile base, finely ciliate, the principal ones 5-9 cm long, 2-3.5 cm wide. Peduncle and pedicel combined 1-3 cm long, jointed at or below the middle, always 1-flowered. Perianth segments rose-color, about 1 cm long, spreading only near the tip; anthers each double-pointed; lobes of the style nearly 1 mm long. Berry red, globose, about 1 cm long. May-July. — Rich woods.

Triantha
FALSE ASPHODEL (TOFIELDIACEAE)

Streptopus lanceolatus

Triantha glutinosa (Michx.) Baker
STICKY FALSE ASPHODEL   *threatened / native / cons* 10

Tofieldia glutinosa (Michx.) Pers.

Perennial herb, from a bulb. Stems erect, nearly leafless, 2-5 dm long, covered with sticky hairs. Leaves 2-4 from base of plant, linear, hairy, 8-20 cm long and to 8 mm wide, sometimes with 1 bractlike leaf near middle of stem. Flowers white, on sticky-hairy stalks 3-6 mm long, in a raceme 2-5 cm long when in flower, becoming longer when fruiting, 2-3 at each node of the raceme, upper flowers opening first; tepals 6, oblong lance-shaped, 4 mm long; stamens 6. Fruit an oblong capsule, 5-6 mm long; seeds about 1 mm long, with a slender tail at each end. June-July. — Sandy or gravelly shores, interdunal wetlands, calcareous fens, rocky shores.

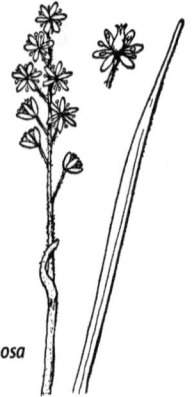
Triantha glutinosa

Trillium
TRILLIUM, WAKE-ROBIN (MELANTHIACEAE)

Perennial herbs from a stout short rhizome, the erect stem bearing a single whorl of 3 ample leaves and a single, large, terminal, sessile or peduncled flower. Perianth segments distinct to the base, the sepals green, the petals white or colored, often of a different shape. Ovary 3-lobed or 3-6-angled or winged, 3-celled with numerous ovules; style short or none; stigmas 3. Fruit a many-seeded berry. A few species are cultivated, especially *Trillium grandiflorum*.

1 Petals white to pink, 3.5–9 cm long, distinctly longer than the sepals, obtuse (occasional small plants with shorter petals—though still longer than sepals—may be recognized by the straight styles and broad obovate petals); stigmatic styles straight (though sometimes spreading) or slightly curved at very tip, uniform in diameter; peduncles held above the leaves. . . . . *T. grandiflorum*
1 Petals white usually less than 2.5 cm long, seldom much longer than sepals; stigmatic styles spreading, thick at base, tapering, and recurved; peduncles usually reflexed and held below the leaves . . . .
. . . . *T. cernuum*

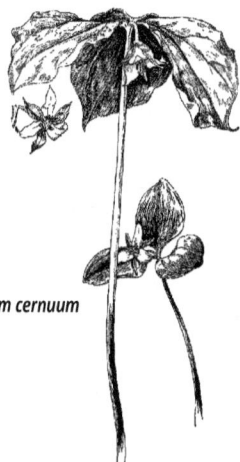

Trillium cernuum L.

WHIP-POOR-WILL-FLOWER   *native / cons* **8**

Stems slender, 2-4 dm tall. Leaves broadly rhombic-obovate, commonly
6-10 cm long at anthesis, acuminate, narrowed from near the middle to
an acute base and obscurely petioled. Peduncle 1-4 cm long, reflexed or
recurved below the leaves. Sepals lance-shaped, acuminate, about equaling
the petals; petals normally white, 1.5-2.5 cm long; anthers 3-7 mm long,
to 1/3 longer than the filaments; ovary white or pinkish. May-June. —
Conifer swamps, bog margins, moist or wet mixed forests, often with
paper birch; thickets along streams; less often in rich hardwood forests.

Trillium cernuum

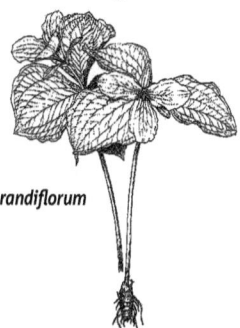

Trillium grandiflorum (Michx.) Salisb.

LARGE-FLOWER WAKEROBIN   *native / cons* **6**

Stems 2-4 dm tall at anthesis. Leaves ovate to rhombic: or subrotund, at
anthesis commonly 8-12 cm long, short-acuminate, narrowed from below
the middle to an acute base. Peduncle erect or declined, usually 5-8 cm
long. Sepals lance-shaped, spreading, 3-5 cm long; petals normally white,
ascending from the base, spreading above, obovate, 4-6 cm long; filaments
nearly as long as the anthers and scarcely wider, the whole stamen 15-25
mm long. — Moist to rather dry deciduous forests, forming large colonies
in beech-maple forests; less common in oak-hickory woods, swamps,
mixed conifer-hardwoods.

Uvularia

BELLWORT (COLCHICACEAE)   *Trillium grandiflorum*

Uvularia grandiflora Sm.

LARGE-FLOWER BELLWORT   *native / cons* **7**

Perennial herb from a slender rhizome, the erect stem forked above the
middle, the lower portion bearing a few bladeless sheaths and up to 4
leaves; leaves sessile or perfoliate, ; flowers yellow or greenish yellow,
nodding. Flowers perfect. Stem at anthesis 2-5 dm tall, at maturity to 1 m
tall, forking above, bearing 0-2 leaves below the fork, 4-8 on the sterile
branch, and several leaves and 1-4 flowers on the fertile branch. Leaves
perfoliate, broadly oval to oblong, to 12 cm long at maturity (reaching
full size after anthesis), minutely pubescent beneath. Flowers yellow, perfect, terminal, but appearing axillary by prolongation of the branches,
nodding; perianth segments 6, 25-50 mm long, acute or acuminate,
smooth within; stamens 6; filaments short. Capsule obovoid, 3-lobed;
seeds subglobose, few in each cell.
April-May. — Rich woods, preferring calcareous soil.

Uvularia grandiflora

Orchidaceae

ORCHID FAMILY

Perennial herbs, from fleshy or tuberous roots, corms or bulbs. Leaves simple, along the stem
and alternate, or mostly at base of plant, stalkless and usually sheathing the stem, parallel-
veined, often somewhat fleshy. Flowers perfect (with both staminate and pistillate parts), irregular, showy in some species, in heads of 1 or 2 flowers at ends of stems, or with several to many
flowers in a spike, raceme or panicle, each flower usually subtended by a bract; sepals 3, green
or colored, sometimes resembling the lateral petals, the lateral sepals free, or joined to form an
appendage below the lip, or joined with the lateral petals to form a hood over the lip (*Spiranthes*);
petals 3 white or colored, the 2 lateral petals alike, the lowest petal different and called the lip;
stamens 1-2, attached to the style and forming a stout column; ovary inferior. Fruit a many-
seeded capsule, opening by 3 or sometimes 6 longitudinal slits, but remaining closed at tip and
base; seeds very small.

One of the world's largest families of vascular plants, with over 700 genera and an estimated 25,000-30,000 species, most of which occur in the tropics.

1   Lip a showy inflated pouch 1–5 cm long. 2
1   Lip showy or inconspicuous, but not an inflated pouch with a small opening, usually flat with or without a slender basal spur (or if somewhat saccate, hardly showy and less than 1 cm long). 4
2   Plants with leafy stems; lip a closed pouch (i.e., open only at base above) . . . . *Cypripedium*
2   Plants with leaves basal; lip split down middle above or open at base about half its length 3
3   Basal leaf single, petiolate, the blade less than 7 cm long, produced in late summer and withering after the plant blooms the following spring; lip ca. 1.5–2 cm long; plants less than 20 cm tall *Calypso*
3   Basal leaves 2, longer, tapered to sheathing bases and not distinctly petiolate, present throughout the summer (but not winter); lip ca. 4–5 cm lon; plants more than 20 cm tall. . . . *Cypripedium acaule*
4   Flower solitary (rarely plants with 2 flowers in a population of 1-flowered ones) 5
4   Flowers 2 or more on one plant 6
5   Leaf linear, at most up to 7 (very rarely 10) mm wide, often poorly developed at flowering time, folded or plicate longitudinally, sheathing stem at base; plant from a small bulbous corm. . *Arethusa*
5   Leaf elliptic or lanceolate, usually over 7 mm wide and well developed at flowering time, flat, arising near middle of stem, sessile but not sheathing at base; plant from slender roots and rhizome . . . . . .
    . . . . *Pogonia*
6   Lip prolonged into a distinct (usually slender and elongate) spur at base 2–40 mm long 7
6   Lip at most somewhat swollen or saccate (but not with a spur 2 mm or more long) 11
7   Leaves cauline. 8
7   Leaves all basal or nearly so, or absent at flowering time (bracts subtending flowers may be leaflike)
    9
8   Spur a thick pouch 2–3 mm long, much shorter than the lip. . . . . *Coeloglossum*
8   Spur slender, sometimes ± clavate, 7–40 mm long, equaling (at most slightly shorter than) to much longer than the lip . . . . *Platanthera*
9   Flowers entirely white and/or green, the lip lanceolate to narrowly linear, entire; lateral petals free .
    . . . . *Platanthera*
9   Flowers with white lip (spotted or not) broadly ovate to oblong, often crenate or lobed; lateral petals connivent or fused with dorsal sepal to form a pink to purple hood. 10
10 Leaves 1; lip less than 1 cm long, spotted, notched at apex and with a lateral lobe on each side. . . . .
    . . . . *Amerorchis*
10 Leaves normally 2; lip over 1 cm long, unspotted, not lobed . . . . *Galearis*
11 Plants lacking green color (except sometimes in fruit), leafless with red, yellow, brown, or purplish stems arising from a coralloid rhizome . . . . *Corallorhiza*
11 Plants with green color, bearing leaves at some time in the year (if leaves absent at flowering time or plants apparently lacking green, arising from tubers, corms, or short rhizomes, not a coralloid mass) 12
12 Leaves a single opposite pair, definitely cauline, not at all sheathing the stem . . . . *Neottia*
12 Leaves solitary, alternate, absent, or basal (or almost basal, with sheathing bases). 13
13 Stem leafy, with 4 or more conspicuous broadly ovate-lanceolate to elliptic leaves; perianth ca. 7–10 mm long; flowers greenish, at least the petals suffused with pink; upper part of stem and axis of inflorescence finely pubescent. . . . . *Epipactis*
13 Stem with the leaves fewer than 4, narrow, and/or basal (or absent); perianth various, but if pinkish then 10 mm or more long and the vegetative parts completely glabrous. 14
14 Perianth 10–12 mm long, white or creamy; inflorescence dense, spike-like . . . . *Spiranthes*
14 Perianth longer or shorter, or not whitish and the inflorescence not spike-like 15
15 Perianth 10 mm or more long, at least in part usually with some shade of pink or purple 16
15 Perianth less than 10 mm long, greenish, white, or yellowish, with no trace of pink or purple 17
16 Flowers 2–3 cm or more wide, the lip uppermost, bearded with a tuft of yellow-tipped hairs; leaf solitary (rarely 2), several times longer than wide . . . . *Calopogon*
16 Flowers less than 1.5 cm broad, the lip lowermost and not bearded; blade of leaf not over 3.5 times longer than wide. . . . . *Liparis*
17 Leaves 1 or 2, sheathing at the base, the scape naked to the inflorescence; flowers on short pedicels, the raceme glabrous and not 1-sided nor noticeably twisted. 18
17 Leaves 3 or more (or withering at flowering time), the stem above them bearing small bracts or scales; flowers sessile or almost so in a narrow spike-like inflorescence, which is 1-sided or spirally twisted, or pubescent (or both). 19
18 Leaves 1 (very rarely 2); perianth less than 4 mm long. . . . . *Malaxis*

18 Leaves 2; perianth over 4 mm long .... *Liparis*
19 Leaves ovate to elliptic, basal or nearly so, present and firm at flowering time, the midvein and/or other veins margined in white or pale green (not always visible in dry plants); lip pouched or saccate at the base.. *Goodyera*
19 Leaves ovate-elliptic to linear and grass-like, sometimes cauline, often withering at flowering time (in wider-leaved species), not marked with whitish; lip not pouched .... *Spiranthes*

Amerorchis
ROUND-LEAF ORCHID

Amerorchis rotundifolia (Banks) Hultén
ROUND-LEAF ORCHID   *threatened / native / cons* **10**
  Galearis rotundifolia (Banks ex Pursh) R. M. Bateman
  Orchis rotundifolia Banks
Perennial herb, roots few from a slender rhizome. Stems leafless, smooth, 15-30 cm long. Leaves single from near base of plant, oval, 4-15 cm long and 2-8 cm wide; usually with 1-2 bladeless sheaths below. Flowers 4 or more, in a raceme 3-8 cm long; sepals white to pale pink; petals white to pink or purple-tinged, the 2 lateral petals joined with the upper sepal to form somewhat of a hood over the column; lip white, with purple spots, 6-10 mm long and 4-7 mm wide, 3-lobed, the terminal lobe largest and notched at tip; spur about 5 mm long, shorter than lip. June-July. — Moist to wet conifer woods (on moss under cedar, balsam fir, tamarack, or black spruce); usually found over underlying limestone and where sphagnum mosses not predominant.

Amerorchis rotundifolia

Arethusa
DRAGON'S-MOUTH

Arethusa bulbosa L.
DRAGON'S-MOUTH   *native / cons* **10**
Perennial herb; roots few, fibrous, from a corm. Stems leafless, smooth, 1-4 dm long. Leaves 1, linear, small and bractlike at flowering time, later expanding to 2 dm long and 3-8 mm wide; lower stem with 2-4 bladeless sheaths. Flowers single at ends of stems, sepals rose-purple, oblong, 2.5-5 cm long; petals joined and more or less hoodlike over the column, lip pink, streaked with rose-purple, 2.5-4 cm long, curved downward near middle. June-July. — Open bogs and conifer swamps (in sphagnum moss), floating mats around bog lakes, calcareous fens; often with grass-pink (*Calopogon tuberosus*) and rose pogonia (*Pogonia ophioglossoides*).

Calopogon
GRASS-PINK

Arethusa bulbosa

Calopogon tuberosus (L.) B.S.P.
GRASS-PINK   *native / cons* **8**
Perennial herb, from a corm. Stems leafless, smooth, 2-7 dm long. Leaves 1 near base of plant, linear, 1-4 dm long and 2-15 mm wide. Flowers pink to purple, 2-15 in a loose raceme, 3-12 cm long; sepals ovate, 1-2.5 cm long; petals oblong, 1-2.5 cm long, the lip located above the lateral petals, 1-2 cm long, bearded on inside with yellow-tipped bristles. — Open bogs and floating mats, openings in conifer swamps, calcareous fens near Lake Michigan shoreline. Distinguished from swamp-pink (*Arethusa bulbosa*) and rose pogonia (*Pogonia ophioglossoides*) by having a raceme of several flowers vs. single flowers in *Arethusa* and *Pogonia*.

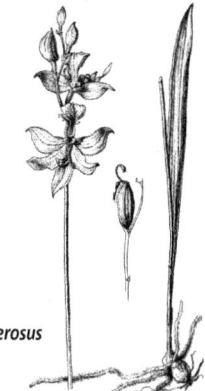

Calopogon tuberosus

Calypso
FAIRY-SLIPPER ORCHID

Calypso bulbosa (L.) Oakes
CALYPSO, FAIRY-SLIPPER   *threatened / native / cons* 10
Perennial herb, from a corm. Stems 0.5-2 dm long, with 2-3 bladeless
sheaths on lower portion. Leaves single from the corm, ovate, 3-5 cm
long and 2-3 cm wide, petioles 1-5 cm long. Flowers 1, nodding at end of
stem; sepals and lateral petals similar, pale purple to pink, lance-shaped,
1-2 cm long and 3-5 mm wide, lip white to pink, streaked with purple,
1.5-2 cm long and 5-10 mm wide, the lip extended to form a white "apron"
with several rows of yellow bristles. May-June. — Mature conifer forests
or mixed forests of conifers and deciduous trees (such as balsam fir, hemlock, and paper birch), usually in shade; soils rich in woody humus.

The single leaf of calypso appears in late August or September, persists
through the winter, and withers after flowering in spring. Between fruiting in June and July and the emergence of the new leaf in late summer
of fall, no aboveground portions of the plant may be visible.

Calypso bulbosa

Coeloglossum
BRACTED ORCHID

Coeloglossum viride (L.) Hartman
BRACTED ORCHID   *native / cons* 6
   Dactylorhiza viridis (L.) R. M. Bateman, A. M. Pridgeon & M. Chase
Habenaria viridis (L.) R. Br.
Stems slender or stout, 2-5 dm tall. Lowest 1 or 2 leaves reduced to bladeless
sheaths; principal foliage leaves obovate, 5-12 cm long, to 5 cm wide, the
upper progressively narrower and shorter and passing gradually into the
bracts. Inflorescence loose or compact, 5-20 cm long; bracts foliaceous,
lance-shaped, exceeding the flowers, the lowest to 5 cm long. Flowers
greenish, often tinged with purple; lip oblong, 6-10 mm long, slightly
widened distally, terminating in 3 teeth, the central one the shortest;
petals lance-shaped, nearly concealed by the incurved sepals; spur pouch-
like, 2-3 mm long. June-Aug. — Moist woods.

Coeloglossum viride

Corallorhiza
CORAL-ROOT

Yellow, brown, or purplish saprophytic herbs, lacking in chlorophyll, and
parasitic on fungi inhabiting their characteristic coral-like rhizomes.
Stems with a few sheathing scales toward the base and a terminal raceme
of small, usually bicolored flowers. Sepals and lateral petals narrow, similar
and nearly equal, spreading or projecting over the column; lateral sepals
united with the base of the column. Lip deflexed, oblong to rotund, often
with two lateral lobes, the margins usually upturned. Fruit a pendent
capsule.

1 Lip with a small lobe or elongate tooth on each side near the base (sometimes difficult to see in dried
   specimens). 2
1 Lip entire, or merely denticulate or erose. 3
2 Sepals and petals 3-nerved; summit of ovary with a low protuberance (like a rudimentary spur) usually
   visible below the base of the lip; lip 4.5–7 mm long . . . . **C. maculata**
2 Sepals and petals 1-nerved (or the latter rarely weakly 3-nerved); summit of ovary without visible protuberance; lip 2.5–4.5 mm long. . . . . **C. trifida**
3 Sepals and petals 3–5-nerved, 8–15 mm long, conspicuously striped with purple, the lip solid purplish
   apically . . . . **C. striata**
3 Sepals and petals 1-nerved (or faintly 3-nerved), less than 6 mm long, not conspicuously striped . . . 4
4 Perianth 3–4.5 mm long, purplish; lip white, spotted with purplish . . . . **C. odontorhiza**
4 Perianth 4–5.5 mm long, yellowish; lip unspotted white (or rarely spotted) . . . . **C. trifida**

Corallorhiza maculata (Raf.) Raf.
SPOTTED CORAL-ROOT   *native / cons* 7
Perennial saprophytic herb. Stems pinkish purple, 2-5 dm tall. Raceme 5-15 cm long, 10-40-flowered. Sepals and lateral petals more or less spotted or suffused with purple, narrowly oblong to oblance-shaped, 6-8 mm long, 3-nerved, the lateral sepals somewhat divergent; spur a prominent, sometimes divergent swelling near the summit of the ovary; lip white, irregularly spotted with purple, 6-8 mm long, bearing 2 conspicuous lateral lobes below the middle and 2 short parallel ridges on the face, the terminal lobe rounded and deflexed, 3-4 mm wide. July-Sept. — Woods.

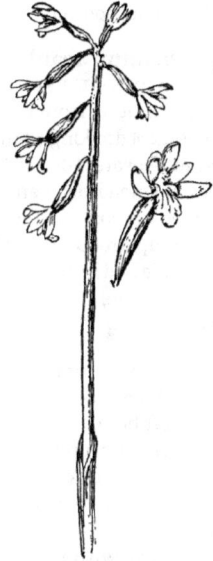

Corallorhiza odontorhiza (Willd.) Poir.
AUTUMN CORAL-ROOT   *native / cons* 8
Perennial saprophytic herb. Stems purple to brown, or greenish above, 1-2 dm tall. Raceme 3-6 cm long, 5-15-flowered. Sepals and lateral petals extending forward, over the column, scarcely spreading, oblong, 3-5 mm long, purplish green to dark purple; lip short-clawed or nearly sessile, expanded above into a rotund blade, 3-4 mm long, white, purple-margined and with 2 purple spots, entire or finely erose. Aug-Sept. — Open woods.

Corallorhiza striata Lindl.
STRIPED CORAL-ROOT   *native / cons* 9
Perennial saprophytic herb. Stems purple or magenta, rather stout, 2-4 dm tall, with 2-4 acute or cuspidate scales. Raceme 5-12 cm long, usually 10-20-flowered. Sepals and lateral petals arching forward, oblong lance-shaped, 10-14 mm long, yellowish white with 3 conspicuous longitudinal stripes of purple, usually also purple on the margin; lip 8-12 mm long, white, heavily striped with purple, or purple throughout. May-July. — Moist or dry woods.

Corallorhiza maculata

Corallorhiza trifida Chatelain
YELLOW CORALROOT   *native / cons* 7
Perennial saprophytic herb, roots absent. Stems yellow-green, smooth, 1-3 dm long, single or in clusters from the coral-like rhizome. Leaves reduced to 2-3 overlapping sheaths on lower stem. Flowers yellow-green, 5-15 in a raceme 3-8 cm long; sepals and lateral petals yellow-green, linear, 3-5 mm long, lip white, sometimes with purple-spots, obovate, 3-5 mm long and 2-3 mm wide. Capsules drooping, 1-1.5 cm long and 3-7 mm wide. May-June. — Moist to wet, mostly conifer woods, swamps (often under white cedar); usually where shaded.

Cypripedium
LADY'S-SLIPPER

Corallorhiza trifida

Erect perennial herbs, from coarse, fibrous roots. Stems unbranched, often clumped, hairy. Leaves 2 or more at base of plant or along stem, broad. Flowers 1 or 2, large and mostly showy at ends of stems, white, pink or yellow; lateral sepals similar to lateral petals, the sepals joined to form a single appendage below the lip; lateral petals free and spreading, lip inflated and pouchlike, projecting forward; stamens 2, 1 on each side of column. Fruit a many-seeded capsule.

1   Lip pouch pink to purple; leaves 2 at base of stem .... *C. acaule*
1   Lip pouch yellow or white; leaves 3 or more on stem 2
2   Pouch yellow, sometimes brown- or purple-dotted 3
2   Pouch white to pink, or pink with white patches 4
3   Sepals and petals red-brown; lateral petals strongly twisted, brown-purple; pouch less than 4 cm long
    .... *C. parviflorum var. makasin*
3   Sepals and petals yellow to brown-green; lateral petals wavy, green with red-brown streaks; pouch more than 4 cm long .... *C. parviflorum var. pubescens*
4   Pouch projected downward into a cone-shaped spur .... *C. arietinum*

4   Pouch not spurred . . . . *C. reginae*

Cypripedium acaule Ait.
PINK LADY'S-SLIPPER   *native / cons* **7**
Perennial herb, from coarse rhizomes; roots long and cordlike. Stems
leafless, 2-4 dm long, glandular-hairy. Leaves 2 at base of plant, opposite,
oval to obovate, 1-2 dm long and 3-10 cm wide, thinly hairy, stalkless.
Flowers 1, nodding at end of stem; sepals and lateral petals yellow-green
to green-brown, the 2 lower sepals joined to form a single sepal below
the lip; lip drooping, pink with red veins, 3-5 cm long, cleft along the
upper side and hiding the opening. May-June. — Forests, typically where
shaded, acidic, and nutrient-poor; sometimes on hummocks in conifer
swamps.   *Cypripedium acaule*

Cypripedium arietinum Ait. f.
RAM'S-HEAD LADY'S-SLIPPER   *threatened / native / cons* **10**
Perennial herb, from a coarse rhizome, roots long and cordlike. Stems
slender, 1-4 dm long, thinly hairy. Leaves 3-5, above middle of stem, stalkless, oval, often folded, 5-10 cm long and 1.5-3 cm wide, finely hairy.
Flowers 1 or sometimes 2 at ends of stems; sepals and lateral petals
similar, green-brown; lip an inflated pouch, 1.5-2.5 cm long, white or
pink-tinged, with prominent red-veins, extended downward to form a
conical pouch. Late May-June. — Conifer swamps, wet forest openings
(often with white cedar); also in drier, sandy, conifer and mixed conifer-
deciduous forests, and on low dunes under conifers.

Cypripedium parviflorum Salisb. var. **makasin** (Farw.) Sheviak
YELLOW LADY'S-SLIPPER   *native / cons* **9**
Cypripedium calceolus var. parviflorum (Salisb.) Fernald
Perennial herb, from rhizomes, roots long and numerous. Stems 1.5-6
dm long, glandular-hairy. Leaves 2-5, alternate along stem, ascending,
oval, 5-18 cm long and 2-7 cm wide, sparsely hairy, stalkless. Flowers 1
(rarely 2) at ends of stems; sepals purple-brown, the lateral sepals joined
below the lip, notched at tip; lateral petals linear, purple-brown, spirally
twisted, 2-5 cm long; lip an inflated pouch, 1.5-3 cm long, yellow, often
with purple veins and spots near opening. May-July. — Conifer swamps,
wet meadows, fens, and moist forests (often under cedar); sphagnum
mosses are usually sparse; sites are shaded or sunny, with organic or
mineral, often calcium-rich soil.
    Our two varieties may be distinguished by the size of the pouch (lip)
and the color of the sepals and petals: in var. makasin, the lip is mostly
2-3 cm long, and the sepals and petals are dark red; in var. pubescens, the
lip is mostly 3-6 cm long and the sepals and petals are yellow-green;
however, intermediate forms may occur.

Cypripedium arietinum

Cypripedium parviflorum Salisb. var. **pubescens** (Willd.) Knight
YELLOW LADY'S-SLIPPER   *native / cons* **9**
Cypripedium calceolus var. pubescens (Willd.) Correll
Perennial herb, from a rhizome, roots long and numerous. Stems 1.5-6
dm long, glandular-hairy. Leaves 3-6, alternate along stem, ascending,
ovate to oval, 8-20 cm long and 3-8 cm wide, sparsely hairy. Flowers 1
(rarely 2) at ends of stems; sepals yellow-green, the lateral sepals joined
below the lip, notched at tip; lateral petals linear, yellow-green, often
streaked with red-brown, usually spirally twisted, 4-8 cm long; lip an inflated pouch, 3-6 cm long, yellow, often with purple veins near opening.
May-July. — Conifer swamps, bogs, fens, especially where soils derived
from limestone; also in wetter hardwood forests.

Cypripedium parviflorum
var. pubescens

Cypripedium reginae Walt.
SHOWY LADY'S-SLIPPER   *native / cons* 9

Perennial herb, from a coarse rhizome, roots many, long and cordlike.
Stems 4-10 dm long, strongly glandular-hairy. Leaves 4-12, alternate along
stem, spreading or ascending, broadly oval, 10-25 cm long, 4-12 cm wide,
abruptly tapered to tip, nearly smooth to hairy, stalkless; reduced to
sheaths at base. Flowers 1 or often 2 at ends of stems, the subtending
bract leaflike, 6-12 cm long; sepals and lateral petals white, the lateral
sepals joined to form an appendage under the lip, rounded at tip; lip an
inflated pouch, 3-5 cm long, white, streaked and spotted with pink or
purple. June-July. — Conifer and hardwood swamps (especially balsam
fir-cedar-tamarack swamps), bogs, calcareous fens, sedge meadows, floating mats, wet openings, wet clayey slopes, ditches; especially where open
and sunny; most abundant in openings in wet forests and swamps not
dominated by sphagnum mosses.

Cypripedium reginae

Showy lady's-slipper is our largest lady's-slipper. Avoid touching plants
as the hairs can be irritating.

Epipactis
HELLEBORINE

Epipactis helleborine (L.) Crantz
HELLEBORINE   *introduced (invasive)*

Perennial herb. Stems erect, to 8 dm tall. Leaves alternate, sessile and
clasping, ovate to lance-shaped, the lower to 10 cm long, the upper progressively shorter and narrower. Flowers in a terminal, many-flowered
raceme 1-3 dm long; bracts linear or narrowly lance-shaped, the lower
surpassing the flowers; sepals and lateral petals ovate-lance-shaped, 10-
14 mm long, acute, dull green, strongly veined with purple; lip greenish
and purple, strongly saccate in the basal half, the terminal lobe broadly
ovate, crested at its base with two elevated swellings. July-Aug. — Native
of Europe; established and spreading in deciduous and mixed woods.

Epipactis helleborine

Galearis
SHOWY ORCHID

Galearis spectabilis (L.) Raf.
SHOWY ORCHID   *native / cons* 6
  Orchis spectabilis L.

Perennial herb. Leaves 2, rather fleshy, narrowly obovate to broadly elliptic,
8-15 cm long. Scape 1-2 dm tall, stout; bracts foliaceous, oblong lance-
shaped, 15-50 mm long. Sepals and lateral petals pink to pale purple, 13-
18 mm long, all connivent; lip white, 15-20 mm long, rhombic-obovate;
spur stout, about equaling the lip. May-June. — Rich woods.

Galearis spectabilis

Goodyera
RATTLESNAKE-PLANTAIN

Perennial herbs from a short rhizome, plants glandular-pubescent on the scape, bracts, ovary,
and sepals. Leaves in a basal cluster, commonly reticulated with white, narrowed to a broad,
petiole-like base. Flowers in a spike-like raceme of white or greenish flowers, atop an erect
scape with several scale-like bracts. Upper sepal and lateral petals coherent by their margins,
forming a concave galea extending forward over the lip. Lateral sepals free, scarcely spreading,
except at the tip. Lip shorter than the galea, conspicuously pouch-like at base, prolonged upwards
into a horizontal or deflexed beak; lateral petals white.

1 Leaf blades with only the midvein outlined above in white or pale green, the largest blades usually 4–
   6 cm long; plants 20–50 cm tall; perianth 6–9 mm long .... ***G. oblongifolia***
1 Leaf blades with white or pale green reticulation throughout (sometimes not on the midvein), the

largest blades often less than 4 cm long; plants 5–30 cm tall; perianth 2.5–5.5 (rarely to 6.5) mm long 2
2 Stem with usually 7–10 cauline bracts (undeveloped leaves); beak of lip (beyond the large pouch) less than 1 mm long, about 1/4 the total length of the lip or usually less; inflorescence densely flowered on all sides. . . . . **G. pubescens**
2 Stem with 2–5 cauline bracts; beak of lip 1–2 mm long, about 1/2 the total length of the lip; pouch shallow or deep; inflorescence strongly one-sided or loosely flowered on all sides. 3
3 Lip deeply pouched, the pouch about as deep as long, the beak often strongly turned downward at maturity; plants mostly 10–20 cm tall; largest leaf blades mostly 1–2 cm long; cauline bracts 2–4 (usually 3) . . . . **G. repens**
3 Lip shallowly pouched, the pouch longer than deep, the beak horizontal or slightly recurved; plants usually 17–25 cm tall; largest leaf blades mostly 2–4 cm long; cauline bracts usually 4–5 . **G. tesselata**

Goodyera oblongifolia Raf.
GREEN-LEAF, WESTERN RATTLESNAKE-PLANTAIN   *native / cons* **8**
Stem stout, 3-4.5 dm tall. Leaf blades ovate-lance-shaped to narrowly elliptic, to 8 cm long, about a third as wide. Raceme 1-sided, densely flowered, 6-12 cm long. Galea 8-10 mm long; lateral sepals lance-shaped, 6-7
mm long, acute; lip ovate in general outline, 6-7 mm long, its beak about
two-thirds as long as the semi-globose body, curved-deflected, broad,
with upturned sides. July-Aug. — Dry woods.

Goodyera pubescens (Willd.) R. Br.
DOWNY RATTLESNAKE-PLANTAIN   *native / cons* **7**
Stem rather stout, 2-4 dm tall. Leaf blades ovate or ovate-lance-shaped,
3-6 cm long, with 5 or 7 white veins and numerous white reticulate veinlets. Raceme dense, many-flowered, 4-10 cm long. Galea broadly elliptic,
very convex, 4-4.5 mm long, upturned at the summit; lateral sepals
broadly ovate to obovate, 3.5-4 mm long, abruptly short-acuminate; lip
subglobose, 3.5-4 mm long, its straight beak less than 1 mm long and
scarcely projecting beyond the ventricose body. July-Aug. — Dry woods.

Goodyera repens (L.) R. Br.
DWARF RATTLESNAKE-PLANTAIN   *native / cons* **9**
Stem slender, 1-3 dm tall, glandular-pubescent. Leaf blades ovate to oblong,
1.5-3 cm long. Raceme loosely flowered, one-sided, 3-6 cm long. Galea 3.5-
5 mm long; lateral sepals slightly shorter, broadly ovate; lip deeply pouch-
like at base, its beak triangular, acute, abruptly deflected. Anther blunt.
July-Aug. — Dry woods.

Goodyera tesselata

Goodyera tesselata Lodd.
CHECKERED RATTLESNAKE-PLANTAIN   *native / cons* **7**
Stem 1.5-4 dm tall. Leaf blades ovate or ovate-lance-shaped, 3-5 cm long.
Raceme loosely flowered, often spiral, 4-10 cm long. Galea strongly convex,
4-5 mm long; lateral sepals ovate, about 4 mm long; lip about 3.5 mm
long, strongly saccate at base, narrowed distally to a beak about as long
as the body, with upturned sides and blunt or rounded at the tip; anther
abruptly narrowed to a conspicuous sharp beak. July-Aug. — Dry woods.

Liparis
WIDE-LIP ORCHID

Liparis loeselii (L.) L. C. Rich.
FEN-ORCHID   *native / cons* **7**
Small, smooth perennial herb, from a bulblike base. Stems erect, 1-2.5
dm long, upper stem somewhat angled in section. Leaves 2 from base of
plant, ascending, sheathing at base, shiny, lance-shaped to oval, 4-15 cm
long and 1-4 cm wide. Flowers 2-15, yellow-green, small, upright, in an
open raceme 2-10 cm long and 1-2 cm wide; sepals narrowly lance-shaped,
4-6 mm long and 1-2 mm wide; lateral petals linear, 3-5 mm long, often

Liparis loeselii

twisted and bent forward under the lip; lip yellow-green, obovate, 4–5 mm long, tipped with a short point. Capsules persistent, short-cylindric, 8–12 mm long. June–Aug. — Conifer swamps, fens, floating mats, streambanks, sandy shores, ditches; soils peaty to mineral, acid to calcium-rich.

Malaxis
ADDER'S-MOUTH ORCHID
Small perennial herbs. Leaves 1-5 from base of plant or single along stem. Flowers green-white, spaced or crowded in slender or cylindric racemes at ends of stems.

1 Flowers evenly spaced in a raceme 5–11 cm long .... *M. monophyllos*
1 Flowers crowded near top of raceme, the raceme 2–5 cm long .... *M. unifolia*

Malaxis monophyllos (L.) Sw.
WHITE ADDER'S-MOUTH ORCHID   *native / cons* **9**
Malaxis brachypoda (Gray) Fern.
Perennial herb, from a bulblike base; roots few, fibrous. Stems smooth, 1-2 dm long. Leaves single, appearing to be attached well above base of stem, the leaf base clasping stem, ovate to oval, 3-7 cm long and 1.5-4 cm wide. Flowers small, green-white, 14-30 or more, in a long, slender, spike-like raceme 4-11 cm long and to 1 cm wide; on stalks 1-2 mm long, the flowers evenly spaced in the raceme; lip heart-shaped, bent downward, 2-3 mm long and 1-2 mm wide, narrowed at middle to form a long, lance-shaped tip, with a pair of lobes at base. June-Aug. — Conifer swamps (white cedar, balsam fir, black spruce), especially in wet depressions and where soils are marly; sphagnum moss hummocks in conifer swamps, wet hardwood forests.
Ours are var. *brachypoda* (Gray) F. Morris & Eames.

Malaxis unifolia Michx.
GREEN ADDER'S-MOUTH ORCHID   *native / cons* **6**
Small perennial herb, from a bulblike base; roots few, fibrous. Stems smooth, 1-3 dm long. Leaves single, attached near middle of stem, ovate, 2-7 cm long and 1-4 cm wide. Flowers small, green, numerous in a cylindric raceme 1.5-6 cm long and 1-2 cm wide, the upper flowers crowded, the lower flowers more widely spaced; lip very small, 1-2 mm long, with 3 teeth at tip. June-Aug. — Sphagnum moss hummocks in swamps, sedge meadows, thickets; also in drier forests including pine plantations.

Malaxis unifolia

Neottia
TWAYBLADE

Neottia cordata (L.) Rich.
HEART-LEAF TWAYBLADE   *native / cons* **10**
Listera cordata (L.) R. Br.
Perennial herb, roots fibrous. Stems 1-3 dm long, glandular-hairy above the leaves, smooth below. Leaves 2, opposite near middle of stem, 1-4 cm long and 1-3 cm wide, stalkless. Flowers green to red-purple, 6-20 in a raceme 3-12 cm long and 1-2 cm wide; lip slender, 3-5 mm long, with 2 teeth on side near base, the tip cleft halfway or more into spreading linear lobes. June-July. — Bogs and conifer swamps, where usually on sphagnum moss hummocks; hemlock groves.
Formerly placed in genus *Listera*.

Neottia cordata

Platanthera
REIN-ORCHID
Perennial herbs, from a cluster of fleshy roots. Stems erect, smooth. Leaves mostly along the stem, upright, reduced to sheaths at base and upward on stem; leaves basal in *P. orbiculata*.

Flowers white or green, several to many in a spike or raceme; upper sepal joined with petals to form a hood over the column; lateral sepals spreading; lip linear to ovate or 3-lobed, entire, toothed or fringed, extended backward into a spur, the spur commonly curved; stamens 1, the anther attached to the top of the short column. Fruit a many-seeded capsule. Previously included in *Habenaria*, that genus now considered tropical, and *Platanthera* occurring in temperate regions.

1 Lip prominently ciliate or fringed 2
1 Lip entire or toothed, but not fringed. 3
2 Flowers pink-purple; divisions of the lip broadly fan-shaped, copiously lacerate-fringed, but the fringe usually cut less than half the distance to the base of the division of the lip . . . . *P. psycodes*
2 Flowers yellowish, cream, or greenish; at least the lateral divisions of the lip more narrowly cuneate, mostly cut into a long fringe more than half their length. . . . . *P. lacera*
3 Leaves all basal, the stem at most with reduced bracts 4
3 Leaves cauline (along the stem) 7
4 Leaves about twice as long as wide, or longer; spur less than 12 mm long 5
4 Leaves less than twice as long as broad, orbicular or almost so; spur 16–40 mm long 6
5 Lip with truncate 3-toothed or crenate tip; spur 7–11 mm long, much exceeding the lip . *P. clavellata*
5 Lip tapered to a pointed or rounded, untoothed tip; spur about equaling lip or at most ca. 2 mm longer . . . . *P. obtusata*
6 Scape naked (rarely with a bract); spurs (16–24 mm long) tapered evenly to rounded tip; lip yellowish green, tending to turn upward near the end . . . . *P. hookeri*
6 Scape with 1–6 bracts between leaves and inflorescence; spurs parallel-sided or even somewhat club-shaped toward tip; lip whitish green, tending to turn downward . . . . *P. orbiculata*
7 Lip truncate and 2–3-toothed or -lobed at tip. . . . . *P. clavellata*
7 Lip tapered, rounded (or almost truncate and obscurely crenulate) but not 2–3-toothed at tip 8
8 Lip much shorter than the spur, broadly rounded (or almosst truncate) at tip, with an erect tubercle near the base and a lateral tooth or projection on each side near the base . . . . *P. flava*
8 Lip 1–2 mm shorter than, about equaling, or slightly longer than the spur, tapered to narrow tip, with neither a tubercle nor lateral teeth (at most, broadly widened basally) 9
9 Flowers pure white, lip strongly expanded basally . . . . *P. dilatata*
9 Flowers green, greenish yellow or greenish white, lip cuneate to strap-shaped, not or only slightly widened at base 10
10 Anther sacs essentially in contact above the rounded stigma (separated at tip by less than 0.3 mm); lips 2.5–5 mm long . . . . *P. aquilonis*
10 Anther sacs separated at tip by ca. 0.4 mm or more, stigma pointed; lips 4–8 mm long . *P. huronensis*

Platanthera aquilonis Sheviak
BOG ORCHID   *native / cons 7*

Perennial herb. Stems 60 cm long. Leaves several, ascending to spreading, gradually reduced to bracts upwards; blade linear-lanceolate, 3-23 cm long and to 4 cm wide. Spikes lax to very dense. Flowers not showy, yellowish green with dull yellowish lip; lateral sepals spreading to reflexed; petals ovate, margins entire; lip descending, 2.5-6 cm long and to 1.5 mm wide, projecting, not thickened at base, margins entire; spur clavate or sometimes rather cylindric, 2-5 mm long, apex usually broadly obtuse. May-Aug. — Moist to wet including moist forests, cedar swamps, riverbanks, wet meadows, fens, ditches and borrow pits. The *Platanthera hyperborea* complex, including *P. dilatata*, *P. aquilonis*, and *P. huronensis*, are often difficult to separate; living rather than dried plants are easiest to identify.

Platanthera aquilonis

Platanthera clavellata (Michx.) Luer
GREEN WOODLAND ORCHID   *native / cons 6*
   Habenaria clavellata (Michx.) Spreng.

Perennial herb. Stems slender, 1-4 dm long. Foliage leaf 1, linear-oblong to oblong lance-shaped, commonly 7-16 cm long, to 3 cm wide, blunt; upper leaves 1 or few, much reduced, the uppermost linear, scale-like. Inflorescence open, 5-15-flowered, 2-6 cm long; bracts narrowly lance-shaped, shorter than the flowers. Flowers divergent from the axis, white or tinged with green or yellow, twisted to one side so that the spur is lateral; lip broadly cuneate, 3-5 mm long, shallowly 3-lobed at the summit; petals and sepals broadly ovate, about equal; spur strongly curved, dilated at the tip, 8-12 mm long. July-Aug. — Acid bogs and wet soils, especially in sphagnum.

Platanthera clavellata

Platanthera dilatata (Pursh) Lindl.
WHITE BOG-ORCHID   *native / cons* **10**
Habenaria dilatata (Pursh) Hook.
Piperia dilatata (Pursh) Szlach. & Rutk.

Perennial herb, strongly clove-scented, roots fleshy. Stems stout or slender, to 1 m long. Leaves 3-6, alternate along stem, upright, lance-shaped, to 10-20 cm long and 1-3 cm wide, with 1-2 small, bractlike leaves above and 1 bladeless sheath at base of stem. Flowers 10-60, bright white, upright, in a raceme 1-2.5 dm long; lateral sepals lance-shaped, 4-9 mm long and 1-3 mm wide; lateral petals similar but joined with upper sepal to form somewhat of a hood over the column; lip lance-shaped, widened at base, 6-8 mm long; spur slender, 4-8 mm long. June-July.— Wet, open bogs and floating mats, conifer swamps, streambanks, shores and seeps; often where sandy or calcium-rich (as in calcareous fens), not in deep sphagnum moss.

Similar to northern bog-orchid (*P. huronensis*) but with white rather than green-tinged flowers as in *P. huronensis*.

Platanthera dilatata

Platanthera flava (L.) Lindl.
PALE GREEN ORCHID   *threatened / native / cons* **7**
Habenaria flava (L.) R. Br.

Perennial herb, roots fleshy. Stems 3-7 dm long. Leaves 2-4, alternate along stem, lance-shaped or oval, to 5-15 cm long and 2-5 cm wide, with 1-3 bractlike leaves above. Flowers 15 or more, green-yellow or green, stalkless, in a raceme 5-15 cm long and 2-4 cm wide; sepals ovate, 2-3 mm long; lip bent downward, 3-6 mm long, the margin irregular, with a tooth near base on each side; spur 4-6 mm long. June-July. — Wet depressions in hardwood swamps, alder thickets, sedge meadows, moist sand prairies; often where calcium-rich, sometimes where disturbed. Ours are var. *herbiola* (R. Br.) Luer.

Platanthera hookeri (Torr.) Lindl.
HOOKER'S ORCHID   *native / cons* **9**
Habenaria hookeri Torr.

Perennial herb; scape 2-4 dm tall, bractless, rising from a few fleshy roots. Leaves 2, basal, commonly broadly elliptic to rotund, 6-12 cm long, usually 2/3 to fully as wide, blunt or rounded, abruptly narrowed to the base. Flowers sessile, yellowish green, ascending; lip triangular-lance-shaped, directed outward, upcurved, 8-12 mm long; lateral petals lance-shaped, incurved and more or less adjacent under the upper sepal; spur 13-24 mm long, directed, downward, tapering to the tip. June-July. — Coniferous or mixed forests, wooded dunes, soils often sandy.

Platanthera huronensis (Nutt.) Lindl.
NORTHERN BOG-ORCHID   *native / cons* **7**
Habenaria hyperborea (L.) R. Br.

Perennial herb, roots fleshy. Stems 2-8 dm long. Leaves 2-7, alternate on stem, linear to oblong, 5-30 cm long and 2-5 cm wide, with 1-3 smaller leaves above. Flowers small, green, erect, many in a raceme 4-25 cm long; lateral sepals ovate and spreading; lateral petals lance-shaped, curved upward and joined with upper sepal to form a loose hood over column;

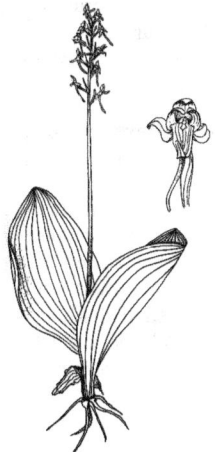

Platanthera hookeri

356 ORCHIDACEAE (Orchid Family)   *monocots*

lip lance-shaped, 3-7 mm long, not abruptly widened at base; spur curved
forward under the lip, about as long as lip, 3-7 mm long. June-Aug. —
Moist to wet forests and swamps, thickets, streambanks, wet meadows,
wet sand along Lake Michigan shoreline, ditches.

Platanthera huronensis

Platanthera lacera (Michx.) G. Don
GREEN FRINGED ORCHID   *native / cons* **7**
  Habenaria lacera (Michx.) R. Br.
Perennial herb, roots fleshy. Stems 3-8 dm long. Leaves 3-7, alternate on
stem, lance-shaped to oval, to 5-15 cm long and 1-4 cm wide; upper leaves
much smaller. Flowers white or green-white, in a usually compact, many-
flowered raceme, 5-20 cm long and 2-5 cm wide; sepals broadly oval, 4-7
mm long, the lateral ones deflexed behind the lip; lateral petals linear,
entire; lip 10-16 mm long and 5-20 mm wide, deeply 3-lobed, each lobe
fringed with a few long segments; spur curved, 1-2 cm long. June-Aug. —
Hummocks in open sphagnum bogs, conifer bogs, swamps, wet meadows,
sandy prairie, thickets, ditches.

Platanthera obtusata (Banks) Lindl.
BLUNT-LEAF ORCHID   *native / cons* **10**
  Habenaria obtusata (Banks) Richards.
Perennial herb, roots fleshy. Stems leafless, slender, 1-3 dm long. Leaves 1
at base of stem, ascending, persistent through flowering, obovate, 5-15
cm long and 1-4 cm wide, blunt-tipped, long-tapered to base. Flowers 4-
20, green-white, in a raceme 3-12 cm long and 1-2 cm wide; lateral sepals
ovate, spreading; petals ascending, widened below middle; lip lance-
shaped, widened at base, 4-6 mm long; spur curved, tapered to a thin tip,
5-8 mm long. June-Aug. — Shaded hummocks in conifer swamps (especially under cedar, black spruce or balsam fir), wet mixed conifer-deciduous forests, alder thickets.

Platanthera obtusata

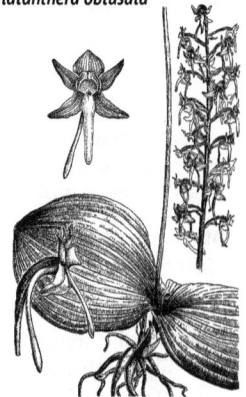

Platanthera orbiculata (Pursh) Lindl.
ROUND-LEAF ORCHID   *native / cons* **9**
  Habenaria orbiculata (Pursh) Torr.
Platanthera macrophylla (Goldie) P.M. Brown
Perennial herb, roots fleshy. Stems 2-6 dm long, leafless apart from 1-6
small bracts. Leaves 2, opposite at base of plant, spreading or lying flat on
ground, more or less round, shiny, 6-15 cm long and 4-15 cm wide. Flowers
green-white, several in a raceme 5-20 cm long and 3-6 cm wide; sepals
ovate, to 1 cm long; petals ovate, 6-7 mm long; lip entire, rounded at tip,
10-15 mm long and 2 mm wide; spur 2-3 cm long, somewhat widened at
tip. Late June-Aug. — Shaded conifer swamps of white cedar, balsam fir,
and black spruce, especially where underlain by marl; also in drier pine
forests.

Platanthera psycodes (L.) Lindl.
LESSER PURPLE FRINGED ORCHID   *native / cons* **7**
  Habenaria psycodes (L.) Spreng.
Perennial herb, roots thick and fleshy. Stems stout, 3-10 dm long. Leaves
4-12, alternate on stem, lance-shaped or oval, the upper much smaller
and narrow. Flowers rose-purple, in a densely flowered, cylindric raceme
4-20 cm long and 3-5 cm wide; sepals oval to obovate, 4-6 mm long; petals
spatula-shaped, finely toothed on margins; lip broad, 8-14 mm wide,
deeply 3-lobed, the lobes fan-shaped, fringed to less than half way to
base; spur curved, about 2 cm long. July-Aug. — Wetland margins, shores,
wet forests, wet meadows, roadside ditches; typically not on sphagnum
moss.

Platanthera orbiculata

Platanthera psycodes

Pogonia
SNAKE-MOUTH

Pogonia ophioglossoides (L.) Ker-Gawl.
ROSE POGONIA, SNAKE-MOUTH ORCHID   *native / cons* **9**

Perennial herb, spreading by surface runners (stolons) which send up a
stem every 10 cm or more apart. Stems slender, smooth, 1.5-4 dm long.
Leaves single, attached about halfway up stem, narrowly oval, 3-10 cm
long and 1-2.5 cm wide, stalkless. Flowers pink to purple, usually 1 at end
of stems; sepals widely spreading, petals oval, hovering over the column;
lip pink with purple veins, 1.5-2 cm long and 5-10 mm wide, fringed at
tip, bearded with yellow bristles. June-July. — Conifer swamps and open
bogs in sphagnum moss, floating sedge mats, sedge meadows, sandy interdunal wetlands.

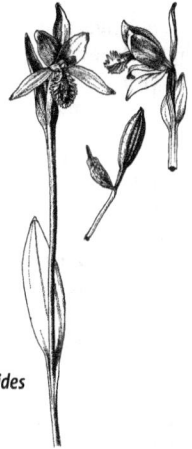

Pogonia ophioglossoides

Spiranthes
LADIES'-TRESSES

Perennial herbs, from a cluster of tuberous roots. Stems slender, erect. Leaves largest at base of
plant, becoming smaller upward on stem, the stem leaves erect and sheathing. Flowers small,
white or creamy, spirally twisted in a densely flowered, spike-like raceme; sepals and lateral
petals similar, the lateral petals joined with all 3 sepals or with only the upper sepal to form a
hood over lip and column; lip folded upward near middle so that margins embrace the column,
curved downward beyond the middle, with a pair of bumps or thickenings at base; anthers 1,
from back of the short column.

1 Leaves widely spreading or lying flat in a basal rosette, short-petioled, sometimes withered at flowering
   time, their blades less than 4.5 cm long, about 2/5 as wide as long or wider; perianth 2.5–5.5 mm
   long . . . . *S. lacera*
1 Leaves ascending, not distinctly petioled, usually present at flowering time, their blades (non-sheathing
   portion) over 4.5 cm long and less than 2/5 as wide as long; flowers usually in 2 or more rows in a
   crowded spike (sometimes one-sided); perianth larger than 5.5 mm long 2
2 Lip fiddle-shaped, strongly constricted behind expanded tip; lateral sepals united for at least half
   their length with dorsal sepal and lateral petals, forming a hood . . . . *S. romanzoffiana*
2 Lip oblong, often erose-margined but not strongly constricted; at least the lateral sepals free (or
   easily separated if connivent when young) . . . . *S. cernua*

Spiranthes cernua (L.) L.C. Rich.
WHITE NODDING LADIES'-TRESSES   *native / cons* **5**
Spiranthes incurva (Jennings) M.C. Pace

Perennial herb, roots fleshy. Stems 1-5 dm long, upper stem short-hairy,
lower stem smooth. Leaves mostly at base of plant, usually present at
flowering time, linear to oblong lance-shaped, 6-25 cm long and 5-15 mm
wide; upper stem leaves 3-5, much smaller and bractlike. Flowers white,
in a spike-like raceme 3-15 cm long, with 2-4 vertical rows of flowers, the
rows spirally twisted; sepals and petals hairy on outside; lateral petals
joined with upper sepal to form a hood; lip white, yellow-green at center,
6-10 mm long and 3-6 mm wide, slightly narrowed at middle, curved
downward, the tip curved inward toward stem, the tip wavy-margined
or with small rounded teeth, the base of lip with a pair of backward-
pointing bumps. Aug-Oct. — Open, usually sandy wetlands such as wet
meadows, lakeshores, moist prairies, ditches and roadsides.

Spiranthes lacera (Raf.) Raf.
NORTHERN SLENDER LADIES'-TRESSES   *native / cons* **8**

Perennial herb. Stems arising from a cluster of thickened roots, very slender, 1-6 dm tall. Basal leaves often present at anthesis, oval to oblong, 2-5
cm long, a to a 1/3 as wide; stem leaves reduced to small scales. Flowers
white, in a very slender twisted raceme 4-10 cm long, glabrous or nearly

Spiranthes cernua

so; lip oblong, about 4 mm long, white with green median area, its tip abruptly deflexed, crisped on the margin. Aug.–Sept. — Dry sandy soil, often with blueberry and bracken fern in open woods of jack pine, red pine, and oak; moist aspen groves, conifer thickets along shores and on dunes.

Spiranthes romanzoffiana Cham.
HOODED LADIES'-TRESSES   *native / cons* **9**
Perennial herb, roots thick and fleshy. Stems 1-4 dm long, upper stem finely hairy. Leaves mostly from base of plant, present at flowering time, upright, linear to narrowly lance-shaped, 5-20 cm long and 3-9 mm wide, the stem leaves becoming smaller and bractlike. Flowers white or cream-colored, in a spike-like raceme 3-10 cm long, with 1-3 vertical rows of flowers, the rows spirally twisted; sepals and lateral petals joined to form a hood over the lip; lip ovate, strongly constricted near middle (violin-shaped), curved downward, the tip ragged and bent inward toward stem, the bumps at base very small. July–Sept. — Open wetlands including wet meadows, fens, lakeshores, open swamps, ditches, seeps; usually in neutral or calcium-rich habitats.

Spiranthes romanzoffiana

Poaceae

GRASS FAMILY
Perennial or annual herbs, clumped or spreading by rhizomes. Stems (culms) usually hollow, with swollen, solid nodes. Leaves linear, parallel-veined, alternate in 2 ranks or rows, sheathing the stem, the sheaths usually split vertically, sometimes joined and tubular as in brome (*Bromus*) and mannagrass (*Glyceria*); with a membranous or hairy ring (ligule) at top of sheath between blade and stem, or the ligule sometimes absent; a pair of projecting lobes (auricles) sometimes present at base of blade.

Flowers (florets) small, usually perfect (with both staminate and pistillate parts), or sometimes either staminate or pistillate, the staminate and pistillate flowers separate on the same or different plants. Florets grouped into spikelets, each spikelet with 1 to many florets, the florets stalkless and alternate along a small stem or axis (rachilla), with a pair of small bracts (glumes) at base of each spikelet (the glumes rarely absent); the glumes usually of different lengths, the lowermost (or first) glume usually smaller, the upper (or second) glume usually longer. Within the spikelet, each floret subtended by 2 bracts, the larger one (lemma) containing the flower, the smaller one (palea) covering the flower; the lemma and palea often enclosing the ripe fruit (grain or caryopsis); stamens usually 3 or sometimes 6, usually exserted when flowering; ovary superior, never enclosed in a sac (as in sedges); styles 2-3-parted, the stigmas often feathery.

Spikelets grouped in a variety of heads, most commonly in branching heads (panicles), or stalked along an unbranched stem (rachis) in a raceme, or the spikelets stalkless along an unbranched stem in a spike; spikelets breaking (disarticulating) either above or below the glumes when mature, the glumes remaining in the head if falling above the glumes, or the glumes falling with the florets if disarticulation is below the glumes.

KEY TO POACEAE GROUPS

1 Tip of plant with a large "tassel" or spike-like raceme bearing staminate florets in pairs, the pistillate florets either sunken in hardened joints of rachis below the staminate portion or in separate "ears" lower on the plant . . . . **Key 1**

1 Tip of plant not as above; upper portion of inflorescence bearing pistillate or bisexual florets, or rarely staminate florets in short one-sided spikes 2

2 Spikelets all unisexual, segregated into different and dissimilar parts of the inflorescence on same plants . . . . **Zizania**

2 Spikelets perfect, or if unisexual, then scattered among bisexual spikelets 3

3 Spikelets forming a simple spike or spikes, directly sessile or subsessile on main axis of inflorescence or at most on secondary branches 4

3 Spikelets not forming simple spikes as above; pediceled and/or on tertiary or further branches of the inflorescence; in some species congested and hence spike-like, but not directly sessile or subsessile (reduced panicle branches usually visible upon removal of some spikelets)................... 5
4 Spike solitary, terminal (its rachis a continuation of the culm), the spikelets on opposite sides of rachis .... *Key 2*
4 Spikes several, one-sided (spikelets in two rows on one side of rachis) .... *Key 3*
5 Spikelets with only 1–2 florets... 6
5 Spikelets with 3 or more florets, including any sterile ones 12
6 Spikelets with an involucre consisting of long subtending bristles....................... *Key 3*
6 Spikelets without an involucre of bristles (although glumes or lemmas may be awned).......... 7
7 Glumes or lemmas (or both) laterally compressed or keeled (lateral nerves, if present, less prominent than midnerve) .... go to couplet 12
7 Glumes and lemmas rounded on back, not keeled (nerves, if present, about equally prominent) . . . 8
8 Glumes very unequal in length, one of them minute, or absent, or at most about half as long as the spikelet. (Note: In Key 3 species, a sterile lemma is present that closely resembles the large second glume opposite it and might easily be misinterpreted as a glume; the true first glume is a small, sometimes minute and membranous, even deciduous, scale at the very base of the spikelet; a reduced palea, often associated with the sterile lemma, will also help to identify the latter as part of a sterile floret and not a glume) 9
8 Glumes equal in length, neither of them much reduced nor absent 10
9 Spikelets disarticulating below the glumes, ± elliptic (less than 3 times as long as wide); a sterile lemma resembling the larger glume present .... *Key 3*
9 Spikelets disarticulating above the glumes, lanceolate (3–10 times as long as wide); no sterile lemma present .... *Key 4*
10 Spikelets paniculate, all (or mostly) 1-flowered, bisexual, the florets all alike (no sterile lemmas or separate sterile pedicels present); spikelets less than 4 mm long, except in *Hesperostipa* with awns over 5 cm long .... *Key 4*
10 Spikelets paniculate or racemose, basically 2-flowered (the lower floret staminate or sterile with often suppressed palea); spikelets 3 mm or more long.. 11
11 Spikelets all alike, not paired with a pedicel bearing a rudimentary, staminate, or no floret .... *Panicum virgatum*
11 Spikelets in pairs, usually of two kinds: one sessile and with a bisexual floret, the other a hairy pedicel with or without a staminate or rudimentary floret (rarely 2 stalked florets with 1 sessile one) .... *Key 1*
12 Spikelets all or mostly containing 1 bisexual floret and no sterile or vestigial ones below it 13
12 Spikelets all or mostly containing 2–several florets, the lower ones sometimes staminate or rudimentary (scale-like or reduced to tiny hairy appendages)... 14
13 Glumes both completely absent; spikelets strongly flattened, appressed and overlapping, the lemmas scabrous or hispid-ciliate .... *Leersia*
13 Glumes (one or both) usually present; spikelets various but not as above.................. *Key 4*
14 Glumes shorter than the lowest floret (excluding awns if present); awn of lemma none, terminal, arising from between terminal teeth and not twisted, or at most subterminal.............. *Key 5*
14 Glumes (at least one of them, not necessarily the first glume) longer than lowest floret; awn of lemma none or arising from between terminal teeth and strongly twisted below, or (the usual condition) inserted on the middle or lower part of the lemma 15
15 Spikelets containing one bisexual awnless floret (the lemma sometimes membranaceous) with two additional, often dissimilar (sometimes awned) staminate, sterile, or vestigial lemmas below it *Key 6*
15 Spikelets usually containing 2 or more bisexual florets (staminate or sterile florets, if present, above the fertile one and/or fertile lemma awned) .... *Key 7*

KEY 1
TRIBE SACCHAREAE
Subtribe Andropogoninae (*Andropogon, Schizachyrium*)
Subtribe Sorghinae (*Sorghastrum*)

1 Inflorescence an open to contracted panicle..................................... *Sorghastrum*
1 Inflorescence of 1 or several narrow or spike-like simple racemes (some spikelets sessile and some pediceled) 2
2 Spike-like simple racemes solitary at the ends of the branches .... *Schizachyrium*
2 Spike-like simple racemes 10–20 at the ends of the branches .... *Andropogon*

KEY 2

TRIBE HORDEEAE
Subtribe Hordeinae (*Elymus, Hystrix, Hordeum, Leymus, Secale*)
TRIBE POEAE
Subtribe Loliinae (*Lolium*)

1 Lemmas smooth and glabrous, except for a spiny-ciliate keel and exposed margin, tapering into a long awn . . . . *Secale*
1 Lemmas smooth to scabrous or pubescent, but not simply with spiny-ciliate keel and margin, awned or awnless 2
2 Larger glumes 3.3–6.5 mm broad with at least 3 prominent nerves, the keel or midnerve not centered . . . . *Leymus*
2 Larger glumes less than 2.5 mm broad, variously nerved (or glumes absent). 3
3 Spikelets mostly 2–3 at each node of the rachis; glumes usually 4–6 (the spikelet arrangement may be obscured by reduction or asymmetric positions of some spikelets, but the basic structure is revealed by the presence of usually 4–6 glumes subtending the entire group of spikelets); glumes usually vestigial or absent in *Elymus hystrix* with mostly 2 easily recognized narrow spikelets at each node . . . 4
3 Spikelets clearly 1 at each node (or most nodes) of the rachis; glumes variously arranged (or absent), but not more than 2 6
4 Spikelets 2 at each node of the rachis (or at some nodes, only 1, rarely 3, but total number of glumes (awn-like or broader) developed at a node not more than 4) . . . . *Elymus*
4 Spikelets basically 3 at each node of the rachis (the lateral 2 in commonest species reduced to bristles), this arrangement most easily recognized by the presence of 6 awn-like or narrowly lanceolate and awn-tipped glumes at a node 5
5 Body of larger lemmas ca. 3.5–6 mm long; rachis of spike readily disintegrating at maturity *Hordeum*
5 Body of larger lemmas ca. 8–12 mm long; rachis not disintegrating . . . . *Elymus*
6 Glumes 1 (except terminal spikelet with 2), the narrow edge of the spikelet against the rachis and lacking a glume . . . . *Lolium*
6 Glumes 2 on all spikelets 7
7 Lemmas with awns strongly divergent or recurved at maturity . . . . *Elymus trachycaulus*
7 Lemmas with awns nearly straight or absent 8
8 Lemmas densely hairy . . . . *Elymus lanceolatus*
8 Lemmas glabrous (rarely slightly pubescent), smooth or scabrous . . . . *Elymus*

KEY 3

TRIBE PANICEAE
Subtribe Anthephorinae (*Digitaria*)
Subtribe Boivinellinae (*Echinochloa*)
Subtribe Cenchrinae (*Setaria*)
Subtribe Panicinae (*Dicanthelium, Panicum*)

1 Spikelets with an involucre consisting of long subtending bristles . . . . *Setaria*
1 Spikelets without an involucre (although glumes or lemmas may be awned) 2
2 Spikelets spiny-hispid and usually also awned; ligule none . . . . *Echinochloa*
2 Spikelets glabrous or pubescent but not coarsely hispid and not awned; ligule present, distinct or nearly absent (of hairs or membranous). 3
3 Inflorescence composed of 1-sided spikes or spike-like racemes, the rachis of each winged or at least flat on the side opposite the spikelets . . . . *Digitaria*
3 Inflorescence an open panicle, not spike-like or distinctly one-sided 4
4 Ligule a membranous collar 1–1.5 mm high, without hairs; base of leaf blade and very summit of sheath without special zone of short pubescence or long hairs or cilia; first glume minute or vestigial; fertile lemma leathery in texture, with thin flat translucent margins . . . . *Digitaria cognata*
4 Ligule usually partly or entirely of short or long hairs; (if ligule membranous, plants not otherwise as above: ligule ca. 0.5 mm long or virtually absent; or summit of sheath or basal margin of blade pubescent or ciliate; and/or first glume more than 0.5 mm long); fertile lemma hard and shiny 5
5 Spikelets at least sparsely pubescent towards their margins . . . . *Dichanthelium*
5 Spikelets glabrous. 6
6 Terminal panicle 8–40 cm long, (smaller in occasional depauperate individuals of annual species); annuals or perennials, but without clear remnants of overwintering basal rosette leaves; flowering and fruiting summer-fall . . . . *Panicum*

6 Terminal panicle 2.5–8 (–12) cm long; tufted perennials with clear remnants of old, dead leaves from the previous year present at the base, these sometimes formed into a clear overwintering rosette; flowering spring and fruiting in late spring–early summer . . . . *Dichanthelium*

KEY 4
TRIBE BRACHYELYTREAE (*Brachyelytrum*)
TRIBE POEAE
  Subtribe Agrostidinae (*Agrostis, Ammophila, Calamagrostis*)
  Subtribe Miliinae (*Milium*)
  Subtribe Phleinae (*Phleum*)
  Subtribe Poinae (*Alopecurus, Apera, Cinna, Ventenata*)
TRIBE STIPEAE
  Subtribe Stipinae (*Hesperostipa, Oryzopsis, Piptatherum*)
TRIBE ZOYSIEAE
  Subtribe Sporobolinae (*Calamovilfa, Sporobolus*)

1 Lemma with awn (or awns) strictly terminal 2
1 Lemma with awn absent or dorsal or subterminal 6
2 Body of lemma 8–23 mm long. 3
2 Body of lemma less than 7 mm long. 4
3 Glumes 9.5–45 mm long. . . . . *Hesperostipa*
3 Glumes rudimentary or one of them up to 5 mm long . . . . *Brachyelytrum*
4 Glumes acute to obtuse, more than 1 mm wide, scarcely if at all keeled, the spikelets nearly terete; lemma rounded on the back, firm and hardened 5
4 Glumes acuminate, not over 1 mm wide, keeled, the spikelets somewhat compressed; lemma keeled, membranous or thin. . . . . *Muhlenbergia*
5 Principal leaf blades basically flat (just the margins involute), basal or nearly so, densely and very finely rough-puberulent, with strong closely spaced veins. . . . . *Oryzopsis*
5 Principal leaf blades involute or, if flat, all cauline and glabrous. . . . . *Piptatherum*
6 Spikelets 10–15 mm long; anthers 5–8 mm long; panicle crowded and spike-like, 10–20 (–28) mm across at the middle . . . . *Ammophila*
6 Spikelets less than 8 mm long (excluding awns); anthers to 4.5 mm long (usually much shorter); panicle various. 7
7 Spikelets sessile or nearly so, crowded in a very dense spike-like panicle (branches of panicle suppressed, scarcely if at all visible without dissection of panicle). 8
7 Spikelets in open or contracted (but not densely spike-like) inflorescences, with evident pedicels and/or panicle branches. 10
8 Glumes awnless . . . . *Alopecurus*
8 Glumes awned 9
9 Plants with scaly rhizomes; glumes gradually tapered into awn; ligule to ca. 1 mm long . . . .
   . . . . *Muhlenbergia*
9 Plants without scaly rhizomes; glumes abruptly rounded or truncate, the awn distinct; ligule over 1 mm long . . . . *Phleum*
10 Spikelets rounded on back, not keeled (neither glumes nor lemma with a midvein more prominent than other nerves), at least 2.5 mm long; lemma shiny, distinctly firmer in texture than the glumes 11
10 Spikelets keeled (glumes and/or lemmas with midvein more prominent than other nerves) or less than 2.5 mm long; lemma no firmer in texture than the glumes. 12
11 Lemmas with appressed pubescence; leaves with blades usually involute; upper ligules not over 3 mm long . . . . *Piptatherum pungens*
11 Lemmas glabrous; leaves with blades broad and flat; upper ligules mostly 4–6 (–8) mm long . *Milium*
12 Ligule a fringe of short hairs; lemmas awnless 13
12 Ligule membranous (at most minutely ciliate at summit of membrane); lemmas awned or awnless 14
13 Lemma (4.7–) 5–6.7 mm long, surrounded with a tuft of long hairs (more than half its length) at its base . . . . *Calamovilfa*
13 Lemma 1.5–5.5 mm long, without long hairs at its base . . . . *Sporobolus*
14 Lemma with long hairs at base (on or near callus) 15
14 Lemma without long hairs at base (at most with hairs on callus less than 0.5 mm long) 16
15 Long hairs at least in part arising from lower portion of lemma; glumes (excluding awn-tips if present) shorter than lemma . . . . *Muhlenbergia*

15 Long hairs restricted to callus at base of floret; glumes slightly exceeding lemma . . . . ***Calamagrostis***
16 Glumes both distinctly shorter than lemma; lemma awnless . . . . ***Muhlenbergia***
16 Glumes (one or both of them) equaling or exceeding the lemma and/or the lemma awned 17
17 Floret raised above base of glumes on a minute stalk; spikelet articulated below the glumes; lemma
    with a small subterminal awn; stamen 1. . . . . ***Cinna***
17 Floret not stalked; spikelets articulated above the glumes; lemma awnless or with dorsal awn; stamens
    3 . . . . ***Agrostis***

KEY 5
TRIBE ARUNDINEAE (*Phragmites*)
TRIBE BROMEAE (*Bromus*)
TRIBE ERAGROSTIDEAE
    Subtribe Eragrostidinae (*Eragrostis*)
TRIBE MELICEAE (*Glyceria, Schizachne*)
TRIBE POEAE
    Subtribe Aveninae (*Graphephorum*)
    Subtribe Coleanthinae (*Sclerochloa*)
    Subtribe Dactylidinae (*Dactylis*)
    Subtribe Holcinae (*Deschampsia*)
    Subtribe Loliinae (*Festuca, Lolium, Vulpia*)

1 Plants tall and stout (usually over 1.5 m tall) with larger leaf blades 1–3.5 cm wide; spikelets ca. 11–
    17 mm long; ligule a densely ciliate brown band; rachilla (above the lowest floret) with silky beard
    about equaling or exceeding the lemmas . . . . ***Phragmites***
1 Plants generally less than 1.5 m tall with narrow leaves less than 1 cm wide; spikelets and ligule various;
    rachilla with beard shorter or absent 2
2 Spikelets sessile or at most very short-pediceled, crowded into dense clusters, these either at the
    ends of elongate panicle branches. . . . . ***Dactylis***
2 Spikelets short- to long-pediceled in an open panicle. 3
3 Callus at base of floret with dense beard of straight hairs 0.5 mm or more long 4
3 Callus glabrous, minutely puberulent, or cobwebby (not bearded with straight hairs) 7
4 Awn of lemma arising near base. . . . . ***Deschampsia***
4 Awn of lemma absent, terminal, subterminal, or arising between terminal teeth. 5
5 Lemmas awnless, weakly 5-nerved; sheaths open . . . . ***Graphephorum***
5 Lemmas with short or long awn; plants with closed sheaths 6
6 Callus with distinct beard, the lemma glabrous or nearly so; grain glabrous . . . . ***Schizachne***
6 Callus lacking a distinct beard, the pubescence like that of the lemma; grain pubescent at the summit
    . . . . ***Bromus***
7 Glumes (at least one of them) and usually also lemmas strongly keeled (lemmas in a few species
    rounded on the back); awns absent or not over 2 mm 8
7 Glumes and lemmas rounded on back, not keeled (or obscurely so toward tip); awns absent or pres-
    ent 11
8 Larger glumes ca. 4.5–7 mm long . . . . ***Dactylis***
8 Larger glumes not over 4.4 mm long 9
9 Larger glumes obovate, broadest above the middle. . . . . ***Key 3***
9 Larger glumes broadest at or below the middle 10
10 Ligule a fringe of hairs; lemmas with 3 prominent nerves, glabrous; spikelets 2–30-flowered . . . .
    . . . . ***Eragrostis***
10 Ligule a membranous scale, the cilia, if any, shorter than the scale; lemmas with 3–5 nerves, glabrous,
    hairy, and/or cobwebby at base; spikelets various . . . . ***Poa***
11 Lemmas usually 2-toothed or minutely 2-lobed at the tip and usually with at least a short awn arising
    from just below or between the teeth (if teeth apparently united, as in some species of Bromus, the
    awn thus subterminal); sheaths closed nearly to their summit . . . . ***Bromus***
11 Lemmas not 2-lobed or 2-toothed at tip, awnless or with strictly terminal awn; sheaths open (or
    closed in *Glyceria*, with prominently nerved and awnless lemmas, and in the youngest shoots of
    Festuca rubra) 12
12 Lemmas acute at the tip, awned 13
12 Lemmas acutish or obtuse, awnless 15
13 Blades of leaves flat (or merely once-folded), at least the larger ones (2.5–) 3–8 mm broad . . . ***Lolium***

13 Blades of leaves strongly involute, less (usually much less) than 3 mm broad 14
14 Plants perennial, usually in dense tufts including numerous dry sheaths of previous years; florets open at anthesis, with 3 anthers... *Festuca*
14 Plants annual, usually in small tufts or solitary; florets cleistogamous, with usually one included anther (rarely 3 anthers) .... *Vulpia*
15 Nerves of lemma prominent, straight and becoming parallel at the tip; sheaths closed for most of their length (but easily splitting open) .... *Glyceria*
15 Nerves of lemma very weak (or if visible, then converging, not parallel, at the tip); sheaths open . 16
16 Blades of leaves strongly involute, usually much less than 3 mm wide.................... *Festuca*
16 Blades of leaves flat (or merely once-folded), at least the larger ones 3–8 mm wide............. 17
17 Larger lemmas 2.5–4.5 mm long; anthers 0.8–1.4 mm long; spikelets mostly containing 2–4 (–5) florets and borne beyond the middle of the primary panicle branches .... *Festuca subverticillata*
17 Larger lemmas 5.5–8 mm long; anthers 2.2–3.5 (–3.8) mm long; spikelets often containing 5 or more florets, borne below as well as above the middle of the primary panicle branches .... *Lolium*

KEY 6
TRIBE POEAE
Subtribe Phalaridinae (*Anthoxanthum, Phalaris*)

1 Glumes very unequal; lower lemmas with prominent dorsal awns, concealing the awnless bisexual floret... *Anthoxanthum*
1 Glumes nearly or quite equal; lower lemmas awnless, small and inconspicuous, only the awnless bisexual floret evident .... *Phalaris*

KEY 7
TRIBE DANTHONIEAE (*Danthonia*)
TRIBE POEAE
Subtribe Aveninae (*Graphephorum, Koeleria, Sphenopholis*)
Subtribe Holcinae (*Deschampsia*)

1 Lemmas all awnless; larger glumes obovate (broadest above the middle), generally shorter than the lowest floret... 2
1 Lemmas with distinct twisted, jointed, or curved awn (sometimes largely hidden by the glumes or absent on some florets of a spikelet); glumes mostly ovate to lanceolate, at least one of them longer than the lowest floret... 4
2 Rachilla and callus prominently bearded with long straight hairs .... *Graphephorum*
2 Rachilla and callus glabrous or at most with short hairs (under 0.5 mm long) 3
3 Axis and branches of inflorescence glabrous, at most scabrous; larger glumes not over 3 mm long .. .. *Sphenopholis*
3 Axis and branches of inflorescence densely short-pubescent; larger glumes 3–4.5 mm long . *Koeleria*
4 Larger glumes 6–27 mm long .... *Danthonia*
4 Larger glume less than 6 mm long .... *Deschampsia*

Agrostis
BENT-GRASS, BENT
Perennial grasses, clumped or spreading by rhizomes or sometimes by stolons. Leaves soft, auricles absent, ligules membranous, sheaths open, usually smooth and glabrous. Head an open panicle. Spikelets small, 1-flowered, breaking above glumes; glumes more or less equal length, 1-veined; floret shorter than glumes; lemma awnless or with a short straight awn; palea small or absent; stamens usually 3.

1 Palea present, about half as long as the lemma or longer; anthers 1–1.5 mm long............... 2
1 Palea absent or vestigial; anthers ca. 0.6 mm long or shorter 3
2 Plants rhizomatous but not stoloniferous, i.e., stems arising from underground rhizomes, straight or curved at the very base, otherwise erect and nearly or quite straight; larger leaf blades mostly 3–7 mm wide; spikelets usually flushed with red or purplish; bases of middle panicle branches mostly meeting the axis of the panicle at an angle of 30–45 degrees (except when very immature)........ .... *A. gigantea*

2 Plants stoloniferous but not rhizomatous, i.e., stems usually decumbent at their bases, the lower nodes often strongly bent and/or rooting, but underground rhizomes absent; larger leaf blades 1.7–3 (–4) mm wide; spikelets pale, greenish; bases of middle panicle branches usually strongly ascending or appressed to axis of panicle, at most diverging about 15 degrees (but panicle branches often spreading distally) . . . . *A. stolonifera*

3 Longest panicle branches less than 6 (–12) cm long and the uppermost leaf blade more than 5 cm long; leaf blades flat, the wider ones 1.5–3.5 mm wide; panicle branches forked about or below the middle, often smooth or only sparingly hispidulous-scabrous; panicle pale, greenish (very rarely red-tinged) . . . . *A. perennans*

3 Longest panicle branches more than 6 cm long or uppermost leaf blade less than 5 cm long (or both conditions); leaf blades usually ± involute, the widest to 1.5 (rarely 3) mm wide; panicle branches often not forked until beyond the middle, copiously hispidulous-scabrous; panicle flushed with red . . . . *A. scabra*

Agrostis gigantea Roth

BLACK BENT *introduced*

Perennial grass, rhizomatous and sod-forming, not stoloniferous. Stems to 10 dm long or sometimes more; auricles absent; ligules membranous, larger (upper) ligules mostly 2.5-6 mm long, higher than wide; leaf blades 3-8 mm wide. Panicle 10-20 cm long, notably suffused with purplish red, at anthesis triangular-ovoid, with widely spreading unequal branches, sometimes later more contracted; at least some of the panicle branches floriferous to the base; panicle branches and often the pedicels scabrous. Spikelets rather crowded, 2-3.5 mm long; glumes scabrous along the keel; lemma 2/3 as long as the glumes, scabrous near tip, usually unawned; callus minutely bearded. — Native of Europe, cultivated and escaped into moist meadows, shores, coastal marshes, and other moist places.

Agrostis gigantea

Agrostis perennans (Walt.) Tuckerman

UPLAND BENT *native / cons* **4**

Tufted perennial grass. Stems 5-10 dm tall; auricles absent; ligules membranous, leaf blades flat, 2-6 mm wide, elongate, the uppermost blade more than 5 cm long. Panicle mostly pale greenish, 10-25 cm long, notably longer than wide, the smooth or sparsely scabrous branches forking near or below the middle, soon divaricate, the longest ones often less than 6 cm long. Spikelets 1.8-2.8 mm long; glumes subequal, scabrous on the midvein; lemma 1.3-2 mm, awnless or rarely with a very short, slender awn near the tip; palea obsolete. — Various habitats, usually in dry soil.

Agrostis scabra Willd.

ROUGH BENT *native / cons* –

Tufted perennial grass. Stems slender, 3-9 dm long; auricles absent; ligules membranous; leaves mostly basal or below the middle of the stem, usually erect, the blade flat or more often involute, 1-2 mm wide. Panicle ovoid or pyramidal, 1-3 dm long, sometimes half as long as the whole plant, very diffuse, more or less reddish, the scabrous filiform branches divaricate, mostly forking well above the middle. Spikelets 1.2-3.2 mm long; glumes scabrous on the midvein; lemma awnless or with a short straight awn; callus short-bearded; palea obsolete or to 0.3 mm long. — Common; widely distributed in many habitats, and variable.

Agrostis scabra

Agrostis stolonifera L.

REDTOP, SPREADING BENT *introduced*

Agrostis alba var. *palustris* (Huds.) Pers.
Agrostis palustris Huds.

Perennial grass, spreading by rhizomes and also sometimes by stolons. Stems erect or more or less horizontal at base, 3-10 dm or more long; auricles absent; ligules membranous, usually splitting at tip, 2-5 mm long; leaf blades ascending, 2-8 mm wide, rough-to-touch. Panicle open, 3-20

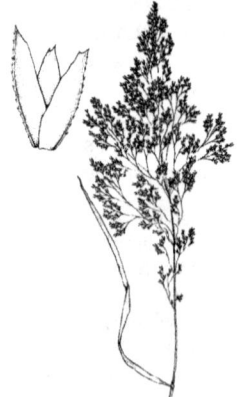

Agrostis stolonifera

cm long, the branches spreading, branched and with spikelets along their entire length. Spikelets 1-flowered, usually purple, 2-4 mm long; glumes lance-shaped, 1.5-2.5 mm long; lemma 2/3 length of glumes, 1-2 mm long; palea present, about half as long as lemma. July-Sept. — Wet meadows, ditches, streambanks and shores; disturbed areas.

Alopecurus
MEADOW-FOXTAIL
Annual or perennial grasses. Stems erect or more or less horizontal at base. Leaves mostly from lower 1/2 of the stems; sheaths open; auricles absent; ligules membranous, entire to lacerate. Heads densely flowered, cylindric, spike-like panicles. Spikelets 1-flowered, flattened, breaking below the glumes; glumes equal length, 3-nerved, often silky hairy on back, awnless; lemma about as long as glumes or shorter, awned from the back, the awn shorter to longer than the glume tips; palea absent. The narrow panicles resemble those of timothy (Phleum).

1 Spikelets (excluding awns) ca. 4–6.5 mm long; awns mostly exserted ca. 3.5–6 mm beyond tips of glumes; anthers ca. 2.4–3.5 mm long .... *A. pratensis*
1 Spikelets not over 3 mm long; awns exserted at most only ca. 1 mm (usually included); anthers less than 2 mm long .... *A. aequalis*

Alopecurus aequalis Sobol.
SHORT-AWN FOXTAIL					*native / cons* **4**
Annual or short-lived perennial grass. Stems single or in small clumps, slender, erect to more or less horizontal, 2-6 dm long, often rooting at the nodes; auricles absent; ligules 2-6.5 mm long, obtuse; leaf blades 1-5 mm wide, finely rough-to-touch above; ligule membranous, rounded to elongate, 2-7 mm long. Panicle erect, spike-like, 2-7 cm long and 3-5 mm wide. Spikelets 1-flowered; glumes 2-3 mm long, blunt-tipped, hairy on the keel and veins; lemma about equaling the glumes, awned from back, the awn straight, to 1.5 mm longer than glume tips. June-Aug. — Shallow water or mud of wet meadows, marshes, ditches, springs, open bogs, fens, shores and streambanks; sometimes where calcium-rich.

Alopecurus pratensis

Alopecurus pratensis L.
FIELD MEADOW-FOXTAIL					*introduced*
Perennial grass, shortly rhizomatous. Stems erect or decumbent at base, 4-8 dm long; auricles absent; ligules 1.5-3 mm long, obtuse to truncate; upper sheaths not or scarcely inflated. Panicle spike-like, 2-8 cm long, 5-10 mm wide, scarcely tapering; glumes 4-5.5 mm long, the keel narrowly winged, conspicuously ciliate, especially above the middle, with hairs 1-1.5 mm long; awn inserted about halfway between the base and middle of the lemma, exserted 2-6 mm. — Native of Eurasia; naturalized in moist meadows, fields, and waste places.

Ammophila
BEACH-GRASS
Ammophila breviligulata
Ammophila breviligulata Fern.
AMERICAN BEACH-GRASS					*native / cons* **10**
  Calamagrostis breviligulata (Fern.) Saarela
Coarse, stiff, perennial grass, from long running rhizomes. Stems stout, erect, glabrous, 5-10 dm tall; sheaths glabrous; ligule membranous, ovate or truncate, 1-3 mm long; leaf blades flat at base, involute above, 4-8 mm wide when unrolled, scabrous above, glabrous beneath. Panicle dense, 1-4 dm long, 1-2.5 cm thick, its base often enclosed in the upper sheath. Spikelets 1-flowered, strongly flattened, articulated above the glumes; glumes about equal, 10-15 mm long, linear-lance-shaped, keeled, the first 1-nerved, the second 3-nerved, scabrous on the keel; lemmas shorter than

the glumes, scaberulous; obscurely 3-5-nerved, awnless, subtended by a tuft of short hairs 1-3 mm long from the callus. — Dunes and dry sandy shores along Lake Michigan; useful as sand-binders in dune control.

Andropogon
BLUESTEM

Andropogon gerardii Vitman
BIG BLUESTEM, TURKEY-FOOT   *native / cons* 4
Perennial grass. Stems stout, 1-3 m tall, forming large bunches or extensive sod; ligules membranous; leaf blades usually 5-10 mm wide, the lower ones and the sheaths sometimes villous. Racemes 2-6, subdigitate, on a long-exserted peduncle, 5-10 cm long. joints of the rachis and pedicels equal, sparsely or usually densely ciliate, densely bearded at the summit. Spikelets of two kinds, in pairs at the joints of the rachis, one sessile and perfect, the other pediceled and staminate, sterile, or abortive; glumes of the fertile spikelet equal or nearly so, leathery, flat to concave on the back, lacking a midnerve, often ciliate; fertile lemma shorter than the glumes, narrow, hyaline, usually ending in a long awn awn 8-15 mm long, twisted below and more or less bent. — Moist or dry soil of prairies, roadsides; in dry open woods, old fields, rarely in fens and sedge meadows.

Andropogon gerardii

Anthoxanthum
SWEET VERNAL GRASS

Anthoxanthum hirtum (Schrank) Y. Schouten & Veldkamp
SWEETGRASS   *native / cons* 7
Hierochloe hirta (Schrank) Borbás
Hierochloe odorata (L.) Beauv.
Perennial grass, from creeping rhizomes; plants nicely sweet-scented, especially when dried. Stems erect, 2-6 dm tall, smooth; sheaths brownish or reddish; ligules membranous, 2-5 mm long; leaf blades of basal and stem leaves 2.5-5.5 mm wide, upper surface glabrous and shiny, undersurface pilose; fertile stem leaves short, 1-4 cm long, leaves on sterile shoots much longer. Head a pyramid-shaped panicle, 5-10 cm long, the branches spreading to drooping. Spikelets 3-flowered, the lower 2 florets staminate, the terminal spikelet perfect, golden brown, or green or purple at base and golden near tips, 5 mm long, breaking above the glumes; glumes ovate, shiny, 4-6 mm long; lemmas 3-4 mm long, the staminate lemma hairy. May-July. — Wet meadows, shores, low prairie; often where sandy.

The fragrance emitted when fresh plants are crushed or burned is from coumarin, an anti-coagulant agent.

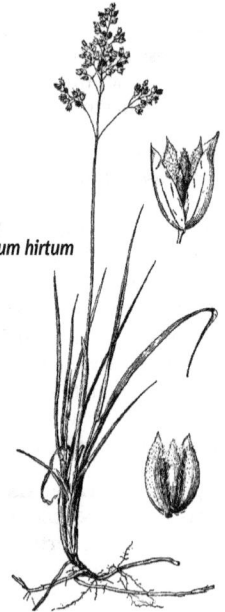
Anthoxanthum hirtum

Brachyelytrum
SHORTHUSK

Brachyelytrum aristosum (Michx.) Beauv. ex Branner & Coville
BEARDED SHORTHUSK   *native / cons* 7
Perennial forest understory grass from knotty rhizomes. Leaves broad, mostly along the stem; sheaths open; auricles absent; ligules short, membranous; lower leaf blades absent or reduced; upper leaf blades flat. Panicles narrow, few-flowered; spikelets readily deciduous, 1-flowered, articulated above the glumes. Glumes minute, subulate to triangular, 1-nerved. Lemma linear-subulate, rounded on the back, sharply 5-nerved, gradually tapering to an elongate awn. Palea 2-keeled; rachilla prolonged into an elongate bristle appressed to the furrow of the palea. — Moist to dry de-

Brachyelytrum aristosum

ciduous forests, lowland forests, moist thickets, sandy pine forests, coniferous swamps.

Previously included within *Brachyelytrum erectum* (Schreb. ex Spreng) Beauv., that species common in Wisconsin but not currently known from Door County.

Bromus
BROME, CHESS, CHEAT-GRASS

Perennial grasses. Leaves generally flat; sheaths closed to near top, usually pubescent; auricles usually absent; ligules membranous, usually erose or lacerate. Head a panicle of drooping spikelets. Spikelets with several to many flowers, breaking above the glumes; glumes shorter than lemmas; lemmas awned or unawned; stamens usually 3.

1   First glume with one distinct nerve; second glume with 3 (–5) nerves 2
1   First glume with 3 (–5) distinct nerves; second glume with 5–7 nerves. 4
2   Awns 10–30 mm long, as long as or longer than their lemmas; apex of lemma beyond insertion of awn 1.5–2.7 mm long; annual weed . . . . *B. tectorum*
2   Awns absent or up to 8 mm long, shorter than their lemmas; apex of lemma less than 1.5 mm long; perennials, mostly native (*B. inermis* introduced) 3
3   Plants with elongate rhizomes; lemmas (at least when fresh) usually flushed with purplish, especially toward the margins, the awns absent or less than 4 (–5.5) mm long; anthers 3.3–4.7 (–6) mm long . .
    . . . . *B. inermis*
3   Plants without elongated rhizomes; lemmas (when fresh) green (very rarely flushed with purple), the larger awns 3–7 (–9) mm long; anthers various. . . . . *B. ciliatus*
4   Larger awns 2–3.5 mm long; primary branches of inflorescence mostly longer than the spikelets; anthers 1.5–2 mm long; ligule less than 0.7 mm long, glabrous on the back; native perennial . . . . . .
    . . . . *B. kalmii*
4   Larger awns 3.5–14 mm long; branches of inflorescence mostly much shorter than the spikelets; anthers 0.5–1.2 mm long; ligule 0.5–2 mm long, pubescent on the back (side next to the blade); introduced annual . . . . *B. hordeaceus*

Bromus ciliatus L.
FRINGED BROME   *native / cons 7*

Perennial grass, rhizomes absent. Stems single or few together, smooth or hairy at nodes, 5-12 dm long; sheaths usually with long hairs; ligule membranous, short, to 2 mm long, ragged across tip; leaf blades flat, 4-10 mm wide, usually with long, soft hairs mainly on upper surface. Panicle loose, open, 1-3 dm long, the branches usually drooping. Spikelets large, 4-10-flowered, 1.5-3 cm long and 5-10 mm wide; glumes usually more or less smooth, lance-shaped, the first glume 4-9 mm long, the second glume 6-10 mm long, often tipped with a short awn; lemma 10-15 mm long, more or less smooth on back, usually long-hairy along lower margins, tipped with an awn 2-6 mm long; palea about as long as body of lemma. July-Sept. — Streambanks, shores, thickets, sedge meadows, fens, marshes; also in moist woods.

Bromus ciliatus

Bromus hordeaceus L.
SOFT CHESS   *introduced*
     Bromus mollis L.

Annual or biennial grass. Stems 4-12 dm tall, glabrous or puberulent; lower sheaths densely pilose; upper sheaths pubescent or glabrous; ligules 1-1.5 mm long, hairy, erose; leaf blades 2-5 mm wide. Panicle rather dense, 5-10 cm long, its branches usually shorter than the spikelets. Spikelets 6-9-flowered, usually softly pubescent, rarely glabrous; first glume 3-nerved, 5-7 mm long; second glume broader, 5-nerved, 7-9 mm long, often awned; lemmas broad, 7-9 mm long, 7-nerved, the teeth triangular, 1.5-2 mm long and 1-1.5 mm wide; awns 6-9 mm long. — Native of Europe; introduced in our range in roadsides, cultivated ground, and waste places.

Bromus hordeaceus

Bromus inermis Leyss.
SMOOTH BROME   *introduced*

Perennial grass, with short to long-creeping rhizomes. Stems 5-10 dm tall; sheaths glabrous; auricles sometimes present; ligules to 3 mm long, glabrous, truncate, erose; leaf blades glabrous, 8-15 mm wide. Panicle spreading at anthesis, later contracted, 1-2 dm long, often with 4-10 branches from a node. Spikelets 15-30 mm long, about 3 mm wide, 7-11-flowered; first glume 1-nerved, 4-8 mm long; second glume 3-nerved, 7-11 mm long; lemmas 10-12 mm long, 3-5-nerved, the outer pair of nerves often inconspicuous, obtuse or retuse, glabrous or scaberulous, awnless or with an awn to 2 mm long. — Native of Europe, cultivated for forage and often escaped.

Bromus inermis (spikelet)

Bromus kalmii Gray
KALM'S BROME   *native / cons* **8**

Perennial grass, not rhizomatous. Stems slender, loosely tufted or solitary, 5-10 dm tall, mostly glabrous, often pubescent at the nodes; sheaths usually villous, varying to glabrous; auricles absent; ligules 0.5-1 mm long, glabrous, truncate, erose; leaf blades 1-2 dm long and 5-10 mm wide, glabrous or pubescent on both sides. Panicle nodding, 5-10 cm long or rarely longer, the relatively few spikelets drooping on slender flexuous pedicels. Spikelets 15-25 mm long, 6-11-flowered, softly villous; first glume 3-nerved, 6-7 mm long; second glume 5-nerved, 7-9 mm long, lemmas 7-nerved, 8-10 mm long, obtuse, the awn 2-3 mm long; the teeth about a third as wide as long. — Dry woods, rocky banks, and sandy or gravelly soil.

Bromus kalmii (spikelet)

Bromus tectorum L.
CHEAT GRASS   *introduced*

Annual grass. Stems tufted, 3-7 dm tall; sheaths and blades softly pubescent, the latter 2-4 mm wide; ligules 2-3 mm long, glabrous, obtuse, lacerate. Panicle 1-2 dm long, repeatedly branched, bearing rather crowded, drooping spikelets to 3 cm long, on slender pedicels; first glume subulate, 1-nerved, 5.5-7 mm long; second glume subulate, 3-nerved, 8-10 mm long; lemmas narrowly lance-shaped, 5-7-nerved, 10-12 mm long, pubescent throughout, usually hirsute toward the tip, acuminate into slender scarious teeth; awn 12-17 mm long; palea conspicuously ciliate. — Native of s Europe, established as a weed in waste ground and on roadsides; now a prominent feature of many sagebrush ecosystems in the western states.

Bromus tectorum

Calamagrostis
REED-GRASS

Perennial grasses, spreading by rhizomes. Stems single or in clumps. Leaves flat or inrolled, green or waxy blue-green, smooth or rough-to-touch; sheaths smooth; ligule large, membranous, usually with an irregular, ragged margin. Head a loose and open or dense and contracted panicle. Spikelets 1-flowered, breaking above glumes; glumes nearly equal, lance-shaped; lemma shorter than glumes, lance-shaped, awned from back, the awn about as long as lemma, the base of lemma (callus) bearded with a tuft of hairs, these shorter to as long as lemma; palea shorter than lemma; stamens 3.

1 Leaf blades rather lax, to 10 mm wide; panicle mostly open with rather loosely ascending to spreading branches at flowering time; lemma nearly or quite smooth, membranous and translucent for at least the apical half; awn nearly or quite smooth, at least on basal half; callus hairs about as long as lemma (occasionally shorter), uniform in length and distribution; palea not over 2 mm long . . **C. canadensis**
1 Leaf blades stiff, to 4 mm wide; panicle mostly narrow and contracted with strongly ascending branches at flowering time; lemma usually firm and prominently scabrous, colorless and translucent only toward the tip; awn distinctly but minutely antrorsely scabrous its entire length (at 20x); callus

hairs generally shorter than lemma, unequal in length or distribution (those immediately below the middle of the lemma shorter than those at the side, or absent; do not confuse the hairy prolongation of the rachilla behind the palea); palea often longer than 2 mm .... **C. stricta**

Calamagrostis canadensis (Michx.) Beauv.
BLUEJOINT   *native / cons* **5**

Perennial grass, from creeping rhizomes. Stems erect, in small clumps, 6-15 dm long, often rooting from lower nodes when partly underwater; leaf blades flat, green to waxy blue-green, 3-8 mm wide, rough-to-touch on both sides; sheaths smooth; ligules 3-7 mm long. Panicle more or less open, 8-20 cm long, the branches upright or spreading. Spikelets 1-flowered, 2-6 mm long; glumes more or less equal, 2-4 mm long, smooth or finely rough-hairy on back; lemma more or less smooth, awned from middle of back, the awn straight, base with dense callus hairs about as long as lemma. June-Aug. — Wet meadows, shallow marshes, calcareous fens, streambanks, thickets.

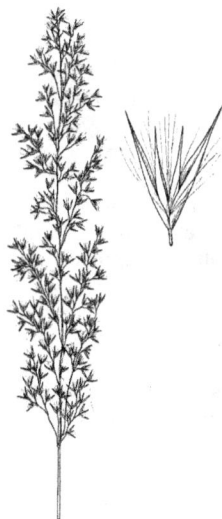

Calamagrostis stricta (Timm) Koel.
SLIM-STEM REED-GRASS   *native / cons* **7**
Calamagrostis inexpansa Gray

Perennial grass, spreading by rhizomes; plants waxy blue-green. Stems erect, 3-12 dm long; leaf blades stiff, often inrolled, 1-4 mm wide when flattened. Panicle narrow, 5-15 cm long, the branches short, upright to erect. Spikelets 1-flowered; glumes 3-6 mm long, smooth or rough-hairy on back; lemma rough-hairy, 2-4 mm long, awned, the awn straight, from near middle of back, base with many callus hairs, half to as long as lemma. June-Sept. — Wet meadows, shallow marshes, shores, streambanks.

Calamagrostis canadensis

Calamovilfa
SAND-REED

Calamovilfa longifolia (Hook.) Scribn.
SAND-REED   *native / cons* **9**
Sporobolus rigidus (Buckl.) P. M. Peterson

Perennial grass, from creeping rhizomes, the rhizomes covered with shiny, scale-like leaves. Stems stout, stiffly erect, to 2 m tall; sheaths much overlapping at base, glabrous except usually more or less villous at the throat; ligule a ring of short hairs 1-2 mm long; leaf blades flat and 3-8 mm wide at base, involute above, tapering to a fine point. Panicle open, 1-4 dm long, with ascending branches. Spikelets 1-flowered, articulated above the glumes; glumes 1-nerved; first glume ovate, 3.5-6 mm long, second glume 4.5-7.5 mm long; lemma glabrous on the back, equaling or slightly shorter than the second glume, awnless, subtended by a conspicuous tuft of hairs from the callus, the hairs about half as long as the lemma. — Dry sandy prairies and dunes. Var. *magna,* primarily of dunes along Lake Michigan, is state threatened, and known from Door County.

Calamovilfa longifolia

Cinna
WOOD-REED

Cinna latifolia (Trev.) Griseb.
DROOPING WOOD-REED   *native / cons* **7**

Perennial grass, with weak rhizomes; Stems single or in small groups, erect, 5-13 dm long, not swollen at base; sheaths smooth to finely roughened; ligules pale, 2-7 mm long; leaf blades 5-15 mm wide, usually rough-to-touch; auricles absent; ligule brown, membranous, with an irregular, jagged margin. Panicle loose, open, pale green, satiny, 1-3.5 dm long, the branches spreading to drooping. Spikelets 1-flowered, laterally compressed, breaking below the glumes; glumes narrowly lance-shaped, 1-

veined, 2-4 mm long; lemma 2-4 mm long, finely rough-hairy on back, usually with an awn to 1.5 mm long from just below the tip, the awn usually longer than the tip. July-Aug. — Wet woods, swamps, springs.

Dactylis
ORCHARD-GRASS

Dactylis glomerata L.
ORCHARD-GRASS   *introduced*
Perennial, densely tufted grass. Stems 5-12 dm tall; sheaths closed for at least 1/2 their length compressed or keeled, scaberulous; auricles absent; ligules membranous, 3-11 mm long, truncate to acuminate; leaf blades flat, elongate, 3-8 mm wide, with a conspicuous midrib and white, scabrous margins. Panicles open, 1-2 dm long, the lower branches naked at base, erect or divergent. Spikelets few-flowered, flat, disarticulating above the glumes and between the lemmas, nearly sessile in dense one-sided clusters, 3-6-flowered; glumes unequal, nearly as long as the lemmas, lance-shaped, 1-3-nerved, keeled, usually ciliate on the keel; lemmas 5-8 mm long, usually ciliate on the keel, awnless or with an awn to 2 mm long. — Introduced from Europe, cultivated for hay or pasture; occasional escape to moist fields, meadows, lawns, and roadsides.

Cinna latifolia

Danthonia
WILD OATGRASS

Danthonia spicata (L.) Beauv.
POVERTY WILD OATGRASS   *native / cons* **4**
Perennial grass. Stems densely tufted, erect, 2-6 dm tall; sheaths glabrous or sparsely pilose; leaves mostly at or near the base, blades usually involute, 1-2 mm wide, seldom more than 10 cm long, glabrous or sparsely pilose; uppermost stem blades erect to ascending; ligule reduced to a tuft of hairs, auricles absent. Panicle contracted, racemiform, 2-5 cm long, the short branches rarely bearing more than 1 spikelet; glumes 8.5-13 mm long; lemmas broadly ovate, sparsely pilose on the back, 3.4-5.2 mm long, including the triangular teeth which are 0.8-1.8 mm long; awn arising between the lemma teeth, 4.5-7 mm long. — Dry woods in sandy or stony soil; occasionally found in marshy or boggy places.

The basal leaves tend to curl and form distinctive tufts.

Deschampsia
HAIRGRASS

Danthonia spicata

Tufted perennial grasses. Leaves usually mainly basal, narrow, flat or involute; sheaths open; auricles absent; ligules membranous. Flowers in panicles; spikelets yellowish or purple, 2-flowered, disarticulating above the glumes and between the lemmas; rachilla hairy, prolonged beyond the base of the upper lemma. Glumes membranous, usually shining, equaling or longer than the lemmas. Lemmas membranous, obtuse or truncate and erose-toothed, rounded on the back, obscurely 5-nerved, the mid-nerve diverging at or below the middle into a short awn, the callus bearded.

1   Leaf blades involute, 1–2 mm wide; ligule 1–2.5 mm long; lemmas minutely scabrous-pubescent, bearing a conspicuously bent awn 1–3 mm longer than the lemmas; palea not bifid at tip . . . .
. . . . ***D. flexuosa***
1   Leaf blades flat or conduplicate, 1–5 mm wide; ligule usually 3–12 mm long; lemmas glabrous, bearing a nearly straight awn shorter than to slightly exceeding the lemmas; palea bifid at the tip . . . .
. . . . ***D. cespitosa***

Deschampsia caespitosa (L.) Beauv.

TUFTED HAIRGRASS   *native / cons* **10**
Stems densely tufted stiff, erect, 3-10 dm long. Leaves mostly from base
of plant, usually shorter than head, 5-30 cm long, usually at least some
flat and 1-4 mm wide, the remainder folded or rolled and to 1 mm wide;
sheaths glabrous; ligules 2-13 mm long, white, translucent. Panicles narrow to open, 1-4 dm long, the panicle branches threadlike, upright to
spreading, the lower branches in groups of 2-5, flowers mostly near branch
tips. Spikelets 2-flowered, purple-tinged, fading to silver with age, 2-5
mm long, breaking above the glumes; glumes shiny, 2-5 mm long, the
first glume slightly shorter than second glume; lemma smooth, 2-4-
toothed across the flat tip, awned from near base on back, the awn shorter
to about as long as lemma. June-July. — Wet meadows, streambanks,
shores, calcium-rich seeps, rocky shores of Lake Michigan.

Deschampsia flexuosa (L.) Trin.
HAIRGRASS   *native / cons* **8**
   Avenella flexuosa (L.) Drej.
Stems densely tufted, 3-10 dm long. Leaves mostly at or near the base;
blades involute, 1-2 mm wide; ligules 1.5-3.5 mm long. Panicle loose and
open, somewhat nodding, to 15 cm long, the lowest branches in fascicles
of 2-5. Spikelets 4.3-6 mm long; first glume 3-4.5 mm long; second glume
acuminate, 3.6-5.3 mm long; lemmas minutely scabrous; awn twisted
below the middle, the distal half somewhat divergent, surpassing the
lemma by 1-3 mm. — Dry woods, fields, and sand hills.

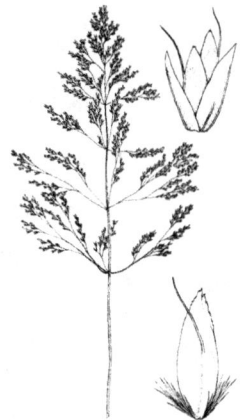

Deschampsia caespitosa

Dichanthelium
PANICGRASS
Perennial grasses, tufted or sometimes rhizomatous, sometimes with hard, corm-like bases.
Stems hollow, usually erect or ascending, sometimes decumbent in the fall, usually branching
from the lower stem nodes in summer and fall, terminating in small panicles that are usually
partly included in the sheaths. Basal rosettes of winter leaves sometimes present. Stem leaves
usually markedly longer and narrower than the rosette blades; ligules of hairs, membranous, or
membranous and ciliate, sometimes absent. Flowers in terminal panicles (vernal) developing
late spring to early summer, and sometimes lateral panicles(autumnal) in late-summer or fall;
disarticulation below the glumes.

   Dichanthelium is often included in genus *Panicum,* the two genera being similar in form. However, molecular data reinforce the separation of *Dichanthelium* as a distinct genus.

1 Lower glumes thinner and more weakly veined than the upper glumes, attached about 0.2 mm below
   the upper glumes, the bases clasping the pedicels; spikelets attenuate basally . . . . . . **D. portoricense**
1 Lower glumes similar in texture and vein prominence to the upper glumes, attached immediately
   below the upper glumes, the bases not clasping the pedicels; spikelets usually not attenuate basally
  2
2 Ligules of hairs; stems arising from caudices; lower florets sterile; stem blades 1–18 mm wide, bases
   usually tapered, rounded, or truncate at the base, sometimes cordate . . . . **D. acuminatum**
2 Ligules with a membranous base, ciliate distally; stems usually arising from slender rhizomes; lower
   florets often staminate; stem blades 5–40 mm wide, often with a cordate base 3
3 Spikelets ellipsoid, not turgid, with pointed apices; stem blades 4–6, cordate at the base; sheaths
   without papillose-based hairs. . . . . **D. latifolium**
3 Spikelets obovoid, turgid, with rounded apices; stem blades 3–4, tapered, rounded or truncate to cordate at the base; sheaths with papillose-based hairs . . . . **D. xanthophysum**

Dichanthelium acuminatum (Sw.) Gould & C. A. Clark
HAIRY PANICGRASS   *native / cons* **5**
   Panicum acuminatum Sw.
Stems densely tufted, erect to prostrate, usually straight and radiating
from the base, glabrous to pilose or villous; sheaths softly pubescent, pa-
pillose-pilose, or glabrous; ligule hairs 2-5 mm long; leaf blades 4-12 cm

long, 5-12 mm wide, glabrous or pubescent on either or both sides. Primary panicle ovoid, with divergent, often flexuous branches, the axis glabrate to villous. Spikelets plumpellipsoid to obovoid, 1-2 mm long, finely pubescent; first glume broadly angular-rotund, usually less than a third as long than the second glume. Autumnal phase spreading or prostrate, copiously branched chiefly from the middle nodes; blades about half as large as the vernal ones; panicles few-flowered, mostly surpassed by the leaves. — Moist or dry situations, open woods, dunes, shores, and prairies.

Dichanthelium latifolium (L.) Gould & C. A. Clark
BROADLEAVED PANICGRASS   native / cons 7
Panicum latifolium L.

Stems tufted, slender, erect, 4-10 dm long, usually glabrous, rarely sparsely puberulent; sheaths pubescent; ligules to 0.7 mm long, membranous, ciliate, the cilia longer than the membranous portion; leaf blades lance-shaped, spreading, glabrous. or nearly so on both sides, ciliate at the cordate base, the larger usually 10-16 cm long, 15-30 mm wide. Primary panicle tardily exsert, ovoid with ascending branches, 6-12 cm long. Spikelets oblong-obovoid, 2.9-3.7 mm long, averaging 3.3 mm, softly villosulous; first glume about half as long, acute; second glume and sterile lemma shorter than the fruit. Autumnal phase sparsely branched from the middle nodes, the leaf blades not much reduced or greatly crowded; panicles small, included at base. — Moist or dry woods and thickets.

Dichanthelium acuminatum

Dichanthelium portoricense (Desv. ex Ham.) Hansen & Wunderlin
BLUNT-GLUMED PANICGRASS   native / cons 7
Panicum columbianum Scribn.

Stems densely tufted, erect or ascending, 2-5 dm tall, often purplish, densely short-pubescent with minute hairs 0.1-0.4 mm long, or toward the tips of the lower internodes sometimes 1 mm long; sheaths similarly pubescent; ligule hairs 0.5-1.5 mm long; leaf blades 3-7 cm long and 3-7 mm wide, glabrous above or with a few widely scattered hairs, minutely puberulent beneath. Primary panicle ovoid, 2-6 cm long, its axis puberulent. Spikelets obovoid, obtuse, 1.4-1.9 mm long, finely pubescent; first glume averaging 2/5 as long, triangular-ovate. Autumnal phase spreading or decumbent, branched early from most of the nodes; blades scarcely reduced; panicles smaller, surpassed by the leaves. — Moist or dry, especially sandy soil.

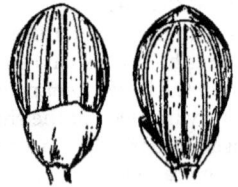
Dichanthelium portoricense

Dichanthelium xanthophysum (Gray) Freckmann
PALE PANICGRASS   native / cons 5
Panicum xanthophysum A. Gray

Stems few or several in loose tufts, erect or ascending, 2-5 dm long, glabrous; sheaths loose, often exceeding the internodes, glabrous to pilose or papillose-pilose; ligules to 0.5 mm long, membranous, ciliate, the cilia longer than the membranous bases; leaf blades yellowish green, erect or nearly so, the larger 10-15 cm long and 10-20 mm wide, glabrous on both sides, slightly narrowed to the rounded, papillose-ciliate base. Primary panicle 5-10 cm long, very narrow, with erect branches. Spikelets obovoid, 3.3-3.8 mm long, minutely puberulent; first glume about half as long, triangular-ovate; second glume slightly shorter than the sterile lemma and fruit. Autumnal phase with 1 or 2 erect branches, bearing scarcely reduced leaf blades equaling or exceeding the shorter panicles. — Dry sandy soil of open woodlands.

Dichanthelium xanthophysum

Digitaria
CRABGRASS

Annual or perennial grasses. Stems spreading, branched from the base. Leaves wide, flat, prostrate, with the tips ascending; sheaths open, ligules membranous. Flowers in several terminal, digitate, spike-like racemes. Spikelets 1-flowered, single or in clusters of 2 or 3 on unequal pedicels on one side of an elongate rachis. First glume minute or lacking; second glume a third to fully as long as the spikelet, conspicuously 5-7-nerved. Fertile lemma cartilaginous with hyaline margins, acute, often shining.

1 Inflorescence an open panicle, neither spike-like nor distinctly one-sided . . . . *D. cognata*
1 Inflorescence composed of 1-sided spikes or spike-like racemes, the rachis of each winged or at least flat on the side opposite the spikelets 2
2 Spikelets ca. 2–2.3 mm long, the fertile lemma dark brown; second glume nearly or fully as long as the floret; sheaths and blades usually nearly or quite glabrous (except around summit of sheath) . . . . *D. ischaemum*
2 Spikelets ca. 2.5–3 mm long, the fertile lemma light or dark grayish; second glume only about half as long as the floret; sheaths and usually blades pilose, at least toward base of plant . . . . *D. sanguinalis*

Digitaria cognata (J.A. Schultes) Pilger
WITCH-GRASS   *native / cons* 2
Leptoloma cognatum (J.A. Schultes) Chase
Perennial grass. Stems tufted, 4-7 dm long; lower sheaths villous to papillose-hirsute; ligules 0.2-1.5 mm long, entire to lacerate; leaf blades narrow, 5-8 cm long. Panicle often a third to half the height of the plant, diffusely branched, purplish, villous in the axils; pedicels 3-8 cm long, 3-angled, scabrous. Spikelets acute, 2.5-3 mm long, nearly glabrous on the second glume and sterile lemma. — Dry, especially sandy soil.

Digitaria ischaemum (Schreb.) Schreb.
SMOOTH CRABGRASS   *introduced*
Annual grass (or sometimes longer-lived). Stems branched and spreading from the decumbent base, rooting at the nodes, 2-5 dm long; sheaths glabrous or sparsely pubescent; ligules 0.6-2.5 mm long; leaf blades to 9 cm long and 3-5 mm wide, glabrous. Racemes 2-5, or rarely to 8, 4-10 cm long; rachis broadly winged, about 1 mm wide. Spikelets elliptic or somewhat obovate, 1.7-2.1 mm long, often purple; first glume lacking or minute and hyaline; second glume and sterile lemma equal and about as long as the spikelet, both more or less pubescent or subtomentose with capitellate hairs, especially in stripes between the nerves; fertile lemma purple-black. — Eurasian weed of lawns, waste places.

Digitaria ischaemum

Digitaria sanguinalis (L.) Scop.
HAIRY CRABGRASS   *introduced*
Annual grass. Stems decumbent or prostrate, much branched, rooting at the nodes, usually 3-6 dm long. leaf blades 4-10 cm long, 5-10 mm wide, pilose. Racemes 3-6 in each of 1-3 whorls, 5-15 cm long; rachis broadly winged, 1 mm wide, scabrous on the margins; pedicels triquetrous, scabrous. Spikelets 2.4-3 mm long; first glume minute, often deciduous, the second half as long as the spikelet; sterile lemma usually scabrous on the 5 strong nerves; fertile lemma greenish brown. — Eurasian; often a troublesome weed of fields, gardens, lawns, and waste ground.

Echinochloa
BARNYARD-GRASS

Echinochloa muricata (Beauv.) Fern.
BARNYARD-GRASS   *native / cons* 1
Weedy annual grass. Stems 1 m or more long; sheaths glabrous; ligules absent; leaf blades 5-30 mm wide. Panicle green to purple, sometimes

Digitaria sanguinalis

strongly purple, 1-3 dm long, panicle branches spreading, hairs on branches absent or to 3 mm long and shorter than spikelets. Spikelets 2-4 mm long (excluding awns); glumes awnless; sterile lemma awnless or with an awn 5-10 mm long; tip of fertile lemma firm, shiny, gradually tapered to the stiff beak, the lemma body and beak not separated by a line of tiny hairs (the beak itself often short-hairy). July-Sept. — Shores, streambanks and ditches, where sometimes in shallow water.

Elymus
Echinochloa muricata

WILD RYE

Tufted perennial grasses. Leaves flat, sheaths open for most of their length, auricles often present, ligules short. Head a densely flowered spike. Spikelets usually 2 at each node of spike, breaking above or below glumes; glumes narrow and awnlike; lemmas tipped with a long awn; stamens 3.

1 Spikelets clearly 1 at each node (or most nodes) of the rachis; glumes not more than 2. 2
1 Spikelets mostly 2-3 at each node of the rachis 4
2 Lemmas densely hairy; leaves with narrow (rarely as much as 4.5 mm wide) often involute blades, the whole plant usually strongly glaucous. . . . . ***E. lanceolatus***
2 Lemmas glabrous (rarely slightly pubescent), smooth or scabrous; leaves various. 3
3 Stems tufted, rhizomes absent; anthers 1-2.2 mm long; rachilla readily disarticulating between the florets when mature (on dry specimens, the florets very easily dislodged and empty glumes often remaining on older plants) . . . . ***E. trachycaulus***
3 Stems from elongate rhizomes; anthers 3-6 mm long; rachilla often not readily disarticulating (florets not easily dislodged on dry specimens except over-ripe ones, empty glumes seldom if ever present) . . . . ***E. repens***
4 Glumes absent or vestigial, or, if present, slenderly awn-like their entire length and at least one much shorter than the others at a node; spikelets horizontally spreading at maturity (ascending when young), well separated, clearly revealing the entire rachis . . . . ***E. hystrix***
4 Glumes present, awn-like to lanceolate, of about equal length; spikelets ascending at maturity, usually concealing much of the rachis. 5
5 Larger paleas (lowest in each spikelet) 8.6-13 mm long; awns of lemmas usually widely spreading at maturity 6
5 Larger paleas 5.5-8.5 mm long; awns of lemmas mostly straight 7
6 Body of glume about twice as long as its awn, or longer; awns of lemmas usually straight at maturity; spike curved to erect. . . . . (go to couplet 7)
6 Body of glume about equaling its awn, or shorter; awns of lemmas ± curved at maturity (straight when young); spike curved to strongly nodding . . . . ***E. canadensis***
7 Glumes less than 1 mm wide, scarcely if at all widened above the base. . . . . ***E. villosus***
7 Glumes, at least the broadest, 1-2 mm wide, clearly expanded and flattened above the base. 8
8 Base of glumes not conspicuously bowed out, but flattened, hardened for less than 1 mm; glumes not thickened above the base on inner face, with very narrow, thin, translucent margins, often slightly overlapping; stem leaves 5-6 . . . . ***E. glaucus***
8 Base of glumes bowed out, terete and hard for 1 mm or more; glumes also thickened, pale, and hardened on inner face for about the basal half or more, with firm margins, not at all overlapping; stem leaves (6-) 7-10 . . . . ***E. virginicus***

Elymus canadensis L.

NODDING WILD RYE   *native / cons 4*

Stems loosely tufted, stout, 1 m or more tall; sheaths often reddish brown; auricles 1.5-4 mm long, brown or purplish black; ligules to 2 mm long, truncate, ciliolate; leaf blades flat, or involute when dry, usually 8-20 mm wide, glabrous to sparsely pilose. Spike 10-15 cm long, usually nodding, often interrupted at base by the elongation of the lower internodes, 1-2 cm thick, excluding the awns. Spikelets 3-7-flowered; glumes, including the awns, usually 15-30 mm long, 3-5-nerved, glabrous, scabrous, or pubescent; lemmas, including the outwardly curved awns, 3-5 cm long,

Elymus canadensis

glabrous, scabrous, or pubescent. — Dry or moist soil, often where sandy or gravelly, usually in full sun. Variable.

Elymus hystrix L.
BOTTLEBRUSH-GRASS   *native / cons* 6
Hystrix patula Moench.

Plants occasionally glaucous, particularly the spikes; stems usually solitary or loosely tufted, not rhizomatous, 6-10 dm tall; sheaths usually glabrous, sometimes pilose, often purplish; auricles usually present, 0.5-3 mm long, brown to black; ligules 1-3 mm long; leaf blades 8-13 mm wide. Spikes 5-12 cm long, the internodes of the flexuous 2-edged rachis 4-10 mm long. Spikelets usually in pairs; glumes varying, even on the same plant, from none to setaceous and to 16 mm long; lemmas 8-11 mm long, tipped with a rough awn 1-4 cm long. — Moist deciduous woods, especially in wet or slightly disturbed areas.

Spikelets soon horizontally divergent, the lemmas easily detached.

Elymus hystrix

Elymus lanceolatus (Scribn. & J.G. Sm.) Gould
STREAMSIDE WILD RYE   *native / cons* –
Agropyron dasystachyum (Hook.) Scribn. & J.G. Sm.

Plants strongly rhizomatous, often glaucous; stems slender, 5-8 dm tall from long rhizomes; leaves often mostly basal; sheaths glabrous or pubescent; auricles usually present on the lower leaves; ligules to 0.5 mm long, erose, sometimes ciliolate; leaf blades involute when dry, 1-3 mm wide. Spikes 7-15 cm long. Spikelets 4-8-flowered, 12-20 mm long; glumes narrowly lance-shaped, villous to glabrous, 3-5-nerved, the first 7-10 mm long, the second somewhat longer; lemmas 9-12 mm long, usually awnless; in the typical variety, moderately hairy, the hairs stiff, shorter than 1 mm. — Subsp. *psammophilus* (Scribn. & J. G. Sm.) Gould, with lemmas densely hairy, the hairs to 1 mm long or longer, is state threatened; it occurs on sandy beaches and dunes along Lake Michigan.

Elymus repens (L.) Gould
QUACK-GRASS   *introduced (invasive)*
Agropyron repens (L.) Beauv.
Elytrigia repens (L.) Nevski

Plants strongly rhizomatous, sometimes glaucous. Stems erect, usually 5-10 dm tall; sheaths pilose or glabrous near base; auricles to 1 mm long; ligules to 1.5 mm long; leaf blades flat, soft, 5-10 mm wide, with numerous slender nerves about 0.2 mm apart. Spikes 6-17 cm long, with numerous ascending, overlapping spikelets; rachis joints usually flat on one side, rounded on the other. Spikelets 10-18 mm long, 4-8-flowered; glumes narrowly oblong to lance-shaped, 8-14 mm long, sharply nerved, acuminate or short-awned; lemmas similar in size and shape, less sharply nerved, acuminate or with an awn to 10 mm long. — Eurasian native, abundant and often a noxious weed in meadows, fields, roadsides, and waste places.

Highly variable in color from green to glaucous, in pubescence, and in presence and length of awns.

Elymus repens

Elymus trachycaulus (Link) Gould ex Shinners
SLENDER WILD RYE   *native / cons* 4
Agropyron caninum (L.) Beauv.
Agropyron trachycaulum (Link) Steud.

Stems loosely tufted, erect, 4-10 dm tall; sheaths usually glabrous, sometimes hirsute or villous; auricles absent or to 1 mm long; ligules 0.2-0.8 mm long, truncate; leaf blades 4-10 mm wide, flat to involute, usually straight and ascending, with numerous fine sharp nerves. Spikes 6-20 cm long. Spikelets erect or ascending, few-flowered, in ours mostly not

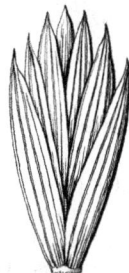

Elymus trachycaulus (spikelet)

imbricate, the tip of one not reaching to the base of the next one above on the same side; glumes 5-7-nerved, acuminate or short-awned; lemmas awnless or with straight awns to 2 cm long (rarely more); rachilla readily disarticulating between the lemmas, leaving the persistent glumes attached. — Dry, open, rocky woods, sandy shores and barrens; rarely in fens and tamarack swamps.

Variable but distinguished by the short anthers (when young), and by the readily disintegrating spikelets (when mature); the rachilla is also nearly always villous.

Elymus villosus Muhl.
HAIRY WILD RYE *native / cons* 6

Plants often persistently deep green. Stems tufted, slender, 5-10 dm tall; sheaths glabrous to pilose; auricles 1-3 mm long, brownish; ligules less than 1 mm long, entire or erose; leaf blades 4-12 mm wide, lax, dark glossy green, softly villous on the upper side. Spikes slightly or strongly nodding, dense, 5-12 cm long; glumes setaceous, not widened above the base, 0.4-1 mm wide, strongly 1-3-nerved, 15-30 mm long, including the awn; lemmas 2-4 cm long, including the straight ascending awn. — Swampy forests and riverbanks; also in drier woods.

Glumes and lemmas usually conspicuously hirsute.

Elymus virginicus L.
VIRGINIA WILD RYE *native / cons* 6

Plants sometimes glaucous, especially in the spikes. Stems tufted, 6-12 dm long; sheaths usually glabrous, rarely hirsute, occasionally reddish or purplish; auricles absent or to 1.8 mm long, pale brown; ligules less than 1 mm long; leaf blades flat, lax, 5-15 mm wide, rough-to-touch on both sides. Spikes erect, 5-15 cm long, the base of spike often covered by top of upper sheath. Spikelets usually 2 at each node, 2-4-flowered, breaking below glumes; glumes firm, 1-2 mm wide, yellowish, bowed-out at base, tapered to a straight awn about 1 cm long; lemmas 6-9 mm long, smooth to hairy, usually with a straight awn to 3 cm long. July-Aug. — Floodplain forests, thickets, streambanks.

Elymus virginicus

Eragrostis
LOVEGRASS

Eragrostis cilianensis (All.) Vign.
STINK-GRASS *introduced*

Annual grass; stems densely tufted, spreading or ascending from a decumbent base, rarely erect, 1-4 dm long; sheaths open, glabrous, occasionally glandular, hairy at tip, the hairs to 5 mm long; ligules to 0.8 mm long, ciliate; leaf blades 5-20 cm long, 2-6 mm wide. Panicle ovoid to subcylindric, 5-15 cm long, the branches spreading, the pedicels usually 1-2 mm long. Spikelets broadly linear, 2.5-3 mm wide, 10-40-flowered; first glume 1.3-1.9 mm long; second glume 1.5-2 mm long; lemmas broadly elliptic-ovate, closely imbricate, 2.1-2.6 mm long, glandular on the keel; grain 0.7 mm long, dull brown. — Native of Europe; a weed of moist ground.

Eragrostis cilianensis

Festuca
FESCUE

Annual or perennial grasses, often densely tufted. Leaves flat to involute, auricles absent; ligules membranous, usually truncate, usually ciliate. Flowers in open or contracted panicles. Spikelets 3-11-flowered, the rachilla disarticulating above the glumes and between the lemmas. Glumes narrow, unequal, 1-3-nerved, usually shorter than the lemmas. Lemmas rounded on back, obscurely 5-nerved, usually awned from the apex. Paleas about equaling the lemmas. Stamens 1 or 3.

1 Blades of leaves flat (or merely once-folded), at least the larger ones 3–8 mm wide; lemmas awnless or rarely with awn less than 1 mm long. . . . . *F. subverticillata*
1 Blades of leaves strongly involute, usually much less than 3 mm wide; lemmas awned or awnless. . 2
2 Margins of lemmas thin and membranous; tip of ovary bristly-pubescent; awns mostly more than 3 mm long, nearly equaling or longer than the bodies of their lemmas; mature panicle open and lax . . . . *F. occidentalis*
2 Margins of lemmas at most very narrowly membranous-bordered, the lemmas firm and thick through-out; summit of ovary glabrous; awns all less than 3 mm long, shorter than the bodies of their lemmas; mature panicle rather narrow, crowded, and compact, the branches strongly ascending or, if spreading, very short 3
3 Sheaths closed in young leaves, the old ones dark reddish brown basally, becoming fibrous by splitting between the prominent pale veins; basal shoots usually arising laterally, the stems thus tending to be strongly curved or bent at the base; anthers mostly 2–3.5 mm long . . . . *F. rubra*
3 Sheaths open most of their length even in young leaves (margins ± overlapping), the old ones mostly pale or drab brown, not becoming fibrous; basal shoots erect, the stems thus nearly or quite straight from the base upwards; anthers various. 4
4 Lower panicle branches often spreading; anthers 2–3 mm long . . . . *F. trachyphylla*
4 Lower panicle branches strongly ascending; anthers less than 2 mm long . . . . *F. saximontana*

Festuca occidentalis Hook.
WESTERN FESCUE   *threatened / native / cons* 10
Stems slender, tufted, 4-8 dm tall, glabrous, shining; sheaths closed for much less than 1/2 their length, glabrous, somewhat persistent or slowly shredding into fibers; collars glabrous; ligules to 0.4 mm long, usually longer at the sides; leaf blades 0.3-0.7 mm wide, conduplicate, upper surface smooth or finely roughened. Panicle narrow, flexuous, more or less secund, 5-20 cm long. Spikelets on slender pedicels, 6-10 mm long, 3-5-flowered; first glume 2.7-3.6 mm long; second glume 3.54.5 mm long; lemmas green or suffused with purple, soft and membranous, the body 4.5-6.5 mm long, the awn two-thirds to fully as long. — Cobble beaches and stabilized dunes along Lake Michigan, dry woods.

Festuca occidentalis

Festuca rubra L.
RED FESCUE   *introduced*
Stems glabrous, 3-10 dm tall, usually loosely tufted, often decumbent at base, frequently rhizomatous; sheaths closed for about 3/4 their length when young, soon disintegrating into loose fibers, usually pubescent, reddish; collars glabrous; ligules 0.1-0.5 mm long; leaf blades usually conduplicate, to 2.5 mm wide, sometimes flat and 1.5-7 mm wide. Panicle 5-20 cm long, narrow with ascending branches, or in some forms loosely spreading. Spikelets 4-7-flowered; first glume subulate, 2.6-4.5 mm long; second glume broader, 3.5-5.5 mm long; lemmas 4.8-6.1 mm long; longest awns 1-3 mm long. Variable, and many horticultural varieties exist. — Widely distributed in n Europe and North America; Wisconsin plants considered adventive.

Festuca rubra

Festuca saximontana Rydb.
ROCKY MOUNTAIN FESCUE   *native / cons* 7
   Festuca brachyphylla J.A. Schultes
Stems very slender, densely tufted, not stoloniferous, glabrous; sheaths closed for about 1/2 their length, usually persistent, rarely slowly shred-ding into fibers, mostly pale or drab brown; collars glabrous; ligules to 0.5 mm long; leaf blades 0.5-1.2 mm wide, conduplicate, upper surface glabrous or sparsely puberulent, undersurface scabrous or puberulent. Panicle 1-10 cm, narrow and spiciform, or somewhat open at anthesis, the first pedical of the lowermost branches usually no more than 5 mm,

from the base. Spikelets 2-4-flowered, the first glume 2-3 mm, 1-nerved, the second 2.5-4.5 mm, 3-nerved; lemmas mostly 3.5-6) mm, with a short awn 1-3 mm. — Dry forests, shores, dunes, and disturbed places. Similar to *F. trachyphylla* in general appearance.

Festuca subverticillata (Pers.) Alexeev
NODDING FESCUE   *native / cons* **4**
Stems few in a tuft, 6-12 dm tall, smooth; sheaths closed for less than 1/3 their length, glabrous or sparsely pilose, shredding into fibers; ligules mostly 0.5-1 mm long; leaf blades 4-10 mm wide, flat or loosely convolute, glabrous or sparsely pilose. Panicle long-exsert, 15-30 cm long; branches slender, elongate, racemiform, eventually widely spreading, bearing spikelets only above the middle. Spikelets relatively remote, the tip of one barely reaching the base of the next, usually 3-flowered, occasionally 4-5-flowered, 4-6 mm long; first glume subulate, averaging 2.8 mm long; second glume ovate, averaging 3.4 mm long; lemmas acute, averaging 3.7 mm long, appressed till maturity. — Moist forests of beech-maple or oak-hickory; occasionally in wet conifer woods. Resembles *F. paradoxa*, but its spikelets are less crowded on the branches.

Festuca subverticillata

Festuca trachyphylla (Hack.) Krajina
HARD FESCUE, SHEEP FESCUE   *introduced*
Festuca ovina auct. p.p. non L.
Stems densely tufted, without rhizomes, 20-80 cm tall, glabrous or with sparse hairs; sheaths closed for less than 1/3 their length, usually glabrous, rarely pubescent, persistent; collars glabrous; ligules to 0.5 mm long; leaf blades about 1 mm in diameter, usually conduplicate, rarely flat. Panicle 3-15 cm long, contracted, with 1-2 branches per node; branches erect or stiffly spreading, lower branches with 2 or more spikelets. Spikelets 5-9 mm long, with 3-7 florets; glumes exceeded by the upper florets, mostly glabrous; lower glumes 2-4 mm long; upper glumes 3-5 mm long; lemmas lance-shaped, usually smooth on the lower portion and scabrous or pubescent upwards, especially on the margins, awns 0.5-2.5 mm long, usually less than 1/2 as long as the lemma body. — Native of Europe, widely introduced as a turf grass and sometimes weedy.

Festuca trachyphylla

Glyceria
MANNA GRASS
Perennial grasses, loosely clumped or spreading by rhizomes. Stems upright, or reclining at base and often rooting at lower nodes. Leaves flat or folded; sheaths closed for most of their length; ligules scarious, erose to lacerate. Head an open panicle. Spikelets 3-flowered, ovate to linear, round in section or somewhat flattened, breaking above the glumes; glumes unequal, shorter than lemmas, 1-veined; lemmas unawned, usually 7-veined; palea about as long as lemma; stamens 3 or 2.

ADDITIONAL SPECIES

Glyceria maxima (Hartman) Holmb. (Reed sweetgrass), native to Eurasia and an aggressive invader of wetlands, was first discovered in the USA in Racine County in 1975. In Wisconsin, the species is now known from several additional locations, including Door County. Spreading by rhizomes, the grass may form large patches and grow to 2.5 m tall. It is similar to large plants of *G. grandis*, but differs in its firmer, more prow-tipped lemmas as well as its larger lemmas and usually larger anthers.

1 Spikelets linear-cylindric, 10 mm long or longer . . . . **G. borealis**
1 Spikelets ovate, 2–7 mm long 2
2 Spikelets 4–7 mm long . . . . **G. grandis**
2 Spikelets 2–4 mm long . . . . **G. striata**

Glyceria borealis (Nash) Batchelder
NORTHERN MANNA GRASS   *native / cons* **8**
Stems erect or reclining at base, often rooting from lower nodes, 6-12 dm
long. Leaves flat or folded, 2-5 mm wide, smooth; sheaths smooth; ligule
3-10 mm long. Panicle 2-4 dm long, with stiff, erect to ascending, branches
to 8-12 cm long, each with several spikelets. Spikelets linear, mostly 6-
12-flowered, 1-1.5 cm long; glumes rounded at tip, 2-3 mm long; lemmas
3-4 mm long, 7-veined. June-Aug. Marshes, ponds, stream, ditches, often
in shallow water or mud.   *Glyceria borealis*

Glyceria grandis S. Wats.
AMERICAN MANNA GRASS   *native / cons* **6**
Stems loosely tufted, erect, stout, 1-1.5 m long and 4-6 mm wide. Leaves
flat, smooth, 6-12 mm wide; sheaths smooth; ligules translucent, 3-6 mm
long. Panicle large, open, much-branched, 2-4 dm long, usually nodding
at tip, branches lax and drooping when mature. Spikelets ovate, purple,
slightly flattened, 5-9-flowered, 4-7 mm long; glumes pale or white, 1-3
mm long; lemmas purple, 2-3 mm long. June-Sept. — Marshes, ditches,
streams, lakes and ponds, open bogs, fens; usually in shallow water or
mud.

Glyceria striata (Lam.) A.S. Hitchc.
FOWL MANNA GRASS   *native / cons* **4**
Plants pale green. Stems loosely tufted erect, slender, 3-10 dm long. Leaves
flat or folded, smooth, 2-6 mm wide; sheaths smooth; ligules 1-3 mm
long. Panicle open, loose, 1-2 dm long, the branches lax, drooping. Spikelets
ovate, often purple, 3-7-flowered, 3-4 mm long; glumes 0.5-1.5 mm long;
lemma 2 mm long, strongly 7-veined. June-Aug. Swamps, thickets, low
areas in forests, wet meadows, springs, streambanks.

Graphephorum
FALSE OAT

Graphephorum melicoides (Michx.) Desv.
PURPLE FALSE OAT   *endangered/ native / cons* **10**
   Trisetum melicoides (Michx.) Vasey
Tufted perennial grass. Stems smooth or finely hairy, 4-9 dm long. Leaves
flat, 3-8 mm wide, sparsely long-hairy; sheaths glabrous or hairy; ligules
membranous, 1.5-3.5 mm long, rounded or truncateragged at tip. Head a
slender, nodding panicle, 10-20 cm long, the branches upright to drooping,
to 6 cm long, the spikelets mostly above middle of branch. Spikelets 2-
flowered, 6-7 mm long, finely hairy; glumes somewhat unequal, 4-7 mm
long, the first glume 1-veined, the second glume 3-veined; lemma unawned; stalk within spikelet (rachilla) and base of lemma white-hairy.
— Mixed forests and ridge and swale ecosystems near Lake Michigan,
shoreline dolomitic sites. The single Door County collection dates from
1938.

Glyceria grandis

Hesperostipa
NEEDLEGRASS

Hesperostipa spartea (Trin.) Barkworth
PORCUPINE GRASS   *native / cons* **9**
   Stipa spartea Trin.
Perennial grass; stems in small tufts, 6-12 dm tall; sheaths glabrous;
ligules membranous, of the upper leaves 4-6 mm long, the lower much
shorter; leaf blades 2-5 mm wide, glabrous beneath, scabrous and usually
also pubescent above, the lower elongate, tapering to a fine point. Panicle

Graphephorum melicoides

narrow, more or less nodding, 1-2 dm long, the few branches each bearing 1-few spikelets; spikelets 1-flowered, articulated above the glumes; glumes about equal, 28-42 mm long, tapering to a very slender point; mature lemma 18-21 mm long, brown, pubescent at base, less so above; awn stout, stiff, 12-20 cm long, twice geniculate near the middle, the central segment usually 1.5-3 cm long. — Sandy, often calcareous places; dune ridges, oak savanna, dry prairies.

The awn is hygroscopic, imparting a twisting motion to the fruit as it winds or unwinds, pushing the sharp basal callus into the soil, and serving to bury the grain.

Hesperostipa spartea

Hordeum
BARLEY

Hordeum jubatum L.
FOXTAIL-BARLEY   *native / cons –*

Tufted perennial grass; plants smooth to densely hairy. Stems erect or reclining at base, 2-7 dm long; sheaths glabrous or pubescent; auricles absent; ligules scarious, less than 1 mm long; leaf blades usually flat, 2-5 mm wide. Head a terminal spike, erect to nodding, 3-10 cm long, appearing bristly due to the long, spreading awns from glumes and lemmas, disarticulating at each joint. Spikelets 1-flowered, 3 at each node, the center spikelet fertile, stalkless, the 2 lateral spikelets sterile, short-stalked, reduced to 1-3 spreading awns; the 3 spikelets at each node falling as a unit; glumes of fertile spikelet awnlike; lemma lance-shaped, tipped by a long awn; the glume and lemma awns 2-7 cm long. June-Sept. — Wet meadows, ditches, shores, shallow marshes, disturbed areas; often where brackish.

Hordeum jubatum

Koeleria
JUNEGRASS

Koeleria macrantha (Ledeb.) J. A. Schultes
PRAIRIE JUNEGRASS   *native / cons 7*
Koeleria pyramidata (Lam.) Beauv.

Perennial tufted grass. Stems tufted, erect, 3-6 dm tall, pubescent below the panicle. Leaves mostly basal; sheaths glabrous or pubescent; ligules membranous, 0.5-2 mm long, erose; blades 1-3 mm wide, flat, or involute when dry, glabrous or pubescent. Panicle spike-like, shining, silvery-green, 5-12 cm long, 1-2 cm thick. Spikelets normally 2-flowered, disarticulating above the glumes and between the lemmas, subsessile, overlapping, more or less scabrous; glumes unequal, obscurely keeled, scariously margined, the first 1-nerved, the second broadest above the middle, 3-5-nerved ; lemma about as long as the glumes, rounded on the back, acute, scarious at margin and tip, obscurely 5-nerved awnless; palea hyaline, nearly as long as the lemma. — Dry soil, prairies, sand hills, open woods. Variable, especially in pubescence.

Koeleria macrantha

Leersia
CUT-GRASS

Leersia oryzoides (L.) Sw.
RICE CUT-GRASS   *native / cons 3*

Loosely tufted perennial grass, from long-creeping rhizomes. Stems weak and sprawling, rooting at nodes, 1-1.5 m long. Leaves flat, 2-3 dm long and 5-10 mm wide, rough-to-touch, the margins fringed with short spines; sheaths open, rough-hairy; ligules membranous, 0.5-1 mm long, flat-topped; auricles absent. Panicle open at end of stem and from leaf axils (these often partly enclosed by leaf sheaths), 1-2 dm long, the branches ascending to spreading. Spikelets 1-flowered, oval, 5 mm long and 1-2 mm wide, compressed, pale green, turning brown with age; glumes absent; lemma covered with bristly hairs. July-Sept. — Muddy or sandy streambanks, shores, swales and marshes; sometimes forming large patches.

Leersia oryzoides

Leymus
LYME GRASS

Leymus arenarius (L.) Hochst.
EUROPEAN LYME GRASS   *introduced*
  Elymus arenarius L.
Perennial grass, weakly tufted, from a stout rhizome, strongly glaucous. Stems stout, erect, 6-12 dm tall, glabrous; sheaths open, crowded and overlapping at base; auricles usually present; ligules membranous, 0.3-2.5 mm long ; leaf blades 8-18 mm wide, minutely scaberulous above. Spike stout, dense, 1-2 dm long, 1-2 cm thick. Spikelets with 2-5 florets, scabrous to villous or almost tomentose; glumes lance-shaped, 17-35 mm long, 3-4 mm wide, strongly 3-5-nerved, overlapping at base; lemmas nearly as long, acuminate, awnless. — Native of Europe, locally introduced on Lake Michigan sandy shores.

Leymus arenarius

Lolium
RYE GRASS
Lolium now includes several taxa previously placed in *Festuca* and *Schedonorus:* the large, broad- and flat-leaved species with awned (or at least sharply pointed) lemmas.

1 Inflorescence an unbranched spike; glumes 1 (except for 2 glumes on the terminal spikelet) . . . .
   . . . . *L. perenne*
1 Inflorescence branched; glumes 2 in all spikelets 2
2 Auricles at top of leaf sheath ciliate, having at least 1 or 2 hairs along the margins; panicle branches at the lowest node usually paired, the shorter with 1–13 spikelets, the longer with 3–19 spikelets; lemmas 5.5–7 mm long, usually scabrous at least distally, unawned or with an awn up to 4 mm long
   . . . . *L. arundinaceum*
2 Auricles glabrous; panicle branches at the lowest node 1 or 2, if paired the shorter with 1–2 (3) spikelets, the longer with 2–6 (9) spikelets; lemmas 7–8.5 mm long, usually smooth, sometimes slightly scabrous distally, unawned or with a mucro to 0.2 mm long . . . . *L. pratense*

Lolium arundinaceum (Schreb.) S.J. Darby-shire
TALL RYE GRASS   *introduced*
  Festuca arundinacea Schreb.
  Schedonorus arundinaceus (Schreb.) Dumort.
Tufted perennial grass, sometimes with rhizomes. Stems erect above a geniculate base, to 1.5 m tall, glabrous; sheaths smooth; auricles ciliate, having at least 1 or 2 hairs along the margins; ligules ca. 1 mm long; leaf blades glabrous or scaberulous, 4-8 mm wide, dilated at base into conspicuous auricles. Panicle erect or nodding at the tip, 1-2 dm long, contracted at least after flowering, the internodes of the branches less than 2x as long as the spikelets. Spikelets 4-11-flowered, usually 7-8-flowered; first glume subulate, 2.5-4.5 mm long; second glume lance-shaped, sharply nerved, 3.5-7 mm long; lemmas 5.5-8 mm long, scarious at the acute tip, occasionally with a short awn. — Native of Europe; cultivated for forage and as a turfgrass; established in fields and meadows.

Lolium arundinaceum

Lolium perenne L.
ENGLISH RYEGRASS   *introduced*
Perennial tufted grass. Stems slender, 3-7 dm tall, glabrous throughout;

sheaths open; ligules membranous, to 4 mm long; leaf blades flat, glossy, 2-4 mm wide. Spike slender, 1-2 dm long, smooth on the back of each joint opposite the spikelet, minutely scabrous on the sharp margin. Spikelets solitary at each node, usually 5-10-flowered, placed edgewise to the rachis, the edge fitting into a concavity in the axis, disarticulating above the glumes and between the lemmas; first glume absent except in the terminal spikelet; second glume (on the side of the spikelet away from the rachis) strongly 3-5-nerved; lemmas awnless, the lowest 5.5-8 mm long, the upper progressively reduced. — Native of Europe; cultivated in meadows and lawns and often included in commercial seed mixes; and escaped and established on roadsides and in waste places.

Lolium pratense (Huds.) S.J. Darbyshire
MEADOW RYE GRASS   *introduced*
   Festuca elatior L. p.p.
   Festuca pratensis Huds.
   Schedonorus pratensis (Huds.) Beauv.
Tufted perennial grass. Stems often basally decumbent, to 1.3 m tall, glabrous; old sheaths brown, decaying to fibers; ligules glabrous to 0.5 mm long; leaf blades lax, 3-5 mm wide, dilated at base into conspicuous auricles. Panicle 1-2.5 dm, erect or nodding at the tip, contracted at least after anthesis, the internodes of the branches less than 2x as long as the spikelets. Spikelets 10-15 mm long, 4-10-flowered; first glume subulate, 2.5-4 mm, 1-veined, the second lance-shaped, 3.5-5 mm, 3-5-veined, with hyaline margins; lemmas 5.5-7 mm long, usually glabrous, 5-veined, the tip hyaline, acute, rarely with a short awn to 2 mm long. — Native of Europe, cultivated for forage and established in fields, meadows, and moist soil.

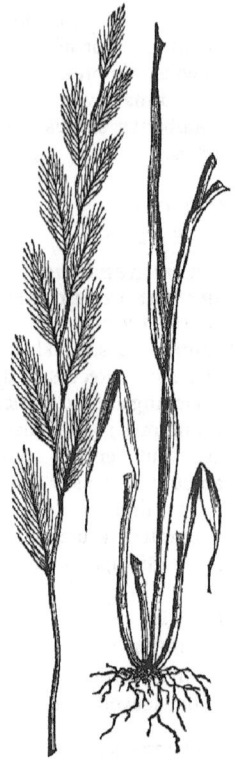

Milium
MILLET GRASS

Milium effusum L.
AMERICAN MILLET GRASS   *native / cons 7*
Perennial rhizomatous grass. Stems erect from a bent base, 6-12 dm tall, glabrous; sheaths open; ligule membranous, 3-9 mm long, obtuse-erose; leaf blades flat, broad, 8-17 mm wide, glabrous, or scaberulous on the margin. Panicle 1-3 dm long, ovoid or pyramidal, the branches in fascicles of 2 or 3, widely spreading and bearing drooping spikelets beyond their middle. Spikelets 1-flowered, disarticulation above the glumes; glumes equal, ovate or elliptic, rounded on the back, scaberulous, about 3 mm long, 3-nerved; lemma about as long as the glumes, awnless, nerveless, obtuse, rounded on the back, at first thin, at maturity firm, white, and shining, its margins partly covering a palea of similar texture. — Rich, moist or dry woods.

Lolium perenne

Milium effusum

Muhlenbergia
MUHLY

Perennial grasses, clumped or with creeping rhizomes. Stems erect or reclining at base, often branching from base. Leaves smooth to hairy, ligules membranous. Head a panicle, usually narrow and spike-like, sometimes open and spreading, at ends of stems and sometimes also from leaf axils. Spikelets 1-flowered, breaking above glumes; glumes usually nearly equal in length, 1-veined, the tip often awned; lemma lance-shaped, 3-veined, sometimes awned, some species with long, soft hairs at lemma base; palea about as long as lemma.

1 Glumes (including prominent awn-tip) 3.5–7.5 mm long, mostly distinctly longer than the body of the lemma; lemma at most short-awned; anthers 0.8–1.3 mm long . . . . **M. glomerata**
1 Glumes generally less than 3.6 mm long (rarely, especially on lower spikelets of panicle, up to 4 mm),

mostly about equaling or shorter than the body of the lemma; lemma awnless to long-awned; anthers not over 0.5 mm long . . . . *M. mexicana*

Muhlenbergia glomerata (Willd.) Trin.
MARSH MUHLY   *native / cons* **9**

Perennial grass, spreading from rhizomes. Stems upright, 3-9 dm long, sometimes with a few branches from base, dull and finely hairy between nodes; sheaths finely roughened, slightly keeled; ligules membranous, to 0.6 mm long, truncate, fringed; leaf blades flat, lax, 2-6 mm wide, usually scabrous. Panicle narrow, crowded, cylindric, 2-10 cm long and 5-10 mm wide, the lower clusters of spikelets often separate from one another. Spikelets 1-flowered, often purple-tinged, 5-6 mm long; glumes nearly equal, longer than the floret, tipped with an awn 1-5 mm long; lemma lance-shaped, 2-3 mm long, with long, soft hairs at base. Aug-Sept. — Swamps, wet meadows, marshes, springs, open bogs, fens, calcareous shores.

Muhlenbergia glomerata

Muhlenbergia mexicana (L.) Trin.
MEXICAN MUHLY   *native / cons* **4**

Perennial grass, from scaly rhizomes. Stems upright, 2-8 dm long, sometimes branched from base; dull and finely hairy between nodes; sheaths smooth or finely roughened, somewhat keeled; ligules membranous, to 1 mm long, truncate, fringed; leaf blades flat, lax, 2-6 mm wide, scabrous or smooth. Panicle narrow, densely flowered, 5-15 cm long and 2-10 mm wide, from ends of stems and leafy branches. Spikelets 1-flowered, green or purple, 2-3 mm long; glumes nearly equal, lance-shaped, 3-4 mm long, about as long as floret, tipped with a short awn about 1 mm long; lemma lance-shaped, 2-3 mm long, unawned or with an awn to 7 mm long. Aug-Sept. — Swamps, floodplain forests, thickets, wet meadows, marshes, springs, fens and streambanks.

Muhlenbergia mexicana

Oryzopsis
MOUNTAIN RICEGRASS

Oryzopsis asperifolia Michx.
WHITE-GRAIN MOUNTAIN RICEGRASS   *native / cons* **6**

Loosely tufted perennial grass. Stems 3-7 dm tall, often widely spreading. Leaves mostly basal; sheaths open, glabrous; auricles absent; basal ligules to 0.7 mm long, sometimes longest at the sides, ciliate; blades of basal leaves 30-90 cm long and 4-9 mm wide, upper surface glaucous; upper stem leaves with greatly reduced blades 3 cm long or less or lacking. Raceme slender, 2-6 cm long, the paired branches each with a single spikelet. Spikelets 1-flowered, articulated above the glumes; glumes equal or nearly so, broad, 7-8.5 mm long, abruptly acute or mucronate; lemma about equaling the glumes, becoming indurate at maturity, pale green or yellowish; awn 6-14 mm long, straight or twisted, articulated with the lemma, readily detached. — Moist or dry open woods, forested dunes.

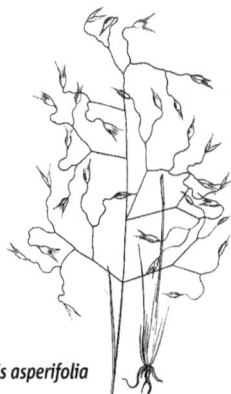

New leaves start to develop in mid-summer, the blades at first erect, then bending downward and remaining green through winter. Sheaths below the level of the duff are usually bright purple. Our other species formerly in this genus are now placed in *Piptatherum.*

Oryzopsis asperifolia

Panicum
PANIC-GRASS

Annual or perennial grasses. Heads narrow to open panicles (ours). Spikelets small, with 1 fertile flower; glumes usually unequal, the first glume membranous, usually very small, second glume green, about as long as spikelet; sterile lemma similar to second glume, enclosing the palea and sometimes a staminate flower, fertile lemma whitish, smooth.

1 Spikelets all or mostly 3 mm or more in length. . . . . *P. virgatum*
1 Spikelets less than 3 mm long . . . . *P. capillare*

Panicum capillare L.

COMMON PANIC-GRASS   *native / cons* 1

Annual grass, hirsute or hispid, hairs papillose-based, often purplish. Stems branched from the base, erect, ascending, or decumbent, to 7 dm long; sheaths rounded; ligules membranous, ciliate, the cilia 0.5-1.5 mm long; leaf blades spreading, 3-18 mm wide. Panicles diffusely branched, sometimes 2/3 as long as the entire plant. Spikelets all or mostly on long pedicels, the first glume to half as long as the sterile lemma; margins of the lemma distinctly inrolled. — Dry or moist soil, often a weed in fields in gardens, widely distributed and variable.

Panicum virgatum L.

SWITCHGRASS   *native / cons* 4

Perennial grass, from hard, scaly rhizomes, often forming large bunches. Stems stout, erect, to 3 m tall; sheaths longer than the lower internodes, shorter than those above, glabrous or pilose, especially on the throat, margins ciliate; ligules a dense zone of silky hairs 2-6 mm long; leaf blades flat, erect to spreading, 2-5 dm long, 2-15 mm wide, undersurface sometimes densely pubescent, margins scabrous. Panicle open, freely branched, pyramidal, usually 2-4 dm long. Spikelets ovoid, soon widened distally by spreading of the glumes and sterile lemma, 2.2-5.6 mm long; first glume half or more as long as the spikelet; second glume and sterile lemma about equal, conspicuously nerved, acute to long-acuminate. — Open woods, prairies, dunes, shores. Not currently known from Door County but common in Wisconsin and to be expected.

Panicum capillare

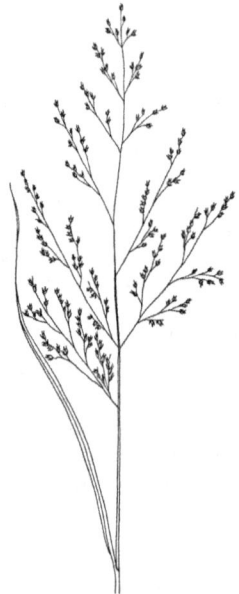

Panicum virgatum

Phalaris

CANARY-GRASS

Annual or perennial grasses. Leaves glabrous, auricles absent; ligule large, membranous. Flowers in dense or spike-like panicles of medium-sized or large spikelets. Spikelets articulated above the glumes, with 1 perfect terminal flower and 1 or 2 minute sterile lemmas below it. Glumes about equal, compressed and keeled, usually winged along the midnerve; lateral nerves usually stronger than the midnerve. Lemmas awnless, shorter than the glumes, the sterile linear, resembling tufts of hairs at the base of a solitary functional floret; the fertile lemma firm or leathery, often shining.

1 Rhizomatous perennial, with elongate lobed panicle (or the lower branches spreading at anthesis); glumes mostly 4–5.7 mm long, the keel not winged . . . . *P. arundinacea*
1 Annual, with very dense, compact, ovoid panicle; glumes mostly 6–8 mm long, the keel prominently winged . . . . *P. canariensis*

Phalaris arundinacea L.

REED CANARY-GRASS   *introduced*

Tall perennial grass, spreading by scaly rhizomes and typically forming large, dense colonies. Stems stout, smooth, 5-20 dm long; sheaths smooth; ligules membranous, 4-9 mm long, truncate, lacerate; leaf blades flat, 5-20 mm wide, surfaces scabrous, margins serrate. Panicle narrow, densely flowered, 5-25 cm long, often purple-tinged, the branches short and up-right. Spikelets 4-6 mm long, breaking above glumes, with 1 fertile flower and 2 small sterile lemmas below; glumes nearly equal, longer than fertile floret, lance-shaped, tapered to tip or short-awned, becoming straw-colored with age, 3-veined; fertile lemma ovate, 3 mm long, shiny; palea as long as lemma. June-July. — Wet meadows, shallow marshes, ditches, shores and streambanks.

Reed canary-grass is an aggressive, highly competitive wetland species, now widely naturalized, often to the detriment of our native flora. Wisconsin populations are likely a mix of native and Eurasian strains, including cultivars developed for forage.

Phalaris canariensis L.

COMMON CANARY-GRASS *introduced*

Annual grass. Stems erect, 3-9 dm tall; ligules membranous, 3-6 mm long, rounded to obtuse, lacerate; leaf blades 2-10 mm wide. Panicle dense, ovoid, usually about 3 cm long; glumes broad, 7-8 mm long, broadly winged on the keel, the midnerve marked by a broad green stripe; fertile lemma about 5 mm long; sterile lemmas 2, linear, about 2.5 mm long. — Native of Europe; introduced and adventive; grown for bird seed.

Distinguished by the exposed, nearly semi-circular ends of the glumes.

Phleum
TIMOTHY

Phleum pratense L.

COMMON TIMOTHY *introduced*

Tufted perennial grass. Stems mostly 5-10 dm tall; sheaths open; auricles absent or inconspicuous; ligules membranous, 2-4 mm long, not ciliate; leaf blades flat, typically 5-8 mm wide, rough-margined. Panicle spike-like and cylindric, usually 5-10 cm long, 6-8 mm thick. Spikelets 1-flowered, strongly flattened, articulated above the glumes; glumes equal, compressed and keeled, hispid-ciliate on the keel, 3-nerved, 2.6-3.2 mm long, rounded to the tip; lemma much shorter than the glumes, thin and delicate, 3-5-nerved, awnless; palea narrow, somewhat shorter than to nearly equaling the lemma. — Introduced from Eurasia as a forage grass, commonly cultivated for hay and pasture; escaped to fields, roadsides, and disturbed places.

Phragmites
REED

Phragmites australis (Cav.) Trin.

COMMON REED *native (see desc.) / cons* 1

Phragmites communis Trin.

Tall, stout perennial reed, from deep, scaly rhizomes, or the rhizomes sometimes exposed and creeping over the soil; often forming large colonies. Stems erect, hollow, 1-4 m long and 5-15 mm wide near base, the internodes often purple; sheaths open, mostly overlapping; ligule membranous, white, 1 mm long, ciliate. leaf blades flat, long, 1-3 cm wide. Head a large, plumelike panicle, purple when young, turning yellow-brown with age, 15-40 cm long, much-branched, the branches angled or curved upward. Spikelets 3-7-flowered, linear, 10-15 mm long, breaking above the glumes; the stem within the spikelet (rachilla) covered with long silky hairs, these longer than the florets and becoming exposed as the lemmas spread after flowering; glumes unequal, the first glume half the length of second glume. Grain seldom maturing. Aug.-Sept. — Fresh to brackish marshes, shores, streams, ditches, occasional in tamarack swamps; sometimes in shallow water.

Two subspecies in Wisconsin, one native (subsp. *americanus*, whose distribution is poorly understood) and one introduced and invasive (subsp. *australis*, the only taxon reported for Door County to date):

Phalaris arundinacea

Phragmites australis

1 Plants rarely forming a monoculture; ligules 1–1.7 mm long; lower glumes 3–6.5 mm long; upper glumes 5.5–11 mm long; lemmas 8–13.5 mm long; leaf sheaths deciduous, exposing stems in winter . . . . subsp. *americanus*

1 Plants invasive and often forming a monoculture; ligules 0.4–0.9 mm long; lower glumes 2.5–5 mm long; upper glumes 4.5–7.5 mm long; lemmas 7.5–12 mm long; leaf sheaths not deciduous, stems not exposed in winter. . . . . subsp. *australis*

Piptatherum
RICEGRASS

Tufted perennial grasses, sometimes rhizomatous. Leaf blades with flat or involute, auricles absent, ligules membranous to hyaline. Spikelets often large, in contracted or open panicles. Spikelets 1-flowered, articulated above the glumes. Glumes equal or nearly so, broad. Lemma about equaling the glumes, becoming indurate at maturity, with a terminal, readily detached awn.

1 Blades flat, mostly 5–18 mm wide; body of lemma 5.5–7 mm long; ligules absent or to 0.5 mm long. . . . . *P. racemosum*

1 Blades involute, less than 2 mm wide; body of lemma 2.5–4 mm long; ligules of upper leaves 1.5–3 mm long. . . . . *P. canadense*

Piptatherum pungens (Torr. ex Spreng.) Dorn
SHORT-AWN MOUNTAIN RICEGRASS   *native / cons* 8
Oryzopsis pungens (Torr.) Hitchc.
Piptatheropsis pungens (Torr. ex Spreng.) Romasch., P. M. Peterson & R. J. Soreng
Stems densely tufted, 2-5 dm tall; leaves mostly basal; sheaths smooth or somewhat scabrous; ligules 0.5-2.5 mm long, truncate to acute; blades 0.5-1.8 mm wide, flat to involute (at least when dry), scaberulous. Panicle 3-8 cm long, usually slender with appressed or strongly ascending branches, or ovoid and open at anthesis; glumes elliptic to obovate, 3.5-4 mm long, very thin, the lateral nerves inconspicuous; lemma gray or pale green; awn 1-2 mm long, straight or slightly bent. — Sandy dry woods, usually with aspen, oak, jack pine, and red pine; dunes and rocky places.

Piptatherum racemosum (Sm.) Barkworth
BLACK-SEED MOUNTAIN RICEGRASS   *native / cons* –
Oryzopsis racemosa (Sm.) Ricker ex Hitchc.
Patis racemosa (Sm.) Romasch., P. M. Peterson & Soreng
Stems 4-10 dm tall, loosely tufted from a knotty rhizome; sheaths usually glabrous; basal leaves more or less absent; upper stem leaves distinctly longer than the lower, the blades usually 1-2 dm long and 8-15 mm wide, scaberulous above, pubescent beneath; ligules 0.3-0.7 mm long, truncate. Panicle sparsely branched, 1-2 dm long, the few straight branches spreading or ascending, bearing the few appressed spikelets toward the ends; glumes herbaceous, narrowly elliptic, 7-9 mm long, 7-nerved; lemma dark brown and shining, somewhat shorter than the glumes; awn 12-22 mm long. — Moist, rich deciduous forests and wooded dunes, sometimes in disturbed places; not often found in dry woods of jack pine or oak.

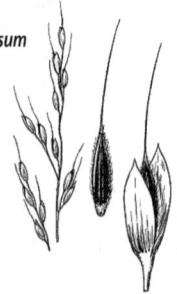

Piptatherum pungens

Poa
BLUEGRASS

Piptatherum racemosum

Annual or perennial grasses, with or without rhizomes or stolons, densely to loosely tufted or the culms solitary. Leaves mostly near base, flat to folded, midrib 2-grooved, the tip keeled similar to the bow of a boat; sheaths partly closed, auricles absent, ligules membranous. Head an open panicle. Spikelets small, with 2 to several flowers breaking above the glumes; glumes nearly equal, the first glume usually 1-veined, the second glume 3-veined; lemmas often with a tuft of distinctive cobwebby hairs at base; palea nearly as long as lemma.

POA GROUP KEY

Some species appear more than once in the keys; 'flag leaf' refers to the uppermost leaf which is often angled outward from the stem.

1 Plants annual or perennial; anthers 0.1–1 mm long in all florets and well developed, or only the upper 1–2 florets with rudimentary anthers . . . . ***Subkey 1***
1 Plants perennial; some anthers1.3–4 mm long, or the florets pistillate and all anthers vestigial and to 0.2 mm long, or longer and poorly developed 2
2 Plants rhizomatous or stoloniferous, rhizomes or stolons usually longer than 5 mm; basal leaves of the erect shoots with well-developed blades; plants densely to loosely tufted or the stems solitary. . . . . ***Subkey 2***
2 Plants neither rhizomatous nor stoloniferous; basal leaves of the erect shoots sometimes without blades; plants densely tufted . . . . ***Subkey 3***

POA SUBKEY 1

Plants annual or perennial. Stems not bulbous at base. Basal leaf sheaths not swollen at the base. Spikelets not bulbiferous, florets developing normally. Anthers 0.1-1 mm long.

1 Plants annual, sometimes surviving for a second season, introduced, weedy species; calluses glabrous; lemmas usually softly puberulent to long-villous on the keel and marginal veins, often also on the lateral veins, glabrous between the veins, non-alpine plants rarely glabrous throughout; palea keels smooth, usually short- to long-villous near the apices, rarely glabrous; panicle branches and glume keels smooth. . . . . ***P. annua***
1 Plants perennial, native, sometimes growing in disturbed habitats; calluses webbed or glabrous, if glabrous, the lmma pubescence not as above or the palea keels at least slightly scabrous near the apices; panicle branches and glume keels smooth of scabrous 2
2 Calluses webbed or glabrous, if webbed, the lemmas hairy on the keel and marginal veins . . . . ***P. nemoralis***
2 Calluses webbed; lemma keels glabrous throughout or, if hairy on the proximal 1/2, the marginal veins glabrous. 3
3 Lemmas hairy only on the keels; branches in whorls of (2)3–5(7) . . . . ***P. alsodes***
3 Lemmas usually glabrous, marginal veins rarely sparsely hairy at the base, hairs to 0.15 mm long; branches 1–3 per node. . . . . ***P. saltuensis***

POA SUBKEY 2

Plants with rhizomes or stolons, densely to loosely tufted or the stems solitary.

1 Stems and nodes strongly compressed; stems usually geniculate; lower stem nodes usually exserted; panicle branches angled, scabrous on the angles; sheaths closed for 1/10–1/5 their length . . . . ***P. compressa***
1 Stems terete to somewhat compressed, nodes not or only weakly compressed; stems geniculate or not; lower stem nodes exserted or not; panicle branches angled or terete, smooth or scabrous; sheath closure varied 2
2 Lemma keels softly puberulent for 3/5 their length, hairs usually sparse, marginal veins glabrous or puberulent to 1/4 their length, intercostal regions smooth and glabrous; lateral veins prominent; calluses webbed; palea keels smooth, muriculate, tuberculate, or scabridulous; lower glumes 1-veined, usually arched to sickle-shaped; ligules 3–10 mm long, acute to acuminate; panicle branches angled, angles densely scabrous; plants usually weakly stoloniferous . . . . ***P. trivialis***
2 Lemmas glabrous or variously pubescent, if as above, the lateral veins faint or moderately prominent or the calluses glabrous or the palea keels distinctly scabrous or hairy or the lower glumes 3-veined; calluses glabrous or hairy; palea keels scabrous at least near the apices; lower glumes 1–3-veined, not arched, not sickle-shaped; ligules 0.5–18 mm long, truncate to acuminate; panicle branches terete or angled, smooth or scabrous; plants stoloniferous or not. 3
3 Sheaths closed for 1/10–1/5 their length; spikelets 3–5 mm long; lemmas glabrous between the keels and marginal veins; panicle branches angled, angles densely scabrous; plants sometimes stoloniferous, sometimes branching above the stem bases; florets bisexual . . . . ***P. palustris***
3 Sheaths closed for 1/5–9/10 their length; spikelets 3.5–12 mm long; lemmas glabrous or hairy between the keels and marginal veins; panicle branches terete or angled, smooth or scabrous; plants rarely stoloniferous, usually rhizomatous, never branching above the stem bases; florets bisexual or unisexual . . . . ***P. pratensis***

POA SUBKEY 3

Plants perennial, loosely to densely tufted, rhizomes and stolons absent. Stems not bulbous at base. Basal sheaths not swollen. Spikelets not bulbiferous, florets developing normally.

1 Calluses usually dorsally webbed . . . . *P. trivialis*
1 Calluses glabrous or with a crown of hairs. 2
2 Lemma lateral veins pronounced, keels pubescent, marginal veins glabrous or softly puberulent at the base, lemmas glabrous elsewhere; lower glumes 1-veined, subulate to narrowly lanceolate, usually arched to sickle-shaped; callus web well-developed . . . . *P. trivialis*
2 Lemma lateral veins obscure to pronounced, keels glabrous throughout or, if pubescent, the marginal veins distinctly pubescent for more than 1/4 their length, lemma lateral veins and intercostal regions glabrous or pubescent, or, if pubescent as in P. trivialis, then the callus web short, scant, poorly developed and the lower glumes 3-veined and lanceolate or broader. 3
3 Sheaths closed for 1/3–3/4 their length . . . . *P. saltuensis*
3 Sheaths closed for up to 1/4 their length 4
4 Spikelets lanceolate; glumes subulate to narrowly lanceolate, gradually tapering to narrowly acuminate tips; ligules to 0.5 mm long, truncate; flag leaf nodes at or above the middle of the stems; flag leaf blades usually longer than their sheaths; rachillas usually hairy, hairs to 0.15 mm long; webs usually short, scanty . . . . *P. nemoralis*
4 Spikelets and glumes not as above or, if so, the ligules 1.5–6 mm long, truncate to acute, and the rachillas glabrous; flag leaf nodes at or above the lower 1/3 of the stem; flag leaf blades longer or shorter than their sheaths; webs short or long, scanty or not 5
5 Panicles 10–30 cm long, branches 4–15 cm long; stems closely spaced to isolated at the base; lower glumes tapering to the apices; lemma keels abruptly inwardly arched beneath the scarious tips; lemma margins distinctly inrolled; rachillas usually muriculate, rarely sparsely hispidulous; web hairs usually longer than 2/3 the length of the lemmas . . . . *P. palustris*
5 Panicles 3–15 cm long, branches 0.4–8 cm long; stems closely spaced at the base; lower glumes abruptly narrowing to the apices, lengths 4.5–6.3 times the widths; lemma keels not abruptly inwardly arched beneath the scarious apices; lemma margins not or slightly inrolled; rachillas usually softly puberulent; web hairs shorter than 1/2(2/3) the length of the lemmas . . . . *P. interior*

Poa alsodes Gray
GROVE BLUEGRASS   *native / cons 5*

Loosely tufted perennial grass, rhizomes absent. Stems slender, 3-8 dm long; sheaths closed for 1/2-7/8 their length; ligules to 2 mm long, smooth or sparsely scabrous, truncate to obtuse; leaf blades 1-4 mm wide, flat, lax. Panicle open, lax, 10-20 cm long, the branches becoming widely spreading, mostly in groups of 4-5, with 1 to few spikelets near tip of branch; base of panicle sometimes remaining enclosed by sheath. Spikelets ovate, 2-3-flowered, 3-5 mm long; glumes nearly equal, 2-4 mm long; lemmas 2-4 mm long, with cobwebby hairs at base. May-July. — Alder thickets, swamp hummocks, most common in moist deciduous or mixed conifer-deciduous forests.

Poa alsodes

Poa annua L.
ANNUAL BLUEGRASS   *introduced*

Annual grass, densely tufted. Stems to 3 dm long, prostrate to ascending; sheaths closed for about 1/3 their length, terete or weakly compressed, smooth; ligules 0.5-3 mm long, glabrous, decurrent, obtuse to truncate; leaf blades 1-10 cm long, 1-4 mm wide, flat or weakly folded, thin, soft, smooth, margins usually slightly scabrous, broadly prow-shaped at tip. Panicle ovoid, 2-8 cm long, with few ascending branches bearing rather crowded spikelets above the middle. Spikelets green, 3-6-flowered, 3-5 mm long; glumes broadly lance-shaped, acute, scarious-margined, indistinctly nerved, the first 1.5-2.4 mm, the second 1.8-2.8 mm long; lemmas thin, elliptic, 5-nerved, obtuse, pubescent on the nerves, not webbed at base. — Native of Eurasia and a widely distributed weedy species of roadsides, lawns, forest trails, clearings, shores, and disturbed places.

Poa annua

Poa compressa L.

CANADA BLUEGRASS   *introduced*

Perennial grass; the shoots usually solitary, sometimes loosely tufted, extensively rhizomatous. Stems erect, 2-7 dm tall, strongly flattened, especially above; sheaths closed for 1/10-1/5 their length, distinctly compressed; ligules 1-3 mm long, scabrous, ciliolate; leaf blades 1.5-4 mm wide, flat. Panicle usually compact and narrow, bluish or grayish green, 2-8 cm long, the branches usually in pairs, bearing spikelets nearly to the base; pedicels of the lateral spikelets 0.5 mm long. Spikelets 3-6-flowered, 4-6 mm long; first glume 1.7-2.4 mm long; second glume 1.8-2.6 mm long; lemmas firm, obscurely nerved, 2-2.8 mm long, slightly pubescent on the nerves below, somewhat webbed at base. — Native of Europe; open, usually dry places, especially in acidic soil.

Poa compressa differs from *P. pratensis* in its flattened, less tufted stems, lemmas with sparse or even absent web at the base, and a more slender panicle with fewer branches at each node.

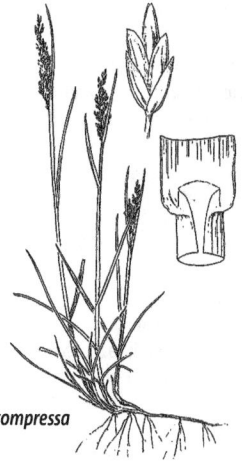

Poa compressa

Poa interior Rydb.

INTERIOR BLUEGRASS   *native / cons −*

Densely tufted perennial grass, green or less often glaucous, rhizomes and stolons absent. Stems to 80 cm long, erect or ascending; sheaths closed for up to 1/5 their length, terete; ligules 0.5-1.5 mm long, scabrous, truncate to obtuse, ciliolate; leaf blades mostly flat, thin, soft, 1-3 mm wide, narrowly prow-shaped at tip. Panicles to 15 cm long; branches to 8 cm long, ascending to widely spreading, angled, the angles scabrous. Spikelets mostly 2-3-flowered, narrowly ovate, laterally compressed 3-6 mm long, usually not glaucous; glumes lance-shaped, distinctly keeled, keels smooth or sparsely scabrous; calluses usually webbed, webs usually scant, less than 1/2 the lemma length, frequently tiny; lemmas 2.4-4 mm long, lance-shaped, distinctly keeled, straight or gradually arched, keels and marginal veins short-villous. — Shallow rocky or sandy soil of outcrops and talus slopes.

Distinguished from *P. nemoralis* by its longer ligules and wider glumes and lemmas; differs from *P. palustris* in having a densely tufted habit, scantly webbed calluses, and lemmas with wider hyaline margins.

Poa nemoralis

Poa nemoralis L.

WOODLAND BLUEGRASS   *introduced*

Densely tufted perennial grass, green or glaucous, rhizomes and stolons absent. Stems slender, 4-8 dm tall; sheaths closed for 1/10-1/5 their length, terete; ligules 0.2-0.8 mm long, sparsely to densely scabrous, truncate; leaf blades 1-3 mm wide, mostly flat, narrowly prow-shaped at tip. Panicle narrowly ovoid, 1-2 dm long, eventually loose and open, the slender branches in fascicles of about 5, bearing spikelets above the middle. Spikelets 2-4-flowered; glumes narrowly lance-shaped, long-acuminate, the first 2.2-3 mm long, conspicuously narrower than the first lemma, the second 2.3-3.3 mm long; lemmas 3-nerved, 2.1-3.1 mm long. — Dry, sandy or rocky soil, forest borders and clearings, old farmsteads.

Poa palustris L.

FOWL BLUEGRASS   *native / cons 5*

Loosely tufted perennial grass, often with stolons. Stems smooth, 4-12 dm long, reclining at base and rooting from lower nodes, lower portion often purple-tinged; sheaths closed for 1/10-1/5 their length, slightly compressed, glabrous or sparsely retrorsely scabrous; ligules 1.5-6 mm long, smooth or sparsely scabrous, tips obtuse to acute, frequently lacerate; leaf blades 1.5-8 mm wide, flat, narrowly prow-shaped at tip. Panicle

Poa palustris

loosely spreading (narrow when emerging from sheath), 1-3 dm long, the branches in mostly widely separated groups along panicle stem (rachis). Spikelets 2-4-flowered, 2-5 mm long and 1-2 mm wide; glumes nearly equal, lance-shaped, 2-3 mm long, often purple; lemma 2-3 mm long, often purple on sides, with cobwebby hairs at base. June-Sept. — Wet meadows, marshes, shores, streambanks, ditches and low prairie; also moist woods.

Poa pratensis L.
KENTUCKY BLUEGRASS   *introduced*

Perennial grass, sometimes glaucous, densely to loosely tufted or the shoots solitary; extensively rhizomatous and sod-forming. Stems 3-10 dm tall; sheaths closed for 1/4-1/2 their length, terete to slightly compressed, glabrous; collars glabrous; ligules mostly 1-2 mm long, ciliolate or glabrous; leaf blades 0.4-4.5 mm wide, flat, folded, or involute, soft and lax to moderately firm, tips usually broadly prow-shaped. Panicle ovoid, rather dense, its branches spreading or ascending, the lower chiefly in fascicles of 4 or 5. Spikelets 3-5-flowered, with very short rachilla-joints; first glume 1.8-2.9 mm long; second glume 2.3-3 mm long; lemmas distinctly 5-nervcd, thinly to densely pubescent on the nerves, webbed at base, glabrous between the nerves, the lowest 2.5-3.5 mm long; anthers 1-1.4 mm long. Moist or dry soil, disturbed places, woods, fields, avoiding acidic soils and heavy shade, often cultivated in lawns and meadows. — Introduced from Europe and naturalized in much of North America.

Poa saltuensis Fern. & Wieg.
OLD-PASTURE BLUEGRASS   *native / cons 7*

Loosely tufted perennial grass; rhizomes and stolons absent. Stems slender, usually weak, 3-10 dm tall; sheaths closed for 1/3-2/3 their length; ligules 0.2-3 mm long, smooth or sparsely scabrous, truncate to obtuse; leaf blades 1-4 mm wide, flat, thin, lax, veins prominent. Panicle loose, more or less nodding, 5-10 cm long, the slender branches bearing a few spikelets beyond the middle, the lower branches usually in pairs, rarely solitary or in 3's. Spikelets ovate, 2-4-flowered, 3-4 mm long; glumes acute, the first lance-shaped to ovate, 1.7-2.6 mm. long, the second ovate, 2-3 mm long; lemmas firm, obscurely nerved, oblong, 2.4-3.2 mm long, glabrous except the webbed base. — Dry or rocky deciduous and mixed woods.

Poa trivialis L.
ROUGH-STALK BLUEGRASS   *introduced*

Short-lived perennial grass, tufted and forming mats from aboveground stolons. Stems slender to stout, erect from a decumbent base, 5-10 dm tall, scabrous below the panicle; sheaths closed for about 1/3-1/2 their length, compressed, usually densely scabrous; ligules 3-10 mm long, scabrous, acute to acuminate; leaf blades 1-5 mm wide, flat, lax, soft, sparsely scabrous over the veins, margins scabrous, tips narrowly prow-shaped. Panicle soon long-exsert, ovoid, the ascending branches in fascicles of 5-8 with numerous crowded spikelets; pedicels scabrous. Spikelets ovate or elliptic, 2- or 3-flowered; glumes lance-shaped, incurved, the first 1.7-2.9 mm., the second 2-3.3 mm long; lemmas thin, narrowly ovate, sharply 5-nerved, acute, 2.3-3.2 mm long, glabrous except the keel and webbed base. — Native of Europe; meadows, moist woods, roadsides, along shaded trails.

Poa pratensis

Poa trivialis

Schizachne
FALSE MELIC GRASS

Schizachne purpurascens (Torr.) Swallen

FALSE MELIC GRASS — *native / cons* **7**

Loosely tufted perennial grass. Stems erect from a short-decumbent base, 3-10 dm tall; sheaths closed almost to the top; ligules membranous, 0.5-1.5 mm long; leaf blades mostly erect, elongate, 1-5 mm wide. Panicle with few drooping branches each bearing 1-3 slender spikelets about 2 cm long. Spikelets 3-5-flowered, usually purplish, disarticulating above the glumes and between the lemmas; glumes purple at base, unequal, 5-8 mm long, 3-5-nerved; fertile lemmas 8-10 mm long, strongly nerved, densely short-bearded at base, bifid for a fourth of their length, with an awn 8-15 mm long between the teeth, awns at length divergent. — Drier, sandy or rocky woods and openings; deciduous forests.

Schizachne purpurascens

Schizachyrium
LITTLE BLUESTEM

Schizachyrium scoparium (Michx.) Nash

LITTLE BLUESTEM — *native / cons* **4**

Andropogon scoparius Michx.

Perennial grass; loosely or densely tufted or with rhizomes, green to purplish, sometimes glaucous. Stems 5-12 dm tall, often freely branched above; sheaths rounded or keeled, glabrous or pubescent; auricles absent; ligules membranous, 0.5-2 mm long; leaf blades 3-7 mm wide. Racemes solitary, usually long-exsert, bearing 5-20 pairs of spikelets on a straight or flexuous, white-ciliate rachis. Sessile spikelets with 2 florets; glumes exceeding the florets; lower florets reduced to hyaline lemmas; upper florets bisexual, lemmas hyaline, bilobed or bifid to 7/8 of their length, awned from the sinuses. Pedicellate spikelets usually shorter than the sessile spikelets, sterile or staminate, with 1 floret. Late summer and early fall. — An important prairie species; in Wisconsin often in drier sandy woods and openings, old fields, sand dunes and shores.

Schizachyrium scoparium

Secale
RYE

Secale cereale L.

RYE — *introduced*

Annual or biennial grass. Stems branched from the base 5-10 dm tall; sheaths open; ligules membranous, truncate, often lacerate; leaf blades 4-12 mm wide, usually glabrous. Spikes densely flowered, 8-15 cm long, often distinctly nodding when mature. Spikelets usually 2-flowered, solitary at each joint of the rachis; glumes linear-subulate, 1-nerved, shorter than the lemmas; lemmas lance-subulate, 5-nerved, with their sides toward the axis, tapering into a long awn 3-8 cm long; disarticulation in the rachis, at the nodes, tardy or the spikes not disarticulating. — An important Eurasian cereal grass. Mostly along roadsides, where planted for erosion control following construction; also on shores, dunes, along railroads, and in old fields; not long-persisting.

Secale cereale

Setaria
BRISTLE GRASS

Tufted annual grasses (ours). Ligules membranous and ciliate or of hairs. Spikelets all alike, with 1 perfect flower, turgid or plano-convex, subtended by an involucre of 1 to many slender bristles, articulated and eventually deciduous above the bristles, aggregated into cylindric, spike-like, terminal panicles. First glume triangular to ovate, 3-5-nerved, half as long as the spikelet or less. Second glume longer, sometimes equaling the spikelet. Sterile lemma equaling the spikelet, several-nerved. Fertile lemma indurate, smooth or transversely rugose.

1 Fertile lemmas mostly ca. 3 mm long, rugose with distinctly transverse ridges, the upper half exposed at maturity; bristles 5 or more per spikelet, becoming orange or golden-brown; sheaths glabrous ..
   .... *S. pumila*
1 Fertile lemmas less than 3 mm long, evenly and finely rugose or reticulate or smooth (without transverse ridges), the upper half largely or entirely concealed at maturity; bristles fewer than 5 per spikelet, pale greenish or purple (rarely yellow) at maturity; sheaths ciliate with long hairs on the margins .... *S. viridis*

Setaria pumila (Poir.) Roem. & Schult.
PEARL-MILLET   *introduced*
  Setaria glauca (L.) Beauv.
Annual grass. Stems usually erect, solitary or tufted, 4-8 dm tall; sheaths glabrous; ligules ciliate; leaf blades 4-10 mm wide, loosely twisted, upper surface with papillose-based hairs near base. Panicles spike-like, usually 5-10 cm long, or sometimes longer, the axis pubescent; bristles 3-10 mm long, yellow or tawny at maturity. Spikelets thick, 3-3.5 mm long; first glume 5-nerved, half as long as the spikelet; second glume 5-nerved or usually 7-nerved, 2/3 as long as the spikelet; fertile lemma transversely rugose. Native of Europe, introduced in lawns, roadsides, railroads, cultivated fields, and disturbed places.

Setaria viridis (L.) Beauv.
GREEN FOXTAIL-GRASS   *introduced*
Annual grass. Stems usually branched and often geniculate at base; sheaths mostly glabrous, margins ciliate upwards; ligules 1-2 mm long, ciliate; leaf blades flat, 4-25 mm wide, scabrous or smooth. Panicle 1-7 cm long, the short branches uniformly spaced on the rachis; bristles 1-3 below each spikelet, upwardly barbed, green, purple, or tawny, usually 2-10 mm long. Spikelets 2-2.5 mm long; sterile lemma usually as long as the finely rugose fertile one. — Native of Eurasia; weedy in gardens, cultivated fields, and disturbed places.

Setaria viridis

Sorghastrum
INDIAN GRASS

Sorghastrum nutans (L.) Nash
YELLOW INDIAN GRASS   *native / cons* 5
Large perennial grass from short scaly rhizomes. Stems in loose tufts, 1-2 m tall, the nodes densely pubescent; sheaths glabrous to hirsute; ligules membranous, 2-6 mm long, usually with thick, pointed auricles; leaf blades 10-70 cm long and to 1 cm wide, usually glabrous. Panicle narrow, 10-25 cm long, the ultimate branches of the panicle bearing short racemes of 1-5 spikelets; terminal spikelet usually with 2 sterile pedicels adjacent the nodes. Spikelets lance-shaped, 6-8 mm long; sessile spikelet perfect, subterete; pediceled spikelet absent, represented by its pedicel only; first glume pale brown, villous, its edges enclosing the margin of the glabrous or ciliate second glume; lemmas hyaline, bifid, awned from the sinuses; the awn 9-15 mm long, twisted below, bent at about a third of its length; sterile pedicel densely villous, 4-5 mm long. — Moist or dry prairies, open woods, fields, shores, and rarely, in marshes; sometimes spreading in disturbed places as along roadsides and railroads.

Sorghastrum nutans

Sphenopholis
WEDGESCALE

Sphenopholis intermedia (Rydb.) Rydb.
SLENDER WEDGESCALE　　　　　　　　*native / cons* **7**
 Sphenopholis obtusata var. *major* (Torr.) K.S. Erdman

Perennial grass. Stems tufted, 3-12 dm tall; sheaths open, smooth or finely roughened, sometimes pubescent; ligules membranous, 1.5-2.5 mm, erose-ciliate, often lacerate; auricles absent; leaf blades flat, soft, 2-5 mm wide, Panicle slender, shining, 8-15 cm long, 1-3 cm wide, more or less lobed or irregular in contour, not spike-like except when very young. Spikelets 2-flowered; first glume subulate, 1.5-2.4 mm long, less than 0.5 mm wide; second glume obovate, broadly acute or apiculate, thin in texture, obscurely nerved, 2-2.7 mm long; lower lemma 2.3-2.9 mm long. — Moist to wet gravelly shores, tamarack swamps, marsh borders, thickets, forest depressions; sometimes in moist woods.

　　Similar in appearance to *Koeleria macrantha,* but differs in its more open panicle, the very narrow first glume, and the essentially glabrous foliage and panicle.

Sphenopholis intermedia

Sporobolus
DROPSEED

Annual or perennial grasses. Leaf blades narrow, often involute; sheaths open, usually glabrous, often ciliate at the top; ligules of short hairs. Panicles open or contracted. Spikelets 1-flowered, articulated above the glumes. Glumes lance-shaped to ovate, 1-nerved, from much shorter than to somewhat longer than the lemma. Lemma rounded on the back, nerveless or 1-nerved, awnless. Palea about as long as the lemma or longer. Fruit differs from a true grain in that the pericarp is free from the seed coat, and often slipping away, at least when moist.

1　Plants perennial. . . . . *S. cryptandrus*
1　Plants annuals or short-lived perennials flowering in the first year 2
2　Lemmas strigose; spikelets 2.3-6 mm long; mature fruits 1.8-2.7 mm long . . . . *S. vaginiflorus*
2　Lemmas glabrous; spikelets 1.6-3 mm long; mature fruits 1.2-1.8 mm long . . . . *S. neglectus*

Sporobolus cryptandrus (Torr.) Gray
SAND-DROPSEED　　　　　　　　*native / cons* **3**

Tufted perennial grass, rhizomes absent, bases not hard and knotty. Stems solitary or in small tufts, 3-10 dm tall, the lower portion usually covered by sheaths; sheath tips with conspicuous tufts of hairs, the hairs to 4 mm long; ligules to 1 mm long; leaf blades 2-6 mm wide, flat or drying involute, 2-6 mm wide, tapering to a long point. Panicle ovoid or pyramidal, 1-2 dm long, at base usually partly included in the upper sheath; branches alternate, soon widely divergent, the branchlets more or less appressed and forming a dense narrow cluster. Spikelets 2-3 mm long; glumes acute, the first half or less as long as the second; lemma about equaling the second glume. — Dry, especially sandy soil, cedar glades, barrens, fields and dunes; often in sandy disturbed areas such as roadsides and railways.

Sporobolus neglectus Nash
SMALL DROPSEED　　　　　　　　*native / cons* **2**

Tufted annual grass; plants delicate, slender, very similar to S. vaginiflorus. Stems 10-45 cm tall, wiry, erect to decumbent; sheaths inflated, mostly glabrous but the tips with small tufts of hairs to 3 mm long; ligules to 0.3 mm long; leaf blades to 2 mm wide, flat to loosely involute. Panicle rarely exsert, usually permanently exceeded by the uppermost blade. Spikelets smaller, the glumes and lemma less acuminate and proportionately wider; first glume 1.5-2.4 mm long; second glume 1.7-2.7 mm

Sporobolus cryptandrus

long; lemma 2-3 mm long, glabrous, about equaling the wide palea. — Dry sterile or sandy soil of roadsides and fields; also along shores and on mudflats.

Sporobolus vaginiflorus (Torr.) Wood
POVERTY-GRASS   *native / cons* 1

Annual grass. Stems tufted, erect to spreading, 15-60 cm tall, very thin and wiry, seldom more than 1 mm in diameter; sheaths often inflated, sometimes with sparse hairs at base, glabrous or the tips with small tufts of hairs to 3 mm long; ligules to 0.3 mm long; leaf blades to 2 mm wide, the lower elongate, the upper progressively shorter to only 1-2 cm long. Panicle slender, 2-5 cm long, eventually exsert; axillary panicles also developed and mostly included in the lower sheaths. Spikelets crowded; glumes and lemma lance-shaped, straight; first glume 2.8-4.1 mm long; second glume 2.9-4.6 mm long; lemma 3-5 mm long, minutely villous; palea equaling or somewhat exceeding the lemma. — Dry sandy or sterile soil as along roadsides (especially where gravelly) and in fields.

Very similar to *S. neglectus* and impossible to distinguish without spikelets; *S. vaginiflorus* differs in having strigose lemmas, sheaths that are sparsely hairy towards the base and, usually, longer spikelets. Both differ from *S. cryptandrus* in their annual habit, and by having nearly equal glumes.

Sporobolus vaginiflorus

Vulpia
SIX-WEEKS FESCUE

Similar to *Festuca* and only marginally distinct, but now customarily treated as a separate genus. Our species a tufted annual, withering by mid-summer, cleistogamous or nearly so, with most commonly only one, less often 2 anthers, these small, to ca 1.5 mm long. Sheaths open, usually glabrous; auricles absent; ligules usually shorter than 1 mm, membranous. Grain linear-cylindric, tapering to both ends.

Vulpia octoflora (Walt.) Rydb.
SIX-WEEKS FESCUE   *native / cons* 3
  Festuca octoflora Walt.

Stems slender, erect or geniculate at base, 1-4 dm tall, glabrous or puberulent; ligules 0.3-1 mm long; leaf blades 0.5-1 mm wide, flat or rolled, glabrous or pubescent. Panicle slender, 3-10 cm long, with a few ascending or rarely spreading branches. Spikelets flattened, 6-10-flowered, rarely with fewer; first glume subulate, 2-4 mm long; second glume lance-shaped, about 1/4 longer; lemma involute, straight, soon diverging and exposing the rachilla, gradually tapering into an erect awn 1-4 mm long. — Dry sandy places, often where disturbed.

Vulpia octoflora

Zizania
WILD RICE

Zizania palustris L.
NORTHERN WILD RICE   *native / cons* 8

Large annual grass; stems to 3 m tall, usually at least partly immersed in water; roots fleshy, yellow. Sheaths open, glabrous or with scattered hairs; ligules 3-15 mm long; leaf blades 0.5-1.7 cm wide. Panicles 25-60 cm long; staminate and pistillate flowers separate on same plant, the staminate flowers on lower panicle branches, pistillate flowers on upper branches; staminate branches ascending or divergent; pistillate branches mostly appressed or ascending, a few sometimes divergent. Staminate spikelets 6-17 mm long, lanceolate, acuminate or awned, the awns to 2 mm long. Pistillate spikelets 8-33 mm long, lanceolate or oblong, leathery or indurate, lustrous, glabrous or with lines of short hairs, tips usually hirsute

Zizania palustris

and abruptly narrowed, awned, the awns to 10 cm long; glumes absent; lemma prominently 3-ribbed, awned; lemmas and paleas remaining clasped at maturity. Grain 6-30 mm long.

Zizania palustris is the source of commercial wild rice (California is the nation's largest producer); in the Great Lakes region, harvesting is most common in Minnesota, especially by Native Americans where large areas of lakes and shallow marshes may be dominated by this plant. The grain is also an excellent food for waterfowl. Many populations are intentional introductions.

Pontederiaceae
PICKERELWEED FAMILY

Heteranthera
MUD-PLANTAIN

Heteranthera dubia (Jacq.) MacM.
GRASS-LEAF MUD-PLANTAIN   *native / cons* 6
Zosterella dubia (Jacq.) Small

Aquatic perennial herb, with lax stems and leaves, or plants sometimes exposed and forming small, leafy rosettes. Stems slender, forked, often rooting at lower nodes, to 1 m long. Leaves alternate, linear, flat, translucent, rounded at tip or tapered to a point, 2-12 cm long and 2-6 mm wide, the midrib and veins inconspicuous; petioles absent. Flowers 1, perfect (with both staminate and pistillate parts), opening on water surface, light yellow, enclosed in a spathe from upper leaf axils, the spathe membranous, 2-5 cm long, surrounding much of the slender perianth tube; perianth tube often curved, 2-8 cm long, the 6 perianth segments linear, 4-6 mm long; stamens 3, all alike. Fruit a many-seeded capsule about 1 cm long. July-Sept. — Shallow water, muddy shores of ponds, lakes, streams and marshes.

Distinguished from the pondweeds (*Potamogeton*) by lack of a leaf midrib.

Heteranthera dubia

Potamogetonaceae
PONDWEED FAMILY

This treatment includes *Zannichellia*, previously included in a separate family.

1 Submersed leaves opposite or whorled, floating leaves absent. . . . . *Zannichellia*
1 Submersed leaves alternate, floating leaves (sometimes present) alternate or opposite. 2
2 Stipules adnate to the leaves for 10–30 mm or more (at least on the larger leaves), adnate for ca. 2/3 of the length of the stipule; leaves all submersed, filiform to narrowly linear (up to 2.5 mm wide). . .
    . . . . *Stuckenia*
2 Stipules free from the leaves or adnate for less than half the length of the stipule (adnate for 5 mm or less except in *P. robbinsii*); leaves submersed or floating, filiform to ovate, oblong, or elliptic . . . .
    . . . . *Potamogeton*

Potamogeton
PONDWEED

Aquatic perennial herbs, with only underwater leaves or with both underwater and floating leaves, from rhizomes or tubers, sometimes reproducing and over-wintering by free-floating winter buds. Stems long, wavy, anchored to bottom by roots and rhizomes. Leaves alternate, or becoming opposite upward in some species, simple, with an open or closed sheath at base. Underwater leaves usually linear and threadlike, sometimes broader, margins often wavy, usually stalkless. Floating leaves, if present, oval or ovate, stalked, with a waxy upper surface. Flowers perfect, regular, green to red, in stalked spikes at ends of stems or from leaf axils, usually raised above water surface, the spikes with few to many small flowers; perianth of 4 sepal-like bracts; stamens 4. Fruit a 4-parted, beaked achene.

KEY TO POTAMOGETON

1 Plants with underwater leaves only, these all alike . . . . *Group 1*
1 Plants with 2 kinds of leaves: broad floating leaves and broad or narrow underwater leaves . *Group 2*

Potamogeton Group 1
Plants with underwater leaves only, these all alike.

1 Leaves broad, lance-shaped to oval or ovate, never linear 2
1 Leaves linear 5
2 Leaf margins wavy-crisped, finely toothed . . . . *P. crispus*
2 Leaf margins flat or sometimes wavy, entire (or rarely finely toothed at tip) 3
3 Base of leaf blade tapered, not clasping stem . . . . *P. illinoensis*
3 Base of leaf blade clasping stem 4
4 Stems whitish; leaves 10–30 cm long; fruit 4–5 mm long . . . . *P. praelongus*
4 Stems green; leaves 5–12 cm long; fruit 2–4 mm long. . . . . *P. richardsonii*
5 Stipules joined with lower part of leaf to form a sheath at least 1 cm long . . . . *P. robbinsii*
5 Stipules free from leaf, or rarely joined to leaf base for only 1–2 mm 6
6 Plants with slender creeping rhizomes. . . . . *P. foliosus*
6 Plants with short rhizomes or rhizomes absent (plants often rooting at lower nodes of stem) 7
7 Leaves without glands at base. . . . . *P. foliosus*
7 At least some of leaves with pair of glands at base. 8
8 Leaves with 5–7 nerves . . . . *P. friesii*
8 Leaves with 3 (rarely 1 or 5) nerves. . . . . *P. pusillus*

Potamogeton Group 2
Plants with 2 kinds of leaves: broad floating leaves and broad or narrow underwater leaves.

1 Underwater leaves linear . . . . *P. natans*
1 Underwater leaves broad, never linear 2
2 Floating leaves with 30–55 nerves; underwater leaves with 30–40 nerves . . . . *P. amplifolius*
2 Floating leaves with fewer than 30 nerves; underwater leaves with less than 30 nerves 3
3 Margins of underwater leaves finely toothed near tip . . . . *P. illinoensis*
3 Margins of underwater leaves entire . . . . *P. gramineus*

Potamogeton amplifolius Tuckerman
LARGE-LEAF PONDWEED   *native / cons 7*
Stems round in section, usually unbranched, to 1 m or more long and 2-4 mm wide. Upper underwater leaves ovate, folded and sickle-shaped, 8-20 cm long and 2-7 cm wide, many-veined; lower underwater leaves lance-shaped, to 2 cm wide, often not folded, usually decayed by fruiting time, many-veined; petioles 1-5 cm long. Floating leaves usually present at flowering time, ovate 5-10 cm long and 3-6 cm wide, many-veined, rounded at tip or abruptly tapered to a sharp tip, rounded at base; petioles 5-15 cm long; stipules open and free of the petioles, 5-12 cm long, long-tapered to a sharp tip. Flowers in dense cylindric spikes, 3-6 cm long in fruit; stalks 6-20 cm long, widening near tip. Achenes green-brown to brown, 4-5 mm long, beak to 1 mm long. July-Aug. — Shallow to deep water of lakes and rivers.

Potamogeton crispus L.
CURLY PONDWEED   *introduced (invasive)*
Stems compressed, with few branches, to 8 dm long and 1-2 mm wide. Leaves all underwater, oblong, 3-9 cm long and 5-10 mm wide, rounded at tip, slightly clasping at base, stalkless, 3-5-veined, margins wavy-

Potamogeton amplifolius

crisped, finely toothed; stipules 4-10 mm long, slightly joined at base, early shredding. Flowers in dense cylindric spikes, 1-2 cm long, appearing bristly in fruit from long achene beaks; on stalks 2-6 cm long. Achenes brown, 2-3 mm long, with a beak 2-3 mm long. April-June. — Shallow to deep water of lakes and rivers; pollution-tolerant.

Potamogeton foliosus Raf.
LEAFY PONDWEED   *native / cons* **6**
Stems compressed, much-branched, to 8 dm long and 1 mm wide. Leaves all underwater, linear, 1-8 cm long and 1-2 mm wide, 1-3-veined, stalkless; stipules free, 0.5-2 cm long, glands usually absent at base of stipules. Flowers in rounded to short-cylindric spikes, 2-7 mm long, with 1-2 whorls of flowers, on stalks 5-12 mm long, widened at tip. Achenes green-brown, 1.5-3 mm long, winged, the beak to 0.5 mm long. June-Aug. — Shallow to deep water of lakes, ponds, rivers and streams.

Potamogeton friesii Rupr.
FLAT-STALK PONDWEED   *native / cons* **8**
Stems compressed, branched, 1-1.5 m long and to 1 mm wide. Leaves all underwater, linear, 3-7 cm long and 1.5-3 mm wide, tip rounded with a short slender point, tapered to the base, 5-7-veined, stalkless, margins flat or becoming rolled under; stipules free, 5-20 mm long, fibrous, often shredding above, 2 glands present at base of stipule. Flowers in cylindric spikes, 8-16 mm long, with 2-5 whorls of flowers, on stalks 1.5-6 cm long. Achenes olive-green to brown, 2-3 mm long, beak flat, short. June-Aug. — Shallow to deep water of lakes, ponds, rivers and streams.

Potamogeton crispus

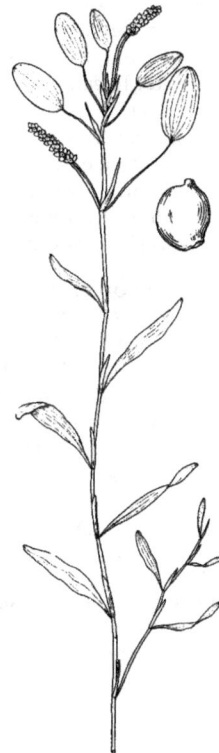

Potamogeton gramineus L.
GRASSY PONDWEED   *native / cons* **7**
Stems slender, slightly compressed, much-branched, to 8 dm long and 1 mm wide. Underwater leaves variable, linear to lance-shaped or oblong lance-shaped, 3-9 cm long and 3-12 mm wide, 3-7-veined, tapered to a stalkless base. Floating leaves usually present, oval, 2-6 cm long and 1-3 cm wide, 11-19-veined, rounded at base; petioles 2-10 cm long, shorter to longer than blade; stipules free, persistent, 1-4 cm long. Flowers in dense, cylindric spikes, 1.5-4 cm long, the stalks thicker than stem, 2-10 cm long. Achenes dull green, 2-3 mm long. June-Aug.
    Shallow to deep water of lakes and ponds.

Potamogeton illinoensis Morong
ILLINOIS PONDWEED   *native / cons* **6**
Stems nearly round in section, usually branched, to 2 m long and 2-5 mm wide. Underwater leaves lance-shaped to obovate, 6-20 cm long, 2-4 cm wide, 9-17-veined, tapered to a broad, flat petiole, 2-4 cm long; stipules free, persistent, 3-8 cm long. Floating leaves sometimes absent, opposite, lance-shaped to oval, 5-14 cm long and 2-6 cm wide, 13- to many-veined, often short-awned from the rounded tip, rounded to wedge-shaped at base; petioles 3-10 cm long, shorter than blades. Flowers in dense cylindric spikes, 2-6 cm long, on stalks 4-20 cm long, usually wider than stem. Achenes olive-green, 3-4 mm long, the beak short, blunt. July-Sept. — Shallow to deep water of lakes and rivers.

Potamogeton gramineus

Potamogeton natans L.
FLOATING PONDWEED   *native / cons* **5**
Stems slightly compressed, usually unbranched, 0.5-2 m long and 1-2 mm wide. Underwater leaves reduced to linear, bladeless, expanded petioles (phyllodes), these often absent by flowering time, 10-30 cm long and

1-2 mm wide. Floating leaves ovate to oval, 4-10 cm long and 2-5 cm wide, usually tipped with a short point, rounded to heart-shaped at base, many-veined; petioles usually much longer than blades, the blade often angled at juncture with petiole; stipules free, 4-10 cm long, persistent or shredding with age. Flowers in dense cylindric spikes, 2-5 cm long, stalks thicker than the stem, 6-14 cm long. Achenes green-brown to brown, 3-5 mm long, with a loose, shiny covering, the beak short. June-Aug. — Usually shallow water (to 2 m deep) of ponds, lakes, rivers and peatlands.

Potamogeton praelongus Wulfen
WHITE-STEM PONDWEED native / cons 8

Stems white-tinged, compressed, branched, to 2-3 m long and 2-4 mm wide, the shorter internodes often zigzagged. Leaves all underwater, lance-shaped, 10-30 cm long and 1-4 cm wide, with 3-5 main veins, rounded and hoodlike at tip, base more or less heart-shaped and clasping stem, stalkless, margins entire and gently wavy; stipules free, white, 1-3 cm long, fibrous at tip. Flowers in dense, cylindric spikes 2-5 cm long; stalks erect, 1-4 dm long, as wide as stem. Achenes green-brown, swollen, 4-5 mm long, the beak rounded, 0.5 mm long. June-Aug. — Shallow to deep water of lakes, streams.

Potamogeton natans

Potamogeton pusillus L.
SLENDER PONDWEED native / cons 7

Stems very slender, round in section, usually freely branched, 2-10 dm long and about 0.5 mm wide; rhizomes more or less absent. Leaves all underwater, linear, 1-7 cm long and 0.5-2 mm wide, tapered to a stalkless base, the midvein broad; stipules free, boat-shaped, brown-green, 4-10 mm long and 2x width of leaf base, soon decaying, glands sometimes present at stipule base. Flowers in short-cylindric spikes 2-10 mm long, the flowers in 1-3 whorls, on slender, upright stalks 1-5 cm long. Achenes green to brown, 1-2 mm long, the beak flat. June-Aug. — Shallow water (to 2 m deep) of lakes and ponds, less often in streams.

Potamogeton richardsonii (Benn.) Rydb.
RED-HEAD PONDWEED native / cons 5

Stems brown to yellow-green, round in section, sparingly to freely branched, mostly 3-10 dm long and 1-2.5 mm wide, the shorter internodes rarely zigzagged. Leaves all underwater, lance-shaped, 5-12 cm long and 1-2.5 cm wide, with 13 or more prominent veins, base heart-shaped and clasping stem, stalkless, margins entire and gently wavy; stipules free, 1-2 cm long, soon shredding into white fibers. Flowers in dense cylindric spikes 1.5-4 cm long, on stalks 2-20 cm long, the stalks strongly curved when in fruit. Achenes green to brown, 2-4 mm long, the beak short. July-Aug. — Shallow to deep water of lakes, streams.

Potamogeton richardsonii

Potamogeton robbinsii Oakes
FERN PONDWEED native / cons 8

Stems few-branched below, much-branched above, to 1 m long; rhizomes not tuberous. Leaves all underwater, crowded in 2 ranks, linear, 4-10 cm long and 3-7 mm wide, tapered to a pointed tip, abruptly narrowed at base, with rounded auricles where joined with stipule, midvein pronounced, margins pale; stipules joined to leaf for 5-15 mm, soon decaying into fibers. Flowers on underwater, cylindric spikes 1-2 cm long, with 3-5 separated whorls of flowers, the inflorescence often branched into 5-20 stalks, 2-5 cm long, at ends of stems. Achenes rarely produced, 3-5 mm long, the beak thick, somewhat curved; reproduction most commonly by stem fragments which root from the nodes. July-Aug. — Shallow to deep water of lakes, ponds and streams.

Potamogeton robbinsii

Stuckenia
FALSE PONDWEED

Stuckenia is a small genus of perennial aquatic herbs, now segregated from *Potamogeton*. In Stuckenia, the stipules are joined to the blade for 2/3 to nearly the entire length of the stipule; in *Potamogeton*, the stipules in most species are free, or if adnate, joined for well less than half the length of the stipule. Also, submersed leaves of *Potamogeton* are translucent, flat, and without grooves or channels; in *Stuckenia*, submersed leaves are opaque, channeled, and turgid.

1 Leaves gradually tapered to tip; rhizomes tuber-bearing; stigmas raised on a tiny style . . **S. pectinata**
1 Leaves rounded, blunt-tipped or tipped with a short, sharp point, stigmas inconspicuous, broad and
not raised . . . . *S. filiformis*

Stuckenia filiformis (Pers) Boerner
THREADLEAF FALSE PONDWEED   *native / cons* **8**
Potamogeton filiformis Pers.
Stems more or less round in section, branched from base, mostly unbranched above, 1-5 dm or more long and 1 mm wide, from a long, tuber-
bearing rhizome. Leaves all underwater, narrowly linear, 5-10 cm long
and 0.2-2 mm wide, 1-veined; stipules 1-3 cm long, joined to base of leaf
blade, forming a tight sheath around stem. Flowers in underwater spikes,
1-5 cm long, with 2-5 separated whorls of flowers, on slender stalks 2-12
cm long. Achenes olive-green, 2-3 mm long, the beak flat, tiny. July-Aug.
— Mostly shallow water (to 1 m) in lakes and rivers.

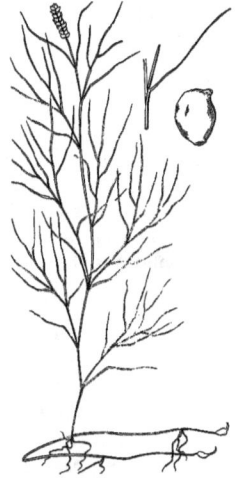

Stuckenia pectinata (L.) Boerner
SAGO FALSE PONDWEED   *native / cons* **3**
Potamogeton pectinatus L.
Stems slender, round in section, 3-10 dm long and 1-2 mm wide much-
branched and forking above, fewer branched near base, from rhizomes
tipped with a white tuber. Leaves all underwater, threadlike to narrowly
linear, 3-12 cm long and 0.5-1.5 mm wide, stalkless; stipules joined to
base of blade for 1-3 cm, forming a sheath around stem. Flowers on underwater, cylindric spikes 1-5 cm long, with 2-5 whorls of flowers, on lax,
threadlike stalks to 15 cm long. Achenes yellow-brown, 3-4 mm long, the
beak to 0.5 mm long; the large fruit an important waterfowl food. June-
Sept. — Shallow to deep water of lakes, ponds and streams; tolerant of
brackish water.

Stuckenia pectinata

Zannichellia
HORNED-PONDWEED

Zannichellia palustris L.
HORNED-PONDWEED   *native / cons* **7**
Perennial aquatic herb, with creeping rhizomes, and often forming extensive underwater mats. Stems slender and delicate, wavy, 0.5-5 dm
long, branched from base. Leaves simple, opposite (or upper leaves appearing whorled), threadlike, 2-8 cm long and 0.5 mm wide, stalkless;
stipules membranous and soon deciduous. Flowers small, produced underwater, either staminate or pistillate, separate on plant but from same
leaf axil, with 1 staminate flower and usually 4 (varying from 1-5) pistillate
flowers at each node, surrounded by a membranous, spathelike bract;
petals and sepals absent; staminate flower a single anther. Fruit a brown
to red-brown, crescent-shaped nutlet, gently wavy on margins, 2-3 mm
long, tipped by a beak 1-2 mm long; the fruit mostly 2-6 per node. June-
Aug. — Submerged in fresh or brackish water of streams, muddy lake
and pond bottoms, marshes and ditches.

Zannichellia palustris

Smilacaceae

GREENBRIER FAMILY

Smilax

GREENBRIER

Perennial herbs (with annual stems), or vining shrubs, climbing by tendrils terminating the stipules, with wide, longitudinally nerved, net-veined, alternate leaves and axillary peduncled umbels of small yellow or greenish yellow flowers. Flowers dioecious, the staminate often the larger. Perianth segments alike, spreading. Stamens in the staminate flower 6; filaments slender or flattened; anthers oblong. Stamens of the pistillate flower reduced to 6 filiform staminodes. Ovary 3-celled, with 1 or 2 ovules in each cell; style none or very short; stigmas solitary or 3, oblong, recurved. Fruit a 1-6-seeded berry. Leaves of all species vary greatly in size and shape.

1 Stems woody and prickly; leaves glabrous beneath (sometimes roughened on main veins) . . . .
. . . . ***S. tamnoides***
1 Stems herbaceous, never prickly; leaves finely puberulent, at least on the veins beneath 2
2 Pistillate (and usually also staminate) flowers fewer than 25 in an umbel; leaves fewer than 20 (usually 7–9) on a plant; stems under 50 cm tall; peduncles usually shorter than the petioles or slightly longer; tendrils completely absent (rarely on upper 2–3 nodes). . . . . ***S. ecirrata***
2 Pistillate and staminate flowers usually more than 25 in an umbel and plants with one or more other exceptions to the above (i.e., leaves more than 20, stems over 50 cm tall, peduncles more than 2 cm longer than petioles, tendrils present on several upper nodes). . . . . ***S. illinoensis***

Smilax ecirrata (Engelm.) S. Wats.

UPRIGHT CARRION-FLOWER   *native / cons* 5

Perennial herb. Stems annual, erect, usually without tendrils, or producing a few tendrils from the upper leaves only, to 8 dm tall. Leaves narrowly to broadly ovate, truncate to cordate at base, convexly narrowed to a short cusp, pubescent beneath. Umbels 1-3, to 25-flowered; peduncles arising from the axils of lance-linear bracts along the lower leafless portion of the stem, or rarely also from the axil of the lowest leaf. Perianth green, the tepals 4-6 mm long. Berries purplish black, globose, ca. 10 mm wide, not glaucous. May-June. — Rich deciduous woods.

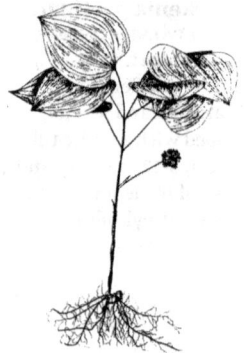

Smilax illinoensis Mangaly

ILLINOIS GREENBRIER   *native / cons* 5

Perennial herb. Stems annual, erect, unbranched, to 1 m long; prickles absent. Leaves narrowly ovate, pubescent and not glaucous on underside, base rounded to truncate, margins convex, tip acute to acuminate; petiole thin, equaling or longer than blade; tendrils few, short. Umbels 3-10, axillary to leaves and bracts, 10-50-flowered; tepals 3.5-4.5 mm long. Berries blue to black, globose. May-June. — Woods, thickets.

Smilax ecirrata

Smilax tamnoides L.

CHINAROOT   *native / cons* 5

Smilax hispida Muhl. ex Torr.

Vine, from a short knotty rhizome. Stems often climbing high (to 7 m or more), usually conspicuously thorny and often densely so; branches nearly terete. Leaves thin, ovate to rotund; commonly 8-12 cm long and 6-10 cm wide at maturity; acute to rounded or cuspidate; at base rounded, truncate, or cordate; not thickened at the margin; minutely serrulate (at 10x), at least near base; 5-7-nerved, the reticulate veinlets not prominently raised. Umbels many, axillary to leaves, to 25-flowered; peduncle often drooping, 1.5-6.5 cm long; perianth green to bronze. Berries black, globose, 6-10 mm wide, not glaucous. May-June. — Moist woods and thickets.

Smilax tamnoides

Typhaceae

CAT-TAIL FAMILY

Family now includes genus Sparganium from former family Sparganiaceae (discontinued under APG III).

1 Pistillate flowers in one to several spherical heads; perianth of greenish sepals; leaves strongly keeled (3-angled in cross-section). . . . . *Sparganium*
1 Pistillate flowers in an elongate densely flowered spike; perianth of white hairs; leaves flat-elliptic in cross-section . . . . *Typha*

Sparganium

BUR-REED

Perennial sedgelike herbs, floating or emergent in shallow water, from rhizomes and forming colonies. Stems stout, usually erect, unbranched, round in section. Leaves long, broadly linear, sheathing stem at base. Flowers crowded in round heads, the heads with either staminate or pistillate flowers; staminate heads few to many, borne above pistillate heads in a unbranched or sparsely branched inflorescence; the pistillate heads 1 to several, from leaf axils or borne above axils on upper stem; sepals and petals reduced to chaffy, spatula-shaped scales, these appressed to the achenes in the mature pistillate heads; staminate flowers with mostly 3-5 stamens; pistillate flowers with a 1-2-chambered pistil, stigmas 1 or 2. Fruit a beaked, nutletlike achene, stalkless or short-stalked.

1 Plants large, about 1 m tall; leaves usually erect; stigmas 2; achenes broadly oblong pyramid-shaped . . . . *S. eurycarpum*
1 Plants smaller, leaves erect or floating; stigmas 1; achenes slender 2
2 Fruiting heads about 1 cm wide; staminate head 1 (often absent by fruiting time); achene beaks less than 1 mm long . . . . *S. natans*
2 Fruiting heads 1.5 cm or more wide; staminate heads 2 or more; achene beaks 2 mm or more long 3
3 Fruiting heads 1.5–2 cm wide; anthers and stigma less than 1 mm long; leaves mostly flat . . . . *S. fluctuans*
3 Fruiting heads larger mostly 2–3 cm wide; anthers and stigma 1–4 mm long; leaves often keeled . . 4
4 Fruiting heads or branches all from leaf axils . . . . *S. americanum*
4 At least some fruiting heads or branches borne above leaf axils 5
5 Leaves floating; achene beak 1–3 mm long. . . . . *S. angustifolium*
5 Leaves usually stiffly erect and emersed; achene beak 3–5 mm long . . . . *S. emersum*

Sparganium americanum Nutt.

AMERICAN BUR-REED   *native / cons* **8**

Perennial herb. Stems stout, erect, mostly unbranched, 3-10 dm long. Leaves linear, flat to somewhat keeled, to 1 m long and 4-12 mm wide; leaflike bracts on upper stem shorter than leaves, widened at base. Inflorescence usually unbranched, or with a few, straight branches; pistillate heads sessile, 2-4 on main stem, sometimes with 1-3 on branches, 2 cm wide when mature; scales widest at tip; staminate heads 3-10 on main stem, sometimes with 1-5 on branches. Achenes widest at middle, tapered to both ends, dull brown, 3-5 mm long, the beak straight, 2-4 mm long. July-Aug. — Marshes, shallow water, streambanks.

Sparganium angustifolium Michx.

NARROW-LEAF BUR-REED   *native / cons* **9**

Sparganium acaule (Beeby) Rydb.
Sparganium chlorocarpum var. *acaule* (Beeby) Fern.
Sparganium emersum var. *angustifolium* (Michx.) Taylor & MacBryde
Sparganium multipedunculatum (Morong) Rydb.

Perennial herb. Stems long and usually floating. Leaves floating, mostly 2-3 mm wide, often wider at base. Inflorescence unbranched; pistillate heads 1-3, shiny, about 2 cm wide, the lowest stalked, the upper pistillate

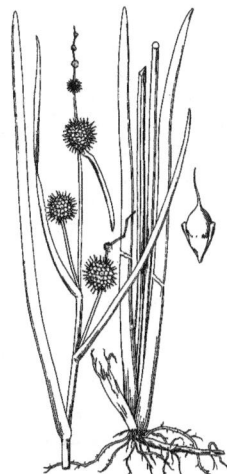

Sparganium americanum

heads sessile; scales spatula-shaped, ragged at tip; staminate heads 2-6, close together above pistillate heads. Achenes spindle-shaped, 5-7 mm long, dull brown except at red-brown base, abruptly contracted to a beak 1 mm long. July-Aug. — Lakes, ponds and shores.

Sparganium emersum Rehmann
NARROW-LEAF BUR-REED   *native / cons* **8**
  Sparganium chlorocarpum Rydb.
  Sparganium simplex Huds.
Perennial herb. Stems usually erect, sometimes lax and trailing in water, 2-6 dm long. Leaves linear, yellow-green, flat to keeled, 3-7 dm long and 3-6 mm wide, usually longer than stems; bracts leaflike, erect, barely widened at base. Inflorescence unbranched, 1-2 dm long; pistillate heads 1-4, sessile or lowest head often stalked, at least 1 head on stem above leaf axils, 1.5-2.5 cm wide when mature; scales spatula-shaped, widest at tip; staminate heads usually 2-5, 1.5-2 cm wide at flowering time. Achenes widest at middle, tapered to both ends, 4-5 mm long, shiny olive-green, the beak 3-5 mm long. June-Aug. — Shallow water or mud of marshes, streams, ditches, open bogs, ponds.

Sparganium emersum

Sparganium eurycarpum Engelm.
BROAD-FRUIT BUR-REED   *native / cons* **5**
  Sparganium californicum Greene
  Sparganium greenei Morong
Perennial herb. Stems stout, branched, 4-10 dm long. Leaves linear, bright green, keeled, 8-10 dm long and 5-12 mm wide; bracts leaflike, slightly widened at base. Inflorescence 1-3 dm long, branched from the bract axils; lower branches with 1 pistillate head and several staminate heads, main stem and upper branches with 6-10 staminate heads; pistillate heads 2-6, 1.5-2.5 cm wide in fruit, scales spatula-shaped; staminate heads numerous, 1-2 cm wide. Achenes oblong pyramid-shaped, 6-8 mm long, the top flattened, 4-7 mm wide, brown to golden-brown, the beak 2-4 mm long. June-Aug. — Usually in shallow water of marshes, streams, ditches, ponds and lakes, often with cat-tails (*Typha*).

Sparganium eurycarpum

Sparganium fluctuans (Morong) B. L. Robins.
FLOATING BUR-REED   *native / cons* **10**
Perennial herb. Stems slender, floating, to 15 dm long. Leaves floating, linear, flat, translucent, 3-10 mm wide, underside with netlike veins; bracts leaflike, short, widened at base. Inflorescence usually branched, the main stem with 2-4 staminate heads, the branches with 1 pistillate head near base and 2-3 staminate heads above; pistillate heads 2-4, 1.5-2 cm wide when mature, scales oblong; staminate heads to 1 cm wide. Achenes obovate, 3-4 mm long, sometimes narrowed near middle, brown, the beak curved, 2-3 mm long. — In shallow water of ponds and lakes.

Sparganium natans L.
ARCTIC BUR-REED   *native / cons* **9**
  Sparganium minimum (Hartman) Wallr.
Perennial herb. Stems usually long and floating, sometimes shorter and upright, 1-3 dm or more long. Leaves linear, dark green, thin, flat, 2-6 mm wide; bracts leaflike, short, somewhat widened at base. Inflorescence unbranched; pistillate heads 2-3, from bract axils, sessile or the lowest sometimes short-stalked, 1 cm wide when mature; scales spatula-shaped, widest at tip; staminate heads usually 1 (rarely 2). Achenes broadly oval, 3-4 mm long, dull green-brown, the beak 1-2 mm long. — Shallow water, pond margins.

Sparganium fluctuans

Typha
CAT-TAIL

Large, familiar, reedlike perennials, from fleshy rhizomes and forming colonies. Stems erect, unbranched, round in section, sheathed for most of length by overlapping leaf sheaths. Leaves mostly near base of plant, alternate in 2 ranks, erect, linear, spongy. Flowers tiny, either staminate or pistillate, separate on same plant; petals and sepals reduced to bristles. Staminate flowers usually of 3-5 stamens, bristles absent or 1-3 or more. Pistillate flowers intermixed with some sterile flowers; pistil 1, raised on a short stalk (gynophore), with numerous bristles near base, the bristles longer than pistil; small bracts (bractlets) also sometimes present, these intermixed with the bristles, slender but with a widened brown tip. Heads with staminate flowers above pistillate in a single, dense, cylindric spike, the staminate and pistillate portions of the spike unalike, contiguous in broad-leaf cat-tail (*T. latifolia*) or separated in narrow-leaf cat-tail (*T. angustifolia*); the mature spike brown and fuzzy in appearance due to the crowded stigmas and gynophore bristles. Fruit a yellow-brown achene, 1-2 mm long, the style persistent, long and slender with an expanded stigma.

A hybrid between *T. angustifolia* and *T. latifolia* is termed *Typha* × *glauca* Godr. Usually larger than either parent, staminate and pistillate portions of hybrid plants are usually separated by a space to 4 cm long. The staminate portion of the spike is light brown, 0.5-2 dm long and about 1 cm wide at flowering time; the pistillate portion is dark brown, 10-20 cm long and 1-2 cm wide. Since *Typha* × *glauca* is sterile, reproduction is vegetative by rhizomes. The hybrid can occur wherever populations of *T. angustifolia* and *T. latifolia* overlap.

1 Staminate and pistillate portions of spike usually separated; leaves to 1 cm wide; stigmas long and slender, pale brown. . . . . *T. angustifolia*
1 Staminate and pistillate portions of spike usually contiguous, not separated; leaves mostly 1–2 mm wide; stigmas broad and flattened, dark brown . . . . *T. latifolia*

Typha angustifolia L.
NARROW-LEAF CAT-TAIL   *introduced*

Perennial emergent herb. Stems erect, 1-2 m long. Leaves upright, flat, 4–10 mm wide. Flowers either staminate or pistillate, on separate portions of the spike, separated by an interval of 2-10 cm; staminate portion 7-20 cm long and 7-15 mm wide, staminate bractlets brown; pistillate portion of spike dark brown, 10-20 cm long and 1-2 cm wide; each flower with 1 bristlelike bractlet, these flat and brown at the widened tip, gynophore hairs brown-tinged at tips; stigmas pale brown, linear, 1 mm long. Fruit 5-7 mm long, subtended by many fine hairs, the hairs slightly widened and brown at tip. June. — Marshes, lakeshores, streambanks, roadside ditches, pond margins, usually in shallow water; more tolerant of brackish conditions than *Typha latifolia*.

Typha latifolia L.
BROAD-LEAF CAT-TAIL   *native / cons* 1

Perennial emergent herb. Stems erect, 1-2.5 m long. Leaves upright, mostly 1-2 cm wide. Flowers either staminate or pistillate, the staminate and pistillate portions of spike normally contiguous, rarely separated by 3-4 mm; staminate portion 5-15 cm long and 1.5-2 cm wide at flowering time, staminate bractlets white; pistillate portion of spike dark brown, 10-15 cm long and 2-3 cm wide when mature, pistillate bractlets absent, gynophore hairs white; stigma lance-shaped, becoming dark brown, less than 1 mm long. Fruit 1 cm long, with many white, linear hairs from base. June. — Marshes, lakeshores, streambanks, ditches, pond margins, usually in shallow water; less tolerant of brackish conditions than *Typha angustifolia*.

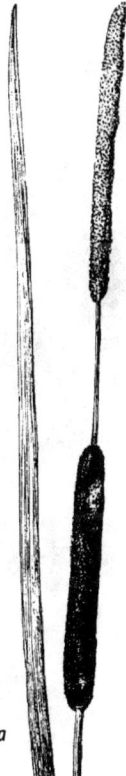

Typha angustifolia

404

Cave Point County Park, Door County.

FAMILY KEYS

SUMMARY OF FAMILY GROUP KEYS

Ferns and Fern Relatives

Fern relatives (horsetails, quillworts, clubmosses); leaves unlobed, awl-shaped, scalelike, or grasslike, and not "fern-like". . . . . GROUP 1
Small ferns, growing on rock or on rock in thin soil. . . . . GROUP 2
Small ferns, growing in soil, not associated with rock outcrops. . . . . GROUP 3
Medium to large ferns, growing on rock, or over rock in thin soil . . . . GROUP 4
Medium to large ferns, growing in soil (not on rock outcrops). . . . . GROUP 5

Seed Plants

Aquatic plants, with all leaves underwater or floating on water surface . . . . . . GROUP 6
Woody plants (trees, shrubs, woody vines, and small evergreen creeping plants). . . . . GROUP 7
Herbaceous plants lacking both green color and developed leaves at flowering time . . . . . GROUP 8
Inflorescence apparently converted to bulblets, tufts of leaves, etc . . . . GROUP 9
Monocots; leaves with parallel veins. . . . . GROUP 10
Flowers in an involucrate head (i.e., the flowers clustered in a head above a whorl of bracts) GROUP 11
Herbaceous plants with single-sex flowers . . . . GROUP 12
Herbaceous dicots, flowers bisexual
    Perianth of 1 series (calyx or corolla present but not both)
        Ovary inferior . . . . GROUP 13
        Ovary superior . . . . GROUP 14
    Perianth of 2 series (both calyx and corolla present)
        Ovaries 2 or more in each flower . . . . GROUP 15
        Ovary inferior . . . . GROUP 16
        Ovary superior
            Stamens more numerous than the petals. . . . . GROUP 17
            Stamens the same number as the petals or fewer, and petals separate . . . . GROUP 18
            Corolla regular and stamens the same number as its lobes, and petals united . . . . GROUP 19
            Corolla either bilaterally symmetrical or stamens fewer than its lobes, or both and petals . united . . . . GROUP 20

FERN AND FERN RELATIVES GROUPS

Includes true ferns and fern relatives (quillworts, clubmosses, spike-mosses).

1 Leaves not "fern-like;" unlobed, variously awl-shaped, scale-like, or grass-like (the fern relatives will key here plus *Asplenium rhizophyllum*). . . . . GROUP 1
1 Leaves "fern-like," variously lobed or divided (true ferns will key here; the following groups are based on size of frond and habitat, either soil or rock) . . . .
  2
2 Plants small, leaf blades (not including the stipe) small, less than 30 cm long or wide (some species will key both here and in the next lead no. 2). . . . .
  3
3 Plants growing on rock, rock walls, or over rock in thin soil . . . . GROUP 2
3 Plants terrestrial, growing in soil, not associated with rock outcrops. . . . . GROUP 3
2 Plants larger, leaf blades medium to large, more than 30 cm long or wide . . . .
  5
4 Plants growing on rock, or over rock in thin soil mats or pockets of soil . . . . . . GROUP 4
4 Plants growing in soil, not associated with rock outcrops. . . . . GROUP 5

GROUP 1

Fern relatives; leaves unlobed, awl-shaped, scalelike, or grasslike, and not "fern-like."

1 Stems obviously jointed; leaves small and scalelike, in a whorl from the joints or nodes, or sometimes absent; spores borne in a terminal conelike strobilus covered with peltate scales (i.e., scales more or less round and attached at middle like an umbrella) . . . . EQUISETACEAE
1 Stems not jointed; leaves scalelike or larger, but if scalelike not in whorls from the nodes; spores borne variously, but if in a terminal strobilus the scales not peltate 2
2 Leaves very numerous and overlapping along creeping, ascending, or erect stems; the leaves usually scalelike or awl-like, 0.5-2 (-3) mm wide, typically sharp- or hair-tipped; sporangia borne in strobili 3
2 Leaves not as above 6
3 Sporangia borne in flattened or 4-sided strobili sessile at tips of leafy branches; spores and sporangia of two sizes, the megasporangia larger and borne at base of strobili . . . . SELAGINELLACEAE
3 Sporangia borne either in axils of normal foliage leaves, or in strobili sessile at tips of leafy branches, or stalked on specialized branches with fewer and smaller leaves; spores and sporangia of one size
. . . . 4
4 Leafy stems erect, simple or dichotomously branched, the ultimate branches upright; sporophylls like the sterile leaves or only slightly smaller, in annual bands along the stem; vegetative reproduction by leafy gemmae near tip of stem (*Huperzia*) . . . . LYCOPODIACEAE
4 Leafy stems prostrate or erect, if erect then generally branched, the ultimate branches spreading (horizontal) or ascending; sporophylls differing from sterile leaves, either broader and shorter, or more spreading, grouped into terminal cones; lacking vegetative reproduction by gemmae. 5
5 Shoots flat-branched, 1-5 mm wide (including the leaves); leaves scalelike, dimorphic, overlapping and appressed to stem, in 4 ranks; strobili on long, branched stalks (*Diphasiastrum*). . . . .
. . . . LYCOPODIACEAE
5 Shoots round-branched, usually 5-8 mm wide (including the leaves), leaves awl-shaped, monomorphic (though sometimes differing in size), separate, spreading or ascending, in 6 ranks; strobili sessile at stem tips (*Lycopodium*) . . . . LYCOPODIACEAE
6 Plants with 1 (-several) leaves, the sterile leaf blade ovate, entire-margined, obtuse, the longer fertile portion with 2 rows of sporangia somewhat embedded in it (*Ophioglossum*) . . . .
. . . . OPHIOGLOSSACEAE
6 Plant with many leaves, generally 5 or more, not divided into separate sterile and fertile segments, the leaves lance-shaped with a long-tapering tip (often with a plantlet which can root to form new plants) (*Asplenium rhizophyllum*) . . . . ASPLENIACEAE

GROUP 2

Small ferns, growing on rock or on rock in thin soil.

1 Fronds pinnatifid or 2-pinnatifid, most of the pinnae not fully divided from one another (the rachis winged by leaf tissue for most or all its length). 2
1 Fronds pinnate, pinnate-pinnatifid, 2-pinnate, or even more divided (rachis naked for most of its length, but often winged in upper portion) 4
2 Fronds 2-pinnatifid, at least the lowermost pinnae deeply lobed (*Phegopteris*) THELYPTERIDACEAE
2 Fronds 1-pinnatifid 3
3 Blades with long, narrow tapering tip, upper portion of blade unlobed or only slightly lobed; sori elongate (*Asplenium*) . . . . ASPLENIACEAE
3 Fronds without a long, narrow tapering tip; blade lobed for most of its length; sori round (Polypodium). . . . . POLYPODIACEAE
4 Fronds 1-pinnate or 1-pinnate-pinnatifid 5
4 Fronds 2-pinnate or more divided 7
5 Sori on the undersurface of the leaf, away from the margins (*Asplenium*) . . . . ASPLENIACEAE
5 Sori on the undersurface of the leaf, along margins and more-or-less hidden beneath either the unmodified inrolled leaf margin or under a modified, reflexed false indusium 6
6 Stipes green to straw-colored for at least the upper 1/3, rachis green; fronds dimorphic, the fertile longer than the sterile and with narrower segments (*Cryptogramma*) . . . . PTERIDACEAE

6   Stipes and rachis dark brown to almost black throughout; fronds similar or somewhat different (Pellaea). . . . . PTERIDACEAE
7   Blade broadly triangular in outline (*Gymnocarpium*) . . . . CYSTOPTERIDACEAE
7   Blade elongate, mostly lance-shaped, generally 4x or more as long as wide, not notably triangular in outline 8
8   Sori on margins, usually more-or-less hidden under the inrolled margin of the pinnule 9
8   Sori not on margins, slightly to strongly hidden by indusia 11
9   Sori round or oblong, distinct and separate along the pinnule margins; fronds bright-green, smooth, herbaceous, delicate, and flexible (*Adiantum pedatum*) . . . . PTERIDACEAE
9   Sori continuous along the pinnule margins; fronds mostly dark-green, often hairy, leathery, tough, and stiff. 10
10 Fronds 2-3-pinnate, more or less densely hairy (*Cheilanthes*). . . . . PTERIDACEAE
10 Fronds 1-2-pinnate, smooth or sparsely and inconspicuously hairy (*Pellaea*) . . . . PTERIDACEAE
11 Blades 3-12 cm long; sori elongate, covered by a flap-like, entire indusium (*Asplenium*) . . . .
. . . . ASPLENIACEAE
11 Blades 4-30 (-50) cm long; sori round, surrounded or covered by an entire, fringed, or divided indusium (*Cystopteris*). . . . . CYSTOPTERIDACEAE

GROUP 3

Small ferns, terrestrial, growing in soil, not associated with rock outcrops.

1   Stipe branched once dichotomously, each branch with 3-7 pinnae in one direction only, the outline of the blade fan-shaped, often wider than long (*Adiantum pedatum*). . . . . PTERIDACEAE
1   Stipe not branched dichotomously, the outline of the blade either longer than wide or triangular and about as wide as long 2
2   Fronds pinnatifid or 2-pinnatifid, most of the pinnae not fully divided from one another (the rachis winged by leaf tissue for most or all of its length). 3
2   Fronds 1-pinnate, 1-pinnate-pinnatifid, 2-pinnate, or even more divided (the rachis naked for most of its length, or often winged in the upper portion) 5
3   Sporangia borne on an erect stalk that arises at or above ground level from stipe of sterile leaf blade (joining stipe of sterile leaf above the rhizome) (*Botrychium, Botrypus*). . OPHIOGLOSSACEAE
3   Sporangia either borne on normal leaf blades or on specialized (fertile) fronds 4
4   Fronds all alike, sori on normal leaf blades (*Phegopteris*) . . . . THELYPTERIDACEAE
4   Fronds of two types; sori on fronds significantly different than normal fronds (*Onoclea sensibilis*) .
. . . . ONOCLEACEAE
5   Fronds broadly triangular in outline, about as broad as long; sporangia borne on an erect stalk that arises at or above ground level from the stipe of the sterile leaf blade (joining the stipe of the sterile leaf above the rhizome) 6
5   Fronds lance-shaped in outline, much longer than broad; sporangia either borne on normal leaf blades, on slightly dimorphic blades, or on an erect stalk that arises at or above ground level from the stipe of the sterile leaf blade (joining the stipe of the sterile leaf above the rhizome). 7
6   Sporangia borne on normal leaf blades (*Gymnocarpium*). . . . . CYSTOPTERIDACEAE
6   Sporangia borne on an erect stalk that arises at or above ground level from the stipe of the sterile leaf blade (joining the stipe of the sterile leaf above the rhizome) (*Sceptridium*) OPHIOGLOSSACEAE
7   Blades 1-8 cm long; sporangia borne on an erect stalk that arises at or above ground level from the stipe of the sterile leaf blade (joining the stipe of the sterile leaf above the rhizome) (*Botrychium*).
. . . . OPHIOGLOSSACEAE
7   Blades 10-30 (-100) cm long; sporangia either on normal leaf blades or on slightly modified blades 8
8   Fronds evergreen, dark green, somewhat leathery (*Polystichum*) . . . . DRYOPTERIDACEAE
8   Fronds light to medium green, herbaceous, deciduous to semi-evergreen 9
9   Sori elongate; leaf blades somewhat dimorphic, the fertile larger and erect, the sterile smaller and prostrate, the larger leaf blades 2-4 (-6.5) cm wide (*Asplenium platyneuron*) . . . . ASPLENIACEAE
9   Sori round; leaf blades monomorphic; the larger leaf blades 5-15 cm wide (*Thelypteris*) . . . .
. . . . THELYPTERIDACEAE

GROUP 4

Medium to large ferns, growing on rock, or over rock in thin soil.

1   Fronds 1-pinnate-pinnatifid or less divided, the pinnae entire, toothed, lobed or pinnatifid 2

1   Fronds 2-pinnate or more divided, the pinnae divided to their midribs 5

2   Sori elongate, the indusium flap-like, attached along the side; leaf blades (if more than 30 cm long) less than 7 cm wide (*Asplenium platyneuron*) . . . . ASPLENIACEAE

2   Sori circular or globular, the indusium peltate, kidney-shaped, or cuplike; leaf blades (if more than 30 cm long) more than 5 cm wide 3

3   Fronds 1-pinnate, the pinnae toothed and each with a slight to prominent lobe near the base on the side towards the leaf tip, dark green, somewhat leathery; indusia peltate (*Polystichum*). . . . . DRYOPTERIDACEAE

3   Fronds 1-pinnate-pinnatifid, the pinnae pinnatifid, generally lacking a prominent basal lobe, light green to dark green, herbaceous to slightly leathery; indusium either kidney-shaped or cuplike 4

4   Vascular bundles in the stipe 3-7 (*Dryopteris*) . . . . DRYOPTERIDACEAE

4   Vascular bundles in the stipe 2, uniting above (*Thelypteris*). . . . . THELYPTERIDACEAE

5   Sori marginal and borne on underside of the false indusium; stipes and rachis shiny black or reddish-black, glabrous except at the very base of the stipe; pinnules fan-shaped or obliquely elongate (*Adiantum*). . . . . PTERIDACEAE

5   Sori not marginal, borne on undersurface of leaf blade (or if marginal, as in *Pteridium*, borne on undersurface of the leaf); stipes darkened only near base (if at all), rachis green, tan, or reddish; pinnules not notably fan-shaped or obliquely elongate 6

6   Blades broadly triangular in outline, about as long as wide (*Pteridium aquilinum*). . . . . DENNSTAEDIACEAE

6   Blades elongate, mostly lanceolate, generally 4x or more as long as wide. 7

7   Vascular bundles (3-) 5 (-7) in the stipe; mostly larger woodland ferns (*Dryopteris*). . . . . DRYOPTERIDACEAE

7   Vascular bundles 2 in the stipe (or joining near the leaf blade into 1); ferns of woods and rocky places. 8

8   Fronds 10-30 cm wide, the tip acute to acuminate; indusium flaplike (*Athyrium angustum*) . . . . ATHYRIACEAE

8   Fronds 4-9 cm wide, the tip long-attenuate; indusium pocketlike or hoodlike (*Cystopteris bulbifera*) . . . . CYSTOPTERIDACEAE

GROUP 5

Medium to large ferns, growing in soil (not on rock outcrops).

1   Blades broadly (about equilaterally) triangular, pentagonal, or flabellate in outline, 0.7 1.3x as long as wide. 2

1   Fronds elongate in outline, mostly ovate, lanceolate, oblanceolate, or narrowly triangular, 1.5-10x or more as long as wide 4

2   Blades fan-shaped in outline, the stipe branched once dichotomously, each branch bearing 3-7 pinnae (*Adiantum pedatum*). . . . . PTERIDACEAE

2   Blades broadly triangular in outline, the stipe not branched dichotomously 3

3   Sporangia in a stalked, specialized, fertile portion of the blade; texture of mature blades somewhat fleshy; plants solitary from a short underground rhizome with thick, mycorrhizal roots; plants of moist woods (*Botrypus virginianus*) . . . . OPHIOGLOSSACEAE

3   Sporangia in marginal, linear sori, indusium absent, protected by the revolute leaf margin and a minute false indusium; texture of mature leaf blades hard and stiff; plants colonial from deep rhizomes; plants of moist to dry woodlands and openings (*Pteridium aquilinum*). DENNSTAEDIACEAE

4   Fronds 2-pinnate or more divided, the pinnae divided to their midribs 5

4   Fronds 1-pinnate-pinnatifid or less divided; the pinnae entire, toothed, lobed or pinnatifid 8

5   Blade divided into sterile and fertile portions; sterile pinnae located below terminal fertile pinnae, the sterile pinnules 30-70 mm long and 8-23 mm wide, finely toothed, tip rounded to somewhat pointed; fertile pinnae greatly reduced in size, the fertile pinnules 7-11 mm long and 2-3 mm wide (*Osmunda regalis*). . . . . OSMUNDACEAE

5 Blade not divided into sterile and fertile portions, the pinnules bearing sporangia only slightly if at all reduced in size, both fertile and sterile pinnules mostly 4-20 mm long and 2-10 mm wide . 6

6 Vascular bundles (3-) 5 (-7) in the stipe (*Dryopteris*) . . . . DRYOPTERIDACEAE

6 Vascular bundles 2 in the stipe (or joining upwards near leaf blade into 1) 7

7 Fronds more than 10 cm wide, the tip acute to acuminate; indusium flaplike; pealike bulblets absent (*Athyrium angustum*) . . . . ATHYRIACEAE

7 Fronds 4-9 cm wide, the tip long-tapering; indusium pocketlike or hoodlike; bulblets often present on upper portion of blade (*Cystopteris bulbifera*) . . . . CYSTOPTERIDACEAE

8 Fronds 1-pinnate or 1-pinnate-pinnatifid, the pinnae fully divided from one another (rachis naked for most of its length, but often winged in upper portion); fronds dimorphic or not (*Onoclea sensibilis*) . . . . ONOCLEACEAE

8 Fronds 1-pinnatifid, most of the pinnae not fully divided from one another (rachis winged by leaf tissue for most or all of its length); fronds dimorphic, the fertile much modified, stiff and/or woody 9

9 Rhizomes long-creeping, fronds scattered, forming patches (*Thelypteris*) . . . THELYPTERIDACEAE

9 Rhizomes short-creeping, the fronds clustered, not forming patches (or rhizomes of both long and short, but fronds borne only in clusters on the short erect rhizomes in *Matteucia*) 10

10 Plants medium to large, fronds typically 60-300 cm tall; fronds either strongly dimorphic, the fertile fronds very unlike the sterile, brown at maturity (*Matteucia* and *Osmunda cinnamomea*) or fertile pinnae very unlike the sterile, brown at maturity, borne as an interruption in the blade, with normal green pinnae above and below (*Osmunda claytoniana*); rachises scaleless, stipes scaleless (except at the base in *Matteucia*) 11

10 Plants mostly smaller, the fronds 30-100 cm tall; fronds not at all or only slightly dimorphic, the fertile differing in various ways, such as having narrower pinnae (as in *Polystichum acrostichoides* and *Thelypteris palustris*) or fertile fronds taller and more deciduous (as in *Asplenium platyneuron* and *Dryopteris cristata*), but not as described in the first lead; rachises and stipes variously scaly or scaleless, but at least the stipe and often also the rachis scaly if the plants over 1 m tall 12

11 Fronds strongly tapering to the base from the broadest point (well beyond the midpoint of the blade), lowermost pinnae much less than 1/2 as long as the largest pinnae (*Matteucia struthiopteris*) . . . . ONOCLEACEAE

11 Fronds slightly if at all tapering to the base, about equally broad through much of their length, lowermost pinnae much more than 1/2 as long as the largest pinnae (*Osmunda*). . OSMUNDACEAE

12 Sori elongate, the indusium elongate, attached along one side as a flap 13

12 Sori roundish; the indusium kidney-shaped or nearly round, attached by a central stalk, or sometimes absent 15

13 Stipe and rachis lustrous brownish black; fertile fronds 2-8 (-12) cm wide (*Asplenium platyneuron*) . . . . ASPLENIACEAE

13 Stipe and rachis green; fertile fronds 10-20 (-30) cm wide 14

14 Fronds 1-pinnate-pinnatifid (the pinnae pinnatifid) (*Deparia acrostichoides*) . . . . ATHYRIACEAE

14 Fronds 1-pinnate (the pinnae entire) (*Diplazium pycnocarpon*) . . . . ATHYRIACEAE

15 Fronds 1-pinnate, the pinnae toothed and each with a slight to prominent lobe near the base on the side towards the leaf tip, dark green, leathery; indusia peltate (round, stalk attached to the center) (*Polystichum acrostichoides*) . . . . DRYOPTERIDACEAE

15 Fronds 1-pinnate-pinnatifid, the pinnae pinnatifid, generally without prominent basal lobe, light green to dark green, herbaceous to somewhat leathery; indusium kidney-shaped 16

16 Vascular bundles in the stipe 4-7 (*Dryopteris*) . . . . DRYOPTERIDACEAE

16 Vascular bundles in the stipe 2, uniting upwards (*Thelypteris*) . . . . THELYPTERIDACEAE

SEED PLANT GROUPS

Includes leads to 15 Group Keys and to several families with specialized features.

1 Plants aquatic, the leaves or plant body entirely submersed or floating on the surface of the water (at most, the inflorescence and bracts, not leaves, held above the surface) .... GROUP 6

1 Plants with at least some leaves or stems above the water, or plants terrestrial 2

2 Plants woody (trees, shrubs, and woody vines), with erect, trailing, or viny above ground stems living through the winter and continuing to grow the next season (leaves may be evergreen or deciduous). .... GROUP 7

2 Plants herbaceous, the perennial parts, if any, below or on the surface of the ground (to which the stems die back each year), not producing woody stems which survive the winter well above ground (therefore, without aerial evergreen leaves although there may be basal winter-green leaves)... 3

3 Plant lacking green color (often wholly parasitic or saprophytic) and the leaves none at flowering time or reduced to tiny scales) .... GROUP 8

3 Plant with green color and the leaves usually developed. 4

4 Inflorescences producing only small bulblets or tufts of little leaves (or modified floral parts), but no flowers or fruit. .... GROUP 9

4 Inflorescences normal 5

5 Perianth parts (2), 3, (4), or 6 (never 5) and leaves (or other green photosynthetic parts when leaves are absent or reduced) parallel-veined (the 3 or more main veins running from base of blade to apex and ± parallel, with or without minute cross-veins), entire, simple .... GROUP 10

5 Perianth parts various (often 5) but leaves netted-veined (or with only the midvein conspicuous), entire or toothed, simple or compound (the main veins, if more than 1, branching and ± reticulate) 6

6 Inflorescence a dense "head" (either a true head or a spadix), of a few to many small sessile flowers on a common receptacle (not merely an elongate spike), subtended by 1 or more small or large bracts .... GROUP 11

6 Inflorescence not an involucrate head, or if head-like the individual flowers short-pediceled and/or the "head" not immediately subtended by 1 or more bracts 7

7 Inflorescence of "false flowers" consisting of small cup-like structures (uniform in texture and not composed of separate parts like bracts or scales) each bearing 1-5 glands on its rim (sometimes with additional petaloid appendages) and including 2 or more stamens and 1 central stalked 3-lobed pistil (which ripens into an exserted, 3-lobed capsule); sap milky (*Euphorbia*) .... EUPHORBIACEAE

7 Inflorescence various, but not composed of such structures; pistil only rarely stalked (and if so, not 3-lobed); sap various 8

8 Anthers and stigma fused into a central structure obscuring the individual reproductive parts; ovaries 2, ripening into follicles, the seeds each with a tuft of hairs (except *Vinca*); sap milky..... .... APOCYNACEAE

8 Anthers and stigmas not fused to each other, of diverse but recognizable structure; ovaries, seeds, and sap various but not combined as above 9

9 Flowers unisexual, containing one or more stamens or pistils, but not both .... GROUP 12

9 Flowers all or mostly bisexual, containing both stamen(s) and pistil(s) (although these may not all be equally mature at the same time) 10

10 Perianth none. 11

10 Perianth present (but not always conspicuous). 12

11 Leaves deeply lobed or compound .... RANUNCULACEAE

11 Leaves unlobed, entire (*Callitriche, Hippuris*) .... PLANTAGINACEAE

12 Perianth of only one series (calyx or corolla) 13

12 Perianth of two series (both calyx and corolla) 14

13 Ovary inferior. .... GROUP 13

13 Ovary (or ovaries) superior. .... GROUP 14

14 Ovaries 2 or more in each flower. .... GROUP 15

14 Ovary 1 in each flower (styles or stigmas may be separate) 15

15 Ovary inferior. .... GROUP 16

15 Ovary superior 16
16 Stamens more numerous than the petals . . . . GROUP 17
16 Stamens the same number as the petals or lobes (not lips) of the corolla, or fewer. 17
17 Petals separate . . . . GROUP 18
17 Petals connate at least at the base 18
18 Corolla regular and the stamens the same number as its lobes . . . . GROUP 19
18 Corolla either bilaterally symmetrical or the stamens fewer than its lobes (not lips), or both con-
   ditions present. . . . . GROUP 20

GROUP 6

Aquatic plants, with all leaves underwater or floating on water surface.

1 Plants without distinct stem and leaves, free-floating at or below surface of water (except where
   stranded by drop in water level), the segments (internodes) small (to 15 mm long, but in most
   species much smaller), often remaining attached where budded from parent plant 2
1 Plants with distinct stem and/or leaves, usually anchored in substrate, mostly larger 3
2 Plant body once to several times equally 2-lobed or 2-forked . RICCIACEAE (a family of liverworts)
2 Plant body not consistently 2-lobed or 2-forked . . . . ARACEAE
3 Plants with floating leaves present (blades, or at least their terminal portions, floating on the
   surface of the water, usually ± smooth and firm in texture, especially compared with submersed
   leaves, or submersed leaves none) 4
3 Plants without any floating leaves, entirely submersed (except sometimes for inflorescences and
   associated bracts) 11
4 Blades of some or all floating leaves on a plant sagittate or deeply lobed at base 5
4 Blades of floating leaves all unlobed (at most subcordate at base), simple, the petiole small or
   absent in ribbon-like leaves. 6
5 Floating blades (at least some of them) sagittate, the tip and lobes acute (Note that plants with
   sagittate leaves extending above the water surface will not key here) (*Sagittaria*). ALISMATACEAE
5 Floating (and any other) blades circular to ± elliptic in outline, rounded at tip with deep sinus at
   base. . . . . NYMPHAEACEAE
6 Floating leaves small (less than 1 cm long), crowded in a terminal rosette; submersed leaves dis-
   tinctly opposite; flowers solitary, axillary (*Callitriche*). . . . . PLANTAGINACEAE
6 Floating leaves larger, not in a rosette; submersed leaves alternate, basal, or absent; flowers mostly
   in a terminal inflorescence 7
7 Leaves narrow and ribbon-like, the blades many times longer than wide, without distinct petiole
   (though in some species a sheath surrounds the stem). 8
7 Leaves (at least floating ones) with ± elliptic blades and distinct petioles 9
8 Leaves ± rounded at tip (even if tapered), the floating portion smooth and shiny, somewhat yellow-green to bright green when fresh, occasionally keeled but midvein scarcely if at all more
   prominent than others; leaf not differentiated into blade and sheath, the submersed portion
   similar to the floating but more evidently with a fine closely checkered pattern; flowers and fruit
   in spherical heads (*Sparganium*) . . . . TYPHACEAE
8 Leaves sharply acute at tip, the floating portion rather dull, ± blue-green when fresh, with midrib;
   leaf including a sheath around stem and a membranous ligule at junction of sheath and blade;
   flowers and fruit in paniculate spikelets . . . . POACEAE
9 Leaves all basal; petals 3, white . . . . ALISMATACEAE
9 Leaves cauline (along the stem), alternate or opposite; petals 4-6, pink or dull and inconspicuous
   10
10 Veins netted; flowers bright pink, in dense ovoid to cylindrical spikes (*Persicaria amphibia*) . . . .
   . . . . POLYGONACEAE
10 Venation parallel; flowers dull, in narrow cylindrical spikes . . . . POTAMOGETONACEAE
11 Leaves (or leaf-like structures) all basal and simple 12
11 Leaves cauline, simple or compound (basal and dissected in one species) 24
12 Leaves flat, widest about the middle or parallel-sided 13
12 Leaves (or similar vegetative stems) filiform or terete or only slightly flattened (especially basally),
   elongate and limp to short and quill-like, less than twice as broad as thick 18

13 Leaf blades not over twice as long as wide juvenile............................ NYMPHAEACEAE
13 Leaf blades more than twice as long as broad 14
14 Leaves stiff and erect or somewhat outcurved, less than 20 cm long..........................15
14 Leaves limp, more than 20 cm long, ribbon-like..16
15 Base of leaf somewhat sheathing, with a membranous ligule (as in a grass) at base of spreading blade (*Pontederia*).. PONTEDERIACEAE
15 Base of leaf not sheathing and with no ligule (*Sagittaria*)...................... ALISMATACEAE
16 Midvein not evident, all veins of essentially equal prominence, with the tiny cross-veins giving a checkered appearance to the leaf, which is thus uniformly marked with minute rectangular cells ca. 1-2 mm long or smaller (*Sparganium*).. TYPHACEAE
16 Midvein (and usually some additional longitudinal veins) evident, the veins not all of equal prominence, not dividing the leaf into minute rectangular cells 17
17 Leaves with the central third (or more) of distinctly different pattern (more densely reticulate) than the two marginal zones; plants dioecious, the staminate flowers eventually liberated from a dense inflorescence submersed at base of plant, the pistillate solitary on a long ± spiraled stalk which reaches the surface of the water; plants without milky juice (*Vallisneria*) ....
.... HYDROCHARITACEAE
17 Leaves ± uniform in venation, not 3-zoned; plants monoecious, with emergent inflorescence of white-petaled flowers (but these scarce on plants with submersed tape-like leaves); plants often with milky juice (*Sagittaria*) .... ALISMATACEAE
18 Major erect structures solitary, spaced along a simple or branched delicate rhizome, consisting either of rather yellowish stems bearing minute alternate bumps as leaves or of filiform leaves mostly buried in the substrate and with a few minute bladder-like organs................... 19
18 Major erect structures solitary to densely tufted, consisting of filiform or quill-like leaves or stems, with neither alternate bumps or bladders 20
19 Leaves merely minute alternate bumps on stem; bladders not present; flowers sessile, inconspicuous, regular (*Myriophyllum tenellum*)..................................... HALORAGACEAE
19 Leaves filiform, mostly buried in substrate (only the green tips, incurled when young, protruding); bladders (minute) usually present on the delicate branching rhizomes and buried leaf bases; flowers short-pediceled, showy (yellow or purple), bilaterally symmetrical (*Utricularia*).........
.... LENTIBULARIACEAE
20 Leaves very limp (retaining no stiffness when removed from water and hence irregularly sinuate, bent, or matted on herbarium specimens) though a stiffer straight stem may also be present, mostly more than 20 cm long, ca. 0.2-1 mm wide 21
20 Leaves usually firm (retaining stiffness when removed from water and hence straight or with an even curve in herbarium specimens), less (in most species much less) than 20 cm long, of various widths... 22
21 Leaves slightly expanded basally for ca. (0.7-) 2-10 cm, sheathing the next inner leaf at least dorsally (usually the sheath continued ventrally as an almost invisible membrane), with tiny ligule or pair of auricles at the summit; rhizome various; inflorescence a lateral spikelet or terminal cyme (*Schoenoplectus subterminalis*)..................................... CYPERACEAE
21 Leaves (actually vegetative stems) terete their entire length, not expanded basally nor sheathing each other, but each separate and closely surrounded at base for ca. (0.6-) 1 cm or more by a very delicate membranous tubular sheath (this may require careful dissection to distinguish); rhizome less than 2 mm in diameter; inflorescence (rare on plants otherwise entirely submersed) a single strictly terminal spikelet (*Eleocharis acicularis*) .... CYPERACEAE
22 Leaves filiform throughout, not broader basally nor sheathing each other, solitary (rarely) or in small tufts along a filiform whitish rhizome, each leaf (actually a vegetative stem) closely surrounded at its base for ca. 6 mm or more by a very delicate membranous tubular sheath (this sometimes requiring careful dissection to distinguish); inflorescence (rare on completely submersed plants) a single terminal spikelet (*Eleocharis acicularis*) .... CYPERACEAE
22 Leaves linear or tapered from base to apex, or if otherwise uniformly filiform then expanded at base or sheathing each other, without individual tubular sheaths as described above; inflorescence various 23
23 Leaves somewhat flattened at least basally, widest at the base, gradually tapered to sharp apex; plants with buried rhizome (*Juncus pelocarpus*)................................. JUNCACEAE

23 Leaves ± terete, scarcely or no wider at base than at middle, of ± uniform width at least to the middle (or even slightly thicker there before tapering to apex); plants with green stolons strongly arching above substrate (*Ranunculus flammula*). . . . . RANUNCULACEAE

24 Leaves compound, dissected, forked, or deeply lobed 25

24 Leaves simple, unlobed, usually entire (toothed in a few species) 31

25 Leaves apparently in a basal rosette, few (*Sium suave*) . . . . APIACEAE

25 Leaves definitely cauline: opposite, whorled, or alternate. 26

26 Leaves all or mostly opposite or whorled 27

26 Leaves definitely all alternate. 29

27 Leaves (or whorled branches) rolled inward at tip when young, bearing tiny stalked bladders; flowers emersed, bilaterally symmetrical, purple or yellow (*Utricularia*). . . . . LENTIBULARIACEAE

27 Leaves not inrolled at tip, without bladders; flowers various but not as above 28

28 Leaves once or twice dichotomously forked, the segments usually sparsely toothed along one edge; flowers inconspicuous, axillary, submersed . . . . CERATOPHYLLACEAE

28 Leaves not dichotomously forked, pectinate (comb-like), the segments entire; flowers emersed (*Myriophyllum*) . . . . HALORAGACEAE

29 Leaves with a definite central axis (following midvein); flowers various (*Proserpinaca*). . . . .
. . . . HALORAGACEAE

29 Leaves with no definite central axis (except sometimes after initially forking at the stem); flowers emersed, with conspicuous corolla 30

30 Petiole present (sometimes very short), ± adnate to a stipular sheath; plants without bladders; flowers regular, white or yellow, with numerous separate carpels forming achenes (*Ranunculus*). .
. . . . RANUNCULACEAE

30 Petioles and stipular sheaths absent; plants with small stalked bladders on leaves or on separate branches; flowers bilaterally symmetrical, yellow, with a single pistil producing a capsule (*Utricularia*). . . . . LENTIBULARIACEAE

31 Leaves much reduced, ± scale-like, not over 7 mm long, never distinctly opposite or whorled . . . 32

31 Leaves much longer or distinctly opposite or whorled (or both conditions). 33

32 Leaves minute, yellowish, merely widely spaced bumps or scales on stem (*Myriophyllum tenellum*)
. . . . HALORAGACEAE

32 Leaves to 7 mm long, green or brownish, loosely overlapping . . . . liverworts, aquatic mosses

33 Leaves alternate, with ligule-like stipules 34

33 Leaves opposite or whorled, without stipules 36

34 Leaf blades ± filiform, terete or at least half as thick as broad, and the stipule adnate to leaf base for 10-30 mm or more, forming a sheath around the stem. . . . . POTAMOGETONACEAE

34 Leaf blades definitely flattened and several times as broad as thick (even if narrow), or stipule little if at all adnate to blade (or both conditions) 35

35 Blades flattened, ribbon-like (up to 5 or even 7.5 mm wide), with no definite midrib (no central vein more prominent than others except rarely toward base); flowers solitary, rare, cleistogamous in axils of submersed leaves or (these almost never on submersed plants) with 6 bright yellow tepals (*Heteranthera*) . . . . PONTEDERIACEAE

35 Blades flattened with a definite midrib or filiform; flowers in spherical or cylindrical spikes, neither cleistogamous nor with showy yellow perianth. . . . . POTAMOGETONACEAE

36 Leaves nearly filiform, not over 0.5 mm wide, very gradually tapered from base to apex but not abruptly expanded basally, perfectly smooth; plants perennial by slender rhizomes; flowers axillary, 1 staminate flower (a single stamen) and (1) 2-several carpels at a node; fruit slightly curved and minutely toothed on convex side (*Zannichellia*). POTAMOGETONACEAE

36 Leaves broader; or if filiform then abruptly expanded basally and with apiculate or toothed margins, the plants annual, and the fruit solitary and ellipsoid 37

37 Leaves definitely whorled 38

37 Leaves opposite (in some species, with bushy axillary tufts of leaves which may give a falsely whorled appearance) 41

38 Whorled structures ("branches") cylindrical, elongate, usually stiff with calcium deposits; plants with distinctive musky odor (*Chara*, a macro-algae) . . . . CHARACEAE

38 Whorled structures (true leaves) flattened, short (not over 20 mm long) or elongate and very limp; plants without odor 39

39 Leaves 6-12 (usually 9) in a whorl, not over 2.5 mm wide, 12-25 times as long as wide; flowers bisexual, apetalous, sessile in axils of emersed leaves or bracts (*Hippuris*) ..... PLANTAGINACEAE

39 Leaves mostly 3-4 (rarely 6) in a whorl, 0.8-5 mm wide, at most 10-13 times as long; flowers bisexual or unisexual, but with petals 40

40 Leaves mostly 3 (rarely 6) in a whorl, very thin (2 cell layers) and delicate; stem round (not angled), smooth; flowers unisexual, with 3 often pink petals, at least the pistillate long-stalked from entirely submersed stem (*Elodea*) .... HYDROCHARITACEAE

40 Leaves mostly 4 in a whorl, stiff and firm; stem 4 sided, often with minutely retrorse-scabrous angles; flowers bisexual, with 3-4 white petals (usually not developed on wholly submersed plants) (*Galium*) .... RUBIACEAE

41 Largest leaves at least 1-4 cm long, with distinct petiole and expanded, entire blade (*Lysimachia nummularia*) .... PRIMULACEAE

41 Largest leaves smaller, or sessile, or toothed (or all of these) 42

42 Leaves large, 3-13 cm long, 5-20 mm wide (*Veronica anagallis-aquatica*) .... PLANTAGINACEAE

42 Leaves small (shorter or narrower than the above, or usually both). 43

43 Leaves linear and bidentate at apex when well submersed, often becoming obovate, ± weakly 3 nerved, and not necessarily bidentate toward summit of stem (or in floating rosettes); fruit solitary in axils, somewhat heart-shaped, of two 2-seeded segments (*Callitriche*) ...... PLANTAGINACEAE

43 Leaves filiform or tapered from base to apex, but essentially uniform on a plant and if linear not bidentate at apex; fruit various. 44

44 Leaves at least 3 times as long as wide, broader at base than at middle; fruit absent or solitary in axils of leaves and ± ellipsoid (*Najas*) .... HYDROCHARITACEAE

44 Leaves less than 3 times as long as wide, often nearly round 45

45 Stems stiffly erect; leaves weakly pinnately veined (with evident midvein), with reddish to blackish shiny dots (these often also on stem) (*Lysimachia terrestris*) .... PRIMULACEAE

45 Stems ± lax; leaves 3-nerved, without dark dots or flecks (though emersed leaves have translucent dots) (*Hypericum boreale*). .... HYPERICACEAE

GROUP 7

Woody plants (trees, shrubs, woody vines, and small evergreen creeping plants such as *Vinca*, *Linnaea*, and *Mitchella*).

1 Leaves scalelike (ca. 4 mm or less long and often appressed/imbricate) or needle-like, evergreen (except in *Larix* in Pinaceae) 2

1 Leaves with expanded (or dissected) blades, neither scale-like nor needle-like, if linear, then herbaceous, not stiff; deciduous or occasionally evergreen; occasionally absent at flowering time 7

2 Plant with leaves scale-like (or less than 3 mm long) 3

2 Plant with leaves needle-like or narrowly linear (over 3 mm long) 4

3 Plants trees or shrubs, fragrant when crushed, producing small dry or berry-like female cones but never flowers or true fruit. .... CUPRESSACEAE

3 Plants a woody parasite, not fragrant, producing flowers and fruit in season (*Arceuthobium*) .... SANTALACEAE

4 Leaves opposite or whorled .... CUPRESSACEAE

4 Leaves alternate or in clusters 5

5 Seed solitary in a red, fleshy, cup-like aril; leaves flattened, with strongly decurrent base, persistent, appearing 2-ranked, all green on both sides (may be yellowish beneath) .... TAXACEAE

5 Seeds borne on scales of a dry woody cone; leaves flattened or not (but if so, not strongly decurrent, readily falling when dry, not 2-ranked, and/or with white lines beneath) 6

6 Leaves evergreen (except *Larix* with leaves spirally arranged), arranged in clusters, spiraled around the stem, or in flattened 2-ranked sprays; cones slightly to very much longer than wide, the cone scales flattened. .... PINACEAE

6 Leaves deciduous, arranged in flattened 2-ranked sprays; cones globular, with peltate cone scales .... CUPRESSACEAE

7   Leaves opposite or whorled or nearly so (evident from scars if leaves not expanded at anthesis) . 8
7   Leaves alternate 37
8   Flowers appearing before leaves are expanded 9
8   Flowers appearing after the leaves have expanded (i.e., leaves present) 18
9   Perianth of both calyx and corolla 10
9   Perianth of only one cycle of parts, or none 12
10 Ovary superior; petals separate; flowers often unisexual (*Acer*) . . . . SAPINDACEAE
10 Ovary inferior; petals united; flowers bisexual 11
11 Flowers numerous in terminal cymes (*Sambucus*) . . . . ADOXACEAE
11 Flowers in pairs on axillary peduncles (*Lonicera*) . . . . CAPRIFOLIACEAE
12 Inflorescence an ament (catkin); bud scale 1 (*Salix purpurea*) . . . . SALICACEAE
12 Inflorescence otherwise, of clustered or pediceled flowers but not an elongate ament; bud scales
     more than 1 13
13 Flowers staminate or bisexual 14
13 Flowers pistillate 16
14 Stamens 2 (-4) (*Fraxinus*) . . . . OLEACEAE
14 Stamens 5 or more 15
15 Calyx lobes 4; stamens 8; buds scurfy-pubescent (*Shepherdia*) . . . . ELAEAGNACEAE
15 Calyx lobes 5; stamens 5-10; buds not scurfy-pubescent (*Acer*) . . . . SAPINDACEAE
16 Ovary with 2 divergent lobes (*Acer*) . . . . SAPINDACEAE
16 Ovary unlobed 17
17 Floral tube with a prominent disk at its summit; buds scurfy-pubescent; young fruit rotund (*Shep-
     herdia*) . . . . ELAEAGNACEAE
17 Floral tube without a prominent disk; buds not scurfy; young fruit strongly flattened (*Fraxinus*). .
     . . . . OLEACEAE
18 Leaves compound 19
18 Leaves simple 23
19 Plant a climbing or trailing vine (*Clematis*) . . . . RANUNCULACEAE
19 Plant erect, not a vine 20
20 Petals none; fruit a samara (winged) 21
20 Petals well developed and conspicuous; fruit various but not a samara 22
21 Ovary 2-lobed; fruit united in pairs; stamens ca. 5-10; leaflets usually 3-5 (*Acer negundo*) . . . .
     . . . . SAPINDACEAE
21 Ovary not lobed; fruits not paired; stamens 2 (-4); leaflets 5-11 (*Fraxinus*) . . . . OLEACEAE
22 Petals united; leaves pinnately compound with 5 or more leaflets; fruit fleshy (*Sambucus*) . . . .
     . . . . ADOXACEAE
22 Petals separate; leaves trifoliolate; fruit dryish . . . . STAPHYLEACEAE
23 Stamens more numerous than the petals or lobes of the corolla (or of the calyx if corolla is absent),
     or flowers strictly pistillate 24
23 Stamens the same number as the lobes or petals of the corolla or fewer 29
24 Petals united . . . . ERICACEAE
24 Petals separate or none 25
25 Stamens usually more than 10; corolla yellow (*Hypericum*) . . . . HYPERICACEAE
25 Stamens 10 or fewer, or flowers strictly pistillate; corolla pink, green, greenish-yellow, or white 26
26 Leaves palmately lobed, toothed; fruit a samara, united in pairs (*Acer*) . . . . SAPINDACEAE
26 Leaves unlobed, entire or toothed; fruit a berry or capsule, not paired 27
27 Plant a bushy shrub, with scurfy or stellate pubescence (*Shepherdia*) . . . . ELAEAGNACEAE
27 Plant barely woody at base, glabrous to somewhat tomentose but not scurfy or stellate pubescent;
     flowers bisexual with showy pink (to white) petals; fruit a capsule 28
28 Leaves evergreen, very shiny, toothed; stigma nearly sessile (*Chimaphila*) . . . . ERICACEAE
28 Leaves deciduous, dull, entire; stigma on an elongate style . . . . LYTHRACEAE
29 Petals separate 30
29 Petals united 32

30 Flowers in terminal inflorescences . . . . CORNACEAE
30 Flowers axillary 31
31 Fruit a red to purple capsule, the seeds enclosed in a red or orange aril; styles unlobed; stamens alternating with the petals (*Euonymus*) . . . . CELASTRACEAE
31 Fruit a dry inconspicuous capsule or fleshy and indehiscent, the seeds not arillate; styles often lobed; stamens opposite the petals . . . . RHAMNACEAE
32 Ovary inferior 33
32 Ovary superior 36
33 Flowers and fruits in dense spherical peduncled heads or paired at the ends of trailing branches; leaves entire, with broad stipules between the petiole bases . . . . RUBIACEAE
33 Flowers and fruits pediceled in small clusters or ± branched inflorescences; leaves entire or toothed, with stipules none or slender and partly adnate to petioles 34
34 Leaves of flowering shoots or flowering portions of shoots entire or somewhat undulate or sinuous, not sharply or regularly toothed . . . . CAPRIFOLIACEAE
34 Leaves with margins lobed, ± regularly toothed, crenate, or finely crenulate, or at least with regular minute gland-like teeth 35
35 Calyx lobes up to 1.5 mm long and broadly triangular to broadly rounded or virtually absent; corolla rotate (flat with very short tube); style very short or essentially absent; fruit fleshy with one pit (*Viburnum*) . . . . ADOXACEAE
35 Calyx lobes (1.6-) 2-6.5 (-7.5) mm long, linear or narrowly lanceolate; corolla tubular; style elongate, conspicuous; fruit dry . . . . CAPRIFOLIACEAE
36 Ovaries 2 (but styles and stigmas united); plants evergreen creeper with blue flowers solitary in the leaf axils (*Vinca*) . . . . APOCYNACEAE
36 Ovary 1; plant erect, with flowers in inflorescences . . . . OLEACEAE
37 Leaves deeply dissected into linear-filiform segments, aromatic (*Artemisia*) . . . . ASTERACEAE
37 Leaves simple, compound (then leaflets broader than linear-filiform), or absent at anthesis, aromatic or not 38
38 Plants dioecious 39
38 Plants not dioecious, the flowers either bisexual or unisexual (if the latter, then both sexes on the same individual) 58
39 Plant a climbing vine (or trailing in absence of support). 40
39 Plant ± erect, not climbing. 45
40 Stems with tendrils. 41
40 Stems without tendrils (aerial roots may be present along stem) 42
41 Leaves entire; stems prickly (at least below); perianth of 6 tepals. . . . . SMILACACEAE
41 Leaves toothed; stems unarmed; perianth of 5 petals and 5 (sometimes vestigial) sepals . . . . VITACEAE
42 Leaves trifoliolate; plants climbing by adventitious roots (*Toxicodendron*) . . . . . . ANACARDIACEAE
42 Leaves simple or with more than 3 leaflets; plants climbing by twining stems 43
43 Leaves pinnately veined, simple (*Celastrus*) . . . . CELASTRACEAE
43 Leaves palmately veined or compound 44
44 Sepals and petals each 6; leaves ± peltate (petiole attached in from margin of the blade), at most somewhat lobed but not toothed . . . . MENISPERMACEAE
44 Sepals (often vestigial) and petals each 5; leaves with marginal petiole, toothed . . . . . . VITACEAE
45 Flowers (at least the male ones) in cylindrical to nearly spherical aments (catkins) 46
45 Flowers not in aments 47
46 Crushed foliage pungently aromatic; twigs resin-dotted; fruit an achene or dry drupe . . . . MYRICACEAE
46 Crushed foliage in most species not at all aromatic; twigs without resinous dots (may be generally shiny); fruit a capsule . . . . SALICACEAE
47 Leaves compound, present at anthesis 48
47 Leaves simple, or unexpanded at anthesis 50
48 Leaves punctate with translucent oil glands; stems sometimes prickly. . . . . RUTACEAE
48 Leaves without translucent oil dots; stems unarmed 49

49 Leaflets nearly or quite entire except for one or more coarse teeth near the base, each with a large gland beneath; fruit a samara (*Ailanthus*).................................... SIMAROUBACEAE

49 Leaflets regularly toothed (glandless) or entire; fruit a small smooth to glandular-pubescent drupe; plant a native shrub .... ANACARDIACEAE

50 Flowers pistillate 51

50 Flowers staminate.. 55

51 Calyx and corolla both present (the former sometimes very small and inconspicuous) 52

51 Calyx and corolla not differentiated, or absent 54

52 Inflorescences terminal, crowded or many-flowered .... ANACARDIACEAE

52 Inflorescences axillary... 53

53 Stigma nearly or quite sessile... AQUIFOLIACEAE

53 Stigma clearly on an elongate style.. RHAMNACEAE

54 Style and stigma 1; leaves entire .... RHAMNACEAE

54 Stigma nearly or quite sessile.. AQUIFOLIACEAE

55 Stamens more numerous than the sepals or the petals (or perianth none).......... LAURACEAE

55 Stamens the same number as the perianth parts (same as the petals if both sepals and petals present).. 56

56 Inflorescences terminal, crowded or many-flowered .... ANACARDIACEAE

56 Inflorescences axillary 57

57 Stamens alternate with the sepals (opposite the petals if any) .... RHAMNACEAE

57 Stamens opposite the sepals (alternate with the petals if any) .... AQUIFOLIACEAE

58 Flowers (at least the staminate) in aments or dense spherical heads (always unisexual and individually inconspicuous) 59

58 Flowers not in aments or heads (often bisexual and/or conspicuous) 62

59 Staminate flowers in dense spherical heads (*Fagus*) .... FAGACEAE

59 Staminate flowers in cylindrical to ellipsoid aments.................................... 60

60 Pistillate flowers solitary or in small clusters; styles 3 or leaves compound.................... 61

60 Pistillate flowers in aments, heads, or cone-like structures (in *Corylus*, the red styles protruding from an ament resembling a leaf bud); styles 2; leaves simple.................... BETULACEAE

61 Leaves pinnately compound; styles 2 .... JUGLANDACEAE

61 Leaves simple (may be deeply lobed); styles usually 3.............................. FAGACEAE

62 Perianth none or apparently of a single series of parts 63

62 Perianth clearly differentiated into calyx and corolla.................................... 71

63 Leaves compound (*Ailanthus*) SIMAROUBACEAE

63 Leaves simple (or absent at anthesis) 64

64 Stamens more numerous than the segments or lobes (if any) of the perianth (or perianth none) 65

64 Stamens the same number as the lobes or segments of the perianth 66

65 Stamens 8; perianth lobes essentially none (*Dirca*) .... THYMELAEACEAE

65 Stamens 5-7 or 9; perianth lobes or segments 5 or 6 (*Rhododendron*) .... ERICACEAE

66 Styles 2, 3, or 5 .... ULMACEAE

66 Style 1 (may be branched above)... 67

67 Plant a vine, climbing or trailing by tendrils.................................... VITACEAE

67 Plant an erect shrub or tree... 68

68 Inflorescences terminal .... CORNACEAE

68 Inflorescences lateral.. 69

69 Leaves beneath and branchlets silvery-scurfy; stamens 4 (*Elaeagnus*).......... ELAEAGNACEAE

69 Leaves and branchlets glabrous or nearly so, not scurfy; stamens 4-6 70

70 Stamens alternating with the sepals .... RHAMNACEAE

70 Stamens opposite the sepals.. AQUIFOLIACEAE

71 Ovaries at least 3, distinct.. ROSACEAE

71 Ovary 1 72

72 Corolla bilaterally symmetrical (or petal only 1); stamens 10 (usually with some of the filaments connate).. FABACEAE

72 Corolla essentially regular; stamens various 73
73 Petals united. 74
73 Petals separate 78
74 Stamens more numerous than the corolla lobes . . . . ERICACEAE
74 Stamens the same number as the corolla lobes 75
75 Stamens adnate to the corolla (and falling with it if the corolla is deciduous); plants vining to
    shrubby; fruit a red berry or drupe. 76
75 Stamens free from the corolla; plant an erect or trailing shrub (not climbing); fruit various 77
76 Flowers white, on short (< 5 mm) pedicels; leaves toothed . . . . AQUIFOLIACEAE
76 Flowers purple (except in rare albinos), pedicels > 7 mm; leaves entire-margined (though sometimes
    lobed) . . . . SOLANACEAE
77 Stigma on a well developed style; fruit a capsule. . . . . ERICACEAE
77 Stigma nearly sessile; fruit a red drupe . . . . AQUIFOLIACEAE
78 Ovary at least partly inferior 79
78 Ovary entirely superior 85
79 Stamens more than the number of petals 80
79 Stamens the same number as the petals 81
80 Style 1 (*Vaccinium*). . . . . ERICACEAE
80 Styles 2-5 . . . . ROSACEAE
81 Petals 4 82
81 Petals 5 83
82 Flowers white, in terminal cymes, blooming in early to mid-summer; fruit fleshy; leaves entire. .
    . . . . CORNACEAE
82 Flowers yellow, in small axillary clusters, blooming in late fall; fruit a capsule; leaves with rounded
    teeth . . . . HAMAMELIDACEAE
83 Flowers in umbels, umbels either solitary or arranged in larger inflorescences. . . . . ARALIACEAE
83 Flowers in racemes, small axillary clusters, or domes or flat-topped corymbs 84
84 Flowers in racemes or small axillary clusters; small shrubs less than 2 m tall GROSSULARIACEAE
84 Flowers in terminal, domed or flat-topped corymbs; large shrubs or small trees more than 2 m tall
    (*Crataegus*) . . . . ROSACEAE
85 Stamens more than twice as many as the petals 86
85 Stamens twice as many as the petals or fewer 88
86 Corolla yellow; fruit a capsule. . . . . CISTACEAE
86 Corolla white to pink; fruit indehiscent 87
87 Inflorescence apparently borne at the middle of a tongue-shaped bract; leaves palmately veined
    (*Tilia*) . . . . MALVACEAE
87 Inflorescence borne normally; leaves pinnately veined . . . . ROSACEAE
88 Leaves compound 89
88 Leaves simple. 95
89 Leaves even-pinnate or even-bipinnate; fruit a large woody legume (pod splitting on 2 sutures) . .
    . . . . FABACEAE
89 Leaves odd-pinnate, trifoliolate, or palmate; fruit a samara, drupe, berry, or 4-5-lobed capsule. . 90
90 Leaflets nearly or quite entire except for one or more coarse teeth near the base, each with a large
    gland beneath; fruit a samara (*Ailanthus*). . . . . SIMAROUBACEAE
90 Leaflets toothed or entire but without large glands; fruit various (a samara only in trifoliolate *Pte-
    lea*) 91
91 Inflorescences terminal 92
91 Inflorescences lateral or axillary 93
92 Fruit a samara, in loose open cymes; leaves trifoliolate, punctate with translucent oil glands (*Pte-
    lea*) . . . . RUTACEAE
92 Fruit a glandular-pubescent drupe, in dense panicles; leaves pinnately compound, without translu-
    cent glands . . . . ANACARDIACEAE
93 Leaflets strongly spiny-toothed; flowers yellow (*Berberis*) . . . . BERBERIDACEAE
93 Leaflets without spines; flowers greenish yellow 94

94 Leaves palmately compound with mostly 5-7 leaflets, if trifoliolate then leaflets sharply toothed or pinnately lobed; plant a vine with tendrils; stamens opposite the petals (i.e., alternate with the sepals) (*Parthenocissus*) . . . . VITACEAE

94 Leaves trifoliolate or pinnately compound with entire or nearly entire leaflets; plant a shrub or vine with adventitious roots (not tendrils); stamens alternating with the petals (i.e., opposite the sepals) (*Toxicodendron*). . . . . ANACARDIACEAE

95 Styles 2, separate to the base; petals 4, yellow, linear . . . . HAMAMELIDACEAE

95 Style 1 or 3 (may be lobed or cleft at summit); petals various 96

96 Stems spiny; flowers yellow, 6-merous (*Berberis*) . . , ... BERBERIDACEAE

96 Stems unarmed; flowers white, pink, or greenish, 4-5-merous. 97

97 Stamens more numerous than the petals; inflorescence an umbel or raceme; plant a low evergreen subshrub . . . . ERICACEAE

97 Stamens the same number as the petals; inflorescence various; plant a bushy shrub, deciduous except in *Rhododendron* 98

98 Leaves evergreen, densely white- or brown-tomentose beneath, revolute (*Rhododendron*). . . . . ERICACEAE

98 Leaves deciduous, glabrous or nearly so, with flat margins. 99

100 Stamens alternating with the sepals (i.e., opposite the petals); style 3-lobed . . . . . RHAMNACEAE

100 Stamens opposite the sepals (i.e., alternating with the petals); style nearly or quite absent . . . . AQUIFOLIACEAE

GROUP 8

Herbaceous plants lacking both green color and developed leaves at flowering time.

1 Plants not anchored in the ground, solely parasitic on and attached to stems of other plants at maturity 2

1 Plants clearly anchored in the ground, not attached to other above-ground plants 3

2 Stem up to 15 mm long, with minute opposite leaves (scale-like); flowers in May, unisexual (plants dioecious), the staminate with stamens adnate to calyx lobes, the pistillate with inferior ovary; parasites on conifers (*Arceuthobium*) . . . . SANTALACEAE

2 Stem elongate, with minute alternate leaves; flowers in late summer, bisexual, the stamens partly adnate to corolla and the ovary superior; parasites on flowering plants (*Cuscuta*) . . . . CONVOLVULACEAE

3 Stem buried in ground; flowers in late winter or earliest spring, crowded in a spadix with a nearly or partly buried hood-like brownish or mottled spathe (green leaves from rhizome appearing after flowering); stamens 4; plant with skunk-like odor (*Symplocarpus*) . . . . ARACEAE

3 Stem or flower stalk above ground; flowers later, solitary or in a few- to many-flowered raceme, umbel, or head; stamens various; plant with odor, if any, not skunk-like. 4

4 Flowers completely 3-merous and regular, in an umbel on a naked peduncle arising from an underground, onion-smelling bulb (*Allium*). . . . . LILIACEAE

4 Flowers not completely 3-merous, regular or bilaterally symmetrical, not in an umbel, on aerial stems 5

5 Inflorescence a single dense, short spike with spirally arranged scales; flowers lacking petals and sepals; stem with tubular sheaths at base (*Eleocharis*) . . . . CYPERACEAE

5 Inflorescence of normal flowers not aggregated into a single dense spike, flowers with at least tiny petals, often showy; stem without tubular sheaths at base (except some Orchidaceae) 6

6 Petals 5, mostly united in a tube, the flower slightly to distinctly bilaterally symmetrical, not spurred; stamens 4 . . . . OROBANCHACEAE

6 Petals 3-5 but not united in a tubular corolla, the flower regular or strongly bilaterally symmetrical (sometimes spurred); stamens various. 7

7 Perianth strongly bilaterally symmetrical; stamens 1-2 8

7 Perianth regular; stamens 4-10 9

8 Sepals and petals 3, the lower petal a definite lip, the others little modified; ovary inferior; plants of various habitats but not aquatic; perianth of various colors. . . . . ORCHIDACEAE

8 Sepals apparently 2 and petals 5, but corolla basically 2-lipped; ovary superior; plants of wet shores, ponds, and bog pools, with perianth yellow (*Utricularia*) . . . . LENTIBULARIACEAE

9   Corolla at least 5 mm long; stamens 8-10; plants of wetlands to dry places . . . . ERICACEAE
9   Corolla less than 5.5 mm long; stamens 4 plant of wet lake shores, nearly or quite aquatic (*Myrio-phyllum tenellum*) . . . . HALORAGACEAE

GROUP 9

Inflorescence apparently converted to bulblets or tufts of leaves.

1   Leaves with flat, net-veined (or dissected) blades 2
1   Leaves terete or slender and parallel-veined 3
2   Leaves with narrow, sparsely toothed leaflets or further dissected; stem hollow; bulblets produced in the axils of broad-based acuminate bracts or leaves, not transversely segmented (*Cicuta bulbifera*). . . . . APIACEAE
2   Leaves simple and entire; stem solid; bulblets otherwise (*Lysimachia terrestris*) . . . PRIMULACEAE
3   Bulblets in a ± spherical head or umbel; plants with odor of onion or garlic (*Allium*) . . LILIACEAE
3   Bulblets not in a distinct umbel or spherical head; plants without strong odor 4
4   Leaves terete, septate (with hard cross-partitions, easily seen on dry specimens or felt by gently pinching a leaf and drawing it between the fingers) (*Juncus*) . . . . JUNCACEAE
4   Leaves flat, neither terete nor septate 5
5   Stem ± triangular and solid (*Scirpus*) . . . . CYPERACEAE
5   Stem terete, with hollow internodes . . . . POACEAE

GROUP 10

Monocots; leaves with parallel veins.

1   Inflorescence a spadix, subtended by a spathe which may be broad and hood-like or elongate; leaves in some species compound or net-veined. 2
1   Inflorescence not a spadix (if flowers in a head, this with neither an elongate fleshy axis nor a conspicuous subtending spathe); leaves simple, rarely net-veined (in *Smilax ecirrata, Trillium*, and some Alismataceae). 3
2   Leaves narrow, sword-like, with ± parallel sides; spathe appearing like a continuation of the leaf-like peduncle (the spadix thus apparently lateral). . . . . ACORACEAE
2   Leaves expanded; spathe clearly differentiated from peduncle . . . . ARACEAE
3   Perianth much reduced: absent, or composed solely of bristles (these small and stiff or elongate and cottony), or of chaffy or scale-like parts, never conspicuously petaloid 4
3   Perianth at least in part of ± conspicuous white or colored petals. 9
4   Individual flowers subtended by 1 or 2 scales; leaves ± elongate, grass-like, usually with a sheath at the base surrounding the stem; fruit a 1-seeded grain or nutlet (achene) .5
4   Individual flowers subtended by no scales or only by bristles, or with a regular perianth of chaffy scales (or tepals); leaves and fruit various 6
5   Each fertile flower subtended by a single scale (others may be at base of spikelet); sheaths of leaves closed (margins connate); stems frequently triangular (but 4-several-angled or terete in many species), usually solid; leaves usually 3-ranked (especially in a species with terete hollow stem); stamens with filament attached to end of anther; fruit a definitely 2- or 3 sided (rarely nearly terete) nutlet . . . . CYPERACEAE
5   Each flower subtended by 2 scales (almost opposite each other, one rarely absent); sheaths often open; stems ± terete (sometimes flattened), never triangular; leaves not clearly 3-ranked (basically 2-ranked); stamens with filament attached near middle of anther (or apparently so because of sagittate anthers); fruit usually a grain neither flattened (2-sided) nor triangular . . . . . . POACEAE
6   Inflorescence composed of separate staminate and pistillate portions, the former consisting of conspicuous stamens, sooner or later withering, leaving only the pistillate portion conspicuous . . . . TYPHACEAE
6   Inflorescence composed of bisexual flowers, without conspicuously separate staminate and pistillate portions. 7
7   Flowers in a branched or umbellate inflorescence, solitary or, more often, clustered into small heads of 2 or more; fruit a 3- to many-seeded capsule . . . . JUNCACEAE
7   Flowers in a single elongate spike or zigzag raceme; fruit indehiscent or a 1-2-seeded follicle . . . 8

8 Spike (truly a spadix) apparently lateral; fruit of each flower indehiscent .... ACORACEAE
8 Spike or raceme terminal; fruit of each flower consisting of 3 or 6 1-2-seeded follicles ....
.... JUNCAGINACEAE
9 Flowers bilaterally symmetrical... 10
9 Flowers regular (radially symmetrical) 12
10 Ovary inferior; fertile stamens 1 or 2, united with the pistil; flowers not blue (almost any other color) .... ORCHIDACEAE
10 Ovary superior; fertile stamens 3 or 6, free; flowers blue (except albinos), at least in part 11
11 Sepals colored like the petals; stamens 6, all fertile; flowers in a dense elongate inflorescence (*Pontederia*) .... PONTEDERIACEAE
11 Sepals greenish, unlike the petals; stamens 6, 3 with imperfect anthers; flowers few (*Commelina*).
.... COMMELINACEAE
12 Sepals and petals of quite different color and/or texture, the former green or brownish........ 13
12 Sepals and petals both colored and petaloid, usually similar in shape (tepals) or the sepals (in *Iris*) of different size and shape... 15
13 Leaves in a single whorl of 3 on the stem .... LILIACEAE
13 Leaves all basal or, if cauline, not in a single whorl of 3 14
14 Petals yellow; flowers in a single compact head less than 12 mm across .... XYRIDACEAE
14 Petals white or pinkish; flowers in a more open or larger inflorescence .... ALISMATACEAE
15 Ovary inferior (flowers bisexual) 16
15 Ovary superior (or flowers unisexual)... 18
16 Stamens 3; leaves equitant .... IRIDACEAE
16 Stamens 6; leaves not equitant 17
17 Ovary only half-inferior, part of it adnate to the perianth, glabrous (at most granular-roughened) (*Anticlea*) .... LILIACEAE
17 Ovary clearly inferior .... LILIACEAE
18 Stamens 3; tepals 6, yellow; plants creeping on wet shores (*Heteranthera*) ..... PONTEDERIACEAE
18 Stamens and tepals 4 or 6, the latter yellow or not; plants erect, of various habitats 19
19 Flowers or inflorescences lateral, arising from the axils of alternate cauline leaves or scales ... 20
19 Flowers or inflorescences terminal on scapes or leafy (simple or branched) stems 22
20 Leaves scale-like, mostly brownish or yellowish, those on the much-branched upper portion of the plant subtending short green filiform branches (often mistaken for leaves) ...... LILIACEAE
20 Leaves broad, flat, green (scale-like leaves or bracts may be present in addition to normal leaves) 21
21 Leaves net-veined with long or short (but distinct) petioles; flowers unisexual, in umbels of several to many .... SMILACACEAE
21 Leaves parallel-veined, sessile, clasping, or perfoliate at base; flowers bisexual, 1-5 at a node .....
.... LILIACEAE
22 Leaves withering before plant flowers (*Allium tricoccum*) .... LILIACEAE
22 Leaves present at flowering time 23
23 Leaves all in one or two whorls on the stem (*Medeola*) .... LILIACEAE
23 Leaves alternate or basal, or if in whorls these more than two or some alternate leaves also present 24
24 Flowers more than 3.5 cm long (various members of the traditional Lily Family will key from this point in the key. Alternately, go to the Liliaceae family key on page 337)..................... 25
24 Flowers less than 3.5 cm long 28
25 Leaves perfoliate (*Uvularia grandiflora*) .... LILIACEAE
25 Leaves not perfoliate... 26
26 Principal leaves cauline, not crowded toward base of plant .... LILIACEAE
26 Principal leaves basal or nearly so; stem leafless above or with very small bracts 27
27 Flowers solitary .... LILIACEAE
27 Flowers 2-many (*Hemerocallis*)... LILIACEAE
28 Flowers in a many (7-60 or more) flowered umbel on an unbranched stem or scape; plants with odor of onion or garlic.. LILIACEAE

28 Flowers solitary or in a raceme, panicle, or corymb, on a simple or branched stem (if in an umbel, then either the stem branched or forked or flowers less than 7) 29
29 Plants with principal leaves clearly cauline; basal leaves (at least of current season) absent or at most apparently one. 30
29 Plants with the principal leaves all basal (at ground level) or nearly so; cauline leaves absent, reduced to bracts, or at most much smaller or fewer than basal leaves. 32
30 Stem forked or branched; perianth over 12 mm long . . . . LILIACEAE
30 Stem unbranched (above the ground and below the inflorescence); perianth usually less than 12 mm long 31
31 Ovary with a single style; fruit a berry; inflorescence a raceme (tepals up to 7 mm long) or if a panicle, the tepals less than 3 mm long; leaves broadly lanceolate to ovate. . . . . LILIACEAE
31 Ovary with 3 styles (one on each lobe); fruit a capsule; inflorescence a panicle; tepals ca 5-13 mm long; leaves very elongate . . . . LILIACEAE
32 Flower solitary (*Erythronium*). . . . . LILIACEAE
32 Flowers 2 or more in an inflorescence 33
33 Tepals united for half or more of their length 34
33 Tepals completely separate or united at base only 36
34 Perianth blue, less than 6 mm long; leaves linear (longest leaves at least 20-40 times as long as wide, up to 8 mm wide); plants bulbous (*Muscari*). . . . . LILIACEAE
34 Perianth white, 5-10 mm long when mature; leaves lanceolate to elliptic (longest leaves less than 20 times as long as wide or over 1 cm wide, or both); plants not bulbous 35
35 Leaves elliptic, the widest 2-6 cm broad; stems up to 35 cm tall, about equaling or shorter than the leaves; flowers nodding, on pedicels longer than the subtending bracts; perianth smooth on outside; fruit a berry (*Convallaria*) . . . . LILIACEAE
35 Leaves narrowly lanceolate or oblanceolate, the widest less than 2 cm broad; stems over 40 cm tall, much surpassing the leaves; flowers ascending on pedicels shorter than the subtending bracts; perianth ± granular-roughened on the outside; fruit a capsule (*Aletris*) . . . . LILIACEAE
36 Ovary with 3 styles (1 on each lobe) in bisexual or pistillate flowers (or flowers all staminate) . . . .
. . . . LILIACEAE
36 Ovary with a single style; flowers bisexual 37
37 Leaves broad (over 3 cm); perianth yellow; fruit a blue berry (*Clintonia*) . . . . LILIACEAE
37 Leaves long and narrow (less than 1.5 cm wide); perianth white to blue; fruit a capsule . . . .
. . . . CARYOPHYLLACEAE

GROUP 11

Flowers in an involucrate head (i.e., the flowers clustered in a head above a whorl of bracts).

1 Flowers on a thick fleshy axis (inflorescence a spadix) subtended by a single large overtopping or enveloping bract (spathe); perianth none or of 4 tepals. 2
1 Flowers not in a spadix overtopped by a spathe; perianth various. 3
2 Leaves narrow, sword-like, with ± parallel sides; spathe appearing like a continuation of the leaf-like peduncle (the spadix thus apparently lateral). . . . . ACORACEAE
2 Leaves expanded; spathe clearly differentiated from peduncle and terminal . . . . ARACEAE
3 Ovary inferior. 4
3 Ovary superior. 6
4 Leaves opposite (very rarely whorled), toothed or pinnatifid; corolla 4-lobed, lilac-purple; stamens 4, separate (*Dipsacus*). . . . . DIPSACACEAE
4 Leaves and corolla not combined as above, e.g., leaves alternate and/or entire or corolla 5-lobed and/or not lilac-purple. . . . . ASTERACEAE
6 Leaves alternate, compound; involucral bract 3-foliolate; flowers strongly bilaterally symmetrical, papilionaceous (as in other legumes) (*Trifolium*) . . . . FABACEAE
6 Leaves opposite, simple; flowers often nearly or quite regular 7
7 Plant with minty odor; ovary deeply 4-lobed, with 1 style; petals united. . . . . LAMIACEAE
7 Plant without minty odor; ovary not lobed, with 2 styles; petals separate (*Petrorhagia, Dianthus*) . .
. . . . CARYOPHYLLACEAE

GROUP 12

Herbaceous plants with single-sex flowers.

1 Leaves compound 2
1 Leaves simple 7
2 Leaves palmately compound (or 3-foliolate) 3
2 Leaves pinnately compound or more than once compound 6
3 Flowers in umbels 4
3 Flowers in spikes or panicles 5
4 Leaves cauline, in a single whorl (*Panax*) .... ARALIACEAE
4 Leaves alternate and basal (*Sanicula*) .... APIACEAE
5 Margins of leaflets entire; flowers at the base of a prolonged fleshy spadix subtended by a single large bract (spathe) (*Arisaema*) .... ARACEAE
5 Margins of leaflets toothed; flowers on normal herbaceous (but not fleshy) pedicels or axes (*Clematis*) .... RANUNCULACEAE
6 Flowers in tight ovoid heads or umbels (*Aralia*) .... ARALIACEAE
6 Flowers in panicles (*Thalictrum*) .... RANUNCULACEAE
7 Plant with leaves all basal 8
7 Plant with leaves all or mostly cauline 9
8 Flowers in dense spikes .... PLANTAGINACEAE
8 Flowers pediceled in panicles (*Rumex*) .... POLYGONACEAE
9 Leaves peltate or pubescent with forked/stellate hairs .... EUPHORBIACEAE
9 Leaves neither peltate nor with forked/stellate hairs 10
10 Leaves opposite or whorled 11
10 Leaves alternate (at least at upper nodes) 16
11 Flowers solitary in axils of leaves; perianth none; stamen 1 .... PLANTAGINACEAE
11 Flowers in axillary or terminal inflorescences 12
12 Leaves hastate, otherwise unlobed but entire to coarsely or irregularly toothed; pistillate flowers and fruit mostly concealed by a pair of bracts with margins ± united at base (*Atriplex*) ....
   .... AMARANTHACEAE
12 Leaves not hastate, in some species deeply lobed, in some closely toothed; pistillate flowers without 2 basal bracts 13
13 Inflorescence terminal; corolla white or colored 14
13 Inflorescence axillary; corolla none or of reduced scales 15
14 Cauline leaves deeply pinnately lobed; style 1; stamens 3-4 (*Valeriana*) .... CAPRIFOLIACEAE
14 Cauline leaves unlobed; styles 3-7; stamens 10 (*Silene*) .... CARYOPHYLLACEAE
15 Plant a vine; leaves deeply 3-7-lobed (*Humulus*) .... CANNABACEAE
15 Plant erect, not a vine; leaves unlobed .... URTICACEAE
16 Flowers with 6 petaloid tepals and 6 stamens or 3 carpels (dioecious); inflorescences on long peduncles from the nodes (not terminal); leaves with several prominent longitudinal veins (including midrib) .... SMILACACEAE
16 Flowers either with other numbers of tepals, stamens, and carpels or the inflorescence terminal (on main stem or branches); leaves various but without several prominent long veins 17
17 Perianth with both calyx and corolla (sometimes very inconspicuous); plants climbing or trailing, with tendrils .... CUCURBITACEAE
17 Perianth absent or of 1 series of parts (tepals); plants erect or prostrate, without tendrils 18
18 Flowers very small, in axillary clusters [plants monoecious; look for pistillate flowers for keying]
   19
18 Flowers small or not, in chiefly terminal inflorescences (spikes, panicles, or racemes on main stem and/or branches) 20
19 Style 1; stamens 4 or 5 .... URTICACEAE
19 Styles (or sessile stigmas) 2-3; stamens various (*Amaranthus, Atriplex*) .... AMARANTHACEAE
20 Flowers consistently 3-merous (tepals 6, stamens 6, carpels 3); stipules united into a sheath (ocrea) surrounding the stem above each node (*Rumex*) .... POLYGONACEAE

20 Flowers not consistently 3-merous (tepals 5 or fewer, stamens usually 5, styles often 2); stipules none . . . . AMARANTHACEAE

GROUP 13

Herbaceous dicots with bisexual flowers, perianth in 1 series, ovary inferior.

1 Stamens more numerous than the 1-4 perianth lobes or parts. 2
1 Stamens the same number as or fewer than the perianth lobes or parts, or perianth 5-merous (or both conditions) 3
2 Perianth with 1-3 (rarely 4) lobes; stamens 6 or 12 . . . . ARISTOLOCHIACEAE
2 Perianth 4-parted; stamens numerous (*Poterium*) . . . . ROSACEAE
3 Leaves all or mostly opposite or whorled 4
3 Leaves alternate or basal 9
4 Inflorescence a dense terminal cluster of flowers (sessile or nearly so) .5
4 Inflorescence of solitary, axillary, or clearly pediceled flowers 7
5 Leaves apparently whorled; bracts below the inflorescence large and white (*Cornus canadensis*) . . .
   . . . . CORNACEAE
5 Leaves clearly opposite; bracts below the inflorescence greenish or inconspicuous 6
6 Heads subtended by several involucral bracts below a receptacle with sessile flowers . . . .
   . . . . DIPSACACEAE
6 Heads not subtended by a distinct involucre, with visible branching structure; flowers sessile but not on a common receptacle (*Valeriana*) . . . . CAPRIFOLIACEAE
7 Leaves compound, in a single whorl (*Panax*) . . . . ARALIACEAE
7 Leaves simple or deeply lobed, opposite or in several whorls (rarely the lower alternate in *Valeriana*) 8
8 Leaves in whorls (*Galium*) . . . . RUBIACEAE
8 Leaves opposite . . . . CAPRIFOLIACEAE
9 Leaves entire, simple and unlobed; flowers in cymes or few-flowered cymules; style 1. . . . .
   . . . . SANTALACEAE
9 Leaves (at least the cauline ones) toothed or crenulate, often deeply lobed or compound; flowers in umbels, axillary, or ovoid to cylindric heads; styles 2, 3, or 5 10
10 Tepals and stamens each 5 11
10 Tepals and stamens each 3 or 4 12
12 Styles 5; fruit fleshy, berry-like (*Aralia*) . . . . ARALIACEAE
12 Styles 2; fruit dry, splitting into 2 achene-like indehiscent parts (mericarps) . . . . APIACEAE
13 Stamens and tepals 3 (*Proserpinaca*) . . . . HALORAGACEAE
13 Stamens and tepals 4 . . . . ROSACEAE

GROUP 14

Herbaceous dicots with bisexual flowers, perianth in 1 series, superior ovary.

1 Ovaries more than 1 in each flower, the carpels separate at least above the middle of the ovaries 2
1 Ovary 1 in each flower (bearing 1 or more styles), the carpels united at least below the styles. . . .4
2 Stipules conspicuous; leaves pinnately compound (*Poterium*). . . . . ROSACEAE
2 Stipules none or leaves simple 3
3 Ovaries united for most of lower half; leaves simple, unlobed (*Penthorum*) . . . . . . PENTHORACEAE
3 Ovaries distinct; leaves of most species lobed or compound . . . . RANUNCULACEAE
4 Leaves bipinnately compound, fruit a legume . . . . FABACEAE
4 Leaves simple or compound (but not bipinnate); fruit not a legume. 5
5 Plants with a solitary large (ca. 3-5 cm wide) white flower between a single usually opposite or subopposite pair of long-petioled cauline eccentrically peltate and deeply lobed leaves (*Podophyllum*). . . . . BERBERIDACEAE
5 Plants with more flowers per stem or, if only one, then leaves not as above 6
6 Stamens more than twice as many as the perianth lobes or parts 7
6 Stamens only twice as many as the perianth lobes or parts, or fewer 10

7 Leaves tubular, open at apex and hence pitcher-like .... SARRACENIACEAE
7 Leaves flat, of normal structure, simple or compound but not hollow 8
8 Perianth small and inconspicuous (stamens more showy); leaves compound with definite flat broad leaflets .... RANUNCULACEAE
8 Perianth well developed, showy; leaves simple or dissected into very narrowly linear segments . 9
9 Leaf blades entire, unlobed except for deeply cordate base; plants aquatic (*Nuphar*) ....
   .... NYMPHAEACEAE
9 Leaf blades deeply lobed or dissected; plants terrestrial .... PAPAVERACEAE
10 Style 1 or none (stigmas may be 2 or more) 11
10 Styles 2 or more 17
11 Stamens more numerous than the perianth divisions.................................... 12
11 Stamens the same number as or fewer than the perianth lobes............................ 13
12 Flowers bilaterally symmetrical; perianth colorful (white, yellow, or pink) ...... PAPAVERACEAE
12 Flowers regular; perianth dull, greenish .... BRASSICACEAE
13 Leaves alternate or basal 14
13 Leaves opposite 16
14 Perianth parts (and stamens) 6, 8, or 9 .... BERBERIDACEAE
14 Perianth parts (and usually stamens) 4 15
15 Leaves simple, entire (*Parietaria*) .... URTICACEAE
15 Leaves pinnately compound with toothed leaflets (*Poterium*) .... ROSACEAE
16 Flowers solitary or few in axils of leaves, sessile or nearly so..................... LYTHRACEAE
16 Flowers in terminal inflorescences (on stems and branches) (*Froelichia*) ...... AMARANTHACEAE
17 Leaves opposite or whorled.. 18
17 Leaves alternate.. 19
18 Leaves opposite.. CARYOPHYLLACEAE
18 Leaves whorled .... MOLLUGINACEAE
19 Plants with a ± membranous stipular sheath (ocrea) surrounding the stem above each node.....
   .... POLYGONACEAE
19 Plants lacking stipules of any kind .... AMARANTHACEAE

GROUP 15

Herbaceous dicots with bisexual flowers, perianth of 2 series, ovaries 2 or more in each flower.

1 Style and/or stigmas united (i.e., 1 in each flower, but style may be branched) 2
1 Style and stigmas separate (1 on each ovary, or scarcely developed) 5
2 Ovaries 2; corolla regular, of united petals; stamens 5; sap in most species milky . APOCYNACEAE
2 Ovaries 4 or more; corolla regular or bilaterally symmetrical, of united or separate petals; stamens 2, 4, 5, or numerous; sap not milky 3
3 Petals separate; ovaries apparently 5 or more; stamens numerous, their filaments connate, at least for much of their length, into a tube around the style; leaves palmately veined (may be deeply lobed) .... MALVACEAE
3 Petals united; ovaries apparently 4; stamens 2, 4, or 5, their filaments not connate (but ± adnate to corolla); leaves mostly pinnately veined... 4
4 Leaves alternate; stamens 5; corolla regular (bilaterally symmetrical only in the very bristly *Echium*); stems not angled (rarely winged) and foliage not aromatic................... BORAGINACEAE
4 Leaves opposite; stamens 2 or 4; corolla bilaterally symmetrical or in a few genera essentially regular; stems usually 4-angled ("square") and foliage often aromatic when bruised (minty or citrus-like) .... LAMIACEAE
5 Sepals (or sepal-like bracts) 3 (*Hepatica*)................................. RANUNCULACEAE
5 Sepals 4 or more.. 6
6 Sepals separate to the base; stamens and petals individually falling from the receptacle after anthesis... RANUNCULACEAE
6 Sepals, petals, and stamens united to form a saucer- or cup-like floral tube ("hypanthium") at the margin of which the stamens and petals are borne 7
7 Carpels as many as, or more than, the petals 8

7 Carpels fewer than the petals 9
8 Leaves succulent, simple, entire, estipulate. . . . . CRASSULACEAE
8 Leaves not succulent, deeply lobed or compound, toothed, stipulate . . . . ROSACEAE
9 Leaves simple, at most shallowly lobed . . . . SAXIFRAGACEAE
9 Leaves clearly compound . . . . ROSACEAE

GROUP 16

Herbaceous dicots with bisexual flowers, perianth of 2 series, ovary inferior.

1 Stamens twice as many as the petals (or nearly so) 2
1 Stamens the same number as the petals or corolla lobes, or fewer 5
2 Style 1 (sometimes very short) 3
2 Styles 2 or more 4
3 Petals spreading; herbaceous plants; fruit a capsule or dry and indehiscent . . . . ONAGRACEAE
3 Petals strongly reflexed; creeping evergreen wetland subshrubs; fruit a berry (*Vaccinium oxycoccos* and *V. macrocarpon*) . . . . ERICACEAE
4 Sepals 2; leaves succulent; styles 3. . . . . PORTULACACEAE
4 Sepals (4-) 5; leaves not succulent; styles 2 . . . . SAXIFRAGACEAE
5 Petals united 6
5 Petals separate 14
6 Stem leaves alternate 7
6 Stem leaves opposite or whorled (rarely the lower alternate in *Valeriana* with pinnate leaves) . . . 8
7 Corolla bilaterally symmetrical (*Lobelia*) . . . . CAMPANULACEAE
7 Corolla regular . . . . CAMPANULACEAE
8 Leaves whorled . . . . RUBIACEAE
8 Leaves opposite 9
9 Stipules present (connate around the stem). . . . . RUBIACEAE
9 Stipules absent. 10
10 Flowers sessile in terminal heads 11
10 Flowers visibly pediceled (even if crowded) or axillary. 12
11 Heads subtended by several involucral bracts below a receptacle with sessile flowers . . . .
. . . . DIPSACACEAE
11 Heads not subtended by a distinct involucre, with visible branching structure; flowers sessile but not on a common receptacle (*Valeriana*) . . . . CAPRIFOLIACEAE
12 Flowers numerous, in rather dense terminal inflorescences (at ends of stem and branches) (*Valeriana*) . . . . CAPRIFOLIACEAE
12 Flowers axillary or on paired pedicels on a peduncle 13
13 Leaves strictly entire, stems erect (*Triosteum*) . . . . CAPRIFOLIACEAE
13 Leaves shallowly toothed on apical half; stems trailing; flowers on paired pedicels on a peduncle (*Linnaea*) . . . . CAPRIFOLIACEAE
14 Stamens and petals each 2 (*Circaea*) . . . . ONAGRACEAE
14 Stamens (fertile) and petals each 4 or 5 (stamens sometimes alternating with staminodia, which may have gland-tipped divisions). 15
15 Petals 4 (*Cornus canadensis*) . . . . CORNACEAE
15 Petals 5. 16
16 Leaves simple; styles 2 or stigmas 4 and sessile; inflorescence various. 17
16 Leaves compound; inflorescence an umbel 18
17 Flowers in panicles, in cymes, or solitary. . . . . SAXIFRAGACEAE
17 Flowers in umbels . . . . APIACEAE
18 Styles 5; fruit berry-like (*Aralia*). . . . . ARALIACEAE
18 Styles 2-3; fruit various 19
19 Leaves alternate or basal; fruit dry, splitting into 2 achene-like indehiscent parts (mericarps). . . .
. . . . APIACEAE
19 Leaves in a single whorl; fruit berry-like (*Panax*). . . . . ARALIACEAE

GROUP 17

Herbaceous dicots with bisexual flowers, perianth of z series, ovary i and superior, stamens more numerous than the petals.

1 Corolla bilaterally symmetrical 2
1 Corolla regular (radially symmetrical) 9
2 Sepals all or partly petal-like in appearance or prolonged into a spur 3
2 Sepals not petal-like in form or appearance, usually green 4
3 Spur none; stamens 6, 7, or 8; leaves entire .... POLYGALACEAE
3 Spur present on one of the sepals... BALSAMINACEAE
4 Sepals 2, separate, usually deciduous early in anthesis; leaves dissected or twice-compound .....
.... PAPAVERACEAE
4 Sepals 4 or more, usually ± connate 5
5 Lower 2 petals forming a laterally compressed "keel" that encloses the stamens; leaves once-compound.. FABACEAE
5 Lower petals not forming a keel nor enclosing the stamens................................... 6
6 Flowers completely 5-merous (sepals and petals 5, stamens 5 or usually 10); pistil long-beaked; corolla pink or purple; leaves deeply lobed or cleft or compound (the main stem leaves opposite and toothed or cleft).. GERANIACEAE
6 Flowers with at least the carpels fewer than 5; corolla and leaves various 7
7 Leaves compound .... FABACEAE
7 Leaves simple, deeply lobed to entire 8
8 Sepals and petals each 4; stamens 6... BRASSICACEAE
8 Sepals and petals each (4) 5-7; stamens 2x as many .... LYTHRACEAE
9 Leaves tubular, open at apex and hence pitcher-like; style greatly expanded, large and umbrella-shaped... SARRACENIACEAE
9 Leaves flat or at most succulent, of usual shapes; style not unusually expanded 10
10 Plants with a solitary large (ca 3-5 cm wide) white flower between a single, usually opposite, long-petioled pair of deeply lobed leaves (Podophyllum) .... BERBERIDACEAE
10 Plants with more flowers per stem or, if only one, then leaves not as above 11
11 Sepals 2 12
11 Sepals 3 or more 13
12 Leaves lobed, compound, or coarsely toothed, not succulent; sap in most species colored (yellow to orange)... PAPAVERACEAE
12 Leaves unlobed, entire, succulent; sap watery............................. PORTULACACEAE
13 Stamens more than 2x as many as the petals.. 14
13 Stamens twice as many as the petals or fewer... 18
14 Leaves compound .... RANUNCULACEAE
14 Leaves simple.. 15
15 Plant truly aquatic, with all leaves basal, the petioles all arising from a rhizome buried under water (except when stranded)..................................... NYMPHAEACEAE
15 Plant terrestrial, with at least some leaves on the stem................................... 16
16 Style 1 (or none, with 3 sessile stigmas) .... CISTACEAE
16 Styles 2 or more, evident... 17
17 Leaves opposite, with translucent dots; petals yellow (Hypericum) .... HYPERICACEAE
17 Leaves alternate, without translucent dots; petals of various colors............... MALVACEAE
18 Stamens fewer than twice as many as the petals 19
18 Stamens exactly twice as many as the petals... 22
19 Style 1 or none; leaves usually alternate, simple or compound, entire or toothed. . BRASSICACEAE
19 Styles 2-5; leaves opposite or whorled, simple and entire 20
20 Petals yellow (Hypericum) .... HYPERICACEAE
20 Petals white, pink, or red 21
21 Stamens 9, in 3 distinct groups of 3 each, with 3 conspicuous glands alternating with the groups (Triadenum)... HYPERICACEAE

21 Stamens various but neither 9 nor in groups . . . . CARYOPHYLLACEAE
22 Petals 3 (*Trillium*) . . . . LILIACEAE
22 Petals 4 or more 23
23 Sepals and petals each 6 or more (*Lythrum*) . . . . LYTHRACEAE
23 Sepals and petals each 4 or 5 24
24 Leaves compound or deeply divided nearly to base of blade 25
24 Leaves simple and entire, toothed, or shallowly lobed 28
25 Leaves opposite 26
25 Leaves alternate 27
26 Leaves uniformly trifoliolate . . . . OXALIDACEAE
26 Leaves palmately compound or lobed (*Geranium*) . . . . GERANIACEAE
27 Styles 5; leaves with 3 obcordate leaflets . . . . OXALIDACEAE
27 Style 1; pinnately compound . . . . FABACEAE
28 Style 1 29
28 Styles 2 or more 31
29 Anthers opening by longitudinal slits, not especially showy; stamens not skewed . LYTHRACEAE
29 Floral tube none, all parts arising directly from the receptacle; petal color various 30
30 Sepals of 2 sizes, the 2 outer ones very much narrower and often shorter than the 3 inner ones (appearing as mere appendages on them); petals yellow . . . . CISTACEAE
30 Sepals all of nearly the same size and shape; petals white, greenish, or pink . . . . ERICACEAE
31 Ovary lobed, with a style on each lobe. 32
31 Ovary unlobed, the styles all arising together 33
32 Leaves not succulent, all or mostly basal, (cauline leaves, if any, few and small or a single pair); lobes of ovary 2 . . . . SAXIFRAGACEAE
32 Leaves succulent, all or mostly cauline; lobes of ovary 4 or 5 . . . . CRASSULACEAE
33 Petals yellow; leaves with translucent dots (*Hypericum*) . . . . HYPERICACEAE
33 Petals white to pink or red (never yellow); leaves without translucent dots. CARYOPHYLLACEAE\\

GROUP 18

Herbaceous dicots with bisexual flowers, perianth of z series, ovary 1 and superior, stamens the same number as the petals or fewer, and petals separate.

1 Leaves compound or dissected 2
1 Leaves entire or toothed to deeply lobed. 4
2 Flowers solitary on leafless peduncles arising from the ground . . . . VIOLACEAE
2 Flowers on leafy stems 3
3 Petals and stamens each 6; leaves 2-3-times compound, with flat, broad leaflets (*Caulophyllum*) . . .
. . . . BERBERIDACEAE
3 Petals and stamens each 5; leaves dissected . . . . GERANIACEAE
4 Leaves opposite or whorled 5
4 Leaves alternate or basal 12
5 Sepals 2 or 3; petals 2-6. 6
5 Sepals and petals each 4-6 (or more) 7
6 Cauline leaves 2; flowers pedunculate (*Claytonia*). . . . . MONTIACEAE
6 Cauline leaves numerous; flowers essentially sessile. . . . . PORTULACACEAE
7 Leaves deeply palmately lobed. . . . . GERANIACEAE
7 Leaves entire or merely toothed. 8
8 Style 1, sometimes very short or the stigma ± sessile 9
8 Styles 2-5. 10
9 Floral tube or disk well developed, with sepals and petals borne at its margin. . . . . . . LYTHRACEAE
9 Floral tube or disk none . . . . GENTIANACEAE
10 Flowers completely 5-merous, including 5 styles; stamens with filaments connate at the base around the ovary; ovary 5- (or 10-) locular . . . . LINACEAE

10 Flowers with styles usually fewer than 5 (and petals sometimes 4); stamens not connate; ovary with 1 locule 11
11 Petals yellow; leaves with translucent dots (*Hypericum*) . . . . HYPERICACEAE
11 Petals white to pink or red; leaves without translucent dots. . . . . CARYOPHYLLACEAE
12 Leaves shallowly to deeply palmately lobed . . . . VIOLACEAE
12 Leaves unlobed or pinnately lobed 13
13 Styles 2 or more 14
13 Style 1 or none 15
14 Leaves essentially all basal; flowers white . . . . DROSERACEAE
14 Leaves cauline; flowers yellow . . . . LINACEAE
15 Floral tube well developed and prolonged, with sepals and petals borne at its margin . . . .
 . . . . LYTHRACEAE
15 Floral tube none or very little developed. 16
16 Corolla bilaterally symmetrical, saccate or spurred at the base . . . . VIOLACEAE
16 Corolla regular, without a spur 17
17 Petals and sepals each 4 . . . . BRASSICACEAE
17 Petals and sepals each 5 18
18 Flowers solitary, terminal; styles essentially none (stigmas 4, nearly sessile); stamens alternating with cleft, gland-tipped staminodia (*Parnassia*) . . . . CELASTRACEAE
18 Flowers in a terminal umbel or raceme; style present; staminodia none. 19
19 Principal leaves all basal; inflorescence a stalked umbel . . . . PRIMULACEAE
19 Principal leaves all or partly cauline; inflorescence a terminal raceme (*Lysimachia*) . . . .
 . . . . PRIMULACEAE

GROUP 19

Herbaceous dicots with bisexual flowers, perianth of 2 series, ovary 1 and superior, corolla regular and stamens the same number as its lobes, and petals united.

1 Leaves all basal. 2
1 Leaves all or mostly cauline. 4
2 Leaves covered with conspicuous stalked glands. . . . . DROSERACEAE
2 Leaves without stalked glands 3
3 Perianth 4-merous; flowers in spikes or heads; corolla scarious . . . . PLANTAGINACEAE
3 Perianth 5-merous; flowers in umbels; corolla petaloid . . . . PRIMULACEAE
4 Ovary deeply 4-lobed, appearing like 4 separate ovaries [and also keyed as such] but with one style arising deep in the midst of the lobes 5
4 Ovary not conspicuously lobed (may be slightly 4- or 2-lobed or notched at apex, where style arises) 6
5 Leaves opposite; stamens 2 or 4; stems 4-angled ("square") and foliage aromatic (minty or citrus-like) . . . . LAMIACEAE
5 Leaves alternate; stamens 5; stem not angled (rarely winged) and foliage not aromatic. . . . .
 . . . . BORAGINACEAE
6 Leaves opposite (or whorled), at least below the inflorescence. 7
6 Leaves alternate, at least below the inflorescence. 12
7 Flowers in dense heads or short spikes; corolla 4-lobed (*Plantago*) . . . . PLANTAGINACEAE
7 Flowers in crowded or more open racemes or other inflorescences; corolla lobes 4-7 8
8 Stamens opposite the corolla lobes (i.e., each stamen arising and oriented above the middle of a lobe) and readily visible . . . . PRIMULACEAE
8 Stamens alternating with the corolla lobes (sometimes hidden in a corolla tube or closed corolla) 9
9 Lobes of corolla 4 . . . . GENTIANACEAE
9 Lobes of corolla 5 10
10 Stigmas 3; ovary with 3 locules . . . . POLEMONIACEAE
10 Stigma 1 (may be 2-lobed); ovary with 2 (or 4 or 5) locules 11

11 Leaves glabrous; ovary 1-locular; fruit a 2-valved capsule .... GENTIANACEAE
11 Leaves strongly clammy-pubescent; ovary 2-locular; fruit a berry or a 2-valved capsule (*Leuco-physalis*).. SOLANACEAE
12 Blades of leaves deeply lobed, dissected, or compound 13
12 Blades of leaves entire, toothed, or at most shallowly lobed (or merely cordate).............. 18
13 Plant a twining or trailing vine... 14
13 Plant, whether erect or prostrate, not a vine... 15
14 Corolla deeply funnel-shaped (or even trumpet-shaped) .... CONVOLVULACEAE
14 Corolla ± flat (rotate) (*Solanum*).. SOLANACEAE
15 Anthers forming a cone around the pistil .... SOLANACEAE
15 Anthers clearly separate.. 16
16 Leaves 3-foliolate.. MENYANTHACEAE
16 Leaves otherwise lobed, compound, or dissected... 17
17 Leaves pinnately compound or pinnately dissected into entire filiform lobes; ovary 3-locular; stigmas or style branches 3; capsule 3-valved .... POLEMONIACEAE
17 Leaves not compound: pinnatifid or bipinnatifid, the segments not both entire and filiform; ovary 1-locular; stigmas or style branches 2; capsule 2-valved...................... BORAGINACEAE
18 Flowers or inflorescences axillary 21
18 Flowers or inflorescences terminal 20
19 Fruit a 4-seeded capsule; corolla large, funnel-shaped; stigmas clearly 2, separate (except in *Ipomoea*, where at most 2-3-lobed) .... CONVOLVULACEAE
19 Fruit a many-seeded berry or capsule; corolla ± flat (rotate) or bell-shaped; stigma 1 SOLANACEAE
20 Anthers separate, at least some of them on hairy filaments; fruit a capsule (*Verbascum*)..........
.... SCROPHULARIACEAE
20 Anthers forming a cone around the pistil, on glabrous filaments; fruit a berry (*Solanum*).........
.... SOLANACEAE

GROUP 20

Herbaceous dicots with bisexual flowers, perianth of z series, ovary 1 and superior, corolla either bilaterally symmetrical or stamens fewer than its lobes, or both and petals united.

1 Fertile (anther-bearing) stamens 5... 2
1 Fertile stamens 2 or 4 3
2 Ovary deeply 4-lobed; plant strongly bristly-hairy; fruit (1-) 4 nutlets (*Echium*)... BORAGINACEAE
2 Ovary not lobed; plants glabrous or with dense pubescence; fruit a capsule (*Verbascum*) ....
.... SCROPHULARIACEAE
3 Corolla with a spur or sac at the base.. 4
4 Calyx 2-parted (*Utricularia*).. LENTIBULARIACEAE
4 Calyx 5-parted .... PLANTAGINACEAE
3 Corolla not prolonged into a spur or sac at the base 5
5 Stem leaves all alternate 6
5 Stem leaves all or mostly opposite or whorled... 9
6 Corolla nearly regular, the lobes equaling or exceeding the tube (*Veronica*) .... PLANTAGINACEAE
6 Corolla bilaterally symmetrical, ± 2-lipped, the lobes (not lips) distinctly shorter than the tube.. 7
7 Bracts of inflorescence contrasting with the leaves, cream, yellow, or red at least apically (*Castilleja*)
.... OROBANCHACEAE
7 Bracts of inflorescence the same color as the leaves, green or purplish-green 8
8 Cauline leaves deeply pinnately lobed (*Pedicularis*) .... OROBANCHACEAE
8 Cauline leaves unlobed (at most shallowly toothed)...................... PLANTAGINACEAE
9 Ovary deeply 4-lobed, appearing like 4 separate ovaries around the base of the single style [and also keyed as such], the fruit (1-) 4 nutlets; plants usually with a 4-angled ("square") stem and often a minty or citrus-like aroma when bruised .... LAMIACEAE
9 Ovary not 4-lobed (at most, somewhat 2-lobed), the fruit a capsule; stem in only a few species 4-angled or with aroma when bruised 10

10 Fertile stamens 2 11

11 Flowers in axillary racemes or spikes. . . . . SCROPHULARIACEAE

11 Flowers in terminal racemes or spikes, or solitary or paired in the axils of the leaves. . . . .
. . . . PLANTAGINACEAE

10 Fertile stamens 4 12

12 Corolla nearly regular, the lobes about equal 13

13 Corolla salverform (trumpet-shaped, with a slender tube of almost uniform diameter) (*Verbena*). .
. . . . VERBENACEAE

13 Corolla funnel-shaped or bell-shaped, with a tube broad toward its summit. 14

14 Corolla pink to purple (white in albinos); calyx and other parts glabrous (at most scabrous) or
with hairs of distinctly different lengths (*Agalinis*). . . . . OROBANCHACEAE

14 Corolla bright yellow; calyx tube, pedicels, and/or stems with hairs of uniform or mixed lengths
(not of 2 distinct lengths and only rarely completely glabrous) (*Mimulus*) . . . . PHRYMACEAE

12 Corolla strongly bilaterally symmetrical. 15

15 Mature flowers and fruit strongly reflexed, nearly sessile and in remote pairs on opposite sides of
a spike-like terminal raceme; calyx with 3 upper teeth bristle-like and 2 lower teeth broadly trian-
gular (*Phryma*) . . . . PHRYMACEAE

15 Mature flowers and fruit not strongly reflexed, and otherwise not as above (long-pediceled, alternate,
and/or crowded); calyx with teeth equal or subequal (never bristle-like) 16

16 Upper lip of corolla apparently absent (the corolla split lengthwise above) or much shorter than
the lower lip and 4-lobed . . . . LAMIACEAE

16 Upper lip of corolla well developed, of 2 lobes (or these ± fused into one), often nearly or quite as
long as the lower lip 17

17 Inflorescence terminal and branched (± paniculate); stamens 4 fertile plus 1 staminodium 18

18 Leaves below the inflorescence distinctly petioled; corolla brownish, less than 12 mm long; staminodium broad (ca. 1-2 mm) and flat at the free apex (mostly adnate to the upper lip), glabrous
(*Scrophularia*) . . . . SCROPHULARIACEAE

18 Leaves below the inflorescence sessile; corolla white to purple-violet, ca. 15-30 (-45) mm long; staminodium slender, elongate (of similar diameter and length as the style), close to lower lip of
corolla, bearded at the apex (*Penstemon*) . . . . PLANTAGINACEAE

17 Inflorescence a spike or raceme (no branched stalks), or flowers all axillary; stamens 4 fertile, in
most genera with no staminodium (or only a very rudimentary one). 19

19 Leaves (especially middle and lower ones) deeply pinnately toothed or lobed ca. one-third or more
the distance to the midrib . . . . OROBANCHACEAE

19 Leaves of main stem toothed or entire but not so deeply pinnately toothed or lobed (uppermost
leaves or bracts may have small basal lobes) 20

20 Sepals separate nearly or quite to the base; flowers in a compact terminal inflorescence (*Chelone*)
. . . . PLANTAGINACEAE

20 Sepals (at least at anthesis) fused ca. 1/3 or more the length of the calyx; flowers solitary in the leaf
axils or in a loose terminal inflorescence 21

21 Flowers (all or many of them, especially lower ones) in the axils of alternate bracts in a distinct
terminal or racemose inflorescence. . . . . OROBANCHACEAE

21 Flowers all solitary in the axils of opposite (or whorled) leaves or bracts 22

22 Lobes less than 1/3 the total length of the calyx, or corolla bright yellow (or both conditions) (*Mimu-
lus*) . . . . PHRYMACEAE

22 Lobes ca. 1/2 or more the total length of the calyx; corolla with whitish to pink or magenta ground
color (plus dark spots and/or yellow markings) . . . . OROBANCHACEAE

REFERENCES

Listed below are some of the more popular and readily available field guides and other resources to the plants and vegetation of Door County and Wisconsin.

Black, M., and E. Judziewicz. 2009. *Wildflowers of Wisconsin and the Great Lakes region: A comprehensive field guide.* Madison: University of Wisconsin Press.

Chadde, Steve W. 2013. *Wetland Plants of Wisconsin.* 2nd Ed. (Available at amazon.com.)

Chadde, Steve W. 2013. *Wisconsin Flora.* (Available at amazon.com.)

Chadde, Steve W. 2013. *Midwest Ferns.* (Available at amazon.com.)

Curtis, J. T. 1959. *The Vegetation of Wisconsin.* Madison: University of Wisconsin Press.

Fassett, N.C. 1951. *Grasses of Wisconsin.* Madison: University of Wisconsin Press.

Fassett, N. C. 1976. *Spring Flora of Wisconsin.* 4th ed. Madison: University of Wisconsin Press.

Gleason, H. A., and A. Cronquist. 1991. *Manual of vascular plants of northeastern United States and adjacent Canada.* 2nd ed. New York: New York Botanical Garden.

Hipp, A. *Field guide to Wisconsin sedges: An introduction to the genus* Carex *(Cyperaceae).* Madison: University of Wisconsin Press.

Holmgren, N. H., and P. K. Holmgren, eds. 1998. *Illustrated companion to Gleason and Cronquist's manual.* New York: New York Botanical Garden.

Judziewicz, Emmet J. 2001. *Flora and vegetation of the Grand Traverse Islands (Lake Michigan), Wisconsin and Michigan.* Michigan Botanist 40(4):81–208.

Judziewicz, Emmet J., Robert W. Freckmann, Lynn G. Clark, and Merel R. Black. 2014. *Field Guide to Wisconsin Grasses.* Madison: University of Wisconsin Press.

Kartesz, J. T. 2003. *A Synonymized Checklist and Atlas with Biological Attributes for the Vascular Flora of the United States, Canada, and Greenland.* Second Edition. In: Kartesz, J.T. Synthesis of the North American Flora, Version 2.0.

Kowal, R. R. 2007. *Keys to the Asteraceae of Wisconsin.* Unpublished report.

Reznicek, A. A., E. G. Voss, and B. S. Walters. 2011. *Michigan Flora Online.* University of Michigan. http://www.michiganflora.net.

Swink, F., and G. Wilhelm. 1994. *Plants of the Chicago region.* 4th ed. Indianapolis: Indiana Academy of Science.

Tryon, R., N. Fassett, D. Dunlop, and M. Diemer. 1953. *The Ferns and Fern Allies of Wisconsin.* Madison: University of Wisconsin Press.

Voss, E.G. and A. A. Reznicek. 2011. *Field Manual of Michigan Flora.* Ann Arbor: University of Michigan Press.

Wetter, M. A., T. S. Cochrane, M. R. Black, H. H. Iltis, and P. E. Berry. 2001. *Checklist of the vascular plants of Wisconsin.* DNR Technical Bulletin 192. Department of Natural Resources, Madison. Available online at: www.botany.wisc.edu/wisflora/.

Wilhelm Gerould, and Laura Rericha. 2017. *Flora of the Chicago Region: A Floristic and Ecological Synthesis.* Indianapolis: Indiana Academy of Science.

Wisconsin Department of Natural Resources. 2015. *The ecological landscapes of Wisconsin: An assessment of ecological resources and a guide to planning sustainable management. Chapter 8, Central Lake Michigan Coastal Ecological Landscape.* Wisconsin Department of Natural Resources, PUB-SS-1131J 2015, Madison.

Wisconsin Department of Natural Resources. 2015. *The ecological landscapes of Wisconsin: An assessment of ecological resources and a guide to planning sustainable management. Chapter 15, Northern Lake Michigan Coastal Ecological Landscape.* Wisconsin Department of Natural Resources, PUB-SS-1131Q 2015, Madison.

ONLINE RESOURCES

A wealth of information about Wisconsin's flora is available online; especially useful is the website of the Wisconsin State Herbarium, UW-Madison: *http://wisflora.herbarium.wisc.edu/index.php*

For species of conservation concern (endangered, threatened), information was obtained from the Wisconsin Natural Heritage Inventory: *dnr.wi.gov/topic/nhi/*

GLOSSARY

abaxial—On the side away from the axis, usually refers to the underside of a leaf (compare with adaxial).

acaulescent—Without an upright, leafy stem.

achene—A one-seeded, dry, indehiscent fruit with the seed coat not attached to the mature wall of the ovary.

acid—Having more hydrogen ions than hydroxyl (OH) ions; a pH less than 7.

acuminate—Tapering to a narrow point, more tapering than acute, less than attenuate.

acute—Gradually tapered to a tip.

adaxial—On the side toward the axis, usually refers to the top side of a leaf (compare with abaxial).

adnate—Fused with a structure different from itself, as when stamens are adnate to petals (compare with connate).

adventive—Not native to and not fully established in a new habitat.

alkaline—Having more hydroxyl ions than hydrogen ions; a pH greater than 7.

alluvial—Deposits of rivers and streams.

alternate—Borne singly at each node, as in leaves on a stem.

ament—Spikelike inflorescence of same-sexed flowers (either male or female); same as catkin.

androgynous—Spike with both staminate and pistillate flowers, the pistillate located at the base, below the staminate (compare with gynaecandrous).

angiosperm—A plant producing flowers and bearing seeds in an ovary.

annual—A plant that completes its life cycle in one growing season, then dies.

anther—Pollen-bearing part of stamen, usually at the end of a stalk called a filament.

anthesis—The period during which a flower is fully open and functional.

anthocyanic—Pigmented with anthocyanins, this usually manifested as a tinging or suffusion of pink, red, or purple.

aphyllopodic—Having basal sheaths without blades; with new shoots arising laterally from parent shoot (compare with phyllopodic).

apiculate—Having an apiculus.

apiculus—An abrupt, very small, projected tip.

appressed—Lying flat to or parallel to a surface.

aquatic—Living in water.

areole—In leaves, the spaces between small veins.

aril—A specialized appendage on a seed, often brightly colored, derived from the seed coat.

aristate—Tipped with a slender bristle.

armed—Bearing a sharp projection such as a prickle, spine, or thorn.

aromatic—Strongly scented.

ascending—Angled upward.

asymmetrical—Not symmetrical.

attenuate—Tapering gradually to a prolonged tip.

auricle—An ear-shaped appendage to a leaf or stipule.

awl-shaped—Tapering gradually from a broad base to a sharp point.

awn—A bristle-like organ.

axil—Angle between a stem and the attached leaf.

barb—Sharp, thorn-like projection.

basal—From base of plant.

basic—A pH greater than 7.

beak—A slender, terminal appendage on a 3-dimensional organ.

beard—Covering of long or stiff hairs.

berry—Fruit with the seeds surrounded by fleshy material.

biennial—A plant that completes its life cycle in two growing season, typically flowering and fruiting in the second year, then dying.

bifid—Cleft into two more or less equal parts.

blade—Expanded, usually flat part of a leaf or petiole.

bloom—A whitish powdery or waxy coating that can be rubbed away.

bog—A wet, acidic, nutrient-poor peatland characterized by sphagnum and other mosses, shrubs and sedges. Technically, a type of peatland raised above its surroundings by peat accumulation and receiving nutrients only from precipitation.

boreal—Far northern latitudes.

brackish—Salty.

bract—An accessory structure at the base of some flowers, usually appearing leaflike.

bractlet—A secondary bract (*Typha*).

branchlets—A small branch.

bristle—A stiff hair.

bud—An undeveloped shoot, inflorescence, or flower, in woody plants often covered by scales and serving as the overwintering stage.

bulb—A group of modified leaves serving as a food-storage organ, borne on a short, vertical, underground stem (compare with corm).

bulbil—A bulb-like structure borne in the leaf axils or in place of flowers.

bulblet—Small bulb borne above ground, as in a leaf axil.

ca.—About, approximately (Latin *circa*).

caducous—Falling off early, as stipules that leave behind a scar.

callosity—A hardened thickening.

callus—A firm, thickened portion of an organ; the firm base of the lemma in the Poaceae.

calcareous fen—An uncommon wetland type associated with seepage areas, and which receive groundwater enriched with primarily calcium and magnesium bicarbonates.

calcium-rich—Refers to wetlands underlain by limestone or receiving water enriched by calcium compounds.

calyx—All the sepals of a flower.

campanulate—Bell-shaped.

capillary—Very fine, hair-like, not-flattened.

capitate—Abruptly expanded at the apex, thereby forming a knob-like tip.

capsule—A dry, dehiscent fruit splitting into 3 or more parts.

carpel—Fertile leaf of an angiosperm, bearing the ovules. A pistil is made up of one or more carpels.

caruncle—An appendage at or near the hilum of some seeds.

caryopsis—The dry, indehiscent seed of grasses.

catkin—Spikelike inflorescence of same-sexed flowers (either male or female); same as ament.

caudex—Firm, hardened, summit of a root mass that functions as a perennating organ.

cauline—Of or pertaining to the aboveground portion of the stem.

cespitose—Growing in a compact cluster with closely spaced stems; tufted, clumped.

chaff—Thin, dry scales; in the Asteraceae, sometimes found as chaffy bracts on the receptacle.

cilia—Hairs found at the margin of an organ.

ciliate—Provided with cilia.

circumboreal—Refers to a species distribution pattern which circles the earth's boreal regions.

clasping—Leaves that partially encircle the stem at the base.

clavate—Widened in the distal portion, like a baseball bat.

claw—The narrow, basal portion of perianth parts.

cleistogamous—Type of flower that remains closed and is self-pollinated.

clumped—Having the stems grouped closely together; tufted.

colony-forming—A group of plants of the same species, produced either vegetatively or by seed.

column—The joined style and filaments in the Orchidaceae.

coma—A tuft of fine hairs, especially at the tip of a seed.

composite—An inflorescence that is made up of many tiny florets crowded together on a receptacle; members of the Aster Family (Asteraceae).

compound leaf—A leaf with two or more leaflets.

concave—Curved inward.

conduplicate—Folded lengthwise into nearly equal parts.

cone—The dry fruit of conifers composed of overlapping scales.

conifer—Cone-bearing woody plants.

connate—Two like parts that are fused (compare with adnate).

connivent—Converging and touching but not actually fused, applies to like organs.

convex—Curved outward.

convolute—Arranged such that one edge is covered and the other is exposed, usually referring to petals in bud.

cordate—With a rounded lobe on each side of a central sinus; heart-shaped.

coriaceous—With a firm, leathery texture.

corm—A short, vertical, enlarged, underground stem that serves as a food storage organ (compare with bulb).

corolla—Collectively, all the petals of a flower.

corymb—An indeterminate inflorescence, somewhat similar to a raceme, that has elongate lower branches that create a more or less flat-topped inflorescence.

costa (plural costae)—A prominent midvein or midrib of a leaflet.

crenate—With rounded teeth.

crenulate—Finely crenate.

crisped—An irregularly crinkled or curled leaf margin.

crown—Persistent base of a plant, especially a grasses.

culm—The stem of a grass or grasslike plant, especially a stem with the inflorescence.

cuneate—Tapering to the base with relatively straight, non-parallel margins; wedge-shaped.

cyme—A type of inflorescence in which the central flowers open first.

deciduous—Not persistent.

decumbent—A stem that is prostrate at the base and curves upward to have an erect or ascending, apical portion.

decurrent—Possessing an adnate line or wing that extends down the axis below the node, usually referring to leaves on a stem.

dehiscent—Splitting open at maturity.

deltate—Triangle-shaped.

dentate—Provided with outward oriented teeth.

depauperate—Poorly developed due to unfavorable conditions.

dicots—One of two main divisions of the Angiosperms (the other being the Monocots); plants having 2 seed leaves (cotyledons), net-venation, and flower parts in 4s or 5s (or multiples of these numbers).

dioecious—Bearing only male or female flowers on a single plant.

dimorphic—Having two forms.

disarticulation—Spikelets breaking either above or below the glumes when mature, the glumes remaining in the head if disarticulation above the glumes, or the glumes falling with the florets if disarticulation is below the glumes.

discoid—In composite flowers (Asteraceae), a head with only disk (tubular) flowers, the ray flowers absent.

disjunct—A population of plants widely separated from its main range.

disk—In the Asteraceae, the central part of the head, composed of tubular flowers.

dissected—Leaves divided into many smaller segments.

disturbed—Natural communities altered by human influences.

divided—Leaves which are lobed nearly to the midrib.

dolomite—A type of limestone consisting of calcium magnesium carbonate.

dorsal—Underside, or back of an organ.

driftless area—Portions of sw Wisconsin, ne Iowa, and se Minnesota that are not covered by glacial drift.

drupe—A fleshy fruit with a single large seed such as a cherry.

echinate—With spines.

eglandular—Without glands.

elliptic—Broadest at the middle, gradually tapering to both ends.

emergent—Growing out of and above the water surface.

emersed leaf—Growing above the water surface or out of water.

endangered—A species in danger of extinction throughout all or most of its range if current trends continue.

endemic—A species restricted to a particular region.

entire—With a smooth margin.

erect—Stiffly upright.

erose—With a ragged edge.

escape—A cultivated plant which establishes itself outside of cultivation.

evergreen—Plant retaining its leaves throughout the year.

excurrent—With the central rib or axis continuing or projecting beyond the organ.

exserted—Extending beyond the mouth of a structure such as stamens extending out from the mouth of the corolla.

falcate—Sickle-shaped

false indusium—A modified tooth or reflexed margin of a fern leaf that covers the sorus.

fen—An open wetland usually dominated by herbaceous plants, and fed by in-flowing, often calcium- and/or magnesium-rich water; soils vary from peat to clays and silts.

fern—Perennial plants with spore-bearing leaves similar to the vegetative leaves and bearing sporangia on their underside, or the spore-bearing leaves much modified.

fibrous—A cluster of slender roots, all with the same diameter.

filament—The stalk of a stamen which supports the anther.

filiform—Thread-like.

flexuous—An elongate axis that arches or bends in alternating directions in a zig-zag fashion.

floating mat—A feature of some ponds where plant roots form a carpet over some or all of the water surface.

floodplain—That part of a river valley that is occasionally covered by flood waters.

floret—A small flower in a dense cluster of flowers; in grasses the flower with its attached lemma and palea.

follicle—A dry, dehiscent fruit that splits along one side when mature.

floricane—the second-year flowering stem of *Rubus* (compare with primocane).

genus—The first part of the scientific name for a plant or animal (plural genera).

glabrate—Nearly glabrous or becoming so.

glabrous—Lacking hairs.

gland—An appendage or depression which produces a sticky or greasy substance.

glandular—Bearing glands.

glaucous—Having a bluish appearance.

glumes—A pair of small bracts at base of each spikelet the lowermost (or first) glume usually smaller the upper (or second) glume usually longer.

grain—The fruit of a grass; the swollen seedlike protuberance on the fruit of some *Rumex*.

gymnosperm—Plants in which the seeds are not produced in an ovary, but usually in a cone.

gynaecandrous—Having both staminate and pistillate flowers on the same spike, the staminate located at the base, below the pistillate (compare with androgynous).

gynophore—The central stalk of some flowers, especially in cat-tails (*Typha*).

halophyte—A plant adapted to growing in a salty substrate.

hastate—More or less triangular in outline with outward-oriented basal lobes.

haustorium—A specialized, root-like connection to a host plant that a parasite uses to extract nourishment.

hardwoods—Loosely used to contrast most deciduous trees from conifers.

herb—A herbaceous, non-woody plant.

herbaceous—Like an herb; also, leaflike in appearance.

hilum—The scar at the point of attachment of a seed.

hirsute—Pubescent with coarse, somewhat stiff, usually curving hairs, coarser than villous but softer than hispid.

hispid—Pubescent with coarse, stiff hairs that may be uncomfortable to the touch, coarser than hirsute but softer than bristly.

hummock—A small, raised mound formed by certain species of sphagnum moss.

humus—Dark, well-decayed organic matter in soil.

hybrid—A cross-breed between two species.

hydric—Wet (compare with mesic, xeric).

hypanthium—A ring, cup, or tube around the ovary; the sepals, petals and stamens are attached to the rim of the hypanthium.

imbricate—Overlapping, as shingles on a roof.

indehiscent—Not splitting open at maturity.

indusium—In ferns, a membranous covering over the sorus (plural indusia).

inferior—The position of the ovary when it is below the point of attachment of the sepals and petals.

inflorescence—A cluster of flowers.

insectivorous—Refers to the insect trapping and digestion habit of some plants as a nutrition supplement.

interdunal swale—Low-lying areas between sand dune ridges.

internode—Portion of a stem between two nodes.

introduced—A non-native species.

invasive—Non-native species causing significant ecological or economic problems.

involucral bract—A single member of the involucre; sometimes called phyllary in composite flowers (Asteraceae).

involucre—A whorl of bracts, subtending a flower or inflorescence.

irregular flower—Not radially symmetric; with similar parts unequal.

joint—A node or section of a stem where the branch and leaf meet.

keel—A central rib like the keel of a boat.

lance-shaped—Broadest near the base, gradually tapering to a narrower tip.

lateral—Borne on the sides of a stem or branch.

lax—Loose or drooping.

leaf axil—The point of the angle between a stem and a leaf.

leaflet—One of the leaflike segments of a compound leaf.

lemma—In grasses, the lower bract enclosing the flower (the upper, smaller bract is the palea).

lens-shaped—Biconvex in shape (like a lentil).

lenticel—Blisterlike openings in the epidermis of woody stems, admitting gases to and from the plant, and often appearing as small oval dots on bark.

ligulate—Having a ligule; in the Asteraceae, the strap-shaped corolla of a ray floret.

ligule—In grasses and grasslike plants, the membranous or hairy ring at top of sheath between the blade and stem.

linear—Narrow and flat with parallel sides.

lip—Upper or lower part of a 2-lipped corolla; also the lower petal in most orchid flowers.

lobed—With lobes; in leaves divisions usually not over halfway to the midrib.

local—Occurring sporadically in an area.

low prairie—Wet and moist herbaceous plant community, typically dominated by grasses.

margin—The outer edge of a leaf.

marl—A calcium-rich clay.

marsh—Wetland dominated by herbaceous plants, with standing water for part or all the growing season, then often drying at the surface.

megaspore—Large, female spores.

mesic—Moist, neither dry nor wet (compare with hydric, xeric).

microspore—Small, male spores.

midrib—The prominent vein along the main axis of a leaf.

mixed forest—A type of forest composed of both deciduous and conifer trees.

moat—The open water area ringing the outer edge of a peatland or floating mat.

monecious—Having male and female reproductive parts in separate flowers on the same plant.

monocots—One of two main divisions of the Angiosperms (the other being the Dicots); plants with a single seed leaf (cotyledon); typically having narrow leaves with parallel veins, and flower parts in 3s or multiples of 3.

muck—An organic soil where the plant remains are decomposed to the point where the type of plants forming the soil cannot be determined.

mucro—A sharp point at termination of an organ or other structure.

naked—Without a covering; a stalk or stem without leaves.

native—An indigenous species.

naturalized—An introduced species that is established and persistent in an ecosystem.

needle—A slender leaf, as in the Pinaceae.

nerve—A leaf vein.

neutral—A pH of 7.

node—The spot on a stem or branch where leaves originate.

nutlet—A small dry fruit that does not split open along a seam.

oblanceolate—Reverse lance-shaped; broadest at the apex, gradually tapering to the narrower base.

oblique—Emerging or joining at an angle other than parallel or perpendicular.

oblong—Broadest at the middle, and tapering to both ends, but broader than elliptic.

obovate—Broadly rounded at the apex, becoming narrowed below.

ocrea—A tube-shaped stipule or pair of stipules around the stem; characteristic of the Smartweed Family (Polygonaceae).

opposite—Leaves or branches which are paired opposite one another on the stem.

organic—Soils composed of decaying plant remains.

oval—Elliptical.

ovary—The lower part of the pistil that produces the seeds.

ovate—Broadly rounded at the base, becoming narrowed above; broader than lanceolate.

palea—The uppermost of the two inner bracts subtending a grass flower (the lower bract is the lemma).

palmate—Divided in a radial fashion, like the fingers of a hand.

panicle—An arrangement of flowers consisting of several racemes.

papilla (plural: papillae)—A short, rounded or cylindrical projections.

pappus—The modified sepals of a composite flower which persist atop the ovary as bristles, scales or awns.

parallel-veined—With several veins running from base of leaf to leaf tip, characteristic of most monocots.

peat—An organic soil formed of partially decomposed plant remains.

peatland—A wetland whose soil is composed primarily of organic matter (mosses, sedges, etc.); a general term for bogs and fens.

peltate—More or less circular, with the stalk attached at a point on the underside.

pepo—A fleshy, many-seeded fruit with a tough rind, as a melon.

perennial—Living for 3 or more years.

perfect—A flower having both male (stamens) and female (pistils) parts.

perianth—Collectively, all the sepals and petals of a flower.

perigynium—A sac-like structure enclosing the pistil in *Carex* (plural perigynia).

petal—An individual part of the corolla, often white or colored.

petiole—The stalk of a leaf.

phyllary—An involucral bract subtending the flower head in composite flowers (Asteraceae).

phyllode—An expanded petiole.

phyllopodic—Having the basal sheaths blade-bearing; with new shoots arising from the center of parent shoot (compare with aphyllopodic).

pinna—The primary or first division in a fern frond or leaf (plural pinnae).

pinnate—Divided once along an elongated axis into distinct segments.

pinnule—The pinnate segment of a pinna.

pistil—The seed-producing part of the flower, consisting of an ovary and one or more styles and stigmas.

pith—A spongy central part of stems and branches.

pollen—The male spores in an anther.

prairie—An open plant community dominated by herbaceous species, especially grasses.

primocane—The first-year, vegetative stem in *Rubus* (compare with floricane).

pro sp. When a taxon is transferred from the non-hybrid category to the hybrid category, the author citation remains unchanged, but may be followed by an indication in parentheses of the original category.

prostrate—Lying flat on the ground.

raceme—A grouping of flowers along an elongated axis where each flower has its own stalk.

rachilla—A small stem or axis.

rachis—The central axis or stem of a leaf or inflorescence.

radiate heads—In composite flowers, heads with both ray and disk flowers (Asteraceae).

ray flower—A ligulate or strap-shaped flower in the Asteraceae, where often the outermost series of flowers in the head.

receptacle—In the Asteraceae, the enlarged summit of the flower stalk to which the sepals, petals, stamens, and pistils are usually attached.

recurved—Curved backward.

regular—Flowers with all the similar parts of the same form; radially symmetric.

rhizome—An underground, horizontal stem.

rib—A pronounced vein or nerve.

rootstock—Similar to rhizome but referring to any underground part that spreads the plant.

rosette—A crowded, circular clump of leaves.

samara—A dry, indehiscent fruit with a well-developed wing.

saprophyte—A plant that lives off of dead organic matter.

scale—A tiny, leaflike structure; the structure that subtends each flower in a sedge (Cyperaceae).

scape—A naked stem (without leaves) bearing the flowers.

section—Cross-section.

secund—Flowers mostly on 1 side of a stalk or branch.

sedge meadow—A community dominated by sedges (Cyperaceae) and occurring on wet, saturated soils.

seep—A spot where water oozes from the ground.

sepal—A segment of the calyx; usually green in color.

sheath—Tube-shaped membrane around a stem, especially for part of the leaf in grasses and sedges.

shrub—A woody plant with multiple stems.

silicle—Short fruit of the Mustard Family (Brassicaceae), normally less than 2x longer as wide.

silique—Dry, dehiscent, 2-chambered fruit of the Mustard Family (Brassicaceae), longer than a silicle.

simple—An undivided leaf.

sinus—The depression between two lobes.

smooth—Without teeth or hairs.

sorus—Clusters of spore containers (plural sori).

spadix—A fleshy axis in which flowers are embedded.

spathe—A large bract subtending or enclosing a cluster of flowers.

spatula-shaped—Broadest at tip and tapering to the base.

sphagnum moss—A type of moss common in peatlands and sometimes forming a continuous carpet across the surface; sometimes forming layers several meters thick; also loosely called peat moss.

spike—A group of unstalked flowers along an unbranched stalk.

spikelet—A small spike; the flower cluster (inflorescence) of grasses (Poaceae) and sedges (Cyperaceae).

sporangium—The spore-producing structure (plural sporangia).

spore—a one-celled reproductive structure that gives rise to the gamete-bearing plant.

sporophyll—A modified, spore-bearing leaf.

spreading—Widely angled outward.

spring—A place where water flows naturally from the ground.

spur—A hollow, pointed projection of a flower.

stamen—The male or pollen-producing organ of a flower.

staminode—An infertile stamen.

stem—The main axis of a plant.

stigma—The terminal part of a pistil which receives pollen.

stipe—A stalk.

stipule—A leaflike outgrowth at the base of a leaf stalk.

stolon—A horizontal stem lying on the soil surface.

style—The stalklike part of the pistil between the ovary and the stigma.

subspecies—A subdivision of the species forming a group with shared traits which differ from other members of the species (subsp.).

subtend—Attached below and extending upward.

succulent—Thick, fleshy and juicy.

superior—Referring to the position of the ovary when it is above the point of attachment of sepals, petals, stamens, and pistils.

swale—A slight depression.

swamp—Wooded wetland dominated by trees or shrubs; soils are typically wet for much of year or sometimes inundated.

talus—Fallen rock at the base of a slope or cliff.

taproot—A main, downward-pointing root.

tendril—A threadlike appendage from a stem or leaf that coils around other objects for support (as in *Vitis*).

tepal—Sepals or petals not differentiated from one another.

terete—Circular in cross-section.

terminal—Located at the end of a stem or stalk.

thallus—A small, flattened plant structure, without distinct stem or leaves.

thicket—A dense growth of woody plants.

threatened—A species likely to become endangered throughout all or most of its range if current trends continue.

translucent—Nearly transparent.

tree—A large, single-stemmed woody plant.

tuber—An enlarged portion of a root or rhizome.

truncate—Abruptly cut-off.

tubercle—Base of style persistent as a swelling atop the achene different in color and texture from achene body.

turion—A specialized type of shoot or bud that overwinters and resumes growth the following year.

umbel—A cluster of flowers in which the flower stalks arise from the same level.

umbelet—A small, secondary umbel in an umbel, as in the Apiaceae.

upright—Erect or nearly so.

urceolate—Constricted at a point just before an opening; urn-shaped.

utricle—A small, one-seeded fruit with a dry, papery outer covering.

valve—A segment of a dehiscent fruit; the wing of the fruit in *Rumex*.

variety—Taxon below subspecies and differing from other varieties within the same subspecies (var.).

vein—A vascular bundle, as in a leaf.

velum—The membranous flap that partially covers the sporangium in *Isoetes*.

venation—The pattern of veins on an organ.

ventral Front side.

ventricose—Inflated or distended.

verrucose—Covered with small, wart-like projections.

verticil—One whorled cycle of organs.

verticillate—Arranged in whorls.

villous—Pubescent with long, soft, bent hairs, the hairs not crimped or tangled.

vine—A trailing or climbing plant, dependent on other objects for support.

viscid—Sticky, glutinous.

whorl—A group of 3 or more parts from one point on a stem.

wing—A thin tissue bordering or surrounding an organ.

woody—Xylem tissue (the vascular tissue which conducts water and nutrients).

xeric—Dry (compare with hydric, mesic).

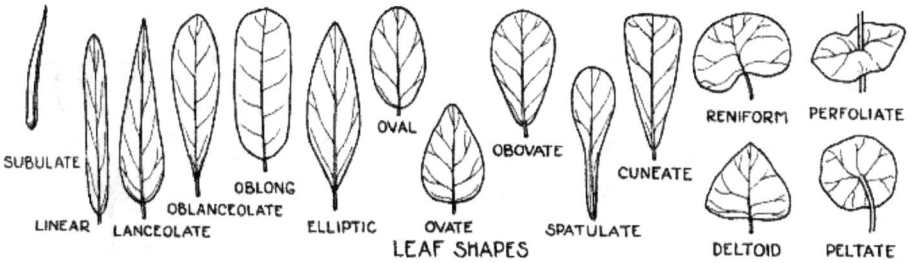

SUBULATE · LINEAR · LANCEOLATE · OBLANCEOLATE · OBLONG · ELLIPTIC · OVAL · OVATE · OBOVATE · SPATULATE · CUNEATE · RENIFORM · PERFOLIATE · DELTOID · PELTATE

LEAF SHAPES

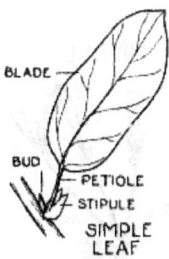

BLADE · BUD · PETIOLE · STIPULE

SIMPLE LEAF

PALMATELY COMPOUND

PINNATELY COMPOUND

BIPINNATELY COMPOUND

DISSECTED

DECOMPOUND

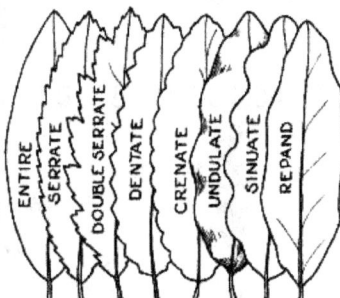

ENTIRE · SERRATE · DOUBLE SERRATE · DENTATE · CRENATE · UNDULATE · SINUATE · REPAND

MARGINS

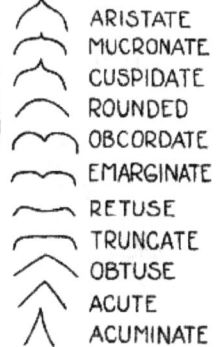

APEX OF BLADE
ARISTATE
MUCRONATE
CUSPIDATE
ROUNDED
OBCORDATE
EMARGINATE
RETUSE
TRUNCATE
OBTUSE
ACUTE
ACUMINATE

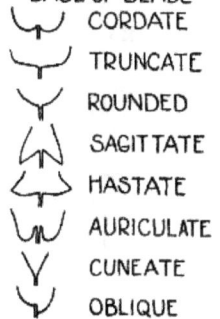

BASE OF BLADE
CORDATE
TRUNCATE
ROUNDED
SAGITTATE
HASTATE
AURICULATE
CUNEATE
OBLIQUE

PINNATELY LOBED

PINNATELY CLEFT

PINNATELY PARTED

PINNATELY DIVIDED

PALMATELY LOBED

PALMATELY CLEFT

PALMATELY PARTED

PALMATELY DIVIDED

Leaf characters (from *A Flora of the Marshes of California* by Herbert L. Mason, 1957.

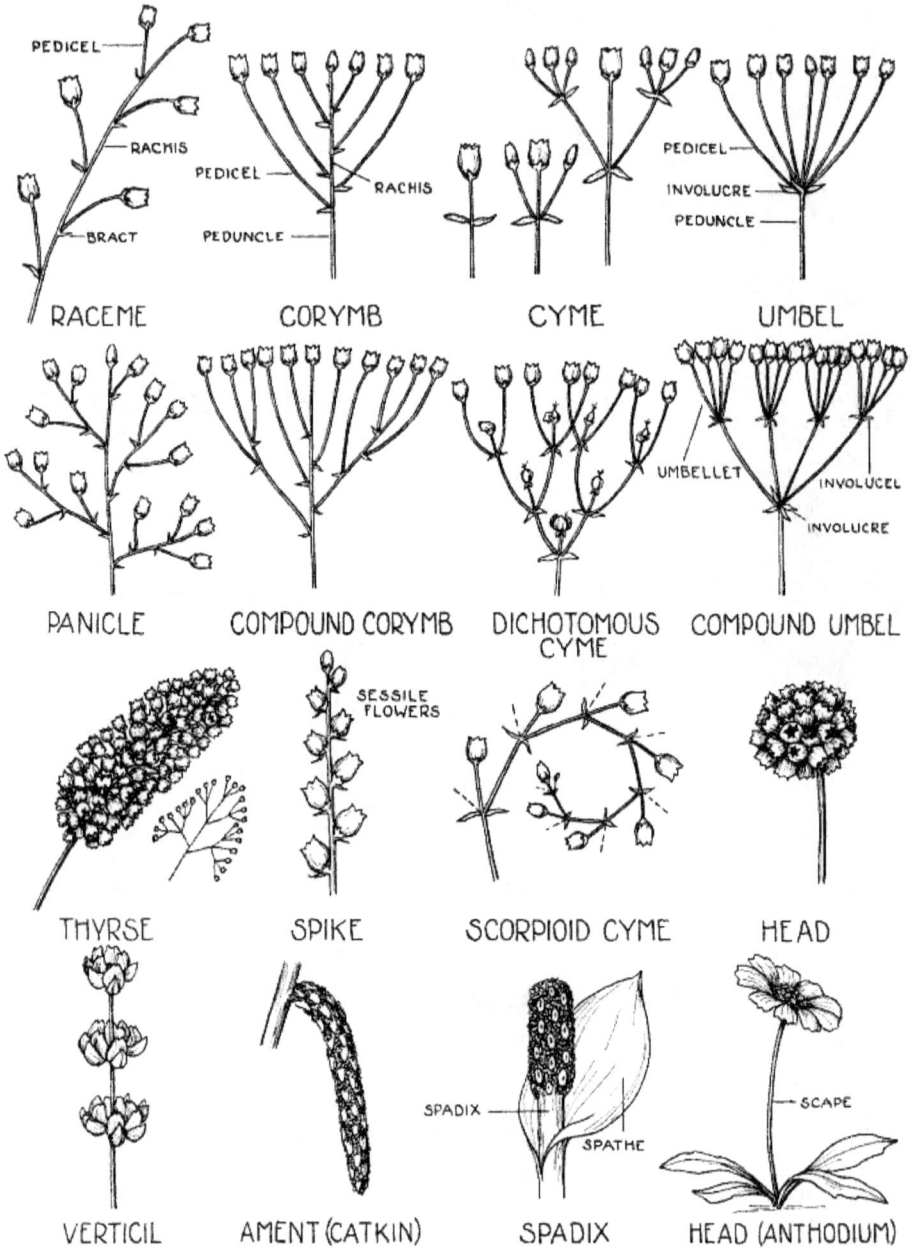

PEDICEL
RACHIS
PEDICEL
BRACT
PEDUNCLE

RACEME

PEDICEL
RACHIS
PEDUNCLE

CORYMB

CYME

PEDICEL
INVOLUCRE
PEDUNCLE

UMBEL

PANICLE

COMPOUND CORYMB

DICHOTOMOUS
CYME

UMBELLET
INVOLUCEL
INVOLUCRE

COMPOUND UMBEL

THYRSE

SESSILE
FLOWERS

SPIKE

SCORPIOID CYME

HEAD

VERTICIL

AMENT (CATKIN)

SPADIX
SPATHE

SPADIX

SCAPE

HEAD (ANTHODIUM)

Types of inflorescences (from *A Flora of the Marshes of California* by Herbert L. Mason, 1957.

INDEX

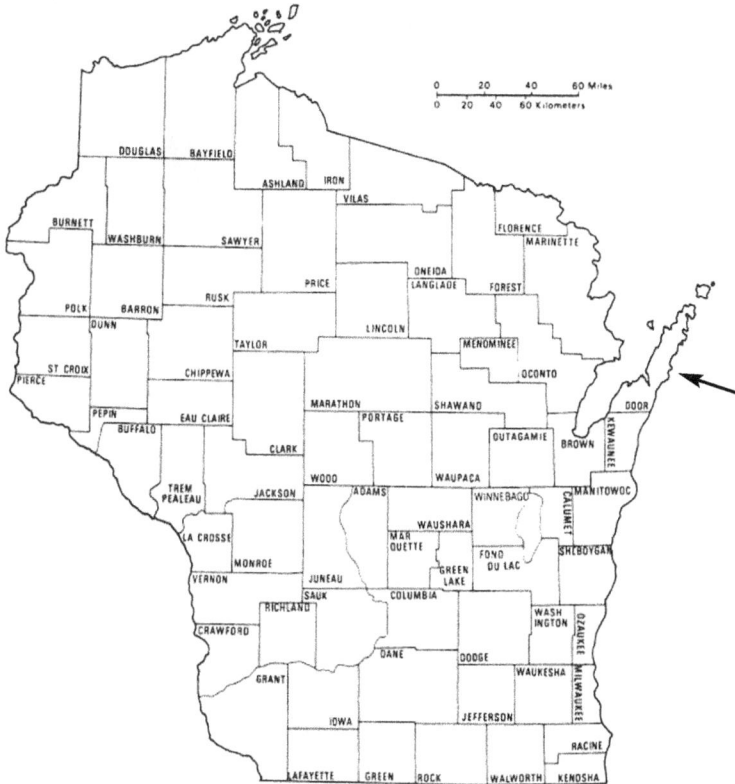

Wisconsin counties; arrow indicates Door County.